CELL MECHANICS AND TUMOR DEVELOPMENT

CELL MECHANICS AND TUMOR DEVELOPMENT

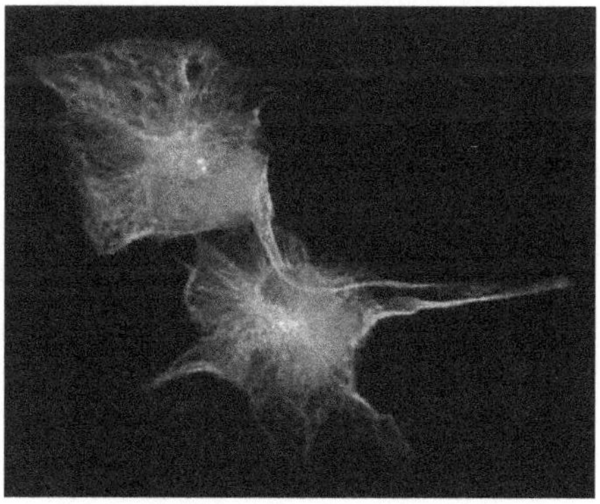

Ronald L Huston
University of Cincinnati, USA

World Scientific

NEW JERSEY • LONDON • SINGAPORE • BEIJING • SHANGHAI • HONG KONG • TAIPEI • CHENNAI • TOKYO

Published by

World Scientific Publishing Co. Pte. Ltd.
5 Toh Tuck Link, Singapore 596224
USA office: 27 Warren Street, Suite 401-402, Hackensack, NJ 07601
UK office: 57 Shelton Street, Covent Garden, London WC2H 9HE

British Library Cataloguing-in-Publication Data
A catalogue record for this book is available from the British Library.

CELL MECHANICS AND TUMOR DEVELOPMENT

ISBN 978-981-120-894-2

For any available supplementary material, please visit
https://www.worldscientific.com/worldscibooks/10.1142/11516#t=suppl

Typeset by Stallion Press
Email: enquiries@stallionpress.com

Acknowledgements

Support for the research, forming the basis for this book, was generously provided via Bardyl Tirana and Jennifer Wei Liao from the Schafer Foundation for Centriole Research.

This support is genuinely appreciated.

Contents

Chapter 1 Introduction

I wrote this book to share what I discovered in reviewing over 750 articles about cellular centrioles, centriole duplication, cell duplication, mitotic spindles, centriole errors, and cancer.

Some might say this is simply an annotated literature review of twenty-first century writings. But it is much more than that. It is a critical analysis whose ultimate objective is treatment of cancer.

That said, the focus initially is yet on only one *tiny* part of a cell: the centrioles. Centrioles are precise geometric structures lying adjacent to the nucleus in all human and animal cells.

"Tiny" means less than 500 nm in size.

Centrioles occur in pairs. They look like perpendicular hollow cylinders.

Figure 1.1 shows a photograph of a centriole pair model — prepared by fellow researcher Roger Adelman.

Chapter II provides a detailed description of centrioles. There we find that unlike other cell parts centrioles do *not* have a membrane cover.

The centriole walls, however, are not smooth. Instead they insist of nine triplets of long hollow cylinders known as "microtubules" (MTs) — positioned side-by-side appearing as a blade, and inclined to the centriole axis.

The centrioles are surrounded by layers of electron dense proteins. The centrioles together with their surroundings is sometimes known as the "centrosome."

The centrosome emits the MTs and, as such, the centromere is known as the microtubule organizing center (MTOC).

Figure 1.1. Model of a centriole pair

Chapter III provides the detailed description of the MTs. We see there that the MTs are long filaments of α- and β-tubulin dimers stacked end-to-end — approximately 400 nm long but only 25 nm in diameter.

Each of the tubulin dimers is only about 5 nm in size. Therefore, with a 25-nm diameter and 10 nm of walls the hollow is only about 15 nm in diameter.

Figure 1.2 shows a photograph of an MT model prepared by Teri Ellis.

When viewed endwise, for an unknown reason the MTs form a ring of 13 filaments, as represented in Figure 1.3.

The base of the MTs is surrounded by a multitude of layers of electron dense proteins giving the MTs a negative charge at its base — and consequently a positive charge at the outer end.

The MTs are *not* fixed on the centriole walls. Instead they are moving with new tubulin inserted at the base and the tubulin discarded at the outer end.

In addition to the centromere there is another organelle known as the chromosome. It consists of concentrated (compacted) DNA.

The centrioles and the DNA are the only parts of a cell which are able to self-duplicate. Interestingly when they duplicate themselves, they do it simultaneously.

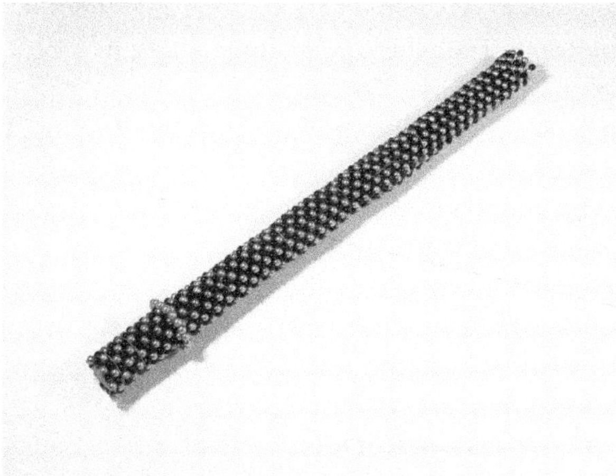

Figure 1.2. Model of a microtubule

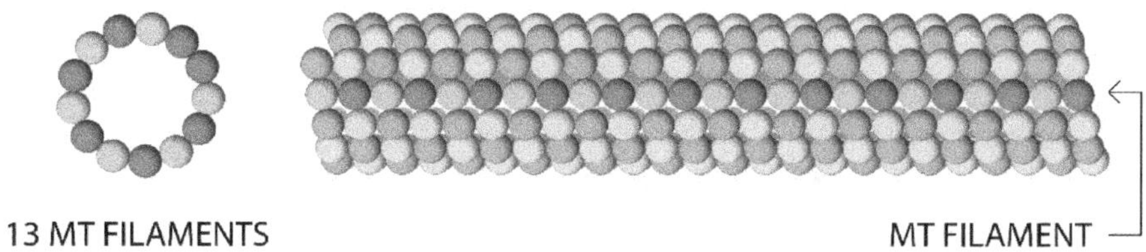

Figure 1.3. End view of microtubule filaments

In Chapter IV, we look at cell duplication, cell separation, and the mitotic spindle. We see that the beginning of cell duplication starts with centriole duplication as discussed in Chapter II. With centriole duplication there are now two pairs of centrioles. The newest of these then leaves the older and moves around the nucleus to the diametrically opposite side.

While the newer centriole is moving along the nucleus two events occur: (1) The new centriole becomes more mature (it is only approximately 80% mature as it departs) and (2) the nucleus elongates, softens, and shrinks in its center.

With the centrioles on opposite sides of the nucleus they emit elongated MTs out around the nucleus where they couple with receptors known as "kinetochores."

The kinetochores, as with the MTs are long strands. They are emitted from the chromosome. Then the chromosome might thus be known as the kinetochore organizing center (KNOC).

When the kinetochores emerge from the chromosome, they seek to find MTs. They approach the MTs from the side of the MT. Upon attaching themselves to the MT shaft they slide onward toward the shaft end and there remain attached to the MT.

The coupled MT/kinetochores then form a cord-like pattern engulfing the nucleus, known as the "mitotic spindle" about the nucleus. The opposite centrioles then pull the nucleus apart forming two bodies and thus the duplicated cells.

Chapter V considers centriole errors leading to the development of tumors and cancer. Although there are many known cancer-causing substances (e.g., smoking), the tumor growth stems from centriolar errors. The most common error is having more than two centrioles per cell — so called "supernumerary" centrioles. This produces premature cell division with ever increasing rapidity.

In this chapter, we also explore new treatments for cancer. The thesis is that cancer has a greater surrounding magnetic field than normal cells. The enhanced magnetic field is thus a "biomarker" suitable for the attraction of nanoparticles. If the ferrous nanoparticles go into a cancer cell, external vibration will cause the cell to head beyond its temperature survival limit.

A series of my papers provide the content for the chapters. These are

R.L. Huston, "On Centrioles, Microtubules, and Cellular Electromagnetism", *Journal of Nanotechnology in Engineering and Medicine*, Vol. 5, No. 3, 2014, pp. 031003, 1–5.

R.L. Huston, "Using the Electromagnetics of Cancer's Centrosome Clusters to Attract Therapeutic Nanoparticles", *Advances in Bioscience and Biotechnology*, Vol. 6, 2015, pp. 172–181.

R.L. Huston, "Mechanics of Centriole Microtubules", *Advances in Bioscience and Biotechnology*, Vol. 7, 2016, pp. 266–277.

R.L. Huston, "A Review of Centriole Activity, and Wrongful Activity, during Cell Division", *Advances in Bioscience and Biotechnology*, Vol. 7, 2016, pp. 169–182.

R.L. Huston, "A Review of Electromagnetic Activity in Cellular Mechanics", *Advances in Bioscience and Biotechnology*, Vol. 7, 2016, pp. 360–371.

R.L. Huston, "Electromagnetic Enhancement of Cancer Cells as Biomarkers for Nanotheranostics", *Journal of Nanotechnology Nanomedicine & Nanobiotechnology*, Vol. 4, No. 1, 2017, pp. 916–920.

R.L. Huston, "Delivering Nanoparticles to Cancer Cells", *Tissue Engineering and Nanotheranostics*, Edited by D. Shi and Q. Lin, World Scientific, Singapore, 2018, Chapter 7, pp. 207–226.

R.L. Huston, "A Real-Life Picture of the Origin of a Tumor", *Advances in Bioscience and Biotechnology*, Vol. 9, 2018, pp. 1–6.

In addition to these papers there are two volumes which are helpful in understanding cell behavior in general. These are:

B. Alberts, *et al.*, *Molecular Biology of the Cell*, 3rd Edition, Garland Publishing, New York, 1994.

J. Hardin, G. Bertoni and I.J. Kleinsmith, *Becker's World of the Cell*, 8th Edition, Pearson, New York, 2015.

Chapter 2

Centrioles

2.1. Introduction

Figure 2.1 shows a planar view of a typical human/animal cell. Such cells are also called "eukaryotic" cells.

From the figure, we see that the nucleus is the largest part of the cell. The other items are called "organelles." Of all the organelles we are focusing on the small ones lying adjacent to the nucleus and known as the centriole.

"Small" means less than 500 nm in size.

Figure 2.2 shows a photograph of a model of a centriole pair, prepared by my colleague Roger Adelman.

From Figure 2.2, we see that centrioles occur in pairs. They look like hollow cylinders. The walls, however, are not smooth. Instead they consist of nine triplets of long hollow cylinders called "microtubules" (MTs) — attached side-by-side to each other, forming the shape of a blade. The centriole walls thus consist of blade-like structures. Also, the blades are inclined to the centriole areas.

The centrioles are radically different from all the other organelles. Centrioles have precise straight lines and curved shapes. They have no membrane cover.

Why focus on centrioles? Because they represent the principal physics of cell duplication and cell separation.

The existence of centrioles has been known for a long time but it was not until the turn of the century that they began to be studied in detail.

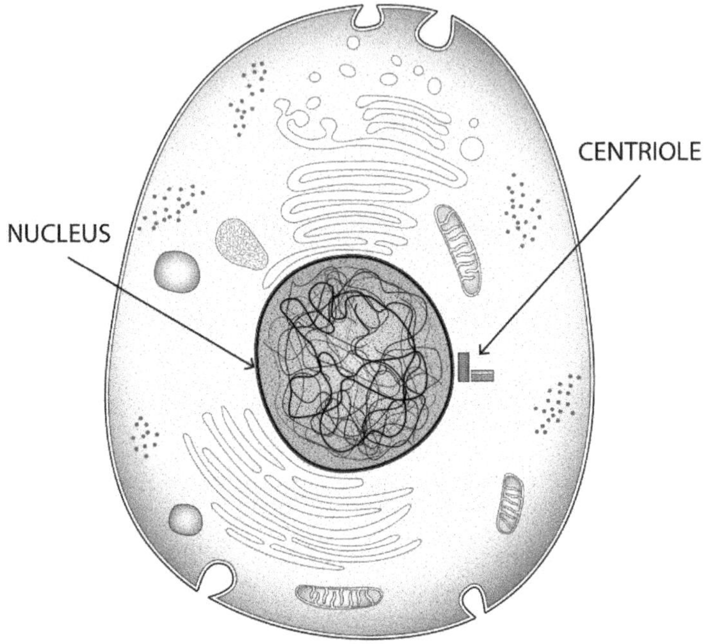

Figure 2.1. A planar (cross-section) view of a human/animal ("eukaryotic") cell

Figure 2.2. A photograph of a model of centrioles

2.2. Centriole Knowledge

Much remains unknown, but this is what is known to date:

1. Centrioles are small orthogonal cylinders adjacent to the nucleus.
2. They are 400–500 nm long and 200 nm in diameter [124, 127].
3. The two cylinders are called "mother" and "daughter" with the daughter being attached to the proximal end side of the mother, with "proximal" meaning toward the nucleus.
4. The daughter is only about 80% as long as the mother.
5. The mother is more fully developed than the daughter [124, 131].
6. The walls of each centriole are composed of nine sets of lengthwise MT triplets, side-by-side forming the appearance of an inclined blade.
7. The proximal ends of the centriole are immersed in a cloud of electron dense proteins known as the "microtubule organizing center" (MTOC) [127, 133].
8. The centrioles, together with the MTOC, are also known as the centrosome.
9. As noted earlier, the centrioles are the only organelles without a membrane cover.
10. The centrioles provide the physical forces for cell division.
11. When cell division is about to occur, each of the centrioles develops a daughter centriole. As the daughter centrioles grow, the original pair of centrioles becomes two pairs, with the mother centriole having a new daughter and the daughter centriole have a daughter of its own [47, 60, 109, 111, 127, 199, 208].
12. Once there are two pairs of centrioles, the pairs begin to separate. The mother–new daughter is then known as the "mature" pair and the daughter–daughter is known as the "immature" pair.
13. Once the centriole pairs are separated, the immature pair migrates to the diametrically opposite side of the nucleus.
14. The two diametrically opposed centriole pairs then pull apart from one another while at the same time the middle of the nucleus begins to soften and shrink.
15. As the centrioles are duplicating, the DNA is also duplicating.

16. If the centriole geometry is disturbed, there may be more than a single pair of centrioles in a cell. This condition is known as "supernumerary centrioles."
17. Similarly, when the centrioles are disturbed, the DNA is also disturbed — a condition known as "aneuploidy."
18. Centriole geometry can be disturbed by more than one daughter growing from the mother centriole.
19. Centriole damage can also be disturbed by loss of perpendicularity of the centriole cylinders [90].
20. Proteins Plk4 (also known as "SAK") and SAS6 are needed for centriole duplication. But an excess of these proteins can produce abnormal geometry.
21. Cancer cells have supernumerary centrioles [1, 4, 6, 8–10, 18, 21, 23, 90, 96, 100, 103–105, 110, 113, 143, 145, 152, 155, 192, 195, 196, 202].
22. Centrioles are the only organelles, aside from the nucleus, that undergo exact duplication.
23. Without a membrane cover about the centrioles, proteins, particles, and chemicals can go in and out of the centrosome without significant resistance [155].
24. There are more than 100 proteins in the pericentriolar materials.
25. It appears that the answers to questions of tumorigenesis are somewhere within the centrosome.
26. There are more than 100 proteins in the pericentriolar region.
27. The daughter centriole arising on the side of the mother centriole is known as "canonical assembly."
28. At the onset of centriole duplication, the protein STIL is recruited to the nascent daughter centriole, which then waiting with the proteins Plk4 and SAS-6, prevents over duplication.
29. Just prior to centriole duplication a protein known as asterless (AsP) attaches itself to one of the nine outside MT blades of the centriole wall. AsP then recruits Plk4 to deposit itself for the AsP. The Plk4 then appears as a dot on the proximal end of the mother centriole.
30. The daughter centriole arises from the Plk4 dot.
31. After the Plk4 dot is established, a cartwheel stimulated by protein SAS-6 begins to grow. The ends of the cartwheel support the development of MTs for the walls of the daughter centriole.

2.3. Bibliography — Mechanics of Centrioles

B. Alberts, *et al.*, *Molecular Biology of the Cell*, 3rd Edition, Garland Publishing, New York, 1994, pp. 803–818.

A. Asnacious and O. Hamant, "The Mechanics Behind Cell Polarity", *Trends in Cell Biology*, Vol. 22, No. 11, 2012, pp. 584–591.

J. Azimzadeh and M. Bornens, "Structure and Duplication of the Centrosome", *Journal of Cell Science*, Vol. 120, No. 13, 2007, pp. 2139–2142. [CrossRef] [PubMed]

W.M. Becker, *The World of the Cell*, Benjamin Cummings Publishing Co., Menlo Park, CA, 1986, pp. 464–468.

J. Beisson and M. Jerka-Dziadesz, "Polarities of the Centriolar Structure: Morphogenetic Consequences", *Biology of the Cell*, Vol. 91, No. (4–5), 1999, pp. 367–378.

M. Bettencourt-Dias and D.M. Glover, "Centrosome Biogenesis and Function: Centrosomics Brings New Understanding", *Nature Reviews. Molecular Cell Biology*, Vol. 8, No. 6, 2007, pp. 451–463. [CrossRef] [PubMed]

M. Bettencourt-Dias and D.M. Glover, "SnapShot: Centriole Biogenesis", *Cell*, Vol. 136, No. 1, 2009, pp. 188–188.

R. Binggeli and R.C. Weinstein, "Membrane Potentials and Sodium Channels: Hypotheses for Growth Regulation and Cancer Formations Based on Changes in Sodium Channels and Gap Junctions", *Journal of Theoretical Biology*, Vol. 123, No. 4, 1986, pp. 377–408.

M. Bornens, "The Centrosome in Cells and Organisms", *Science*, Vol. 335, No. 6067, 2012, pp. 422–425. [CrossRef] [PubMed]

T. Boveri, *The Origin of Malignant Tumors*, Williams and Wilkins, Baltimore, MD, 1929.

R. Brinkley, "Managing the Centrosome Numbers Game: From Chaos to Stability in Cancer Cell Division", *Trends in Cell Biology*, Vol. 11, No. 1, 2001, pp. 18–21.

J.A. Brown and J.A. Tuszynski, "A Review of the Ferroelectric Model of Microtubules", *Ferroelectrics*, Vol. 220, No. 1, 1999, pp. 141–155.

J.Y. Chan, "A Clinical Overview of Centrosome Amplification in Human Cancers", *International Journal of Biological Sciences*, Vol. 7, 2011, pp. 1122–1144.

M. Cifra, *et al.*, "Electric Field Generated by Axial Longitudinal Vibration Modes of Microtubule", *Biosystems*, Vol. 100, No. 2, 2010, pp. 122–131. [CrossRef] [PubMed]

J. Cuzick, *et al.*, "Electropotential Measurements as a New Diagnostic Modality for Breast Cancer", *Lancet*, Vol. 352, No. 9125, 1998, pp. 359–363.

A.B. D'Assoro, W.L. Lingle, and J.L. Salisbury, "Centrosome Amplification and the Development of Cancer", *Oncogene*, Vol. 21, 2002, pp. 6146–6153.

R.J. Davies, *et al.*, "Detection of the Cancer-Prone Colon, Using Transepithelial Impedance Analysis", *Archives of Surgery*, Vol. 124, 1989, pp. 480–484.

M.B.A. Djarngoz, *et al.*, "Directional Movement of Rat Prostate Cancer Cells in Direct-Current Electric Field: Involvement of Voltage-Gated Na^+ Channel Activity", *Journal of Cell Science*, Vol. 114, No. 14, 2001, pp. 2597–2705.

S. Doxey, "Re-Evaluating Centrosome Function", *Nature Reviews: Molecular Cell Biology*, Vol. 2, No. 9, 2001, pp. 688–689. [CrossRef] [PubMed]

A. Duensing, *et al.*, "Centriole Overduplication through the Concurrent Formation of Multiple Daughter Centrioles at Single Material Templates", *Nature Oncogene*, Vol. 26, No. 43, 2007, pp. 6280–6288.

S.K. Dutcher, "The Tubulin Family: Alpha to Eta", *Current Opinion in Cell Biology*, Vol. 13, No. 1, 2001, pp. 49–54.

K. Fukasawa, "Centrosome Amplifications, Chromosome Instability and Cancer Development", *Cancer Letters*, Vol. 230, 2005, pp. 6–19.

K. Fukasawa, *et al.*, "Abnormal Centrosome Amplification in the Absence of P53", *Science*, Vol. 271, 1966, pp. 1744–1747.

N.J. Ganem, S.A. Godinho, and D. Pellman, "A Mechanism Linking Extra Centrosomes to Chromosomal Instability", *Nature*, Vol. 460, No. 7252, 2009, pp. 278–282.

S.A. Godinho, M. Kwan, and D. Pellman, "Centrosomes and Cancer: How Cancer Cells Divide with Too Many Centrosomes", *Cancer Metastasis Reviews*, Vol. 28, No. 1–2, 2009, pp. 85–98.

T.M. Goepfert, *et al.*, "Centrosome Amplification and Overexpression of Aurora A Are Early Events in Rat Mammary Carcinogenesis", *Cancer Research*, Vol. 62, No. 14, 2002, pp. 4115–4122.

S.R. Hameroff, "A New Theory of the Origin of Cancer: Quantum Coherent Entanglement, Centrioles, Mitosis, and Differentiation", *Bio Systems*, Vol. 77, No. 1–3, 2004, pp. 119–136.

W.G. Hol, "The Role of the Alpha-Helix Dipole in Protein Function and Structure", *Progress in Biophysics and Molecular Biology*, Vol. 45, No. 3, 1985, pp. 149–155.

J. Kleyleon-Sohn, *et al.*, "Plk4-Induced Centriole Biogenesis in Human Cells", *Developmental Cell*, Vol. 13, No. 2, 2007, pp. 190–202. [CrossRef] [PubMed]

T. Kobayashi and B.D. Cynlacht, "Regulating the Transition from Centriole to Basal Body", *Journal of Cell Biology*, Vol. 193, No. 3, 2011, pp. 435–444.

A. Kramer, K. Neben, and A.D. Ho, "Centrosome Replication, Genomic Instability and Cancer", *Leukemia*, Vol. 16, No. 5, 2002, pp. 767–775. [CrossRef] [PubMed]

K. Kukasawa, "Oncogenes and Tumor Suppressors Take on Centrosomes", *Nature Reviews Cancer*, Vol. 7, 2007, pp. 911–924.

M. Kwon, *et al.*, "Mechanisms to Suppress Multiplanar Divisions in Cancer Cells with Extra Chromosomes", *Genes & Development*, Vol. 22, No. 16, 2008, pp. 2189–203. [CrossRef] [PubMed]

B.M.H. Lange and K. Gull, "Structure and Function of the Centriole in Animal Cells: Progress and Questions", *Trends in Cell Biology*, Vol. 6, No. 9, 1996, pp. 348–352. [CrossRef] [PubMed]

W.L. Lingle, *et al.*, "Centrosome Amplification Drives Chromosomal Instability in Breast Tumor Development", *Proceedings of the National Academy of Sciences United States of America*, Vol. 99, No. 4, 2002, pp. 1978–1983.

W.L. Lingle and J.L. Salisbury, "The Role of the Centrosome in the Development of Malignant Tumors", *Current Topics in Developmental Biology*, Vol. 49, No. 1, 1999, pp. 313–339.

H. Loffler, *et al.*, "DNA Damage-Induced Centrosome Amplification Occurs via Excessive Formation of Centriolar Satellites", *Oncogene*, Vol. 32, 2013, pp. 2963–2972.

H. Loffler, *et al.*, "Structure Meets Function–Centrosomes, Genome Maintenance and the DNA Damage Response", *Experimental Cell Research*, Vol. 312, No. 14, 2006, pp. 2633–2640. [CrossRef] [PubMed]

J. Luders and T. Stearns, "Microtubule-Organizing Centres: A Re-Evaluation", *Nature Reviews. Molecular Cell Biology*, Vol. 8, No. 2, 2007, pp. 161–167. [CrossRef]

R. Mahen and A.R. Venkitaraman, "Pattern Formation in Centrosome Assembly", *Current Opinion in Cell Biology*, Vol. 24, No. 1, 2012, pp. 19–23.

V. Marthiens, M. Piel, and R. Basto, "Never Tear Us Apart — The Importance of Centrosome Clustering", *Journal of Cell Science*, Vol. 125, 2012, pp. 3281–3292.

J. Marx, "Do Centrosome Abnormalities Lead to Cancer?", *Science*, Vol. 292, No. 5516, 2001, pp. 426–429. [CrossRef] [PubMed]

D. McCaig, *et al.*, "Controlling Cell Behavior Electrically: Current Views and Future Potential", *Physiological Reviews*, Vol. 86, No. 3, 2005, pp. 943–978.

P.G. McKean, S. Vaughan, and K. Gull, "The Extended Tubulin Superfamily", *Journal of Cell Science*, Vol. 144, No. 15, 2001, pp. 2723–2733.

E.A. Nigg, "Centrosome Duplication: Of Rules and Licenses", *Trends in Cell Biology*, Vol. 17, No. 5, 2008, pp. 215–221. [CrossRef]

E.A. Nigg and J.W. Raft, "Centrioles, Centrosomes, and Cilia in Health and Disease", *Cell*, Vol. 139, No. 4, 2009, pp. 663–678.

E.A. Nigg and T. Stearns, "The Centrosome Cycle: Centriole Biogenesis, Duplication and Inherent Asymmetries", *Nature Cell Biology*, Vol. 13, No. 10, 2011, pp. 1154–1160.

G.A. Pihan, *et al.*, "Centrosome Abnormalities and Chromosome Instability Occur Together in Pre-Invasive Carcinomas", *Cancer Research*, Vol. 63, No. 6, 2003, pp. 1398–1404.

J. Pokorny, "Electrodynamic Activity of Healthy and Cancer Cells", *Journal of Physics: Conference Series*, Vol. 329, 2011, p. 012007.

J. Pokorny, *et al.*, "Cancer Physics: Diagnosis Based on Damped Cellular Elastoelectrical Vibrations in Microtubules", *European Biophysics Journal*, Vol. 40, No. 6, 2011, pp. 747–759.

J. Pokorny, J. Hasek, and F. Jelinek, "Electromagnetic Field of Microtubules: Effects on Transfer of Mass Particles and Electrons", *Journal of Biological Physics*, Vol. 31, No. 3–4, 2005a, pp. 501–514. [CrossRef] [PubMed]

J. Pokorny, J. Hasek, and F. Jelinek, "Endogenous Electric Field and Organization of Living Matter", *Electromagnetic Biology and Medicine*, Vol. 24, No. 3, 2005b, pp. 183–197. [CrossRef]

W. Portet, *et al.*, "Elastic Vibrations in Seamless Microtubules", *European Biophysics Journal*, Vol. 34, No. 7, 2005, pp. 912–920. [CrossRef] [PubMed]

X.S. Qian, J.Q. Zhang, and C.Q. Ru, "Wave Propagation in Orthotropic Microtubules", *Journal of Applied Physiology*, Vol. 101, No. 8, 2007, p. 084702. [CrossRef]

J.W. Raff, "Centrosomes and Cancer: Lessons from a TACC", *Trends in Cell Biology*, Vol. 12, No. 5, 2002, pp. 222–225.

A. Ridley and R. Heald, "Cell Structure and Dynamics — Editorial Overview", *Current Opinion in Cell Biology*, Vol. 23, No. 1, 2011, pp. 1–3. [CrossRef]

J.L. Salisbury, *et al.*, "Centrosomes and Cancer", *Biology of the Cell*, Vol. 91, No. 6, 1999, pp. 451–460. [CrossRef]

M.V. Sataric, J.A. Tuszynski, and J. Pokorny, "Kinlike Excitations as an Energy-Transfer Mechanism in Microtubules", *Physical Review E*, Vol. 28, 1993, pp. 569–597.

P.W. Schafer, "Centrioles of a Human Cancer: Intercellular Order and Intracellular Disorder", *Science*, Vol. 164, No. 3885, 1969, pp. 1300–1303. [CrossRef] [PubMed]

P.W. Schafer, "Centrioles: Intercellular Order in Normal and Malignant Cells", *Journal of Thoracic and Cardiovascular Surgery*, Vol. 63, No. 3, 1975, pp. 472–477.

P.W. Schafer and J.A. Chandler, "Electron Probe X-Ray Microanalysis of a Normal Centriole", *Science*, Vol. 170, No. 3963, 1970, pp. 1204–1209. [CrossRef] [PubMed]

L. Schoeckel, *et al.*, "Cleavage of Cohesion Rings Coordinates the Separation of Centrioles and Chromatides", *Nature Cell Biology*, Vol. 13, No. 8, 2011, pp. 966–972.

G. Sinder and J.L. Nordberg, "The Good, the Bad and the Ugly: The Practical Consequences of Centrosome Amplification", *Current Opinion in Cell Biology*, Vol. 16, 2004, pp. 49–64.

P. Strand and P. Gonczy, "Mechanisms of Procentriole Formation", *Trends Cell in Biology*, Vol. 18, No. 8, 2008, pp. 389–396. [CrossRef][PubMed]

S. Suresh, "Biomechanics and Biophysics of Cancer Cells", *Acta Materialia*, Vol. 55, No. 12, 2007, pp. 3989–4014.

A.-M. Tassin and M. Bornens, "Centrosome Structure and Microtubules Nucleation in Animal Cells", *Biology of the Cell*, Vol. 91, No. 4–5, 1999, pp. 343–354. [CrossRef] [PubMed]

M.-C. Tsai and J. Ahringer, "Microtubules Are Involved in Anterior–Posterior Axis Formation in *C. elegans* Embryos", *Journal of Cell Biology*, Vol. 179, No. 3, 2007, pp. 397–402. [CrossRef] [PubMed]

B.D. Vitre and D.W. Cleveland, "Centrosomes, Chromosomes Instability (CIN) and Aneuploidy", *Current Opinion in Cell Biology*, Vol. 24, 2012, pp. 809–815.

J. Vulprecht, *et al.*, "STIL Is Required for Centriole Duplication in Human Cells", *Journal of Cell Science*, Vol. 125, No. 5, 2012, pp. 1353–1362.

F. Wang, *et al.*, "Dual Surface-Functionalized Janus Nanocomposites of Polystyrene/Fe_3O_4@SiO_2 for Simultaneous Tumor Cell Targeting and Stimulus-Induced Drug Release", *Advanced Materials*, Vol. 25, No. 25, 2013, pp. 3485–3489. [CrossRef] [PubMed]

C. Wong and T. Stearns, "Centrosome Number is Controlled by a Centrosome Intrinsic Block to Reduplication", *Nature Cell Biology*, Vol. 5, 2003, pp. 539–544.

Y. Zhau and Q. Zhan, "Electric Oscillation and Coupling of Chromatics Regulate Chromosome Packaging and Transcription in Eukaryotic Cells", *Theoretical Biology & Medical Modelling*, Vol. 9, No. 1, 2012, pp. 27–38.

D. Zyss and F. Gergely, "Centrosome Function in Cancer: Guilty or Innocent?", *Trends in Cell Biology*, Vol. 19, No. 7, 2009, pp. 334–346.

2.4. Annotated Literature Review

F.G. Agrican, E. Schiebel, and B.R. Mardin, "Separate to Operate: Control of Centrosome Positioning and Separation", *Philosophical Transactions of the Royal Society of London. Series B, Biological Sciences*, Vol. 369, 2014, p. 10.

Review Notes: The authors present a review and discussion as how centriole pairs separate. No mention of electromagnetic forces.

B. Alberts, *et al.*, *Molecular Biology of the Cell*, 3rd Edition, Garland Publishing, New York, 1994.

Review Notes: Centrioles arise by the duplication of existing centrioles (p. 819). Daughter centrioles begin to grow near the base of each parent centriole and at a right angle to it.

C. Arquint, *et al.*, "Cell-Cycle-Regulated Expression of STIL Controls Centriole Number in Human Cells", *Journal of Cell Science*, Vol. 125, 2012, pp. 1342–1352.

Review Notes: STIL is required for centriole duplication, but an overexpression produces multiple daughter centrioles around each mother. STIL is recruited to nascent daughter centrioles at the onset of centriole duplication. STIL cooperates with SAS-6 and Plk4 in the control of centriole number.

J. Azimzadeh and M. Bornens, "Structure and Duplication of the Centrosome", *Journal of Cell Science*, Vol. 120, 2007, pp. 2139–2142.

Review Notes: Centriole elongation reaches full length during cell cycle. Centrosome separation is believed to require disassembly of the linkage connecting the two centriole pairs, but not the centrioles within a pair.

J. Azimzadeh and W.F. Marshall, "Building the Centriole", *Current Biology*, 2010, Vol. 20, No. 18, pp. R816–R825.

Review Notes: The proximal end is negative and the distal end is positive.

R. Bahtz, *et al.*, "GCP6 Is a Substrate of Plk4 and Required for Centriole Duplication", *Journal of Cell Science*, Vol. 125, 2011, pp. 486–496.

Review Notes: This paper focuses on the protein GCP6 and its role in centriole construction.

— GCP6 is needed by Plk4 (SAK) but it is not clear how.

— GCP6 is part of the γ-TuRC.

— GCP6 is localized to the perimeter of the centrosome — at a distance from the centrioles.

F. Balzarotti, *et al.*, "Nanometer Resolution Imaging and Tracking of Fluorescent Molecules with Minimal Photon Fluxes", *Science*, Vol. 355, 2017, pp. 606–612.

Review Notes: The authors introduce "MINFLUX" which localized photon emitters in space. It minimizes the fluorescence photons needed for precise resolution. MINFLUX attains ~1 nm precision.

R. Basto, *et al.*, "Flies without Centrioles", *Cell*, Vol. 125, No. 7, 2006, pp. 1375–1386.

Review Notes: Flies can live without centrioles but they are not normal and soon die.

M. Bettencourt-Dias and D.M. Glover, "Centrosome Biogenesis and Function: Centrosomics Bring New Understanding", *Nature Reviews Molecular Cell Biology*, Vol. 8, 2007, pp. 451–463.

Review Notes: The centrosome is the microtubular organizing center (MTOC). The centrosome is surrounded by electron dense material. Centrioles are surrounded by electron dense material. There are as many as 300 proteins in the pericentriolar material (PCM). Overexpression of SAK/Plk4 leads to centrosome amplification.

M. Bettencourt-Dias, *et al.*, "Centrosomes and Cilia in Human Disease", *Trends in Genetics*, Vol. 27, 2011, pp. 307–315.

Review Notes: Centrosome abnormalities are common in cancer. Centrosome clustering is observed in tumors.

S. Blachon, *et al.*, "Drosophila Asterless and Vertebrate CEP152 Are Orthologs Essential for Centriole Duplication", *Genetics*, Vol. 180, No. 4, 2008, pp. 2081–2094.

Review Notes: Cep152 is the ortholog of asterless. Centrosome amplification is a common feature of cancer cells. Asterless is required for initiation of centriole duplication.

S.P. Blagden and D.M. Glover, "Polar Expeditions — Provisioning the Centrosome for Mitosis", *Nature Cell Biology*, Vol. 5, No. 6, 2003, pp. 505–511.

Review Notes: Centrosomes enlarge dramatically after mitotic entry.

M. Bornens, "Centrosome Composition and Microtubule Anchoring Mechanisms", *Current Opinion in Cell Biology*, Vol. 14, No. 1, 2002, pp. 25–34.

Review Notes: This paper discusses how the centrioles develop the microtubule organizing center (MTOC) and thus the centrosome.

M. Bornens, "Organelle Positioning and Cell Polarity", *Nature Reviews Molecular Cell Biology*, Vol. 9, 2008, pp. 874–886.

Review Notes: "Cell polarity" is a structural rather than an electrical phenomenon. Non-random positioning of each organelle is preserved and transmitted during cell division.

M. Bornens, "The Centrosome in Cells and Organisms", *Science*, Vol. 335, 2012, pp. 422–426.

Review Notes: The centrosome transmits polarity to daughter centrioles. The daughter centriole is only ~80% as long as the more mature mother centriole.

M. Bornens and P. Gönczy, "Centrosomes Back in the Limelight", *Philosophical Transactions of the Royal Society of London. Series B, Biological Sciences*, Vol. 369, 2014, p. 15.

Review Notes: Centrioles are responsible for the assembly of microtubules (MTs). Centrosome is sometimes called the "microtubule organizing center" (MTOC).

"Diplosome" — a centriole and a developing centriole

"Procentriole" — a developing centriole

"Basal body" — mother centriole

Centrosome, MTOC, and "pericentriole material" (PCM) are all the same — "electron dense material." In the MT blades: ABC they are close to each other and thus provides stability to one another. This helps establish the stability of the MTs. The centriole pair is anchored to the nucleus.

B.R. Brinkley, "Managing the Centrosome Numbers Game: From Chaos to Stability in Cancer Cell Division", *Trends in Cell Biology*, Vol. 11, No. 1, 2001, pp. 18–21.

Review Notes: Daughter centrioles always form perpendicular to the mother and at the proximal end. It appears that regulatory constraints, as opposed to structure constraints, dictate the length and form of the daughter centrioles. Centriole duplication is coupled with DNA replication — but not in tumor generation. Supernumerary centrioles may be due to an uncoupling with DNA regulations. Excess centrosomes are associated with the loss of one or more chromosomes (aneuploidy).

D.A. Brito, S.M. Gouveia, and M. Bettencourt-Dias, "Deconstructing the Centriole: Structure and Number Control", *Current Opinion in Cell Biology*, Vol. 24, No. 1, 2012, pp. 4–13.

Review Notes: Centriole assembly is "tightly coordinated" with DNA duplication.

G.J. Brouhard and L.M. Rice, "The Contribution of αβ-tubulin Curvature to Microtubule Dynamics", *The Journal of Cell Biology*, Vol. 207, No. 3, 2014, pp. 323–334.

Review Notes: The β-tubulin determines curvature at the distal ends of MTs.

C.W. Brownlee and G.C. Rogers, "Show Me Your License, Please: Deregulation of Centriole Duplication Mechanisms that Promote Amplification", *Cellular and Molecular Life Sciences*, Vol. 70, No. 6, 2013, pp. 1021–1034.

Review Notes: The mechanics of centriole biogenesis remain poorly understood. Centriole amplification and DNA replication are similar.

C.W. Brownlee, *et al.*, "The Protein Phosphatase 2A Regulatory Submit Twins Stabilizes Plk4 to Induce Centriole Amplification", *Journal of Cell Biology*, Vol. 195, No. 2, 2011, pp. 231–243.

Review Notes: Supernumerary centrioles can cause aneuploidy and tumorigenesis.

Z. Carvalho-Santos, *et al.*, "Stepwise Evolution of the Centriole-Assembly Pathway", *Journal of Cell Science*, Vol. 123, 2010, pp. 1414–1426.

Review Notes: Centrioles are found in eukaryotic cells.

Z. Carvalho-Santos, *et al.*, "Tracing the Origins of Centrioles, Cilia and Flagella", *Journal of Cell Biology*, Vol. 194, No. 2, 2011, pp. 165–175.

Review Notes: Centrioles may be visualized by electron microscopes.

J.Y. Chan, "A Clinical Overview of Centrosome Amplification in Human Cancers", *International Journal of Biological Sciences*, Vol. 7, No. 8, 2011, pp. 1122–1144.

Review Notes: Twenty-first century has a surge in centriole research. Centrosome amplification is a hallmark of cancer.

P. Chang, *et al.*, "ε-Tubulin Is Required for Centriole Duplication and Microtubule Organization", *Nature Cell Biology*, Vol. 5, 2003, pp. 71–76.

Review Notes: γ-Tubulin is a component of the sub-distal appendages of the centriole. γ-Tubulin is required for centriole duplication and organization of PCM.

C.W. Chang, *et al.*, "CEP295 Interacts with Microtubules and Is Required for Centriole Elongation", *Journal of Cell Science*, Vol. 129, 2016, pp. 2501–2513.

Review Notes: The protein Cep295 interacts with the MTs to elongate them. Too much → long centrioles, too little → short centrioles.

P.L. Chavali, M. Putz, and F. Gergely, "Small Organelle, Big Responsibility: The Role of Centrosomes in Development and Disease", *Philosophical Transactions of the Royal Society of London. Series B, Biological Sciences*, Vol. 369, 2014, p. 12.

Review Notes: Identifies various disease, aside from tumorigenesis, that can occur with centrosome dysfunction.

I.M. Cheeseman and A. Desai, "Molecular Architecture of the Kinetochore-Microtubule Interface", *Nature Reviews Molecular Cell Biology*, Vol. 19, 2008, pp. 33–46.

Review Notes: Kinetochores have more than 80 components (proteins). These proteins are organized. Minor errors in the structure can lead to aneuploidy and tumorigenesis. Major errors cause cell death.

F. Chen, P.W. Tillberg, and E.S. Boyden, "Expansion Microscopy", *Science*, Vol. 347, No. 6221, 2015, pp. 543–548.

Review Notes: Physically expanding a cell could improve imaging.

K. Cheng and C. Zou, "Physics Models of Centriole Replication", *Medical Hypotheses*, Vol. 67, No. 3, 2006, pp. 572–577.

Review Notes: Externally applied current can induce an electric field in the centriole. Centriole replication may be accelerated by an electric current. This may be useful in wound healing. The mother centriole has a strong positive charge. The daughter's charge continues to grow. When the mother and daughter both have strong positive charges, separation occurs.

K. Cheng and C. Zou, "Theoretical Models of Cytokinesis and Mitosis (Without Astens)", *American Journal of Plant Physiology*, Vol. 1, No. 1, 2006, pp. 21–27.

Review Notes: Plants do not have centrioles. Instead, it is believed that growth occurs due to electromagnetic forces.

O. Cizmecioglu, *et al.*, "CEP152 Acts as a Scaffold for Recruitment of Plk4 and CPAP to the Centrosome", *Journal of Cell Biology*, Vol. 191, No. 4, 2010, pp. 731–739.

Review Notes: There is a correlation between excess centrosomes, aneuploidy, and cancer. Asterless failure prevents the recruitment of Plk4 to the centrosome Plk4 which appears to regulate itself.

P.T. Conduit, "Microtubule Organization: A Complex Relation", *Journal of Cell Biology (JCB)*, Vol. 213, 2016, pp. 809–812.

Review Notes: Paper argues that the γ-tubulin ring complex (γ-TuRC) varies between different species.

I. Cunha-Ferreira, *et al.*, "The SCF/Slimb Ubiquitin Ligase Limits Centrosome Amplification through Degradation of SAK/PLK4", *Current Biology*, Vol. 19, 2009, pp. 43–49.

Review Notes: Abnormal centrioles lead to cancer. An overexpression of SAK/Plk4 leads to cancer.

I. Cunha-Ferreira, *et al.*, "Regulation of Autophosphorylation Controls PLK4 Self-Destruction and Centriole Number", *Current Biology*, Vol. 23, No. 22, 2013, pp. 2245–2254.

Review Notes: Plk4 needs to be at a certain level before centriole duplication occurs, but then Plk4 self-destructs before too many centrioles are formed.

I. Cunha-Ferreira, I. Bento, and M. Bettencourt-Dias, "From Zero to Many: Control of Centriole Number in Development and Disease", *Traffic*, Vol. 10, No. 5, 2009, pp. 482–498.

Review Notes: The centrosome is electron dense. An overexpression of SAK/Plk4 and SAS-6 causes the formation of "flower" (multiple daughter) phenomena.

A. Dammermann, *et al.*, "SAS-4 Is Recruited to a Dynamic Structure in Newly Forming Centrioles that Is Stabilized by the γ-Tubulin-Mediated Addition of Centriolar Microtubules", *Journal of Cell Biology*, Vol. 180, No. 4, 2008, pp. 771–785.

Review Notes: The small size of centrioles makes it difficult to observe the kinematics.

A. Dart, "Extra! Extra! Read All About It", *Nature Reviews Cancer*, Vol. 17, February 2017, pp. 143. doi:10.1038/nrc2017.12

Review Notes: "extra centrosomes were sufficient to initiate tumorigenesis in mammals." This quote is based upon findings in the article: M.S. Levine, *et al.*, "Centrosome Amplification Is Sufficient to Promote Spontaneous Tumorigenesis in Mammals", *Developmental Cell*, Vol. 12, 2016, p. 022, 2017.

M. Delattre, C. Canard, and P. Gönczy, "Sequential Protein Recruitment in *C. elegans* Centriole Formation", *Current Biology*, Vol. 16, 2006, pp. 1844–1849.

Review Notes: Centriole formation is a poorly understood process. But just before centriole formation, the original pair splits slightly.

M. Delattre and P. Gönczy, "The Arithmetic of Centrosome Biogenesis", *Journal of Cell Science*, Vol. 117, 2004, pp. 1619–1630.

Review Notes: SAS-4 and ZYG-1 govern daughter formation. PCM is electron dense. As mitosis begins the two centriole pairs lose their orthogonal arrangement and split slightly while maintaining a flexible connection.

S.J. Doxsey, "The Centrosome — A Tiny Organelle with Big Potential", *Nature Genetics*, Vol. 20, No. 2, 1998, pp. 104–106.

Review Notes: Centrosome ~1 nm in diameter. Tumorigenesis questions are likely to have answers in small places like the centrosome.

S.J. Doxsey, "Centrosomes as Command Centres for Cellular Control", *Nature Cell Biology*, Vol. 3, No. 5, 2001a, pp. E105–E108.

Review Notes: There is a direct link between centrosome activity and DNA replication.

S.J. Doxsey, "Re-evaluating Centrosome Function", *Nature Reviews Molecular Cell Biology*, Vol. 2, No. 9, 2001b, pp. 688–698.

Review Notes: Centrioles and chromosomes (DNA) are the only organelles able to self-duplicate. Centrioles elongate at the distal end.

S.J. Doxsey, D. McCollum, and W. Theurkauf, "Centrosomes in Cellular Regulation", *Annual Review of Cell and Developmental Biology*, Vol. 21, 2005, pp. 411–434.

Review Notes: Centrosomes bind more than 100 regulatory proteins.

S. Duensing, "A Tentative Classification of Centrosome Abnormalities in Cancer", *Cell Biology International*, Vol. 29, 2005, pp. 352–359.

Review Notes: Centrosome anomalies are found in malignant tumors.

A. Duensing, *et al.*, "Centriole Overduplication through the Concurrent Formation of Multiple Daughter Centrioles at Single Maternal Templates", *Oncogene*, Vol. 26, No. 43, 2007, pp. 6280–6288.

Review Notes: Abnormal centrosome members are detected in virtually all cancers. A single maternal centriole can generate multiple daughter centrioles. Multiple daughter centrioles arise from a presence of Plk4 and E/cyclin-dependent kinase2 (CDK2). Figure 2 shows a picture of multiple daughter centrioles in a flower-type configuration.

N.S. Dzhindzhev, *et al.*, "Asterless Is a Scaffold for the Onset of Centriole Assembly", *Nature*, Vol. 457, 2010, pp. 614–718.

Review Notes: Asterless provides a molecular platform for Plk4. The amino terminus interacts with the Polo box of Plk4. Too much or too little Plk4 causes a failure of centriole assembly. Asterless is Cep152.

F. Eckerdt, *et al.*, "Identification of a Polo-like Kinase 4-Dependent Pathway for De Novo Centriole Formation", *Current Biology*, Vol. 21, No. 5, 2011, pp. 428–432.

Review Notes: Supernumerary centrioles are a key cause of cancer.

M.W. Elting, *et al.*, "Force on Spindle Microtubule Minus Ends Moves Chromosomes", *Journal of Cell Biology*, Vol. 206, No. 2, 2014, pp. 245–256.

Review Notes: Proteins dynein and NuMA exert tension forces to stabilize the minus ends of MTs.

E.N. Firat-Karalar and T. Stearns, "The Centriole Duplication Cycle", *Philosophical Transactions of the Royal Society of London. Series B, Biological Sciences*, Vol. 369, 2014, p. 17.

Review Notes: Defects in centriole number are associated with cancer. This is a review paper on how centrioles are assembled and controlled. Centriole proteins at the site of assembly: Plk4, SASS6, and STIL. Plk4 localizes at a dog ant and then recruits SASS6. It is not clear what targets Plk4 to the site on the mother centriole. It is an asymmetric event. The concentration of Plk4 is critical to limiting duplication to one site. Once the Plk4 is established, a cartwheel with ninefold symmetry based on SASS6 is assembled. At the base of the ninefold is a stalk which keeps the centrioles connected. γ-tubulin forms on the ends of the c to form MTs. There is a triplet of MTs A, B, and C (A near center and longest). The kinetics of the protein reactions and interactions appear to be unknown. Separase is required for disengagement. Authors present a model for centriole development. One of the steps is that Plk4 is limiting.

A.M. Fry, "Solving the Centriole Disengagement Puzzle", *Nature Cell Biology*, Vol. 17, No. 1, 2015, pp. 3–5.

Review Notes: Tumor cells have excess chromosomes.

J. Fu and D.M. Glover, "Structured Illumination of the Interface between Centriole and Pericentriolar Material", *Open Biology*, Vol. 2, 2012, pp. 1–30.

Review Notes: Centrosome behavior is still poorly understood. The centrosome increases in size upon entry into mitosis. Asterless is a scaffold for Plk4 and SAS-6. SAS-6 assembles the cartwheel structure. Plk1 produces a dramatic increase in PCM. Also, Plk1 recruits tubulin.

K. Fukasawa, "Introduction", *Oncogene*, Vol. 21, 2002, pp. 6140–6145.

Review Notes: This is an introduction paper to a series of reviews. What is new is the idea that multiple centrosomes can occur in four ways:

(1) Over duplication of centrioles (the "flower")

(2) Failure to undergo cytokinesis (i.e., cell dividing)

(3) Separation of the centriole pair

(4) Centrosome occurring without centriole

K. Fukasawa, "Oncogenes and Tumour Suppressors Take on Centrosomes", *Nature Reviews Cancer*, Vol. 7, 2007, pp. 911–924.

Review Notes: Centrosome abnormalities are present in solid tumors. Centrosome instability accelerates tumor progression. The understanding of centrosome function is likely to be an aid in cancer prevention and control.

B.J. Galleta, *et al.*, "Asterless Is Required for Centriole Length Control and Sperm Development", *Journal of Cell Biology*, Vol. 213, No. 4, 2016, pp. 435–474.

Review Notes: Asterless (Asl) is an orthologue of Cep152. Asl is the Drosophila melanogaster orthologue of Cep152. Authors show that Asterless (Asl) is required for centriole length control but not for PCM recruitment.

N.J. Ganem, S.A. Godinho, and D. Pellman, "A Mechanism Linking Extra Centrosomes to Chromosomal Instability", *Nature*, Vol. 460, No. 7252, 2009, pp. 278–282.

Review Notes: The earliest event in cartwheel assembly is likely to be the recruitment of Plk4 on the surface of the proximal end of the mother. Centrioles maintain their geometry through numerous cell divisions.

D. Gisselsson, *et al.*, "Centrosomal Abnormalities, Multipolar Mitoses, and Chromosomal Instability in Head and Neck Tumours with Dysfunctional Telomeres", *British Journal of Cancer*, Vol. 87, 2002, pp. 202–207.

Review Notes: Confirms previous findings.

S.A. Godinho, M. Kwon, and D. Pellman, "Centrosomes and Cancer: How Cancer Cells Divide with too Many Centrosomes", *Cancer Metastasia Review*, Vol. 28, 2009, pp. 85–98.

Review Notes: Cancer cells will die if the clustering of centrosomes can be prevented.

P. Gönczy, "Towards a Molecular Architecture of Centriole Assembly", *Nature Reviews Molecular Cell Biology*, Vol. 13, 2012, pp. 425–435.

Review Notes: This paper provides a general description of centrioles with an extensive list of references.

P. Gönczy, "Centrosomes and Cancer: Revisiting a Long-Standing Relationship", *Nature Reviews Cancer*, Vol. 15, 2015, pp. 639–642.

Review Notes: Centrioles recruit PCM. This material forms concentric layers around the centriole. Extra centrosomes cluster together.

M. Gottardo, G. Callaini, and M.G. Riparbelli, "Procentriole Assembly without Centriole Disengagement — A Paradox of Male Gametogenesis", *Journal of Cell Science*, Vol. 127, 2014, pp. 3434–3439.

Review Notes: Procentrioles appear in male butterflies even with the centrioles still being engaged. But they do not grow and develop.

P. Guichard, *et al.*, "Procentriole Assembly Revealed by Cryo-Electron Tomography", *EMBO Journal*, Vol. 29, No. 9, 2010, pp. 1565–1572. doi:10.1038/emboj.2010.45

Review Notes: The A MT forms a template for B and C MTs.

S.R. Hameroff, "A New Theory of the Origin of Cancer: Quantum Coherent Entanglement, Centrioles, Mitosis, and Differentiation", *Biosystems*, Vol. 77, No. 1–3, 2004, pp. 119–136.

Review Notes: Centrioles can recruit PCM to form MTs from the beginning of the cell cycle. Mother centrioles and daughter centrioles have different abilities to recruit MT material.

Daughter centrioles become mature during mitosis.

Plk1 is needed for centriole disagreement.

M. Hilbert, *et al.*, "SAS-6 Engineering Reveals Interdependence between Cartwheel and Microtubules in Determining Centriole Architecture", *Nature Cell Biology*, Vol. 18, No. 4, 2016, pp. 393–403. doi:10.1038/ncd3329

Review Notes: Paper discusses SAS-6 activity.

M. Hiraki, *et al.*, "Bld10p Constitutes the Cartwheel-Spoke Tip and Stabilizes the 9-Fold Symmetry of the Centriole", *Current Biology*, Vol. 17, No. 20, pp. 1776–1783.

Review Notes: Protein Bld10 is a tip agent that connects the MTs to the cartwheel.

M. Hirono, "Cartwheel Assembly", *Philosophical Transactions of the Royal Society of London. Series B, Biological Sciences*, Vol. 369, 2014, p. 16.

Review Notes: The initial stage of centriole assembly is the continued assembly — nine spokes. The central part is SAS6-Bld12p. The distal part is Cep135/Bld10p.

A.J. Holland, *et al.*, "Polo-like Kinase 4 Kinase Activity Limits Centrosome Overduplication by Autoregulating Its Own Stability", *Journal of Cell Biology*, Vol. 188, No. 2, 2010, pp. 191–198.

Review Notes: Centrosome duplication is controlled by centriole replication. Plk4 plays a key role in initiating centriole duplication. Overexpression of Plk4 promotes centriole overduplication and for the formation of extra centrosomes. Plk4 is regulated by its own kinase activity.

A.J. Holland, *et al.*, "The Autoregulated Instability of Polo-like Kinase 4 Limits Centrosome Duplication to Once per Cycle", *Genes and Development*, Vol. 26, No. 24, 2012, pp. 2684–2689.

Review Notes: Plk autoregulation guards against duplication instability.

S.A. Hormeño, *et al.*, "Single Centrosome Manipulation Reveals Its Electric Charge and Associated Dynamic Structure", *Biophysical Journal*, Vol. 97, No. 4, 2009, pp. 1022–1030.

Review Notes: Precise physical properties of the centrosome are yet to be determined.

R.L. Huston, "Mechanics of Centriole Microtubules", *Advances in Biosciences and* Biotechnology, Vol. 7, 2016, pp. 266–277.

Review Notes: MT review paper on MTs.

R. Ibrahim, *et al.*, "Electron Tomography Study of Isolated Human Centrioles", *Microscopy Research and Technique*, Vol. 72, No. 1, 2009, pp. 42–48.

Review Notes: Paper presents a model of centriole appendages.

A. Jord, *et al.*, "Calibrated Mitotic Oscillator Drives Motile Ciliogenesis", *Science*, Vol. 358, 2017, pp. 803–806.

Review Notes: Authors state that the vibratory process needed to drive cell division forward also performs other functions.

C. Kasbek, *et al.*, "Preventing the Degradation of Mps1 at Centrosomes Is Sufficient to Cause Centrosome Reduplication in Human Cells", *Molecular Biology of the Cell*, Vol. 18, 2007, pp. 4457–4469.

Review Notes: Supernumerary centrosomes promote the assembly of abnormal mitotic spindles. A prolonged S-phase produces extra centrosomes.

K. Kawamura, *et al.*, "Induction of Centrosome Amplification and Chromosome Instability in Human Bladder Cancer Cells by p53 Mutation and Cyclin E Overexpression", *Cancer Research*, Vol. 64, 2005, pp. 4800–4809.

Review Notes: Human cancers have centrosome amplification and chromosomal instability (CIN). Loss of protein P53 induces centrosome instability.

D. Keller, *et al.*, "Mechanisms of HsSAS-6 Assembly Promoting Centriole Formation in Human Cells", *Journal of Cell Biology*, Vol. 204, No. 5, 2014, pp. 697–712.

Review Notes: SAS-6 imparts the ninefold symmetry of the centrioles.

A. Khodjakov and C.L. Rieder, "Centrosomes Enhance the Fidelity of Cytokinesis in Vertebrates and Are Required for Cell Cycle Progression", *Journal of Cell Biology*, Vol. 153, No. 1, 2001, pp. 237–242.

Review Notes: Cells without centrioles do not survive. Cancer cells have extra centrosomes. Normal cells require Plk4 to survive. Cancer cells do not need Plk4 to multiply.

M. Kirkham, *et al.*, "SAS-4 Is a *C. elegans* Centriolar Protein that Controls Centrosome Size", *Cell*, Vol. 112, 2003, pp. 575–587.

Review Notes: SAS-4 is like tubulin in that it is incorporated into centrioles during their duplication. SAS-4 is necessary for centriole duplication. The centriole size is related to the levels of SAS-4. Partial depletion of SAS-4 leads to defective centrioles.

D. Kitagawa, *et al.*, "Structural Basis of the 9-Fold Symmetry of Centrioles", *Cell*, Vol. 144, No. 3, 2011, pp. 364–375.

Review Notes: "The mechanisms directing centriole formation are incompletely understood and represent a fundamental open question in biology" (2011). The ninefold symmetry probably comes from a geometric characteristic of a protein.

J.E. Klebba, *et al.*, "Polo-like Kinase 4 Autodestructs by Generating Its Slimb-Binding Phosphodegron", Current *Biology*, Vol. 23, 2013, pp. 2255–2261.

Review Notes: Plk4 generates its own phosphodegron without needing assistance from other kinases.

J.E. Klebba, *et al.*, "Autoinhibition and Relief Mechanism for Polo-like Kinase 4", *Proceedings of the Academy of* Sciences *United States of America*, Vol. 112, No. 7, 2015a, pp. E657–E666.

Review Notes: Plk4 is an enzyme — a master regulator of cell division. It induces centriole amplification. It is also self-regulating, but overexpressed in cancer.

J.E. Klebba, *et al.*, "Two Polo-like Kinase Binding Domains in Asterless Perform Distinct Roles in Regulating Kinase Stability", *Journal of Cell Biology*, Vol. 208, No. 4, 2015b, pp. 401–414.

Review Notes: This paper explores how Asterless (Asl) and Plk4 interact. Overexpression of Asl recruits Plk4, but then regulates Plk4 for stability, in new centriole development. Asterless is a binding partner of Pk4. Asterless (Asl) acts as a scaffolding protein. Plk4 is primarily regulated by protein turnover and it promotes its own destruction to suppress centriole overduplication. Plk4 forms a single spot on each centriole enabling the spawning of a single daughter centriole during the S-phase. Asterless is believed to target Plk4 toward the centriole.

J. Kley-Sohn, *et al.*, "Plk4-Induced Centriole Biogenesis in Human Cells", *Developmental Cell*, Vol. 13, No. 2, 2007, pp. 190–202.

Review Notes: The centrosomes induce the creation of new centrosomes. The ninefold symmetry ("cartwheel") has been proposed to serve as a scaffold for the assembly of centrioles.

T. Kobayashi and B.D. Bynlacht, "Regulating the Transition from Centriole to Basal Body", *Journal of Cell Biology*, Vol. 193, No. 3, 2011, pp. 435–444.

Review Notes: Centriole defects promote cancer.

G. Kohlmaier, *et al.*, "Overly Long Centrioles and Defective Cell Division upon Excess of the SAS-4-Related Protein CPAP", *Current Biology*, Vol. 19, 2009, pp. 1012–1018.

Review Notes: Discusses centriole length with a focus on CPAP — a protein related to SAS-4. CPAP controls the centriole length.

D. Kong, *et al.*, "Centriole Maturation Requires Regulated Plk1 Activity during Two Consecutive Cell Cycles", *Journal of Cell Biology*, Vol. 206, 2014, pp. 855–865.

Review Notes: Plk1 is required for maturation of the centrioles. This is a poorly understood process.

N. Korzeniewski, *et al.*, "Cullin 1 Functions as a Centrosomal Suppressor of Centriole Multiplication by Regulating Polo-like Kinase 4 Protein Levels", *Cancer Research*, Vol. 69, No. 16, 2009, pp. 6668–6675.

Review Notes: Tumors have an abnormal number of centrioles. Multiple daughters can form from a single centriole. These are called "centriole flowers." Following Plk4 recruitment SAS-6 is necessary for centriole development.

N. Korzeniewski, *et al.*, "Daughter Centriole Elongations Is Controlled by Proteolysis", *Molecular Biology of the Cell*, Vol. 21, 2010, pp. 2942–2951.

Review Notes: Centrioles can be elongated by chemicals.

N. Korzeniewski, M. Hohenfellner, and S. Duensing, "CANDI Promotes PLK4-Mediated Centriole Overduplication and Is Frequently Disrupted in Prostate Cancer", *Neoplasia*, Vol. 14, No. 9, 2012, pp. 799–806.

Review Notes: Cancer cells have aberrant centrosome numbers. Centriole duplication arises from Plk1 and separase (polo-like kinase 1). The right regulation of Plk4 is needed for healthy cell mitosis. A single maternal centriole assembles multiple daughters concurrently. "...CAND1 is a centrosomal protein that controls the stability of Plk4, a master regulator of centriole biogenesis. CAND1 was found to synergize with Plk4 to stimulate centriole overduplication..."

A. Kramer, K. Neben, and A.D. Ho, "Centrosome Replication, Genomic Instability and Cancer", *Leukemia*, Vol. 16, No. 5, 2002, pp. 767–775.

Review Notes: Errors in centrosome duplication cause cancer. Numerical chromosome imbalance known as "aneuploidy" is a cause of cancer. There are approximately 100 proteins in the centrosome.

M. Kwon, "Using Cell Culture Models of Centrosome Amplification to Study Centrosome Clustering in Cancer", *Methods in Molecular Biology*, Vol. 1413, 2016, pp. 367–392.

Review Notes: Full article not found. The presence of extra centrosomes imposes the tendency for catastrophic multipolar division.

B.G. Lambrus, *et al.*, "p53 Protects against Genome Instability Following Centriole Duplication Failure", *Journal of Cell Biology*, Vol. 210, No. 1, 2015, pp. 63–77.

Review Notes: p53 is a protector against irregular cell development in the absence of Plk4. Centrosome aberrations can lead to cancer.

F. Lampert and S. Westermann, "A Blueprint for Kinetochores–New Insight into the Molecular Mechanics of Cell Division", *Nature Reviews Molecular Cell Biology*, Vol. 12, 2011, pp. 407–412.

Review Notes: Kinetochore assembly is restricted to centrosomes which are chromosomal domains.

B.M. Lange and K. Gull, "A Molecular Marker for Centriole Maturation in the Mammalian Cell Cycle", *Journal of Cell Biology*, Vol. 130, No. 4, 1995, pp. 919–927.

Review Notes: The paper provides excellent description of centrioles. The acquisition of connexin makes a daughter centriole mature during mitosis. It is not known if there are any unique proteins in the centrioles.

S. Lawo, *et al.*, "Subdiffraction Imaging of Centrosomes Reveals Higher-Order Organizational Features of Pericentriolar Material", *Nature Cell Biology*, Vol. 14, No. 11, 2012, pp. 1148–1158. doi:10.1038/ncb2591

Review Notes: The centrosome is not simply a cloud of proteins. Instead it is structured layers of matter. The exact architecture of PCM is not known but it appears to have an intricate mode of assembly. The paper purports to present the higher-order PCM organization in regulating centrosome assembly. Kinetochores and centrioles have no membrane cover. The centrosome is like a vapor cover over the centriole. That is, the centrosome takes the place of the membrane. The PCM is largely featureless in EM imaging.

B. Leber, *et al.*, "Proteins Required for Centrosome Clustering in Cancer Cells", *Science Translational Medicine*, Vol. 2, No. 33, 2010, pp. 1–11.

Review Notes: Centrosomal clustering leads division and multiplication of cancer cells, but the mechanisms for centriole clustering are not completely understood.

K. Lee and K. Rhee, "Separase-Dependent Cleavage of Pericentrin B Is Necessary and Sufficient for Centriole Disengagement during Mitosis", *Cell*, Vol. 11, 2012, pp. 2476–2485.

Review Notes: Protein: Pericentrin B is necessary and sufficient for disengagement during mitosis.

M. Lee and V. Vasioukhin, "Cell Polarity and Cancer — Cell and Tissue Polarity as a Non-Canonical Tumor Suppressor", *Journal of Cell Science*, Vol. 121, 2008, pp. 1141–1150.

Review Notes: Regarding cell polarity: The apical and basal sections of a cell are separated by a physical barrier called the "apical junction complex" (AJC).

S. Leidel, *et al.*, "SAS-6 Defines a Protein Family Required for Centrosome Duplication in *C. elegans* and in Human Cells", *Nature Cell Biology*, Vol. 7, No. 2, 2005, pp. 115–125.

Review Notes: SAS-6 is recruited to the centrioles for daughter formation — to the mother first.

M.S. Levine, *et al.*, "Centrosome Amplification Is Sufficient to Promote Spontaneous Tumorigenesis in Mammals", *Developmental Cell*, Vol. 40, 2017, pp. 313–322.

Review Notes: Supernumerary centrioles promote tumorigenesis. Centrosome amplification results in aneuploidy. Overexpression of Plk4 promotes centrosome amplification.

T.C. Lin, A. Neuner, and E. Schiebel, "Targeting of γ-Tubulin Complexes to Microtubule Organizing Centers: Conservation and Divergence", *Trends in Cell Biology*, Vol. 25, No. 5, 2015, pp. 296–307.

Review Notes: The PCM is a highly structured consecutive layers of protein.

W.L. Lingle, *et al.*, "Centrosome Hypertrophy in Human Breast Tumors: Implications for Genomic Stability and Cell Polarity", *Proceedings of the National Academy of Sciences*, Vol. 95, 1998, pp. 2950–2955.

Review Notes: Numerous centrioles are seen in cancer tissues.

W.L. Lingle and J.L. Salisbury, "Altered Centrosome Structure Is Associated with Abnormal Mitoses in Human Breast Tumors", *American Journal of Pathology*, Vol. 155, No. 6, 1999, pp. 1941–1951.

Review Notes: Defects in chromosomes can lead to cancer.

J. Lončarek, P. Herget, and A. Khodjakov, "Centriole Reduplication during Prolonged Interphase Requires Procentriole Maturation Governed by Plk1", *Current Biology*, Vol. 20, 2010, pp. 1277–1282.

Review Notes: Multiple rounds of duplication is called "reduplication." Mechanisms for reduplication are poorly understood (2010). If the daughter centriole develops full length before separation this may lead to reduplication. Delayed disengagement promotes reduplication.

J. Loncarek and A. Khodjakov, "Ab ovo or de nous? Mechanisms of Centriole Duplication", *Molecules and Cells*, Vol. 27, No. 2, 2009, pp. 135–142.

Review Notes: In Latin "centrosome" means "central body." PCM is electron dense. Centrioles have the same shape in all mammals. Centrioles duplicate from pre-existing centrioles. During duplication, the mother centriole maintains the PCM. Once duplication begins the cartwheel protein occurs. The developed mother daughter centriole pair is called a "diplosome." Early in mitosis the diplosome breaks down. As with DNA overduplication is blocked. It is not clear only one daughter forms near the wall of the mother. There is a symmetry breaking mechanism — not clear what it is.

J. Lončarek, *et al.*, "Control of Daughter Centriole Formation by the Pericentriolar Material", *Nature Cell Biology*, Vol. 10, 2008, pp. 322–328.

Review Notes: Methods preventing centriole overduplication are poorly understood. Physical removal of the daughter induces a reduplication of the mother in the S-phase of cell duplication. Author proposes that the pericentriolar material cloud associated with the mother centriole restricts the number of daughters that can form simultaneously. It is for large, multiple daughters can form. As long as the daughter is attached to the mother, formation of additional daughters is prevented. Supernumerary centrioles cause geometric instability. The attachment of the daughter centriole to the wall of the mother inhibits the formation of additional daughters. Quote: "A single mother centriole is capable of repeatedly duplicating at least four times in the same cell cycle. However, continuous association of the daughter centriole with its mother prevents the mother centriole from reduplicating."

R. Mahen and A.R. Venkitaraman, "Pattern Formation in Centrosome Assembly", *Current Opinion in Cell Biology*, Vol. 24, No. 1, 2012, pp. 14–23.

Review Notes: The number of proteins on the centrosome may exceed 1000.

M.R. Mahjoub and T. Stearns, "Supernumerary Centrosomes Nucleate Extra Cilia and Compromise Primary Cilium Signaling", *Current Biology*, Vol. 22, No. 17, 2012, pp. 1628–1634.

Review Notes: Plk4 (aka SAK) overexpression induces formation of multiple centrioles.

M.S. Markov, "Magnetic Field Theory: A Review", *Electromagnetic Biology and Medicine*, Vol. 26, 2007, pp. 1–23.

Review Notes: Quote: Little research has been done on cellular magnetism. Magnetic fields can increase cell growth, as in wound healing.

W.F. Marshall, "Centriole Assembly: The Origin of Nin-ness", *Current Biology*, Vol. 17, No. 24, 2007a, pp. R1057–R1059.

Review Notes: Speculation about nine-ness: the number 9 arises due to the inner diameter of the centriole being fixed and thus 9 blades or 3 are needed along the circumference.

W.F. Marshall, "What is the Function of Centrioles?" *Journal of Cellular Biochemistry*, Vol. 100, No. 4, 2007b, pp. 916–922.

Review Notes: Centrioles form centrosomes and cilia.

W.F. Marshall and J.L Rosenbaum, "How Centrioles Work: Lessons from Green Yeast", *Current Opinion in Cell Biology*, Vol. 12, No. 1, 2000, pp. 119–125.

Review Notes: Centrioles seen over a century ago. Centrioles are needed to form the centrosome. Pre-existing centrioles are the templates for making new centrioles.

G. Marteil, *et al.*, "Over-Elongation of Centrioles in Cancer Promotes Centriole Amplification and Chromosome Missegregation", *Nature Communications*, Vol. 9, 2018, p. 1258.

Review Notes: Title problem is discussed. Long centrioles can lead to centriole fragmentation and tumorigenesis.

V. Marthiens, *et al.*, "Centrosome Amplification Causes Microcephaly", *Nature Cell Biology*, Vol. 15, 2013, pp. 731–742.

Review Notes: Multiple centrioles cause cancer.

K. Matsuo, *et al.*, "Kendrin Is a Novel Substrate for Separase Involved in the Licensing of Centriole Duplication", *Current Biology*, Vol. 22, 2012, pp. 915–921.

Review Notes: Authors state that kendrin regulates separase in centriole disengagement.

V. Mennella, *et al.*, "Amorphous No More: Subdiffraction View of the Pericentriolar Material Architecture", *Trends in Cell Biology*, Vol. 24, No. 3, 2014, pp. 188–197.

Review Notes: The centrosome is electron dense material. The centrosome increases in size at the onset of mitosis. This is called "centrosome maturation." Little is known about the pericentriolar material.

P. Meraldi and E.A. Nigg, "The Centrosome Cycle", *FEBS Letters*, Vol. 521, No. 1–3, 2002, pp. 9–13.

Review Notes: G1 → S-phase → G2 → Mitosis. Centrioles are duplicated in S-phase and they mature in G2 phase.

A. Meunier and N. Spassky, "Centriole Continuity: Out with the New, in with the Old", *Current Opinion in Cell Biology*, Vol. 38, 2016, pp. 60–67. doi:10.1016/j.ceb.2016.02.007

Review Notes: Centrioles govern cell reproduction.

T.C. Moyer, *et al.*, "Binding of STIL to Plk4 Activates Kinase Activity to Promote Centriole Assembly", *Journal of Cell Biology*, Vol. 209, No. 6, 2015, pp. 863–878.

Review Notes: Plk4 primes the binding of STIL to SAS6. Also, supernumerary centrosomes promote cancer.

E.A. Nigg, "Mitotic Kinases as Regulators of Cell Division and Its Checkpoints", *Nature Reviews Molecular Cell Biology*, Vol. 2, 2001, pp. 21–31.

Review Notes: DNA replication and centrosome duplication occur at the same time.

E.A. Nigg, L. Cajanek, and C. Arquint, "The Centrosome Duplication Cycle in Health and Disease", *FEBS Letters*, Vol. 588, 2014, pp. 2366–2372.

Review Notes: Centrosome aberrations have long been implicated in cancer. The process is not yet fully understood. This is a short review of the gene products involved.

E.A. Nigg and J.W. Raft, "Centrioles, Centrosomes, and Cilia in Health and Disease", *Cell*, Vol. 139, No. 4. 2009, pp. 663–678.

Review Notes: Centriole existence has been known for over 50 years but they are just now becoming understood. There is a connection between extra centrioles and tumor growth.

E.A. Nigg and T. Steams, "The Centrosome Cycle: Centriole Biogenesis, Duplication and Inherent Asymmetries", *Nature Cell Biology*, Vol. 13, No. 10, 2011, pp. 1154–1160.

Review Notes: Healthy centrioles duplicate only once per cell cycle. Aberrations in centrioles are associated with cancer. Excess Plk4 triggers the formation of supernumerary centrioles.

J.K. Pagan, *et al.*, "Degradation of Cep68 and PCNT Cleavage Mediate Cep215 Removal from the PCM to Allow Centriole Separation, Disengagement and Licensing", *Nature Cell Biology*, Vol. 17, 2015, pp. 31–43.

Review Notes: Tumor cells have excess centrioles.

M. Paintrand, *et al.*, "Centrosome Organization and Centriole Architecture: Their Sensitivity to Divalent Cations", *Journal of Structural Biology*, Vol. 108, No. 2, 1992, pp. 107–128.

Review Notes: More detailed description of the centrosome is lacking.

C.G. Pearson and M. Winey, "Plk3/SAK/ZYG-1 in the Regulation of Centriole Duplication", *F1000 Biology Reports*, Vol. 2, 2010, p. 58.

Review Notes: Defects in centrioles are found in cancer. The first sign of daughter centriole is the electron dense cartwheel (~25 nm in diameter) at the proximal end of the mother centriole?

N. Peel, *et al.*, "Overexpressing Centriole-Replication Proteins in Vivo Induces Centriole Overduplication and De Novo Formation", *Current Biology*, Vol. 17, 2007, pp. 834–843.

Review Notes: Abnormal centriole numbers produce cancer. Overexpression of SAK (Plk4) leads to centriole-like structures.

L. Pelletier, *et al.*, "Centriole Assembly in Caenorhabditis elegans", *Nature*, Vol. 444, 2006, pp. 619–623.

Review Notes: The mechanisms underlying daughter centriole formation remain unclear. The formation is dependent on SAS-5 and SAS-6 proteins with the spoke end for the MTs dependent on SAS-4.

M. Piel, *et al.*, "The Respective Contribution of the Mother and Daughter Centrioles to Centrosome Activity and Behavior in Vertebrate Cells", *Journal of Cell Biology*, Vol. 149, No. 2, 2000, pp. 317–330.

Review Notes: Centrioles maintain their perpendicularity through mitosis. Centrin is a small protein concentrating within the distal end of the centriole. During mitosis there are centriole pairs on opposite sides of the nucleus and the daughter centriole oscillates and the mother remains in place while the daughter migrates away.

M. Piel, *et al.*, "Centrosome-Dependent Exit of Cytokinesis in Animal Cells", *Science*, Vol. 291, 2001, pp. 1550–1553.

Review Notes: In movement about the nucleus the centriole speed was as fast as 10 nm/min.

S.L. Prosser and L. Pelletier, "Centrosome Biology: The Ins and Outs of Centrosome Assembly", *Current Biology*, Vol. 25, 2015, pp. R656–R659.

Review Notes: Upon mitosis the centrosome increases dramatically in size. The centrosome is not a featureless cloud but instead an organized structure.

J.W. Raff and R. Basto, "Centrosome Amplification and Cancer", *Developmental Cell*, Vol. 40, 2017, pp. 217–218.

Review Notes: Enlarged centrosome (i.e., supernumerary centrioles) leads to cancer.

J. Reina and C. Gonzalez, "When Fate Follows Age: Unequal Centrosomes in Asymmetric Cell Division", *Philosophical Transactions of the Royal Society of London. Series B, Biological Sciences*, Vol. 369, 2014, p. 13.

Review Notes: The paper is focused on centrosome asymmetry in cells that divide by "self-renewing asymmetric division" (SRAD).

C.L. Rieder, S. Faruki, and A. Khodjakov, "The Centrosome in Vertebrates: More than a Microtubule-Organizing Center", *Trends in Cell Biology*, Vol. 11, No. 10, 2001, pp. 413–419.

Review Notes: The centrosome, in addition to being an MTOC, provides structural foundation.

A. Rodrigues-Martins, *et al.*, "Revisiting the Role of the Mother Centriole in Centriole Biogenesis", *Science*, Vol. 315, No. 5827, 2007, pp. 1046–1050.

Review Notes: SAK aka Plk4 is a necessary regulator of centriole formation but overexpression leads to centriole amplification. SAK itself may be regulated by the cell itself. "The mother centriole is not a template but a platform for construction of the daughter centriole."

A. Rodrigues-Martins, *et al.*, "From Centriole Biogenesis to Cellular Function: Centrioles Are Essential for Cell Division at Critical Developmental States", *Cell Cycle*, Vol. 7, No. 1, 2008, pp. 11–16.

Review Notes: Cancers are associated with abnormalities in centrosome structure. Centriole biogenesis involves SAK/Plk4. SAS-6 organizes the nine pericentriolar units. The daughter centriole arising on the side of the mother centriole is called "canonical assembly." SAK/Plk4 is an upstream regulator of canonical centriole formation. An overexpression of SAK/Plk4 leads to formation of multiple centrosomes.

M.D. Rose, S. Biggins, and L.L. Satterwhite, "Unravelling the Tangled Web at the Microtubule-Organizing Center", *Current Opinion in Cell Biology*, Vol. 5, No. 1, 1993, pp. 105–115.

Review Notes: The MTOC produces tubulin and a reeling in and out of the MTs.

J.L. Salisbury, *et al.*, "Centrosomes and Cancer", *Biology of the Cell*, Vol. 91, No. 6, 1999, pp. 451–460.

Review Notes: Centrosome disruption can lead to disruption in cell tissue. Tumor cells have supernumerary centrioles and multiple centrosomes.

S. Santaguida and A. Musacchio, "The Life and Miracles of Kinetochores", *EMBO Journal*, Vol. 28, 2009, pp. 2511–2531.

Review Notes: Kinetochores have four functions:

(1) Connect to centromeric chromatin

(2) Connect to MTs

(3) Monitor the state of the kinetochores/MT attachment

(4) Correct improper attachments

W. Saunders, "Centrosome Amplification and Spindle Multipolarity in Cancer Cells", *Seminars in Cancer Biology*, Vol. 15, 2005, pp. 25–32.

Review Notes: Centrosomal defects lead to tumorigenesis. The paper reviews how centrosomal defects lead to CIN. Centrosomal defects occur early in premalignant tissue. Loss of p53 does not promote centrosome amplification. DNA damage occurs with changes in centrosome number. Centrosome amplification in the absence of cytokinesis is rare.

P.W. Schafer, "Centrioles of a Human Cancer: Intracellular Order and Intracellular Disorder", *Science*, Vol. 164, 1969, pp. 1300–1303.

Review Notes: An initial foundation paper.

P.W. Schafer, "Centrioles: Intercellular Order in Normal and Malignant Cells", *Journal of Thoracic and Cardiovascular Surgery*, Vol. 63, No. 3, 1972, pp. 472–477.

Review Notes: This is a classical Paul Schafer paper.

P.W. Schafer and J.A. Chandler, "Electron Probe X-Ray Microanalysis of a Normal Centriole", *Science*, Vol. 170, 1970, pp. 1204–1205.

Review Notes: An initial foundation paper.

L. Schöckel, *et al.*, "Cleavage of Cohesion Rings Coordinates the Separation of Centrioles and Chromatids", *Nature Cell Biology*, Vol. 13, No. 8, 2011, pp. 966–972.

Review Notes: Supernumerary centrioles contribute to the development of cancer.

S. Sen, S. Subramanial, and D.E. Discher, "Indentation and Adhesive Probing of a Cell Membrane with AFM: Theoretical Model and Experiments", *Biophysical Journal*, Vol. 89, 2005, pp. 3203–3213.

Review Notes: The paper presents a modeling of cell membrane response to induction by an atomic force microscopy (AFM) probe.

J.E. Sillibourne, *et al.*, "Autophosphorylation of Polo-like Kinase 4 and Its Role in Centriole Duplication", *Molecular Biology of the Cell*, Vol. 21, No. 4, 2010, pp. 547–561.

Review Notes: Plk4 is a regulator of centriole duplication and has a single polo box.

L.K. Slevin, *et al.*, "The Structure of the Plk4 Polo Box Reveals Two Tandem Polo Boxes Required for Centriole Duplication", *Structure*, Vol. 20, No. 11, 2012, pp. 1905–1917.

Review Notes: Proteins required for centriole formation: Plk4/SAK, Asterless/Cep152, SAS-6, SAS-6/Ana2/STIL, Cep135, and SAS-9/CPAP. Asterless (Asl) is a scaffolding protein which recruits Plk4 to the site on the daughter for assembly. SAS-6 forms the ninefold symmetric cartwheel scaffold for the daughter centriole base.

G. Sluder, "One to Only Two: A Short History of the Centrosome and Its Duplication", *Philosophical Transactions of the Royal Society of London. Series B, Biological Sciences*, Vol. 369, 2014, p. 8.

Review Notes: Paper reviews various theories as to why only one procentriole occurs with a mother centriole. The author states that the answer is still unknown.

G. Sluder, F.J. Miller, and C.L. Rieder, "The Reproduction of Centrosomes: Nuclear versus Cytoplasmic Controls", *Journal of Cell Biology*, Vol. 103, 1986, pp. 1873–1881.

Review Notes: Centriole duplication appears to be driven by elements of the cytoplasm instead of by the nucleus.

G. Sluder and J.L. Nordberg, "The Good, the Bad and The Ugly: The Practical Consequences of Centrosome Amplification", *Current Opinion in Cell Biology*, Vol. 16, No. 1, 2004, pp. 49–54.

Review Notes: The presence of more than two centrosomes during mitosis is characteristic of cancer. There is a survival of the fittest of cancer cells (via *clustering* of centrosomes?).

M.H. Song, *et al.*, "Centrioles: Some Self-Assembly Required", *Current Opinions in Cell Biology*, Vol. 20, 2008, pp. 688–693.

Review Notes: The regulators for centriole duplication are the same as those for chromosome separation.

K.F. Sonnen, *et al.*, "3D-Structured Illumination Microscopy Provides Novel Insight into Architecture of Human Centrosomes", *Biology Open*, Vol. 1, 2012, pp. 965–976.

Review Notes: The daughter centriole arises from a dot of Plk4 on the proximal end of the mother. It is not clear what mechanisms lead to the concentration of Plk4 or whether the choice of the site is random or somehow regulated. How does Plk4 attract SAS-6 to form a cartwheel structure? At the proximal end of the centriole SAS6, Plk4, and STIL localize. Plk4 forms a dot before SAS6. Plk4 is an early marker for new centriole formation. "...definitive information about the spatial organization of various components remains scarce." PCM undergoes a drastic expansion at the beginning of mitosis. Plk4 is an early marker (the first) for new centrioles. It falls onto a ring stained by Cep152. Also Plk4 colocalizes with SAS-6. The site for MT nucleation is determined on the mother centriole. It is not clear whether the nucleation site is chosen at random or is somehow predetermined.

K.F. Sonnen, *et al.*, "Human Cep192 and Cep152 Cooperate in Plk4 Recruitment and Centriole Duplication", *Journal of Cell Science*, Vol. 126, 2014, pp. 3223–3233.

Review Notes: Plk4 is a regulator of centriole duplication and it depends upon Asterless. Cep192 recruits Cep152 and Plk4. Plk4 attraction to centrioles depends upon electrostatic interaction.

P. Strnad, *et al.*, "Regulated HsSAS-6 Levels Ensure Formation of a Single Procentriole per Centriole during Centrosome Duplication Cycle", *Developmental Cell*, Vol. 13, No. 2, 2007, pp. 203–213.

Review Notes: To form a new centriole we need centrin-3, Plk4, and SAS-6.

P. Strnad and P. Gönczy, "Mechanisms of Procentriole Formation", *Trends in Cell Biology*, Vol. 18, No. 8, 2008, pp. 389–396.

Review Notes: Centrosome determines cell polarity. Centrioles and PCM.

This is a primary site of MT nucleation — aka MTOC (microtubule organizing center). At the onset of procentriolar assembly, the cartwheel is the first ninefold structure to appear. Then there is the appearance of MTs at the tips of the spokes. The cartwheel is the structural scaffold onto which the centriole MTs assemble. It is speculated that SAS-6 dictates the ninefold symmetry of the centriole.

A. Suzuki, *et al.*, "How the Kinetochore Couples Microtubule Force and Centromere Stretch to Move Chromosomes", *Nature Cell Biology*, Vol. 18, No. 4, 2016, pp. 382–392. doi:10.1038/ncb3323

Review Notes: Kinetochores pull the chromosome apart.

C.J.C. Tang, *et al.*, "The Human Microcephaly Protein STIL Interacts with CPAP and Is Required for Procentrioles Formation", *EMBP Journal*, Vol. 30, 2011, pp. 4790–4804.

Review Notes: This paper focuses on STIL and CPAP for procentriole formation. STIL makes an asymmetric location on the daughter cell — required for new centriole formation. "Overexpression of STIL induces a formation of multiple procentrioles around the parent centriole." Conversely, too little STIL prevents normal centriole development. CPAP induces centriole elongation. Quote: "... our results indicate that the two MCPH-associated proteins STIL and CPAP interact with each other and are required for procentriole formation, implying a central role of centriole biogenesis in MCPH." MCPH is "microcephaly."

A.M. Tassin and M. Bornens, "Centrosome Structure and Microtubule Nucleation in Animal Cells", *Biology of the Cell*, Vol. 91, No. 4–5, 1999, pp. 343–354.

Review Notes: Centriole is the order, organizer, and power within the centrosome. The distal end of the centriole is positive.

M.F. Tsou and T. Stearns, "Controlling Centrosome Number: Licenses and Blocks", *Current Opinion* in *Cell Biology*, Vol. 18, No. 1. 2006a, pp. 74–78.

Review Notes: If the centrioles remain intact duplication is prevented.

M.F. Tsou and T. Stearns, "Mechanism Limiting Centrosome Duplication to Once Per Cell Cycle", *Nature*, Vol. 442, No. 7105, 2006b, pp. 847–951.

Review Notes: Human cancers exhibit centrosome amplification.

M.F.B. Tsou, *et al.*, "Polo Kinase and Separase Regulate the Mitotic Licensing of Centriole Duplication in Human Cells", *Developmental Cell*, Vol. 17, 2009, pp. 344–354.

Review Notes: A new daughter occurs during the S-phase cycle of cell duplication. The physical engagement between a mother and daughter centriole creates a block to further centriole assembly.

Centriole disengagement is poorly understood. The orthogonal configuration of centrioles prevents overduplication of centrioles.

Y. Uetake, *et al.*, "Cell Cycle Progression and De Novo Centriole Assembly After Centrosomal Removal in Untransformed Human Cells", *Journal of Cell Biology*, Vol. 176, 2007, pp. 173–182.

Review Notes: Centrioles can assemble de nova after centrosome removal.

B.C.M. van de Weerdt and R.H. Medema, "Polo-like Kinases: A Team in Control of the Division", *Cell* Cycle, Vol. 5, No. 8, 2006, pp. 853–864.

Review Notes: Plk1 promotes mitotic entry and is involved in chromosome maturation and separation. Plk4 is involved in centriole replication.

H. Varmark, *et al.*, "Asterless Is a Centriolar Protein Required for Centrosome Function and Embryo Development in Drosophila", *Current Biology*, Vol. 17, No. 20, 2007, pp. 1735–1745.

Review Notes: Asterless (Asl) is needed for stable centrosome development.

B. Vitre, *et al.*, "Chronic Centrosome Amplification without Tumorigenesis", *Proceedings of the National Academy of Sciences United States of America*, Vol. 112, November 2015, pp. E6321–E6330. doi: 10.1073/pnas.1519388112

Review Notes: Under absence of p. 53 and overexpression of Plk4 supernumerary centrioles occur without producing a tumor.

I.A. Vorobjev and Y.S. Chentsov, "Centrioles in the Cell Cycle. 1. Epithelial Cells", *Journal of Cell Biology*, Vol. 93, No. 3, 1982, pp. 938–949.

Review Notes: Mother cells have left the mitotic cycle. Replication of centrioles does not occur.

J. Vulprecht, *et al.*, "STIL Is Required for Centriole Duplication in Human Cells", *Journal of Cell Science*, Vol. 125, 2012, pp. 1353–1362.

Review Notes: STIL is a centrosomal protein whose overexpression results in excess centriole formation. An extra centrosome number leads to cancer. Too little STIL leads to a loss of centriole formation. STIL recruits SAS-6.

X. Wan, *et al.*, "Protein Architecture of the Human Kinetochore Microtubule Attachment Site", *Cell*, Vol. 137, 2009, pp. 672–684.

Review Notes: Robust spindle MT plus-end attachment is "end-on," and MTs bound to kinetochores are known as kinetochore microtubules (kMTs).

C. Wiese and Y. Zheng, "A New Function for the Gamma-Tubulin Ring Complex as a Microtubule Minus-End Cap", *Nature Cell Biology*, Vol. 2, 2000, pp. 358–364.

Review Notes: The ring has a lock washer shape.

M. Winey and E. O'Toole, "Centriole Structure", *Philosophical Transactions of the Royal Society of London. Series B, Biological Sciences*, Vol. 369, 2014, p. 14.

Review Notes: Centrioles are among the largest protein-based structures found in cell types. They are ~250 nm in diameter and 500 nm long. Paper is a review of ultrastructural observations about centrioles and associated structures. The paper largely repeats what is already known about centrioles from earlier papers.

J.B. Woodruff, O. Wueseke, and A.A. Hyman, "Pericentriolar Material Structure and Dynamics", *Philosophical Transactions of the Royal Society of London. Series B, Biological Sciences*, Vol. 369, 2014, p. 10.

Review Notes: PCM concentrates tubulin and serves as the primary organizing center for MTs. The PCM has no membrane cover. Instead the proteins contribute the separation. PCM appears as an amorphous mass surrounding the centrioles. PCM proteins can self-assemble. PCM forms toroids about the centrioles.

W.J. Wong, *et al.*, "The Conversion of Centrioles to Centrosomes: Essential Coupling of Duplication with Segregation", *Journal of Cell Biology*, Vol. 183, 2011, pp. 727–739.

Review Notes: Centrioles cannot duplicate without PCM. "They acquire PCM by becoming modified with a Plk1 dependence early in mitosis."

J. Xiao and T. Ha, "Flipping Nanoscopy on Its Head", *Science*, Vol. 355, 2017, pp. 582–584.

Review Notes: This is a preview article for the article by Balzarotti, *et al.* and other articles [See 680]. The article states that with the naked eye the smallest object we can see is approximately the size of the human hair. With a magnifying glass we can see 100 times smaller; with light microscopy we can see ~300 times smaller (~250 nm). With fluorescence nanoscopy we can see structures 20–30 nm across. However, most proteins are less than 5 nm across. MINFLUX.

D. Zyss and F. Gergely, "Centrosome Function in Cancer: Guilty or Innocent?" *Trends in Cell Biology*, Vol. 19, No. 7, 2009, pp. 334–346.

Review Notes: Cancer cells have a wide range of centrosome abnormalities. Centrosome dysfunction results in cancer.

2.5. First Author Alphabetical Listing of References

F.G. Agrican, E. Schiebel, and B.R. Mardin, "Separate to Operate: Control of Centrosome Positioning and Separation", *Philosophical Transactions of the Royal Society of London. Series B, Biological Sciences*, Vol. 369, 2014, p. 10.

B. Alberts, *et al.*, *Molecular Biology of the Cell*, 3rd Edition, Garland Publishing, New York, 1994.

C. Arquint, *et al.*, "Cell-Cycle-Regulated Expression of STIL Controls Centriole Number in Human Cells", *Journal of Cell Science*, Vol. 125, 2012, pp. 1342–1352.

J. Azimzadeh and M. Bornens, "Structure and Duplication of the Centrosome", *Journal of Cell Science*, Vol. 120, 2007, pp. 2139–2142.

J. Azimzadeh and W.F. Marshall, "Building the Centriole", *Current Biology*, Vol. 20, No. 18, 2010, pp. R816–R825.

R. Bahtz, *et al.*, "GCP6 Is a Substrate of Plk4 and Required for Centriole Duplication", *Journal of Cell Science*, Vol. 125, 2011, pp. 486–496.

F. Balzarotti, *et al.*, "Nanometer Resolution Imaging and Tracking of Fluorescent Molecules with Minimal Photon Fluxes", *Science*, Vol. 355, 2017, pp. 606–612.

R. Basto, *et al.*, "Flies without Centrioles", *Cell*, Vol. 125, No. 7, 2006, pp. 1375–1386.

M. Bettencourt-Dias and D.M. Glover, "Centrosome Biogenesis and Function: Centrosomics Brings New Understanding", *Nature Reviews Molecular Cell Biology*, Vol. 8, 2007, pp. 451–463.

M. Bettencourt-Dias, *et al.*, "Centrosomes and Cilia in Human Disease", *Trends in Genetics*, Vol. 27, 2011, pp. 307–315.

S. Blachon, *et al.*, "Drosophila Asterless and Vertebrate CEP152 Are Orthologs Essential for Centriole Duplication", *Genetics*, Vol. 180, No. 4, 2008, pp. 2081–2094.

S.P. Blagden and D.M. Glover, "Polar Expeditions — Provisioning the Centrosome for Mitosis", *Nature Cell Biology*, Vol. 5, No. 6, 2003, pp. 505–511.

M. Bornens, "Centrosome Composition and Microtubule Anchoring Mechanisms", *Current Opinion in Cell Biology*, Vol. 14, No. 1, 2002, pp. 25–34.

M. Bornens, "Organelle Positioning and Cell Polarity", *Nature Reviews Molecular Cell Biology*, Vol. 9, 2008, pp. 874–886.

M. Bornens, "The Centrosome in Cells and Organisms", *Science*, Vol. 335, 2012, pp. 422–426.

M. Bornens and P. Gönczy, "Centrosomes Back in the Limelight", *Philosophical Transactions of the Royal Society of London. Series B, Biological Sciences*, Vol. 369, 2014, p. 15.

B.R. Brinkley, "Managing the Centrosome Numbers Game: From Chaos to Stability in Cancer Cell Division", *Trends in Cell Biology*, Vol. 11, No. 1, 2001, pp. 18–21.

D.A. Brito, S.M. Gouveia, and M. Bettencourt-Dias, "Deconstructing the Centriole: Structure and Number Control", *Current Opinion in Cell Biology*, Vol. 24, No. 1, 2012, pp. 4–13.

G.J. Brouhard and L.M. Rice, "The Contribution of αβ-Tubulin Curvature to Microtubule Dynamics", *Journal of Cell Biology*, Vol. 207, No. 3, 2014, pp. 323–334.

C.W. Brownlee, *et al.*, "The Protein Phosphatase 2A Regulatory Submit Twins Stabilizes Plk4 to Induce Centriole Amplification", *Journal of Cell Biology*, Vol. 195, No. 2, 2011, pp. 231–243.

C.W. Brownlee and G.C. Rogers, "Show Me Your License, Please: Deregulation of Centriole Duplication Mechanisms that Promote Amplification", *Cellular and Molecular Life Sciences*, Vol. 70, No. 6, 2013, pp. 1021–1034.

Z. Carvalho-Santos, *et al.*, "Stepwise Evolution of the Centriole-Assembly Pathway", *Journal of Cell Science*, Vol. 123, 2010, pp. 1414–1426.

Z. Carvalho-Santos, *et al.*, "Tracing the Origins of Centrioles, Cilia and Flagella", *Journal of Cell Biology*, Vol. 194, No. 2, 2011, pp. 165–175.

J.Y. Chan, "A Clinical Overview of Centrosome Amplification in Human Cancers", *International Journal of Biological Sciences*, Vol. 7, No. 8, 2011, pp. 1122–1144.

P. Chang, *et al.*, "ε-Tubulin Is Required for Centriole Duplication and Microtubule Organization", *Nature Cell Biology*, Vol. 5, 2003, pp. 71–76.

C.W. Chang, *et al.*, "CEP295 Interacts with Microtubules and Is Required for Centriole Elongation", *Journal of Cell Science*, Vol. 129, 2016, pp. 2501–2513.

P.L. Chavali, M. Putz, and F. Gergely, "Small Organelle, Big Responsibility: The Role of Centrosomes in Development and Disease", *Philosophical Transactions of the Royal Society of London. Series B, Biological Sciences*, Vol. 369, 2014, p. 12.

I.M. Cheeseman and A. Desai, "Molecular Architecture of the Kinetochore-Microtubule Interface", *Nature Reviews Molecular Cell Biology*, Vol. 19, 2008, pp. 33–46.

F. Chen, P.W. Tillberg, and E.S. Boyden, "Expansion Microscopy", *Science*, Vol. 347, No. 6221, 2015, pp. 543–548.

K. Cheng and C. Zou, "Physics Models of Centriole Replication", *Medical Hypotheses*, Vol. 67, 2006a, pp. 572–577.

K. Cheng and C. Zou, "Theoretical Models of Cytokinesis and Mitosis (without Astens)", *American Journal of Plant Physiology*, Vol. 1, No. 1, 2006b, pp. 21–27.

O. Cizmecioglu, *et al.*, "CEP152 Acts as a Scaffold for Recruitment of Plk4 and CPAP to the Centrosome", *Journal of Cell Biology*, Vol. 191, No. 4, 2010, pp. 731–739.

P.T. Conduit, "Microtubule Organization: A Complex Relation", *Journal of Cell Biology (JCB)*, Vol. 213, 2016, pp. 809–812.

I. Cunha-Ferreira, *et al.*, "The SCF/Slimb Ubiquitin Ligase Limits Centrosome Amplification through Degradation of SAK/PLK4", *Current Biology*, Vol. 19, 2009, pp. 43–49.

I. Cunha-Ferreira, *et al.*, "Regulation of Autophosphorylation Controls PLK4 Self-Destruction and Centriole Number", *Current Biology*, Vol. 23, No. 22, 2013, pp. 2245–2254.

I. Cunha-Ferreira, I. Bento, and M. Bettencourt-Dias, "From Zero to Many: Control of Centriole Number in Development and Disease", *Traffic*, Vol. 10, No. 5, 2009, pp. 482–498.

A. Dammermann, *et al.*, "SAS-4 Is Recruited to a Dynamic Structure in Newly Forming Centrioles that Is Stabilized by the γ-Tubulin-Mediated Addition of Centriolar Microtubules", *Journal of Cell Biology*, Vol. 180, No. 4, 2008, pp. 771–785.

A. Dart, "Extra! Extra! Read All about it", *Nature Reviews Cancer*, Vol. 17, February 2017, p. 143. doi:10.1038/nrc2017.12

M. Delattre, C. Canard, and P. Gönczy, "Sequential Protein Recruitment in *C. elegans* Centriole Formation", *Current Biology*, Vol. 16, 2006, pp. 1844–1849.

M. Delattre and P. Gönczy, "The Arithmetic of Centrosome Biogenesis", *Journal of Cell Science*, Vol. 117, 2004, pp. 1619–1630.

S.J. Doxsey, "The Centrosome — A Tiny Organelle with Big Potential", *Nature Genetics*, Vol. 20, No. 2, 1998, pp. 104–106.

S.J. Doxsey, "Centrosomes as Command Centres for Cellular Control", *Nature Cell Biology*, Vol. 3, No. 5, 2001a, pp. E105–E108.

S.J. Doxsey, "Re-evaluating Centrosome Function", *Nature Reviews Molecular Cell Biology*, Vol. 2, No. 9, 2001b, pp. 688–698.

S. Doxsey, D. McCollum, and W. Theurkauf, "Centrosomes in Cellular Regulation", *Annual Review of Cell and Developmental Biology*, Vol. 21, 2005, pp. 411–434.

S. Duensing, "A Tentative Classification of Centrosome Abnormalities in Cancer", *Cell Biology International*, Vol. 29, 2005, pp. 352–359.

A. Duensing, *et al.*, "Centriole Overduplication through the Concurrent Formation of Multiple Daughter Centrioles at Single Maternal Templates", *Oncogene*, Vol. 26, No. 43, 2007, pp. 6280–6288.

N.S. Dzhindzhev, *et al.*, "Asterless Is a Scaffold for the Onset of Centriole Assembly", *Nature*, Vol. 457, 2010, pp. 614–718.

F. Eckerdt, *et al.*, "Identification of a Polo-like Kinase 4-Dependent Pathway for De Novo Centriole Formation", *Current Biology*, Vol. 21, No. 5, 2011, pp. 428–432.

M.W. Elting, *et al.*, "Force on Spindle Microtubule Minus Ends Moves Chromosomes", *Journal of Cell Biology*, Vol. 206, No. 2, 2014, pp. 245–256.

E.N. Firat-Karalar and T. Stearns, "The Centriole Duplication Cycle", *Philosophical Transactions of the Royal Society of London. Series B, Biological Sciences*, Vol. 369, 2014, p. 17.

A.M. Fry, "Solving the Centriole Disengagement Puzzle", *Nature Cell Biology*, Vol. 17, No. 1, 2015, pp. 3–5.

J. Fu and D.M. Glover, "Structured Illumination of the Interface between Centriole and Pericentriolar Material", *Open Biology*, Vol. 2, 2012, pp. 1–30.

K. Fukasawa, "Introduction", *Oncogene*, Vol. 21, 2002, pp. 6140–6145.

K. Fukasawa, "Oncogenes and Tumour Suppressors Take on Centrosomes", *Nature Reviews Cancer*, Vol. 7, 2007, pp. 911–924.

B.J. Galleta, *et al.*, "Asterless Is Required for Centriole Length Control and Sperm Development", *Journal of Cell Biology*, Vol. 213, No. 4, 2016, pp. 435–474.

N.J. Ganem, S.A. Godinho, and D. Pellman, "A Mechanism Linking Extra Centrosomes to Chromosomal Instability", *Nature*, Vol. 460, No. 7252, 2009, pp. 278–282.

D. Gisselsson, *et al.*, "Centrosomal Abnormalities, Multipolar Mitoses, and Chromosomal Instability in Head and Neck Tumours with Dysfunctional Telomeres", *British Journal of Cancer*, Vol. 87, 2002, pp. 202–207.

S.A. Godinho, M. Kwon, and D. Pellman, "Centrosomes and Cancer: How Cancer Cells Divide with too Many Centrosomes", *Cancer Metastasia Review*, Vol. 28, 2009, pp. 85–98.

P. Gönczy, "Towards a Molecular Architecture of Centriole Assembly", *Nature Reviews Molecular Cell Biology*, Vol. 13, 2012, pp. 425–435.

P. Gönczy, "Centrosomes and Cancer: Revisiting a Long-Standing Relationship", *Nature Reviews Cancer*, Vol. 15, 2015, pp. 639–642.

M. Gottardo, G. Callaini, and M.G. Riparbelli, "Procentriole Assembly without Centriole Disengagement — A Paradox of Male Gametogenesis", *Journal of Cell Science*, Vol. 127, 2014, pp. 3434–3439.

P. Guichard, *et al.*, "Procentriole Assembly Revealed by Cryo-Electron Tomography", *The EMBO Journal*, Vol. 29, No. 9, 2010, pp. 1565–1572. doi:10.1038/emboj.2010,45

S.R. Hameroff, "A New Theory of the Origin of Cancer: Quantum Coherent Entanglement, Centrioles, Mitosis, and Differentation", *Biosystems*, Vol. 77, No. 1–3, 2004, pp. 119–136.

M. Hilbert, *et al.*, "SAS-6 Engineering Reveals Interdependence between Cartwheel and Microtubules in Determining Centriole Architecture", *Nature Cell Biology*, Vol. 18, No. 4, 2016, pp. 393–403. doi:10.1038/ncd3329

M. Hiraki, "Bld10p Constitutes the Cartwheel-Spoke Tip and Stabilizes the 9-Fold Symmetry of the Centriole", *Current Biology*, Vol. 17, No. 20, pp. 1776–1783.

M. Hirono, "Cartwheel Assembly", *Philosophical Transactions of the Royal Society of London. Series B, Biological Sciences*, Vol. 369, 2014, p. 16.

A.J. Holland, *et al.*, "The Autoregulated Instability of Polo-like Kinase 4 Limits Centrosome Duplication to Once Per Cycle", *Genes and Development*, Vol. 26, No. 24, 2012, pp. 2684–2689.

A.J. Holland, *et al.*, "Polo-like Kinase 4 Kinase Activity Limits Centrosome Overduplication by Autoregulating Its Own Stability", *Journal of Cell Biology*, Vol. 188. No. 2, 2010, pp. 191–198.

S.A. Hormeño, *et al.*, "Single Centrosome Manipulation Reveals Its Electric Charge and Associated Dynamic Structure", *Biophysical Journal*, Vol. 97, No. 4, 2009, pp. 1022–1030.

R.L. Huston, "Mechanics of Centriole Microtubules", *Advances in Biosciences and Biotechnology*, Vol. 7, 2016, pp. 266–277.

R. Ibrahim, *et al.*, "Electron Tomography Study of Isolated Human Centrioles", *Microscopy Research and Technique*, Vol. 72, No. 1, 2009, pp. 42–48.

A. Jord, *et al.*, "Calibrated Mitotic Oscillator Drives Motile Ciliogenesis", *Science*, Vol. 358, 2017, pp. 803–806.

C. Kasbek, *et al.*, "Preventing the Degradation of Mps1 at Centrosomes Is Sufficient to Cause Centrosome Reduplication in Human Cells", *Molecular Biology of the Cell*, Vol. 18, 2007, pp. 4457–4469.

K. Kawamura, *et al.*, "Induction of Centrosome Amplification and Chromosome Instability in Human Bladder Cancer Cells by p. 53 Mutation and Cyclin E Overexpression", *Cancer Research*, Vol. 64, 2005, pp. 4800–4809.

D. Keller, *et al.*, "Mechanisms of HsSAS-6 Assembly Promoting Centriole Formation in Human Cells", *Journal of Cell Biology*, Vol. 204, No. 5, 2014, pp. 697–712.

A. Khodjakov and C.L. Rieder, "Centrosomes Enhance the Fidelity of Cytokinesis in Vertebrates and Are Required for Cell Cycle Progression", *Journal of Cell Biology*, Vol. 153, No. 1, 2001, pp. 237–242.

M. Kirkham, *et al.*, "SAS-4 Is a *C. elegans* Centriolar Protein that Controls Centrosome Size", *Cell*, Vol. 112, 2003, pp. 575–587.

D. Kitagawa, *et al.*, "Structural Basis of the 9-Fold Symmetry of Centrioles", *Cell*, Vol. 144, No. 3, 2011, pp. 364–375.

J.E. Klebba, *et al.*, "Polo-like Kinase 4 Autodestructs by Generating Its Slimb-Binding Phosphodegron", *Current Biology*, Vol. 23, 2013, pp. 2255–2261.

J.E. Klebba, *et al.*, "Autoinhibition and Relief Mechanism for Polo-like Kinase 4", *Proceedings of the Academy of Sciences United States of America*, Vol. 112, No. 7, 2015a, pp. E657–E666.

J.E. Klebba, *et al.*, "Two Polo-like kinase Binding Domains in Asterless Perform Distinct Roles in Regulating Kinase Stability", *Journal of Cell Biology*, Vol. 208, No. 4, 2015b, pp. 401–414.

J. Kley-Sohn, *et al.*, "Plk4-Induced Centriole Biogenesis in Human Cells", *Developmental Cell*, Vol. 13, No. 2, 2007, pp. 190–202.

T. Kobayashi and B.D. Bynlacht, "Regulating the Transition from Centriole to Basal Body", *Journal of Cell Biology*, Vol. 193, No. 3, 2011, pp. 435–444.

G. Kohlmaier, *et al.*, "Overly Long Centrioles and Defective Cell Division upon Excess of the SAS-4-Related Protein CPAP", *Current Biology*, Vol. 19, 2009, pp. 1012–1018.

D. Kong, *et al.*, "Centriole Maturation Requires Regulated Plk1 Activity during Two Consecutive Cell Cycles", *Journal of Cell Biology*, Vol. 206, 2014, pp. 855–865.

N. Korzeniewski, *et al.*, "Cullin 1 Functions as a Centrosomal Suppressor of Centriole Multiplication by Regulating Polo-like Kinase 4 Protein Levels", *Cancer Research*, Vol. 69, No. 16, 2009, pp. 6668–6675.

N. Korzeniewski, *et al.*, "Daughter Centriole Elongations Is Controlled by Proteolysis", *Molecular Biology of the Cell*, Vol. 21, 2010, pp. 2942–2951.

N. Korzeniewski, M. Hohenfellner, and S. Duensing, "CANDI Promotes PLK4-Mediated Centriole Overduplication and Is Frequently Disrupted in Prostate Cancer", *Neoplasia*, Vol. 14, No. 9, 2012, pp. 799–806.

A. Kramer, K. Neben, and A.D. Ho, "Centrosome Replication, Genomic Instability and Cancer", *Leukemia*, Vol. 16, No. 5, 2002, pp. 767–775.

M. Kwon, "Using Cell Culture Models of Centrosome Amplification to Study Centrosome Clustering in Cancer", *Methods in Molecular Biology*, Vol. 1413, 2016, pp. 367–392.

B.G. Lambrus, *et al.*, "p. 53 Protects Against Genome Instability Following Centriole Duplication Failure", *Journal of Cell Biology*, Vol. 210, No. 1, 2015, pp. 63–77.

F. Lampert and S. Westermann, "A Blueprint for Kinetochores — New Insight into the Molecular Mechanics of Cell Division", *Nature Reviews Molecular Cell Biology*, Vol. 12, 2011, pp. 407–412.

B.M. Lange and K. Gull, "A Molecular Marker for Centriole Maturation in the Mammalian Cell Cycle", *Journal of Cell Biology*, Vol. 130, No. 4, 1995, pp. 919–927.

S. Lawo, *et al.*, "Subdiffraction Imaging of Centrosomes Reveals Higher-Order Organizational Features of Pericentriolar Material", *Nature Cell Biology*, Vol. 14, No. 11, 2012, pp. 1148–1158.

B. Leber, *et al.*, "Proteins Required for Centrosome Clustering in Cancer Cells", *Science Translational Medicine*, Vol. 2, No. 33, 2010, pp. 1–11.

K. Lee and K. Rhee, "Separase-Dependent Cleavage of Pericentrin B Is Necessary and Sufficient for Centriole Disengagement during Mitosis", *Cell*, Vol. 11, 2012, pp. 2476–2485.

S. Leidel, *et al.*, "SAS-6 Defines a Protein Family Required for Centrosome Duplication in *C. elegans* and in Human Cells", *Nature Cell Biology*, Vol. 7, No. 2, 2005, pp. 115–125.

M.S. Levine, *et al.*, "Centrosome Amplification Is Sufficient to Promote Spontaneous Tumorigenesis in Mammals", *Developmental Cell*, Vol. 40, 2017, pp. 313–322.

T.C. Lin, A. Neuner, and E. Schiebel, "Targeting of γ-Tubulin Complexes to Microtubule Organizing Centers: Conservation and Divergence", *Trends in Cell Biology*, Vol. 25, No. 5, 2015, pp. 296–307.

W.L. Lingle, *et al.*, "Centrosome Hypertrophy in Human Breast Tumors: Implications for Genomic Stability and Cell Polarity", *Proceedings of the National Academy of Sciences*, Vol. 95, 1998, pp. 2950–2955.

W.L. Lingle and J.L. Salisbury, "Altered Centrosome Structure Is Associated with Abnormal Mitoses in Human Breast Tumors", *American Journal of Pathology*, Vol. 155, No. 6, 1999, pp. 1941–1951.

J. Lončarek, P. Herget, and A. Khodjakov, "Centriole Reduplication during Prolonged Interphase Requires Procentriole Maturation Governed by Plk1", *Current Biology*, Vol. 20, 2010, pp. 1277–1282.

J. Lončarek, *et al.*, "Control of Daughter Centriole Formation by the Pericentriolar Material", *Nature: Cell Biology*, Vol. 10, No. 3, 2008, pp. 322–328.

J. Lončarek and A. Khodjakov, "Ab ovo or de nous? Mechanisms of Centriole Duplication", *Molecules and Cells*, Vol. 27, No. 2, 2009, pp. 135–142.

R. Mahen and A.R. Venkitaraman, "Pattern Formation in Centrosome Assembly", *Current Opinion in Cell Biology*, Vol. 24, No. 1, 2012, pp. 14–23.

M.R. Mahjoub and T. Stearns, "Supernumerary Centrosomes Nucleate Extra Cilia and Compromise Primary Cilium Signaling", *Current Biology*, Vol. 22, No. 17, 2012, pp. 1628–1634.

M.S. Markov, "Magnetic Field Theory: A Review", *Electromagnetic Biology and Medicine*, Vol. 26, 2007, pp. 1–23.

W.F. Marshall, "Centriole Assembly: The Origin of Nine-ness", *Current Biology*, Vol. 17, No. 24, 2007a, pp. R1057–R1059.

W.F. Marshall, "What Is the Function of Centrioles?" *Journal of Cellular Biochemistry*, Vol. 100, No. 4, 2007b, pp. 916–922.

W.F. Marshall and J.L Rosenbaum, "How Centrioles Work: Lessons from Green Yeast", *Current Opinion in Cell Biology*, Vol. 12, No. 1, 2000, pp. 119–125.

G. Marteil, *et al.*, "Over-Elongation of Centrioles in Cancer Promotes Centriole Amplification and Chromosome Missegregation", *Nature Communications*, Vol. 9, 2018, pp. 1258.

V. Marthiens, *et al.*, "Centrosome Amplification Causes Microcephaly", *Nature Cell Biology*, Vol. 15, 2013, pp. 731–742.

K. Matsuo, *et al.*, "Kendrin Is a Novel Substrate for Separase Involved in the Licensing of Centriole Duplication", *Current Biology*, Vol. 22, 2012, pp. 915–921.

V. Mennella, *et al.*, "Amorphous No More: Subdiffraction View of the Pericentriolar Material Architecture", *Trends in Cell Biology*, Vol. 24, No. 3, 2014, pp. 188–197.

P. Meraldi and E.A. Nigg, "The Centrosome Cycle", *FEBS Letters*, Vol. 521, No. 1–3, 2002, pp. 9–13.

A. Meunier and N. Spassky, "Centriole Continuity: Out with the New, In with the Old", *Current Opinion in Cell Biology*, Vol. 38, 2016, pp. 60–67. doi:10.1016/j.ceb.2016.02.007

T.C. Moyer, *et al.*, "Binding of STIL to Plk4 Activates Kinase Activity to Promote Centriole Assembly", *Journal of Cell Biology*, Vol. 209, No. 6, 2015, pp. 863–878.

E.A. Nigg, "Mitotic Kinases as Regulators of Cell Division and Its Checkpoints", *Nature Reviews Molecular Cell Biology*, Vol. 2, 2001, pp. 21–31.

E.A. Nigg, L. Cajanek, and C. Arquint, "The Centrosome Duplication Cycle in Health and Disease", *FEBS Letters*, Vol. 588, 2014, pp. 2366–2372.

E.A. Nigg and J.W. Raft, "Centrioles, Centrosomes, and Cilia in Health and Disease", *Cell*, Vol. 139, No. 4. 2009, pp. 663–678.

E.A. Nigg and T. Steams, "The Centrosome Cycle: Centriole Biogenesis, Duplication and Inherent Asymmetries", *Nature Cell Biology*, Vol. 13, No. 10, 2011, pp. 1154–1160.

J.K. Pagan, *et al.*, "Degradation of Cep68 and PCNT Cleavage Mediate Cep215 Removal from the PCM to Allow Centriole Separation,

Disengagement and Licensing", *Nature Cell Biology*, Vol. 17, 2015, pp. 31–43.

M. Paintrand, *et al.*, "Centrosome Organization and Centriole Architecture: Their Sensitivity to Divalent Cations", *Journal of Structural Biology*, Vol. 108, No. 2, 1992, pp. 107–128.

C.G. Pearson and M. Winey, "Plk3/SAK/ZYG-1 in the Regulation of Centriole Duplication", *F1000 Biology Reports*, Vol. 2, 2010, p. 58.

N. Peel, *et al.*, "Overexpressing Centriole-Replication Proteins In Vivo Induces Centriole Overduplication and De Novo Formation", *Current Biology*, Vol. 17, 2007, pp. 834–843.

L. Pelletier, *et al.*, "Centriole Assembly in Caenorhabditis Elegans", *Nature*, Vol. 444, 2006, pp. 619–623.

M. Piel, *et al.*, "The Respective Contributions of the Mother and Daughter Centrioles to Centrosome Activity and Behavior in Vertebrate Cells", *Journal of Cell Biology*, Vol. 149, No. 2, 2000, pp. 317–330.

M. Piel, *et al.*, "Centrosome-Dependent Exit of Cytokinesis in Animal Cells", *Science*, Vol. 291, 2001, pp. 1550–1553.

S.L. Prosser and L. Pelletier, "Centrosome Biology: The Ins and Outs of Centrosome Assembly", *Current Biology*, Vol. 25, 2015, pp. R656–R659.

J.W. Raff and R. Basto, "Centrosome Amplification and Cancer", *Developmental Cell*, Vol. 40, 2017, pp. 217–218.

J. Reina and C. Gonzalez, "When Fate Follows Age: Unequal Centrosomes in Asymmetric Cell Division", *Philosophical Transactions of the Royal Society of London. Series B, Biological Sciences*, Vol. 369, 2014, p. 13.

C.L. Rieder, S. Faruki, and A. Khodjakov, "The Centrosome in Vertebrates: More than a Microtubule-Organizing Center", *Trends in Cell Biology*, Vol. 11, No. 10, 2001, pp. 413–419.

A. Rodrigues-Martins, *et al.*, "Revisiting the Role of the Mother Centriole in Centriole Biogenesis", *Science*, Vol. 315, No. 5827, 2007, pp. 1046–1050.

A. Rodrigues-Martins, *et al.*, "From Centriole Biogenesis to Cellular Function: Centrioles Are Essential for Cell Division at Critical Developmental States", *Cell Cycle*, Vol. 7, No. 1, 2008, pp. 11–16.

M.D. Rose, S. Biggins, and L.L. Satterwhite, "Unravelling the Tangled Web at the Microtubule-Organizing Center", *Current Opinion in Cell Biology*, Vol. 5, No. 1, 1993, pp. 105–115.

J.L. Salisbury, *et al.*, "Centrosomes and Cancer", *Biology of the Cell*, Vol. 91, No. 6, 1999, pp. 451–460.

S. Santaguida and A. Musacchio, "The Life and Miracles of Kinetochores", *EMBO Journal*, Vol. 28, 2009, pp. 2511–2531.

W. Saunders, "Centrosome Amplification and Spindle Multipolarity in Cancer Cells", *Seminars in Cancer Biology*, Vol. 15, 2005, pp. 25–32.

P.W. Schafer, "Centrioles of a Human Cancer: Intercellular Order and Intracellular Disorder", *Science*, Vol. 164, 1969, pp. 1300–1303.

P.W. Schafer and J.A. Chandler, "Electron Probe X-ray Microanalysis of a Normal Centriole", *Science*, Vol. 170, 1970, pp. 1204–1205.

P.W. Schafer, "Centrioles: Intercellular Order in Normal and Malignant cells", *Journal of Thoracic and Cardiovascular Surgery*, Vol. 63, No. 3, 1972, pp. 472–477.

L. Schöckel, *et al.*, "Cleavage of Cohesion Rings Coordinates the Separation of Centrioles and Chromatids", *Nature Cell Biology*, Vol. 13, No. 8, 2011, pp. 966–972.

S. Sen, S. Subramanial, and D.E. Discher, "Indentation and Adhesive Probing of a Cell Membrane with AFM: Theoretical Model and Experiments", *Biophysical Journal*, Vol. 89, 2005, pp. 3203–3213.

J.E. Sillibourne, *et al.*, "Autophosphorylation of Polo-like Kinase 4 and Its Role in Centriole Duplication", *Molecular Biology of the Cell*, Vol. 21, No. 4, 2010, pp. 547–561.

L.K. Slevin, *et al.*, "The Structure of the Plk4 Polo Box Reveals Two Tandem Polo Boxes Required for Centriole Duplication", *Structure*, Vol. 20, No. 11, 2012, pp. 1905–1917.

G. Sluder, "One to Only Two: A Short History of the Centrosome and Its Duplication", *Philosophical Transactions of the Royal Society of London. Series B, Biological Sciences*, Vol. 369, 2014, p. 8.

G. Sluder, F.J. Miller, and C.L. Rieder, "The Reproduction of Centrosomes: Nuclear versus Cytoplasmic Controls", *Journal of Cell Biology*, Vol. 103, 1986, pp. 1873–1881.

G. Sluder and J.L. Nordberg, "The Good, The Bad and The Ugly: The Practical Consequences of Centrosome Amplification", *Current Opinion in Cell Biology*, Vol. 16, No. 1, 2004, pp. 49–54.

M.H. Song, *et al.*, "Centrioles: Some Self-Assembly Required", *Current Opinions in Cell Biology*, Vol. 20, 2008, pp. 688–693.

K.F. Sonnen, *et al.*, "3D-Structured Illumination Microscopy Provides Novel Insight into Architecture of Human Centrosomes", *Biology Open*, Vol. 1, 2012, pp. 965–976.

K.F. Sonnen, *et al.*, "Human Cep192 and Cep152 Cooperate in Plk4 Recruitment and Centriole Duplication", *Journal of Cell Science*, Vol. 126, 2014, pp. 3223–3233.

P. Strnad, *et al.*, "Regulated HsSAS-6 Levels Ensure Formation of a Single Procentriole per Centriole during Centrosome Duplication Cycle", *Developmental Cell*, Vol. 13, No. 2, 2007, pp. 203–213.

P. Strnad and P. Gönczy, "Mechanisms of Procentriole Formation" *Trends in Cell Biology*, Vol. 18, No. 8, 2008, pp. 389–396.

A. Suzuki, *et al.*, "How the Kinetochore Couples Microtubule Force and Centromere Stretch to Move Chromosomes", *Nature Cell Biology*, Vol. 18, No. 4, 2016, pp. 382–392. doi:10.1038/ncb3323

C.J.C. Tang, *et al.*, "The Human Microcephaly Protein STIL Interacts with CPAP and Is Required for Procentrioles Formation", *EMBP Journal*, Vol. 30, 2011, pp. 4790–4804.

A.M. Tassin and M. Bornens, "Centrosome Structure and Microtubule Nucleation in Animal Cells", *Biology of the Cell*, Vol. 91, No. 4–5, 1999, pp. 343–354.

M.F.B. Tsou, *et al.*, "Polo Kinase and Separase Regulate the Mitotic Licensing of Centriole Duplication in Human Cells", *Developmental Cell*, Vol. 17, 2009, pp. 344–354.

M.F. Tsou and T. Stearns, "Controlling Centrosome Number: Licenses and Blocks", *Current Opinion in Cell Biology*, Vol. 18, No. 1. 2006a, pp. 74–78.

M.F. Tsou and T. Stearns, "Mechanism Limiting Centrosome Duplication to Once Per Cell Cycle", *Nature*, Vol. 442, No. 7105, 2006b, pp. 847–951.

Y. Uetake, *et al.*, "Cell Cycle Progression and de novo Centriole Assembly after Centrosomal Removal in Untransformed Human Cells", *Journal of Cell Biology*, Vol. 176, 2007, pp. 173–182.

B.C.M. van de Weerdt and R.H. Medema, "Polo-like Kinases: A Team in Control of the Division", *Cell* Cycle, Vol. 5, No. 8, 2006, pp. 853–864.

H. Varmark, *et al.*, "Asterless Is a Centriolar Protein Required for Centrosome Function and Embryo Development in Drosophila", *Current Biology*, Vol. 17, No. 20, 2007, pp. 1735–1745.

B. Vitre, *et al.*, "Chronic Centrosome Amplification without Tumorigenesis", *Proceedings of the National Academy of Sciences United States of America*, Vol. 112, November 2015, pp. E6321–E6330. doi: 10.1073/pnas.1519388112

I.A. Vorobjev and Y.S. Chentsov, "Centrioles in the Cell Cycle. 1. Epithelial Cells", *Journal of Cell Biology*, Vol. 93, No. 3, 1982, pp. 938–949.

J. Vulprecht, *et al.*, "STIL Is Required for Centriole Duplication in Human Cells", *Journal of Cell Science*, Vol. 125, 2012, pp. 1353–1362.

X. Wan, *et al.*, "Protein Architecture of the Human Kinetochore Microtubule Attachment Site", *Cell*, Vol. 137, 2009, pp. 672–684.

C. Wiese and Y. Zheng, "A New Function for the Gamma-Tubulin Ring Complex as a Microtubule Minus-End Cap", *Nature Cell Biology*, Vol. 2, 2000, pp. 358–364.

M. Winey and E. O'Toole, "Centriole Structure", *Philosophical Transactions of the Royal Society of London. Series B, Biological Sciences*, Vol. 369, 2014, p. 14.

W.J. Wong, *et al.*, "The Conversion of Centrioles to Centrosomes: Essential Coupling of Duplication with Segregation", *Journal of Cell Biology*, Vol. 183, 2011, pp. 727–739.

J.B. Woodruff, O. Wueseke, and A.A. Hyman, "Pericentriolar Material Structure and Dynamics", *Philosophical Transactions of the Royal Society of London. Series B, Biological Sciences*, Vol. 369, 2014, p.10.

J. Xiao and T. Ha, "Flipping Nanoscopy on Its Head", *Science*, Vol. 355, 2017, pp. 582–584.

D. Zyss and F. Gergely, "Centrosome Function in Cancer: Guilty or Innocent?" *Trends in Cell Biology*, Vol. 19, No. 7, 2009, pp. 334–346.

Chapter 3

Microtubules

3.1. Introduction

As with the beams in a large building, microtubules form the framework of cells. In biology texts 2, 3, and 4, microtubules appear prominently in illustrations of cell divisions (mitosis) as in Figure 3.1.

Unlike structural beams of a building, microtubules change length, as needed, for cellular application. For example, just as the microtubules are long in mitosis, they are short as they form the walls of centrioles (see Chapter II).

These beginning paragraphs are based upon the paper: R.L. Huston, "Mechanics of Centriole Microtubules", *Advances in Bioscience and Biotechnology*, Vol. 7, 2010, pp. 266–277.

J. Harden, G. Bertoni, and L.J. Kleinsmith, *Beecher's World of the Cell*, 8th Edition, Pearson Publishing, 2015.

J.W. Sherman and V.G. Sherman, *Biology: A Human Approach*, Oxford University Press, New York, 1983.

B. Alberts, D. Bray, J. Lewis, M. Raft, K. Roberts, and J.D. Watson, *Molecular Biology of the Cell*, 3rd Edition, Garland Publishing, Inc., New York, 1994.

Physically, microtubules are relatively simple. They are long, hollow cylinders, with approximately 25 nm in diameter and whose length may be as much as 100 times the diameter, or more. Figure 3.2 is a photograph of a simple microtubule model where the dark and light portions represent α and β-tubulin dimers forming the microtubule walls. The lengthwise

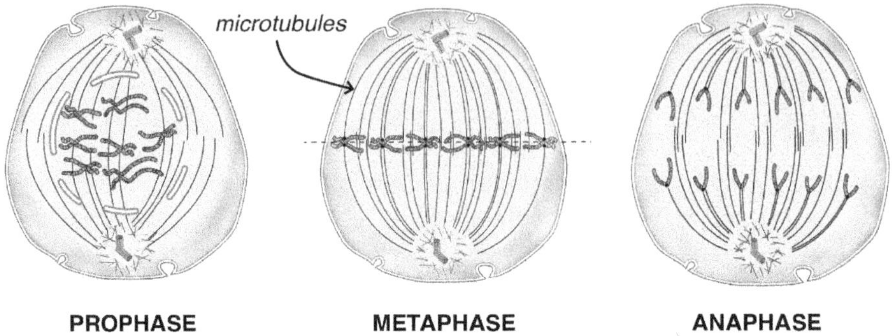

Figure 3.1. Microtubules dividing a cell

Figure 3.2. Photograph of a simplistic microtubule model

Figure 3.3. Microtubule end and side view portion

alignment of these dimers end-6-end form what is known as a "filament." Therefore, the microtubule walls are simply strands of filaments.

Figure 3.3 shows an end view of a microtubule and a short side view. A cross-section shows that there are approximately 13 filaments forming the microtubule wall.

Observe in Figure 3.3 that the filaments are not parallel to each other, but instead are spiral about the microtubule axis. Therefore, the cross-section end shown in Figure 3.2 can be misleading if it is viewed as being planar.

3.2. Applications

Microtubules have two applications in cellular dynamics with short and long lengths.

The short length application forms the walls or "blades" of the centrioles as described in Chapter II. Recall that these blades are aligned triples of microtubules, nine in all, about the centriole. Recall further that these triplets known as A, B, and C, decrease in length in the outward radial direction.

With a centriole cylinder being approximately 410–500 nm long, the corresponding microtubules have similar lengths.

The other microtubule application is the composition of the mitotic spindle as in Figure 3.1. Here the microtubule length is approximately half the length of the entire cell.

3.3. Tubulin

Tubulin is a semi-solid protein used in the development and construction of microtubules. To this end there are three prominent kinds: α, β, and γ. They may be modeled as being approximately cubical with a 5-nm side.

The α- and β-tubulin are usually linked together forming a "dimer." Electronically, α has a negative charge with β being positive. The dimer therefore has a negative to positive charge from the α end to the β end.

The α, β-dimers are connected end-to-end (as in flashlight batteries) and in that way they form the filaments for the microtubules.

The γ-tubulin is used to aid the microtubules emerging out of the centrosome — which is also known as the microtubules organizing center (MTOC).

3.4. The γ-Tubulin Ring Complex (γ-TuRC)

γ-Tubulin plays the major role in extruding the microtubules out of the centrosome (the MTOC). The extrusion is perhaps best described by an illustration as in Figure 3.4 where γ-tubulin is an overlapping ring (like a lock washer). The ring in turn contains 13 pores through which the microtubules are extruded. The pores are sometimes known as "γ-tubulin small complexes" (γ-TuSC).

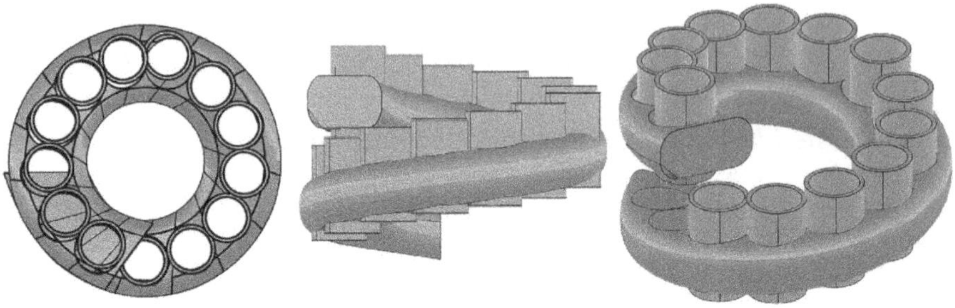

Figure 3.4. A drawing of a γ-tubulin ring complex

3.5. Kinetochores

During mitosis, as the microtubules grow toward the center of the nucleus, they are met by a series of similar strands known as "kinetochores." Kinetochores arise from the chromosome. Their primary purpose is to seek out and link to the microtubules. On occasion more than one kinetochore will engulf one microtubule along the side of the microtubule. When this happens, the kinetochores slide along the microtubule toward the microtubule end with one of the kinetochores linking to the microtubule end and the others fall away.

The connected kinetochore/microtubules then form the mitosis strands.

3.6. Centrosomes and Chromosomes

The centrosome is a large collection of proteins grouped together whose function is to produce microtubules. Thus the centrosome is often known as the MTOC.

Alternatively, the chromosome is a long twisted strand of DNA molecules tightly coiled. As with the centrioles, the chromosome is divided during mitosis (cell division).

3.7. Microtubule Movements and Length Changes

Of all cell organisms, microtubules are the most dynamic in that they are continually moving. They have continuous movement. When a dimer is inserted through the γ-tubulin ring complex as shown in Figure 3.4, the microtubule gets longer. Alternatively, at the distal end the dimers of the microtubule break away and fall into the cytoplasm — thus shortening the length of the microtubule.

The ongoing lengthening and shortening of the microtubules are sometimes referred to as: (1) "growth", (2) "shrinkage", (3) "catastrophe", and (4) "rescue."

"Growth" is elongation, "shrinkage" is shortening, "catastrophe" is the transition from growth to shrinkage, and "rescue" is the transition from shrinkage to growth.

Interestingly, growth and shrinkage can occur simultaneously. When the growth rate is less than the shrinkage, there is "catastrophe." When the growth rate is greater than the shrinkage, there is "rescue."

In addition to all the movement, the individual dimers oscillate at high frequency, approximately 10^7–10^{10} Hz.

3.8. Magnetic Field Development

The tip end of the microtubule has a positive charge. With the microtubule end continually moving, we have a moving charge — a source of electricity. If there is an electric field, there will also be a magnetic field. The magnetic field can then serve as a biomarker for detecting geometric and numeral differences. As an example, defective or excessive microtubules could be a sign of tumors and/or cancer.

3.9. Bibliography — Mechanics of Centriole Microtubules

B. Alberts, *et al.*, *Molecular Biology of the Cell*, 3rd Edition, Garland Publishing, New York, 1994.

D'A.B. Assoro, W.L. Lingle, and J.L. Salisbury, "Centrosome Amplification and the Development of Cancer", *Oncogene*, Vol. 21, 2002, pp. 6146–6153. doi:10.1038/sj.onc.1205772.

J. Azimzadeh and M. Bornens, "Structure and Duplication of the Centrosome", *Journal of Cell Science*, Vol. 120, 2007, pp. 2139–2142. doi:10.1242/jcs.005231.

F. Barenz, D. Mayilo, and O.J. Gruss, "Centriolar Satellites: Busy Orbits Around the Centrosome", *European Journal of Cell Biology*, Vol. 90, 2011, pp. 983–989. doi:10.1016/j.ejcb.2011.07.007.

F.P. Beer and E.R. Johnston Jr., *Mechanics of Materials*, 2nd Edition, McGraw-Hill, New York, 1994.

M. Bettencourt-Dias and D.M. Glover, "Centrosome Biogenesis and Function: Centrosomics Brings New Understanding", *Nature Reviews Molecular Cell Biology*, Vol. 8, 2007, pp. 451–463. doi:10.1038/nrm2180.

M. Bettencourt-Dias and D.M. Glover, "SnapShot: Centriole Biogenesis", *Cell*, Vol. 136, 2009, pp. 188.e1–188.e2. doi:10.1016/j.cell.2008.12.035

A.D. Bicek, *et al.*, "Analysis of Microtubule Curvature", *Methods in Cell Biology*, Vol. 83, 2007, pp. 237–268. doi:10.1016/S0091-679X(07)83010-X.

M. Bornens, "Organelle Positioning and Cell Polarity", *Nature Reviews Molecular Cell Biology*, Vol. 9, 2008, pp. 874–886. doi:10.1038/nrm2524.

M. Bornens, "The Centrosome in Cells and Organisms", *Science*, Vol. 335, 2012, pp. 422–426. doi:10.1126/science.1209037.

A. Bouissou, *et al.*, "γ-Tubulin Ring Complexes Regulate Microtubule Plus End Dynamics", *The Journal of Cell Biology*, Vol. 187, 2009, pp. 327–334. doi:10.1083/jcb.200905060.

B.R. Brinkley, "Managing the Centrosome Numbers Game: From Chaos to Stability in Cancer Cell Division", *Trends in Cell Biology*, Vol. 11, 2001, pp. 18–21. doi:10.1016/S0962-8924(00)01872-9.

L. Cassimeris, *et al.*, "XMAP215 Is a Long Thin Molecule that Does Not Increase Microtubule Stiffness", *Journal of Cell Science*, Vol. 114, 2001, pp. 3025–3033.

P. Chang, *et al.*, "ε-Tubulin Is Required for Centriole Duplication and Microtubule Organization", *Nature Cell Biology*, Vol. 5, 2003, pp. 71–76. doi:10.1038/ncb900.

M. Cifra, D. Havelka, and O. Kucera, "Electric Oscillations Generated by Collective Vibration Modes of Microtubules", *Proceedings of SPIE*, Vol. 7376, 2010, p. 73760N.

M. Cifra, *et al.*, "Electric Field Generated by Axial Longitudinal Vibration Modes of Microtubule", *Biosystems*, Vol. 100, 2010, pp. 122–131. doi:10.1016/j.biosystems.2010.02.007.

I. Cunha-Ferreira, I. Bento, and M. Bettencourt-Dias, "From Zero to Many: Control of Centriole Number in Development and Disease", *Traffic*, Vol. 10, 2009, pp. 482–498. doi:10.1111/j.1600-0854.2009.00905.x.

M. Delattre and P. Gonczy, "The Arithmetic of Centrosome Biogenesis", *Journal of Cell Science*, Vol. 117, 2004, pp. 1619–1630. doi:10.1242/jcs.01128.

J.G. DeLuca, "Kinetochore-Microtubule Dynamics and Attachment Stability", *Methods in Cell Biology*, Vol. 97, 2010, pp. 53–79. doi:10.1016/S0091-679X(10)97004-0.

M. Dogterom, *et al.*, "Force Generation by Dynamic Microtubules", *Current Opinion in Cell Biology*, Vol. 17, 2005, pp. 67–74. doi:10.1016/j.ceb.2004.12.011.

A. Duensing, *et al.*, "Centriole Overduplication through the Concurrent Formation of Multiple Daughter Centrioles at Single Material Templates", *Oncogene*, Vol. 26, 2007, pp. 6280–6288. doi:10.1038/sj.onc.1210456.

S.K. Dutcher, "The Tubulin Fraternity: Alpha to Eta", *Current Opinion in Cell Biology*, Vol. 13, 2001, pp. 49–54. doi:10.1016/S0955-0674(00)00173-3.

K. Fukasawa, "Oncogenes and Tumour Suppressors Take on Centrosomes", *Nature Reviews Cancer*, Vol. 7, 2007, pp. 911–924. doi:10.1038/nrc2249.

N.J. Ganem, S.A. Godinho, and D. Pellman, "A Mechanism Linking Extra Centrosomes to Chromosomal Instability", *Nature*, Vol. 460, 2009, pp. 278–282. doi:10.1038/nature08136.

S.A. Godinho, M. Kwon, and D. Pellman, "Centrosomes and Cancer: How Cancer Cells Divide with Too Many Centrosomes", *Cancer and Metastasis Reviews*, Vol. 28, 2009, pp. 85–98. doi:10.1007/s10555-008-9163-6.

T.M. Goepfert, *et al.*, "Centrosome Amplification and Overexpression of Aurora A Are Early Events in Rat Mammary Carcinogenesis", *Cancer Research*, Vol. 62, 2002, pp. 4115–4122.

D. Guest, *Biology Smart, the Princeton Review*, Random House, New York, 1996, pp. 133–135.

J. Hardin, G. Bertoni, and L.J. Kleinsmith, *Becker's World of the Cell*, 8th Edition, Benjamin Cummings, Boston, 2015.

D. Havelka and M. Cifra, "Calculation of the Electromagnetic Field around a Microtubule", *Acta Polytechnica*, Vol. 49, 2009, pp. 58–63.

D. Havelka, *et al.*, "High-Frequency Electric Field and Radiation Characteristics of Cellular Microtubule Network", *Journal of Theoretical Biology*, Vol. 286, 2011, pp. 31–40. doi:10.1016/j.jtbi.2011.07.007.

A.J. Holland, *et al.*, "Polo-Like Kinase 4 Kinase Activity Limits Centrosome Overduplication by Autoregulating Its Own Stability", *The Journal of Cell Biology*, Vol. 188, 2010, pp. 191–198. doi:10.1083/jcb.200911102.

R.L. Huston, "On Centrioles, Microtubules, and Cellular Electromagnetism", *Journal of Nanotechnology in Engineering and Medicine*, Vol. 5, 2014, p. 021003.

R.L. Huston, "Using the Electromagnetics of Cancer's Centrosome Clusters to Attract Therapeutic Nanoparticles", *Advances in Bioscience and Biotechnology*, Vol. 6, 2015, pp. 172–181. doi:10.4236/abb.2015.63017.

R.L. Huston, "A Review of Centriole Activity, and Wrongful Activity, during Cell Division", *Advances in Bioscience and Biotechnology*, Vol. 7, 2016, pp. 169–182. doi:10.4236/abb.2016.73015.

R.L. Huston, "Delivering Nanoparticles to Cancer Cells", *Tissue Engineering and Nanotheranostics*, Vol. 10, No. 7, 2018, p. 215.

Y. F. Inclán and E. Nogales, "Structural Models for the Self-Assembly and Microtubule Interactions of γ-, δ- and ε-Tubulin", *Journal of Cell Science*, Vol. 114, 2000, pp. 413–422.

M.E. Janson, M.E. de Dood, and M. Dogterom, "Dynamic Instability of Microtubules Is Regulated by Force", *The Journal of Cell Biology*, Vol. 161, 2003, pp. 1029–1034. doi:10.1083/jcb.200301147.

A. Kis, *et al.*, "Nanomechanics of Microtubules", *Physical Review Letters*, Vol. 89, 2002, p. 248101. doi:10.1103/PhysRevLett.89.248101.

S.L. Kline-Smith and C.E. Wolczak, "The Microtubule-Destabilizing Kinesin XKCM1 Regulates Microtubule Dynamic Instability in Cells", *Molecular Biology of the Cell*, Vol. 13, 2002, pp. 2718–2731. doi:10.1091/mbc.E01-12-0143.

T. Kobayashi and B.D. Dynlacht, "Regulating the Transition from Centriole to Basal Body", *Journal of Cell Biology*, Vol. 193, 2011, pp. 435–444. doi:10.1083/jcb.201101005.

J.M. Kollman, *et al.*, "Microtubule Nucleating γ-TuSC Assembles Structures with 13-Fold Microtubule-Like Symmetry", *Nature*, Vol. 466, 2010, pp. 879–882. doi:10.1038/nature09207.

N. Korzeniewski, M. Hohenfellner, and S. Duensing, "CANDI Promotes PLK4-Mediated Centriole Overduplication and is Frequently Disrupted in Prostate Cancer", *NEOPLASIA*, Vol. 14, 2012, pp. 799–806. doi:10.1593/neo.12580.

A. Kramer, K. Neben, and A.D. Ho, "Centrosome Replication, Genomic Instability and Cancer", *Leukemia*, Vol. 16, 2002, pp. 767–775. doi:10.1038/sj.leu.2402454.

M. Kwon, *et al.*, "Mechanisms to Suppress Multipolar Divisions in Cancer Cells with Extra Centrosomes", *Genes & Development*, Vol. 22, 2008, 2189–2203.

B.M. Lange and K. Gull, "Structure and Function of the Centriole in Animal Cells: Progress and Questions", *Trends in Cell Biology*, Vol. 6, 1996, pp. 348–352. doi:10.1016/0962-8924(96)10033-7.

B. Leber, *et al.*, "Proteins Required for Centrosome Clustering in Cancer Cells", *Science Translational Medicine*, Vol. 2, 2010, pp. 33–38. doi:10.1126/scitranslmed.3000915.

W.L. Lingle and J.L. Salisbury, "The Role of the Centrosome in the Development of Malignant Tumors", *Current Topics in Developmental Biology*, Vol. 49, 2000, pp. 313–339. doi:10.1016/S0070-2153(99)49015-5.

T.-C. Lin, A. Neuner, and E. Schiebel, "Targeting of γ-Tubulin Complexes to Microtubule Organizing Centers: Conservation and Divergence", *Trends in Cell Biology*, Vol. 25, 2015, pp. 296–307. doi:10.1016/j.tcb.2014.12.002.

W.L. Lingle, *et al.*, "Centrosome Hypertrophy in Human Breast Tumors: Implications for Genomic Stability and Cell Polarity", *Proceedings of the National Academy of Sciences of the United States of America*, Vol. 95, 1998, pp. 2950–2955. doi:10.1073/pnas.95.6.2950.

W.L. Lingle, *et al.*, "Centrosome Amplification Drives Chromosomal Instability in Breast Tumor Development", *Proceedings of the National*

Academy of Sciences of the United States of America, Vol. 99, 2002, pp. 1978–1983. doi:10.1073/pnas.032479999.

W.L. Lingle and J.L. Salisbury, "Altered Centrosome Structure is Associated with Abnormal Mitoses in Human Breast Tumors", *American Journal of Pathology*, Vol. 155, 1999, pp. 1941–1951. doi:10.1016/S0002-9440(10)65513-7.

H. Loffler, *et al.*, "DNA Damage-Induced Centrosome Amplification Occurs via Excessive Formation of Centriolar Satellites", *Oncogene*, Vol. 32, 2013, pp. 2963–2972. doi:10.1038/onc.2012.310.

J. Loncarek and A. Khodjakov, "Ab ovo or de nous? Mechanisms of Centriole Duplication", *Molecules and Cells*, Vol. 27, 2009, pp. 135–142. doi:10.1007/s10059-009-0017-z.

R. Mahen and A.R. Venkitaraman, "Pattern Formation in Centrosome Assembly", *Current Opinion in Cell Biology*, Vol. 24, 2012, pp. 14–23. doi:10.1016/j.ceb.2011.12.012.

H. Maiato and E. Logarinho, "Mitotic Spindle Multipolarity without Centrosome Amplifications", *Nature Cell Biology*, Vol. 16, 2014, pp. 386–394. doi:10.1038/ncb2958.

E.N. Marieb, *Human Anatomy and Physiology*, 3rd Edition, Benjamin/Cummings Publishing, Redwood City, 1991, Chapter 3, pp. 60–101.

V. Marthiens, *et al.*, "Centrosome Amplification Causes Microcephaly", *Nature Cell Biology*, Vol. 15, 2013, pp. 731–740. doi:10.1038/ncb2746.

J. Marx, "Do Centrosome Abnormalities Lead to Cancer?" *Science*, Vol. 292, 2001, pp. 426–429. doi:10.1126/science.292.5516.426.

V. Mennella, *et al.*, "Amorphous No More: Subdiffraction View of the Pericentriolar Material Architecture", *Trends in Cell Biology*, Vol. 24, 2014, pp. 188–197. doi:10.1016/j.tcb.2013.10.001.

M. Moritz, *et al.*, "Structure of the Gamma-Tubulin Ring Complex: A Template for Microtubule Nucleation", *Nature Cell Biology*, Vol. 2, 2000, pp. 365–370. doi:10.1038/35014058.

E.A. Nigg, "Centrosome Aberrations: Cause or Consequence of Cancer Progression?" *Nature Reviews Cancer*, Vol. 2, 2002, pp. 815–825. doi:10.1038/nrc924.

E.A. Nigg, "Origins and Consequences of Centrosome Aberrations in Human Cancers", *International Journal of Cancer*, Vol. 119, 2006, pp. 2717–2723. doi:10.1002/ijc.22245.

E.A. Nigg and J.W. Raft, "Centrioles, Centrosomes, and Cilia in Health and Disease", *Cell*, Vol. 139, 2009, pp. 663–678. doi:10.1016/j.cell.2009.10.036.

E.A. Nigg and T. Stearns, "The Centrosome Cycle: Centriole Biogenesis, Duplication and Inherent Asymmetries", *Nature Cell Biology*, Vol. 13, 2011, pp. 1154–1160. doi:10.1038/ncb2345.

B.R. Oakley, V. Paolillo, and Y. Zheng, "γ-Tubulin Complexes in Microtubule Nucleation and Beyond", *Molecular Biology of the Cell*, Vol. 26, 2015, pp. 2957–2962. doi:10.1091/mbc.E14-11-1514.

G.A. Pihan, *et al.*, "Centrosome Defects and Genetic Instability in Malignant Tumors", *Cancer Research*, Vol. 58, 1998, pp. 3974–3985.

G.A. Pihan, *et al.*, "Centrosome Abnormalities and Chromosome Instability Occur Together in Pre-Invasive Carcinomas", *Cancer Research*, Vol. 63, 2003, pp. 1398–1404.

J. Pokorny, "Viscous Effects on Polar Vibrations in Microtubules", *Electromagnetic Biology and Medicine*, Vol. 22, 2003, pp. 15–29. doi:10.1081/JBC-120020349.

J. Pokorny, "Physical Aspects of Biological Activity and Cancer", *AIP Advances*, Vol. 2, 2012, p. 011207. doi:10.1063/1.3699057.

J. Pokorny, J. Hasek, and F. Jelínek, "Electromagnetic Field of Microtubules: Effects on Transfer of Mass Particles and Electrons", *Journal of Biological Physics*, Vol. 31, 2005a, pp. 501–514. doi:10.1007/s10867-005-1286-1.

J. Pokorny, J. Hasek, and F. Jelínek, "Endogenous Electric Field and Organization of Living Matter", *Electromagnetic Biology and Medicine*, Vol. 24, 2005b, pp. 185–197. doi:10.1080/15368370500379566.

J. Pokorny, J.Pokorny, and J. Kobilková, "Postulates on Electromagnetic Activity in Biological Systems and Cancer", *Integrative Biology*, Vol. 5, 2013, pp. 1439–1446. doi:10.1039/c3ib40166a.

S. Portet, *et al.*, "Elastic Vibrations in Seamless Microtubules", *European Biophysics Journal*, Vol. 34, 2005, pp. 912–920. doi:10.1007/s00249-005-0461-4.

B. Raynaud-Messina and A. Merdes, "γ-Tubulin Complexes and Microtubule Organization", *Current Opinion in Cell Biology*, Vol. 19, 2007, pp. 24–30. doi:10.1016/j.ceb.2006.12.008.

A. Rodrigues-Martins, *et al.*, "From Centriole Biogenesis to Cellular Function: Centrioles are Essential for Cell Division at Critical Developmental Stages", *Cell Cycle*, Vol. 7, 2008, pp. 11–16. doi:10.4161/cc.7.1.5226.

E. Schiebel, "γ-Tubulin Complexes: Binding to the Centrosome, Regulation and Microtubule Nucleation", *Current Opinion in Cell Biology*, Vol. 12, 2000, pp. 113–118. doi:10.1016/S0955-0674(99)00064-2.

L. Schockel, *et al.*, "Cleavage of Cohesion Rings Coordinates the Separation of Centrioles and Chromatids", *Nature Cell Biology*, Vol. 13, 2011, pp. 966–972. doi:10.1038/ncb2280.

M.M. Shawki and A. Farid, "Low Electric Field Parameters Required to Induce Death of Cancer Cells", *Electromagnetic Biology & Medicine*, Vol. 33, 2014, pp. 159–163. doi:10.3109/15368378.2013.800105.

G. Sluder and J.J. Nordberg, "The Good, the Bad and the Ugly: The Practical Consequences of Centrosome Amplification", *Current Opinion in Cell Biology*, Vol. 16, 2004, pp. 49–54. doi:10.1016/j.ceb.2003.11.006.

P. Strnad and P. Gonczy, "Mechanisms of Procentriole Formation", *Trends in Cell Biology*, Vol. 18, 2008, pp. 389–396. doi:10.1016/j.tcb.2008.06.004.

A.-M. Tassin and M. Bornens, "Centrosome Structure and Microtubule Nucleation in Animal Cells", *Biology of the Cell*, Vol. 91, 1999, pp. 343–354. doi:10.1111/j.1768-322X.1999.tb01092.x.

N. Teixidó-Travesa, J. Roig, and J. Lüders, "The Where, When and How of Microtubule Nucleation — One Ring to Rule Them All", *Journal of Cell Science*, Vol. 125, 2012, pp. 4445–4456. doi:10.1242/jcs.106971.

M.F. Tsou and T. Stearns, "Mechanism Limiting Centrosome Duplication to Once per Cell Cycle", *Nature*, Vol. 442, 2006, pp. 947–951. doi:10.1038/nature04985.

J. Vulprecht, *et al.*, "STIL Is Required for Centriole Duplication in Human Cells", *Journal of Cell Science*, Vol. 125, 2012, pp. 1353–1362. doi:10.1242/jcs.104109.

C.E. Walczak, "Microtubule Dynamics and Tubulin Interacting Proteins", *Current Opinion in Cell Biology*, Vol. 12, 2000, pp. 52–56. doi:10.1016/S0955-0674(99)00056-3.

C. Wiese and Y. Zheng, "A New Function for the γ-Tubulin Ring Complex as a Microtubule Minus-End Cap", *Nature Cell Biology*, Vol. 2, 2000, pp. 358–364. doi:10.1038/35014051.

Z. Yang, *et al.*, "Extra Centrosomes and/or Chromosomes Prolong Mitosis in Human Cells", *Nature Cell Biology*, Vol. 10, 2008, pp. 748–751.

3.10. Annotated Literature Review

S. Afreen and D. Varma, "Cell Divisions: Molecular Pathways for KMN Kinetochore Recruitment", *Current Biology*, Vol. 24, 2015, pp. R328–R347.

Review Notes: Two ways are postulated as the ways the kinetochore attaches to the mitotic spindle.

T. Afrin, *et al.*, "Buckling of Microtubules on Elastic Media via Breakable Bonds", *Biochemical and Biophysical Research Communications*, Vol. 480, 2016, pp. 132–138.

Review Notes: Authors use computer simulation to study microtubule buckling behavior.

A. Akhmanova and C.C. Hoogenraad, "Microtubule Minus-End-Targeting Proteins", *Current Biology*, Vol. 25, 2015, pp. R162–R171.

Review Notes: Paper explores MT minus end regulation, but does not seem to come to firm conclusions. Paper states the better understanding of MT dynamics is needed for new cancer treatment strategies. It also states that plus end dynamics are better understood than minus end dynamics.

T.K. Akhshi, D. Wernike, and A. Piekny, "Microtubules and Actin Crosstalk in Cell Migration and Division", *Cytoskeleton*, Vol. 71, 2014, pp. 1–23.

Review Notes: MTs interact with actin to form the shape, migration, and division of the cell. Actin is an abundant protein. The monomeric form is called globular or G-actin; the polymeric form is filamentous or F-actin.

B. Alberts, *et al.*, *Molecular Biology of the Cell*, 3rd Edition, Garland Publishing, New York, 1994.

Review Notes: For MT: the fastest growing and is defined as "positive" (pp. 805–806). Due to their spiral geometry, an MT may point either CW or CCW (p. 805). Picture of growing/shrinking MT (p. 810).

H. Aldaz, *et al.*, "Insights into Microtubule Nucleation from the Crystal Structure of Human ϒ-Tubulin", *Nature*, Vol. 435, 2005, pp. 523–527.

Review Notes: New MTs require γ-tubulin.

C.L. Asbury, J.F. Tien, and T.N. Davis, "Kinetochores' Gripping Feat: Conformational Wave or Biased Diffusion?" Trends in Cell Biology, Vol. 21, 2011, pp. 38–46.

Review Notes: (2011) The authors present two models for microtubule-kinetochore coupling: wave and biased diffusion.

C. Aumeier, *et al.*, "Self-Repair Promotes Microtubule Rescue", *Nature Cell Biology*, Vol. 18, 2016, pp. 1054–1064.

Review Notes: Authors state that microtubule rescue, after catastrophe, can occur by α, β-dimers filling the gaps left by the catastrophe. Microtubules can obtain rescue by incorporating α, β-dimers from the centrosome.

N.A. Baker, *et al.*, "Electrostatics of Nanosystems: Application to Microtubules and the Ribosome", *Proceedings of the National Academy of Sciences of the United States of America*, Vol. 98, 2001, pp. 10037–10041.

Review Notes: Paper referred from David Lemmon.

H. Berg, *et al.*, "Bioelectromagnetic Field Effects on Cancer Cells and Mice Tumors", *Electromagnetic Biology & Medicine*, Vol. 29, No. 4, 2010, pp. 132–143.

Review Notes: Low frequency EM fields are successful in treating EM tumors.

A.D. Bicek, *et al.*, "Analysis of Microtubule Curvature", *Methods in Cell Biology*, Vol. 83, 2007, Chapter 10, pp. 237–268.

Review Notes: The paper proposes ways to measure curvature of MTs.

$$EI = (1 \text{ to } 200) \times 10^{-24}\ \text{Nm}^2 \text{ (in vitro)}$$

$$I = 10^{-32}\, m^4 \text{ (I calculate } 1.668 \times 10^{-32}\, m^4\text{)}$$

$$E \approx 10^9\ \text{Pa, where Pa is N/m}^2.$$

D.G. Booth and W.C. Earnshaw, "Ki-67 and the Chromosome Periphery Compartment in Mitosis", *Trends in Cell Biology*, Vol. 27, No. 12, 2017, pp. 906–913.

Review Notes: Chromosome periphery is a complex network of proteins. Mitotic chromosomes have five major regions, one of which is the kinetochore.

B.P. Bouchet and A. Akhmanova, "Microtubules in 3D Cell Motility", *Journal of Cell Science*, Vol. 130, 2017, pp. 39–50.

Review Notes: Authors discuss the important effects of microtubules on 3D cell functions.

A. Bouissou, *et al.*, "Gamma-Tubulin Ring Complexes Regulate Microtubule Plus End Dynamics", *Journal of Cell Biology*, Vol. 187, No. 3, 2009, pp. 327–334.

Review Notes: γ-TuRC is formed of γ-TuSC.

C.P. Brangwynne, *et al.*, "Microtubules can Bear Enhanced Compressive Loads in Living Cells Because of Lateral Reinforcement", *Journal of Cell Biology*, Vol. 173, No. 5, 2006, pp. 733–741.

Review Notes: MT can withstand compressive loads due to adjacent and surrounding tissue.

C.P. Brangwynne, *et al.*, "Microtubules can bear Enhanced Compressive Loads in Living Cells because of Lateral Reinforcement", *Journal of Cell Biology*, Vol. 173, No. 5, 2006, pp. 733–761. doi:10.1083/jcb.200601060

Review Notes: MTs can withstand higher compressive loads in a cell than in vitro due to surrounding lateral reinforcement.

G.J. Brouhard and L.M. Rice, "The Contribution of $\alpha\beta$-Tubulin Curvature to Microtubule Dynamics", *Journal of Cell Biology*, Vol. 207, No. 3, 2014, pp. 323–334.

Review Notes: The $\forall$, $\exists$-tubulin determines curvature at the distal ends of MTs.

M. Casenghi, *et al.*, "Polo-Like Kinase 1 Regulates Nip a Centrosome Protein Involved in Microtubule Nucleation", *Developmental Cell*, Vol. 5, 2003, pp. 113–125.

Review Notes: Nip is a centrosomal Plk1 substrate which is controlled by Plk1 in the development of MTs.

L. Cassimeris, *et al.*, "XMAP215 Is a Long Thin Molecule that Does Not Increase Microtubule Stiffness", *Journal of Cell Science*, Vol. 114, 2001, pp. 3025–3033.

Review Notes: EI for an MT $\approx 1.7 \times 10^{-23}$ Nm^2.

B.T. Castle and D.J. Odde, "Optical Control of Microtubule Dynamics in Time and Space", *Cell*, Vol. 162, 2015, pp. 243–245.

Review Notes: Small molecular inhibitors of MT development are used in chemotherapy.

N. Caudron, *et al.*, "A Reassessment of the Factors Affecting Microtubule Assembly and Disassembly In Vitro", *Journal of Molecular Biology*, Vol. 297, 2000, pp. 211–220.

Review Notes: MT nucleation rate remains constant during polymer assembly.

C.-W. Chang, *et al.*, "CEP295 Interacts with Microtubules and Is Required for Centriole Elongation", *Journal of Cell Science*, Vol. 129, 2016, pp. 2501–2513.

Review Notes: The protein CEP 295 interacts with the microtubules to elongate them. Too much → long centrioles; too little → short centrioles.

P. Chang and T. Stearns, "δ-Tubulin and ε-Tubulin: Two New Human Centrosomal Tubulins Reveal New Aspects of Centrosome Structure and Function", *Nature Cell Biology*, Vol. 2, 2000, pp. 30–35.

Review Notes: Of the five tubulins ∀, ∃, γ, ∗, and γ, ∀ and ∃ make up the MT, γ is at the base for nucleation, ∗ is associated with the centriole and γ with the pericentriolar material. (In 424 ∗ is associated with the triplets.)

I.M. Cheeseman and A. Desai, "Molecular Architecture of the Kinetochore-Microtubule Interface", *Nature Reviews Molecular Cell Biology*, Vol. 9, 2008, pp. 33–46.

Review Notes: Kinetochores appear to be more complex than microtubules. Kinetochores have more than 80 compounds. The connection, however, is not clear.

C. Ciferri, *et al.*, "Implications for Kinetochore-Microtubule Attachment from the Structure of an Engineered Ndc80 Complex", *Cell*, Vol. 133, 2008, pp. 427–439.

Review Notes: The NDc80 has four subunits.

M. Cifra, J.Z. Fields, and A. Farhadi, "Electromagnetic Cellular Interactions", *Progress in Biophysics and Molecular Biology*, Vol. 105, 2011, pp. 223–246.

Review Notes: Quote: "A static electric field is generated by a static charged particle. Both an electric field and a magnetic field are generated if a charged particle moves at a constant velocity. The term electromagnetic field is reserved for a condition where a charged particle is accelerated. In most cases the acceleration of a charged particle takes place in the form of an oscillation. Therefore, the electric and magnetic fields also oscillate."

M. Cifra, *et al.*, "Electric Field Generated by Axial Longitudinal Vibration Modes of Microtubule", *Biosystems*, Vol. 100, No. 2, 2010, pp. 122–131.

Review Notes: Calculation of electric field generation via axial longitudinal vibration of MTs. There are several hundred MTs in a cell (Kis *et al.* 2002). The authors find $E > 100$ MPa and $G = 1.4 \pm 0.4$ MPa (orthotropic). Energy supply to MTs from: (1) GTP (hydrolysis of guanosine triphosphate), (2) energy transfer by motor (moving) proteins, and (3) energy released from mitochondria as "wasted" energy in the citric acid cycle. Paper considers axial longitudinal vibration of MT. Hetrodimers have an electric dipole moment of over 1000 Debye (10^{-26} cm). Frequencies of MT vibrations span from kHz to GHz.

P.T. Conduit, "Microtubule Organization: A Complex Relation", *Journal of Cell Biology*, Vol. 213, 2016, pp. 809–812.

Review Notes: Paper argues that the γ-tubulin ring complex (γ-TuRC) varies between different species.

J.R. Cooper, *et al.*, "Catalysis of the Microtubule On-Rate Is the Major Parameter Regulating the Deploymerase Activity of MCAK", Nature Structural and Molecular Biology, Vol. 17, 2010, pp. 77–82.

Review Notes: The focus of the paper is on the role of MCAK in coupling with the microtubule ends. MCAK is a kinesin 13 (a catalysis). The acronym is *mitotic-centromere-associated-kinesin*. MCAK is a regulator with a positive charge. The positive charges in the mech of MCAK depolymerizes microtubules. The mechanism of MCAK is not known. Several models are proposed: (a quote).

(i) A loose tether that allows the molecule to diffuse along the microtubule lattice.

(ii) An obstructive element that prevents tight binding to the lattice.

(iii) A "pry-bar" that destabilizes lateral interactions at the microtubule end.

A facilitator of cross-linking between protofilament peels at the depolymerizing ends of microtubules.

J.G. DeLuca, "Kinetochore-Microtubule Dynamics and Attachment Stability", *Methods in Cell Biology*, Vol. 97, 2010, Chapter 4, pp. 53–79.

Review Notes: Kinetochores are large protein structures built on centrosomeric DNA. Attachments between MTs and kinetochores must be stable to ensure successful mitosis.

M. Dogterom, *et al.*, "Force Generation by Dynamic Microtubules", *Current Opinion in Cell Biology*, Vol. 17, 2005, pp. 67–74.

Review Notes: MTs can push and pull — a closely regulated process. The pushing and pulling seem to be attributed to adding and subtracting tubulin. "Catastrophe" refers to a switch from growing to shrinking. MTs interact with kinetochores and provide for the separation of the chromosomes.

K.H. Downing and E. Nogales, "New Insights into Microtubule Structure and Function from the Atomic Model of Tubulin", *European Biophysics Journal*, Vol. 27, No. 5, 1998, pp. 431–436.

Review Notes: ∀ ∃-dimer tubulin and connections between dimers form the protofilament — has a drawing (Figure 3.1).

S. Dumont and T.J. Mitchison, "Force and Length in the Mitotic Spindle", *Current Biology*, Vol. 19, 2009, pp. R749–R761.

Review Notes: Little is known about the underlying mechanical principles that govern the assembly and function of the spindle. Most animal spindles can be thought of as the superposition of kinetochore, nonkinetochore, and astral microtubules. This is a review paper about the interplay of forces and assembly of spindle dynamics.

M.W. Elting, *et al.*, "Force on Spindle Microtubule Minus Ends Moves Chromosomes", *Journal of Cell Biology*, Vol. 206, No. 2, 2014, pp. 245–256.

Review Notes: Dynein and NuMA exert tension forces to stabilize the minus ends of MTs.

M.K. Gardner, "Cell Biology: Microtubule Collisions to the Rescue", *Current Biology*, Vol. 26, 2016, pp. R1287–R1289.

Review Notes: Colliding microtubules can pressure their lengths.

F. Gittes, *et al.*, "Flexural Rigidity of Microtubules and Actin Filaments Measured from Thermal Fluctuations in Shape", *Journal of Cell Biology*, Vol. 120, No. 4, 1993, pp. 923–934.

Review Notes: $EI = 2.2 \times 10^{-23}$ Nm^2 (same as 76).

S.A. Godinho, M. Kwon, and D. Pellman, "Centrosomes and Cancer: How Cancer Cells Divide with too Many Centrosomes", *Cancer Metastasis Review*, Vol. 28, 2009, pp. 85–98.

Review Notes: Mother centrioles contain a set of distal appendages for anchoring MTs — daughter centrioles do not have this appendage.

E.L. Grishchuk, *et al.*, "Force Production by Disassembling Microtubules", *Nature*, Vol. 438, No. 17, 2005, pp. 385–388.

Review Notes: (2005) The paper proposes that MT forces on organelles are formed by the curved filaments coming off the positive end of the MT attaching to an organelle.

S.R. Hameroff, "A New Theory of the Origin of Cancer: Quantum Coherent Entanglement, Centrioles, Mitosis, and Differentiation", *Biosystems*, Vol. 77, No. 1–3, 2004, pp. 119–136.

Review Notes: Microtubules are ferromagnetic. "It has been demonstrated that it is possible to orient arrays of MTs by the application of an electric field …" (See also 38).

L. Hamon, P.A. Curmi, and D. Pastré, "High-Resolution Imaging of Microtubules and Cytoskeleton Structures by Atomic Force Microscopy", *Methods in Cell Biology*, Vol. 95, 2010, pp. 157–174.

Review Notes: AFM can image in a liquid environment.

D. Havelka and M. Cifra, "Calculation of the Electromagnetic Field Around a Microtubule", *Acta Polytechnica*, Vol. 49, 2009, pp. 58–63.

Review Notes: "Tubulin dimers have a strong electric dipole moment." Tubulin dimer dipole moment: over 1,000 Debye = 10^{-26} cm. MT oscillation frequency: from MHz to GHz.

D. Havelka, *et al.*, "High-Frequency Electric Field and Radiation Characteristics of Cellular Microtubule Network", *Journal of Theoretical Biology*, Vol. 286, 2011, pp. 31–40.

Review Notes: MT are electrically polar. Vibrations generate an oscillating electric field.

S.M. Hinshaw and S.C. Harrison, "Kinetochore Function from the Bottom Up", *Trends in Cell Biology*, Vol. 28, No. 1, 2018, pp. 22–33.

Review Notes: It is now known that kinetochores transmit chromosomal DNA to microtubules. Kinetochores enable the process of connecting chromosomal DNA to microtubules. Cancer cells exhibit defects in this process.

J. Howard and A.A. Hyman, "Dynamics and Mechanics of the Microtubule Plus End", *Nature*, Vol. 422, 2003, pp. 753–758.

Review Notes: MTs grow and shrink. Polarity of tubulin dimer reflects the polarity of the MT. In mitosis the "pushing" from the two ends of the spindle centers the nucleus.

S. Hormeño, *et al.*, "Single Centrosome Manipulation Reveals Its Electric Charge and Associated Dynamic Structure", *Biophysical Journal*, Vol. 97, No. 4, 2009, pp. 1022–1030.

Review Notes: Centrosome and MTs have same polarity. Surface charge density of a centrosome is : e^3 $e = 1.6 \times 10^{-9}$ C. This is much smaller than the charge of an MT.

Y.F. Inclán and E. Nogales, "Structural Models for the Self-Assembly and Microtubule Interactions of Gamma-, Delta- and Epsilon-Tubulin", *Journal of Cell Science*, Vol. 114, 2000, pp. 413–422.

Review Notes: ∀∃-tubulin heterodimers self-assemble to form the MTs. γ-tubulin nucleates the MT. γ-tubulin is believed to be capable of longitudinal assembly. ∗-tubulin may seed the triplets.

C. Janke, "The Tubulin Code: Molecular Components, Readout Mechanisms, and Functions", *Journal of Cell Biology*, Vol. 206, No. 4, 2014, pp. 461–472.

Review Notes: Microtubules look alike in almost all species. How tubulin heterogeneity controls MT functions is still relatively unknown.

M.E. Janson, M.E. deDood, and M. Dogterom, "Dynamic Instability of Microtubules Is Regulated by Force", *Journal of Cell Biology*, Vol. 161, No. 6, 2003, pp. 1029–1034.

Review Notes: MTs alternating between elongating and shortening (apparently at random) is a process called "dynamic instability." The time of elongation is known as "catastrophe time." Force reduces the catastrophe time by limiting the rate of tubulin addition.

D. Job, O. Valiron, and B. Oakley, "Microtubule Nucleation", *Current Opinion in Cell Biology*, Vol. 15, 2003, pp. 111–117.

Review Notes: (2001) MT assembly is usually seeded by γ-tubulin ring complexes γ-TuRC: γ-tubulin ring complex; γ-TuSC: γ-tubulin small complex.

A.P. Joglekar, K. Bloom, and E.D. Salmon, "In Vivo Protein Architecture of the Eukaryotic Kinetochore with Nanometer Scale Accuracy", *Current Biology*, Vol. 19, 2009, pp. 694–699.

Review Notes: The authors use two-color fluorescence microscopy to image the proteins for budding yeast kinetochore-microtubule connection. The claim is that budding yeast connection is the same as in eukaryotic cells.

A.P. Joglekar, *et al.*, "Molecular Architecture of a Kinetochore–Microtubule Attachment Site", *Nature Cell Biology*, Vol. 8, 2006, pp. 581–585.

Review Notes: Proposes a ring model for kinetochore connection to a microtubule (2006).

M.A. Jordan and L. Wilson, "Microtubules as a Target for Anticancer Drugs", *Nature Reviews: Cancer*, Vol. 4, No. 4, 2004, pp. 253–265.

Review Notes: There may be a variety of types of MTs. Current MT drugs reduce the MT mass. But the drugs also work at lower concentrations.

K. Kamath, E. Oroudjev, and M.A. Jordan, "Determination of Microtubule Dynamic Instability in Living Cells", *Methods in Cell Biology*, Vol. 97, 2010, pp. 1–14.

Review Notes: MTs are dynamically unstable exhibiting periods of rapid growth and shortening. ∃-tubulin is at the + end.

S. Kasas, *et al.*, "Oscillation Modes of Microtubules", *Biology of the Cell*, Vol. 96, No. 9, 2004, pp. 697–700.

Review Notes: Shows various ways an MT could oscillate.

T.J. Keating and G.G. Borisky, "Immunostructural Evidence for the Template Mechanism of Microtubule Nucleation", *Nature Cell Biology*, Vol. 2, 2000, pp. 352–357.

Review Notes: Authors indicate that MT nucleation follows a pointed template (helix) as opposed to a protofilament model (2000).

I.A. Kent and T.P. Lele, "Microtubule Based Force Generation", *Wiley Interdisciplinary Reviews: Nanomedicine and Nanobiotechnology*, Vol. 9, No. 3, 2016, p. e1428.

Review Notes: This is a review paper on how MTs and kinetochore link together. MTs in dynamic instability undergo cycles of polymerization and depolymerization controlled by GTP-processes ("rescue" and "catastrophe"). GTP must be bound to β-tubulin for individual dimers to be added. There are different views of how MTs and kinetochore connect to each other: one is via a ring from the kinetochore encircling the MT; the other is to have MT-like strings immersing with the catastrophe dimers of an MT.

A. Kis, *et al.*, "Nanomechanics of Microtubules", *Physical Review Letters*, Vol. 89, No. 24, 2002, pp. 248101-1–248101-4.

Review Notes: Microtubules are made of α-tubulin and β-tubulin. Microtubules help define cell morphology. They are building blocks for cilia and flagella. They are stiff in length but work in shear. 100G < E. Temperature may affect MT properties. Testing may lead to misleading results. This paper claims to report elastic moduli for the first time. MTs are anisotropic and inhomogeneous. MTs are similar to carbon nanotube ropes.

S.L. Kline-Smith and C.E. Wolczak, "The Microtubule-Destabilizing Kinesin XKCM1 Regulates Microtubule Dynamic Instability in Cells", *Molecular Biology of the Cell*, Vol. 13, 2002, pp. 2718–2731.

Review Notes: MT dynamics can be described by four parameters: (1) growth, (2) shrinkage, (3) catastrophe (transition from growth to shrinkage), and (4) rescue (transition from shrinkage to growth (edited quote)).

J.M. Kollman, *et al.*, "Microtubule Nucleating γ-TuSC Assembles Structures with 13-Fold Microtubule-Like Symmetry", *Nature*, Vol. 466, 2010, pp. 879–883.

Review Notes: 13-fold structure allows protofilaments to move straight without being twisted by three or four symmetries.

B. Lacroix, *et al.*, "Microtubule Dynamics Scale with Cell Size to Set Spindle Length and Assembly Timing", Developmental Cell, Vol. 45, 2018, pp. 496–511.

Review Notes: Paper states what might be expected: That is, the microtubule growth rate scales with the cell size.

B.M. Lange and K. Gull, "Structure and Function of the Centriole in Animal Cells: Progress and Questions", *Trends in Cell Biology*, Vol. 6, No. 9, 1996, pp. 348–352.

Review Notes: Centriole barrels contain only a few proteins namely: ∀- and ∃-tubulin. Inside the cylinder there is also γ-tubulin. γ-tubulin also occurs within the proximal end. (Some believe γ-tubulin is responsible for centriole duplication and/or growth of MTs.)

R. Leguy, *et al.*, "Monomeric ϒ-Tubulin Nucleates Microtubules", *Journal of Biological Chemistry*, Vol. 275, No. 29, 2000, pp. 21975–21980.

Review Notes: (2000) γ-tubulin is required for MT nucleation but the mechanism is not understood (a near quote).

T.-C. Lin, A. Neuner, and E. Schiebel, "Targeting of ϒ-Tubulin Complexes to Microtubule Organizing Centers: Conservation and Divergence", *Trends in Cell Biology*, Vol. 25, No. 5, 2015, pp. 296–307.

Review Notes: MT nucleation depends upon γ-tubulin. The γ-tubulin small complex (γ-TuSC) then forms a scaffold for the ∀/∃-tubulin dimers to initiate polymerization.

R. Mahen and A.R. Venkitaraman, "Pattern Formation in Centrosome Assembly", *Current Opinion in Cell Biology*, Vol. 24, No. 1, 2012, pp. 14–23.

Review Notes: Microtubules (at the base) are surrounded by a region of dense proteins (electron-dense material?).

H. Maiato, *et al.*, "The Dynamic Kinetochore-Microtubule Interface", *Journal of Cell Science*, Vol. 117, 2004, pp. 5461–5477.

Review Notes: (2005) Paper says exact structure of kinetochore is unknown. Very good picture model of MT kinetochore connection. Paper says that MT ends begin to explore cytoplasm after nucleus softening.

E.N. Marieb, *Human Anatomy and Physiology*, 3rd Edition, Benjamin/Cummings, Redwood City, CA, 1991, Chapter 3, pp. 60–101.

Review Notes: 2010 AFM studies of MTs are rare.

Provides procedures for AFM studies.

G. Marteil, *et al.*, "Over-Elongation of Centrioles in Cancer Promotes Centriole Amplification and Chromosome Missegregation", *Nature Communications*, Vol. 9, 2018, pp. 1258.

Review Notes: The authors relate centriole elongation to tumorigenesis. The causes are not well understood. Centriole elongation promotes deregulation of events.

P.G. McKean, S. Vaughan, and K. Gull, "The Extended Tubulin Superfamily", *Journal of Cell Science*, Vol. 114, Pt. 15, 2001, pp. 2723–2733.

Review Notes: ∀ ∃-dimer defines polarity: Negative (proximal) and positive (distal).

F.J. McNally, "Mechanisms of Spindle Positioning", *Journal of Cell Biology*, Vol. 200, 2013, pp. 131–159.

Review Notes: The author discusses mitosis when the nucleus is not centered.

F.J. McNally and A. Roll-Mecak, "Microtubule-Severing Enzymes: From Cellular Functions to Molecular Mechanism", *Journal of Cell Biology*, Vol. 217, No. 12, 2018, pp. 4057–4069.

Review Notes: There are microtubule enzymes that cause internal breaks in the MT. This can lead to disorders. The mechanism is not well-understood.

V. Mennella, *et al.*, "Subdiffraction-Resolution Fluorescence Microscopy Reveals a Domain of the Centrosome Critical for Pericentriolar Material Organization", *Nature Cell Biology*, Vol. 14, No. 11, 2012, pp. 1159–1168.

Review Notes: The primary role of the pericentriolar material (PCM) is to anchor the microtubules through the γ-tubulin ring complex γTuRC. The PCM is *not* an amorphous material.

A. Mershin, *et al.*, "Tubulin Dipole Moment, Dielectric Constant and Quantum Behavior: Computer Simulations, Experimental Results and Suggestions", *Biosystems*, Vol. 77, 2004, pp. 73–85.

Review Notes: (2004) The electric dipole moment has only been calculated via computer simulations (not measured). (2004) There exists little experimental evidence about the electrical properties of tubulin and MTs. MTs are not steady-state systems; energy is continually being put in and taken out.

T. Mitchison and M. Kirschner, "Dynamic Instability of Microtubule Growth", Nature, Vol. 312, 1984a, pp. 237–242.

Review Notes: Paper requested by Roger Adelman. The essence of the paper suggests that microtubules are frequently shrinking and elongating, but rarely both at the same time.

T. Mitchinson and M. Kirachner, "Microtubule Assembly Nucleated by Isolated Centrosomes", *Nature*, Vol. 312, 1984b, pp. 232–237.

Review Notes: This is a companion paper to paper 608.

M. Moritz and D.A. Agard, "Υ-Tubulin Complexes and Microtubule Nucleation", *Current Opinion in Structural Biology*, Vol. 11, 2001, pp. 174–181.

Review Notes: γ-Tubulin is required for initiating MTs. γ-Tubulin is a protein required for initiating the polymerization of MTs.

M. Moritz, *et al.*, "Reconstruction of Centrosome Microtubule Nucleation in Drosophila", *Methods in Cell Biology*, Vol. 67, 1998, pp. 141–148.

Review Notes: γ-Tubulin ring complex is required for Drosophila centrosomes to nucleate microtubules.

T. Müller-Reichert, R. Kiewisz, and S. Redamann, "Mitotic Spindles Revisited — New Insights from 3D Electron Microscopy", *Journal of Cell Science*, Vol. 131, 2018, p. jcs211383.

Review Notes: "The mitosis spindle functions to ensure faithful segregation of chromosomes during cell division." — from the abstract. This is a review paper on the imaging of mitotic spindles based primarily on advances in 3D electron microscopy.

D.J. Needleman, *et al.*, "Synchrotron X-ray Diffraction Study of Microtubules Buckling and Bundling under Osmotic Stress: A Probe of Interprotofilament Interactions", *Physical Review Letters*, Vol. 193, No. 19, 2004, pp. 198104-1–198104-4.

Review Notes: MT buckle at ~600 Pa.

E.A. Nigg, L. Cajanek, and C. Arquint, "The Centrosome Duplication Cycle in Health and Disease", FEBS Letters, Vol. 588, 2014, pp. 2366–2372.

Review Notes: Centrosome aberrations have long been implicated in cancer. The process is not yet fully understood. This is a short review of the gene products involved.

E. Nogales, "Structural Insights into Microtubule Function", *Annual Review of Biochemistry*, Vol. 69, 2000, pp. 277–302.

Review Notes: This paper is in a review journal, but it is not a traditional literature review paper. Instead it is more of a review of the ∀- and ∃-tubulin structure. There are detailed color drawings of the tubulin in the figures. MTs exist in all eukaryotes. The tubulin dimers are connected head to tail. MTs are highly dynamic and can switch stochastically between growing and shrinking phases.

E. Nogales, *et al.*, "High-Resolution Model of the Microtubule", *Cell*, Vol. 96, No. 1, 1999, pp. 79–88.

Review Notes: The inside and outside of the MT surfaces are different: the outside is smoother with shallow undulation and is rather flat. Alternatively, the inside surface is deeply corrugated.

E. Nogales, S.G. Wolf, and K.H. Downing, "Structure of the αβ-Tubulin Dimer by Electron Crystallography", *Nature*, Vol. 391, 1998, pp. 199–203.

Review Notes: Good pictures of ∀- and ∃-tubulin. The tubulins are almost the same.

C.B. O'Connell and A.L. Khodjakov, "Cooperative Mechanisms of Mitotic Spindle Formation", *Journal of Cell Science*, Vol. 120, 2007, pp. 1717–1722.

Review Notes: Paper discusses interaction of proteins to speed the kinetochore-microtubule connection.

B.R. Oakley, V. Paolillo, and Y. Zheng, "Gamma-Tubulin Complexes in Microtubule Nucleation and Beyond", *Molecular Biology of the Cell*, Vol. 26, 2015, pp. 2957–2962.

Review Notes: γ-TuRC is a 13 γ-tubulin molecule — containing as many as 5 γ-TuSC.

D. Passeri, *et al.*, "Magnetic Force Microscopy", *Biomatter*, Vol. 4, No. 1, 2014. doi:10.4161/biom.29507

Review Notes: "Magnetic force microscopy (MFM) is an atomic force microscopy (AFM)-based technique in which an AFM tip with a magnetic coating is used to probe local magnetic fields with the typical AFM spatial resolution."

J. Paz and J. Lüders, "Microtubule-Organizing Centers: Towards a Minimal Parts List", *Trends in Cell Biology*, Vol. 28, No. 3, 2018, pp. 176–186.

Review Notes: The authors try to identify the proteins essential for the centrosome, the microtubule organizing center (MTOC). The microtubule originates from the gamma-tubulin ring complex (γ-TuRC). Approximately 13 molecules of γ-tubulin form a helix which becomes the template for microtubule assembly.

J. Paz and J. Lüders, "Microtubule-Organizing Centers: Towards a Minimal Parts List", *Trends in Cell Biology*, Vol. 28, No. 3, 2018, pp. 176–187.

Review Notes: The objective is to find the minimal number of proteins needed to make a microtubule. There are hundreds of proteins in the centrosome. These proteins have other functions besides microtubule development.

S. Petry and R.D. Vale, "Microtubule Nucleation at the Centrosome and Beyond", *Nature Cell Biology*, Vol. 17, No. 9, 2015, pp. 1089–1093.

Review Notes: (2015) This is a review of factors and behavior of γ-tubulin — a key MT protein. MTs are nucleated from the centrosome (the MTOC). The MTs are nucleated by the γ-TuRC (γ-tubulin ring complex). The γ-TuRC and the γ-TuSC form a cone at the negative end of MT (from papers 461, 473, 478). MTs can originate outside of the MTOC, for example, at the kinetochore.

A. Pitaval, *et al.*, "Microtubule Stabilization Drives 3D Centrosome Migration to Initiate Primary Ciliogenesis", *Journal of Cell Biology*, Vol. 216, 2017, p. 3713.

Review Notes: The paper title is discussed but no apparent mention of electromagnetic forces.

J. Pokorný, "Endogenous Electromagnetic Forces in Living Cells: Implications for Transfer of Reaction Components", *Electro- and Magnetobiology*, Vol. 20, No. 1, 2001, pp. 59–73.

Review Notes: Electric field near a microtubule can be 10^8 v/m or more. Force F on a dipole is: $F = \nabla(pgE)$ where p is the dipole moment and E is the electric field intensity. Microtubules are composed of subunits that are strong electric dipoles.

J. Pokorný, "Viscous Effects on Polar Vibrations in Microtubules", *Electromagnetic Biology and Medicine*, Vol. 22, No. 1, 2003, pp. 15–29.

Review Notes: The subunits of microtubules: tubulin heterodimers — have electric dipole moments.

"The charges in the microtubule are at most 2 nm from the surface with ~ exponential dependence of the electric potential in the charge layer may be assumed." (See Figure 3.1.)

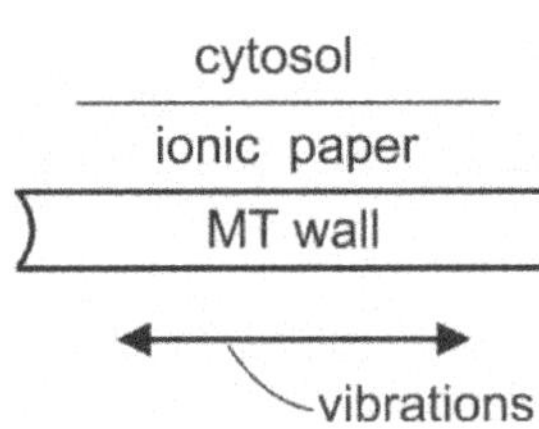

J. Pokorný, "Electrodynamic Activity of Healthy and Cancer Cells", *Journal of Physics: Conference Series*, Vol. 329, No. 1, 2011, p. 012007.

Review Notes: (2011) A very impressive paper that asserts that electromagnetic fields are developed by oscillating MT and that the higher the frequency the greater is the generated field. The paper also states that the electromagnetic field has cellular effects that go beyond the chemical effects. Microtubules can vibrate in kHz to GHz regions.

J. Pokorný, "Physical Aspects of Biological Activity and Cancer", *AIP Advances*, Vol. 2, 2012, pp. 011207-1–011207-11.

Review Notes: Dipole moment of ∀, ∃-tubulin ~1,000 Debye (10^{-26} cm). Induced moment from movement is ~200–400 Debye. This also produces polarization changes.

J. Pokorný, J. Hašek, and F. Jelínek, "Electromagnetic Field of Microtubules: Effects on Transfer of Mass Particles and Electrons", *Journal of Biological Physics*, Vol. 31, 2005, pp. 501–514.

Review Notes: MTs can generate electric fields.

J. Pokorný, *et al.*, "Vibration in Microtubules", *Journal of Biological Physics*, Vol. 23, 1997, pp. 171–179.

Review Notes: Highest MT vibration frequency ~10^8 Hz. MTs have 13–14 protofilaments; ∀, ∃-tubulin each have molecular weight of 55,000 daltons. 1 dalton is the mass of a proton ~1.66×10^{-27} kg — lowest vibration frequency ~10^6 Hz. Half-life of MT ~10 min. Diameter of tubulin ~ 4 nm.

J. Pokorný, J. Pokorný, and J. Kobilková, "Postulates on Electromagnetic Activity in Biological Systems and Cancer", *Integrative Biology*, Vol. 5, No. 12, 2013, pp. 1439–1446.

Review Notes: Longitudinal vibration of MTs generate an endogenome electromagnetic field. MT vibration ~10–30 MHz and 100–200 MHz.

J. Pokorný, *et al.*, "Cancer Physics: Diagnosis Based on Damped Cellular Elastoelectrical Vibrations in Microtubules", *European Biophysics Journal*, Vol. 40, No. 6, 2011, pp. 747–759.

Review Notes: MT shear modulus: $G = 5$ N/m^2, i.e., $G = 5$ Pa.

T.D. Pollard, "Nine Unanswered Questions about Cytokinesis", Journal of Cell Biology, Vol. 916, 2017 Archive, pp. 3007 ff.

Review Notes: The author lists nine questions related to mitosis which are not yet fully understood:

1. What proteins participate?
2. What specifies the site of the furrow?

3. How does the contractile assemble?
4. How is the ring attached to the plasma membrane?
5. What triggers constriction of the contractile ring?
6. How do rings produce force to form the furrow?
7. How does the ring disassemble as it contracts?
8. How does the cell membrane expand in the furrow?
9. How do daughter cells separate?

S. Portet, *et al.*, "Elastic Vibrations in Seamless Microtubules", *European Biophysics Journal*, Vol. 34, No. 7, 2005, pp. 912–920.

Review Notes: Protofilament — a column of an MT (as opposed to a helix of dimers). Typically, there are 13 protofilaments in an MT. Uses a model of a lattice of dimers. Rupture strength of MT ~ 0.4–0.5 Pa. MT: $E = 10^6$–10^9 Pa.

A.F. Powers, *et al.*, "The Ndc80 Kinetochore Complex Forms Load Bearing Attachments to Dynamic Microtubule Tips via Biased Diffusion", *Cell*, Vol. 136, 2009, pp. 865–875.

Review Notes: Authors present models for the kinetochore/microtubule connection. The focus is on the activity of Ndc80 (not defined) in making the coupling.

M.E. Presenti, J.R. Weir, and A. Musacchio, "Progress in the Structural and Functional Characterization of Kinetochores", *Current Opinion in Structural Biology*, Vol. 37, 2016, pp. 152–163.

Review Notes: "Kinetochores are macromolecular complexes built on a specialized chromatin domain called the centromere. Kinetochores provide a site of attachment for spindle microtubules during mitosis" (from the abstract). Similar spelled words: centromere — kinetochores; chromosome — nucleosis (DNA); centrosome — centriole surrounding.

X.S. Qian, J.Q. Zhang, and C.Q. Ru, "Wave Propagation in Orthotropic Microtubules", *Journal of Applied Physics*, Vol. 101, 2007, pp. 084702-1–084702-7.

Review Notes: MTs are orthotropic. This anisotropy lowers torsional and radial wave velocity, but it does not affect longitudinal

velocities. This would not occur if MTs were isotropic. MTs are the most rigid of structures. MT: 25 nm OD; 15 nm ID. 13 parallel protofilaments bonded laterally. Helical structure. Lots of recent studies of MT. Shear modulus and elastic modulus in circumferential direction are lower than in the longitudinal direction. Authors neglect helix effect.

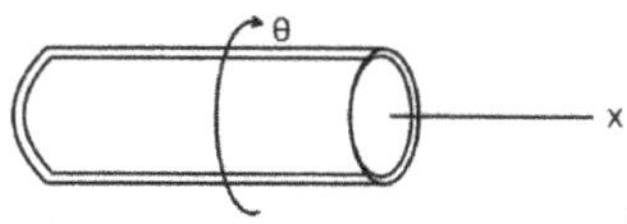

$E_x = 0.5$ to 2 GPa $\upsilon = 0.3$

$E_\theta = 1$ to 4 MPa $\gamma = 1.47$ G = 1 MPa

Wave velocities given but no frequencies.

Torsional wave → 26.1 m/s; longitudinal wave → 824.8 m/s.

B. Raynaud-Messina and A. Merdes, "Gamma-Tubulin Complexes and Microtubule Organization", *Current Opinion in Cell Biology*, Vol. 19, No. 1, 2007, pp. 24–30.

Review Notes: Interesting paper (somewhat speculative) about the base structure of an MT. The focus is on γ-tubulin: Two types γ-tubulin small complex, γ-TuSC and γ-tubulin ring complex, γ-TuRC. γ-TuRC forms the MT base and it only contacts the ∀-tubulin. The ∃-tubulin is on the positive side of the dimer and the ∀-tubulin on the negative side.

L.M. Rice, "A New Look for the Growing Microtubule End?", Journal of Cell Biology, Vol. 217, No. 8, 2018, p. 2609.

Review Notes: The αβ-tubulin constructs are different outside the microtubule within. If we understood the biochemistry of the αβ-tubulin conformational cycle we could predict the end structure. Conversely if we knew the end structure we could infer the biochemistry. The end structure and the biochemistry are both unknown.

R.M. Rios, "The Centrosome-Golgi Apparatus Nexus", *Philosophical Transactions of the Royal Society of London. Series B, Biological Sciences*, Vol. 369, 2014, 17 pages.

Review Notes: The author states that microtubule nucleation can occur at either the centrioles or at the Golgi apparatus.

M.D. Rose, S. Biggins, and L.L. Satterwhite, "Unravelling the Tangled Web at the Microtubule-Organizing Center", *Current Opinion in Cell Biology*, Vol. 5, No. 1, 1993, pp. 105–115.

Review Notes: Two views for the MTOC: (1) a tubulin producer (like a spider web) and (2) a capston (a reeling in/out) of the MTs. (Author says both.)

C. Sacristan, *et al.*, "Dynamic Kinetochore Size Regulation Promotes Microtubule Capture and Chromosome Biorientation in Mitosis", Nature Cell Biology, Vol. 20, 2018, pp. 800–810.

Review Notes: The kinetochore expands to capture a microtubule.

J. Salogiannis and S.L. Reck-Peterson, "Hitchhiking: A Non-Canonical Mode of Microtubule-Based Transport", Trends in Cell Biology, Vol. 27, 2017, pp. 141–150.

Review Notes: Authors suggest that organelles move within a cell by attaching themselves to microtubules and other recruited molecular motors.

A.D. Sanchez and J.L. Feldman, "Microtubule Organizing Centers: From the Centrosome to Non-Centrosomal Sites", Current Opinion in Cell Biology, Vol. 44, 2017, pp. 93–101.

Review Notes: Generally, microtubules are developed in the centrosome also known as (aka) the microtubule organizing center (MTOC) but on occasion they can be developed elsewhere.

M. Sanhaji, *et al.*, "Mitotic Centromere-Associated Kinesin (MCAK): A Potential Cancer Drug Target", *Oncotarget*, Vol. 2, 2011, pp. 935–947. doi:10.18632/ONCOTARGET.416

Review Notes: The paper focuses on mitotic centromere-associated kinesin (MCAK). Restriction of MCAK can present cancer cells from forming and mutating. MCAK is found throughout the cell but especially at the centromere and centrosome. MCAK is positively charged.

S. Santaguida and A. Musacchio, "The Life and Miracles of Kinetochores", *EMBO Journal*, Vol. 28, 2009, pp. 2511–2531.

Review Notes: Kinetochores have four functions:

(1) Connect to centromeric chromatin

(2) Connect to MTs

(3) Monitor the state of the kinetochores/MT attachment

(4) Correct improper attachments.

"The primary function of kinetochores is to create load-bearing attachments between chromosomes and microtubules in a dividing mother cell."

M.V. Satarić and J.A. Tuszyński, "Relationship between the Nonlinear Ferroelectric and Liquid Crystal Models for Microtubules", *Physical Review*, Vol. 67, 2003, pp. 011901-1–011901-11.

Review Notes: MT still not clearly understood. "Travelling kink" discussed. Experimental work called for.

M.V. Sataric, *et al.*, "Modelling the Role of Intrinsic Electric Fields in Microtubules as an Additional Control Mechanism of Bi-Directional Intracellular Transport", *Cell Biochemistry and Biophysics*, Vol. 52, No. 2, 2008, pp. 113–124.

Review Notes: Paper presents a model of how alternating electric fields control microtubule-based transport.

M. Sataric, B.M. Sataric, and J.A. Tuszynski, "Nonlinear Model of Microtubule Dynamics", *Electromagnetic Biology and Medicine*, Vol. 24, 2005, pp. 255–264.

Review Notes: Proposes ferromagnetic properties for MTs.

L. Schaedel, *et al.*, "Microtubules Self-Repair in Response to Mechanical Stress", *Nature Materials*, Vol. 14, 2015, pp. 1156–1165.

Review Notes: This paper is a reference in paper 627. Damaged microtubules are able to incorporate new tubulin dimers and thereby recover their stiffness. The hollow cross-section of microtubules gives them stiffness. Microtubules soften under mechanical stress, but softened microtubules can self-repair. They have self-heating properties. Figure 6 provides a good graphic model of microtubule damage ("catastrophe") and self-repair ("rescue").

H.T. Schek III, *et al.*, "Microtubule Assembly Dynamics at the Nanoscale", *Current Biology*, Vol. 17, No. 17, 2007, pp. 1445–1455.

Review Notes: MTs have variable growth rates, and sometimes shorten as much as five tubulin layers.

E. Schiebel, "Υ-Tubulin Complexes: Binding to the Centrosome, Regulation and Microtubule Nucleation", *Current Opinion in Cell Biology*, Vol. 12, 2000, pp. 113–118.

Review Notes: (2000) MT assembly is initiated by γ-tubulin. γ-tubulin and other proteins (Spc97p and Spc98p) form a core of the forming MT.

J. Scrofani, *et al.*, "Microtubule Nucleation in Mitosis by a Ran GTP-Dependent Protein Complex", *Current Biology*, Vol. 25, 2015, pp. 131–140.

Review Notes: MTs can be developed from chromosomes. The development occurs via γTuRC, but the exact mechanism remains unknown. MTs from the chromosomes are essential for the success of the mitotic spindle.

M.M. Shawki and A. Farid, "Low Electric Field Parameters Required to Induce Death of Cancer Cells", *Electromagnetic Biology & Medicine*, Vol. 33, No. 2, 2014, pp. 159–163.

Review Notes: Low electric fields kill cancer cells.

B. Short, "Mapping the Kinetochore Delta", *Journal of Cell Biology*, Vol. 212, 2016, p. 259.

Review Notes: Kinetochores are not thin flat plates. Measurements have been based on single-molecule, high-resolution colocalization (SHREC). Researchers believe that is not accurate enough.

V. Sikirzhytski, *et al.*, "Microtubules Assemble near Most Kinetochores during Early Prometaphase in Human Cells", *Journal of Cell Biology*, Vol. 217, No. 8, 2018, p. 2641.

Review Notes: Kinetochores are formed by lateral microtubules which assemble quickly into end-on shape.

A. Singh, *et al.*, "Polarized Microtubule Dynamics Directs Cell Mechanics and Coordinates Forces during Epithelial Morphogenesis", *Nature Cell Biology*, Vol. 20, 2018, pp. 1126–1133.

Review Notes: In early cell development in Drosophila wing epithelium development, each cell has its own microtubule that can bear compressive forces.

K.C. Slep, "A Microtubule Dynamics Reconstitutional Convention", *Journal of Cell Biology*, Vol. 205, No. 3, 2016, pp. 305–315.

Review Notes: Paper summarizes results of Moriwaki and Goshima, "Five Factors Can Reconstitute All Three Phases of Microtubule Polymerizative Dynamics", *The Journal of Cell Biology* (2016).

K.F. Sonnen, *et al.*, "3D-Structured Illumination Microscopy Provides Novel Insight into Architecture of Human Centrosomes", *Biology Open*, Vol. 1, 2012, pp. 965–976.

Review Notes: At the beginning of mitosis MT nucleation is stimulated by recruitment and exchange of PCM components. PCM undergoes a drastic expansion at the beginning of mitosis. Plk4 is an early marker (the first) for new centrioles. It falls onto a ring stained by Cep152. Also, Plk4 colocalizes with SAS-6. The site for MT nucleation is determined on the mother centriole. It is not clear whether the nucleation site is chosen at random or is somehow predetermined.

R. Stracke, *et al.*, "Analysis of the Migration Behavior of Single Microtubules in Electric Fields", *Biochemical and Biophysical Research Communications*, Vol. 293, 2002, pp. 602–609.

Review Notes: Net charge on a tubulin dimer is 0.19e (lower than that calculated earlier).

P. Strnad and P. Gönczy, "Mechanisms of Procentriole Formation", *Trends in Cell Biology*, Vol. 18, No. 8, 2008, pp. 389–396.

Review Notes: Centrosome: centrioles and pericentriolar material (PCM). This is the primary site of microtubule nucleation — aka MTOC (microtubule organizing center).

P. Strnad, *et al.*, "Regulated HsSAS-6 Levels Ensure Formation of a Simple Procentriole per Centriole during Centrosome Duplication Cycle", *Developmental Cell*, Vol. 13, No. 2, 2007, pp. 203–213.

Review Notes: To form a new centriole we need centrin-3, Plk4, and SAS-6.

S. Suresh, "Biomechanics and Biophysics of Cancer Cells", *Acta Biomaterialia*, Vol. 3, No. 4, 2007, pp. 413–438.

Review Notes: Young's modulus $E = 1.96$ Pa. Good pictures.

T.U. Tanaka and A. Desai, "Kinetochore-Microtubule Interactions: The Means to the End", *Current Opinion in Cell Biology*, Vol. 20, 2008, pp. 53–63.

Review Notes: Kinetochore components comprise the direct interface with spindle microtubules. Kinetochores initially attach to the surface of a single microtubule. The kinetochore is then "transported" poleward. Ndc80 is the important constituent that interacts with the MT.

M.E. Tanenbaum and R.H. Medema, "Mechanisms of Centrosome Separation and Bipolar Spindle Assembly", *Developmental Cell*, Vol. 19, 2010, pp. 797–806.

Review Notes: Claims that centrosome separation and spindle forces are due to MT compressive forces.

A.-M. Tassin and M. Bornens, "Centrosome Structure and Microtubule Nucleation in Animal Cells", *Biology of the Cell*, Vol. 91, No. 4–5, 1999, pp. 343–354.

Review Notes: MTs within the centriole are highly stable, but this exceptional stability is "yet to be elucidated."

N. Teixido-Travesa, J. Roig, and J. Luders, "The Where, When and How of Microtubule Nucleation–One Ring to Rule Them All", *Journal of Cell Science*, Vol. 125, 2012, pp. 4445–4456.

Review Notes: (2012) γ-tubulin behavior is poorly understood. γ-tubulin provides a *template* for MT nucleation (good pictures).

J.F. Tien, *et al.*, "Cooperation of the Dam1 and Ndc80 Kinetochore Complexes Enhances Microtubule Coupling and Is Regulated by Aurora B", *Journal of Cell Biology*, Vol. 189, No. 4, 2010, pp. 713–23. doi:10.1083/jcb.200910142.

Review Notes: Not clear how microtubules and kinetochores couple. Dam1 is from the kinetochores and Ndc80 is from budding yeast. The study shows that Dam1 and Ndc80 are compatible.

I.M. Tolić, "Mitotic Spindle: Kinetochore Fibers Hold on Tight to Interpolar Bundles", *European Biophysics Journal*, Vol. 47, 2018, pp. 191–203.

Review Notes: Pulling microtubules separate duplicated chromosomes. The microtubules are attached to the chromosomes via kinetochores. Kinetochores appear to be attached laterally to the microtubules along the length of the microtubules.

M.C. Tsai and J. Ahringer, "Microtubules Are Involved in Anterior-Posterior Axis Formation in *C. elegans* Embryos", *Journal of Cell Biology*, Vol. 179, No. 3, 2007, pp. 397–402.

Review Notes: Difficult to read. The conclusion seems to be that MT carry current from polarity in the centrosome. Tubulin is apparently a source of polarity from embryo testing.

J.A. Tuszyński, *et al.*, "Molecular Dynamics Simulations of Tubulin Structure and Calculations of Electrostatic Properties of Microtubules", *Mathematical and Computer Modelling*, Vol. 41, No. 10, 2005, pp. 1055–1070.

Review Notes: Computer modeling of tubulin → MT.

MTs appear to be zipped up lattices of filament.

Each tubulin monomer has more than 400 amino acids.

Centriole MT probably differs from mitotic spindle MT.

∀ and ∃-tubulin are very similar.

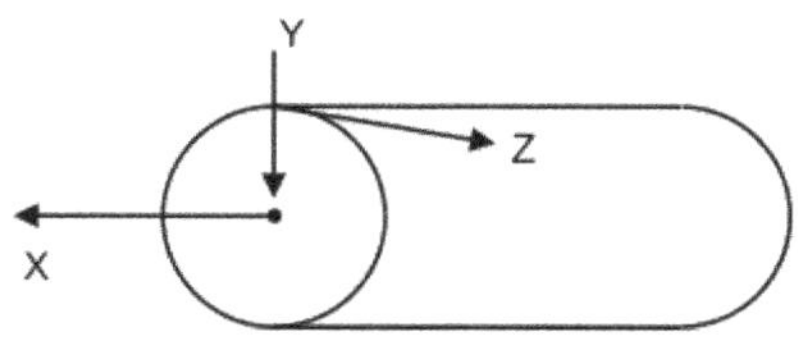

MT dissection

J.A. Tuszyński, *et al.*, "Anisotropic Elastic Properties of Microtubules", *European Physics Journal E Soft Matter*, Vol. 17, No. 1, 2005, pp. 29–35.

Review Notes: Good drawings of tubulin.

Table of elastic properties.

$E = 1.2$ to 2.7×10^9 Pa [8–11]

$E = 1$ to 5×10^8 Pa [5, 12]

$G = 1.4 \times 10^6$ Pa

$v = 0.3$ to 0.4

$$\mathrm{EI} = (16 \text{ to } 45) \times 10^{-24}\ \mathrm{Nm}^2 \ \text{(see 77)}.$$

O. Valiron, N. Caudron, and D. Job, "Microtubule Dynamics", *Cellular and Molecular Life Sciences*, Vol. 58, 2001, pp. 2069–2084.

Review Notes: (2001) "The precise ways in which microtubules achieve their cellular functions is not fully understood yet." "MT arrays in cells are dynamic, capable of assembly, disassembly, and rearrangement on a time scale of seconds to minutes." Dynamics in vitro and in vivo are similar.

A. Vemu, *et al.*, "Severing Enzymes Amplify Microtubule Arrays through Lattice GTP-Tubulin Incorporation", *Science*, Vol. 361, 2018, p. 768.

Review Notes: The breaking up of microtubules is a poorly understood phenomenon.

V.A. Volkov, *et al.*, "Long Tethers Provide High-Force Coupling of the Dam1 Ring to Shortening Microtubules", *Proceedings of the National Academy of Sciences of the United States of America*, Vol. 110, 2013, pp. 7708–7713.

Review Notes: Provides some evidence and a theoretical basis for ring attachment between kinetochores and microtubules.

K. Vukušić, *et al.*, "Microtubule Sliding within the Bridging Fiber Pushes Kinetochore Fibers Apart to Segregate Chromosomes", *Developmental Cell*, Vol. 43, 2017, pp. 11–23.

Review Notes: "The forces that drive chromosome segregation in mitosis in human cells remain poorly understood" — from the abstract. Fiber crosslinking is thought to be a principal factor.

C.E. Walczak, "Microtubule Dynamics and Tubulin Interacting Proteins", *Current Opinion in Cell Biology*, Vol. 12, 2000, pp. 52–56.

Review Notes: (2015) Similar to 489 in the four phases.

X. Wan, *et al.*, "Protein Architecture of the Human Kinetochore Microtubule Attachment Site", *Cell*, Vol. 137, 2009, pp. 672–684.

Review Notes: Robust spindle microtubule (MT) plus-end attachment is "end-on," and MTs bound to kinetochores are known as kinetochore microtubules (kMTs).

C.Y. Wang, C.Q. Ru, and A. Mioduchowski, "Vibration of Microtubules as Orthotropic Shells", *Physica E: Low-Dimensional Systems and Nanostructures*, Vol. 35, No. 1, 2006, pp. 48–56.

Review Notes: "Circumferential" modulus ~ 0.1 "longitudinal module ⇒ orthotropic shell model — (same as 62). Isotropic model overestimates frequencies except in longitudinal and bending modes. (Specifically in radial modes is the major difference.)

S. Werner, A. Pimenta-Marques, and M. Battencourt-Dias, "Maintaining Centrosomes and Cilia", Journal of Cell Science, Vol. 130, 2017, pp. 3789–3800. doi:10.1242/jcs.203505

Review Notes: Centrosomes are thought to be stable. This is a review paper which briefly summarizes centrosome assembly.

C. Wiese and Y. Zheng, "A New Function for the ϒ-Tubulin Ring Complex as a Microtubule Minus-End Cap", *Nature Cell Biology*, Vol. 2, 2000, pp. 358–364.

Review Notes: The focus of the paper is on the behavior of the γ-TuRC in nucleating MTs. γ-tubulin is required for MT nucleation. "γ-TuRC resembles a lock washer ..." "like the first turn of the MT helix". Two models are discussed for MT nucleation: (1) "template" and (2) "protofilament." In template the γ-TuRC is the first turn in an assembled helix. In protofilament, the γ-TuRC uncurls the dimers longitudinally. The authors state that the in vitro results favor the template model. γ-TuRC is a lock washer shaped protein complex.

C. Wiese and Y. Zheng, "Microtubule Nucleation: ϒ-Tubulin and Beyond", *Journal of Cell Science*, Vol. 119, 2006, pp. 4143–4153.

Review Notes: γ-Tubulin plays key role in MT formation. Necessary but not sufficient. γ-Tubulin discovered ~1989.

S.M. Wignall and A.M. Villeneuve, "Lateral Microtubule Bundles Promote Chromosome Alignment during Acentrosomal Oocyte Meiosis", *Nature Cell Biology*, Vol. 11, No. 7, 2009, pp. 839–844.

Review Notes: This paper discusses the assembly of MTs about the chromosomes.

J. Wu, *et al.*, "How Dynein and Microtubules Rotate the Nucleus", Journal of Cellular Physiology, Vol. 226, 2016, pp. 2666–2674.

Review Notes: Authors present a model showing that the microtubule rotates the nucleus.

Z. Yue, *et al.*, "Mechanisms Mitigating Problems Associated with Multiple Kinetochores on One Microtubule in Early Mitosis", *Journal of Cell Science*, Vol. 130, 2017, pp. 2266–2276.

Review Notes: The kinetochore interacts with the lateral side of the microtubule and then is subsequently tethered at the microtubule end. Multiple kinetochores could occasionally interact with a single microtubule. Although multiple kinetochores can interact with the lateral side of microtubule only one kinetochore can form a sustained attachment. Kinetochores can slide along a microtubule. This sliding enables only one kinetochore to be attached to the end of the microtubule.

J. Zhang, W.-H. Guo, and Y.-L. Wang, "Microtubules Stabilize Cell Polarity by Localizing Rear Signals", *Proceedings of the National Academy of Sciences of the United States of America*, Vol. 111, No. 46, 2014, pp. 16383–16388.

Review Notes: MTs do not generate, but instead maintain cell polarity.

W. Zimmerman, *et al.*, "Mitosis-Specific Anchoring of γ-Tubulin Complexes by Pericentrin Controls Spindle Organization and Mitotic Entry", *Molecular Biology of the Cell*, Vol. 15, 2004, pp. 3672–3657.

Review Notes: Author states that the pericentrin provides the anchor for the γ-tubulin ring complex (γ-TuRC).

3.11. First Author Alphabetical Listing of References

S. Afreen and D. Varma, "Cell Divisions: Molecular Pathways for KMN Kinetochore Recruitment", *Current Biology*, Vol. 24, 2015, pp. R328–R347.

T. Afrin, *et al.*, "Buckling of Microtubules on Elastic Media via Breakable Bonds", *Biochemical and Biophysical Research Communications*, Vol. 480, 2016, pp. 132–138.

A. Akhmanova and C.C. Hoogenraad, "Microtubule Minus-End-Targeting Proteins", *Current Biology*, Vol. 25, 2015, pp. R162–R171.

T.K. Akhshi, D. Wernike, and A. Piekny, "Microtubules and Actin Crosstalk in Cell Migration and Division", *Cytoskeleton*, Vol. 71, 2014, pp. 1–23.

B. Alberts, *et al.*, *Molecular Biology of the Cell*, 3rd Edition, Garland Publishing, New York, 1994.

H. Aldaz, *et al.*, "Insights into Microtubule Nucleation from the Crystal Structure of Human ϒ-Tubulin", *Nature*, Vol. 435, 2005, pp. 523–527.

C.L. Asbury, J.F. Tien, and T.N. Davis, "Kinetochores' Gripping Feat: Conformational Wave or Biased Diffusion?" *Trends in Cell Biology*, Vol. 21, 2011, pp. 38–46.

C. Aumeier, *et al.*, "Self-Repair Promotes Microtubule Rescue", *Nature Cell Biology*, Vol. 18, No. 10, 2016, pp. 1054–1064.

N.A. Baker, *et al.*, "Electrostatics of Nanosystems: Application to Microtubules and the Ribosome", *Proceedings of the National Academy of Sciences of the United States of America*, Vol. 98, 2001, pp. 10037–10041.

H. Berg, *et al.*, "Bioelectromagnetic Field Effects on Cancer Cells and Mice Tumors", *Electromagnetic Biology & Medicine*, Vol. 29, No. 4, 2010, pp. 132–143.

A.D. Bicek, *et al.*, "Analysis of Microtubule Curvature", *Methods in Cell Biology*, Vol. 83, 2007, Chapter 10, pp. 237–268.

D.G. Booth and W.C. Earnshaw, "Ki-67 and the Chromosome Periphery Compartment in Mitosis", *Trends in Cell Biology*, Vol. 27, No. 12, 2017, pp. 906–913.

B.P. Bouchet and A. Akhmanova, "Microtubules in 3D Cell Motility", *Journal of Cell Science*, Vol. 130, 2017, pp. 39–50.

A. Bouissou, *et al.*, "Gamma-Tubulin Ring Complexes Regulate Microtubule Plus End Dynamics", *Journal of Cell Biology*, Vol. 187, No. 3, 2009, pp. 327–334.

C.P. Brangwynne, *et al.*, "Microtubules can Bear Enhanced Compressive Loads in Living Cells because of Lateral Reinforcement", *Journal of Cell Biology*, Vol. 173, No. 5, 2006, pp. 733–761. doi:10.1083/jcb.200601060

G.J. Brouhard and L.M. Rice, "The Contribution of αβ-tubulin Curvature to Microtubule Dynamics", *Journal of Cell Biology*, Vol. 207, No. 3, 2014, pp. 323–334.

M. Casenghi, *et al.*, "Polo-Like Kinase 1 Regulates Nip a Centrosome Protein Involved in Microtubule Nucleation", *Developmental Cell*, Vol. 5, 2003, pp. 113–125.

L. Cassimeris, *et al.*, "XMAP215 Is a Long Thin Molecule that Does Not Increase Microtubule Stiffness", *Journal of Cell Science*, Vol. 114, 2001, pp. 3025–3033.

B.T. Castle and D.J. Odde, "Optical Control of Microtubule Dynamics in Time and Space", *Cell*, Vol. 162, 2015, pp. 243–245.

N. Caudron, *et al.*, "A Reassessment of the Factors Affecting Microtubule Assembly and Disassembly In Vitro", *Journal of Molecular Biology*, Vol. 297, 2000, pp. 211–220.

C.-W. Chang, *et al.*, "CEP295 Interacts with Microtubules and Is Required for Centriole Elongation", Journal of Cell Science, Vol. 129, 2016, pp. 2501–2513.

P. Chang and T. Stearns, "δ-Tubulin and ε-Tubulin: Two New Human Centrosomal Tubulins Reveal New Aspects of Centrosome Structure and Function", *Nature Cell Biology*, Vol. 2, 2000, pp. 30–35.

I.M. Cheeseman and A. Desai, "Molecular Architecture of the Kinetochore-Microtubule Interface", *Nature Reviews Molecular Cell Biology*, Vol. 9, 2008, pp. 33–46.

C. Ciferri, *et al.*, "Implications for Kinetochore-Microtubule Attachment from the Structure of an Engineered Ndc80 Complex", *Cell*, Vol. 133, 2008, pp. 427–439.

M. Cifra, J.Z. Fields, and A. Farhadi, "Electromagnetic Cellular Interactions", *Progress in Biophysics and Molecular Biology*, Vol. 105, 2011, pp. 223–246.

M. Cifra, *et al.*, "Electric Field Generated by Axial Longitudinal Vibration Modes of Microtubule", *Biosystems*, Vol. 100, No. 2, 2010, pp. 122–131.

J.R. Cooper, *et al.*, "Catalysis of the Microtubule On-Rate Is the Major Parameter Regulating the Deploymerase Activity of MCAK", *Nature Structural and Molecular Biology*, Vol. 17, 2010, pp. 77–82.

P.T. Conduit, "Microtubule Organization: A Complex Relation", *Journal of Cell Biology*, Vol. 213, 2016, pp. 809–812.

J.G. DeLuca, "Kinetochore-Microtubule Dynamics and Attachment Stability", *Methods in Cell Biology*, Vol. 97, 2010, Chapter 4, pp. 53–79.

M. Dogterom, *et al.*, "Force Generation by Dynamic Microtubules", *Current Opinion in Cell Biology*, Vol. 17, 2005, pp. 67–74.

K.H. Downing and E. Nogales, "New Insights into Microtubule Structure and Function from the Atomic Model of Tubulin", *European Biophysics Journal*, Vol. 27, No. 5, 1998, pp. 431–436.

S. Dumont and T.J. Mitchison, "Force and Length in the Mitotic Spindle", *Current Biology*, Vol. 19, 2009, pp. R749–R761.

M.W. Elting, *et al.*, "Force on Spindle Microtubule Minus Ends Moves Chromosomes", *Journal of Cell Biology*, Vol. 206, No. 2, 2014, pp. 245–256.

M.K. Gardner, "Cell Biology: Microtubule Collisions to the Rescue", *Current Biology*, Vol. 26, 2016, pp. R1287–R1289.

F. Gittes, *et al.*, "Flexural Rigidity of Microtubules and Actin Filaments Measured from Thermal Fluctuations in Shape", *Journal of Cell Biology*, Vol. 120, No. 4, 1993, pp. 923–934.

S.A. Godinho, M. Kwon, and D. Pellman, "Centrosomes and Cancer: How Cancer Cells Divide with too Many Centrosomes", *Cancer Metastasis Review*, Vol. 28, 2009, pp. 85–98.

L. Grishchuk, *et al.*, "Force Production by Disassembling Microtubules", *Nature*, Vol. 438, No. 17, 2005, pp. 385–388.

S.R. Hameroff, "A New Theory of the Origin of Cancer: Quantum Coherent Entanglement, Centrioles, Mitosis, and Differentiation", *Biosystems*, Vol. 77, No. 1–3, 2004, pp. 119–136.

L. Hamon, P.A. Curmi, and D. Pastré, "High-Resolution Imaging of Microtubules and Cytoskeleton Structures by Atomic Force Microscopy", *Methods in Cell Biology*, Vol. 95, 2010, pp. 157–174.

D. Havelka and M. Cifra, "Calculation of the Electromagnetic Field Around a Microtubule", *Acta Polytechnica*, Vol. 49, 2009, pp. 58–63.

D. Havelka, *et al.*, "High-Frequency Electric Field and Radiation Characteristics of Cellular Microtubule Network", *Journal of Theoretical Biology*, Vol. 286, 2011, pp. 31–40.

S.M. Hinshaw and S.C. Harrison, "Kinetochore Function from the Bottom Up", *Trends in Cell Biology*, Vol. 28, No. 1, 2018, pp. 22–33.

S.A. Hormeño, *et al.*, "Single Centrosome Manipulation Reveals Its Electric Charge and Associated Dynamic Structure", *Biophysical Journal*, Vol. 97, No. 4, 2009, pp. 1022–1030.

J. Howard and A.A. Hyman, "Dynamics and Mechanics of the Microtubule Plus End", *Nature*, Vol. 422, 2003, pp. 753–758.

Y.F. Inclán and E. Nogales, "Structural Models for the Self-Assembly and Microtubule Interactions of Gamma-, Delta- and Epsilon-Tubulin", *Journal of Cell Science*, Vol. 114, 2000, pp. 413–422.

C. Janke, "The Tubulin Code: Molecular Components, Readout Mechanisms, and Functions", *Journal of Cell Biology*, Vol. 206, No. 4, 2014, pp. 461–472.

M.E. Janson, M.E. deDood, and M. Dogterom, "Dynamic Instability of Microtubules is Regulated by Force", *Journal of Cell Biology*, Vol. 161, No. 6, 2003, pp. 1029–1034.

D. Job, O. Valiron, and B. Oakley, "Microtubule Nucleation", *Current Opinion in Cell Biology*, Vol. 15, 2003, pp. 111–117.

A.P. Joglekar, K. Bloom, and E.D. Salmon, "In Vivo Protein Architecture of the Eukaryotic Kinetochore with Nanometer Scale Accuracy", *Current Biology*, Vol. 19, 2009, pp. 694–699.

A.P. Joglekar, *et al.*, "Molecular Architecture of a Kinetochore — Microtubule Attachment Site", *Nature Cell Biology*, Vol. 8, 2006, pp. 581–585.

M.A. Jordan and L. Wilson, "Microtubules as a Target for Anticancer Drugs", *Nature Reviews: Cancer*, Vol. 4, No. 4, 2004, pp. 253–265.

K. Kamath, E. Oroudjev, and M.A. Jordan, "Determination of Microtubule Dynamic Instability in Living Cells", *Methods in Cell Biology*, Vol. 97, 2010, pp. 1–14.

S. Kasas, *et al.*, "Oscillation Modes of Microtubules", *Biology of the Cell*, Vol. 96, No. 9, 2004, pp. 697–700.

T.J. Keating and G.G. Borisky, "Immunostructural Evidence for the Template Mechanism of Microtubule Nucleation", *Nature Cell Biology*, Vol. 2, 2000, pp. 352–357.

I.A. Kent and T.P. Lele, "Microtubule Based Force Generation", *Wiley Interdisciplinary Reviews: Nanomedicine and Nanobiotechnology*, Vol. 9, No. 3, 2016, p. e1428.

A. Kis, *et al.*, "Nanomechanics of Microtubules", *Physical Review Letters*, Vol. 89, No. 24, 2002, pp. 248101-1–248101-4.

S.L. Kline-Smith and C.E. Wolczak, "The Microtubule-Destabilizing Kinesin XKCM1 Regulates Microtubule Dynamic Instability in Cells", *Molecular Biology of the Cell*, Vol. 13, 2002, pp. 2718–2731.

J.M. Kollman, *et al.*, "Microtubule Nucleating γ-TuSC Assembles Structures with 13-Fold Microtubule-like Symmetry", *Nature*, Vol. 466, 2010, pp. 879–883.

B. Lacroix, *et al.*, "Microtubule Dynamics Scale with Cell Size to Set Spindle Length and Assembly Timing", *Developmental Cell*, Vol. 45, 2018, pp. 496–511.

B.M. Lange and K. Gull, "Structure and Function of the Centriole in Animal Cells: Progress and Questions", *Trends in Cell Biology*, Vol. 6, No. 9, 1996, pp. 348–352.

R. Leguy, *et al.*, "Monomeric Υ-Tubulin Nucleates Microtubules", *Journal of Biological Chemistry*, Vol. 275, No. 29, 2000, pp. 21975–21980.

T.-C. Lin, A. Neuner, and E. Schiebel, "Targeting of Υ-Tubulin Complexes to Microtubule Organizing Centers: Conservation and Divergence", *Trends in Cell Biology*, Vol. 25, No. 5, 2015, pp. 296–307.

R. Mahen and A.R. Venkitaraman, "Pattern Formation in Centrosome Assembly", *Current Opinion in Cell Biology*, Vol. 24, No. 1, 2012, pp. 14–23.

H. Maiato, *et al.*, "The Dynamic Kinetochore-Microtubule Interface", *Journal of Cell Science*, Vol. 117, 2004, pp. 5461–5477.

E.N. Marieb, *Human Anatomy and Physiology*, 3rd Edition, Benjamin/ Cummings Publishing, Redwood City, CA, 1991, Chapter 3, pp. 60–101.

G. Marteil, *et al.*, "Over-Elongation of Centrioles in Cancer Promotes Centriole Amplification and Chromosome Missegregation", *Nature Communications*, Vol. 9, 2018, pp. 1258.

P.G. McKean, S. Vaughan, and K. Gull, "The Extended Tubulin Superfamily", *Journal of Cell Science*, Vol. 114, Pt. 15, 2001, pp. 2723–2733.

F.J. McNally, "Mechanisms of Spindle Positioning", *Journal of Cell Biology*, Vol. 200, 2013, pp. 131–159.

F.J. McNally and A. Roll-Mecak, "Microtubule-Severing Enzymes: From Cellular Functions to Molecular Mechanism", *Journal of Cell Biology*, Vol. 217, No. 12, pp. 4057–4069.

V. Mennella, *et al.*, "Subdiffraction-Resolution Fluorescence Microscopy Reveals a Domain of the Centrosome Critical for Pericentriolar Material Organization", *Nature Cell Biology*, Vol. 14, No. 11, 2012, pp. 1159–1168.

A. Mershin, *et al.*, "Tubulin Dipole Moment, Dielectric Constant and Quantum Behavior: Computer Simulations, Experimental Results and Suggestions", *Biosystems*, Vol. 77, 2004, pp. 73–85.

T. Mitchison and M. Kirschner, "Dynamic Instability of Microtubule Growth", *Nature*, Vol. 312, 1984a, pp. 237–242.

T. Mitchison and M. Kirachner, "Microtubule Assembly Nucleated by Isolated Centrosomes", *Nature*, Vol. 312, 1984b, pp. 232–237.

M. Moritz and D.A. Agard, "ϒ-Tubulin Complexes and Microtubule Nucleation", *Current Opinion in Structural Biology*, Vol. 11, 2001, pp. 174–181.

M. Moritz, *et al.*, "Reconstruction of Centrosome Microtubule Nucleation in Drosophila", *Methods in Cell Biology*, Vol. 67, 1998, pp. 141–148.

T. Müller-Reichert, R. Kiewisz, and S. Redamann, "Mitotic Spindles Revisited–New Insights from 3D Electron Microscopy", *Journal of Cell Science*, Vol. 131, 2018.

D.J. Needleman, *et al.*, "Synchrotron X-ray Diffraction Study of Microtubules Buckling and Bundling under Osmotic Stress: A Probe of Interprotofilament Interactions", *Physical Review Letters*, Vol. 193, No. 19, 2004, pp. 198104-1–198104-4.

E.A. Nigg, L. Cajanek, and C. Arquint, "The Centrosome Duplication Cycle in Health and Disease", *FEBS Letters*, Vol. 588, 2014, pp. 2366–2372.

E. Nogales, "Structural Insights into Microtubule Function", *Annual Review of Biochemistry*, Vol. 69, 2000, pp. 277–302.

E. Nogales, *et al.*, "High-Resolution Model of the Microtubule", *Cell*, Vol. 96, No. 1, 1999, pp. 79–88.

B.R. Oakley, V. Paolillo, and Y. Zheng, "Gamma-Tubulin Complexes in Microtubule Nucleation and Beyond", *Molecular Biology of the Cell*, Vol. 26, 2015, pp. 2957–2962.

C.B. O'Connell and A.L. Khodjakov, "Cooperative Mechanisms of Mitotic Spindle Formation", *Journal of Cell Science*, Vol. 120, 2007, pp. 1717–1722.

D. Passeri, *et al.*, "Magnetic force microscopy", *Biomatter*, Vol. 4, No. 1, 2014. doi:10.4161/biom.29507

J. Paz and J. Lüders, "Microtubule-Organizing Centers: Towards a Minimal Parts List", *Trends in Cell Biology*, Vol. 28, No. 3, 2018, pp. 176–187.

S. Petry and R.D. Vale, "Microtubule Nucleation at the Centrosome and Beyond", *Nature Cell Biology*, Vol. 17, No. 9, 2015, pp. 1089–1093.

A. Pitaval, *et al.*, "Microtubule Stabilization Drives 3D Centrosome Migration to Initiate Primary Ciliogenesis", *Journal of Cell Biology*, Vol. 216, 2017, p. 3713.

J. Pokorný, "Endogenous Electromagnetic Forces in Living Cells: Implications for Transfer of Reaction Components", *Electro- and Magnetobiology*, Vol. 20, No. 1, 2001, pp. 59–73.

J. Pokorný, "Viscous Effects on Polar Vibrations in Microtubules", *Electromagnetic Biology and Medicine*, Vol. 22, No. 1, 2003, pp. 15–29.

J. Pokorný, "Electrodynamic Activity of Healthy and Cancer Cells", *Journal of Physics: Conference Series*, Vol. 329, No. 1, 2011, p. 012007.

J. Pokorný, "Physical Aspects of Biological Activity and Cancer", *AIP Advances*, Vol. 2, 2012, pp. 011207-1–011207-11.

J. Pokorný, J. Hašek, and F. Jelínek, "Electromagnetic Field of Microtubules: Effects on Transfer of Mass Particles and Electrons", *Journal of Biological Physics*, Vol. 31, 2005, pp. 501–514.

J. Pokorný, *et al.*, "Vibration in Microtubules", *Journal of Biological Physics*, Vol. 23, 1997, pp. 171–179.

J. Pokorný, J. Pokorný, and J. Kobilková, "Postulates on Electromagnetic Activity in Biological Systems and Cancer", *Integrative Biology*, Vol. 5, No. 12, 2013, pp. 1439–1446.

J. Pokorný, *et al.*, "Cancer Physics: Diagnosis Based on Damped Cellular Elastoelectrical Vibrations in Microtubules", *European Biophysics Journal*, Vol. 40, No. 6, 2011, pp. 747–759.

T.D. Pollard, "Nine Unanswered Questions About Cytokinesis", *Journal of Cell Biology*, Vol. 916, 2017 Archive, pp. 3007 ff.

S. Portet, *et al.*, "Elastic Vibrations in Seamless Microtubules", *European Biophysics Journal*, Vol. 34, No. 7, 2005, pp. 912–920.

A.F. Powers, *et al.*, "The Ndc80 Kinetochore Complex Forms Load Bearing Attachments to Dynamic Microtubule Tips via Biased Diffusion", *Cell*, Vol. 136, 2009, pp. 865–875.

M.E. Presenti, J.R. Weir, and A. Musacchio, "Progress in the Structural and Functional Characterization of Kinetochores", *Current Opinion in Structural Biology*, Vol. 37, 2016, pp. 152–163.

X.S. Qian, J.Q. Zhang, and C.Q. Ru, "Wave Propagation in Orthotropic Microtubules", *Journal of Applied Physics*, Vol. 101, 2007, pp. 084702-1–084702-7.

B. Raynaud-Messina and A. Merdes, "Gamma-Tubulin Complexes and Microtubule Organization", *Current Opinion in Cell Biology*, Vol. 19, No. 1, 2007, pp. 24–30.

L.M. Rice, "A New Look for the Growing Microtubule End?", *Journal of Cell Biology*, Vol. 217, No. 8, 2018, p. 2609.

R.M. Rios, "The Centrosome-Golgi Apparatus Nexus", Philosophical Transactions of the Royal Society of London. Series B, *Biological Sciences*, Vol. 369, 2014, 17 pages.

M.D. Rose, S. Biggins, and L.L. Satterwhite, "Unravelling the Tangled Web at the Microtubule-Organizing Center", *Current Opinion in Cell Biology*, Vol. 5, No. 1, 1993, pp. 105–115.

C. Sacristan, *et al.*, "Dynamic Kinetochore Size Regulation Promotes Microtubule Capture and Chromosome Biorientation in Mitosis", *Nature Cell Biology*, Vol. 20, 2018, pp. 800–810.

J. Salogiannis and S.L. Reck-Peterson, "Hitchhiking: A Non-Canonical Mode of Microtubule-Based Transport", *Trends in Cell Biology*, Vol. 27, 2017, pp. 141–150.

A.D. Sanchez and J.L. Feldman, "Microtubule Organizing Centers: From the Centrosome to Non-Centrosomal Sites", *Current Opinion in Cell Biology*, Vol. 44, 2017, pp. 93–101.

M. Sanhaji, *et al.*, "Mitotic Centromere-Associated Kinesin (MCAK): A Potential Cancer Drug Target", *Oncotarget*, Vol. 2, 2011, pp. 935–947. doi:10.18632/oncotarget.416

S. Santaguida and A. Musacchio, "The Life and Miracles of Kinetochores", *EMBO Journal*, Vol. 28, 2009, pp. 2511–2531.

M.V. Satarić, *et al.*, "Modelling the Role of Intrinsic Electric Fields in Microtubules as an Additional Control Mechanism of Bi-Directional Intracellular Transport", *Cell Biochemistry and Biophysics*, Vol. 52, No.2, 2008, pp. 113–124.

M. Satarić, B.M. Satarić, and J.A. Tuszynski, "Nonlinear Model of Microtubule Dynamics", *Electromagnetic Biology and Medicine*, Vol. 24, 2005, pp. 255–264.

M.V. Satarić and J.A. Tuszyński, "Relationship between the Nonlinear Ferroelectric and Liquid Crystal Models for Microtubules", *Physical Review*, Vol. 67, 2003, pp. 011901-1–011901-11.

L. Schaedel, *et al.*, "Microtubules Self-Repair in Response to Mechanical Stress", *Nature Materials*, Vol. 14, 2015, pp. 1156–1165.

H.T. Schek III, *et al.*, "Microtubule Assembly Dynamics at the Nanoscale", *Current Biology*, Vol. 17, No. 17, 2007, pp. 1445–1455.

E. Schiebel, "ϒ-Tubulin Complexes: Binding to the Centrosome, Regulation and Microtubule Nucleation", *Current Opinion in Cell Biology*, Vol. 12, 2000, pp. 113–118.

J. Scrofani, *et al.*, "Microtubule Nucleation in Mitosis by a Ran GTP-Dependent Protein Complex", *Current Biology*, Vol. 25, 2015, pp. 131–140.

M.M. Shawki and A. Farid, "Low Electric Field Parameters Required to Induce Death of Cancer Cells", *Electromagnetic Biology & Medicine*, Vol. 33, No. 2, 2014, pp. 159–163.

B. Short, "Mapping the Kinetochore Delta", *Journal of Cell Biology*, Vol. 212, 2016, p. 259.

V. Sikirzhytski, *et al.*, "Microtubules Assemble Near Most Kinetochores during Early Prometaphase in Human Cells", *Journal of Cell Biology*, Vol. 217, No. 8, 2018, p. 2641.

A. Singh, *et al.*, "Polarized Microtubule Dynamics Directs Cell Mechanics and Coordinates Forces during Epithelial Morphogenesis", *Nature Cell Biology*, Vol. 20, 2018, pp. 1126–1133.

Y.M. Sirenko, M.A. Stroscio, and K.W. Kim, "Elastic Vibrations of Microtubules in a Fluid", *Physical Review E*, Vol. 53, No. 1, 1996, pp. 1003–1010.

K.C. Slep, "A Microtubule Dynamics Reconstitutional Convention", *Journal of Cell Biology*, Vol. 205, No. 3, 2016, pp. 305–315.

K.F. Sonnen, *et al.*, "3D-structured Illumination Microscopy Provides Novel Insight into Architecture of Human Centrosomes", *Biology Open*, Vol. 1, 2012, pp. 965–976.

R. Stracke, *et al.*, "Analysis of the Migration Behavior of Single Microtubules in Electric Fields", *Biochemical and Biophysical Research Communications*, Vol. 293, 2002, pp. 602–609.

P. Strnad and P. Gönczy, "Mechanisms of Procentriole Formation", *Trends in Cell Biology*, Vol. 18, No. 8, 2008, pp. 389–396.

P. Strnad, *et al.*, "Regulated HsSAS-6 Levels Ensure Formation of a Simple Procentriole per Centriole during Centrosome Duplication Cycle", *Developmental Cell*, Vol. 13, No. 2, 2007, pp. 203–213.

S. Suresh, "Biomechanics and Biophysics of Cancer Cells", *Acta Biomaterialia*, Vol. 3, No. 4, 2007, pp. 413–438.

T.U. Tanaka and A. Desai, "Kinetochore-Microtubule Interactions: The Means to the End", *Current Opinion in Cell Biology*, Vol. 20, 2008, pp. 53–63.

M.E. Tanenbaum and R.H. Medema, "Mechanisms of Centrosome Separation and Bipolar Spindle Assembly", *Developmental Cell*, Vol. 19, 2010, pp. 797–806.

A.-M. Tassin and M. Bornens, "Centrosome Structure and Microtubule Nucleation in Animal Cells", *Biology of the Cell*, Vol. 91, No. 4–5, 1999, pp. 343–354.

N. Teixido-Travesa, J. Roig, and J. Luders, "The Where, When and How of Microtubule Nucleation–One Ring to Rule Them All", *Journal of Cell Science*, Vol. 125, 2012, pp. 4445–4456.

J.F. Tien, *et al.*, "Cooperation of the Dam1 and Ndc80 Kinetochore Complexes Enhances Microtubule Coupling and Is Regulated by Aurora B", *Journal of Cell Biology*, Vol. 189, No. 4, 2010, pp. 713–723. doi:10.1083/jcb.200910142.

I.M. Tolić, "Mitotic Spindle: Kinetochore Fibers Hold on Tight to Interpolar Bundles", *European Biophysics Journal*, Vol. 47, 2018, pp. 191–203.

M.C. Tsai and J. Ahringer, "Microtubules Are Involved in Anterior-Posterior Axis Formation in *C. elegans* Embryos", *Journal of Cell Biology*, Vol. 179, No. 3, 2007, pp. 397–402.

J.A. Tuszyński, *et al.*, "Molecular Dynamics Simulations of Tubulin Structure and Calculations of Electrostatic Properties of Microtubules", *Mathematical and Computer Modelling*, Vol. 41, No. 10, 2005, pp. 1055–1070.

J.A. Tuszyński, *et al.*, "Anisotropic Elastic Properties of Microtubules", *European Physics Journal E Soft Matter*, Vol. 17, No. 1, 2005, pp. 29–35.

O. Valiron, N. Caudron, and D. Job, "Microtubule Dynamics", *Cellular and Molecular Life Sciences*, Vol. 58, 2001, pp. 2069–2084.

A. Vemu, *et al.*, "Severing Enzymes Amplify Microtubule Arrays through Lattice GTP-Tubulin Incorporation", *Science*, Vol. 361, 2018, p. 768.

V.A. Volkov, *et al.*, "Long Tethers Provide High-Force Coupling of the Dam1 Ring to Shortening Microtubules", *Proceedings of the National Academy of Sciences*, Vol. 110, 2013, pp. 7708–7713.

K. Vukušić, *et al.*, "Microtubule Sliding within the Bridging Fiber Pushes Kinetochore Fibers Apart to Segregate Chromosomes", Developmental *Cell*, Vol. 43, 2017, pp. 11–23.

C.E. Walczak, "Microtubule Dynamics and Tubulin Interacting Proteins", *Current Opinion in Cell Biology*, Vol. 12, 2000, pp. 52–56.

X. Wan, *et al.*, "Protein Architecture of the Human Kinetochore Microtubule Attachment Site", *Cell*, Vol. 137, 2009, pp. 672–684.

C.Y. Wang, C.Q. Ru, and A. Mioduchowski, "Vibration of Microtubules as Orthotropic Shells", *Physica E: Low-dimensional Systems and Nanostructures*, Vol. 35, No. 1, 2006, pp. 48–56.

C.Y. Wang and L.C. Zhang, "Circumferential Vibration of Microtubules with Long Axial Wavelength", *Journal of Biomechanics*, Vol. 41, 2008, pp. 1892–1896.

S. Werner, A. Pimenta-Marques, and M. Battencourt-Dias, "Maintaining Centrosomes and Cilia", *Journal of Cell Science*, Vol. 130, 2017, pp. 3789–3800. doi:10.1242/jcs.203505.

C. Wiese and Y. Zheng, "A New Function for the ϒ-Tubulin Ring Complex as a Microtubule Minus-end Cap", *Nature Cell Biology*, Vol. 2, 2000, pp. 358–364.

C. Wiese and Y. Zheng, "Microtubule Nucleation: ϒ-Tubulin and beyond", *Journal of Cell Science*, Vol. 119, 2006, pp. 4143–4153.

S.M. Wignall and A.M. Villeneuve, "Lateral Microtubule Bundles Promote Chromosome Alignment during Acentrosomal Oocyte Meiosis", *Nature Cell Biology*, Vol. 11, No. 7, 2009, pp. 839–844.

J. Wu, *et al.*, "How Dynein and Microtubules Rotate the Nucleus", *Journal of Cellular Physiology*, Vol. 226, 2016, pp. 2666–2674.

Z. Yue, *et al.*, "Mechanisms Mitigating Problems Associated with Multiple Kinetochores on One Microtubule in Early Mitosis", *Journal of Cell Science*, 2017, Vol. 130, pp. 2266–2276.

J. Zhang, W.-H. Guo, and Y.-L. Wang, "Microtubules Stabilize Cell Polarity by Localizing Rear Signals", *Proceedings of the National Academy of Sciences of the United States of America*, Vol. 111, No. 46, 2014, pp. 16383–16388.

W. Zimmerman, *et al.*, "Mitosis-Specific Anchoring of γ-Tubulin Complexes by Pericentrin Controls Spindle Organization and Mitotic Entry", *Molecular Biology of the Cell*, Vol. 15, 2004, pp. 3672–3657.

Chapter 4

Cell Duplication and Separation

4.1. Introduction

Cell duplication is the most fascinating subject in all of biology. Duplication produces not only growth, but also determines organ geometry, limb shape, wound healing, recovery from illness, and enhancement of fitness.

The study of cell duplication dates back to the beginning of biology studies. But as extensively as duplication has been studied, comprehensive understanding of the process is still elusive.

In this chapter, we focus on two aspects of duplication: centrioles and DNA. These are the only parts of the body which duplicate themselves, and remarkably, when they duplicate, they do it at the *same time*.

In the balance of this chapter we describe the duplication. This in turn brings us to two parallel active organelles: the centromere and the chromosome.

Since most of the duplication is conducted by microtubules most of our discussion uses material from Chapter III. Therefore, we use the same references herein. Also, we are using the same permissions from *Advances in Bioscience and Biotechnology*.

4.2. Overall Cell Description Focusing on the Centrioles

Figure 4.1 presents a sketch of a typical eukaryotic (human and animal) cell.

Observe that the cell is "pear-shaped," but not symmetrical. The nucleus occupies the largest space. The other cell parts perform various functions depending upon the cell type.

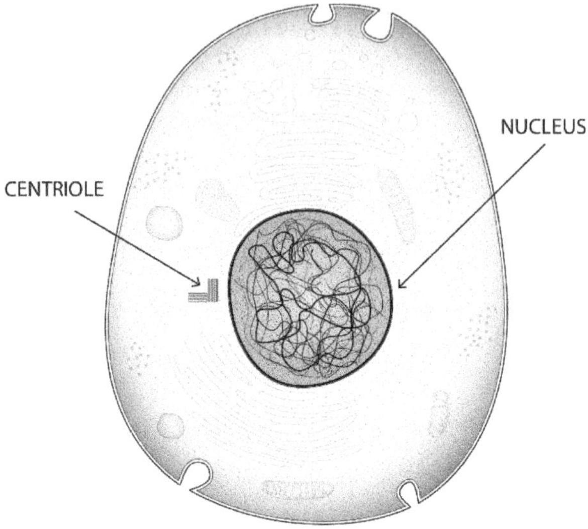

Figure 4.1. Cross-section sketch of a human/animal cell

The centrioles shown in Figure 4.1 are the principal players in causing cell separation. Chapter II provides the details. From afar, the centrioles appear to be two hollow perpendicular cylinders, joined at their base ends. A closer look shows the centrioles to be compared at microtubules — as discussed in Chapter III.

The centrioles have no membrane cover. But they are surrounded by numerous electron dense proteins. The centrioles and their surrounding proteins form what is known as the microtubule organizing center (MTOC).

The largest set of proteins about the centrioles is sometimes known as "pericentriolar material" or PCM. The PCM and the immersed centrioles form what is known as the "centromere."

4.3. Duplication of the Centriole

Consider again the centriole pair, adjacent to the nucleus as in Figure 4.1. When the cell is ready to divide, the pair of centrioles self-duplicate into two pairs. It happens this way: a protein known as "asterless" (AsP) attaches itself to one of the nine outside blades of the centriole. Next, AsP recruits

Plk4 to deposit itself atop the AsP and to become the base for the ensuing structure. While this is happening, yet another protein SAS-6 attaches itself atop the Plk4 to fulfill the intended role of being the "base."

After SAS-6 is established as the base upon one of the nine centriole blades, it expands symmetrically and sends out nine spokes which become platforms for a new ("daughter") centriole. This platform then attracts γ-tubulin to grow the microtubules for the ensuing centriole.

Both members of the centriole pair produce a daughter centriole. The original centrioles then separate from each other. The end result is two centriole pairs.

4.4. Movement of the Newly Created Centriole

The newer centriole pair is only about 80% mature. Next, this, new pair separates itself from the mother pair and begins to migrate around the nucleus to the diametrically opposite side of the nucleus. While this migration is occurring, the new centriole pair becomes mature.

Figures 4.2, 4.3, 4.4, and 4.5 provide a modeling sketch.

Once there are centriole pairs on opposite sides of the nucleus, the centrioles send out microtubules across the nearby nucleus boundary. At the same time the chromosome sends out kinetochores to meet and connect with the advancing microtubules.

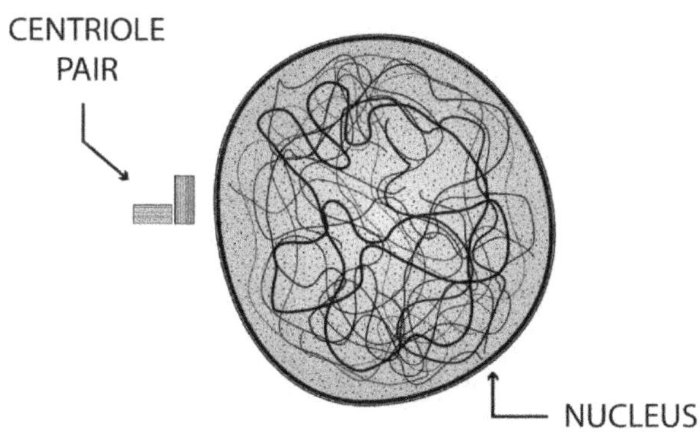

Figure 4.2. The nucleus with its adjacent centriole pair

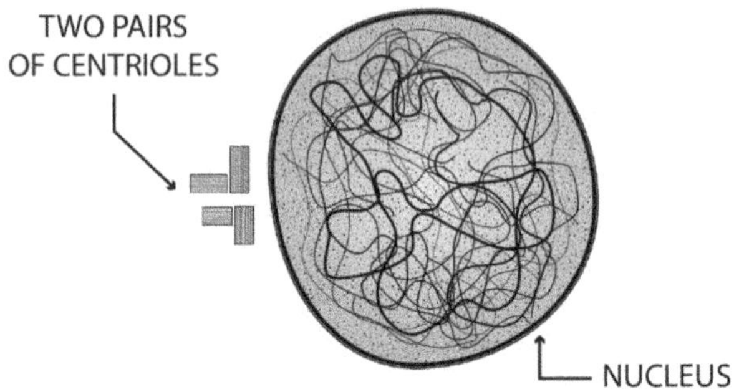

Figure 4.3. Nucleus with two pairs of centrioles

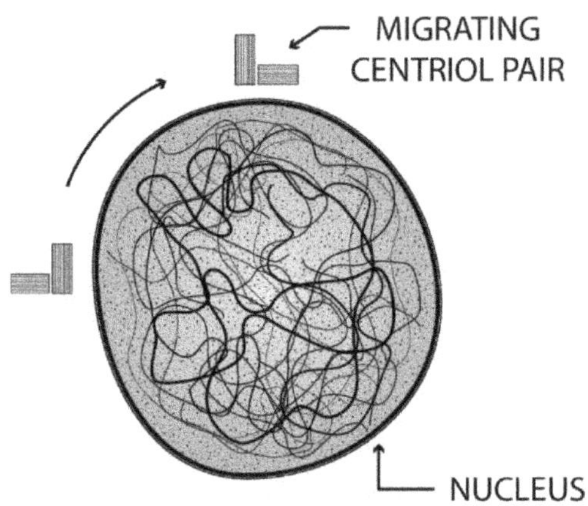

Figure 4.4. Migration of new centriole pair about the nucleus

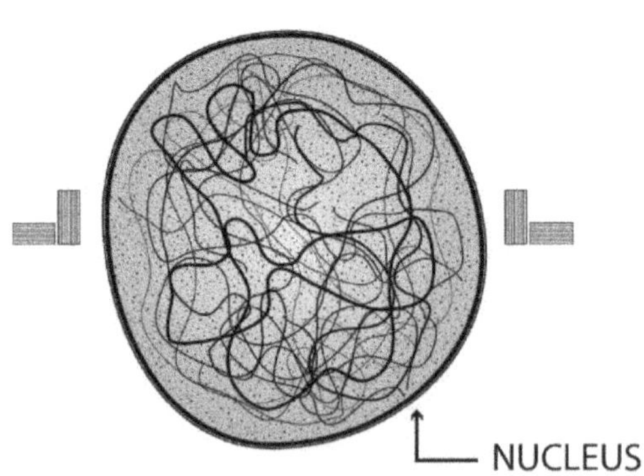

Figure 4.5. The nucleus with diametrically opposite centriole pairs

4.5. The Role of the Kinetochores

The kinetochores seek out the microtubules. They are perpendicular to the microtubules. On occasion, more than one kinetochore will be attached to a microtubule. If this occurs, the kinetochores slide up the shaft of the microtubule with one of the kinetochores anchoring to the microtubule end.

This process results in each half of the nucleus being covered by a set of microtubule/kinetochore strands. Figure 4.6 depicts this condition.

Finally, while this is happening the nucleus itself, in preparation for division, softens in its midsection as represented in Figure 4.7. As the nucleus continues to soften and shrink, separation occurs and the single cell becomes more cells.

The separation mechanics is accomplished via the action of the microtubules and the kinetochores.

4.6. Kinetochores and Microtubules

It may be instructive to look at the microtubules and the kinetochores in greater detail. Chapter III is devoted almost entirely to microtubules so the focus here is upon the kinetochores.

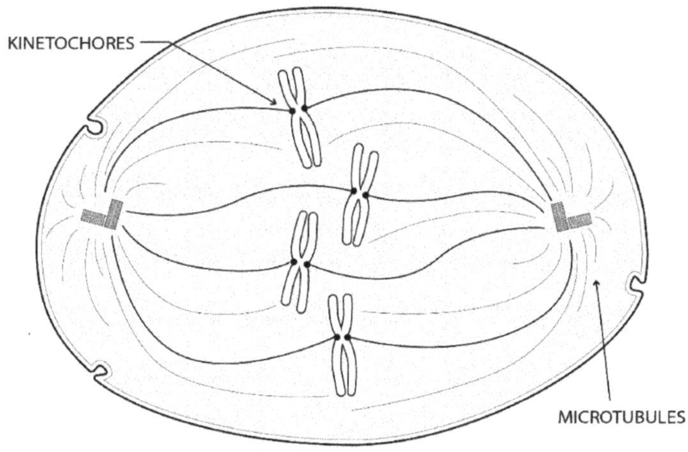

Figure 4.6. The nucleus covered by microtubule/kinetochore strands

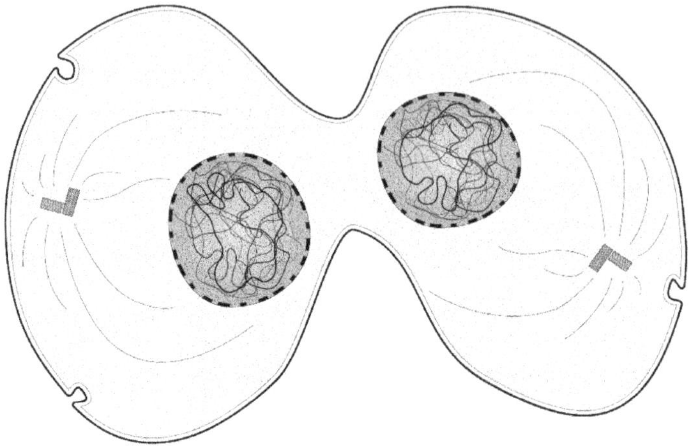

Figure 4.7. Softening of the nucleus

Kinetochores are similar to microtubules in that they are long strands with attachment ends. Although they complement each other their behavior is somewhat different. The microtubules are more robust and the kinetochores seek out the microtubule. The kinetochores approach the microtubule shaft from a perpendicular direction. On occasion, several kinetochores may latch onto one microtubule. When this occurs, the kinetochores slide along the microtubule shaft until they reach the ends, at which case the stronger kinetochore knots to the microtubule end, and the others fall away.

The microtubules originate from an organelle known as the centrosome. From afar the centrosome appears to be a mound of electron dense proteins. But a closer look shows them to be a layered organization of proteins.

As we saw in Chapter III, the microtubules are composed of tubulin from within the centrosome. Thus, the centrosome is also known as the MTOC.

Alternatively, the kinetochores have a similar origin.

4.7. Chromosomes

Figure 4.8 provides a representation of an organelle known as the "chromosome." As seen it has four major "arms" with each arm composed of concentrated DNA.

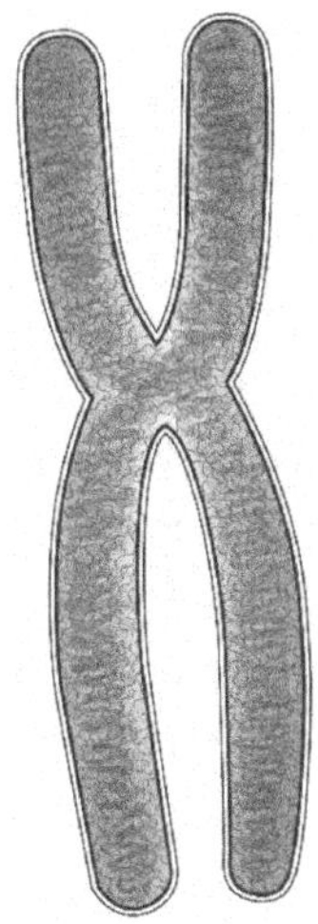

Figure 4.8. Chromosome

The DNA has the remarkability to expand to great lengths or to shrink to infinitesimal size. In the chromosome the DNA is shrunk to its lower limit.

Just as the microtubules emerge from the centrosome, the kinetochores emerge from within the chromosome. Thus, as the centrosome is known as the MTOC, the centrosome may be called the kinetochore organizing center.

From these observations we see parallel between the centrosome and the chromosome and thus also between the microtubules and the kinetochores.

In the following chapter, we describe ways the cell separation and division process can go awry.

4.8. Summary

Cell division via duplication and then separation is a "built-in" feature of eukaryotic (human and animal) cells. When the time for separation occurs, there is simultaneous activity in two analogous organelles: the centrosome and the chromosome. Within the centrosome, the centrioles duplicate themselves and within the chromosome the DNA duplicates.

The centrioles and the DNA are the only two parts of the cell that can duplicate each other — and they do it simultaneously.

Once the centrioles are duplicated, the newer of the duplicated pair migrates around the nucleus to the diametrically opposite side. With the centrioles on opposite sides of the nucleus, the centrosome with its concentrated DNA remains in the center.

While this is happening, the nucleus begins to soften and shrink in its center, between the opposing centrioles. As the shrinking is occurring, the centrosome which contains the centrioles sends out shafts of microtubules about the nucleus and toward its center. Simultaneously, containing the concentrated DNA, sends out shafts of kinetochores toward the oncoming microtubules.

The kinetochores approach the microtubule shafts from a perpendicular direction and then grasp the microtubule shafts. Although several kinetochores may grasp a single microtubule shaft, they eventually move forward and attach themselves at the tip of the microtubule. Once this attachment is made the diametrically opposite centrioles, within their centrosome exert tension on the microtubules and their connected kinetochores, pulling the nucleus apart.

Just prior to this separation the nucleus is covered with an array of microtubules and kinetochores, half on each side.

Eventually the two sides of the nucleus separate with each side moving toward the opposite sided centrioles with their surrounding centrosomes.

Meanwhile the chromosomes with its concentrated DNA remains in the center.

Once the two halves of the nucleus are separated, the remaining cell parts (the "cytoplasm") divide equally between the two halves — thus forming two new cells.

4.9. Bibliography — Cellular Duplication and Separation

B. Alberts, *et al.*, *Molecular Biology of the Cell*, 3rd Edition, Garland Publishing, New York, 1994.

J. Azimzadeh and M. Bornens, "Structure and Duplication of the Centrosome", *Journal of Cell Science*, Vol. 120, 2007, pp. 2139–2142. doi:10.1242/jcs.005231.

F. Barenz, D. Mayilo, and O.J. Gruss, "Centriolar Satellites: Busy Orbits around the Centrosome", *European Journal of Cell Biology*, Vol. 90, 2011, pp. 983–989. doi:10.1016/j.ejcb.2011.07.007.

F.P. Beer and E.R. Johnston Jr., *Mechanics of Materials*, 2nd Edition, McGraw-Hill, New York, 1994.

M. Bettencourt-Dias and D.M. Glover, "Centrosome Biogenesis and Function: Centrosomics Brings New Understanding", *Nature Reviews Molecular Cell Biology*, Vol. 8, 2007, pp. 451–463. doi:10.1038/nrm2180.

M. Bettencourt-Dias and D.M. Glover, "SnapShot: Centriole Biogenesis", *Cell*, Vol. 136, 2009, pp. 188.e1–188.e2. doi:10.1016/j.cell.2008.12.035.

A.D. Bicek, *et al.*, "Analysis of Microtubule Curvature", *Methods in Cell Biology*, Vol. 83, 2007, pp. 237–268. doi:10.1016/S0091-679X(07)-83010-X.

M. Bornens, "Organelle Positioning and Cell Polarity", *Nature Reviews Molecular Cell Biology*, Vol. 9, 2008, pp. 874–886. doi:10.1038/nrm2524.

M. Bornens, "The Centrosome in Cells and Organisms", *Science*, Vol. 335, 2012, pp. 422–426. doi:10.1126/science.1209037.

A. Bouissou, *et al.*, "γ-Tubulin Ring Complexes Regulate Microtubule Plus End Dynamics", *Journal of Cell Biology*, Vol. 187, 2009, pp. 327–334. doi:10.1083/jcb.200905060.

B.R. Brinkley, "Managing the Centrosome Numbers Game: From Chaos to Stability in Cancer Cell Division", *Trends in Cell Biology*, Vol. 11, 2001, pp. 18–21. doi:10.1016/S0962-8924(00)01872-9.

L. Cassimeris, *et al.*, "XMAP215 Is a Long Thin Molecule that Does Not Increase Microtubule Stiffness", *Journal of Cell Science*, Vol. 114, 2001, pp. 3025–3033.

P. Chang, *et al.*, "ε-Tubulin is Required for Centriole Duplication and Microtubule Organization", *Nature Cell Biology*, Vol. 5, 2003, pp. 71–76. doi:10.1038/ncb900.

M. Cifra, D. Havelka, and O. Kucera, "Electric Oscillations Generated by Collective Vibration Modes of Microtubules", *Proceedings of SPIE*, Vol. 7376, 2010, pp. 73760N.

M. Cifra, *et al.*, "Electric Field Generated by Axial Longitudinal Vibration Modes of Microtubule", *Biosystems*, Vol. 100, 2010, pp. 122–131. doi:10.1016/j.biosystems.2010.02.007.

I. Cunha-Ferreira, I. Bento, and M. Bettencourt-Dias, "From Zero to Many: Control of Centriole Number in Development and Disease", *Traffic*, Vol. 10, 2009, pp. 482–498. doi:10.1111/j.1600-0854.2009.00905.x.

A.B. D'Assoro, W.L. Lingle, and J.L. Salisbury, "Centrosome Amplification and the Development of Cancer", *Oncogene*, Vol. 21, 2002, pp. 6146–6153. doi:10.1038/sj.onc.1205772.

M. Delattre and P. Gonczy, "The Arithmetic of Centrosome Biogenesis", *Journal of Cell Science*, Vol. 117, 2004, pp. 1619–1630. doi:10.1242/jcs.01128.

J.G. DeLuca, "Kinetochore-Microtubule Dynamics and Attachment Stability", *Methods in Cell Biology*, Vol. 97, 2010, pp. 53–79. doi:10.1016/S0091-679X(10)97004-0.

M. Dogterom, *et al.*, "Force Generation by Dynamic Microtubules", *Current Opinion in Cell Biology*, Vol. 17, 2005, pp. 67–74. doi:10.1016/j.ceb.2004.12.011.

A. Duensing, *et al.*, "Centriole Overduplication through the Concurrent Formation of Multiple Daughter Centrioles at Single Material Templates", *Oncogene*, Vol. 26, 2007, pp. 6280–6288. doi:10.1038/sj.onc.1210456.

S.K. Dutcher, "The Tubulin Fraternity: Alpha to Eta", *Current Opinion in Cell Biology*, Vol. 13, 2001, pp. 49–54. doi:10.1016/S0955-0674(00)00173-3.

K. Fukasawa, "Oncogenes and Tumour Suppressors Take on Centrosomes", *Nature Reviews Cancer*, Vol. 7, 2007, pp. 911–924. doi:10.1038/nrc2249.

N.J. Ganem, S.A. Godinho, and D. Pellman, "A Mechanism Linking Extra Centrosomes to Chromosomal Instability", *Nature*, Vol. 460, 2009, pp. 278–282. doi:10.1038/nature08136.

S.A. Godinho, M. Kwon, and D. Pellman, "Centrosomes and Cancer: How Cancer Cells Divide with Too Many Centrosomes", *Cancer and Metastasis Reviews*, Vol. 28, 2009, pp. 85–98. doi:10.1007/s10555-008-9163-6.

T.M. Goepfert, *et al.*, "Centrosome Amplification and Overexpression of Aurora a are Early Events in Rat Mammary Carcinogenesis", *Cancer Research*, Vol. 62, 2002, pp. 4115–4122.

D. Guest, *Biology Smart, the Princeton Review*, Random House, New York, 1996, pp. 133–135.

J. Hardin, G. Bertoni, and L. J. Kleinsmith, *Becker's World of the Cell*, 8th Edition, Benjamin Cummings, Boston, MA, 2015.

D. Havelka and M. Cifra, "Calculation of the Electromagnetic Field around a Microtubule", *Acta Polytechnica*, Vol. 49, 2009, pp. 58–63.

D. Havelka, *et al.*, "High-Frequency Electric Field and Radiation Characteristics of Cellular Microtubule Network", *Journal of Theoretical Biology*, Vol. 286, 2011, pp. 31–40. doi:10.1016/j.jtbi.2011.07.007.

A.J. Holland, *et al.*, "Polo-like Kinase 4 Kinase Activity Limits Centrosome Overduplication by Autoregulating Its Own Stability", *The Journal of Cell Biology*, Vol. 188, 2010, pp. 191–198. doi:10.1083/jcb.200911102.

R. Huston, "Delivering Nanoparticles to Cancer Cells", *Tissue Engineering and Nanotheranostics*, Vol. 10, No. 7, 2018, p. 215.

R.L. Huston, "On Centrioles, Microtubules, and Cellular Electromagnetism", *Journal of Nanotechnology in Engineering and Medicine*, Vol. 5, 2014, pp. 021003.

R.L. Huston, "Using the Electromagnetics of Cancer's Centrosome Clusters to Attract Therapeutic Nanoparticles", *Advances in Bioscience and Biotechnology*, Vol. 6, 2015, pp. 172–181. doi:10.4236/abb.2015.63017.

R.L. Huston, "A Review of Centriole Activity, and Wrongful Activity, during Cell Division", *Advances in Bioscience and Biotechnology*, Vol. 7, 2016, pp. 169–182. doi:10.4236/abb.2016.73015.

Y.F. Inclán and F. Nogales, "Structural Models for the Self-Assembly and Microtubule Interactions of γ-, δ- and ε-Tubulin", *Journal of Cell Science*, Vol. 114, 2000, pp. 413–422.

M.E. Janson, M.E. de Dood, and M. Dogterom, "Dynamic Instability of Microtubules Is Regulated by Force", *The Journal of Cell Biology*, Vol. 161, 2003, pp. 1029–1034. doi:10.1083/jcb.200301147.

A. Kis, *et al.*, "Nanomechanics of Microtubules", *Physical Review Letters*, Vol. 89, 2002, pp. 248101. doi:10.1103/PhysRevLett.89.248101.

S.L. Kline-Smith and C.E. Walczak, "The Microtubule-Destabilizing Kinesin XKCM1 Regulates Microtubule Dynamic Instability in Cells", *Molecular Biology of the Cell*, Vol. 13, 2002, pp. 2718–2731. doi:10.1091/mbc.E01-12-0143.

T. Kobayashi and B.D. Dynlacht, "Regulating the Transition from Centriole to Basal Body", *Journal of Cell Biology*, Vol. 193, 2011, pp. 435–444. doi:10.1083/jcb.201101005.

J.M. Kollman, *et al.*, "Microtubule Nucleating γ-TuSC Assembles Structures with 13-Fold Microtubule-like Symmetry", *Nature*, Vol. 466, 2010, pp. 879–882. doi:10.1038/nature09207.

N. Korzeniewski, M. Hohenfellner, and S. Duensing, "CANDI Promotes PLK4-Mediated Centriole Overduplication and is Frequently Disrupted in Prostate Cancer", *NEOPLASIA*, Vol. 14, 2012, pp. 799–806. doi:10.1593/neo.12580.

A. Kramer, K. Neben, and A.D. Ho, "Centrosome Replication, Genomic Instability and Cancer," *Leukemia*, Vol. 16, 2002, pp. 767–775. doi:10.1038/sj.leu.2402454

M. Kwon, *et al.*, "Mechanisms to Suppress Multipolar Divisions in Cancer Cells with Extra Centrosomes," *Genes & Development*, Vol. 22, 2008, pp. 2189–2203.

B.M. Lange and K. Gull, "Structure and Function of the Centriole in Animal Cells: Progress and Questions," *Trends in Cell Biology*, Vol. 6, 1996, pp. 348–352. doi:10.1016/0962-8924(96)10033-7.

B. Leber, *et al.*, "Proteins Required for Centrosome Clustering in Cancer Cells," *Science Translational Medicine*, Vol. 2, 2010, pp. 33–38. doi:10.1126/scitranslmed.3000915.

T.-C. Lin, A. Neuner, and E. Schiebel, "Targeting of γ-Tubulin Complexes to Microtubule Organizing Centers: Conservation and Divergence", *Trends in Cell Biology*, Vol. 25, 2015, pp. 296–307. doi:10.1016/j.tcb.2014.12.002.

W.L. Lingle, *et al.*, "Centrosome Amplification Drives Chromosomal Instability in Breast Tumor Development", *Proceedings of the National Academy of Sciences of the United States of America*, Vol. 99, 2002, pp. 1978–1983. doi:10.1073/pnas.032479999.

W.L. Lingle, *et al.*, "Centrosome Hypertrophy in Human Breast Tumors: Implications for Genomic Stability and Cell Polarity", *Proceedings of the National Academy of Sciences of the United States of America*, Vol. 95, 1998, pp. 2950–2955. doi:10.1073/pnas.95.6.2950.

W.L. Lingle and J.L. Salisbury, "Altered Centrosome Structure is Associated with Abnormal Mitoses in Human Breast Tumors", *American Journal of Pathology*, Vol. 155, 1999, pp. 1941–1951. doi:10.1016/S0002-9440(10)65513-7.

W.L. Lingle and J.L. Salisbury, "The Role of the Centrosome in the Development of Malignant Tumors", *Current Topics in Developmental Biology*, Vol. 49, 2000, pp. 313–339. doi:10.1016/S0070-2153(99)49015-5.

H. Loffler, *et al.*, "DNA Damage-Induced Centrosome Amplification Occurs via Excessive Formation of Centriolar Satellites", *Oncogene*, Vol. 32, 2013, pp. 2963–2972. doi:10.1038/onc.2012.310.

J. Loncarek and A. Khodjakov, "Ab ovo or de nous? Mechanisms of Centriole Duplication", *Molecules and Cells*, Vol. 27, 2009, pp. 135–142. doi:10.1007/s10059-009-0017-z.

R. Mahen and A.R. Venkitaraman, "Pattern Formation in Centrosome Assembly", *Current Opinion in Cell Biology*, Vol. 24, 2012, pp. 14–23. doi:10.1016/j.ceb.2011.12.012

H. Maiato and E. Logarinho, "Mitotic Spindle Multipolarity without Centrosome Amplifications", *Nature Cell Biology*, Vol. 16, 2014, pp. 386–394. doi:10.1038/ncb2958.

E.N. Marieb, *Human Anatomy and Physiology*, 3rd Edition, Benjamin/Cummings Publishing, Redwood City, 1991, Chapter 3, pp. 60–101.

V. Marthiens, *et al.*, "Centrosome Amplification Causes Microcephaly", *Nature Cell Biology*, Vol. 15, 2013, pp. 731–740. doi:10.1038/ncb2746.

J. Marx, "Do Centrosome Abnormalities Lead to Cancer?" *Science*, Vol. 292, 2001, pp. 426–429. doi:10.1126/science.292.5516.426.

V. Mennella, *et al.*, "Amorphous No More: Subdiffraction View of the Pericentriolar Material Architecture", *Trends in Cell Biology*, Vol. 24, 2014, pp. 188–197. doi:10.1016/j.tcb.2013.10.001.

M. Moritz, *et al.*, "Structure of the Gamma-Tubulin Ring Complex: A Template for Microtubule Nucleation", *Nature Cell Biology*, Vol. 2, 2000, pp. 365–370. doi:10.1038/35014058.

E.A. Nigg, "Centrosome Aberrations: Cause or Consequence of Cancer Progression?" *Nature Reviews Cancer*, Vol. 2, 2002, pp. 815–825. doi:10.1038/nrc924.

E.A. Nigg, "Origins and Consequences of Centrosome Aberrations in Human Cancers", *International Journal of Cancer*, Vol. 119, 2006, pp. 2717–2723. doi:10.1002/ijc.22245.

E.A. Nigg and J.W. Raft, "Centrioles, Centrosomes, and Cilia in Health and Disease", *Cell*, Vol. 139, 2009, pp. 663–678. doi:10.1016/j.cell.2009.10.036.

E. A. Nigg and T. Stearns, "The Centrosome Cycle: Centriole Biogenesis, Duplication and Inherent Asymmetries", *Nature Cell Biology*, Vol. 13, 2011, pp. 1154–1160. doi:10.1038/ncb2345.

B.R. Oakley, V. Paolillo, and Y. Zheng, "γ-Tubulin Complexes in Microtubule Nucleation and Beyond", *Molecular Biology of the Cell*, Vol. 26, 2015, pp. 2957–2962. doi:10.1091/mbc.E14-11-1514.

G.A. Pihan, *et al.*, "Centrosome Defects and Genetic Instability in Malignant Tumors", *Cancer Research*, Vol. 58, 1998, pp. 3974–3985.

G.A. Pihan, *et al.*, "Centrosome Abnormalities and Chromosome Instability Occur Together in Pre-Invasive Carcinomas", *Cancer Research*, Vol. 63, 2003, pp. 1398–1404.

J. Pokorny, "Viscous Effects on Polar Vibrations in Microtubules", *Electromagnetic Biology and Medicine*, Vol. 22, 2003, pp. 15–29. doi:10.1081/JBC-120020349.

J. Pokorny, "Physical Aspects of Biological Activity and Cancer", *AIP Advances*, Vol. 2, 2012, p. 011207. doi:10.1063/1.3699057.

J. Pokorny, J. Hasek, and F. Jelínek, "Electromagnetic Field of Microtubules: Effects on Transfer of Mass Particles and Electrons", *Journal of Biological Physics*, Vol. 31, 2005a, pp. 501–514. doi:10.1007/s10867-005-1286-1.

J. Pokorny, J. Hasek, and F. Jelínek, "Endogenous Electric Field and Organization of Living Matter', *Electromagnetic Biology and Medicine*, Vol. 24, 2005b, pp. 185–197. doi:10.1080/15368370500379566.

J. Pokorny, J. Pokorny, and J. Kobilková, "Postulates on Electromagnetic Activity in Biological Systems and Cancer", *Integrative Biology*, Vol. 5, 2013, pp. 1439–1446. doi:10.1039/c3ib40166a.

S. Portet, *et al.*, "Elastic Vibrations in Seamless Microtubules", *European Biophysics Journal*, Vol. 34, 2005, pp. 912–920. doi:10.1007/s00249-005-0461-4.

B. Raynaud-Messina and A. Merdes, "γ-Tubulin Complexes and Microtubule Organization", *Current Opinion in Cell Biology*, Vol. 19, 2007, pp. 24–30. doi:10.1016/j.ceb.2006.12.008.

A. Rodrigues-Martins, *et al.*, "From Centriole Biogenesis to Cellular Function: Centrioles Are Essential for Cell Division at Critical Developmental Stages", *Cell Cycle*, Vol. 7, 2008, pp. 11–16. doi:10.4161/cc.7.1.5226.

E. Schiebel, "γ-Tubulin Complexes: Binding to the Centrosome, Regulation and Microtubule Nucleation", *Current Opinion in Cell Biology*, Vol. 12, 2000, pp. 113–118. doi:10.1016/S0955-0674(99)00064-2.

L. Schockel, *et al.*, "Cleavage of Cohesion Rings Coordinates the Separation of Centrioles and Chromatids", *Nature Cell Biology*, Vol. 13, 2011, pp. 966–972. doi:10.1038/ncb2280.

M.M. Shawki and A. Farid, "Low Electric Field Parameters Required to Induce Death of Cancer Cells", *Electromagnetic Biology & Medicine*, Vol. 33, 2014, pp. 159–163. doi:10.3109/15368378.2013.800105.

G. Sluder and J.J. Nordberg, "The Good, the Bad and the Ugly: The Practical Consequences of Centrosome Amplification", *Current Opinion in Cell Biology*, Vol. 16, 2004, pp. 49–54. doi:10.1016/j.ceb.2003.11.006.

P. Strnad and P. Gonczy, "Mechanisms of Procentriole Formation", *Trends in Cell Biology*, Vol. 18, 2008, pp. 389–396. doi:10.1016/j.tcb.2008.06.004.

A.-M. Tassin and M. Bornens, "Centrosome Structure and Microtubule Nucleation in Animal Cells", *Biology of the Cell*, Vol. 91, 1999, pp. 343–354. doi:10.1111/j.1768-322X.1999.tb01092.x.

N. Teixidó-Travesa, J. Roig, and J. Lüders, "The Where, When and How of Microtubule Nucleation — One Ring to Rule Them All", *Journal of Cell Science*, Vol. 125, 2012, pp. 4445–4456. doi:10.1242/jcs.106971.

M.S. Tsou and T. Stearns, "Mechanism Limiting Centrosome Duplication to Once per Cell Cycle", *Nature*, Vol. 442, 2006, pp. 947–951. doi:10.1038/nature04985.

J. Vulprecht, *et al.*, "STIL Is Required for Centriole Duplication in Human Cells", *Journal of Cell Science*, Vol. 125, 2012, pp. 1353–1362. doi:10.1242/jcs.104109.

C.E. Walczak, "Microtubule Dynamics and Tubulin Interacting Proteins", *Current Opinion in Cell Biology*, Vol. 12, 2000, pp. 52–56. doi:10.1016/S0955-0674(99)00056-3.

C. Wiese and Y. Zheng, "A New Function for the γ-Tubulin Ring Complex as a Microtubule Minus-End Cap", *Nature Cell Biology*, Vol. 2, 2000, pp. 358–364. doi:10.1038/35014051.

Z. Yang, *et al.*, "Extra Centrosomes and/or Chromosomes Prolong Mitosis in Human Cells", *Nature Cell Biology*, Vol. 10, 2008, pp. 748–751.

Chapter 5 Centriole Errors, Tumorigenesis, and Treatment Options

5.1. Introduction

"Cancer" is the most dreaded word a patient hears in diagnosis from his or her treating physician. Survivability depends upon the location and extent of the disease.

In this final chapter, I review common ways cell development can go awry and lead either to cell death or tumorigenesis.

5.2. Centriole Errors

Chapter II provides a detailed description of centrioles: their development and geometry. The centrioles and their counterpart, the DNA, are the only two parts of a cell that can self-duplicate — and remarkably, they do it at the same time.

There is a close connection between the centrioles and the DNA. But it is not clear what the connection is or how signals are communicated. In any event, if that connection is disrupted, or if there is an error in DNA development, the centrioles will be negatively affected.

But aside from DNA malfunction, errors can occur in the development of centrioles themselves.

The most common and deadliest error is the occurrence of more than one pair of centrioles in a non-dividing cell. A second but less common error is a centriole with an extra cylinder, or non-perpendicular cylinders. The development of centrioles depends upon the correct supply of Plk4. Too much produces multiple (or "supernumerary") centrioles. Too little produces weakened centrioles.

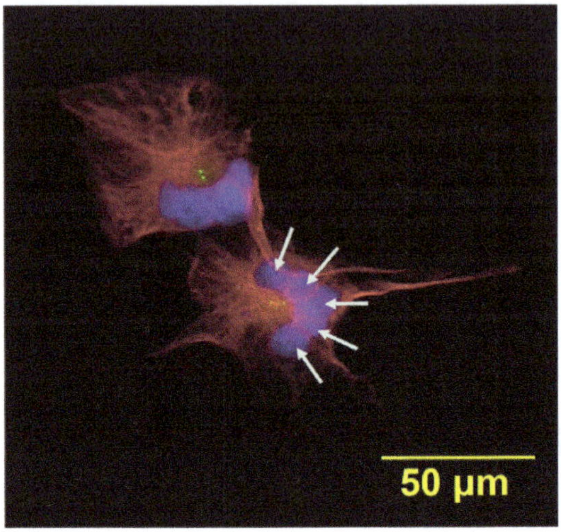

Figure 5.1. A created malignancy. Used with permission from World Scientific Publishing

If there are supernumerary centrioles, the cell division will be disrupted and cells with multiple nuclei will occur. This leads to tumorigenesis. Figure 5.1 presents an illustration of this process.

We (my assistant, Christy Petersen, and I) created the supernumerary centriole cells by placing nominal breast cells in a solution rich in Plk4. The result was a cell with supernumerary centrioles and multiple nuclei as in Figure 5.1.

5.3. Tumorigenesis

Figure 5.1 shows the result of a cell with multiple nuclei. These nuclei are the result of a cell with supernumerary centrioles. If allowed to continue, the cell will develop with closely packed adjacent cells forming a mass of cells, readily identified as a tumor.

If the tumor is malignant, cancerous cells can break away from it and get into the blood vessels. Once cancer cells are in the blood vessels, they can deposit themselves in organs and bones — that is, cells of the tumor can be spread to other parts of the body — so-called metastasis.

5.4. Prevention and Treatment

With a focus on cancer the best result is to physically remove the tumor before metastasis.

If complete physical removal is not possible, the next step is radiation. But the difficulty with radiation is that it is harmful to adjacent tissue and may well be a source of cancer itself.

A third choice is chemotherapy, with chemicals of various kinds are injected into the blood stream. The concept is that the chemicals are like a poison to the system which can be survived by normal cells but not cancer cells — which are presumed to be weaker.

The difficulty with chemotherapy, aside from numerous side effects, is that all of a tumor may not be reached by the poisonous chemical. The blood vessels supporting the tumor often also become malignant and then produce back pressure, keeping the chemical from reaching all parts of the tumor. An apparent success is tumor reduction, but the remaining untouched cells can then begin to grow new tumors.

A fourth, but not yet fully tested, is that tumor cells are believed to have a higher magnetic field than normal cells. If this is the case carbon crated ferrous (iron) nanoparticles will be attached to the tumor.

With ferrous nanoparticles within the tumor, they can be vibrated by external magnetic oscillations. This vibration can cause a temperature elevation beyond the tolerance of the tumor cells thus killing the tumor cells.

Much research is needed before this approach can be tested.

Alphabetical Listing of References

I.R. Adams and J.V. Kilmartin, "Spindle Pole Body Duplication: A Model for Centrosome Duplication", *Trends in Cell Biology*, Vol. 10, No. 8, 2000, pp. 329–335.

AFM Workshop, *Atomic Force Microscopes*, 2012, Chapter 4, p. 75H.

S. Afreen and D. Varma, "Cell Divisions: Molecular Pathways for KMN Kinetochore Recruitment", *Current Biology*, Vol. 24, 2015, pp. R328–R347.

R. Afrin, T. Yamada, and A. Ikai, "Analogies of Force Curves Obtained on the Live Cell Membrane Using Chemically Modified AFM Probes", *Ultramicroscopy*, Vol. 100, 2004, pp 187–195.

F.G. Agrican, E. Schiebel, and B.R. Mardin, "Separate to Operate: Control of Centrosome Positioning and Separation", *Philosophical Transactions Royal Society of London. Series B*, Vol. 369, 2014, 10 pages.

D.K. Ahirwar, *et al.*, "Non-Contact Method for Directing Electrotaxis", *Scientific Reports,* Vol. 5, 2015, pp. 11005.

A. Akhmanova and C.C. Hoogenraad, "Microtubule Minus-End-Targeting Proteins", *Current Biology*, Vol. 25, 2015, pp. R162–R171.

T.K. Akhshi, D. Wernike, and A. Piekny, "Microtubules and Actin Crosstalk in Cell Migration and Division", *Cytoskeleton*, Vol. 71, 2014, pp. 1–23.

N. Akhtar and C.H. Streuli, "An Integrin-ILK-Microtubule Network Orients Cell Polarity and Lumen Formation in Glandular Epithelium", *Nature Cell Biology*, Vol. 15, No. 1, 2013, pp. 17–27.

C.R. Alarcón and S.F. Tavazoie, "Cancer: Endothelial-Cell Killing Promotes Metastasis", *Nature News & Views*, Vol. 536, 2016, pp. 154–155.

B. Alberts, *et al., Molecular Biology of the Cell*, 3rd Edition, Garland Publishing, New York, 1994.

H. Aldaz, *et al.*, "Insights into Microtubule Nucleation from the Crystal Structure of Human ϒ-Tubulin", *Nature*, Vol. 435, 2005, pp. 523–527.

K. Al-Romaih, *et al.*, "Chromosomal Instability in Osteosarcoma and Its Association with Centrosome Abnormalities", *Cancer Genetics and Cytogenetics*, Vol. 144, No. 2, 2003, pp. 91–99.

V. Aranda, M.E. Nolan, and S.K. Muthuswamy, "Par Complex in Cancer: A Regulator of Normal Cell Polarity Joins the Dark Side", *Oncogene*, Vol. 27, 2008, pp. 6878–6887.

M. Arbi, *et al.*, "Controlling Centriole Numbers: Geminin Family Member as Master Regulators of Centriole Amplification and Multiciliogenesis", *Springer Chromosome*, Vol. 127, 2018, pp. 151–174.

V. Archambault and D.M. Glover, "Polo-like Kinases: Conservation and Divergence in Their Functions and Regulation", *Nature Reviews: Molecular Cell Biology*, Vol. 10, 2009, pp. 265–275.

C. Arquint, *et al.*, "Cell-Cycle-Regulated Expression of STIL Controls Centriole Number in Human Cells", *Journal of Cell Science*, Vol. 125, 2012, pp. 1342–1352.

M. Arruebo, *et al.*, "Magnetic Nanoparticles for Drug Delivery", *Nanotoday*, Vol. 2, 2007, pp. 22–32.

H. Asakawa, *et al.*, "Submolecular-Scale Imaging of α-Helices and C-Terminal Domains of Tubulins by Frequency Modulation Atomic Force Microscopy in Liquid", *Biophysical Journal*, Vol. 101, No. 5, 2011, pp. 1270–1276.

C.L. Asbury and T.N. Davis, "Insights into the Kinetochore", *Structure*, Vol. 16, 2008, pp. 834–836.

C.L. Asbury, J.F. Tien, and T.N. Davis, "Kinetochores' Gripping Feat: Conformational Wave or Biased Diffusion?" *Trends in Cell Biology*, Vol. 21, 2011, pp. 38–46.

A. Asnacios and O. Hamant, "The Mechanics Behind Cell Polarity", *Trends in Cell Biology*, Vol. 22, No. 11, 2012, pp. 584–591.

M. Assenheimer, *et al.*, "The T-SCAN Technology: Electrical Impedence as a Diagnostic Tool for Breast Cancer Detection", *Physiological Measurement*, Vol. 22, 2001, pp. 1–8.

C. Aumeier, *et al.*, "Self-Repair Promotes Microtubule Rescue", *Nature Cell Biology*, Vol. 18, No. 10, 2016, pp. 1054–1064.

T. Avidor-Reiss and J. Gopalakrishnan, "Building a Centriole", *Current Opinion in Cell Biology*, Vol. 25, No. 1, 2013, pp. 72–77.

S. Awakumova, *et al.*, "Biotechnological Approaches Toward Nanoparticle Biofunctionalization", *Trends in Biotechnology*, Vol. 32, 2014, pp. 11–20.

Y. Aylon and M. Oren, "Tumor Suppression by p53: Bring in the Hippo", *Cancer Cell*, Vol. 32, 2017, pp. 397–399.

S. Azarmi, W.H. Roa, and R. Löbenberg, "Targeted Delivery of Nanoparticles for the Treatment of Lung Diseases", *Advanced Drug Delivery Reviews*, Vol. 60, 2008, pp. 863–875.

J. Azimzadeh and M. Bornens, "Structure and Duplication of the Centrosome", *Journal of Cell Science*, Vol. 120, 2007, pp. 2139–2142.

J. Azimzadeh and W.F. Marshall, "Building the Centriole", *Current Biology*, Vol. 20, No. 18, 2010, pp. R816–R825.

H. Bae, *et al.*, "Carbon-Coated Iron Oxide Nanoparticles as Contrast Agents in Magnetic Resonance Imaging", *Springer Open Journal, Nanoscale Research Letters*, Vol. 7, 2012, p. 44.

A.D. Baffet, *et al.*, "Meeting Report–Building a Centrosome", *Journal of Cell Science*, Vol. 126, 2013, pp. 3259–3262.

R. Bahtz, *et al.*, "GCP6 Is a Substrate of Plk4 and Required for Centriole Duplication", *Journal of Cell Science*, Vol. 125, 2011, pp. 486–496.

N.A. Baker, *et al.*, "Electrostatics of Nanosystems: Application to Microtubules and the Ribosome", *Proceedings of the National Academy of Sciences of the United States of America*, Vol. 98, 2001, pp. 10037–10041.

F. Balzarotti, *et al.*, "Nanometer Resolution Imaging and Tracking of Fluorescent Molecules with Minimal Photon Fluxes", *Science*, Vol. 355, 2017, pp. 606–612.

F. Bärenz, D. Mayilo, and O.J. Gruss, "Centriolar Satellites: Busy Orbits Around the Centrosome", *European Journal of Cell Biology*, Vol. 90, No. 12, 2011, pp. 983–989.

T.C. Barnes, "Healing Rate of Human Skin Determined by Measurement of the Electrical Potential of Experimental Abrasions: A Study of Treatment with Petrolatum and with Petrolatum Containing Yeast and Liver Extracts", *The American Journal of Surgery*, Vol. 69, 1945, pp. 82–88.

F.A. Barr, H.H.W. Silljé, and E.A. Nigg, "Polo-like Kinases and the Orchestration of Cell Division", *Nature Reviews Molecular Cell Biology*, Vol. 5, 2004, pp. 429–441.

N. Barvitenko, *et al.*, "Integration of Intracellular Signaling: Biological Analogues of Wires, Processors and Memories Organized by a Centrosome 3D Reference System", *BioSystems*, Vol. 173, 2018, pp. 191–206.

R. Basto, *et al.*, "Flies without Centrioles", *Cell*, Vol. 125, No. 7, 2006, pp. 1375–1386.

R. Basto, *et al.*, "Centrosome Amplification Can Initiate Tumorigenesis in Flies", *Cell*, Vol. 133, No. 6, 2008, pp. 1032–1042.

R.O. Becker and J.A. Spadaro, "Electrical Stimulation of Partial Limb Regeneration in Mammals", *Bulletin of the New York Academy of Medicine*, Vol. 48, No. 4, 1972, pp. 627–641.

F.P. Beer and E.R. Johnston, Jr., *Mechanics of Materials*, 2nd Edition, McGraw Hill, New York, 1992.

J. Beisson and M-Jerka-Dziadosz, "Polarities of the Centriolar Structure: Morphogenetic Consequences", *Biology of the Cell*, Vol. 91, No. 4–5, 1999, pp. 367–378.

J. Beisson and M. Wright, "Basal Body/Centriole Assembly and Continuity", *Current Opinion in Cell Biology*, Vol. 15, No. 1, 2003, pp. 96–104.

H. Berg, *et al.*, "Bioelectromagnetic Field Effects on Cancer Cells and Mice Tumors", *Electromagnetic Biology & Medicine*, Vol. 29, No. 4, 2010, pp. 132–143.

H. Berg, "Bioelectric and Biomagnetic Methods for Cancer Research and Therapy–A Survey", *Electromagnetic Biology and Medicine*, Vol. 24, 2005, pp. 423–440.

G. Bergers and D. Hanahan, "Modes of Resistance to Anti-Angiogenic Therapy", *Nature Reviews Cancer*, Vol. 8, 2008, pp. 592–603.

N. Bertrand, *et al.*, "Cancer Nanotechnology: The Impact of Passive and Active Targeting in the Era of Modern Cancer Biology", *Advanced Drug Delivery Reviews*, Vol. 66, 2014, pp. 2–25.

S. Beseme, *et al.*, "Transcriptional Changes in Cancer Cells Induced by Exposure to a Healing Method", *Dose Response*, 2018, pp. 1–8.

M. Bettencourt-Dias, *et al.*, "Genome-Wide Survey of Protein Kinases required for Cell Cycle Progression", *Nature*, Vol. 432, 2004, pp. 980–987.

M. Bettencourt-Dias, *et al.*, "SAK/PLK4 Is Required for Centriole Duplication and Flagella Development", *Current Biology*, Vol. 15, No. 24, 2005, pp. 2199–2207.

M. Bettencourt-Dias and D.M. Glover, "Centrosome Biogenesis and Function: Centrosomics Brings New Understanding", *Nature Reviews Molecular Cell Biology*, Vol. 8, 2007, pp. 451–463.

M. Bettencourt-Dias and D.M. Glover, "SnapShot: Centriole Biogenesis", *Cell*, Vol. 136, No. 1, 2009, p. 188.

M. Bettencourt-Dias, *et al.*, "Centrosomes and Cilia in Human Disease", *Trends in Genetics*, Vol. 27, No. 8, 2011, pp. 307–315.

A. Bianco, K. Kostarelos, and M. Prato, "Applications of Carbon Nanotubes in Drug Delivery", *Current Opinion in Chemical Biology*, Vol. 9, No. 6, 2005, pp. 674–679.

A.D. Bicek, *et al.*, "Analysis of Microtubule Curvature", *Methods in Cell Biology*, Vol. 83, 2007, Chapter 10, pp. 237–268.

D. Bienkowska and C.R. Cowan, "Centrosomes Can Initiate a Polarity Axis from Any Position Within One-Cell *C. elegans* Embryos", *Current Biology*, Vol. 22, No. 7, 2012, pp. 583–589.

D. Bilder, M. Li, and N. Perrimon, "Cooperative Regulation of Cell Polarity and Growth by Drosophila Tumor Suppressors", *Science*, Vol. 289, 2000, pp. 113–116.

R. Binggeli and R.C. Weinstein, "Membrane Potentials and Sodium Channels: Hypotheses for Growth Regulation and Cancer Formation Based on Changes in Sodium Channels and Gap Junctions", *Journal of Theoretical Biology*, Vol. 123, No. 4, 1986, pp. 377–401.

S. Biswas and V.P. Torchilin, "Nanopreparations for Organelle-Specific Delivery in Cancer", *Advanced Drug Delivery Reviews*, Vol. 66, 2014, pp. 26–41.

S. Blachon, *et al.*, "Drosophila Asterless and Vertebrate CEP152 Are Orthologs Essential for Centriole Duplication", *Genetics*, Vol. 180, No. 4, 2008, pp. 2081–2094.

R. Blackwell, *et al.*, "Contributions of Microtubule Dynamic Instability and Rotational Diffusion to Kinetochore Capture", *Biophysical Journal*, Vol. 112, 1917, pp. 552–563.

S.P. Blagden and D.M. Glover, "Polar Expeditions–Provisioning the Centrosome for Mitosis", *Nature Cell Biology*, Vol. 5, No. 6, 2003, pp. 505–511.

K.J. Böhm, *et al.*, "Movement and Alignment of Microtubules in Electric Fields and Electric-Dipole-Moment Estimates", *Electromagnetic Biology and Medicine*, Vol. 24, 2005, pp. 319–330.

B. Bohunicky and S.A. Mousa, "Biosensors: The New Wave in Cancer Diagnosis", *Nanotechnology Science and Applications*, Vol. 4, 2001, pp. 1–10.

D.G. Booth and W.C. Earnshaw, "Ki-67 and the Chromosome Periphery Compartment in Mitosis", *Trends in Cell Biology*, Vol. 27, No. 12, 2017, pp. 906–913.

F. Bordeleau, *et al.*, "Matrix Stiffening Promotes a Tumor Vasculature Phenotype", *Proceedings of the National Academy of Sciences*, Vol. 14, 2017, pp. 492–497.

F. Borel, *et al.*, "Multiple Centrosomes Arise from Tetraploidy Checkpoint Failure and Mitotic Centrosome Clusters in p53 and RB Pocket Protein-Compromised Cells", *Proceedings National Academy of Science*, Vol. 99., No. 15, 2002, pp. 9819–9824.

J.A. Borgia, *et al.*, "*Correlation of Angiogenesis Biomarkers with Early Metastatic Progression in NSCLC as Determined using a Multiplexed Immunoassay Kit*", EMD Millipore, Merck KGGA, Darmstadt, Germany, 2017.

M. Bornens, "Centrosome Composition and Microtubule Anchoring Mechanisms", *Current Opinion in Cell Biology*, Vol. 14, No. 1, 2002, pp. 25–34.

M. Bornens, "Organelle Positioning and Cell Polarity", *Nature Reviews Molecular Cell Biology*, Vol. 9, 2008, pp. 874–886.

M. Bornens, "The Centrosome in Cells and Organisms", *Science*, Vol. 335, 2012, pp. 422–426.

M. Bornens and P. Gonczy, "Centrosomes Back in the Limelight", *Philosophical Transactions Royal Society, London B, Biological Sciences*, Vol. 369, 2014, 15 pages.

C.A.K. Borrebaeck, "Precision Diagnostics: Moving Towards Protein Biomarker Signatures of Clinical Utility in Cancer", *Nature Reviews Cancer*, Vol. 17, 2017, pp. 199–204.

F. Bosveld, A. Ainslie, and Y. Bellaïche, "Sequential Activities of Dynein, Mud and Asp in Centrosome-Spindle Coupling Maintain Centrosome Number upon Mitosis", *Journal of Cell Science*, Vol. 130, 2017, pp. 3557–3567.

B.P. Bouchet and A. Akhmanova, "Microtubules in 3D Cell Motility", *Journal of Cell Science*, Vol. 130, 2017, pp. 39–50.

A. Bouissou, *et al.*, "Gamma-Tubulin Ring Complexes Regulate Microtubule Plus End Dynamics", *Journal of Cell Biology*, Vol. 187, No. 3, 2009, pp. 327–334.

C. Boutin, "To Give Cancer a Deadly Fever, NIST Explores Better Nanoparticle Design."

C.P. Brangwynne, *et al.*, "Microtubules Can Bear Enhanced Compressive Loads in Living Cells Because of Lateral Reinforcement", *Journal of Cell Biology*, Vol. 173, No. 5, 2006, pp. 733–761. doi:10.1083/jcb.200601060

I. Brigger, C. Dubernet, and P. Couvreur, "Nanoparticles in Cancer Therapy and Diagnosis", *Advanced Drug Delivery Reviews*, Vol. 54, 2002, pp. 631–651.

B.R. Brinkley, "Managing the Centrosome Numbers Game: From Chaos to Stability in Cancer Cell Division", *Trends in Cell Biology*, Vol. 11, No. 1, 2001, pp. 18–21.

D.A. Brito, S.M. Gouveia, and M. Bettencourt-Dias, "Deconstructing the Centriole: Structure and Number Control", *Current Opinion in Cell Biology*, Vol. 24, No. 1, 2012, pp. 4–13.

G.J. Brouhard and L.M. Rice, "The Contribution of $\alpha\beta$-Tubulin Curvature to Microtubule Dynamics", *Journal of Cell Biology*, Vol. 207, No. 3, 2014, pp. 323–334.

G. Brouhard and D. Sept, "Microtubules: Sizing Up the GTP Cap", *Current Biology*, Vol. 22, No. 18, 2012, pp. R802–R803.

S.C. Brown, *et al.*, "Nanoparticle Characterization for Cancer Nanotechnology and Other Biological Applications", *Methods in Molecular Biology*, Vol. 624, 2010, pp. 39–65.

J.A. Brown and J.A. Tuszynski, "A Review of the Ferroelectric Model of Microtubules", *Ferroelectrics*, Vol. 220, No. 1, 1999, pp. 141–155.

C.W. Brownlee, *et al.*, "The Protein Phosphatase 2A Regulatory Submit Twins Stabilizes Plk4 to Induce Centriole Amplification", *Journal of Cell Biology*, Vol. 195, No. 2, 2011, pp. 231–243.

C.W. Brownlee and G.C. Rogers, "Show Me Your License, Please: Deregulation of Centriole Duplication Mechanisms that Promote Amplification", *Cellular and Molecular Life Sciences*, Vol. 70, No. 6, 2013, pp. 1021–1034.

F. Bunz, *et al.*, "Targeted Inactivation of p53 in Human Cells Does Not Result in Aneuploidy", *Cancer Research*, Vol. 62, No. 4, 2002, p. 1129.

J. Burclaff and J.C. Mills, "Cell Biology: Healthy Skin Rejects Cancer", *Nature*, Vol. 548, August 2017, pp. 289–290.

D.T. Butcher, T. Alliston, and V.M. Weaver, "A Tense Situation: Forcing Tumor Progression", *Nature Reviews Cancer*, Vol. 9, 2009, pp. 108–122.

J.D. Byrne, T. Betancourt, and L. Branon-Peppas, "Active Targeting Schemes for Nanoparticle Systems in Cancer Therapeutics", *Advanced Drug Delivery Reviews*, Vol. 60, 2008, pp. 1615–1626.

W. Cai, *et al.*, "Applications of Gold Nanoparticles in Cancer Nanotechnology", *Nanotechnology Science and Applications*, Vol. 1, 2008, pp. 17–32.

N.A. Campbell, *Biology*, Benjamin/Cummings, Menlo Park, CA, 1987, p. 231.

P. Carmeliet and R.K. Jain, "Angiogenesis in Cancer and Other Diseases", *Nature*, Vol. 407, No. 6801, 2000, pp. 249–257.

J. Cartwright, "An Electrical Misunderstanding", *Physics of Cancer, Physics World*, Vol. 26, 2013, pp. 25–26.

Z. Carvalho-Santos, *et al.*, "Tracing the Origins of Centrioles, Cilia and Flagella", *Journal of Cell Biology*, Vol. 194, No. 2, 2011, pp 165–175.

Z. Carvalho-Santos, *et al.*, "Stepwise Evolution of the Centriole-Assembly Pathway", *Journal of Cell Science*, Vol. 123, 2010, pp. 1414–1426.

M. Casenghi, *et al.*, "Polo-like Kinase 1 Regulates Nip a Centrosome Protein Involved in Microtubule Nucleation", *Developmental Cell*, Vol. 5, 2003, pp. 113–125.

L. Cassimeris, *et al.*, "XMAP215 Is a Long Thin Molecule that Does Not Increase Microtubule Stiffness", *Journal of Cell Science*, Vol. 114, 2001, pp. 3025–3033.

B.T. Castle and D.J. Odde, "Optical Control of Microtubule Dynamics in Time and Space", *Cell*, Vol. 162, 2015, pp. 243–245.

N. Caudron, *et al.*, "A Reassessment of the Factors Affecting Microtubule Assembly and Disassembly In Vitro", *Journal of Molecular Biology*, Vol. 297, 2000, pp. 211–220.

J.J. Chalmers, *et al*, "Femtogram Resolution of Iron Content on a Per Cell Basis: Ex Vivo Storage of Human Red Blood Cells Leads to Loss of Hemoglobin", *Analytical Chemistry*, Vol. 14, 2017, pp. 3702–3709.

J.Y. Chan, "A Clinical Overview of Centrosome Amplification in Human Cancers", *International Journal of Biological Sciences*, Vol. 7, No. 8, 2011, pp. 1122–1144.

G.K. Chan, S.-T. Lin, and T.J. Yen, "Kinetochore Structure and Function", *Trends in Cell Biology*, Vol. 15, 2005, pp. 587–598.

P. Chang, *et al.*, "ε-Tubulin Is required for Centriole Duplication and Microtubule Organization", *Nature Cell Biology*, Vol. 5, 2003, pp. 71–76.

C.-W. Chang, *et al.*, "CEP295 Interacts with Microtubules and Is Required for Centriole Elongation", *Journal of Cell Science*, Vol. 129, 2016, pp. 2501–2513.

P. Chang and T. Stearns, "δ-Tubulin and ε-Tubulin: Two New Human Centrosomal Tubulins Reveal New Aspects of Centrosome Structure and Function", *Nature Cell Biology*, Vol. 2, 2000, pp. 30–35.

P.L. Chavali, M. Putz, and F. Gergely, "Small Organelle, Big Responsibility: The Role of Centrosomes in Development and Disease", *Philosophical Transactions Royal Society of London, Series B*, Vol. 369, 2014, 12 pages.

I.M. Cheeseman and A. Desai, "Molecular Architecture of the Kinetochore-Microtubule Interface", *Nature Reviews, Molecular Cell Biology*, Vol. 19, 2008, pp. 33–46.

J.J. Chen and L. Galluzzi, "Fighting Resistant Cancers with Iron", *Trends in Cell Biology*, Vol. 28, No. 2, 2018, pp. 74–78.

P.C. Chen, S.C. Mwakwari, and A.K. Oyelere, "Gold Nanoparticles: From Nanomedicine to Nanosensing", *Nanotechnology Science and Applications*, Vol. 1, 2008, pp. 45–65.

F. Chen, P.W. Tillberg, and E.S. Boyden, "Expansion Microscopy", *Science*, Vol. 347, No. 6221, 2015, pp. 543–548.

Z. Cheng, *et al.*, "Multifunctional Nanoparticles: Cost versus Benefit of Adding Targeting and imaging Capabilities", *Science*, Vol. 338, No. 6109, 2012, pp. 903–910.

K. Cheng and C. Zou, "Physics Models of Centriole Replication", *Medical Hypotheses*, Vol. 67, No. 3, 2006, pp.572–577.

K. Cheng and C. Zou, "Theoretical Models of Cytokinesis and Mitosis (Without Asters)", *American Journal of Plant Physiology*, Vol. 1, No. 1, 2006, pp. 21–27.

K. Cho, *et al.*, "Therapeutic Nanoparticles for Drug Delivery in Cancer", *Clinical Cancer Research*, Vol. 14, 2008, pp. 1310–1316.

J. Chomoucka, *et al.*, "Magnetic Nanoparticles and Targeted Drug Delivery", *Pharmacological Research*, Vol. 62, No. 2, 2010, pp. 144–149.

E.K.-H. Chow and D. Ho, "Cancer Nanomedicine: From Drug Delivery to Imaging", *Science Translational Medicine*, Vol. 5, No. 216, 2013, pp. 1–12.

C. Ciferri, *et al.*, "Implications for Kinetochore-Microtubule Attachment from the Structure of an Engineered Ndc80 Complex", *Cell*, Vol. 133, 2008, pp. 427–439.

M. Cifra, "Electrodynamic Eigenmodes in Cellular Morphology", *Biosystems*, Vol. 109, No. 3, 2012, pp. 356–366.

M. Cifra, J.Z. Fields, and A. Farhadi, "Electromagnetic Cellular Interactions", *Progress in Biophysics and Molecular Biology*, Vol. 105, 2011, pp. 223–246.

M. Cifra, D. Havelka, and O. Kucera, "Electric Oscillations Generated by Collective Vibration Modes of Microtubules", *Proceedings of SPIE, Laser Applications in Life Sciences*, Edited by M. Kinnamen and R. Myllyla, Vol. 7376, 2010.

M. Cifra, *et al.*, "Electric Field Generated by Axial Longitudinal Vibration Modes of Microtubule", *Biosystems*, Vol. 100, No. 2, 2010, pp. 122–131.

O. Cizmecioglu, *et al.*, "CEP152 Acts as a Scaffold for Recruitment of Plk4 and CPAP to the Centrosome", *Journal of Cell Biology*, Vol. 191, No. 4, 2010, pp. 731–739.

P.T. Conduit, "Microtubule Organization: A Complex Relation", *Journal of Cell Biology (JCB)*, Vol. 213, 2016, pp. 809–812.

J.R. Cooper, *et al.*, "Catalysis of the Microtubule On-Rate Is the Major Parameter Regulating the Deploymerase Activity of MCAK", *Nature Structural and Molecular Biology*, Vol. 17, 2010, pp. 77–82.

C.R. Cowan and A.A. Hyman, "Centrosomes Direct Cell Polarity Independently of Microtubule Assembly in *C. elegans* Embryos", *Nature*, Vol. 431, No. 7004, 2004, pp. 92–96.

L. Cucullo, *et al.*, "Very Low Intensity Alternating Current Decreases Cell Proliferation", *Glia*, Vol. 51, No. 1, 2005, pp. 65–72.

A.G. Cuenca, *et al.*, "Emerging Implications of Nanotechnology on Cancer Diagnostics and Therapeutics", *Cancer*, Vol. 107, 2006, pp. 459–466.

I. Cunha-Ferreira, I. Bento, and M. Bettencourt-Dias, "From Zero to Many: Control of Centriole Number in Development and Disease", *Traffic*, Vol. 10, No. 5, 2009, pp. 482–498.

I. Cunha-Ferreira, *et al.*, "Regulation of Autophosphorylation Controls PLK4 Self-Destruction and Centriole Number", *Current Biology*, Vol. 23, No. 22, 2013, pp. 2245–2254.

I. Cunha-Ferreira, *et al.*, "The SCF/Slimb Ubiquitin Ligase Limits Centrosome Amplification through Degradation of SAK/PLK4", *Current Biology*, Vol. 19, 2009, pp. 43–49.

J. Cuzick, *et al.*, "Electropotential Measurements as a New Diagnostic Modality for Breast Cancer", *The Lancet*, Vol. 352, No. 9125, 1998, pp. 359–363.

W. Dai, Q. Wang, and F. Traganos, "Polo-like Kinases and Centrosome Regulation", *Oncogene*, Vol. 21, No. 40, pp. 6195–6200.

M.G. Damiano, *et al.*, "Templated High Density Lipoprotein Nanoparticles as Potential Therapies and for Molecular Delivery", *Advanced Drug Delivery Reviews*, Vol. 65, 2013, pp. 649–662.

A. Dammermann, P.S. Maddox, A. Desai, and K. Oegema, "SAS-4 Is Recruited to a Dynamic Structure in Newly Forming Centrioles that Is Stabilized by the ϒ-Tubulin-Mediated Addition of Centriolar Microtubules", *Journal of Cell Biology*, Vol. 180, No. 4, 2008, pp. 771–785.

A. Dammermann, *et al.*, "Centriole Assembly Requires Both Centriolar and Pericentriolar Material Proteins", *Developmental Cell*, Vol. 7, No. 6, 2004, pp. 815–829.

A. Dart, "Extra! Extra! Read all about it", *Nature Reviews Cancer*, published online 23 February, 2017. doi:10.1038/nrc2017.12.

A.B. D'Assoro, *et al.*, "Amplified Centrosomes in Breast Cancer: A Potential Indicator of Tumor Aggressiveness", *Breast Cancer Research and Treatment*, Vol. 75, 2002, pp. 25–34.

A.B. D'Assoro, W.L. Lingle, and J.L. Salisbury, "Centrosome Amplification and the Development of Cancer", *Oncogene*, Vol. 21, No. 40, 2002, pp. 6146–6153.

M.E. Davis, Z. Chen, and D.M. Shin, "Nanoparticle Therapeutics: An Emerging Treatment Modality for Cancer", *Nature Reviews Drug Discovery*, Vol. 7, 2008, pp. 771–782.

R.J. Davies, *et al.*, "Detection of the Cancer-Prone Colon, Using Transepithelial Impedance Analysis", *Archives of Surgery*, Vol. 124, No. 4, 1989, pp. 480–484.

T. Davoli, *et al.*, "Tumor Aneuploidy Correlates with Markers of Immune Evasion and with Reduced Response to Immunotherapy", *Science*, Vol. 355, 2017, 37 pages.

M. Delattre, C. Canard, and P. Gonczy, "Sequential Protein Recruitment in *C. elegans* Centriole Formation", *Current Biology*, Vol. 16, 2006, pp. 1844–1849.

M. Delattre and P. Gönczy, "The Arithmetic of Centrosome Biogenesis", *Journal of Cell Science*, Vol. 117, 2004, pp. 1619–1630.

J.G. DeLuca, "Kinetochore-Microtubule Dynamics and Attachment Stability", *Methods in Cell Biology*, Vol. 97, 2010, Chapter 4, pp. 53–79.

S.B. Dev and G.A. Hofmann, "Electrochemotherapy — A Novel Method of Cancer Treatment", *Cancer Treatment Reviews*, Vol. 21, No. 1, 1994, pp. 105–115.

M.B.A. Djamgoz, *et al.*, "Directional Movement of Rat Prostate Cancer Cells in Direct-Current Electric Field: Involvement of Voltagegated Na+ Channel Activity", *Journal of Cell Science*, Vol. 114, Pt. 14, 2001, pp. 2697–2705.

J. Dobson, "Magnetic Nanoparticles for Drug Delivery", *Drug Development Research*, Vol. 67, 2006, pp. 55–60.

J. Dobson, "Remote Control of Cellular Behaviour with Magnetic Nanoparticles", *Nature Nanotechnology*, Vol. 3, 2008, pp. 139–143.

H. Dodson, *et al.*, "Centrosome Amplification Induced by DNA Damage Occurs During a Prolonged G2 Phase and Involves ATM", *The EMBO Journal*, Vol. 23, No. 19, 2004, pp. 3864–3873.

M. Dogterom, *et al.*, "Force Generation by Dynamic Microtubules", *Current Opinion in Cell Biology*, Vol. 17, 2005, pp. 67–74.

K.H. Downing and E. Nogales, "New Insights into Microtubule Structure and Function from the Atomic Model of Tubulin", *European Biophysics Journal*, Vol. 27, No. 5, 1998, pp. 431–436.

S.J. Doxsey, "The Centrosome — A Tiny Organelle with Big Potential", *Nature Genetics*, Vol. 20, No. 2, 1998, pp. 104–106.

S.J. Doxsey, "Centrosomes as Command Centres for Cellular Control", *Nature Cell Biology*, Vol. 3, No. 5, 2001, pp. E105–E108.

S.J. Doxsey, "Duplicating Dangerously: Linking Centrosome Duplication and Aneuploidy", *Molecular Cell*, Vol. 10, 2002, pp. 439–440.

S.J. Doxsey, "Re-evaluating Centrosome Function", *Nature Reviews Molecular Cell Biology*, Vol. 2, No. 9, 2001, pp. 688–698.

S.J. Doxsey, D. McCollum, and W. Theurkauf, "Centrosomes in Cellular Regulation", *Annual Review of Cell and Developmental Biology*, Vol. 21, 2005, pp. 411–434.

S.J. Doxsey, W. Zimmerman, and K. Mikule, "Centrosome Control of the Cell Cycle", *Trends in Cell Biology*, Vol. 15, No. 6, 2005, pp. 303–311.

S. Duensing, "A Tentative Classification of Centrosome Abnormalities in Cancer", *Cell Biology International*, Vol. 29, 2005, pp. 352–359.

A. Duensing and S. Duensing, "Guilt by Association? P53 and the Development of Aneuploidy in Cancer", *Biochemical and Biophysical Research Communications*, Vol. 331, No. 3, 2005, pp. 694–700.

A. Duensing, *et al.*, "Centriole Overduplication through the Concurrent Formation of Multiple Daughter Centrioles at Single Material Templates", *Oncogene*, Vol. 26, No. 43, 2007, pp. 6280–6288.

A. Duensing, *et al.*, "Centrosome Overduplication Chromosomal Instability, and Human Papillomavirus Oncoproteins", *Environmental and Molecular Mutagenesis*, Vol. 40, 2009, pp. 741–747.

M.J. Duffy, *et al.*, "p53 as a Target for the Treatment of Cancer", *Cancer Treatment Reviews*, Vol. 40, 2014, pp. 1153–1160.

S. Dumont and T.J. Mitchison, "Force and Length in the Mitotic Spindle", *Current Biology*, Vol. 19, 2009, pp. R749–R761.

P. Dupuis-Williams, *et al.*, "Functional Role of ε-Tubular in the Assembly of the Centriolar Microtubule Scaffold", *Journal of Cell Biology*, Vol. 158, No. 7, 2002, pp. 1183–1193.

S.K. Dutcher, "The Tubulin Fraternity: Alpha to Eta", Current *Opinion in Cell Biology*, Vol. 13, No. 1, 2001, pp. 49–54.

S.K. Dutcher, "Long-Lost Relatives Reappear: Identification of New Members of the Tubulin Superfamily", *Current Opinion in Microbiology*, Vol. 6, 2003, pp. 634–640.

N.S. Dzhindzhev, *et al.*, "Asterless Is a Scaffold for the Onset of Centriole Assembly", *Nature*, Vol. 457, 2010, pp. 714–718.

F. Eckerdt, *et al.*, "Identification of a Polo-like Kinase 4-Dependent Pathway for De Novo Centriole Formation", *Current Biology*, Vol. 21, No. 5, 2011, pp. 428–432.

S.I. Ellenbroek, S. Iden, and J.G. Collard, "Cell Polarity Proteins and Cancer", *Seminars in Cancer Biology*, Vol. 22, No. 3, 2012, pp. 208–215.

M.W. Elting, *et al.*, "Force on Spindle Microtubule Minus Ends Moves Chromosomes", *Journal of Cell Biology*, Vol. 206, No. 2, 2014, pp. 245–256.

M. Falco, *et al.*, "Tumour Biomarkers: Homeostasis as a Novel Prognostic Indicator", *Open Biology*, Vol. 6, 2016, pp. 1–31.

A.H. Faraji and P. Wipf, "Nanoparticles in Cellular Drug Delivery", *Bioorganic & Medicinal Chemistry*, Vol. 17, 2009, pp. 2950–2962.

N.H. Farina, *et al.*, "Nanoparticle-Based Targeted Cancer Strategies for Non-Invasive Prostate Cancer Intervention", *Journal of Cell Physiology*, Vol. 233, 2018, pp. 6408–6417.

R.H. Farnsworth, *et al.*, "Vascular Remodeling in Cancer", *Oncogene*, Vol. 33, 2014, pp. 3496–3505.

M. Ferrari, "Cancer Nanotechnology: Opportunities and Challenges", *Nature Reviews Cancer*, Vol. 5, 2005, pp. 161–171.

N. Ferrara, "Molecular and Biological Properties of Vascular Endothelial Growth Factor", *Journal of Molecular Medicine*, Vol. 77, No. 7, 1999, pp. 527–543.

E.N. Firat-Karalar and T. Stearns, "The Centriole Duplication Cycle", *Philosophical Transactions of the Royal Society of London. Series B, Biological Sciences*, Vol. 369, 2014, p. 20130460.

H.A. Fisk, C.P. Mattison, and M. Winey, "Human Mps1 Protein Kinase Is Required for Centrosome Duplication and Normal Mitotic Progression", *Proceedings of the National Academy of Sciences*, Vol. 100, 203, pp. 14875–14880.

N.A. Fonseca, *et al.*, "Bridging Cancer Biology and the Patient's Needs with Nanotechnology-Based Approaches", *Cancer Treatment Reviews*, Vol. 40, No. 5, 2014, pp. 626–635.

S.K. Frandsen, *et al.*, "Normal and Malignant Cells Exhibit Differential Responses to Calcium Electroporation", *Cancer Research, Therapeutics, Targets and Chemical Biology*, Vol. 77, 2017, pp. 4389–4401. doi:10.1158/0008-5472.CAN-16-1611

S.K. Frandsen, *et al.*, "Difference in Membrane Repair Capacity Between Cancer Cell Lines and a Normal Cell Line", *Journal of Membrane Biology*, Vol. 249, 2016, pp. 569–576.

A.M. Fry, "Solving the Centriole Disengagement Puzzle", *Nature Cell Biology*, Vol. 17, No. 1, 2015, pp. 3–5.

J. Fu and D.M. Glover, "Structured Illumination of the Interface between Centriole and Peri-Centriolar Material", *Open Biology*, Vol. 2, 2012, pp. 1–30.

K. Fukasawa, "Introduction", *Oncogene*, Vol. 21, 2002, pp. 6140–6145.

K. Fukasawa, "Centrosome Amplification, Chromosome Instability and Cancer Development", *Cancer Letters*, Vol. 230, 2005, pp. 6–19.

K. Fukasawa, "Oncogenes and Tumour Suppressors Take on Centrosomes", *Nature Reviews Cancer*, Vol. 7, 2007, pp. 911–924.

K. Fukasawa, *et al.*, "Abnormal Centrosome Amplification in the Absence of P53", *Science*, Vol. 271, 1996, pp. 1744–1747.

E.E.M. Furlong and M. Levine, "Developmental Enhancers and Chromosome Topology", *Science*, Vol. 361, 2018, pp. 1341–1345.

B.J. Galletta, *et al.*, "Asterless Is Required for Centriole Length Control and Sperm Development", *Journal of Cell Biology*, Vol. 213, No. 4, 2016, pp. 435–474.

N.J. Ganem, S.A. Godinho, and D. Pellman, "A Mechanism Linking Extra Centrosomes to Chromosomal Instability", *Nature*, Vol. 460, No. 7252, 2009, pp. 278–282.

N.J. Ganem, Z. Storchova, and D. Pellman, "Tetraploidy, Aneuploidy and Cancer", *Current Opinion in Genetics & Development*, Vol. 17, 2007, pp. 157–162.

P. Ganesan and D. Narayanasamy, "Lipid Nanoparticles: Different Preparation Techniques, characterization, Hurdles, and Strategies for the Production of Solid Lipid Nanoparticles and Nanostructured Lipid Carriers for Oral Drug Delivery", *Sustainable Chemistry and Pharmacy*, Vol. 6, 2017, pp. 37–56.

G.G. Ganev, "Pathological Significance of Cilia and Centrioles in Bone and Soft Tissue Tumors", *Journal of Orthopaedic Science*, Vol. 2, 1997, pp. 137–145.

A. Garci-Bennett, M. Nees, and B. Fadeel, "In Search of the Holy Grail: Folate-Targeted Nanoparticles for Cancer Therapy", *Biochemical Pharmacology*, Vol. 81, No. 8, 2011, pp. 976–984.

M.K. Gardner, "Cell Biology: Microtubule Collisions to the Rescue", *Current Biology*, Vol. 26, 2016, pp. R1287–R1289.

B. Garipcan, *et al.*, "Image Analysis of Endothelial Microstructure and Endothelial Cell Dimensions of Human Arteries–A Preliminary Study", *Advanced Engineering Materials*, Vol. 13, 2011, pp. 854–857.

C. Garzan-Coral, "A Force-Generating Machinery Maintains the Spindle at the Cell Center during Mitosis", *Science*, Vol. 352, 2016, pp. 1124–1127.

I. Gasic and T.J. Mitchison, "Autoregulation and Repair in Microtubule Homeostasis", *Current Opinion in Cell Biology*, Vol. 56, 2019, pp. 80–87.

M.P. Gavilan, *et al.*, "The Dual Role of the Centrosome Is Organizing the Microtubule Network in Interphase", *EMBO Reports*, Vol. 19, 2018, pp. 1–21.

J.H. Gibcus, *et al.*, "A Pathway for Mitotic Chromosome Formation", *Science*, Vol. 359, No. 652, 2018.

S. Gilberto and M. Peter, "Dynamic Ubiquitin Signaling in Cell Cycle Regulation", *Journal of Cell Biology*, Vol. 216, 2017, p. 2259.

P.P. Girard, *et al.*, "Cellular Chemomechanics at Interfaces: Sensing, Integration and Response", *Soft Matter*, Vol. 3, 2007, pp. 307–326.

D. Gisselsson, *et al.*, "Centrosomal Abnormalities, Multipolar Mitoses, and Chromosomal Instability in Head and Neck Tumours with Dysfunctional Telomeres", *British Journal of Cancer*, Vol. 87, 2002, pp. 202–207.

F. Gittes, *et al.*, "Flexural Rigidity of Microtubules and Actin Filaments Measured from Thermal Fluctuations in Shape", *Journal of Cell Biology*, Vol. 120, No. 4, 1993, pp. 923–934.

D.R. Glenn, *et al.*, "Single Cell Magnetic Imaging Using a Quantum Diamond Microscope", *Nature Methods*, Vol. 12, 2015, pp. 736–738.

S.A. Godinho, M. Kwon, and D. Pellman, "Centrosomes and Cancer: How Cancer Cells Divide with too Many Centrosomes", *Cancer Metastasis Review*, Vol. 28, 2009, pp. 85–98.

S.A. Godinho and B. Pellman, "Causes and Consequences of Centrosome Abnormalities in Cancer", *Philosophical Transactions of the Royal Society of London. Series B, Biological Sciences*, Vol. 369, 2014, 13 pages.

T.M. Goepfert, *et al.*, "Centrosome Amplification and Overexpression of Aurora A Are Early Events in Rat Mammary Carcinogenesis", *Cancer Research*, Vol. 62, No. 14, 2002, pp. 4115–4122.

B. Goldstein and I.G. Macara, "The PAR Proteins: Fundamental Players in Animal Cell Polarization", *Developmental Cell*, Vol. 13, No. 5, 2007, pp. 609–622.

E.R. Gomes and G.G. Gundersen, "Real-Time Centrosome Reorientation During Fibroblast Migration", *Methods in Enzymology*, Vol. 406, 2006, pp. 579–592.

P. Gönczy, "Towards a Molecular Architecture of Centriole Assembly", *Nature Reviews Molecular Cell Biology*, Vol. 13, 2012, pp. 425–435.

P. Gönczy, "Centrosomes and Cancer: Revisiting a Long-Standing Relationship", *Nature Reviews Cancer*, Vol. 15, 2015, pp. 639–642.

C. Gonzalez, "Spindle Orientation, Asymmetric Division and Tumour Suppression in Drosophila Stem Cells", *Nature Review*, Vol. 8, 2007, pp. 462–472.

R. Goodman and M. Blank, "Insights into Electromagnetic Interaction Mechanisms", *Journal of Cellular Physiology*, Vol. 192, 2002, pp. 16–22.

J. Gopalakrishnan, *et al.*, "Self-Assembling SAS-6 Multimer Is a Core Centriole Building Block", *Journal of Biological Chemistry*, Vol. 285, No. 12, 2010, pp. 8759–8770.

M. Gottardo, G. Callaini, and M.G. Riparbelli, "Procentriole Assembly Without Centriole Disengagement — A Paradox of Male Gametogenesis", *Journal of Cell Science*, Vol. 127, 2014, pp. 3434–3439.

J.R. Gray, C.H. Frith, and J.D. Parker, "In Vivo Enhancement of Chemotherpay with Static Electric or Magnetic Fields", *Bioeletromagnetics*, Vol. 21, No. 8, 2000, pp. 575–583.

D.R. Green, "Cancer and Apoptosis: Who Is Built to Last?" *Cancer Cell*, Vol. 31, 2017, pp. 2–4.

E.L. Grishchuk, *et al.*, "Force Production by Disassembling Microtubules", *Nature*, Vol. 438, No. 17, 2005, pp. 385–388.

S.R. Grobmyer and B.M. Moudqil (Eds.), *Cancer Nanotechnology Methods and Protocols*, Humana Press, Springer, New York, 2010.

D. Guest, *Biology Smart*, The Princeton Review, Random House, New York, 1996, pp. 133–135.

P. Guichard, *et al.*, "Cell-Free Reconstitution Reveals Centriole Cartwheel Assembly Mechanisms", *Nature Communications*, Vol. 8, 2017, p. 14813.

P. Guichard, *et al.*, "Procentriole Assembly Revealed by Cryo-Electron Tomography", *The EMBO Journal*, Vol. 29, No. 9, 2010, pp. 1565–1572. doi:10.1038/emboj.2010.45

G.D. Gupta and L. Pelletier, "Centrosome Biology: Polymer-Based Centrosome Maturation", *Current Biology*, Vol. 27, 2017, pp. R833–R852.

R. Habedanck, *et al.*, "The Polo Kinase Plk4 Functions in Centriole Duplication", *Nature Cell Biology*, Vol. 7, No. 11, 2005, pp. 1140–1146.

B. Haley and E. Frenkel, "Nanoparticles for Drug Delivery in Cancer Treatment", *Urologic Oncology: Seminars and Original Investigations*, Vol. 26, 2008, pp. 57–64.

S.R. Hameroff, "A New Theory of the Origin of Cancer: Quantum Coherent Entanglement, Centrioles, Mitosis, and Differentiation", *Biosystems*, Vol. 77, No. 1–3, 2004, pp 119–136.

L. Hamon, P.A. Curmi, and D. Pastré, "High-Resolution Imaging of Microtubules and Cytoskeleton Structures by Atomic Force Microscopy", *Methods in Cell Biology*, Vol. 95, 2010, pp. 157–174.

K.-H. Han and A.B. Frazier, "Diamagnetic Capture Mode Magnetophoretic Microseparator for Blood Cells", *Journal of Microelectromechanical Systems*, Vol. 14, No. 6, 2005, pp 1412–1431.

D. Hanahan and R.A. Weinberg, "The Hallmarks of Cancer", *Cell*, Vol. 100, 2000, pp. 57–70.

D. Hanahan and R.A. Weinberg, "Hallmarks of Cancer: The Next Generation", *Cell*, Vol. 144, 2011, pp. 646–674.

E.L. Hansen, *et al.*, "Dose-Dependent ATP Depletion and Cancer Cell death following Calcium Electroporation, Relative Effect of Calcium Concentration and Electric Field Strength", *Plos One*, Vol. 10, 2015, p. e0122973. doi:10.1371/journal.pone.0122973

J. Hardin, G. Bertoni, and L.J. Kleinsmith, *Becker's World of the Cell*, 8th Edition, Pearson, Washington, DC, 2012.

P.A. Hardy and H. Zacharias, "Reappraisal of the Hansemann-Boveri Hypothesis on the Origin of Tumors", *Cell Biology International*, Vol. 29, No. 12, 2005, pp. 983–992.

C.C. Harris, *et al.*, "Metabolism of Benzo(a)pyrene, N-nitrosodimethylamine, and N-nitrosopyrrolidine and Identification of the Major Carcinogen-DNA Adducts Formed in Cultured Human Esophagus", *Cancer Research*, Vol. 39, No. 11, 1979, pp. 4401–4406.

S. Hashimoto, *et al.*, "The Measurement of Small Magnetic Signals from Magnetic Nanoparticles Attached to the Cell Surface and Surrounding Living Cells Using a General-Purpose SQUID Magnetometer", *Physics in Medicine and Biology*, Vol. 54, 2009, pp. 2571–2583.

E.M. Hatch, *et al.*, "Cep 152 Interacts with Plk4 and Is Required for Centriole Duplication", *Journal of Cell Biology*, Vol. 191, No. 4, 2010, pp. 721–729.

H.J. Hathaway, *et al.*, "Detection of Breast Cancer Cells Using Targeted Magnetic Nanoparticles and Ultra-Sensitive Magnetic Field Sensors", *Breast Cancer Research*, Vol. 13, No. 3, 2011, p. R108.

D. Havelka and M. Cifra, "Calculation of the Electromagnetic Field Around a Microtubule", *Acta Polytechnica*, Vol. 49, 2009, pp. 58–63.

D. Havelka, *et al.*, "High-Frequency Electric Field and Radiation Characteristics of Cellular Microtubule Network", *Journal of Theoretical Biology*, Vol. 286, 2011, pp. 31–40.

C. He, *et al.*, "Effects of Particle Size and Surface Charge on Cellular Uptake and Biodistribution of Polymeric Nanoparticles", *Biomaterials*, Vol. 31, 2010, pp. 3657–3666.

M. Hilbert, *et al.*, "SAS-6 Engineering Reveals Interdependence between Cartwheel and Microtubules in Determining Centriole Architecture", *Nature, Cell Biology*, Vol. 18, No. 4, 2016, pp. 393–403. doi:10.1038/ncb3329

E.H. Hinchcliffe and G. Sluder, "'It Takes Two to Tango': Understanding How Centrosome Duplication Is Regulated Throughout the Cell Cycle", *Genes & Development*, Vol. 15, 2001, pp. 1167–1181.

L. Hinck and I. Näthke, "Changes in Cell and Tissue Organization in Cancer of the Breast and Colon", *Current Opinion in Cell Biology*, Vol. 26, 2014, pp. 87–95.

S.M. Hinshaw and S.C. Harrison, "Kinetochore Function from the Bottom Up", *Trends in Cell Biology*, Vol. 28, No. 1, 2018, pp. 22–78.

M. Hiraki, *et al.*, "Bld10p Constitutes the Cartwheel-Spoke Tip and Stabilizes the 9-Fold Symmetry of the Centriole", *Current Biology*, Vol. 17, No. 20, 2007, pp. 1776–1783.

M. Hirono, "Cartwheel Assembly", *Philosophical Transactions of the Royal Society of London. Series B, Biological Sciences*, Vol. 369, 2014, 16 pages.

S.N. Hochwald, "Molecular-Targeted Therapy for Cancer and Nanotechnology", *Methods in Molecular Biology*, Vol. 624, 2010, pp. 11–23.

W.G. Hol, "The Role of the Alpha-Helix Dipole in Protein Function and Structure", *Progress in Biophysics and Molecular Biology*, Vol. 45, No. 3, 1985, pp. 149–195.

A.J. Holland and D.W. Cleveland, "Boveri Revisited: Chromosomal Instability, Aneuploidy and Tumorigenesis", *Nature Reviews Molecular Cell Biology*, Volume 10, 2009, pp. 478–487.

A.J. Holland and D.W. Cleveland, "Losing Balance: The Origin and Impact of Aneuploidy in Cancer", *EMBO Reports*, Vol. 13, No. 6, 2012, pp. 501–514.

A.J. Holland, *et al.*, "The Autoregulated Instability of Polo-like Kinase 4 Limits Centrosome Duplication to Once Per Cell Cycle", *Genes and Development*, Vol. 26, No. 24, 2012, pp. 2684–2689.

A.J. Holland, *et al.*, "Polo-like Kinase 4 Kinase Activity Limits Centrosome Overduplication by Autoregulating Its Own Stability", *Journal of Cell Biology*, Vol. 188, No. 2, 2010, pp. 191–198.

M.C. Hollander and A.J. Fornace, Jr., "Genomic Instability Centrosome Amplification, Cell Cycle Checkpoints and Gadd45a", *Oncogene*, Vol. 21, No. 40, 2002, pp. 6228–6233.

S.A. Hormeño, *et al.*, "Single Centrosome Manipulation Reveals its Electric Charge and Associated Dynamic Structure", *Biophysical Journal*, Vol. 97, No. 4, 2009, pp. 1022–1030.

J. Howard and A.A. Hyman, "Dynamics and Mechanics of the Microtubule Plus End", *Nature*, Vol. 422, 2003, pp. 753–758.

S. Hughes, J. Dobson, and A.J. El Haj, "Magnetic Targeting of Mechanosensors in Bone Cells for Tissue Engineering Applications", *Journal of Biomechanics*, Vol. 40, 2007, pp. S96–S104.

K.R. Hurley, *et al.*, "Characterization of Magnetic Nanoparticles in Biological Matrices", *Analytical Chemistry*, Vol. 87, 2015, pp. 11611–11619.

R.L. Huston, "On Centrioles, Microtubules, and Cellular Electromagnitism", *Journal of Nanotechnology in Engineering and Medicine, American Society of Mechanical Engineers (ASME)*, Vol. 5, 2014, pp. 021003-1–021003-5.

R.L. Huston, "Using the Electromagnetics of Cancer's Centrosome Clusters to Attract Therapeutic Nanoparticles", *Advances in Bioscience and Biotechnology*, Vol. 6, 2015, pp. 172–181.

R.L. Huston, "Mechanics of Centriole Microtubules", *Advances in Biosciences and Biotechnology*, Vol. 7, 2016a, pp. 266–277.

R.L. Huston, "A Review of Centriole Activity, and Wrongful Activity, During Cell Division", *Advances in Bioscience and Biotechnology*, Vol. 7, 2016b, pp. 169–182.

R.L. Huston, "A Review of Electromagnetic Activity in Cellular Mechanics", *Advances in Bioscience and Biotechnology*, Vol. 7, 2016c, pp. 360–371.

R. Ibrahim, *et al.*, "Electron Tomography Study of Isolated Human Centrioles", *Microscopy Research Technique*, Vol. 72, No. 1, 2009, pp. 42–48.

D.P. Ignacio, V.C. Cottman, and A.T. Dawes, "Centriole Biogenesis: Symmetry Breaking and Site Selection", *Trends in Cell Biology*, Vol. 29, 2019, pp. 3–5.

Y.F. Inclán and E. Nogales, "Structural Models for the Self-Assembly and Microtubule Interactions of Gamma-, Delta- and Epsilon-tubulin", *Journal of Cell Science*, Vol. 114, 2000, pp. 413–422.

J.R. Jackson, *et al.*, "Targeted Anti-Mitotic Therapies: Can We Impose on Tubulin Agents?" *Nature Reviews: Cancer*, Vol. 7, No. 2, 2007, pp.107–117.

S.C. Jana, J.F. Bazan, and M. Bettencourt-Dias, "Polo Boxes Come Out of the Crypt: A New View of PLK Function and Evolution", *Structure*, Vol. 20, No. 11, 2012, pp. 1801–1804.

S.C. Jana, G. Marteil, and M. Bettencourt-Dias, "Mapping Molecules to Structure: Unveiling Secrets of Centriole and Cilia Assembly with Near-Atomic Resolution", *Current Opinion in Cell Biology*, Vol. 26, 2014, pp. 96–106.

C. Janke, "The Tubulin Code: Molecular Components, Readout Mechanisms, and Functions", *Journal of Cell Biology*, Vol. 206, No. 4, 2014, pp. 461–472.

M.E. Janson, M.E. deDood, and M. Dogterom, "Dynamic Instability of Microtubules Is Regulated by Force", *Journal of Cell Biology*, Vol. 161, No. 6, 2003, pp. 1029–1034.

F. Jiang, *et al.*, "Centrosomal Abnormality Is Common in and a Potential Biomarker for Bladder Cancer", *International Journal of Cancer*, Vol. 106, No. 5, 2003, pp. 661–665.

X. Jin, J.J. Chalmers, and M. Zborowski, "Iron Transport in Cancer Cell Culture Suspensions Measured by Cell Magnetophoresis", *Analytical Chemistry*, Vol. 84, 2012, pp 4520–4526.

D. Job, O. Valiron, and B. Oakley, "Microtubule Nucleation", *Current Opinion in Cell Biology*, Vol. 15, 2003, pp. 111–117.

A.P. Joglekar, K. Bloom, and E.D. Salmon, "In Vivo Protein Architecture of the Eukaryotic Kinetochore with Nanometer Scale Accuracy", *Current Biology*, Vol. 19, 2009, pp. 694–699.

A.P. Joglekar, *et al.*, "Molecular Architecture of a Kinetochore–Microtubule Attachment Site", *Nature Cell Biology*, Vol. 8, 2006, pp. 581–585.

C.N. Johnson, *et al.*, "Magnetic Relaxometry with an Atomic Magnetometer and SQUID Sensors on Targeted Cancer Cells", *Journal of Magnetism and Magnetic Materials*, Vol. 324, 2012, pp. 2613–2619.

A Jord, *et al.*, "Calibrated Mitotic Oscillator Drives Motile Ciliogenesis", *Science*, Vol. 358, 2017, pp. 803–806.

A. Jordan, *et al.*, "The Effect of Thermotherapy Using Magnetic Nanoparticles on Rat Malignant Glioma", *Journal of Neuro-Oncology*, Vol. 78, 2006, pp. 7–14.

M.A. Jordan and L. Wilson, "Microtubules as a Target for Anticancer Drugs", *Nature Reviews: Cancer*, Vol. 4, No. 4, 2004, pp. 253–265.

H. Josephs and R.L. Huston, *Dynamics of Mechanical Systems*, CRC Press, Boca Raton, FL, 2002.

N.W.S. Kam, *et al.*, "Carbon Nanotubes as Multifunctional Biological Transporters and Near-Infrared Agents for Selective Cancer Cell Destruction", *Proceedings of the National Academy of Sciences (PNAS)*, Vol. 102, 2005, pp. 11600–11605.

K. Kamath, E. Oroudjev, and M.A. Jordan, "Determination of Microtubule Dynamic Instability in Living Cells", *Methods in Cell Biology*, Vol. 97, 2010, pp. 1–14.

J. Kamenz and S. Hauf, "Time to Split Up: Dynamics of a Chromosome Separation", *Trends in Cell Biology*, Vol. 27, No. 1, 2017, pp. 42–54.

S. Kanno, *et al.*, "Establishment of a Simple and Practical Procedure Applicable to Therapeutic Angiogenesis", *Circulation*, Vol. 99, No. 20, 1999, pp. 2682–2687.

M. Karki, N. Keyhaninejad, and C.B. Shuster, "Precocious Centriole Disengagement and Centrosome Fragmentation Induced by Mitotic Delay", *Nature Communications*, Vol. 8, 2017, p. 15803.

P.P. Karmali, *et al.*, "Targeting of Albumin-Embedded Paclitaxel Nanoparticles to Tumors. Nanomedicine, *Nanotechnology, Biology, and Medicine*, Vol. 5, 2009, pp. 73–82. doi:10.1016/j.nano.2008.07.007

S. Kasas, *et al.*, "Oscillation Modes of Microtubules", *Biology of the Cell*, Vol. 96, No. 9, 2004, pp. 697–700.

C. Kasbek, *et al.*, "Preventing the Degradation of Mps1 at Centrosomes Is Sufficient to Cause Centrosome Reduplication in Human Cells", *Molecular Biology of the Cell*, Vol. 18, 2007, pp. 4457–4469.

K. Kawamura, *et al.*, "Induction of Centrosome Amplification and Chromosome Instability in Human Bladder Cancer Cells by p53 Mutation and Cyclin E Overexpression", *Cancer Research*, Vol. 64, 2004, pp. 4800–4809.

K. Kawamura, *et al.*, "Chromosome Hyperamplification and Chromosomal Instability in Bladder Cancer", *European Urology*, Vol. 43, No. 5, 2003, pp. 505–515.

T.J. Keating and G.G. Borisky, "Immunostructural Evidence for the Template Mechanism of Microtubule Nucleation", *Nature Cell Biology*, Vol. 2, 2000, pp. 352–357.

D. Keller, *et al.*, "Mechanisms of HsSAS-6 Assembly Promoting Centriole Formation in Human Cells", *Journal of Cell Biology*, Vol. 204, No. 5, 2014, pp. 697–712.

I.A. Kent and T.P. Lele, "Microtubule Based Force Generation", *Wiley Interdisciplinary Reviews: Nanomedicine and Nanobiotechnology*, 2016.

A. Khodjakov and C.L. Rieder, "Centrosomes Enhance the Fidelity of Cytokinesis in Vertebrates and Are Required for Cell Cycle Progression", *Journal of Cell Biology*, Vol. 153, No. 1, 2001, pp. 237–242.

D.-H. Kim, *et al.*, "Biofunctionalized Magnetic-Vortex Microdiscs for Targeted Cancer-Cell Destruction", *Nature Materials*, Vol. 9, 2010, pp. 165–171.

K. Kim, *et al.*, "Ultrahigh-Speed Rotating Nanoelectromechanical System Devices Assembled from Nanoscale Building Blocks", *Nature Communications*, Vol. 15, 2014, p. 3632.

M. Kirkham, *et al.*, "SAS-4 Is a *C. elegens* Centriolar Protein that Controls Centrosome Size", *Cell*, Vol. 112, 2003, pp. 575–587.

E.D. Kirson, *et al.*, "Alternating Electric Fields Arrest Cell Proliferation in Animal Tumor Models and Human Brain Tumors", *Proceedings of the National Academy of Sciences*, Vol. 104, No. 24, 2007, pp. 10152–10157.

E.D. Kirson, *et al.*, "Disruption of Cancer Cell Replication by Alternating Electric Fields", *Cancer Research*, Vol. 64, No. 9, 2004, pp 3288–3295.

A. Kis, *et al.*, "Nanomechanics of Microtubules", *Physical Review Letters*, Vol. 89, No. 24, 2002, pp. 248101-1–248101-4.

D. Kitagawa, *et al.*, "Structural Basis of the 9-Fold Symmetry of Centrioles", *Cell*, Vol. 144, No. 3, 2011, pp. 364–375.

M. Klagsbrun and P.A. D'Amore, "Regulators of Angiogenesis", *Annular Review of Physiology*, Vol. 53, 1991, pp. 217–239.

J.E. Klebba, *et al.*, "Polo-like Kinase 4 Autodestructs by Generating Its Slimb-Binding Phosphodegron", *Current Biology*, Vol. 23, 2013, pp. 2255–2261.

J.E. Klebba, *et al.*, "Autoinhibition and Relief Mechanism for Polo-like Kinase 4", *Proceedings of the Academy of Sciences*, Vol. 112, No. 7, 2015, pp. E657–E666.

J.E. Klebba, *et al.*, "Two Polo-like Kinase 4 Binding Domains in Asterless Perform Distinct Roles in Regulating Kinase Stability", *Journal of Cell Biology*, Vol. 208, No. 4, 2015, pp. 401–414.

J. Kleylein-Sohn, *et al.*, "Plk4-Induced Centriole Biogenesis in Human Cells", *Developmental Cell*, Vol. 13, No. 2, 2007, pp. 190–202.

S.L. Kline-Smith and C.E. Wolczak, "The Microtubule-Destabilizing Kinesin XKCM1 Regulates Microtubule Dynamic Instability in Cells", *Molecular Biology of the Cell*, Vol. 13, 2002, pp. 2718–2731.

D.A. Klos Dehring, *et al.*, "Deuterosome-Mediated Centriole Biogenesis", *Developmental Cell*, Vol. 27, No. 1, 2013, pp. 103–112.

J. A. Knoblich, "Asymmetric Cell Division: Recent Developments and their Implication for Tumour Biology", *Nature Reviews/Molecular Cell Biology*, Vol. 11, 2010, pp. 849–860.

M.A. Ko, *et al.*, "Plk4 Haploinsufficiency Causes Mitotic Infidelity and Carcinogenesis", Nature Genetics, Vol. 37, No. 6, 2005, pp. 883–888.

T. Kobayashi and B.D. Dynlacht, "Regulating the Transition from Centriole to Basal Body", *Journal of Cell Biology*, Vol. 193, No. 3, 2011, pp. 435–444.

R.S. Kochanski and G.G. Borisy, "Mode of Centriole Duplication and Distribution", *Journal of Cell Biology*, Vol. 110, No. 5, 1990, pp. 1599–1605.

G. Kohlmaier, *et al.*, "Overly Long Centrioles and Defective Cell Division upon Excess of the SAS-4-Related Protein CPAP", *Current Biology*, Vol. 19, 2009, pp. 1012–1018.

J.M. Kollman, *et al.*, "Microtubule Nucleating γ-TuSC Assembles Structures with 13-fold Microtubule-like Symmetry", *Nature*, Vol. 466, 2010, pp. 879–883.

D. Kong, *et al.*, "Centriole Maturation Requires Regulated Plk1 Activity during Two Consecutive Cell Cycles", *Journal of Cell Biology*, Vol. 206, 2014, pp. 855–865.

N. Korzeniewski, *et al.*, "Daughter Centriole Elongation Is Controlled by Proteolysis", *Molecular Biology of the Cell*, Vol. 21, 2010, pp. 2942–2951.

N. Korzeniewski, M. Hohenfellner, and S. Duensing, "CANDI Promotes PLK4-Mediated Centriole Overduplication and Is Frequently Disrupted in Prostate Cancer", *Neoplasia*, Vol. 14, No. 9, 2012, pp. 799–806.

N. Korzeniewski, *et al.*, "Cullin 1 Functions as a Centrosomal Suppressor of Centriole Multiplication by Regulating Polo-like Kinase 4 Protein Levels", *Cancer Research*, Vol. 69, No. 16, 2009, pp. 6668–6675.

A. Krämer, J. Lukas, and J. Bartek, "Checking Out the Centrosome", *Cell Cycle*, Vol. 3, No. 11, 2004, pp. 1390–1393.

A. Krämer, K. Neben, and A.D. Ho, "Centrosome Replication, Genomic Instability and Cancer", *Leukemia*, Vol. 16, No. 5, 2002, pp. 767–775.

O. Kučera and M. Cifra, "Cell-to-Cell Signaling through Light: Just a Ghost of Chance?", *Cell Communication and Signaling*, Vol. 11, 2013, pp. 1–8.

O. Kučera, M. Cifra, and J. Pokorný, "Technical Aspects of Measurement of Cellular Electromagnetic Activity", *European Biophysics Journal*, Vol. 39, 2010, pp. 1465–1470.

O. Kučera and D. Havelka, "Mechano-Electrical Vibrations of Microtubules — Link to Subcellular Morphology", *Biosystems*, Vol. 19, No. 3, 2012, pp. 346–355.

J.F. Kukowska-Latallo, *et al.*, "Nanoparticle Targeting of Anticancer Drug Improves Therapeutic Response in Animal Model of Human Epithelial Cancer", *Cancer Research*, Vol. 65, No. 12, 2005, pp. 5317–5324.

Z. Kuncic and S. Lacombe, "Nanoparticle Radio-Enhancement Principles, Progress and Application to Cancer Treatment", *Physics in Medicine & Biology*, Vol. 63, 2018, p. 02TR01.

L. Kurzawa, *et al.*, "Dissipation of Contractile Forces: The Missing Piece in Cell Mechanics", *Molecular Biology of the Cell*, Vol. 28, No. 14, 2017, pp. 1825–1832. doi:10.1091/mbc.e16-09-0672

M. Kwon, "Using Cell Culture Models of Centrosome Amplification to Study Centrosome Clustering in Cancer", *Methods in Molecular Biology*, Vol. 1413, 2016, pp. 367–392.

M. Kwon, *et al.*, "Mechanisms to Suppress Multipolar Divisions in Cancer Cells with Extra Chromosomes", *Genes & Development*, Vol. 22, No. 16, 2008, pp. 2189–2203.

K.R. Lacey, P.K. Jackson, and T. Stearns, "Cyclin-Dependent Kinase Control of Centrosome Duplication", *Proceedings National Academy of Sciences*, Vol. 96, 1999, pp. 2817–2822.

B. Lacroix, *et al.*, "Microtubule Dynamics Scale with Cell Size to Set Spindle Length and Assembly Timing", *Developmental Cell*, Vol. 45, 2018, pp. 496–511.

B.G. Lambrus, *et al.*, "p53 Protects Against Genome Instability Following Centriole Duplication Failure", *Journal of Cell Biology*, Vol. 210, No. 1, 2015, pp. 63–77.

T. Lammers, W.E. Hennink, and G. Storm, "Tumour-Targeted Nanomedicines: Principles and Practice", *British Journal of Cancer*, Vol. 99, 2008, pp. 392–397.

F. Lampert and S. Westermann, "A Blueprint for Kinetochores — New Insight into the Molecular Mechanics of Cell Division", *Nature Reviews: Molecular Cell Biology*, Vol. 12, 2011, pp. 407–412.

B.M. Lange and K. Gull, "A Molecular Marker for Centriole Maturation in the Mammalian Cell Cycle", *Journal of Cell Biology*, Vol. 130, No. 4, 1995, pp. 919–927.

B.M. Lange and K. Gull, "Structure and Function of the Centriole in Animal Cells: Progress and Questions", *Trends in Cell Biology*, Vol. 6, No. 9, 1996, pp. 348–352.

L. Lau, *et al.*, "STED Microscopy with Optimized Labeling Density Reveals 9-Fold Arrangement of a Centriole Protein", *Biophysical Journal*, Vol. 120, No. 12, 2012, pp. 2926–2935. doi:10.1016/j.bpj.2012.05.015

K. Lauber and M. Herrmann, "Tumor Biology: With a Little Help from my Dying Friends", *Current Biology*, Vol. 25, No. 5, 2015, pp. R198–R201.

S. Lawo, *et al.*, "Subdiffraction Imaging of Centrosomes Reveals Higher-Order Organizational Features of Pericentriolar Material", *Nature Cell Biology*, Vol. 14, No. 11, 2012, pp. 1148–1158. doi:10.1038/ncb2591

B. Leber, *et al.*, "Proteins Required for Centrosome Clustering in Cancer Cells", *Science Translational Medicine*, Vol. 2, No. 33, 2010, pp. 1–11.

N. Lecland and J. Lüders, "Imaging and Quantifying the Dynamics of γ-Tubulin at Microtubule Minus Ends in Mitotic Spindles", *The Mitotic Spindle — Methods and Protocols*, Edited by P. Chang and R. Ohi, Humana Press, Springer Science and Business Media, New York, 2018, Chapter 5.

J.-H. Lee, *et al.*, "Artificially Engineered Magnetic Nanoparticles for Ultra-Sensitive Molecular Imaging", *Nature Medicine*, Vol. 13, 2007, pp. 95–99.

K. Lee and K. Rhee, "Separase-Dependent Cleavage of Pericentrin B Is Necessary and Sufficient for Centriole Disengagement During Mitosis", *Cell Cycle*, Vol. 11, 2012, pp. 2476–2485.

M. Lee and V. Vasioukhin, "Cell Polarity and Cancer–Cell and Tissue Polarity as a Non-Canonical Tumor Suppressor", *Journal of Cell Science*, Vol. 121, 2008, pp. 1141–1150.

B.K. Lee, Y. Yun, and K. Park, "PLA Micro- and Nano-Particles", *Advanced Drug Delivery Reviews*, Vol. 107, 2016, pp. 176–191.

R. Leguy, *et al.*, "Monomeric ϒ-Tubulin Nucleates Microtubules", *Journal of Biological Chemistry*, Vol. 275, No. 29, 2000, pp. 21975–21980.

S. Leidel, *et al.*, "SAS-6 Defines a Protein Family Required for Centrosome Duplication in *C. elegans* and in Human Cells", *Nature Cell Biology*, Vol. 7, No. 2, 2005, pp. 115–125.

S. Leidel and P. Gönczy, "Centrosome Duplication and Nematodes: Recent Insights from an Old Relationship", *Developmental Cell*, Vol. 9, No. 3, 2005, pp. 317–325.

D. LeSage, *et al.*, "Optical Magnetic Imaging of Living Cells", *Nature*, Vol. 496, 2013, pp. 486–489.

G.C. Leung, *et al.*, "The SAK Polo-Box Comprises a Structural Domain Sufficient for Mitotic Subcellular Localization", *Nature Structural Biology*, Vol. 9, No. 10, 2002, pp. 719–724.

M.S. Levine, *et al.*, "Centrosome Amplification Is Sufficient to Promote Spontaneous Tumorigenesis in Mammals, *Developmental Cell*, Vol. 40, 2017, pp. 313–322.

S.D. Li and L. Huang, "Stealth Nanoparticles: High Density but Sheddable PEG Is a Key for Tumor Targeting", *Journal of Controlled Release*, Vol. 145, 2010, pp. 178–181.

M.M. Lin, *et al.*, "Development of Superparamagnetic Iron Oxide Nanoparticles (SPIONS) for Translation to Clinical Applications", *IEEE Transactions on Nanobioscience*, Vol. 7, No. 4, 2008, pp. 298–305.

Y. Lin and L. Pelletier, "A Magic Bullet for Targeting Cancers with Supernumerary Centrosomes", *The EMBO Journal*, Vol. 38, 2018, p. e101134.

C.-Y. Lin, *et al.*, "Cancer Stem Cell Markers: What Is Their Diagnostic Value?", *Expert Opinion Medical Diagnosis*, Vol. 4, No. 6, pp. 474–481.

T.-C. Lin, A. Neuner, and E. Schiebel, "Targeting of ϒ-Tubulin Complexes to Microtubule Organizing Centers: Conservation and Divergence", *Trends in Cell Biology*, Vol. 25, No. 5, 2015, pp. 296–307.

W.L. Lingle, *et al.*, "Centrosome Amplification Drives Chromosomal Instability in Breast Tumor Development", *Proceedings of the National Academy of Sciences*, Vol. 99, No. 4, 2002, pp. 1978–1983.

W.L. Lingle, *et al.*, "Centrosome Hypertrophy in Human Breast Tumors: Implications for Genomic Stability and Cell Polarity", *Proceedings of the National Academy of Sciences*, Vol. 95, 1998, pp. 2950–2955.

W.L. Lingle and J.L. Salisbury, "Altered Centrosome Structure Is Associated with Abnormal Mitoses in Human Breast Tumors", *American Journal of Pathology*, Vol. 155, No. 6, 1999, pp. 1941–1951.

W.L. Lingle and J.L. Salisbury, "The Role of the Centrosome in the Development of Malignant Tumors", *Current Topics in Developmental Biology*, Vol. 49, 2000, pp. 313–339.

M. Liong, *et al.*, "Multifunctional Inorganic Nanoparticles for Imaging, Targeting, and Drug Delivery", *ACS Nano*, Vol. 2, No. 5, 2008, pp. 889–896.

Z. Liu, *et al.*, "Drug Delivery with Carbon Nanotubes for In Vivo Cancer Treatment", *Cancer Research*, Vol. 68, 2008, pp. 6652–6660.

H. Löffler, *et al.*, "DNA Damage-Induced Centrosome Amplification Occurs Via Excessive Formation of Centriolar Satellites", *Oncogene*, Vol. 32, No. 24, 2013, pp. 2963–2972.

H. Löffler, *et al.*, "Structure Meets Function — Centrosomes, Genome Maintenance and the DNA Damage Response", *Experimental Cell Research*, Vol. 312, No. 14, 2006, pp. 2633–2640.

J. Lončarek and M. Bettencourt-Dias, "Building the Right Centriole for Each Cell Type", *Journal of Cell Biology*, Vol. 217, No. 3, 2018, p. 823.

J. Lončarek, P. Herget, and A. Khodjakov, "Centeriole Reduplication during Prolonged Interphase Requires Procentriole Maturation Governed by Plk1", *Current Biology*, Vol. 20, 2010, pp. 1277–1282.

J. Lončarek, *et al.*, "Control of Daughter Centriole Formation by the Pericentriolar Material", *Nature: Cell Biology*, Vol. 10, No. 3, 2008, pp. 322–328.

J. Lončarek, A. Khodjakov, "Ab ovo or de nous? Mechanisms of Centriole Duplication", *Molecules and Cells*, Vol. 27, No. 2, 2009, pp. 135–142.

G. Long, *et al.*, "Histological and Finite Element Analysis of Cell Death due to Irreversible Electroporation", *Technology in Cancer Research and Treatment*, Vol. 13, No. 6, 2014, pp. 561–569.

C.A.M. Lopez, *et al.*, "Centrosome Amplification Arises Before Neoplasia and Increases Upon P53 Loss in Tumorigenesis", *Journal of Cell Biology*, Vol. 217, No. 7, 2018 Archive, p. 2353.

P.S. Low, W.A. Henne, and D.D. Doomeweerd, "Discovery and Development of Folic-Acid-Based Receptor Targeting for Imaging and Therapy of Cancer and Inflammatory Diseases", *Accounts of Chemical Research*, Vol. 41, No. 1, 2008, pp. 120–129.

D.M. Lowery, D. Lim, and M.B. Yaffe, "Structure and Function of Polo-like Kinases", *Oncogene*, Vol. 24, No. 2, 2005, pp. 248–259.

A.S. Lübbe, *et al.*, "Clinical Experiences with Magnetic Drug Targeting: A Phase I Study with 4′-Epidoxorubicin in 14 Patients with Advanced Solid Tumors", *Cancer Research*, Vol. 56, 1996, pp. 4686–4693.

J. Lüders, "XMAP215 Joins Microtubule Nucleation Team", *Nature Cell Biology*, Vol. 20, 2018, pp. 507–520.

J. Lüders and T. Stearns, "Microtubule-Organizing Centres: A Re-Evaluation", *Nature Reviews, Molecular Cell Biology*, Vol. 8, No. 2, 2007, pp. 161–167.

S.M. Maghimi, A.C. Hunter, and J.C. Murray, "Nanomedicine: Current Status and Future Prospects", *The FASEB Journal–Review*, Vol. 19, 2005, pp. 311–330.

R. Mahen and A.R. Venkitaraman, "Pattern Formation in Centrosome Assembly", *Current Opinion in Cell Biology*, Vol. 24, No. 1, 2012, pp. 14–23.

M.R. Mahjoub and T. Stearns, "Supernumerary Centrosomes Nucleate Extra Cilia and Compromise Primary Cilium Signaling", *Current Biology*, Vol. 22, No. 17, 2012, pp. 1628–1634.

H. Maiato, *et al.*, "The Dynamic Kinetochore-Microtubule Interface", *Journal of Cell Science*, Vol. 117, 2004, pp. 5461–5477.

H. Maiato and E. Logarinho, "Mitotic Spindle Multipolarity without Centrosome Amplifications", *Nature Cell Biology*, Vol. 16, No. 5, 2014, pp. 386–394.

V. Mailänder and K. Landfester, "Interaction of Nanoparticles with Cells", *Biomacromolecules*, Vol. 10, 2009, pp. 2379–2400.

I.J. Majoros, *et al.*, "Poly (amidoamine) Dendrimer-Based Multifunctional Engineered Nanodevice for Cancer Therapy", *Journal of Medicinal Chemistry*, Vol. 48, 2005, pp. 5892–5899.

J.K. Mando and I.M. Chassman, "The Kinetochore-Microtubule Interface at a Glance", *Journal of Cell Science*, Vol. 131, 2018, p. 7.

R.J. Mannix, *et al.*, "Nanomagnetic Actuation of Receptor-Mediated Signal Transduction", *Nature Nanotechnology*, Vol. 3, 2008, pp. 36–40.

E.N. Marieb, *Human Anatomy and Physiology*, 3rd Edition, Benjamin/Cummings Publishing, Redwood City, CA, 1991, Chapter 3, pp. 60–101.

M.S. Markov, "Magnetic Field Theory: A Review", *Electromagnetic Biology and Medicine*, Vol. 26, 2007, pp. 1–23.

W. Marshall, "Centrioles Take Center Stage", *Current Biology*, Vol. 11, 2001, pp. R487–R496.

W.F. Marshall, "Centriole Assembly: The Origin of Nine-ness", *Current Biology*, Vol. 17, No. 24, 2007a, pp. R1057–R1059.

W.F. Marshall, "What Is the Function of Centrioles?" *Journal of Cellular Biochemistry*, Vol. 100, No. 4, 2007b, pp. 916–922.

W.F. Marshall and J.L. Rosenbaum, "How Centrioles Work: Lessons from Green Yeast", *Current Opinion in Cell Biology*, Vol. 12, No. 1, 2000, pp. 119–125.

G. Marteil, *et al*, "Over-Elongation of Centrioles in Cancer Promotes Centriole Amplification and Chromosome Missegregation", *Nature Communications*, Vol. 9, 2018, p. 1258.

V. Marthiens, M. Piel, and R. Basto, "Never Tear us Apart — The Importance of Centrosome Clustering", *Journal of Cell Science*, Vol. 125, 2012, pp. 3281–3292.

V. Marthiens, *et al.*, "Centrosome Amplification Causes Microcephaly", *Nature Cell Biology*, Vol. 15, 2013, pp. 731–742.

M. Mashreghi, *et al.*, "Angiogenesis Biomarkers and their Targeting Ligands as Potential Targets for Tumor Angiogenesis", *Journal of Cell Physiology*, Vol. 233, 2018, pp. 2949–2965.

J. Marx, "Do Centrosome Abnormalities Lead to Cancer?" *Science*, Vol. 292, No. 5516, 2001, pp. 426–429.

Y. Matsumura and H. Maeda, "A New Concept for Macromolecular Therapeutics in Cancer Chemotherapy: Mechanism of Tumoritropic Accumulation of Proteins and the Antitumor Agent Smancs", *Cancer Research*, Vol. 46, 1986, pp. 6382–6392.

Y. Matsumoto, *et al.*, "Vascular Bursts Enhance Permeability of Tumour Blood Vessels and Improve Nanoparticle Delivery", *Nature Nanotechnology*, Vol. 11, 2016, pp. 533–539.

K. Matsuo, *et al.*, "Kendrin Is a Novel Substrate for Separase Involved in the Licensing of Centriole Duplication", *Current Biology*, Vol. 22, 2012, pp. 915–921.

T. Mayor, *et al.*, "The Centrosomal Protein C-Nap1 Is Required for Cell Cycle-regulated Centrosome Cohesion", *Journal of Cell Biology*, Vol. 151, 2000, pp. 837–846.

L.M. McCaffrey, *et al.*, "Loss of the Par3 Polarity Protein Promotes Breast Tumorigenesis and Metastasis", *Cancer Cell*, Vol. 22, No. 5, 2012, pp. 601–614.

C.D. McCaig, *et al.*, "Controlling Cell Behavior Electrically: Current views and Future Potential", *Physiological Review*, Vol. 85, No. 3, 2005, pp. 943–978.

P.G. McKean, S. Vaughan, and K. Gull, "The Extended Tubulin Superfamily", *Journal of Cell Science*, Vol. 114, Pt. 15, 2001, pp. 2723–2733.

F.J. McNally, "Mechanisms of Spindle Positioning", *Journal of Cell Biology*, Vol. 200, 2013, pp. 131–159.

F.J. McNally and A. Roll-Mecak, "Microtubule-Severing Enzymes: From Cellular Functions to Molecular Mechanism", *Journal of Cell Biology*, Vol. 217, No. 12, pp. 4057–4069.

W.J. Meggs, "Electric Fields Determine the Spatial Organization of Microtubules and Actin Filaments", *Medical Hypothesis*, Vol. 26, No. 3, 1988, pp. 165–170.

V. Mennella, *et al.*, "Amorphous No More: Subdiffraction View of the Pericentriolar Material Architecture", *Trends in Cell Biology*, Vol. 24, No. 3, 2014, pp. 188–197.

V. Mennella, *et al.*, "Subdiffraction-Resolution Fluorescence Microscopy Reveals a Domain of the Centrosome Critical for Pericentriolar Material Organization", *Nature Cell Biology*, Vol. 14, No. 11, 2012, pp. 1159–1168.

P. Meraldi, R. Honda, and E.A. Nigg, "Aurora-A Overexpression Reveals Tetraploidization as a Major Route to Centrosome Amplification in p53-/-Cells", *The EMBO Journal*, Vol. 21, 2002, pp. 483–492.

P. Meraldi and E.A. Nigg, "The Centrosome Cycle", *FEBS Letters*, Vol. 521, No. 1–3, 2002, pp. 9–13.

A. Mershin, *et al.*, "Tubulin Dipole Moment, Dielectric Constant and Quantum Behavior: Computer Simulations, Experimental Results and Suggestions", *Biosystems*, Vol. 77, 2004, pp. 73–85.

M.A. Messerli and D.M. Graham, "Extracellular Electrical Fields Direct Wound Healing and Regeneration", *Biological Bulletin*, Vol. 221, No. 1, 2011, pp. 79–92.

A. Meunier and N. Spassky, "Centriole Continuity: Out with the New, In with the Old", *Current Opinion in Cell Biology*, Vol. 38, 2016, pp. 60–67. doi:10.1016/j.ceb.2016.02.007

C.J. Meyer, *et al.*, "Mechanical Control of Cyclic AMP Signalling and Gene Transcription Through Integrins", *Nature Cell Biology*, Vol. 2, No. 9, 2000, pp. 666–668.

M.D.D. Miell and A.F. Straight, "In Vitro Kinetochore Assembly", *The Mitotic Spindle — Methods and Protocols*, Edited by P. Chang and R. Ohi, Humana Press, Springer Science and Business Media, New York, 2018, Chapter 8.

T. Mitchison and M. Kirschner, "Dynamic Instability of Microtubule Growth", *Nature*, Vol. 312, 1984a, pp. 237–242.

T. Mitchinson and M. Kirschner, "Microtubule Assembly Nucleated by Isolated Centrosomes", *Nature*, Vol. 312, 1984b, pp. 232–237.

M.M. Mogensen, *et al.*, "Microtuble Minus-End Anchorage at Centrosomal and Non-centrosomal Sites: The Role of Ninein", *Journal of Cell Science*, Vol. 113, 2000, pp. 3013–3023.

S.M. Moghimi, A.C. Hunter, and J.C. Murray, "Nanomedicine: Current Status and Future Prospects", *FASEB Journal*, Vol. 19, 2005, pp. 311–330.

M. Moritz and D.A. Agard, "ϒ-Tubulin Complexes and Microtubule Nucleation", *Current Opinion in Structural Biology*, Vol. 11, 2001, pp. 174–181.

M. Moritz, *et al.*, "Reconstruction of Centrosome Microtubule Nucleation in Drosophila", *Methods in Cell Biology*, Vol. 67, 1998, pp. 141–148.

M. Moritz, *et al.*, "Structure of the γ-Tubulin Ring Complex: A Template for Microtubule Nucleation", *Nature Cell Biology*, Vol. 2, No. 6, 2000, pp. 365–370.

T.C. Moyer, *et al.*, "Binding of STIL to Plk4 Activates Kinase Activity to Promote Centriole Assembly", *Journal of Cell Biology*, Vol. 209, No. 6, 2015, pp. 863–878.

D. J. Muller, *et al.*, "Force Probing Surfaces of Living Cells to Molecular Resolution", *Nature, Chemical Biology*, Vol. 5, 2009, pp. 383–390.

T. Müller-Reichert, R. Kiewisz, and S. Redamann, "Mitotic Spindles Revisited — New Insights from 3D Electron Microscopy", *Journal of Cell Science*, Vol. 131, 2018, p. 211383.

L.L. Munn, "Aberrant Vascular Architecture in Tumors and Its Importance in Drug-Based Therapies", *Drug Discovery Today*, Vol. 8, 2003, pp. 396–403.

L.E. Muñoz, *et al.*, "Nanoparticles Size-Dependently Initiate Self-Limiting NETosis-driven Inflammation", *Proceedings National Academy of Sciences*, Vol. 113, 2016, pp. E5856–E5865.

E.A. Murphy, *et al.*, "Nanoparticle-Mediated Drug Delivery to Tumor Vasculature Suppresses Metastasis", *Proceedings National Academy of Science,* Vol. 105, 2008, pp. 9343–9348.

M.E. Mycielska and M.B. Djamgoz, "Cellular Mechanisms of Direct-Current Electric Field Effects: Galvanotaxis and Metastatic Disease", *Journal of Cell Science*, Vol. 177, Pt. 9, 2004, pp. 1631–1639.

F.-D. Naal, *et al.*, "High Hydrostatic Pressure-Induced Cell Death in Human Chondrocytes and Chondrosarcoma Cells", *Anticancer Research*, Vol. 25, 2005, pp. 1977–1982.

R. Nadkarni, S. Berkley, and C. Fradin, "A Comparison of Methods to Measure the Mangetic Moment of Magnetotactic Bacteria through Analysis of Their Trajectories in External Magnetic Fields", *Plos One*, Vol. 8, 2013, p. 1.

T. Nakajima, *et al.*, "Centrosome Aberration Accompanied with p53 Mutation can Induce Genetic Instability in Hepatocellular Carcinoma", *Modern Pathology*, Vol. 17, No. 6, 2004, pp. 722–727.

Y. Nakazawa, *et al.*, "SAS-6 Is a Cartwheel Protein that Establishes the 9-Fold Symmetry of the Centriole", *Current Biology*, Vol. 17, No. 24, 2007, pp. 2169–2174.

H.-J. Nam, R.M. Naylor, and J.M. van Deursen, "Centrosome Dynamics as a Source of Chromosomal Instability", *Trends in Cell Biology*, Vol. 25, No. 2, 2015, pp. 65–73.

H.-J. Nam and J.M. van Deursen, "Cyclin B2 and p53 Control Proper Timing of Centrosome Separation", *Nature Cell Biology*, Vol. 16, No.6, 2014, pp. 538–549.

D.J. Needleman, *et al.*, "Synchrotron X-ray Diffraction Study of Microtubules Buckling and Bundling under Osmotic Stress: A Probe of Interprotofilament Interactions", *Physical Review Letters*, Vol. 193, No. 19, 2004, pp. 198104-1–198104-4.

W.J. Nelson, "Adaptation of Core Mechanisms to Generate Cell Polarity", *Nature*, Vol. 422, 2003, pp. 766–774.

T. Neuberger, *et al.*, "Superparamagnetic Nanoparticles for Biomedical Applications: Possibilities and Limitations of a New Drug Delivery System", *Journal of Magnetism and Magnetic Materials*, Vol. 293, 2005, pp. 483–496.

S. Nie, *et al.*, "Nanotechnology Applications in Cancer", *Annual Review of Biomedical Engineering*, Vol. 9, 2007, pp. 257–288.

E.A. Nigg, "Mitotic Kinases as Regulators of Cell Division and Its Checkpoints", *Nature Reviews Molecular Cell Biology*, Vol. 2, 2001, pp. 21–31.

E.A. Nigg, "Centrosome Aberrations: Cause or Consequence of Cancer Progression?" *Nature Reviews Cancer*, Vol. 2, No. 11, 2002, pp. 815–825.

E.A. Nigg, "Origins and Consequences of Centrosome Aberrations in Human Cancers", *International Journal of Cancer*, Vol. 119, No. 12, 2006, pp. 2717–2723.

E.A. Nigg, "Centrosome Duplication: Of Rules and Licenses", *Trends in Cell Biology*, Vol. 17, No. 5, 2007, pp. 215–221.

E.A. Nigg, L. Cajanek, and C. Arquint, "The Centrosome Duplication Cycle in Health and Disease", *FEBS Letters*, Vol. 588, 2014, pp. 2366–2372.

E.A. Nigg and J.W. Raft, "Centrioles, Centrosomes, and Cilia in Health and Disease", *Cell*, Vol. 139, No. 4, 2009, pp. 663–678.

E.A. Nigg and T. Stearns, "The Centrosome Cycle: Centriole Biogenesis, Duplication and Inherent Asymmetries", *Nature Cell Biology*, Vol. 13, No. 10, 2011, pp. 1154–1160.

E. Nogales, "Structural Insights into Microtubule Function", *Annual Review of Biochemistry*, Vol. 69, 2000, pp. 277–302.

E. Nogales, *et al.*, "High-Resolution Model of the Microtubule", *Cell*, Vol. 96, No. 1, 1999, pp. 79–88.

E. Nogales, S.G. Wolf, and K.H. Downing, "Structure of the $\alpha\beta$ Tubulin Dimer by Electron Crystallography", *Nature*, Vol. 391, 1998, pp. 199–203.

C. Nuñez, S.V. Estévez, and M. del Pilar Chantada, "Inorganic Nanoparticles in the Diagnosis and Treatment of Breast Cancer", *Journal of Biological Inorganic Chemistry*, Vol. 23, 2018, pp. 332–343.

B.R. Oakley, V. Paolillo, and Y. Zheng, "Gamma-Tubulin Complexes in Microtubule Nucleation and Beyond", *Molecular Biology of the Cell*, Vol. 26, 2015, pp. 2957–2962.

M. Oberringer, *et al.*, "Centrosomes Multiplication Accompanies a Transient Clustering of Polyploid Cells during Tissue Repair", *Molecular Cell Biology Research Communications*, Vol. 2, 1999, pp. 190–196.

C.B. O'Connell and A.L. Khodjakov, "Cooperative Mechanisms of Mitotic Spindle Formation", *Journal of Cell Science*, Vol. 120, 2007, pp. 1717–1722.

M. Olivotto, *et al.*, "Electric Fields at the Plasma Membrane Level: A Neglected Element in the Mechanisms of Cell Signalling", *Bio Essays*, Vol. 18, No. 6, 1996, pp. 495–504.

D.P. O'Neal, *et al.*, "Photo-Thermal Tumor Ablation in Mice Using near Infrared-Absorbing Nanoparticles", *Cancer Letters*, Vol. 209, 2004, pp. 171–176.

T. Osaka, *et al.*, "Effect of Surface Charge of Magnetite Nanoparticles on their Internalization into Breast Cancer and Umbilical Vein Endothelial Cells", *Colloids and Surfaces B: Biointerfaces*, Vol. 71, 2009, pp. 325–330.

J. Owen, Q. Pankhurst, and E. Stride, "Magnetic Targeting and Ultrasound Mediated Drug Delivery: Benefits, Limitations and Combination", *International Journal of Hyperthermia*, Vol. 28, No. 4, 2012, pp. 362–373.

J.K. Pagan, *et al.*, "Degradation of Cep68 and PCNT Cleavage Mediate Cep215 Removal from the PCM to Allow Centriole Separation, Disengagement and Licensing", *Nature Cell Biology*, Vol. 17, 2015, pp. 31–43.

M. Paintrand, *et al.*, "Centrosome Organization and Centriole Architecture: Their Sensitivity to Divalent Cations", *Journal of Structural Biology*, Vol. 108, No. 2, 1992, pp. 107–128.

C. Panbianco and M. Gotta, "Coordinating Cell Polarity with Cell Divisions in Space and Time", *Trends in Cell Biology*, Vol. 21, No. 11, 2011, pp. 672–680.

M. Papetti and I.M. Herman, "Mechanisms of Normal and Tumor-derived Angiogenesis", *American Journal of Cell Physiology*, Vol. 282, No. 5, 2002, pp. C947–C970.

J.H. Park, *et al.*, "Magnetic Iron Oxide Nanoworms for Tumor Targeting and Imaging", *Advanced Materials*, Vol. 20, 2008, pp. 1630–1635.

J.H. Park, *et al.*, "Biodegradable Luminescent Porous Silicon Nanoparticles for In Vivo Applications", *Nature Materials*, Vol. 8, 2009a, pp. 331–336.

J.H. Park, *et al.*, "Systematic Surface Engineering of Magnetic Nanoworms for in Vivo Tumor Targeting", *Small*, Vol. 5, 2009b, pp. 694–700.

J.G. Parvani, *et al.*, "Silencing β3 Integrin by Targeted ECO/siRNA Nanoparticles Inhibits EMT and Metastasis of Triple-Negative Breast Cancer", Cancer Research, Vol. 75, 2015, pp. 2316–25. doi:10.1158/0008-5472

D. Passeri, *et al.*, "Magnetic Force Microscopy", *Biomatter*, Vol. 4, 2014, p. 1. doi:10.4161/biom.29507

J. Paz and J. Lüders, "Microtubule-Organizing Centers: Towards a Minimal Parts List", *Trends in Cell Biology*, Vol. 28, No. 3, 2018, pp. 176–187.

C.G. Pearson and M. Winey, "Plk4/SAK/ZYG-1 in the Regulation of Centriole Duplication", *F1000 Biology Reports*, Vol. 2, 2010, p. 58.

N. Peel, *et al.*, "Overexpressing Centriole-Replication Proteins In Vivo Induces Centriole Overduplication and De Novo Formation", *Current Biology*, Vol. 17, 2007, pp. 834–843.

D. Peer, *et al.*, "Nanocarriers as an Emerging Platform for Cancer Therapy", *Nature Nanotechnology*, Vol. 2, 2007, pp. 751–760. doi:10.1038/nnano.2007.387.

L. Pelletier, *et al.*, "Centriole Assembly in *Caenorhabditis elegans*", *Nature*, Vol. 444, 2006, pp. 619–623.

A.E. Pelling, *et al.*, "Local Nanomechanical Motion of the Cell Wall of Saccharomyces Cerevisiae", *Science*, Vol. 305, No. 5687, 2004, pp. 1147–1150.

X.-H. Peng, *et al.*, "Targeted Magnetic Iron Oxide Nanoparticles for Tumor Imaging and Therapy", *International Journal of Nanomedicine*, Vol. 3, 2008, pp. 311–321.

T.J. Pereira, *et al.*, "Inducible Fluorescent Speckle Microscopy", *Journal of Cell Biology*, Vol. 212, 2016, pp. 245–255.

M. Petronczki, *et al.*, "Polo-like Kinase 1 Triggers the Initiation of Cytokinesis in Human Cells by Promoting Recruitment of the RhoGEF Ect2 to the Central Spindle", *Developmental Cell*, Vol. 12, 2007, pp. 713–725.

M. Petronczki, P. Lénárt, and J.-M. Peters, "Polo on the Rise — From Mitotic Entry to Cytokinesis with Plk1", *Developmental Cell*, Vol. 14, 2008, pp. 646–659.

S. Petry and R.D. Vale, "Microtubule Nucleation at the Centrosome and Beyond", *Nature Cell Biology*, Vol. 17, No. 9, 2015, pp. 1089–1093.

D.B. Pettigrew, *et al.*, "Co-oriented Asymmetric, Neurtite Outgrowth from Neuronal Explants in Tissue Culture."

M. Piel, *et al.*, "The Respective Contribution of the Mother and Daughter Centrioles to Centrosome Activity and Behavior in Vertebrate Cells", *Journal of Cell Biology*, Vol. 149, No. 2, 2000, pp. 317–330.

M. Piel, *et al.*, "Centrosome-Dependent Exit of Cytokinesis in Animal Cells", *Science*, Vol. 291, 2001, pp. 1550–1553.

G.A. Pihan, *et al.*, "Centrosome Defects and Genetic Instability in Malignant Tumors", *Cancer Research*, Vol. 58, No. 17, 1998, pp. 3974–3985.

G.A. Pihan, *et al.*, "Centrosome Abnormalities and Chromosome Instability Occur Together in Pre-invasive Carcinomas", *Cancer Research*, Vol. 63, No. 6, 2003, pp. 1398–1404.

A. Pitaval, *et al.*, "Microtubule Stabilization Drives 3D Centrosome Migration to Initiate Primary Ciliogenesis", *Journal of Cell Biology*, Vol. 216, 2017, p. 3713.

H.A. Pohl, "Cellular Spin Resonance", *Journal of Theoretical Biology*, Vol. 93, 1981, pp. 207–213.

H.A. Pohl and T. Branden, "Cellular Spin Resonance of Aging Yeast and Mouse Sarcoma Cells", *Journal of Biological Physics*, Vol. 19, 1982, pp. 17–30.

J. Pokorný, "Endogenous Electromagnetic Forces in Living Cells: Implications for Transfer of Reaction Components", *Electro- and Magnetobiology*, Vol. 20, No. 1, 2001, pp. 59–73.

J. Pokorný, "Viscous Effects on Polar Vibrations in Microtubules", *Electromagnetic Biology and Medicine*, Vol. 22, No. 1, 2003, pp. 15–29.

J. Pokorný, "Biophysical Cancer Transformation Pathway", *Electromagnetic Biology and Medicine*, Vol. 28, No. 2, 2009, pp. 105–123.

J. Pokorný, "Electrodynamic Activity of Healthy and Cancer Cells", *Journal of Physics: Conference Series*, Vol. 329, No. 1, 2011, p. 012007.

J. Pokorný, "Physical Aspects of Biological Activity and Cancer", *AIP Advances*, Vol. 2, 2012, pp. 011207-1–011207-11.

J. Pokorný, *et al.*, "Biophysical Aspects of Cancer — Electromagnetic Mechanism", *Indian Journal of Experimental Biology*, Vol. 46, No. 5, 2008, pp. 310–321.

J. Pokorný, *et al.*, "Cancer Physics: Diagnosis Based on Damped Cellular Elastoelectrical Vibrations in Microtubules", *European Biophysics Journal*, Vol. 40, No. 6, 2011, pp. 747–759.

J. Pokorný, J. Hašek, and F. Jelínek, "Electromagnetic Field of Microtubules: Effects on Transfer of Mass Particles and Electrons", *Journal of Biological Physics*, Vol. 31, 2005a, pp. 501–514.

J. Pokorný, J. Hašek, and F. Jelínek, "Endogenous Electric Field and Organization of Living Matter", *Electromagnetic Biology and Medicine*, Vol. 24, No. 3, 2005b, pp. 185–197.

J. Pokorný, *et al.*, "Vibration in Microtubules", *Journal of Biological Physics*, Vol. 23, 1997, pp. 171–179.

J. Pokorný, F. Jelinek, and V. Trkal, "Electric Field around Microtubules", *Bioelectrochemistry and Bioenergetics*, Vol 45, No. 2, 1998, pp. 239–245.

J. Pokorný, J. Pokorný, and J. Kobilková, "Postulates on Electromagnetic Activity in Biological Systems and Cancer", *Integrative Biology*, Vol. 5, No. 12, 2013, pp. 1439–1446.

T.D. Pollard, "Nine Unanswered Questions about Cytokinesis", *Journal of Cell Biology, ICB Home*, Vol. 916, 2017 Archive, p. 3007 ff.

S. Portet, *et al.*, "Elastic Vibrations in Seamless Microtubules", *European Biophysics Journal*, Vol. 34, No. 7, 2005, pp. 912–920.

P. Pouponneau, *et al.*, "Co-encapsulation of Magnetic Nanoparticles and Doxorubicin into Biodegradable Microcarriers for Deep Tissue Targeting by Vascular MRI Navigation", *Biomaterials*, Vol.32, No. 13, 2011, pp. 3481–3486.

A.F. Powers, *et al.*, "The Ndc80 Kinetochore Complex Forms Load Bearing Attachments to Dynamic Microtubule Tips via Biased Diffusion", *Cell*, Vol. 136, 2009, pp. 865–875.

M.E. Presenti, J.R. Weir, and A. Musacchio, "Progress in the Structural and Functional Characterization of Kinetochores", *Current Opinion in Structural Biology*, Vol. 37, 2016, pp. 152–163.

S.L. Prosser and L. Pelletier, "Centrosome Biology: The Ins and Outs of Centrosome Assembly", *Current Biology*, Vol. 25, 2015, pp. R656–R659.

X.S. Qian, J.Q. Zhang, and C.Q. Ru, "Wave Propagation in Orthotropic Microtubules", *Journal of Applied Physics*, Vol. 101, 2007, pp. 084702-1–084702-7.

N.J. Quintyne, *et al.*, "Spindle Multipolarity Is Prevented by Centrosomal Clustering", *Science*, Vol. 307, No. 5706, 2005, pp. 127–129.

J.W. Raff, "Centrosomes and Cancer: Lessons from a TACC", *Trends in Cell Biology*, Vol. 12, No. 5, 2002, pp. 222–225.

J.W. Raff and R. Basto, "Centrosome Amplification and Cancer: A Question of Sufficiency", *Developmental Cell*, Vol. 40, 2017, pp. 217–218.

H. Rajagopalan and C. Lengauer, "Aneuploidy and Cancer", *Nature*, Vol. 432, 2004, pp. 338–341.

N. Ramakrishnan, *et al.*, "Biophysically Inspired Model for Functionalized Nanocarrier Adhesion to Cell Surface: Roles of Protein Expression and Mechanical Factors", *Royal Society Open Science*, Vol. 3, June 29, 2016, pp. 1–40.

B. Raynaud-Messina and A. Merdes, "Gamma-Tubulin Complexes and Microtubule Organization", *Current Opinion in Cell Biology*, Vol. 19, No. 1, 2007, pp. 24–30.

J. Reina and C. Gonzalez, "When Fate Follows Age: Unequal Centrosomes in Asymmetric Cell Division", *Philosophical Transactions of the Royal Society of London. Series B, Biological Science*, Vol. 369, 2014, 13 pages.

L.M. Rice, "A New Look for the Growing Microtubule End?", *Journal of Cell Biology*, Vol. 217, No. 8, 2018, p. 2609.

A. Ridley and R. Heald, "Cell Structure and Dynamics — Editorial Overview", *Current Opinion in Cell Biology*, Vol. 23, No. 1, 2011, pp. 1–3.

C.L. Rieder, S. Faruki, and A. Khodjakov, "The Centrosome in Vertebrates: More Than a Microtubule-Organizing Center", *Trends in Cell Biology*, Vol. 11, No. 10, 2001, pp. 413–419.

R.M. Rios, "The Centrosome-Golgi Apparatus Nexus", *Philosophical Transactions of the Royal Society of London. Series B, Biological Science*, Vol. 369, 2014, 17 pages.

W. Risau, "Mechanisms of Angiogenesis", *Nature*, Vol. 386, 1997, pp. 671–674.

K.R. Robinson, "The Responses of Cells to Electrical Fields: A Review", *Journal of Cell Biology*, Vol. 101, No. 6, 1985, pp. 2023–2027.

A. Rodrigues-Martins, *et al.*, "DSAS-6 Organizes a Tube-like Centriole Precursor, and Its Absence Suggests Modularity in Centriole Assembly", *Current Biology*, Vol. 17, No. 17, 2007, pp. 1465–1472.

A. Rodrigues-Martins, *et al.*, "Revisiting the Role of the Mother Centriole in Centriole Biogenesis", *Science*, Vol. 316, No. 5827, 2007, pp. 1046–1050.

A. Rodrigues-Martins, *et al.*, "From Centriole Biogenesis to Cellular Function: Centrioles are Essential for Cell Division at Critical Developmental Stages", *Cell Cycle*, Vol. 7, No. 1, 2008, pp. 11–16.

J. Roostalu and T. Surrey, "Microtubule Nucleation: Beyond the Template", *Nature Reviews Molecular Cell Biology*, Vol. 18, 2017, pp. 702–710.

M.D. Rose, S. Biggins, and L.L. Satterwhite, "Unravelling the Tangled Web at the Microtubule-Organizing Center", *Current Opinion in Cell Biology*, Vol. 5, No. 1, 1993, pp. 105–115.

J.M. Rosenholm, *et al.*, "Targeting of Porous Hybrid Silica Nanoparticles to Cancer Cells", ACS Nano, Vol. 3, No. 1, pp. 197–206. doi:10.1021/nn800781r

C. Rossi, *et al.*, "New Perspectives in Cell Communication: Bioelectromagnetic Interactions", *Seminars in Cancer Biology*, Vol. 21, No. 3, 2011, pp. 207–214.

E. Russlahti, S.N. Bhatia, and M.J. Sailor, "Targeting of Drugs and Nanoparticles to Tumors", *Journal of Cell Biology*, Vol. 188, 2010, pp. 759–768.

H.I. Saavedra, *et al.*, "Inactivation of E2F3 Results in Centrosome Amplification", *Cancer Cell*, Vol. 3, 2003, pp. 333–346.

D. Sabino, *et al.*, "Moesin Is a Major Regulator of Centrosome Behavior in Epithelial Cells with Extra Centrosomes", Current Biology, Vol. 25, No. 7, 2015, pp. 879–889.

C. Sacristan, *et al.*, "Dynamic Kinetochore Size Regulation Promotes Microtubule Capture and Chromosome Biorientation in Mitosis", *Nature Cell Biology*, Vol. 20, 2018, pp. 800–810.

O. Sadik, "The Two Faces of Nanotechnology", *American Scientist*, Vol. 105, 2017, pp. 208–211.

S.J. Sahl, S.W. Hell, and S. Jacobs, "Fluorescence Nanoscopy in Cell Biology", *Nature Reviews Molecular Cell Biology*, Vol. 18, 2017, pp. 685–701.

J.L. Salisbury, *et al.*, "Centrosomes and Cancer", *Biology of the Cell*, Vol. 91, No. 6, 1999, pp 451–460.

J. Salogiannis and S.L. Reck-Peterson, "Hitchhiking: A Non-Canonical Mode of Microtubule-based Transport", *Trends in Cell Biology*, Vol. 27, 2017, pp. 141–150.

E. Samiyoshi and A. Sugimoto, "Cell Polarity: Centrosomes Release Signals for Polarization", *Current Biology*, Vol. 22, No. 8, 2012, pp. R281–R283.

A.D. Sanchez and J.L. Feldman, "Microtubule Organizing Centers: From the Centrosome to Non-Centrosomal Sites", *Current Opinion in Cell Biology*, Vol. 44, 2017, pp. 93–101.

M. Sanhaji, *et al.*, "Mitotic Centromere-Associated Kinesin (MCAK): A Potential Cancer Drug Target", *Oncotarget*, Vol. 2, 2011, pp. 935–947. doi:10.18632/oncotarget.416

S. Santaguida and A. Musacchio, "The Life and Miracles of Kinetochores", *The EMBO Journal*, Vol. 28, 2009, pp. 2511–2531.

N. Sanvicens and M.P. Marco, "Multifunctional Nanoparticles — Properties and Prospects for the use in Human Medicine", *Trends in Biotechnology*, Vol. 26, No. 8, 2008, pp. 425–433.

T. Sasaki, *et al.*, "Magnetic Nanoparticles for Improving Cell Invasion in Tissue Engineering", *Journal of Biomedical Materials Research, Part A*, Vol. 86, 2007, pp. 969–979.

M.V. Sataric, *et al.*, "Modelling the Role of Intrinsic Electric Fields in Microtubules as an Additional Control Mechanism of Bi-Directional Intracellular Transport", *Cell Biochemistry and Biophysics*, Vol. 52, No. 2, 2008, pp. 113–124.

M. Sataric, B.M. Sataric, and J.A. Tuszynski, "Nonlinear Model of Microtubule Dynamics", *Electromagnetic Biology and Medicine*, Vol. 24, 2005, pp. 255–264.

M.V. Satarić and J.A. Tuszyński, "Relationship Between the Nonlinear Ferroelectric and Liquid Crystal Models for Microtubules", *Physical Review*, Vol. 67, 2003, pp. 011901-1–011901-11.

M.V. Sataric, J.A. Tuszynski, and R.B. Zakula, "Kinklike Excitations as an Energy-Transfer Mechanism in Microtubules", *Physical Review E*, Vol. 48, No. 1, 1993, pp. 589–597.

N. Sato, *et al.*, "Radiation-Induced Centrosome Overduplication and Multiple Mitotic Spindles in Human Tumor Cells", *Experimental Cell Research*, Vol. 255, No. 2, 2000, pp. 321–326.

N. Sato, *et al.*, "A Possible Role for Centrosome Overduplication in Radiation-Induced Cell Death", *Oncogene*, Vol. 19, 2000, pp. 5281–5290.

W. Saunders, "Centrosomal Amplification and Spindle Multipolarity in Cancer Cells", *Seminars in Cancer Biology*, Vol. 15, 2005, pp. 25–32.

L. Schaedel, *et al.*, "Microtubules Self-Repair in Response to Mechanical Stress", *Nature Materials*, Vol. 14, 2015, pp. 1156–1165.

P.W. Schafer, "Centrioles of a Human Cancer: Intercellular Order and Intracellular Disorder", *Science*, Vol. 164, 1969, pp. 1300–1303.

P.W. Schafer and J.A. Chandler, "Electron Probe X-ray Microanalysis of a Normal Centriole", *Science*, Vol. 170, 1970, pp. 1204–1205.

P.W. Schafer, "Centrioles: Intercellular Order in Normal and Malignant Cells", *Journal of Thoracic and Cardiovascular Surgery*, Vol. 63, No. 3, 1972, pp. 472–477.

P.W. Schafer, 1980. Personal communication.

H.T. Schek III, *et al.*, "Microtubule Assembly Dynamics at the Nanoscale", *Current Biology*, Vol. 17, No. 17, 2007, pp. 1445–1455.

L. Schermelleh, R. Heintzmann, and H. Leonhardt, "A Guide to Super-Resolution Fluorescence Microscopy", *Journal of Cell Biology*, Vol. 190, No. 2, 2010, pp. 165–175. doi:10.1083/jcb.201002018.

R. Scherz-Shouval, *et al.*, "The Reprogramming of Tumor Stroma by HSF1 Is a Potent Enabler of Malignancy", *Cell*, Vol. 158, No. 3, 2014, pp. 564–578.

E. Schiebel, "ϒ-Tubulin Complexes: Binding to the Centrosome, Regulation and Microtubule Nucleation", *Current Opinion in Cell Biology*, Vol. 12, 2000, pp. 113–118.

L. Schöckel, *et al.*, "Cleavage of Cohesion Rings Coordinates the Separation of Centrioles and Chromatids", *Nature Cell Biology*, Vol. 13, No. 8, 2011, pp. 966–972.

J.M. Scholey, G.C. Rogers, and D.J. Sharp, "Mitosis, Microtubules, and the Matrix", *Journal of Cell Biology*, Vol. 154, No. 2, 2001, pp. 261–266.

F. Scholkmann, D. Fels, and M. Cifra, "Non-Chemical and Non-Contact Cell-to-Cell Communication: A Short Review", *American Journal of Translational Research*, Vol. 5, No. 6, 2013, pp. 586–593.

A. Schroeder, *et al.*, "Treating Metastatic Cancer with Nanotechnology", *Nature Reviews Cancer*, Vol. 12, 2012, pp. 39–50.

J. Scrofani, *et al.*, "Microtubule Nucleation in Mitosis by a Ran GTP-Dependent Protein Complex", *Current Biology*, Vol. 25, 2015, pp. 131–140.

S. Seaman, *et al.*, "Eradication of Tumors through Simultaneous Ablation of CD276/B7-H3-Positive Tumor Cells and Tumor Vasculature", *Cancer Cell*, Vol. 31, 2017, pp. 501–515.

S. Sen, S. Subramanial, and D.E. Discher, "Indentation and Adhesive Probing of a Cell Membrane with AFM: Theoretical Model and Experiments", *Biophysical Journal*, Vol. 89, 2005, pp. 3203–3213.

S. Sengupta and R. Sasisekharan, "Exploiting Nanotechnology to Target Cancer", *British Journal of Cancer*, Vol. 96, 2007, pp. 1315–1319.

R.F. Service, "Nanotechnology Takes Aim at Cancer", *Science*, Vol. 310, 2005, pp. 1132–1134.

Y. Shamay, *et al.*, "P-Selection Is a Nanotherapeutic Delivery Target in the Tumor", *Science Translational Medicine*, Vol. 29, 2016, p. 345ra87.

B. Short, "Mapping the Kinetochore Delta", *Journal of Cell Biology*, Vol. 212, 2016, p. 259.

J.E. Sillibourne, *et al.*, "Autophosphorylation of Polo-like Kinase 4 and Its Role in Centriole Duplication", *Molecular Biology of The Cell*, Vol. 21, No. 4, 2010, pp. 547–561.

M.M. Shawki and A. Farid, "Low Electric Field Parameters Required to Induce Death of Cancer Cells", *Electromagnetic Biology & Medicine*, Vol. 33, No. 2, 2014, pp. 159–163.

D. Shi, N.M. Bedford, and H.-S. Cho, "Engineered Multifunctional Nano-carriers for Cancer Diagnosis and Therapeutics", *Small*, Vol. 7, No. 18, 2011, pp. 2549–1567.

J. Shi, *et al.*, "Cancer Nanomedicine: Progress, Challenges and Opportunities", *Nature Reviews Cancer*, 17, 2017, pp. 20–33.

K. Shin, V.C. Fogg, and B. Margolis, "Tight Junction and Cell Polarity", *Annual Review of Cell Development Biology*, Vol. 22, 2006, pp. 207–235.

K. Shinmura, *et al.*, "Direct Evidence for the Role of Centrosomally Localized p53 in the Regulation of Centrosome Duplication", *Oncogene*, Vol. 26, No. 20, 2007, pp. 2939–2944.

V. Sikirzhytski, *et al.*, "Microtubules Assemble Near Most Kinetochores During Early Prometaphase in Human Cells", *Journal of Cell Biology*, Vol. 217, No. 8, 2018, p. 2641.

J.E. Sillibourne and M. Bornens, "Polo-like Kinase 4: The Odd One Out of the Family", *Cell Division*, Vol. 5, 2010, p. 25.

J.E. Sillibourne, *et al.*, "Assessing the Localization of Centrosomal Proteins by PALM/STORM Nanoscopy", *Cytoskeleton*, Vol. 68, 2011, pp. 619–627.

A.V. Singh, "Biotechnological Applications of Supersonic Cluster Beam-Deposited Nanostructured Thin Films: Bottom-Up Engineering to Optimize Cell-Protein-Surface Interactions", *Journal of Biomedical Materials Research*, Vol. 101A, 2013, pp. 2984–3008.

A. Singh, *et al.*, "Polarized Microtubule Dynamics Directs Cell Mechanics and Coordinates Forces During Epithelial Morphogenesis", *Nature Cell Biology*, Vol. 20, 2018, pp. 1126–1133.

Y.M. Sirenko, M.A. Stroscio, and K.W. Kim, "Elastic Vibrations of Microtubules in a Fluid", *Physical Review E*, Vol. 53, No. 1, 1996, pp. 1003–1010.

A. Skibinski and C. Kuperwasser, "The Origin of Breast Tumor Heterogeneity", *Oncogene*, Vol. 34, 2015, pp. 5309–5316.

K.C. Slep, "A Microtubule Dynamics Reconstitutional Convention", *Journal of Cell Biology*, Vol. 205, No. 3, 2016, pp. 305–315.

L.K. Slevin, *et al.*, "The Structure of the Plk4 Cryptic Polo Box Reveals Two Tandem Polo Boxes Required for Centriole Duplication", *Structure*, Vol. 20, No. 11, 2012, pp. 1905–1917.

G. Sluder, "One to Only Two: A Short History of the Centrosome and Its Duplication", *Philosophical Transactions of the Royal Society. Series B, London*, Vol. 369, 2014, 8 pages.

G. Sluder, F.J. Miller, and C.L. Rieder, "The Reproduction of Centrosomes: Nuclear versus Cytoplasmic Controls", *The Journal of Cell Biology*, Vol. 103, 1986, pp. 1873–1881.

G. Sluder and J.J. Nordberg, "The Good, The Bad and The Ugly: The Practical Consequences of Centrosome Amplification", *Current Opinion in Cell Biology*, Vol. 16, No. 1, 2004, pp. 49–54.

M.H. Song, *et al.*, "Centrioles: Some Self-Assembly Required", *Current Opinions in Cell Biology*, Vol. 20, 2008, pp. 688–693.

K.F. Sonnen, *et al.*, "3D-Structured Illumination Microscopy Provides Novel Insight into Architecture of Human Centrosomes", *Biology Open*, Vol. 1, 2012, pp. 965–976.

K.F. Sonnen, *et al.*, "Human Cep192 and Cep152 Cooperate in Plk4 Recruitment and Centriole Duplication", *Journal of Cell Science*, Vol. 126, 2014, pp.3223–3233.

T. Stearns, "Centrioles, in Absentia", *Science*, Vol. 348, No. 6239, 2015, pp. 1091–1092.

R. Stracke, *et al.*, "Analysis of the Migration Behavior of Single Microtubules in Electric Fields", *Biochemical and Biophysical Research Communications*, Vol. 293, 2002, pp. 602–609.

K. Strebhardt and A. Ullrich, "Targeting Polo-like Kinase 1 for Cancer Therapy", *Nature Reviews Cancer*, Vol. 6, 2006, pp. 321–330.

P. Strnad, *et al.*, "Regulated HsSAS-6 Levels Ensure Formation of a Simple Procentriole per Centriole During Centrosome Duplication Cycle", *Developmental Cell*, Vol. 13, No. 2, 2007, pp. 203–213.

P. Strnad and P. Gönczy, "Mechanisms of Procentriole Formation", *Trends in Cell Biology*, Vol. 18, No. 8, 2008, pp. 389–396.

V.M. Stucke, *et al.*, "Human Mps1 Kinase Is Required for the Spindle Assembly Checkpoint but not for Centrosome Duplication", *The EMBO Journal*, Vol. 21, 2002, pp. 1723–1732.

K.N. Sugahara, *et al.*, "Tissue-Penetrating Delivery of Compounds and Nanoparticles into Tumors", *Cancer Cell*, Vol. 16, 2009, pp. 510–520.

C. Sun, J.S. Lee, and M. Zhang, "Magnetic Nanoparticles in MR Imaging and Drug Delivery", *Advanced Drug Delivery Reviews*, Vol. 60, 2008, pp. 1252–1265.

J. Sun, M. Zborowski, and J. Chalmens, "Quantification of Both the Presence, and Oxidation State, of Mn in Bacillus Atrophaeus Spores and its Imparting of Magnetic Susceptibility to the Spores", *Biotechnology and Bioengineering*, Vol. 108, No. 5, 2011, pp. 1119–1129.

S. Suresh, "Biomechanics and Biophysics of Cancer Cells", *Acta Biomater*, Vol. 3, No. 4, 2007, pp. 413–438.

A. Suzuki, *et al.*, "How the Kinetochore Couples Microtubule Force and Centromere Stretch to Move Chromosomes", *Nature, Cell Biology*, Vol. 18, No. 4, 2016, pp. 382–392. doi:10.1038/ncb3323.

C.J. Swallow, *et al.*, "Sak/Plk4 and Mitotic Fidelity", *Oncogene*, Vol. 24, 2005, pp. 306–312.

K. Sztandera, M. Gorzkiewicz, and B. Klajnert-Maculewicz, "Gold Nanoparticles in Cancer Treatment", *Molecular Particles in Cancer Treatment*, Vol. 16, 2019, pp. 1–23.

M. Takeda, M. Sami, and Y.-C. Wang, "A Homeostatic Apical Microtubule Network Shortens Cells for Epithelial Folding via a Basal Polarity Shift", *Nature Cell Biology*, Vol. 20, 2018, pp. 36–45.

T.U. Tanaka and A. Desai, "Kinetochore-Microtubule Interactions: The Means to the End", *Current Opinion in Cell Biology*, Vol. 20, 2008, pp. 53–63.

M.E. Tanenbaum and R.H. Medema, "Mechanisms of Centrosome Separation and Bipolar Spindle Assembly", *Developmental Cell*, Vol. 19, 2010, pp. 797–806.

C.-J.C. Tang, *et al.*, "The Human Microcephaly Protein STIL Interacts with CPAP and is Required for Procentriole Formation", *The EMBP Journal*, Vol. 30, 2011, pp. 4790–4804.

P. Tarapore, *et al.*, "Direct Regulation of the Centrosome Duplication Cycle by the p53-p21Waf1/Cip1 Pathway", *Oncogene*, Vol. 20, 2001, pp. 3173–3184.

P. Tarapore and K. Fukasawa, "Loss of p53 and Centrosome Hyperamplification", *Oncogene*, Vol. 21, No. 40, 2002, pp. 6234–6240.

A.-M. Tassin and M. Bornens, "Centrosome Structure and Microtubule Nucleation in Animal Cells", *Biology of the Cell*, Vol. 91, No. 4–5, 1999, pp. 343–354.

N. Teixido-Travesa, J. Roig, and J. Luders, "The Where, When and How of Microtubule Nucleation — One Ring to Rule them all", *Journal of Cell Science*, Vol. 125, 2012, pp. 4445–4456.

J.F. Tien, *et al.*, "Cooperation of the Dam1 and Ndc80 Kinetochore Complexes Enhances Microtubule Coupling and is Regulated by Aurora B", *Journal of Cell Biology*, Vol. 189, No. 4, 2010, pp. 713–723. doi:10.1083/jcb.200910142

I.M. Tolić, "Mitotic Spindle: Kinetochore Fibers Hold on Tight to Interpolar Bundles", *European Biophysics Journal*, Vol. 47, 2018, pp. 191–203.

M.C. Tsai and J. Ahringer, "Microtubules are Involved in Anterior-Posterior Axis Formation in *C. elegans* Embryos", *Journal of Cell Biology*, Vol. 179, No. 3, 2007, pp. 397–402.

M.F. Tsou and T. Stearns, "Mechanism Limiting Centrosome Duplication to Once Per Cell Cycle", *Nature*, Vol. 442, No. 7105, 2006a, pp. 947–951.

M.F. Tsou and T. Stearns, "Controlling Centrosome Number: Licenses and Blocks", *Current Opinion in Cell Biology*, Vol. 18, No. 1, 2006b, pp. 74–78.

M-F.B. Tsou, *et al.*, "Polo Kinase and Separase Regulate the Mitotic Licensing of Centriole Duplication in Human Cells, *Developmental Cell*, Vol. 17, 2009, pp. 344–354.

J.A. Tuszyński, *et al.*, "Anisotropic Elastic Properties of Microtubules", *European Physics Journal E Soft Matter*, Vol. 17, No. 1, 2005a, pp. 29–35.

J.A. Tuszyński, *et al.*, "Molecular Dynamics Simulations of Tubulin Structure and Calculation of Electrostatic Properties of Microtubules", *Mathematical and Computer Modelling*, Vol. 41, No. 10, 2005b, pp. 1055–1070.

J.A. Tuszyński, *et al.*, "Ferroelectic Behavior in Microtubule Dipole Lattices: Implication for Information Processing, Signaling and Assembly/Disassembly", *Journal of Theoretical Biology*, Vol. 174, No. 4, 1995, pp. 371–380.

K. M. Tyner, R. Kopelman, and M. Philbert, "'Nanosized Voltmeter' Enables Cellular-Wide Electric Field Mapping", *Biophysical Journal*, Vol. 93, No. 4, 2007, pp. 1163–1174.

Y. Uetake, *et al.*, "Cell Cycle Progression and de novo Centriole Assembly after Centrosomal Removal in Untransformed Human Cells", *Journal of Cell Biology*, Vol. 176, 2007, pp. 173–182.

O. Valiron, N. Caudron, and D. Job, "Microtubule Dynamics", *Cellular and Molecular Life Sciences*, Vol. 58, 2001, pp. 2069–2084.

J.M. Valles, Jr., "Model of Magnetic Field-Induced Mitotic Apparatus Reorientation in Frog Eggs", *Biophysical Journal*, Vol. 82, 2002, pp. 1260–1265.

E. Valsami-Jones and I. Lynch, "How Safe Are Nanomaterials?" *Science*, Vol. 350, No. 6259, 2015, pp. 388–389.

M. van Breugel, *et al.*, "Structures of SAS-6 Suggest Its Organization in Centrioles", *Science*, Vol. 331, No. 6021, 2011, pp. 1196–1199.

B.C.M. van de Weerdt and R.H. Medema, "Polo-like Kinases: A Team in Control of the Division", *Cell Cycle*, Vol. 5, No. 8, 2006, pp. 853–864.

M.A. van Vugt and R.H. Medema, "Getting In and Out of Mitosis with Polo-like Kinase-1", *Oncogene*, Vol. 24, 2005, pp. 2844–2859.

H. Varmark, *et al.*, "Asterless Is a Centriolar Protein Required for Centrosome Function and Embryo Development in Drosophila", *Current Biology*, Vol. 17, No. 20, 2007, pp. 1735–1745.

V.G. Vasishto, "Sequentially Programmed Magnetic Field Therapy in the Management of Recurrent Anaplastic Astrocytoma: A Case Study Report and Literature Review", *Case Reports in Oncology*, Vol. 3, No. 2, 2010, pp. 189–194.

S. Vaughan and H.R. Dawe, "Common Themes in Centriole and Centrosomes Movements", *Trends in Cell Biology*, Vol. 21, No.1, 2011, pp. 57–66.

O. Veiseh, J.W. Gunn, and M. Zhang, "Design and Fabrication of Magnetic Nanoparticles for Targeted Drug Delivery and Imaging", *Advanced Drug Delivery Reviews*, Vol. 62, 2010, pp. 284–304.

A. Vemu, *et al.*, "Severing Enzymes Amplify Microtubule Arrays through Lattice GTP-Tubulin Incorporation", *Science*, Vol. 361, 2018, p. 768.

V.J. Venditto and F.C. Szoka Jr., "Cancer Nanomedicines: So Many Papers and So Few Drugs!" *Advanced Drug Delivery Reviews*, Vol. 65, 2013, pp. 80–88.

S. Veno and S. Yamaguchi-Sek, "Magnetic Control of Biological Cell Growth", *Biomagnetics Principles and Applications of Biomagnetic Stimulation and Imaging*, Edited by S. Veno and M.Sekino, CRC Press, 2016, Chapter 7, pp. 186–212.

B. Vitre, *et al.*, "Chronic Centrosome Amplification without Tumorigenesis", *Proceedings of the National Academy of Sciences of the United States of America*, Vol. 112, 2015, pp. E6321–E6330. doi:10.1073/pnas.1519388112

B.D. Vitre and D.W. Cleveland, "Centrosomes, Chromosome Instability (CIN) and Aneuploidy", *Current Opinion in Cell Biology*, Vol. 24, No. 7, 2012, pp. 809–815.

E. Vladimirou, *et al.*, "Nonautonomous Movement of Chromosomes in Mitosis", *Developmental Cell*, Vol. 27, No. 1, 2013, pp. 60–71.

V.A. Volkov, *et al.*, "Long Tethers Provide High-Force Coupling of the Dam1 Ring to Shortening Microtubules", *Proceedings of the National Academy of Sciences of the United States of America*, Vol. 110, 2013, pp. 7708–7713.

G. von Maltzahn, *et al.*, "Nanoparticles that Communicate *In Vivo* to Amplify Tumour Targeting", *Nature Materials*, Vol. 10, 2011, pp. 545–552.

I.A. Vorobjev and Y.S. Chentsov, "Centrioles in the Cell Cycle. 1. Epithelial Cells", *Journal of Cell Biology*, Vol. 93, No. 3, 1982, pp. 938–949.

K. Vukušić, *et al.*, "Microtubule Sliding Within the Bridging Fiber Pushes Kinetochore Fibers Apart to Segregate Chromosomes", *Developmental Cell*, Vol. 43, 2017, pp. 11–23.

J. Vulprecht, *et al.*, "STIL Is Required for Centriole Duplication in Human Cells", *Journal of Cell Science*, Vol. 125, 2012, pp. 1353–1362.

C.E. Walczak, "Microtubule Dynamics and Tubulin Interacting Proteins", *Current Opinion in Cell Biology*, Vol. 12, 2000, pp. 52–56.

K. Walter, "Intelligent Nanoparticles Could Treat Cancer", *Rand Magazine*, Internet October 24, 2017.

K. Walter (Digital Reporter), "Gold Nanoparticles Could Enable More Targeted Drugs", 2018, from an article published in *Nature Communications*.

X. Wan, *et al.*, "Protein Architecture of the Human Kinetochore Microtubule Attachment Site", *Cell*, Vol. 137, 2009, pp. 672–684.

A.Z. Wang, R. Langer, and O.C. Farokhzad, "Nanoparticle Delivery of Cancer Drugs", *Annual Review of Medicine*, Vol. 63, 2012, pp. 185–198.

C.Y. Wang, C.Q. Ru, and A. Mioduchowski, "Vibration of Microtubules as Orthotropic Shells", *Physica E: Low-Dimensional Systems and Nanostructures,* Vol. 35, No. 1, 2006, pp. 48–56.

C.Y. Wang and L.C. Zhang, "Circumferential Vibration of Microtubules with Long Axial Wavelength", *Journal of Biomechanics,* Vol. 41, 2008, pp. 1892–1896.

F. Wang, *et al.*, "Dual Surface-Functionalized Janus Nanocomposites of Polystyrene/ Fe_3O_4@SiO_2 for Simultaneous Tumor Cell Targeting and Stimulus-Induced Drug Release", *Advanced Materials, Communications,* Vol. 25, 2013, pp. 1–5. www.advmat.de.

L. Wang, P. Park, and C-Y Lin, "Characterization of Stem Cell Attributes in Human Osteosarcoma Cell Lines", *Cancer Biology and Therapy,* Vol. 8, No. 6, 2009, pp. 543–552.

W.J. Wang, *et al.*, "The Conversion of Centrioles to Centrosomes: Essential Coupling of Duplication with Segregation", *Journal of Cell Biology,* Vol. 193, No. 4, 2011, pp. 727–739.

S. Warnke, *et al.*, "Polo-like Kinase-2 Is Required for Centriole Duplication in Mammalian Cells", Current Biology, Vol. 14, No. 13, 2004, pp. 1200–1207.

S. Watanuki, *et al.*, "Characterization of Centriole Duplication in Human Epidermis, Bowen's Disease, and Squamous Cell Carcinoma", *Journal of Dermatological Science,* Vol. 91, 2018, pp. 9–18.

J.D. Watson and E.H.C. Crick, "Molecular Structure of Nucleic Acids: A Structure for Deoxyribose Nucleic Acid", *Nature,* Vol. 171, 1953, pp. 737–738.

B.A.A. Weaver and D.W. Cleveland, "Does Aneuploidy Cause Cancer?" *Current Opinion in Cell Biology,* Vol. 18, 2006, pp. 658–667.

D.S. Weiss, R. Kirsner, and W.H. Eaglstein, "Electrical Stimulation and Wound Healing", *Archives of Dermatology,* Vol. 126, No. 2, 1990, pp. 222–225.

M.E. Werner, *et al.*, "Actin and Microtubules Drive Differential Aspects of Planar Cell Polarity in Multiciliated Cells", *Journal of Cell Biology,* Vol. 195, No. 1, 2011, pp. 19–26.

S. Werner, A. Pimenta-Marques, and M. Battencourt-Dias, "Maintaining Centrosomes and Cilia", *Journal of Cell Science,* Vol. 130, 2017, pp. 3789–3800. doi:10.1242/jcs.203505

C. Wiese and Y. Zheng, "A New Function for the ϒ-Tubulin Ring Complex as a Microtubule Minus-End Cap", *Nature Cell Biology,* Vol. 2, 2000, pp. 358–364.

C. Wiese and Y. Zheng, "Microtubule Nucleation: ϒ-Tubulin and Beyond", *Journal of Cell Science,* Vol. 119, 2006, pp. 4143–4153.

S.M. Wignall and A.M. Villeneuve, "Lateral Microtubule Bundles Promote Chromosome Alignment during Acentrosomal Oocyte Meiosis", *Nature Cell Biology,* Vol. 11, No. 7, 2009, pp. 839–844.

K.A. Williams, *et al.*, "Carbon Nanotubes with DNA Recognition", *Nature*, Vol. 420, 2002, p. 761.

M. Winey and E. O'Toole, "Centriole Structure", *Philosophical Transactions of Royal Society of the London. Series B, Biological Sciences*, Vol. 369, 2014, 14 pages.

C. Wong and T. Stearns, "Centrosome Number Is Controlled by a Centrosome-Intrinsic Block to Reduplication", *Nature Cell Biology*, Vol. 5, 2003, pp. 539–544.

C. Wong, *et al.*, "Multistage Nanoparticle Delivery System for Deep Penetration into Tumor Tissue", *Proceedings National Academy of Sciences*, Vol. 108, No. 6, 2011, pp. 2426–2431.

L. Wong, *et al.*, "Prospective Identification of Tumorigenic Osteosarcoma Cancer Stem Cells in OS99-1 Cells Based on High Aldehyde Dehydrogenase Activity", *International Journal of Cancer*, Vol. 128, 2010, pp. 294–303.

V.L. Wong, *et al.*, "Reversible Centriole Depletion with an Inhibitor of Polo-like Kinase 4", *Science*, Vol. 348, No. 6239, 2015, pp. 1155–1160.

E.C. Woodhouse, R.F. Chuaqui, and L.A. Liotta, "General Mechanisms of Metastasis", *Cancer*, Vol. 80, 1997, pp. 1529–1537.

J.B. Woodruff, O. Wueseke, and A.A. Hyman, "Pericentriolar Material Structure and Dynamics", *Philosophical Transactions Royal Society of London. Series B*, Vol. 369, 2014, 10 pages.

J. Wu, *et al.*, "How Dynein and Microtubules Rotate the Nucleus", *Journal of Cellular Physiology*, Vol. 226, 2016, pp. 2666–2674.

J. Xiao and T. Ha, "Flipping Nanoscopy on its Head", *Science*, Vol. 355, 2017, pp. 582–584.

K. Xias, *et al.*, "The Effect of Surface Charge on In Vivo Biodistribution of PEG-Oligocholic Acid Based Micellar Nanoparticles", *Biomaterials*, Vol. 32, 2011, pp. 3435–3446.

C. Xu, "Modification of Superparamagnetic Nanoparticles for Biomedical Applications", Ph.D. Dissertation, Department of Chemistry, Brown University, Providence, RI, 2009.

B. Xue, *et al.*, "Loss of Par3 Promotes Breast Cancer Metastasis by Compromising Cell-Cell Cohesion", *Nature Cell Biology*, Vol. 15, No. 2, 2013, pp. 189–200.

R. Yang and J. Feldman, "SPD-2/CEP192 and CDK Are Limiting for Microtubule-Organizing Center Function at the Centrosome", *Current Biology*, Vol. 24, 2015, pp. 1924–1931.

Z. Yang, *et al.*, "Extra Centrosomes and/or Chromosomes Prolong Mitosis in Human Cells", *Nature Cell Biology*, Vol. 10, 2008, pp. 748–751.

T.Y. Yoo and D. J. Needleman, "Studying Kinetochores In Vivo Using FLIM-FRET", *The Mitotic Spindle — Methods and Protocol*, Edited by P. Chang and R. Ohi, Humana Press, Springer Science and Business Media, New York, 2018, Chapter 11.

Z. Yue, *et al.*, "Mechanisms Mitigating Problems Associated with Multiple Kinetochores on One Microtubule in Early Mitosis", *Journal of Cell Science*, 2017, Vol. 130, pp. 2266–2276.

M. Zanic, "Measuring the Effects of Microtubule-Associated Proteins on Microtubule Dynamics In Vitro", *The Mitotic Spindle — Methods and Protocols*, Edited by P. Chang and R. Ohi, Humana Press, Springer Science and Business Media, New York, 2018, Chapter 4.

J. Zhang, W.-H. Guo, and Y.-L. Wang, "Microtubules Stabilize Cell Polarity by Localizing Rear Signals", *Proceedings National Academy of Sciences*, Vol. 111, No. 462014, 2014, pp. 16383–16388.

L. Zhang, *et al.*, "Nanoparticles in Medicine: Therapeutic Applications and Developments", *Clinical Pharmacology and Therapeutics*, Vol. 83, No. 5, 2008, pp. 761–769.

L. Zhang, *et al.*, "27T Ultra-High Static Magnetic Field Changes Orientation and Morphology of Mitotic Spindles in Human Cells", *Elife*, Vol. 6, 2017, p. e22911.

P. Zhang, *et al.*, "Electrical Stimulation Inhibits Neointimal Hyperplasia after Abdominal Aorta Balloon Injury Through the PTEN/p27Kip1 Pathway", *Acta Biochimica et Biophysica Sinica* (Shanghai), Vol. 42, No. 11, 2010, pp. 807–815. doi:10.1093/abbs/gmq089

R. Zhang, *et al.*, "Enhancement Effect of Nano Fe_3O_4 to the Drug Accumulation of Doxorubicin in Cancer Cells", *Material Science and Engineering C*, Vol. 29, 2009, pp. 1697–1701.

Y. Zhang, *et al.*, "Self-Assembled Coatings on Individual Monodisperse Magnetite Nanoparticles for Efficient Intracellular Uptake", *Biomedical Microdevices*, Vol. 6, 2004, pp. 33–40.

Y. Zhang, N. Kohler, and M. Zhang, "Surface Modification of Superparamagnetic Magnetite Nanoparticles and their Intracellular Uptake", *Biomaterials*, Vol. 23, 2002, pp. 1553–1561.

Y. Zhao, "Centrosome Functions as a Molecular Dynamo in the Living Cell", *Advances in Bioscience and Biotechnology*, Vol. 6, 2015, pp. 452–455.

M. Zhao, J.V. Forrester, and C.D. McCaig, "A Small, Physiological Electric Field Orients Cell Division", *Proceedings National Academy of Sciences*, Vol. 96, No. 9, 1999, pp. 4942–4946.

M. Zhao, *et al.*, "Electrical Stimulation Directly Induced Pre-angiogenic Responses in Vascular Endothelial Cells by Signaling Through VEGF Receptors", *Journal of Cell Science*, Vol. 117, Pt. 3, 2004, pp. 397–405.

Y. Zhao and Q. Zhan, "Electric Fields Generated by Synchronized, Oscillation of Microtubules, Centrosomes and Chromosomes Regulate the Dynamics of Mitosis and Meiosis", *Theoretical Biology and Medical Modeling*, Vol. 9, 2012a, p. 26. doi:10.1186/1742-4682-9-26

Y. Zhao and Q. Zhan, "Electric Oscillation and Coupling of Chromatin Regulate Chromosome Packaging and Transcription in Eukaryotic cells", *Theoretical Biology and Medical Modelling*, Vol. 9, 2012b, pp. 27–38.

Y. Zheng, *et al.*, "Nucleation of Microtubule Assembly by a γ-Tubulin-Containing Ring Complex", *Nature*, Vol. 378, 1995, pp. 578–583.

B. Zhivotovsky and S. Orrenius, "Calcium and Cell Death Mechanisms: A Perspective from the Cell Death Community", *Cell Calcium*, Vol. 50, 2011, pp. 211–221.

F. Zhou and S. Roy, "SnapShot: Motile Cilia", *Cell*, Vol. 162, 2015, pp. 224–224.e1. doi:10.1016/j.cell.2015.06.

Z. Zhou, *et al.*, "MRI Detection of Breast Cancer Micrometastases with a Fibronectin-Targeting Contrast Agent", *Nature Communications*, Vol. 6, 2015, p. 7984. doi:10.1038/ncomms8984

A. Zielichowska, *et al.*, "Applications of Calcium Electroporation to Effective Apoptosis Induction in Fibrosarcoma Cells and Stimulation of Normal Muscle Cells", *Bioelectrochemistry*, Vol. 109, 2016, pp. 70–78.

W. Zimmerman, *et al.*, "Mitosis-Specific Anchoring of γ-Tubulin Complexes by Pericentrin Controls Spindle Organization and Mitotic Entry", *Molecular Biology of the Cell*, Vol. 15, 2004, pp. 3672–3657.

L. Zitvogel and G. Kroemer, "A p53 Regulated Immune Checkpoint Relevant to Cancer", *Science*, Vol. 349, No. 6247, 2015, pp. 476–477.

C. Zou, *et al.*, "Centrobin: A Novel Daughter Centriole-Associated Protein that is Required for Centrole Duplication", *Journal of Cell Biology*, Vol. 171, No. 3, 2005, pp. 437–445.

D. Zyss and F. Gergely, "Centrosome Function in Cancer: Guilty or Innocent?", *Trends in Cell Biology*, Vol. 19, No. 7, 2009, pp. 334–346.

Chronological List of Literature

1. A. Duensing, *et al.*, "Centriole Overduplication through the Concurrent Formation of Multiple Daughter Centrioles at Single Material Templates", *Oncogene*, Vol. 26, No. 43, 2007, pp. 6280–6288.
2. Z. Carvalho-Santos, *et al.*, "Tracing the Origins of Centrioles, Cilia and Flagella", *Journal of Cell Biology*, Vol. 194, No. 2, 2011, pp. 165–175.
3. S.R. Hameroff, "A New Theory of the Origin of Cancer: Quantum Coherent Entanglement, Centrioles, Mitosis, and Differentiation", *Biosystems*, Vol. 77, No. 1–3, 2004, pp 119–136.
4. L. Schöckel, *et al.*, "Cleavage of Cohesion Rings Coordinates the Separation of Centrioles and Chromatids", *Nature Cell Biology*, Vol. 13, No. 8, 2011, pp. 966–972.
5. E.A. Nigg and T. Stearns, "The Centrosome Cycle: Centriole Biogenesis, Duplication and Inherent Asymmetries", *Nature Cell Biology*, Vol. 13, No. 10, 2011, pp. 1154–1160.
6. E.A. Nigg and J.W. Raft, "Centrioles, Centrosomes, and Cilia in Health and Disease", *Cell*, Vol. 139, No. 4, 2009, pp. 663–678.
7. J. Azimzadeh and W.F. Marshall, "Building the Centriole", *Current Biology*, Vol. 20, No. 18, 2010, pp. R816–R825.
8. T. Kobayashi and B.D. Dynlacht, "Regulating the Transition from Centriole to Basal Body", *Journal of Cell Biology*, Vol. 193, No. 3, 2011, pp. 435–444.
9. N.J. Ganem, S.A. Godinho, and D. Pellman, "A Mechanism Linking Extra Centrosomes to Chromosomal Instability", *Nature*, Vol. 460, No. 7252, 2009, pp. 278–282.
10. E.A. Nigg, "Centrosome Aberrations: Cause or Consequence of Cancer Progression?" *Nature Reviews Cancer*, Vol. 2, No. 11, 2002, pp. 815–825.
11. K.A. Williams, *et al.*, "Carbon Nanotubes with DNA Recognition", *Nature*, Vol. 420, 2002, p. 761.

12. Z. Cheng, *et al.*, "Multifunctional Nanoparticles: Cost versus Benefit of Adding Targeting and imaging Capabilities", *Science*, Vol. 338, No. 6109, 2012, pp. 903–910.
13. N. Korzeniewski, M. Hohenfellner, and S. Duensing, "CANDI Promotes PLK4-Mediated Centriole Overduplication and Is Frequently Disrupted in Prostate Cancer", *Neoplasia*, Vol. 14, No. 9, 2012, pp. 799–806.
14. E.D. Kirson, *et al.*, "Alternating Electric Fields Arrest Cell Proliferation in Animal Tumor Models and Human Brain Tumors", *Proceedings of the National Academy of Sciences*, Vol. 104, No. 24, 2007.
15. V.G. Vasishto, "Sequentially Programmed Magnetic Field Therapy in the Management of Recurrent Anaplastic Astrocytoma: A Case Study Report and Literature Review", *Case Reports in Oncology*, Vol. 3, No. 2, 2010, pp. 189–194.
16. D. Zyss and F. Gergely, "Centrosome Function in Cancer: Guilty or Innocent?", *Trends in Cell Biology*, Vol. 19, No. 7, 2009, pp. 334–346.
17. W.J. Meggs, "Electric Fields Determine the Spatial Organization of Microtubules and Actin Filaments", *Medical Hypothesis*, Vol. 26, No. 3, 1988, pp. 165–170.
18. W.F. Marshall, "What Is the Function of Centrioles?" *Journal of Cellular Biochemistry*, Vol. 100, No. 4, 2007, pp. 916–922.
19. J.W. Raff, "Centrosomes and Cancer: Lessons from a TACC", *Trends in Cell Biology*, Vol. 12, No. 5, 2002, pp. 222–225.
20. P. Meraldi and E.A. Nigg, "The Centrosome Cycle", *FEBS Letters*, Vol. 521, No. 1–3, 2002, pp. 9–13.
21. J.L. Salisbury, *et al.*, "Centrosomes and Cancer", *Biology of the Cell*, Vol. 91, No. 6, 1999, pp. 451–460.
22. V.G. Vasishto, "Sequentially Programmed Magnetic Field Therapy in the Management of Recurrent Anaplastic Astrocytoma: A Case Study Report and Literature Review", *Case Reports in Oncology*, Vol. 3, No. 2, 2010, pp. 189–194.
23. J. Howard and A.A. Hyman, "Dynamics and Mechanics of the Microtubule Plus End", *Nature*, Vol. 422, 2003, pp. 753–758.
24. P. Strnad, *et al.*, "Regulated HsSAS-6 Levels Ensure Formation of a Simple Procentriole per Centriole During Centrosome Duplication Cycle", *Developmental Cell*, Vol. 13, No. 2, 2007, pp. 203–213.
25. S.A. Hormeño, *et al.*, "Single Centrosome Manipulation Reveals Its Electric Charge and Associated Dynamic Structure", *Biophysical Journal*, Vol. 97, No. 4, 2009, pp. 1022–1030.

26. M.V. Sataric, *et al.*, "Modelling the Role of Intrinsic Electric Fields in Microtubules as an Additional Control Mechanism of Bi-Directional Intracellular Transport", *Cell Biochemistry and Biophysics*, Vol. 52, No. 2, 2008, pp. 113–124.
27. J. Pokorný, "Electrodynamic Activity of Healthy and Cancer Cells", *Journal of Physics: Conference Series*, Vol. 329, No. 1, 2011, p. 012007.
28. A. Garci-Bennett, M. Nees, and B. Fadeel, "In Search of the Holy Grail: Folate-Targeted Nanoparticles for Cancer Therapy", *Biochemical Pharmacology*, Vol. 81, No. 8, 2011, pp. 976–984.
29. C. Rossi, *et al.*, "New Perspectives in Cell Communication: Bioelectromagnetic Interactions", *Seminars in Cancer Biology*, Vol. 21, No. 3, 2011, pp. 207–214.
30. Y. Zhao and Q. Zhan, "Electric Fields Generated by Synchronized, Oscillation of Microtubules, Centrosomes and Chromosomes Regulate the Dynamics of Mitosis and Meiosis", *Theoretical Biology and Medical Modelling*, Vol. 9, No. 26, 2012. doi: 10.1186/1742-4682-9-26
31. K. M. Tyner, R. Kopelman, and M. Philbert, "'Nanosized Voltmeter' Enables Cellular-Wide Electric Field Mapping", *Biophysical Journal*, Vol. 93, No. 4, 2007, pp. 1163–1174.
32. Y. Zhao and Q. Zhan, "Electric Oscillation and Coupling of Chromatin Regulate Chromosome Packaging and Transcription in Eukaryotic cells", *Theoretical Biology and Medical Modelling*, Vol. 9, 2012, pp. 27–38.
33. AFM Workshop, *Atomic Force Microscopes*, 2012, Chapter 4, p. 75H.
34. W.G. Hol, "The Role of the Alpha-Helix Dipole in Protein Function and Structure", *Progress in Biophysics and Molecular Biology*, Vol. 45, No. 3, 1985, pp. 149–195.
35. M.V. Sataric, J.A. Tuszynski, and R.B. Zakula, "Kinklike Excitations as an Energy-Transfer Mechanism in Microtubules", *Physical Review E*, Vol. 48, No. 1, 1993, pp. 589–597.
36. J.A. Brown and J.A. Tuszynski, "A Review of the Ferroelectric Model of Microtubules", *Ferroelectrics*, Vol. 220, No. 1, 1999, pp. 141–155.
37. J. Pokorný, F. Jelinek, and V. Trkal, "Electric Field around Microtubules", *Bioelectrochemistry and Bioenergetics*, Vol. 45, No. 2, 1998, pp. 239–245.
38. J.A. Tuszynski, *et al.*, "Molecular Dynamics Simulations of Tubulin Structure and Calculation of Electrostatic Properties of Microtubules", *Mathematical and Computer Modelling*, Vol. 41, No. 10, 2005, pp. 1055–1070.
39. J. Pokorný, "Electrodynamic Activity of Healthy and Cancer Cells", *Journal of Physics: Conference Series*, Vol. 329, No. 1, 2011, p. 012007, paper no. 012007, 19th International Fröhlick's Symposium.

40. A. Kis, *et al.*, "Nanomechanics of Microtubules", *Physical Review Letters*, Vol. 89, No. 24, 2002, pp. 248101-1–248101-4.
41. W.G. Hol, "The Role of the Alpha-Helix Dipole in Protein Function and Structure", *Progress in Biophysics and Molecular Biology*, Vol. 45, No. 3, 1985, pp. 149–195.
42. M.V. Satarić and J.A. Tuszyński, "Relationship Between the Nonlinear Ferroelectric and Liquid Crystal Models for Microtubules", *Physical Review*, Vol. 67, 2003, pp. 011901-1–011901-11.
43. J.A. Tuszyński, *et al.*, "Ferroelectric Behavior in Microtubule Dipole Lattices: Implication for Information Processing, Signaling and Assembly/Disassembly", *Journal of Theoretical Biology*, Vol. 174, No. 4, 1995, pp. 371–380.
44. M. Cifra, D. Havelka, and O. Kucera, "Electric Oscillations Generated by Collective Vibration Modes of Microtubules", *Proceedings of SPIE, Laser Applications in Life Sciences*, Edited by M. Kinnamen and R. Myllyla, Vol. 7376, 2010.
45. B. Alberts, *et al.*, *Molecular Biology of the Cell*, 3rd Edition, Garland Publishing, New York, 1994.
46. H.T. Schek III, *et al.*, "Microtubule Assembly Dynamics at the Nanoscale", *Current Biology*, Vol. 17, No. 17, 2007, pp. 1445–1455.
47. E. Samiyoshi and A. Sugimoto, "Cell Polarity: Centrosomes Release Signals for Polarization", *Current Biology*, Vol. 22, No. 8, 2012, pp. R281–R283.
48. G. Brouhard and D. Sept, "Microtubules: Sizing Up the GTP Cap", *Current Biology*, Vol. 22, No. 18, 2012, pp. R802–R803.
49. M. Cifra, "Electrodynamic Eigenmodes in Cellular Morphology", *Biosystems*, Vol. 109, No. 3, 2012, pp. 356–366.
50. O. Kučera and D. Havelka, "Mechano-Electrical Vibrations of Microtubules—Link to Subcellular Morphology", *Biosystems*, Vol. 19, No. 3, 2012, pp. 346–355.
51. P.P. Girard, *et al.*, "Cellular Chemomechanics at Interfaces: Sensing, Integration and Response", *Soft Matter*, Vol. 3, 2007, pp. 307–326.
52. P.W. Schafer, "Centrioles of a Human Cancer: Intercellular Order and Intracellular Disorder", *Science*, Vol. 164, 1969, pp. 1300–1303.
53. P.W. Schafer and J.A. Chandler, "Electron Probe X-ray Microanalysis of a Normal Centriole", *Science*, Vol. 170, 1970, pp. 1204–1205.
54. C.C. Harris, *et al.*, "Metabolism of Benzo(a)pyrene, N-nitrosodimethylamine, and N-nitrosopyrrolidine and Identification of the Major Carcinogen-DNA

Adducts Formed in Cultured Human Esophagus", *Cancer Research*, Vol. 39, No. 11, 1979, pp. 4401–4406.
55. G. Long, *et al.*, "Histological and Finite Element Analysis of Cell Death due to Irreversible Electroporation", *Technology in Cancer Research and Treatment*, Vol. 13, No. 6, 2014, pp. 561–569.
56. P.W. Schafer, "Centrioles: Intercellular Order in Normal and Malignant Cells", *Journal of Thoracic and Cardiovascular Surgery*, Vol. 63, No. 3, 1972, pp. 472–477.
57. J. Pokorný, *et al.*, "Cancer Physics: Diagnosis Based on Damped Cellular Elastoelectrical Vibrations in Microtubules", *European Biophysics Journal*, Vol. 40, No. 6, 2011, pp. 747–759.
58. W.F. Marshall and J.L. Rosenbaum, "How Centrioles Work: Lessons from Green Yeast", *Current Opinion in Cell Biology*, Vol. 12, No. 1, 2000, pp. 119–125.
59. J. Pokorný, *et al.*, "Biophysical Aspects of Cancer—Electromagnetic Mechanism", *Indian Journal of Experimental Biology*, Vol. 46, No. 5, 2008, pp. 310–321.
60. X.S. Qian, J.Q. Zhang, and C.Q. Ru, "Wave Propagation in Orthotropic Microtubules", *Journal of Applied Physics*, Vol. 101, 2007, pp. 084702-1–084702-7.
61. S. Suresh, "Biomechanics and Biophysics of Cancer Cells", *Acta Biomater*, Vol. 3, No. 4, 2007, pp. 413–438.
62. C.Y. Wang, C.Q. Ru, and A. Mioduchowski, "Vibration of Microtubules as Orthotropic Shells", *Physica E: Low-Dimensional Systems and Nanostructures*, Vol. 35, No. 1, 2006, pp. 48–56.
63. C.P. Brangwynne, *et al.*, "Microtubules Can Bear Enhanced Compressive Loads in Living Cells Because of Lateral Reinforcement", *Journal of Cell Biology*, Vol. 173, No. 5, 2006, pp. 733–741.
64. J. Pokorný, J. Hašek, and F. Jelínek, "Electromagnetic Field of Microtubules: Effects on Transfer of Mass Particles and Electrons", *Journal of Biological Physics*, Vol. 31, 2005, pp. 501–514.
65. M. Cifra, *et al.*, "Electric Field Generated by Axial Longitudinal Vibration Modes of Microtubule", *Biosystems*, Vol. 100, No. 2, 2010, pp. 122–131.
66. A.E. Pelling, *et al.*, "Local Nanomechanical Motion of the Cell Wall of Saccharomyces Cerevisiae", *Science*, Vol. 305, No. 5687, 2004, pp. 1147–1150.
67. E.D. Kirson, *et al.*, "Disruption of Cancer Cell Replication by Alternating Electric Fields", *Cancer Research*, Vol. 64, No. 9, 2004, pp. 3288–3295.

68. L. Cucullo, *et al.*, "Very Low Intensity Alternating Current Decreases Cell Proliferation", *Glia*, Vol. 51, No. 1, 2005, pp. 65–72.
69. M.R. Mahjoub and T. Stearns, "Supernumerary Centrosomes Nucleate Extra Cilia and Compromise Primary Cilium Signaling", *Current Biology*, Vol. 22, No. 17, 2012, pp. 1628–1634.
70. S. Portet, *et al.*, "Elastic Vibrations in Seamless Microtubules", *European Biophysics Journal*, Vol. 34, No. 7, 2005, pp. 912–920.
71. D.J. Needleman, *et al.*, "Synchrotron X-ray Diffraction Study of Microtubules Buckling and Bundling under Osmotic Stress: A Probe of Interprotofilament Interactions", *Physical Review Letters*, Vol. 193, No. 19, 2004, pp. 198104-1–198104-4.
72. J.A. Tuszyński, *et al.*, "Anisotropic Elastic Properties of Microtubules", *European Physics Journal E Soft Matter*, Vol. 17, No. 1, 2005, pp. 29–35.
73. F. Gittes, *et al.*, "Flexural Rigidity of Microtubules and Actin Filaments Measured from Thermal Fluctuations in Shape", *Journal of Cell Biology*, Vol. 120, No. 4, 1993, pp. 923–934.
74. J. Cartwright, "An Electrical Misunderstanding", *Physics of Cancer, Physics World*, Vol. 26, 2013, pp. 25–26.
75. J.M. Scholey, G.C. Rogers, and D.J. Sharp, "Mitosis, Microtubules, and the Matrix", *Journal of Cell Biology*, Vol. 154, No. 2, 2001, pp. 261–266.
76. M. Bornens, "The Centrosome in Cells and Organisms", *Science*, Vol. 335, 2012, pp. 422–426.
77. W.F. Marshall, "Centriole Assembly: The Origin of Nine-ness", *Current Biology*, Vol. 17, No. 24, 2007, pp. R1057–R1059.
78. S. Vaughan and H.R. Dawe, "Common Themes in Centriole and Centrosomes Movements", *Trends in Cell Biology*, Vol. 21, No.1, 2011, pp. 57–66.
79. B. Goldstein and I.G. Macara, "The PAR Proteins: Fundamental Players in Animal Cell Polarization", *Developmental Cell*, Vol. 13, No. 5, 2007, pp. 609–622.
80. J.A. Tuszyński, *et al.*, "Molecular Dynamics Simulations of Tubulin Structure and Calculation of Electrostatic Properties of Microtubules", *Mathematical and Computer Modelling*, Vol. 41, No. 10, 2005, pp. 1055–1070.
81. R. Basto, *et al.*, "Flies without Centrioles", *Cell*, Vol. 125, No. 7, 2006, pp. 1375–1386.
82. M.C. Tsai and J. Ahringer, "Microtubules Are Involved in Anterior-Posterior Axis Formation in *C. elegans* Embryos", *Journal of Cell Biology*, Vol. 179, No. 3, 2007, pp. 397–402.

83. J. Pokorný, J. Hašek, and F. Jelínek, "Endogenous Electric Field and Organization of Living Matter", *Electromagnetic Biology and Medicine*, Vol. 24, No. 3, 2005, pp. 185–197.
84. J. Pokorný, "Viscous Effects on Polar Vibrations in Microtubules", *Electromagnetic Biology and Medicine*, Vol. 22, No. 1, 2003, pp. 15–29.
85. K.H. Downing and E. Nogales, "New Insights into Microtubule Structure and Function from the Atomic Model of Tubulin", *European Biophysics Journal*, Vol. 27, No. 5, 1998, pp. 431–436.
86. A. Duensing, *et al.*, "Centriole Overduplication through the Concurrent Formation of Multiple Daughter Centrioles at Single Material Templates", *Oncogene*, Vol. 26, No. 43, 2007, pp. 6280–6288.
87. N. Korzeniewski, M. Hohenfellner, and S. Duensing, "CANDI Promotes PLK4-Mediated Centriole Overduplication and Is Frequently Disrupted in Prostate Cancer", *Neoplasia*, Vol. 14, No. 9, 2012, pp. 799–806.
88. D. Bienkowska and C.R. Cowan, "Centrosomes Can Initiate a Polarity Axis from Any Position within One-Cell *C. elegans* Embryos", *Current Biology*, Vol. 22, No. 7, 2012, pp. 583–589.
89. C.R. Cowan and A.A. Hyman, "Centrosomes Direct Cell Polarity Independently of Microtubule Assembly in *C. elegans* Embryos", *Nature*, Vol. 431, No. 7004, 2004, pp. 92–96.
90. D. Kitagawa, *et al.*, "Structural Basis of the 9-Fold Symmetry of Centrioles", *Cell*, Vol. 144, No. 3, 2011, pp. 364–375.
91. E.A. Nigg, "Origins and Consequences of Centrosome Aberrations in Human Cancers", *International Journal of Cancer*, Vol. 119, No. 12, 2006, pp. 2717–2723.
92. S.A. Godinho, M. Kwon, and D. Pellman, "Centrosomes and Cancer: How Cancer Cells Divide with too Many Centrosomes", *Cancer Metastasis Review*, Vol. 28, 2009, pp. 85–98.
93. S.R. Grobmyer and B.M. Moudqil (Eds.), *Cancer Nanotechnology Methods and Protocols*, Humana Press, Springer, New York, 2010.
94. S.C. Brown, *et al.*, "Nanoparticle Characterization for Cancer Nanotechnology and Other Biological Applications", *Methods in Molecular Biology*, Vol. 624, 2010, pp. 39–65.
95. S.N. Hochwald, "Molecular-Targeted Therapy for Cancer and Nanotechnology", *Methods in Molecular Biology*, Vol. 624, 2010, pp. 11–23.
96. M. Oberringer, *et al.*, "Centrosomes Multiplication Accompanies a Transient Clustering of Polyploid Cells during Tissue Repair", *Molecular Cell Biology Research Communications*, Vol. 2, 1999, pp. 190–196.

97. M.A. Jordan and L. Wilson, "Microtubules as a Target for Anticancer Drugs", *Nature Reviews: Cancer*, Vol. 4, No. 4, 2004, pp. 253–265.
98. J.R. Jackson, *et al.*, "Targeted Anti-Mitotic Therapies: Can We Impose on Tubulin Agents?" *Nature Reviews: Cancer*, Vol. 7, No. 2, 2007, pp.107–117.
99. G. Sluder and J.J. Nordberg, "The Good, The Bad and The Ugly: The Practical Consequences of Centrosome Amplification", *Current Opinion in Cell Biology*, Vol. 16, No. 1, 2004, pp. 49–54.
100. B.R. Brinkley, "Managing the Centrosome Numbers Game: From Chaos to Stability in Cancer Cell Division", *Trends in Cell Biology*, Vol. 11, No. 1, 2001, pp. 18–21.
101. A.B. D'Assoro, W.L. Lingle, and J.L. Salisbury, "Centrosome Amplification and the Development of Cancer", *Oncogene*, Vol. 21, No. 40, 2002, pp. 6146–6153.
102. I.R. Adams and J.V. Kilmartin, "Spindle Pole Body Duplication: A Model for Centrosome Duplication", *Trends in Cell Biology*, Vol. 10, No. 8, 2000, pp. 329–335.
103. M.C. Hollander and A.J. Fornace, Jr., "Genomic Instability Centrosome Amplification, Cell Cycle Checkpoints and Gadd45a", *Oncogene*, Vol. 21, No. 40, 2002, pp. 6228–6233.
104. W. Dai, Q. Wang, and F. Traganos, "Polo-like Kinases and Centrosome Regulation", *Oncogene*, Vol. 21, No. 40, pp. 6195–6200.
105. J. Kleylein-Sohn, *et al.*, "Plk4-Induced Centriole Biogenesis in Human Cells", *Developmental Cell*, Vol. 13, No. 2, 2007, pp. 190–202.
106. R. Basto, *et al.*, "Centrosome Amplification Can Initiate Tumorigenesis in Flies", *Cell*, Vol. 133, No. 6, 2008, pp. 1032–1042.
107. E.A. Nigg, "Centrosome Duplication: Of Rules and Licenses", *Trends in Cell Biology*, Vol. 17, No. 5, 2007, pp. 215–221.
108. M. Bornens, "Centrosome Composition and Microtubule Anchoring Mechanisms", *Current Opinion in Cell Biology*, Vol. 14, No. 1, 2002, pp. 25–34.
109. G.A. Pihan, *et al.*, "Centrosome Abnormalities and Chromosome Instability Occur Together in Pre-Invasive Carcinomas", *Cancer Research*, Vol. 63, No. 6, 2003, pp. 1398–1404.
110. J. Lončarek, *et al.*, "Control of Daughter Centriole Formation by the Pericentriolar Material", *Nature: Cell Biology*, Vol. 10, No. 3, 2008, pp. 322–328.
111. R.S. Kochanski and G.G. Borisy, "Mode of Centriole Duplication and Distribution", *Journal of Cell Biology*, Vol. 110, No. 5, 1990, pp. 1599–1605.

112. M. Piel, *et al.*, "The Respective Contribution of the Mother and Daughter Centrioles to Centrosome Activity and Behavior in Vertebrate Cells", *Journal of Cell Biology*, Vol. 149, No. 2, 2000, pp. 317–330.
113. M.C. Tsai and J. Ahringer, "Microtubules Are Involved in Anterior-Posterior Axis Formation in *C. elegans* Embryos", *Journal of Cell Biology*, Vol. 179, No. 3, 2007, pp. 397–402.
114. M. Bornens, "Centrosome Composition and Microtubule Anchoring Mechanisms", *Current Opinion in Cell Biology*, Vol. 14, No. 1, 2002, pp. 25–34.
115. A. Rodrigues-Martins, *et al.*, "DSAS-6 Organizes a Tube-like Centriole Precursor, and Its Absence Suggests Modularity in Centriole Assembly", *Current Biology*, Vol. 17, No. 17, 2007, pp. 1465–1472.
116. J. Gopalakrishnan, *et al.*, "Self-Assembling SAS-6 Multimer Is a Core Centriole Building Block", *Journal of Biological Chemistry*, Vol. 285, No. 12, 2010, pp. 8759–8770.
117. J. Azimzadeh and M. Bornens, "Structure and Duplication of the Centrosome", *Journal of Cell Science*, Vol. 120, 2007, pp. 2139–2142.
118. Y. Nakazawa, *et al.*, "SAS-6 is a Cartwheel Protein that Establishes the 9-Fold Symmetry of the Centriole", *Current Biology*, Vol. 17, No. 24, 2007, pp. 2169–2174.
119. E.R. Gomes and G.G. Gundersen, "Real-Time Centrosome Reorientation During Fibroblast Migration", *Methods in Enzymology*, Vol. 406, 2006, pp. 579–592.
120. S. Doxsey, "Re-evaluating Centrosome Function", *Nature Reviews Molecular Cell Biology*, Vol. 2, No. 9, 2001, pp. 688–698.
121. W.J. Wang, *et al.*, "The Conversion of Centrioles to Centrosomes: Essential Coupling of Duplication with Segregation", *Journal of Cell Biology*, Vol. 193, No. 4, 2011, pp. 727–739.
122. W.J. Meggs, "Electric Fields Determine the Spatial Organization of Microtubules and Actin Filaments", *Medical Hypothesis*, Vol. 26, No. 3, 1988, pp. 165–170.
123. B.M. Lange and K. Gull, "Structure and Function of the Centriole in Animal Cells: Progress and Questions", *Trends in Cell Biology*, Vol. 6, No. 9, 1996, pp. 348–352.
124. J. Beisson and M. Wright, "Basal Body/Centriole Assembly and Continuity", *Current Opinion in Cell Biology*, Vol. 15, No. 1, 2003, pp. 96–104.
125. P.G. McKean, S. Vaughan, and K. Gull, "The Extended Tubulin Superfamily", *Journal of Cell Science*, Vol. 114, Pt. 15, 2001, pp. 2723–2733.
126. S.K. Dutcher, "The Tubulin Fraternity: Alpha to Eta", Current *Opinion in Cell Biology*, Vol. 13, No. 1, 2001, pp. 49–54.

127. B.M. Lange and K. Gull, "A Molecular Marker for Centriole Maturation in the Mammalian Cell Cycle", *Journal of Cell Biology,* Vol. 130, No. 4, 1995, pp. 919–927.
128. J. Beisson and M-Jerka-Dziadosz, "Polarities of the Centriolar Structure: Morphogenetic Consequences", *Biology of the Cell*, Vol. 91, No. 4–5, 1999, pp. 367–378.
129. A.-M. Tassin and M. Bornens, "Centrosome Structure and Microtubule Nucleation in Animal Cells", *Biology of the Cell*, Vol. 91, No. 4–5, 1999, pp. 343–354.
130. R. Mahen and A.R. Venkitaraman, "Pattern Formation in Centrosome Assembly", *Current Opinion in Cell Biology*, Vol. 24, No. 1, 2012, pp. 14–23.
131. A. Ridley and R. Heald, "Cell Structure and Dynamics—Editorial Overview", *Current Opinion in Cell Biology*, Vol. 23, No. 1, 2011, pp. 1–3.
132. H. Asakawa, *et al.*, "Submolecular-Scale Imaging of α-Helices and C-Terminal Domains of Tubulins by Frequency Modulation Atomic Force Microscopy in Liquid", *Biophysical Journal,* Vol. 101, No. 5, 2011, pp. 1270–1276.
133. B. Raynaud-Messina and A. Merdes, "Gamma-Tubulin Complexes and Microtubule Organization", *Current Opinion in Cell Biology*, Vol. 19, No. 1, 2007, pp. 24–30.
134. K.R. Robinson, "The Responses of Cells to Electrical Fields: A Review", *Journal of Cell Biology*, Vol. 101, No. 6, 1985, pp. 2023–2027.
135. C.D. McCaig, *et al.*, "Controlling Cell Behavior Electrically: Current views and Future Potential", *Physiological Review*, Vol. 85, No. 3, 2005, pp. 943–978.
136. D.A. Brito, S.M. Gouveia, and M. Bettencourt-Dias, "Deconstructing the Centriole: Structure and Number Control", *Current Opinion in Cell Biology*, Vol. 24, No. 1, 2012, pp. 4–13.
137. S.A. Hormeño, *et al.*, "Single Centrosome Manipulation Reveals its Electric Charge and Associated Dynamic Structure", *Biophysical Journal*, Vol. 97, No. 4, 2009, pp. 1022–1030.
138. R. Binggeli and R.C. Weinstein, "Membrane Potentials and Sodium Channels: Hypotheses for Growth Regulation and Cancer Formation Based on Changes in Sodium Channels and Gap Junctions", *Journal of Theoretical Biology*, Vol. 123, No. 4, 1986, pp. 377–401.
139. M.F. Tsou and T. Stearns, "Mechanism Limiting Centrosome Duplication to Once Per Cell Cycle", *Nature*, Vol. 442, No. 7105, 2006, pp. 947–951.
140. K. M. Tyner, R. Kopelman, and M. Philbert, "'Nanosized Voltmeter' Enables Cellular-Wide Electric Field Mapping", *Biophysical Journal*, Vol. 93, No. 4, 2007, pp. 1163–1174.

141. M. Bettencourt-Dias and D.M. Glover, "SnapShot: Centriole Biogenesis", *Cell*, Vol. 136, No. 1, 2009, p. 188.
142. J. Lončarek, *et al.*, "Control of Daughter Centriole Formation by the Pericentriolar Material", *Nature: Cell Biology*, Vol. 10, No. 3, 2008, pp. 322–328.
143. J. Cuzick, *et al.*, "Electropotential Measurements as a New Diagnostic Modality for Breast Cancer", *The Lancet*, Vol. 352, No. 9125, 1998, pp. 359–363.
144. J. Kleylein-Sohn, *et al.*, "Plk4-Induced Centriole Biogenesis in Human Cells", *Developmental Cell*, Vol. 13, No. 2, 2007, pp. 190–202.
145. E. Vladimirou, *et al.*, "Nonautonomous Movement of Chromosomes in Mitosis", *Developmental Cell*, Vol. 27, No. 1, 2013, pp. 60–71.
146. D.A. Klos Dehring, *et al.*, "Deuterosome-Mediated Centriole Biogenesis", *Developmental Cell*, Vol. 27, No. 1, 2013, pp. 103–112.
147. M.A. Messerli and D.M. Graham, "Extracellular Electrical Fields Direct Wound Healing and Regeneration", *Biological Bulletin*, Vol. 221, No. 1, 2011, pp. 79–92.
148. H. Löffler, *et al.*, "DNA Damage-Induced Centrosome Amplification Occurs Via Excessive Formation of Centriolar Satellites", *Oncogene*, Vol. 32, No. 24, 2013, pp. 2963–2972.
149. N. Akhtar and C.H. Streuli, "An Integrin-ILK-Microtubule Network Orients Cell Polarity and Lumen Formation in Glandular Epithelium", *Nature Cell Biology*, Vol. 15, No. 1, 2013, pp. 17–27.
150. M.B.A. Djamgoz, *et al.*, "Directional Movement of Rat Prostate Cancer Cells in Direct-Current Electric Field: Involvement of Voltagegated Na+ Channel Activity", *Journal of Cell Science*, Vol. 114, Pt. 14, 2001, pp. 2697–2705.
151. M.E. Mycielska and M.B. Djamgoz, "Cellular Mechanisms of Direct-Current Electric Field Effects: Galvanotaxis and Metastatic Disease", *Journal of Cell Science*, Vol. 177, Pt. 9, 2004, pp. 1631–1639.
152. J.R. Gray, C.H. Frith, and J.D. Parker, "In Vivo Enhancement of Chemotherapy with Static Electric or Magnetic Fields", *Bioeletromagnetics*, Vol. 21, No. 8, 2000, pp. 575–583.
153. T. Avidor-Reiss and J. Gopalakrishnan, "Building a Centriole", *Current Opinion in Cell Biology*, Vol. 25, No. 1, 2013, pp. 72–77.
154. J. Gopalakrishnan, *et al.*, "Self-Assembling SAS-6 Multimer Is a Core Centriole Building Block", *Journal of Biological Chemistry*, Vol. 285, No. 12, 2010, pp. 8759–8770.
155. M. van Breugel, *et al.*, "Structures of SAS-6 Suggest Its Organization in Centrioles", *Science*, Vol. 331, No. 6021, 2011, pp. 1196–1199.

156. D. Kitagawa, *et al.*, "Structural Basis of the 9-Fold Symmetry of Centrioles", *Cell*, Vol. 144, No. 3, 2011, pp. 364–375.
157. S. Kanno, *et al.*, "Establishment of a Simple and Practical Procedure Applicable to Therapeutic Angiogenesis", *Circulation*, Vol. 99, No. 20, 1999, pp. 2682–2687.
158. M. Zhao, *et al.*, "Electrical Stimulation Directly Induced Pre-angiogenic Responses in Vascular Endothelial Cells by Signaling Through VEGF Receptors", *Journal of Cell Science*, Vol. 117, Pt. 3, 2004, pp. 397–405.
159. M. Papetti and I.M. Herman, "Mechanisms of Normal and Tumor-derived Angiogenesis", *American Journal of Cell Physiology*, Vol. 282, No. 5, 2002, pp. C947–C970.
160. R. Habedanck, *et al.*, "The Polo Kinase Plk4 Functions in Centriole Duplication", *Nature Cell Biology*, Vol. 7, No. 11, 2005, pp. 1140–1146.
161. S. Leidel, *et al.*, "SAS-6 Defines a Protein Family Required for Centrosome Duplication in *C. elegans* and in Human Cells", *Nature Cell Biology*, Vol. 7, No. 2, 2005, pp. 115–125.
162. M.E. Werner, *et al.*, "Actin and Microtubules Drive Differential Aspects of Planar Cell Polarity in Multiciliated Cells", *Journal of Cell Biology*, Vol. 195, No. 1, 2011, pp. 19–26.
163. P. Carmeliet and R.K. Jain, "Angiogenesis in Cancer and Other Diseases", *Nature*, Vol. 407, No. 6801, 2000, pp. 249–257.
164. S. Doxsey, W. Zimmerman, and K. Mikule, "Centrosome Control of the Cell Cycle", *Trends in Cell Biology*, Vol. 15, No. 6, 2005, pp. 303–311.
165. M. Zhao, J.V. Forrester, and C.D. McCaig, "A Small, Physiological Electric Field Orients Cell Division", *Proceedings National Academy of Sciences*, Vol. 96, No. 9, 1999, pp. 4942–4946.
166. A. Asnacios and O. Hamant, "The Mechanics Behind Cell Polarity", *Trends in Cell Biology*, Vol. 22, No. 11, 2012, pp. 584–591.
167. C. Panbianco and M. Gotta, "Coordinating Cell Polarity with Cell Divisions in Space and Time", *Trends in Cell Biology*, Vol. 21, No. 11, 2011, pp. 672–680.
168. M. Olivotto, *et al.*, "Electric Fields at the Plasma Membrane Level: A Neglected Element in the Mechanisms of Cell Signalling", *Bio Essays*, Vol. 18, No. 6, 1996, pp. 495–504.
169. M. Klagsbrun and P.A. D'Amore, "Regulators of Angiogenesis", *Annular Review of Physiology*, Vol. 53, 1991, pp. 217–239.
170. E.M. Hatch, *et al.*, "Cep 152 Interacts with Plk4 and Is Required for Centriole Duplication", *Journal of Cell Biology*, Vol. 191, No. 4, 2010, pp. 721–729.

171. D.-H. Kim, *et al.*, "Biofunctionalized Magnetic-Vortex Microdiscs for Targeted Cancer-Cell Destruction", *Nature Materials*, Vol. 9, 2010, pp. 165–171.
172. H. Löffler, *et al.*, "Structure Meets Function—Centrosomes, Genome Maintenance and the DNA Damage Response", *Experimental Cell Research*, Vol. 312, No. 14, 2006, pp. 2633–2640.
173. A. Krämer, J. Lukas, and J. Bartek, "Checking Out the Centrosome", *Cell Cycle*, Vol. 3, No. 11, 2004, pp. 1390–1393.
174. S.B. Dev and G.A. Hofmann, "Electrochemotherapy—A Novel Method of Cancer Treatment", *Cancer Treatment Reviews*, Vol. 21, No. 1, 1994, pp. 105–115.
175. N. Ferrara, "Molecular and Biological Properties of Vascular Endothelial Growth Factor", *Journal of Molecular Medicine*, Vol. 77, No. 7, 1999, pp. 527–543.
176. B.D. Vitre and D.W. Cleveland, "Centrosomes, Chromosome Instability (CIN) and Aneuploidy", *Current Opinion in Cell Biology*, Vol. 24, No. 7, 2012, pp. 809–815.
177. J. Dobson, "Remote Control of Cellular Behaviour with Magnetic Nanoparticles", *Nature Nanotechnology*, Vol. 3, 2008, pp. 139–143.
178. E. Nogales, *et al.*, "High-Resolution Model of the Microtubule", *Cell*, Vol. 96, No. 1, 1999, pp. 79–88.
179. E.A. Nigg and T. Stearns, "The Centrosome Cycle: Centriole Biogenesis, Duplication and Inherent Asymmetries", *Nature Cell Biology*, Vol. 13, No. 10, 2011, pp. 1154–1160.
180. C.J. Meyer, *et al.*, "Mechanical Control of Cyclic AMP Signalling and Gene Transcription Through Integrins", *Nature Cell Biology*, Vol. 2, No. 9, 2000, pp. 666–668.
181. P.A. Hardy and H. Zacharias, "Reappraisal of the Hansemann-Boveri Hypothesis on the Origin of Tumors", *Cell Biology International*, Vol. 29, No. 12, 2005, pp. 983–992.
182. E.C. Woodhouse, R.F. Chuaqui, and L.A. Liotta, "General Mechanisms of Metastasis", *Cancer*, Vol. 80, 1997, pp. 1529–1537.
183. R.J. Mannix, *et al.*, "Nanomagnetic Actuation of Receptor-Mediated Signal Transduction", *Nature Nanotechnology*, Vol. 3, 2008, pp. 36–40.
184. S. Hughes, J. Dobson, and A.J. El Haj, "Magnetic Targeting of Mechanosensors in Bone Cells for Tissue Engineering Applications", *Journal of Biomechanics*, Vol. 40, 2007, pp. S96–S104.
185. T. Neuberger, *et al.*, "Superparamagnetic Nanoparticles for Biomedical Applications: Possibilities and Limitations of a New Drug Delivery

System", *Journal of Magnetism and Magnetic Materials*, Vol. 293, 2005, pp. 483–496.

186. R.J. Davies, *et al.*, "Detection of the Cancer-Prone Colon, Using Transepithelial Impedance Analysis", *Archives of Surgery*, Vol. 124, No. 4, 1989, pp. 480–484.
187. W. Risau, "Mechanisms of Angiogenesis", *Nature*, Vol. 386, 1997, pp. 671–674.
188. W.L. Lingle, *et al.*, "Centrosome Amplification Drives Chromosomal Instability in Breast Tumor Development", *Proceedings of the National Academy of Sciences*, Vol. 99, No. 4, 2002, pp. 1978–1983.
189. A. Rodrigues-Martins, *et al.*, "From Centriole Biogenesis to Cellular Function: Centrioles are Essential for Cell Division at Critical Developmental Stages", *Cell Cycle*, Vol. 7, No. 1, 2008, pp. 11–16.
190. A. Krämer, K. Neben, and A.D. Ho, "Centrosome Replication, Genomic Instability and Cancer", *Leukemia*, Vol. 16, No. 5, 2002, pp. 767–775.
191. N.J. Ganem, S.A. Godinho, and D. Pellman, "A Mechanism Linking Extra Centrosomes to Chromosomal Instability", *Nature*, Vol. 460, No. 7252, 2009, pp. 278–282.
192. J. Vulprecht, *et al.*, "STIL Is Required for Centriole Duplication in Human Cells", *Journal of Cell Science*, Vol. 125, 2012, pp. 1353–1362.
193. C. Arquint, *et al.*, "Cell-Cycle-Regulated Expression of STIL Controls Centriole Number in Human Cells", *Journal of Cell Science*, Vol. 125, 2012, pp. 1342–1352.
194. J. Lüders and T. Stearns, "Microtubule-Organizing Centres: A Re-Evaluation", *Nature Reviews, Molecular Cell Biology*, Vol. 8, No. 2, 2007, pp. 161–167.
195. P. Strnad and P. Gönczy, "Mechanisms of Procentriole Formation", *Trends in Cell Biology*, Vol. 18, No. 8, 2008, pp. 389–396.
196. H. Rajagopalan and C. Lengauer, "Aneuploidy and Cancer", *Nature*, Vol. 432, 2004, pp. 338–341.
197. M. Kwon, *et al.*, "Mechanisms to Suppress Multipolar Divisions in Cancer Cells with Extra Chromosomes", *Genes & Development*, Vol. 22, No. 16, 2008, pp. 2189–2203.
198. L. Pelletier, *et al.*, "Centriole Assembly in *Caenorhabditis elegans*", *Nature*, Vol. 444, 2006, pp. 619–623.
199. M. Bettencourt-Dias and D. Glover, "Centrosome Biogenesis and Function: Centrosomics Brings New Understanding", *Nature Reviews Molecular Cell Biology*, Vol. 8, 2007, pp. 451–463.

200. G.A. Pihan, *et al.*, "Centrosome Abnormalities and Chromosome Instability Occur Together in Pre-Invasive Carcinomas", *Cancer Research*, Vol. 63, No. 6, 2003, pp. 1398–1404.
201. A.J. Holland, *et al.*, "Polo-like Kinase 4 Kinase Activity Limits Centrosome Overduplication by Autoregulating Its Own Stability", *Journal of Cell Biology*, Vol. 188, No. 2, 2010, pp. 191–198.
202. W.L. Lingle and J.L. Salisbury, "The Role of the Centrosome in the Development of Malignant Tumors", *Current Topics in Developmental Biology*, Vol. 49, 2000, pp. 313–339.
203. T.M. Goepfert, *et al.*, "Centrosome Amplification and Overexpression of Aurora A Are Early Events in Rat Mammary Carcinogenesis", *Cancer Research*, Vol. 62, No. 14, 2002, pp. 4115–4122.
204. S. Doxsey, "The Centrosome—A Tiny Organelle with Big Potential", *Nature Genetics*, Vol. 20, No. 2, 1998, pp. 104–106.
205. J. Marx, "Do Centrosome Abnormalities Lead to Cancer?" *Science*, Vol. 292, No. 5516, 2001, pp. 426–429.
206. F. Wang, *et al.*, "Dual Surface-Functionalized Janus Nanocomposites of Polystyrene/Fe_3O_4@SiO_2 for Simultaneous Tumor Cell Targeting and Stimulus-Induced Drug Release", *Advanced Materials, Communications*, Vol. 25, 2013, pp. 1–5. www.advmat.de.
207. D. Shi, N.M. Bedford, and H.-S. Cho, "Engineered Multifunctional Nanocarriers for Cancer Diagnosis and Therapeutics", *Small*, Vol. 7, No. 18, 2011, pp. 2549–1567.
208. E.K.-H. Chow and D. Ho, "Cancer Nanomedicine: From Drug Delivery to Imaging", *Science Translational Medicine*, Vol. 5, No. 216, 2013, pp. 1–12.
209. P. Pouponneau, *et al.*, "Co-encapsulation of Magnetic Nanoparticles and Doxorubicin into Biodegradable Microcarriers for Deep Tissue Targeting by Vascular MRI Navigation", *Biomaterials*, Vol. 32, No. 13, 2011, pp. 3481–3486.
210. Z. Liu, *et al.*, "Drug Delivery with Carbon Nanotubes for In vivo Cancer Treatment", *Cancer Research*, Vol. 68, 2008, pp. 6652–6660.
211. A. Bianco, K. Kostarelos, and M. Prato, "Applications of Carbon Nanotubes in Drug Delivery", *Current Opinion in Chemical Biology*, Vol. 9, No. 6, 2005, pp. 674–679.
212. G. von Maltzahn, *et al.*, "Nanoparticles that Communicate *in vivo* to Amplify Tumour Targeting", *Nature Materials*, Vol. 10, 2011, pp. 545–552.
213. N.A. Fonseca, *et al.*, "Bridging Cancer Biology and the Patient's needs with Nanotechnology-Based Approaches", *Cancer Treatment Reviews*, Vol. 40, No. 5, 2014, pp. 626–635.

214. A.Z. Wang, R. Langer, and O.C. Farokhzad, "Nanoparticle Delivery of Cancer Drugs", *Annual Review of Medicine*, Vol. 63, 2012, pp. 185–198.
215. L. Hinck and I. Näthke, "Changes in Cell and Tissue Organization in Cancer of the Breast and Colon", *Current Opinion in Cell Biology*, Vol. 26, 2014, pp. 87–95.
216. V. Marthiens, *et al.*, "Centrosome Amplification Causes Microcephaly", *Nature Cell Biology*, Vol. 15, 2013, pp. 731–742.
217. S.I. Ellenbroek, S. Iden, and J.G. Collard, "Cell Polarity Proteins and Cancer", *Seminars in Cancer Biology*, Vol. 22, No. 3, 2012, pp. 208–215.
218. W.J. Nelson, "Adaptation of Core Mechanisms to Generate Cell Polarity", *Nature*, Vol. 422, 2003, pp. 766–774.
219. K. Shin, V.C. Fogg, and B. Margolis, "Tight Junction and Cell Polarity", *Annual Review of Cell Development Biology*, Vol. 22, 2006, pp. 207–235.
220. B. Xue, *et al.*, "Loss of Par3 Promotes Breast Cancer Metastasis by Compromising Cell-Cell Cohesion", *Nature Cell Biology*, Vol. 15, No. 2, 2013, pp. 189–200.
221. A.D. Baffet, *et al.*, "Meeting Report–Building a Centrosome", *Journal of Cell Science*, Vol. 126, 2013, pp. 3259–3262.
222. L.M. McCaffrey, *et al.*, "Loss of the Par3 Polarity Protein Promotes Breast Tumorigenesis and Metastasis", *Cancer Cell*, Vol. 22, No. 5, 2012, pp. 601–614.
223. D. Bilder, M. Li, and N. Perrimon, "Cooperative Regulation of Cell Polarity and Growth by Drosophila Tumor Suppressors", *Science*, Vol. 289, 2000, pp. 113–116.
224. H. Maiato and E. Logarinho, "Mitotic Spindle Multipolarity without Centrosome Amplifications", *Nature Cell Biology*, Vol. 16, No. 5, 2014, pp. 386–394.
225. M.M. Mogensen, *et al.*, "Microtuble Minus-End Anchorage at Centrosomal and Non-Centrosomal Sites: The Role of Ninein", *Journal of Cell Science*, Vol. 113, 2000, pp. 3013–3023.
226. W.L. Lingle and J.L. Salisbury, "Altered Centrosome Structure Is Associated with Abnormal Mitoses in Human Breast Tumors", *American Journal of Pathology*, Vol. 155, No. 6, 1999, pp. 1941–1951.
227. M.D. Rose, S. Biggins, and L.L. Satterwhite, "Unravelling the Tangled Web at the Microtubule-Organizing Center", *Current Opinion in Cell Biology*, Vol. 5, No. 1, 1993, pp. 105–115.
228. Z. Yang, *et al.*, "Extra Centrosomes and/or Chromosomes Prolong Mitosis in Human Cells", *Nature Cell Biology*, Vol. 10, 2008, pp. 748–751.

229. F. Bärenz, D. Mayilo, and O.J. Gruss, "Centriolar Satellites: Busy Orbits Around the Centrosome", *European Journal of Cell Biology*, Vol. 90, No. 12, 2011, pp. 983–989.
230. N.J. Ganem, S.A. Godinho, and D. Pellman, "A Mechanism Linking Extra Centrosomes to Chromosomal Instability", *Nature*, Vol. 460, No. 7252, 2009, pp. 278–282.
231. S.C. Jana, G. Marteil, and M. Bettencourt-Dias, "Mapping Molecules to Structure: Unveiling Secrets of Centriole and Cilia Assembly with Near-Atomic Resolution", *Current Opinion in Cell Biology*, Vol. 26, 2014, pp. 96–106.
232. K.F. Sonnen, *et al.*, "3D-Structured Illumination Microscopy Provides Novel Insight into Architecture of Human Centrosomes", *Biology Open*, Vol. 1, 2012, pp. 965–976.
233. M. Bettencourt-Dias, *et al.*, "Centrosomes and Cilia in Human Disease", *Trends in Genetics*, Vol. 27, No. 8, 2011, pp.307–315.
234. R. Ibrahim, *et al.*, "Electron Tomography Study of Isolated Human Centrioles", *Microscopy Research Technique*, Vol. 72, No. 1, 2009, pp. 42–48.
235. J. Lončarek and A. Khodjakov, "Ab ovo or de nous? Mechanisms of Centriole Duplication", *Molecules and Cells*, Vol. 27, No. 2, 2009, pp. 135–142.
236. P. Gönczy, "Towards a Molecular Architecture of Centriole Assembly", *Nature Reviews Molecular Cell Biology*, Vol. 13, 2012, pp. 425–435.
237. A. Dammermann, *et al.*, "Centriole Assembly Requires Both Centriolar and Pericentriolar Material Proteins", *Developmental Cell*, Vol. 7, No. 6, 2004, pp. 815–829.
238. B. Leber, *et al.*, "Proteins Required for Centrosome Clustering in Cancer Cells", *Science Translational Medicine*, Vol. 2, No. 33, 2010, pp. 1–11.
239. G.A. Pihan, *et al.*, "Centrosome Defects and Genetic Instability in Malignant Tumors", *Cancer Research*, Vol. 58, No. 17, 1998, pp. 3974–3985.
240. W.L. Lingle, *et al.*, "Centrosome Hypertrophy in Human Breast Tumors: Implications for Genomic Stability and Cell Polarity", *Proceedings of the National Academy of Sciences*, Vol. 95, 1998, pp. 2950–2955.
241. M. Bettencourt-Dias, *et al.*, "SAK/PLK4 Is Required for Centriole Duplication and Flagella Development", *Current Biology*, Vol. 15, No. 24, 2005, pp. 2199–2207.
242. I. Cunha-Ferreira, I. Bento, and M. Bettencourt-Dias, "From Zero to Many: Control of Centriole Number in Development and Disease", *Traffic*, Vol. 10, No. 5, 2009, pp. 482–498.

243. K. Fukasawa, *et al.*, "Abnormal Centrosome Amplification in the Absence of P53", *Science*, Vol. 271, 1996, pp. 1744–1747.
244. F. Borel, *et al.*, "Multiple Centrosomes Arise from Tetraploidy Checkpoint Failure and Mitotic Centrosome Clusters in p53 and RB Pocket Protein-Compromised Cells", *Proceedings National Academy of Science*, Vol. 99, No. 15, 2002, pp. 9819–9824.
245. G. Sluder and J.J. Nordberg, "The Good, The Bad and The Ugly: The Practical Consequences of Centrosome Amplification", *Current Opinion in Cell Biology*, Vol. 16, No. 1, 2004, pp. 49–54.
246. P. Dupuis-Williams, *et al.*, "Functional Role of ε-Tubular in the Assembly of the Centriolar Microtubule Scaffold", *Journal of Cell Biology*, Vol. 158, No. 7, 2002, pp. 1183–1193.
247. C. Zou, *et al.*, "Centrobin: A Novel Daughter Centriole-Associated Protein that is Required for Centrole Duplication", *Journal of Cell Biology*, Vol. 171, No. 3, 2005, pp. 437–445.
248. M. Bornens, "Organelle Positioning and Cell Polarity", *Nature Reviews Molecular Cell Biology*, Vol. 9, 2008, pp. 874–886.
249. K. Fukasawa, "Oncogenes and Tumour Suppressors Take on Centrosomes", *Nature Reviews Cancer*, Vol. 7, 2007, pp. 911–924.
250. N.J. Quintyne, *et al.*, "Spindle Multipolarity Is Prevented by Centrosomal Clustering", *Science*, Vol. 307, No. 5706, 2005, pp. 127–129.
251. H. Dodson, *et al.*, "Centrosome Amplification Induced by DNA Damage Occurs During a Prolonged G2 Phase and Involves ATM", *The EMBO Journal*, Vol. 23, No. 19, 2004, pp. 3864–3873.
252. K. Shinmura, *et al.*, "Direct Evidence for the Role of Centrosomally Localized p53 in the Regulation of Centrosome Duplication", *Oncogene*, Vol. 26, No. 20, 2007, pp. 2939–2944.
253. P. Tarapore and K. Fukasawa, "Loss of p53 and Centrosome Hyperamplification", *Oncogene*, Vol. 21, No. 40, 2002, pp. 6234–6240.
254. M.F. Tsou and T. Stearns, "Controlling Centrosome Number: Licenses and Blocks", *Current Opinion in Cell Biology*, Vol. 18, No. 1, 2006, pp. 74–78.
255. S. Leidel and P. Gönczy, "Centrosome Duplication and Nematodes: Recent Insights from an Old Relationship", *Developmental Cell*, Vol. 9, No. 3, 2005, pp. 317–325.
256. M. Delattre and P. Gönczy, "The Arithmetic of Centrosome Biogenesis", *Journal of Cell Science*, Vol. 117, 2004, pp. 1619–1630.
257. B.R. Brinkley, "Managing the Centrosome Numbers Game: From Chaos to Stability in Cancer Cell Division", *Trends in Cell Biology*, Vol. 11, No. 1, 2001, pp. 18–21.

258. K. Fukasawa, "Centrosome Amplification, Chromosome Instability and Cancer Development", *Cancer Letters*, Vol. 230, 2005, pp. 6–19.
259. C. Wong and T. Stearns, "Centrosome Number Is Controlled by a Centrosome-Intrinsic Block to Reduplication", *Nature Cell Biology*, Vol. 5, 2003, pp. 539–544.
260. T. Nakajima, *et al.*, "Centrosome Aberration Accompanied with p53 Mutation can Induce Genetic Instability in Hepatocellular Carcinoma", *Modern Pathology*, Vol. 17, No. 6, 2004, pp. 722–727.
261. N. Sato, *et al.*, "Radiation-Induced Centrosome Overduplication and Multiple Mitotic Spindles in Human Tumor Cells", *Experimental Cell Research*, Vol. 255, No. 2, 2000, pp. 321–326.
262. K. Kim, *et al.*, "Ultrahigh-Speed Rotating Nanoelectromechanical System Devices Assembled from Nanoscale Building Blocks", *Nature Communications*, Vol. 15, 2014, p. 3632.
263. W. Marshall, "Centrioles Take Center Stage", *Current Biology*, Vol. 11, 2001, pp. R487–R496.
264. J.Y. Chan, "A Clinical Overview of Centrosome Amplification in Human Cancers", *International Journal of Biological Sciences*, Vol. 7, No. 8, 2011, pp. 1122–1144.
265. K. Al-Romaih, *et al.*, "Chromosomal Instability in Osteosarcoma and its Association with Centrosome Abnormalities", *Cancer Genetics and Cytogenetics*, Vol. 144, No. 2, 2003, pp. 91–99.
266. F. Jiang, *et al.*, "Centrosomal Abnormality Is Common in and a Potential Biomarker for Bladder Cancer", *International Journal of Cancer*, Vol. 106, No. 5, 2003, pp. 661–665.
267. K. Kawamura, *et al.*, "Chromosome Hyperamplification and Chromosomal Instability in Bladder Cancer", *European Urology*, Vol. 43, No. 5, 2003, pp. 505–515.
268. A.B. D'Assoro, *et al.*, "Amplified Centrosomes in Breast Cancer: A Potential Indicator of Tumor Aggressiveness", *Breast Cancer Research and Treatment*, Vol. 75, 2002, pp. 25–34.
269. K. Kawamura, *et al.*, "Induction of Centrosome Amplification and Chromosome Instability in Human Bladder Cancer Cells by p53 Mutation and Cyclin E Overexpression", *Cancer Research*, Vol. 64, 2004, pp. 4800–4809.
270. C.W. Brownlee and G.C. Rogers, "Show Me Your License, Please: Deregulation of Centriole Duplication Mechanisms that Promote Amplification", *Cellular and Molecular Life Sciences*, Vol. 70, No. 6, 2013, pp. 1021–1034.
271. V. Marthiens, M. Piel, and R. Basto, "Never Tear us Apart—The Importance of Centrosome Clustering", *Journal of Cell Science*, Vol. 125, 2012, pp. 3281–3292.

272. A.J. Holland and D.W. Cleveland, "Boveri Revisited: Chromosomal Instability, Aneuploidy and Tumorigenesis", *Nature Reviews Molecular Cell Biology*, Vol. 10, 2009, pp. 478–487.
273. M. Bettencourt-Dias and D.M. Glover, "Centrosome Biogenesis and Function: Centrosomics Brings New Understanding", *Nature Reviews Molecular Cell Biology*, Vol. 8, 2007, pp. 451–463.
274. P. Tarapore, *et al.*, "Direct Regulation of the Centrosome Duplication Cycle by the p53-p21Waf1/Cip1 Pathway", *Oncogene*, Vol. 20, 2001, pp. 3173–3184.
275. F. Bunz, *et al.*, "Targeted Inactivation of p53 in Human Cells Does Not Result in Aneuploidy", *Cancer Research*, Vol. 62, No. 4, 2002, pp. 1129
276. K. Fukasawa, "Introduction", *Oncogene*, Vol. 21, 2002, pp. 6140–6145.
277. W. Saunders, "Centrosomal Amplification and Spindle Multipolarity in Cancer Cells", *Seminars in Cancer Biology*, Vol. 15, 2005, pp. 25–32.
278. K.R. Lacey, P.K. Jackson, and T. Stearns, "Cyclin-Dependent Kinase Control of Centrosome Duplication", *Proceedings National Academy of Sciences*, Vol. 96, 1999, pp. 2817–2822.
279. J.H. Park, *et al.*, "Magnetic Iron Oxide Nanoworms for Tumor Targeting and Imaging", *Advanced Materials*, Vol. 20, 2008, pp. 1630–1635.
280. N. Sato, *et al.*, "A Possible Role for Centrosome Overduplication in Radiation-Induced Cell Death", *Oncogene*, Vol. 19, 2000, pp. 5281–5290.
281. D. Gisselsson, *et al.*, "Centrosomal Abnormalities, Multipolar Mitoses, and Chromosomal Instability in Head and Neck Tumours with Dysfunctional Telomeres", *British Journal of Cancer*, Vol. 87, 2002, pp. 202–207.
282. P. Meraldi, R. Honda, and E.A. Nigg, "Aurora-A Overexpression Reveals Tetraploidization as a Major Route to Centrosome Amplification in p53-/-cells", *The EMBO Journal*, Vol. 21, 2002, pp. 483–492.
283. C. Kasbek, *et al.*, "Preventing the Degradation of Mps1 at Centrosomes Is Sufficient to Cause Centrosome Reduplication in Human Cells", *Molecular Biology of the Cell*, Vol. 18, 2007, pp. 4457–4469.
284. J. Lončarek, P. Herget, and A. Khodjakov, "Centeriole Reduplication during Prolonged Interphase Requires Procentriole Maturation Governed by Plk1", *Current Biology*, Vol. 20, 2010, pp. 1277–1282.
285. M-F.B. Tsou, *et al.*, "Polo Kinase and Separase Regulate the Mitotic Licensing of Centriole Duplication in Human Cells, *Developmental Cell*, Vol. 17, 2009, pp. 344–354.
286. J. Lončarek, *et al.*, "Control of Daughter Centriole Formation by the Pericentriolar Material", *Nature: Cell Biology*, Vol. 10, No. 3, 2008, pp. 322–328.

287. A. Rodrigues-Martins, *et al.*, "Revisiting the Role of the Mother Centriole in Centriole Biogenesis", *Science*, Vol. 316, No. 5827, 2007, pp. 1046–1050.
288. Y. Uetake, *et al.*, "Cell Cycle Progression and de novo Centriole Assembly after Centrosomal Removal in Untransformed Human Cells", *Journal of Cell Biology*, Vol. 176, 2007, pp. 173–182.
289. D. Hanahan and R.A. Weinberg, "The Hallmarks of Cancer", *Cell*, Vol. 100, 2000, pp. 57–70.
290. D. Hanahan and R.A. Weinberg, "Hallmarks of Cancer: The Next Generation", *Cell*, Vol. 144, 2011, pp. 646–674.
291. M.F. Tsou and T. Stearns, "Mechanism Limiting Centrosome Duplication to Once Per Cell Cycle", *Nature*, Vol. 442, No. 7105, 2006, pp. 947–951.
292. K. Strebhardt and A. Ullrich, "Targeting Polo-like Kinase 1 for Cancer Therapy", *Nature Reviews Cancer*, Vol. 6, 2006, pp. 321–330.
293. T. Mayor, *et al.*, "The Centrosomal Protein C-Nap1 Is Required for Cell Cycle-regulated Centrosome Cohesion", *Journal of Cell Biology*, Vol. 151, 2000, pp. 837–846.
294. M. Petronczki, *et al.*, "Polo-like Kinase 1 Triggers the Initiation of Cytokinesis in Human Cells by Promoting Recruitment of the RhoGEF Ect2 to the Central Spindle", *Developmental Cell*, Vol. 12, 2007, pp. 713–725.
295. F.A. Barr, H.H.W. Sillјé, and E.A. Nigg, "Polo-like Kinases and the Orchestration of Cell Division", *Nature Reviews Molecular Cell Biology*, Vol. 5, 2004, pp. 429–441.
296. E.H. Hinchcliffe and G. Sluder, "'It Takes Two to Tango': Understanding How Centrosome Duplication Is Regulated Throughout the Cell Cycle", *Genes & Development*, Vol. 15, 2001, pp. 1167–1181.
297. S. Doxsey, "Duplicating Dangerously: Linking Centrosome Duplication and Aneuploidy", *Molecular Cell*, Vol. 10, 2002, pp. 439–440.
298. H.A. Fisk, C.P. Mattison, and M. Winey, "Human Mps1 Protein Kinase Is required for Centrosome Duplication and Normal Mitotic Progression", *Proceedings of the National Academy of Sciences*, Vol. 100, 203, pp. 14875–14880.
299. H.I. Saavedra, *et al.*, "Inactivation of E2F3 Results in Centrosome Amplification", *Cancer Cell*, Vol. 3, 2003, pp. 333–346.
300. V.M. Stucke, *et al.*, "Human Mps1 Kinase Is Required for the Spindle Assembly Checkpoint but not for Centrosome Duplication", *The EMBO Journal*, Vol. 21, 2002, pp. 1723–1732.
301. S. Awakumova, *et al.*, "Biotechnological Approaches Toward Nanoparticle Biofunctionalization", *Trends in Biotechnology*, Vol. 32, 2014, pp. 11–20.

302. S.D. Li and L. Huang, "Stealth Nanoparticles: High Density but Sheddable PEG Is a Key for Tumor Targeting", *Journal of Controlled Release*, Vol. 145, 2010, pp. 178–181.
303. J. Chomoucka, *et al.*, "Magnetic Nanoparticles and Targeted Drug Delivery", *Pharmacological Research*, Vol. 62, No. 2, 2010, pp. 144–149.
304. M.E. Davis, Z. Chen, and D.M. Shin, "Nanoparticle Therapeutics: An Emerging Treatment Modality for Cancer", *Nature Reviews Drug Discovery*, Vol. 7, 2008, pp. 771–782.
305. V. Mailänder and K. Landfester, "Interaction of Nanoparticles with Cells", *Biomacromolecules*, Vol. 10, 2009, pp. 2379–2400.
306. X.-H. Peng, *et al.*, "Targeted Magnetic Iron Oxide Nanoparticles for Tumor Imaging and Therapy", *International Journal of Nanomedicine*, Vol. 3, 2008, pp. 311–321.
307. R. Zhang, *et al.*, "Enhancement Effect of Nano Fe_3O_4 to the Drug Accumulation of Doxorubicin in Cancer Cells", *Material Science and Engineering C*, Vol. 29, 2009, pp. 1697–1701.
308. M. Arruebo, *et al.*, "Magnetic Nanoparticles for Drug Delivery", *Nanotoday*, Vol. 2, 2007, pp. 22–32.
309. A.H. Faraji and P. Wipf, "Nanoparticles in Cellular Drug Delivery", *Bioorganic & Medicinal Chemistry*, Vol. 17, 2009, pp. 2950–2962.
310. A. Schroeder, *et al.*, "Treating Metastatic Cancer with Nanotechnology", *Nature Reviews Cancer*, Vol. 12, 2012, pp. 39–50.
311. S. Nie, *et al.*, "Nanotechnology Applications in Cancer", *Annual Review of Biomedical Engineering*, Vol. 9, 2007, pp. 257–288.
312. S. Sengupta and R. Sasisekharan, "Exploiting Nanotechnology to Target Cancer", *British Journal of Cancer*, Vol. 96, 2007, pp. 1315–1319.
313. R.H. Farnsworth, *et al.*, "Vascular Remodeling in Cancer", *Oncogene*, Vol. 33, 2014, pp. 3496–3505.
314. I. Brigger, C. Dubernet, and P. Couvreur, "Nanoparticles in Cancer Therapy and Diagnosis", *Advanced Drug Delivery Reviews*, Vol. 54, 2002, pp. 631–651.
315. P. Carmeliet and R.K. Jain, "Angiogenesis in Cancer and Other Diseases", *Nature*, Vol. 407, No. 6801, 2000, pp. 249–257.
316. S.M. Maghimi, A.C. Hunter, and J.C. Murray, "Nanomedicine: Current Status and Future Prospects", *The FASEB Journal–Review*, Vol. 19, 2005, pp. 311–330.
317. G. Bergers and D. Hanahan, "Modes of Resistance to Anti-Angiogenic Therapy", *Nature Reviews Cancer*, Vol. 8, 2008, pp. 592–603.

318. A.G. Cuenca, *et al.*, "Emerging Implications of Nanotechnology on Cancer Diagnostics and Therapeutics", *Cancer*, Vol. 107, 2006, pp. 459–466.
319. L.L. Munn, "Aberrant Vascular Architecture in Tumors and its Importance in Drug-Based Therapies", *Drug Discovery Today*, Vol. 8, 2003, pp. 396–403.
320. N.W.S. Kam, *et al.*, "Carbon Nanotubes as Multifunctional Biological Transporters and Near-Infrared Agents for Selective Cancer Cell Destruction", *Proceedings of the National Academy of Sciences (PNAS)*, Vol. 102, 2005, pp. 11600–11605.
321. A.S. Lübbe, *et al.*, "Clinical Experiences with Magnetic Drug Targeting: A Phase I Study with 4'-Epidoxorubicin in 14 Patients with Advanced Solid Tumors", *Cancer Research*, Vol. 56, 1996, pp. 4686–4693.
322. D.P. O'Neal, *et al.*, "Photo-Thermal Tumor Ablation in Mice using near Infrared-Absorbing Nanoparticles", *Cancer Letters*, Vol. 209, 2004, pp. 171–176.
323. M. Ferrari, "Cancer Nanotechnology: Opportunities and Challenges", *Nature Reviews Cancer*, Vol. 5, 2005, pp. 161–171.
324. S. Azarmi, W.H. Roa, and R. Löbenberg, "Targeted Delivery of Nanoparticles for the Treatment of Lung Diseases", *Advanced Drug Delivery Reviews*, Vol. 60, 2008, pp. 863–875.
325. M.H. Song, *et al.*, "Centrioles: Some Self-Assembly Required", *Current Opinions in Cell Biology*, Vol, 20, 2008, pp. 688–693.
326. A. Rodrigues-Martins, *et al.*, "Revisiting the Role of the Mother Centriole in Centriole Biogenesis", *Science*, Vol. 316, No. 5827, 2007, pp. 1046–1050.
327. M. Lee and V. Vasioukhin, "Cell Polarity and Cancer–Cell and Tissue Polarity as a Non-Canonical Tumor Suppressor", *Journal of Cell Science*, Vol. 121, 2008, pp. 1141–1150.
328. V. Mennella, *et al.*, "Amorphous No More: Subdiffraction View of the Pericentriolar Material Architecture", *Trends in Cell Biology*, Vol. 24, No. 3, 2014, pp. 188–197.
329. M. Bettencourt-Dias, *et al.*, "Centrosomes and Cilia in Human Disease", *Trends in Genetics*, Vol. 27, No. 8, 2011, pp. 307–315.
330. V. Aranda, M.E. Nolan, and S.K. Muthuswamy, "Par Complex in Cancer: A Regulator of Normal Cell Polarity Joins the Dark Side" *Oncogene*, Vol. 27, 2008, pp. 6878–6887.
331. M. Bettencourt-Dias and D.M. Glover, "Centrosome Biogenesis and Function: Centrosomics Brings New Understanding", *Nature Reviews Molecular Cell Biology*, Vol. 8, 2007, pp. 451–463.

332. G. Kohlmaier, *et al.*, "Overly Long Centrioles and Defective Cell Division upon Excess of the SAS-4-Related Protein CPAP", *Current Biology*, Vol. 19, 2009, pp. 1012–1018.
333. M. Kirkham, *et al.*, "SAS-4 Is a *C. elegens* Centriolar Protein that Controls Centrosome Size", *Cell*, Vol. 112, 2003, pp. 575–587.
334. S. Biswas and V.P. Torchilin, "Nanopreparations for Organelle-Specific Delivery in Cancer", *Advanced Drug Delivery Reviews*, Vol. 66, 2014, pp. 26–41.
335. N. Bertrand, *et al.*, "Cancer Nanotechnology: The Impact of Passive and Active Targeting in the Era of Modern Cancer Biology", *Advanced Drug Delivery Reviews*, Vol. 66, 2014, pp. 2–25.
336. V.J. Venditto and F.C. Szoka Jr., "Cancer Nanomedicines: So Many Papers and So Few Drugs!" *Advanced Drug Delivery Reviews*, Vol. 65, 2013, pp. 80–88.
337. M.G. Damiano, *et al.*, "Templated High Density Lipoprotein Nanoparticles as Potential Therapies and for Molecular Delivery", *Advanced Drug Delivery Reviews*, Vol. 65, 2013, pp. 649–662.
338. C. He, *et al.*, "Effects of Particle Size and Surface Charge on Cellular Uptake and Biodistribution of Polymeric Nanoparticles", *Biomaterials*, Vol. 31, 2010, pp. 3657–3666.
339. K. Xias, *et al.*, "The Effect of Surface Charge on In Vivo Biodistribution of PEG-Oligocholic Acid based Micellar Nanoparticles", *Biomaterials*, Vol. 32, 2011, pp. 3435–3446.
340. O. Veiseh, J.W. Gunn, and M. Zhang, "Design and Fabrication of Magnetic Nanoparticles for Targeted Drug Delivery and Imaging", *Advanced Drug Delivery Reviews*, Vol. 62, 2010, pp. 284–304.
341. T. Lammers, W.E. Hennink, and G. Storm, "Tumour-Targeted Nanomedicines: Principles and Practice", *British Journal of Cancer*, Vol. 99, 2008, pp. 392–397.
342. M.E. Davis, Z. Chen, and D.M. Shin, "Nanoparticle Therapeutics: An Emerging Treatment Modality for Cancer", *Nature Reviews Drug Discovery*, Vol. 7, 2008, pp. 771–782.
343. T. Sasaki, *et al.*, "Magnetic Nanoparticles for Improving Cell Invasion in Tissue Engineering", *Journal of Biomedical Materials Research, Part A*, Vol. 86, 2007, pp. 969–979.
344. K. Cho, *et al.*, "Therapeutic Nanoparticles for Drug Delivery in Cancer", *Clinical Cancer Research*, Vol. 14, 2008, pp. 1310–1316.
345. T. Osaka, *et al.*, "Effect of Surface Charge of Magnetite Nanoparticles on Their Internalization into Breast Cancer and Umbilical Vein Endothelial Cells", *Colloids and Surfaces B: Biointerfaces*, Vol. 71, 2009, pp. 325–330.

346. C. Sun, J.S. Lee, and M. Zhang, "Magnetic Nanoparticles in MR Imaging and Drug Delivery", *Advanced Drug Delivery Reviews*, Vol. 60, 2008, pp. 1252–1265.
347. J. Dobson, "Magnetic Nanoparticles for Drug Delivery", *Drug Development Research*, Vol. 67, 2006, pp. 55–60.
348. J.H. Park, *et al.*, "Systematic Surface Engineering of Magnetic Nanoworms for In Vivo Tumor Targeting", *Small*, Vol. 5, 2009b, pp. 694–700.
349. N. Sanvicens and M.P. Marco, "Multifunctional Nanoparticles—Properties and Prospects for the use in Human Medicine", *Trends in Biotechnology*, Vol. 26, No. 8, 2008, pp. 425–433.
350. D. Peer, *et al.*, "Nanocarriers as an Emerging Platform for Cancer Therapy", *Nature Nanotechnology*, Vol. 2, 2007, pp. 751–760. doi:10.1038/nnano.2007.387
351. O. Kučera and M. Cifra, "Cell-to-Cell Signaling through Light: Just a Ghost of Chance?", *Cell Communication and Signaling*, Vol. 11, 2013, pp. 1–8.
352. F. Scholkmann, D. Fels, and M. Cifra, "Non-Chemical and Non-Contact Cell-to-Cell Communication: A Short Review", *American Journal of Translational Research*, Vol. 5, No. 6, 2013, pp. 586–593.
353. D. Havelka, *et al.*, "High-Frequency Electric Field and Radiation Characteristics of Cellular Microtubule Network", *Journal of Theoretical Biology*, Vol. 286, 2011, pp. 31–40.
354. J. Pokorný, J. Pokorný, and J. Kobilková, "Postulates on Electromagnetic Activity in Biological Systems and Cancer", *Integrative Biology*, Vol. 5, No. 12, 2013, pp. 1439–1446.
355. J. Pokorný, "Physical Aspects of Biological Activity and Cancer", *AIP Advances*, Vol. 2, 2012, pp. 011207-1–011207-11.
356. J. Pokorný, *et al.*, "Vibration in Microtubules", *Journal of Biological Physics*, Vol. 23, 1997, pp. 171–179.
357. M.M. Shawki and A. Farid, "Low Electric Field Parameters Required to Induce Death of Cancer Cells", *Electromagnetic Biology & Medicine*, Vol. 33, No. 2, 2014, pp. 159–163.
358. H. Berg, *et al.*, "Bioelectromagnetic Field Effects on Cancer Cells and Mice Tumors", *Electromagnetic Biology &* Medicine, Vol. 29, No. 4, 2010, pp. 132–143.
359. M. Cifra, J.Z. Fields, and A. Farhadi, "Electromagnetic Cellular Interactions", *Progress in Biophysics and Molecular Biology*, Vol. 105, 2011, pp. 223–246.
360. J. Pokorný, "Biophysical Cancer Transformation Pathway", *Electromagnetic Biology and Medicine*, Vol. 28, No. 2, 2009, pp. 105–123.

361. O. Kučera, M. Cifra, and J. Pokorný, "Technical Aspects of Measurement of Cellular Electromagnetic Activity", *European Biophysics Journal*, Vol. 39, 2010, pp. 1465–1470.
362. K. Cheng and C. Zou, "Physics Models of Centriole Replication", *Medical Hypotheses*, Vol. 67, No. 3, 2006, pp.572–577.
363. G. Sluder, F.J. Miller, and C.L. Rieder, "The Reproduction of Centrosomes: Nuclear versus Cytoplasmic Controls", *The Journal of Cell Biology*, Vol. 103, 1986, pp. 1873–1881.
364. K. Cheng and C. Zou, "Theoretical Models of Cytokinesis and Mitosis (Without Asters)", *American Journal of Plant Physiology*, Vol. 1, No. 1, 2006, pp. 21–27.
365. T.C. Barnes, "Healing Rate of Human Skin Determined by Measurement of the Electrical Potential of Experimental Abrasions: A Study of Treatment with Petrolatum and with Petrolatum Containing Yeast and Liver Extracts", *The American Journal of Surgery*, Vol. 69, 1945, pp. 82–88.
366. D.S. Weiss, R. Kirsner, and W.H. Eaglstein, "Electrical Stimulation and Wound Healing", *Archives of Dermatology*, Vol. 126, No. 2, 1990, pp. 222–225.
367. A.M. Fry, "Solving the Centriole Disengagement Puzzle", *Nature Cell Biology*, Vol. 17, No. 1, 2015, pp. 3–5.
368. H.-J. Nam and J.M. van Deursen, "Cyclin B2 and p53 Control Proper Timing of Centrosome Separation", *Nature Cell Biology*, Vol. 16, No. 6, 2014, pp. 538–549.
369. M. Gottardo, G. Callaini, and M.G. Riparbelli, "Procentriole Assembly without Centriole Disengagement—A Paradox of Male Gametogenesis", *Journal of Cell Science*, Vol. 127, 2014, pp. 3434–3439.
370. G.J. Brouhard and L.M. Rice, "The Contribution of αβ-Tubulin Curvature to Microtubule Dynamics", *Journal of Cell Biology*, Vol. 207, No. 3, 2014, pp. 323–334.
371. M.W. Elting, *et al.*, "Force on Spindle Microtubule Minus Ends Moves Chromosomes", *Journal of Cell Biology*, Vol. 206, No. 2, 2014, pp. 245–256.
372. J. Zhang, W.-H. Guo, and Y.-L. Wang, "Microtubules Stabilize Cell Polarity by Localizing Rear Signals", *Proceedings National Academy of Sciences*, Vol. 111, No. 462014, 2014, pp. 16383–16388.
373. R. Scherz-Shouval, *et al.*, "The Reprogramming of Tumor Stroma by HSF1 is a Potent Enabler of Malignancy", *Cell*, Vol. 158, No. 3, 2014, pp. 564–578.
374. M.J. Duffy, *et al.*, "p53 as a Target for the Treatment of Cancer", *Cancer Treatment Reviews*, Vol. 40, 2014, pp. 1153–1160.

375. D. Kong, *et al.*, "Centriole Maturation Requires Regulated Plk1 Activity during Two Consecutive Cell Cycles", *Journal of Cell Biology*, Vol. 206, 2014, pp. 855–865.
376. C. Janke, "The Tubulin Code: Molecular Components, Readout Mechanisms, and Functions", *Journal of Cell Biology*, Vol. 206, No. 4, 2014, pp. 461–472.
377. R.O. Becker and J.A. Spadaro, "Electrical Stimulation of Partial Limb Regeneration in Mammals", *Bulletin of the New York Academy of Medicine*, Vol. 48, No. 4, 1972, pp. 627–641.
378. I.A. Vorobjev and Y.S. Chentsov, "Centrioles in the Cell Cycle. 1. Epithelial Cells", *Journal of Cell Biology*, Vol. 93, No. 3, 1982, pp. 938–949.
379. M. Paintrand, *et al.*, "Centrosome Organization and Centriole Architecture: Their Sensitivity to Divalent Cations", *Journal of Structural Biology*, Vol. 108, No. 2, 1992, pp. 107–128.
380. D. Sabino, *et al.*, "Moesin Is a Major Regulator of Centrosome Behavior in Epithelial Cells with Extra Centrosomes", *Current Biology*, Vol. 25, No. 7, 2015, pp. 879–889.
381. L. Hamon, P.A. Curmi, and D. Pastré, "High-Resolution Imaging of Microtubules and Cytoskeleton Structures by Atomic Force Microscopy", *Methods in Cell Biology*, Vol. 95, 2010, pp. 157–174.
382. E.N. Marieb, *Human Anatomy and Physiology*, 3rd Edition, Benjamin/Cummings Publishing, Redwood City, CA, 1991, Chapter 3, pp. 60–101.
383. D. Guest, *Biology Smart*, The Princeton Review, Random House, New York, 1996, pp. 133–135.
384. J.E. Klebba, *et al.*, "Autoinhibition and Relief Mechanism for Polo-like Kinase 4", *Proceedings of the Academy of Sciences*, Vol. 112, No. 7, 2015, pp. E657–E666.
385. J.E. Klebba, *et al.*, "Two Polo-like Kinase 4 Binding Domains in Asterless Perform Distinct Roles in Regulating Kinase Stability", *Journal of Cell Biology*, Vol. 208, No. 4, 2015, pp. 401–414.
386. K.F. Sonnen, *et al.*, "Human Cep192 and Cep152 Cooperate in Plk4 Recruitment and Centriole Duplication", *Journal of Cell Science*, Vol. 126, 2014, pp. 3223–3233.
387. J.E. Sillibourne, *et al.*, "Autophosphorylation of Polo-like Kinase 4 and Its Role in Centriole Duplication", *Molecular Biology of The Cell*, Vol. 21, No. 4, 2010, pp. 547–561.
388. N.S. Dzhindzhev, *et al.*, "Asterless Is a Scaffold for the Onset of Centriole Assembly", *Nature*, Vol. 457, 2010, pp. 714–718.

389. O. Cizmecioglu, *et al.*, "CEP152 Acts as a Scaffold for Recruitment of Plk4 and CPAP to the Centrosome", *Journal of Cell Biology*, Vol. 191, No. 4, 2010, pp. 731–739.
390. A. Duensing, *et al.*, "Centrosome Overduplication Chromosomal Instability, and Human Papillomavirus Oncoproteins", *Environmental and Molecular Mutagenesis*, Vol. 40, 2009, pp. 741–747.
391. C.G. Pearson and M. Winey, "Plk4/SAK/ZYG-1 in the Regulation of Centriole Duplication", *F1000 Biology Reports*, Vol. 2, 2010, p. 58.
392. Z. Carvalho-Santos, *et al.*, "Stepwise Evolution of the Centriole-Assembly Pathway", *Journal of Cell Science*, Vol. 123, 2010, pp. 1414–1426.
393. L.K. Slevin, *et al.*, "The Structure of the Plk4 Cryptic Polo Box Reveals Two Tandem Polo Boxes Required for Centriole Duplication", *Structure*, Vol. 20, No. 11, 2012, pp. 1905–1917.
394. S. Blachon, *et al.*, "Drosophila Asterless and Vertebrate CEP152 Are Orthologs Essential for Centriole Duplication", *Genetics*, Vol. 180, No. 4, 2008, pp. 2081–2094.
395. S.C. Jana, J.F. Bazan, and M. Bettencourt-Dias, "Polo Boxes Come Out of the Crypt: A New View of PLK Function and Evolution", *Structure*, Vol. 20, No. 11, 2012, pp. 1801–1804.
396. D.M. Lowery, D. Lim, and M.B. Yaffe, "Structure and Function of Polo-like Kinases", *Oncogene*, Vol. 24, No. 2, 2005, pp. 248–259.
397. G.C. Leung, *et al.*, "The SAK Polo-Box Comprises a Structural Domain Sufficient for Mitotic Subcellular Localization", *Nature Structural Biology*, Vol. 9, No. 10, 2002, pp. 719–724.
398. I. Cunha-Ferreira, *et al.*, "Regulation of Autophosphorylation Controls PLK4 Self-Destruction and Centriole Number", *Current Biology*, Vol. 23, No. 22, 2013, pp. 2245–2254.
399. J.E. Sillibourne and M. Bornens, "Polo-like Kinase 4: The Odd One Out of the Family", *Cell Division*, Vol. 5, 2010, p. 25.
400. A.J. Holland, *et al.*, "The Autoregulated Instability of Polo-like Kinase 4 Limits Centrosome Duplication to Once Per Cell Cycle", *Genes and Development*, Vol. 26, No. 24, 2012, pp. 2684–2689.
401. H.-J. Nam, R.M. Naylor, and J.M. van Deursen, "Centrosome Dynamics as a Source of Chromosomal Instability", *Trends in Cell Biology*, Vol. 25, No. 2, 2015, pp. 65–73.
402. C. Wiese and Y. Zheng, "Microtubule Nucleation: ϒ-Tubulin and Beyond", *Journal of Cell Science*, Vol. 119, 2006, pp. 4143–4153.
403. M.E. Tanenbaum and R.H. Medema, "Mechanisms of Centrosome Separation and Bipolar Spindle Assembly", *Developmental Cell*, Vol. 19, 2010, pp. 797–806.

404. A.J. Holland and D.W. Cleveland, "Losing Balance: The Origin and Impact of Aneuploidy in Cancer", *EMBO Reports*, Vol. 13, No. 6, 2012, pp. 501–514.
405. F. Eckerdt, *et al.*, "Identification of a Polo-like Kinase 4-Dependent Pathway for De Novo Centriole Formation", *Current Biology*, Vol. 21, No. 5, 2011, pp. 428–432.
406. M. Hiraki, *et al.*, "Bld10p Constitutes the Cartwheel-Spoke Tip and Stabilizes the 9-Fold Symmetry of the Centriole", *Current Biology*, Vol. 17, No. 20, 2007, pp. 1776–1783.
407. N. Korzeniewski, *et al.*, "Cullin 1 Functions as a Centrosomal Suppressor of Centriole Multiplication by Regulating Polo-like Kinase 4 Protein Levels", *Cancer Research*, Vol. 69, No. 16, 2009, pp.6668–6675.
408. I. Cunha-Ferreira, *et al.*, "The SCF/Slimb Ubiquitin Ligase Limits Centrosome Amplification through Degradation of SAK/PLK4", *Current Biology*, Vol. 19, 2009, pp. 43–49.
409. A. Duensing and S. Duensing, "Guilt by Association? P53 and the Development of Aneuploidy in Cancer", *Biochemical and Biophysical Research Communications*, Vol. 331, No. 3, 2005, pp. 694–700.
410. B.C.M. van de Weerdt and R.H. Medema, "Polo-like Kinases: A Team in Control of the Division", *Cell Cycle*, Vol. 5, No. 8, 2006, pp. 853–864.
411. N. Peel, *et al.*, "Overexpressing Centriole-Replication Proteins In Vivo Induces Centriole Overduplication and De Novo Formation", *Current Biology*, Vol. 17, 2007, pp. 834–843.
412. J.E. Klebba, *et al.*, "Polo-like Kinase 4 Autodestructs by Generating Its Slimb-Binding Phosphodegron", *Current Biology*, Vol. 23, 2013, pp. 2255–2261.
413. M. Piel, *et al.*, "Centrosome-Dependent Exit of Cytokinesis in Animal Cells", *Science*, Vol. 291, 2001, pp. 1550–1553.
414. M. Piel, *et al.*, "The Respective Contribution of the Mother and Daughter Centrioles to Centrosome Activity and Behavior in Vertebrate Cells", *Journal of Cell Biology*, Vol. 149, No. 2, 2000, pp. 317–330.
415. P. Chang, *et al.*, "ε-Tubulin Is Required for Centriole Duplication and Microtubule Organization", *Nature Cell Biology*, Vol. 5, 2003, pp. 71–76.
416. M. Moritz and D.A. Agard, "ϒ-Tubulin Complexes and Microtubule Nucleation", *Current Opinion in Structural Biology*, Vol. 11, 2001, pp. 174–181.
417. E. Nogales, "Structural Insights into Microtubule Function", *Annual Review of Biochemistry*, Vol. 69, 2000, pp. 277–302.
418. P. Chang and T. Stearns, "δ-Tubulin and ε-Tubulin: Two New Human Centrosomal Tubulins Reveal New Aspects of Centrosome Structure and Function", *Nature Cell Biology*, Vol. 2, 2000, pp. 30–35.

419. E. Schiebel, "ϒ-Tubulin Complexes: Binding to the Centrosome, Regulation and Microtubule Nucleation", *Current Opinion in Cell Biology*, Vol. 12, 2000, pp. 113–118.
420. S.J. Doxsey, "Centrosomes as Command Centres for Cellular Control", *Nature Cell Biology*, Vol. 3, No. 5, 2001, pp. E105–E108.
421. J. Lončarek and A. Khodjakov, "Ab ovo or de nous? Mechanisms of Centriole Duplication", *Molecules and Cells*, Vol. 27, No. 2, 2009, pp. 135–142.
422. T.-C. Lin, A. Neuner, and E. Schiebel, "Targeting of ϒ-Tubulin Complexes to Microtubule Organizing Centers: Conservation and Divergence", *Trends in Cell Biology*, Vol. 25, No. 5, 2015, pp. 296–307.
423. S. Kasas, *et al.*, "Oscillation Modes of Microtubules", *Biology of the Cell*, Vol. 96, No. 9, 2004, pp. 697–700.
424. N.A. Baker, *et al.*, "Electrostatics of Nanosystems: Application to Microtubules and the Ribosome", *Proceedings of the National Academy of Sciences of the United States of America*, Vol. 98, 2001, pp. 10037–10041.
425. C.Y. Wang and L.C. Zhang, "Circumferential Vibration of Microtubules with Long Axial Wavelength", *Journal of Biomechanics*, Vol. 41, 2008, pp. 1892–1896.
426. R. Goodman and M. Blank, "Insights into Electromagnetic Interaction Mechanisms", *Journal of Cellular Physiology*, Vol. 192, 2002, pp. 16–22.
427. Y.M. Sirenko, M.A. Stroscio, and K.W. Kim, "Elastic Vibrations of Microtubules in a Fluid", *Physical Review E*, Vol. 53, No. 1, 1996, pp. 1003–1010.
428. M. Casenghi, *et al.*, "Polo-like Kinase 1 Regulates Nip a Centrosome Protein Involved in Microtubule Nucleation", *Developmental Cell*, Vol. 5, 2003, pp. 113–125.
429. H. Varmark, *et al.*, "Asterless is a Centriolar Protein Required for Centrosome Function and Embryo Development in Drosophila", *Current Biology*, Vol. 17, No. 20, 2007, pp. 1735–1745.
430. R.L. Huston, "On Centrioles, Microtubules, and Cellular Electromagnetism", *Journal of Nanotechnology in Engineering and Medicine, American Society of Mechanical Engineers (ASME)*, Vol. 5, 2014, 021003-1 to 021003-5.
431. R.L. Huston, "Using the Electromagnetics of Cancer's Centrosome Clusters to Attract Therapeutic Nanoparticles", *Advances in Bioscience and Biotechnology*, Vol. 6, 2015, pp. 172–181.
432. A. Mershin, *et al.*, "Tubulin Dipole Moment, Dielectric Constant and Quantum Behavior: Computer Simulations, Experimental Results and Suggestions", *Biosystems*, Vol. 77, 2004, pp. 73–85.

433. R. Stracke, *et al.*, "Analysis of the Migration Behavior of Single Microtubules in Electric Fields", *Biochemical and Biophysical Research Communications*, Vol. 293, 2002, pp. 602–609.
434. E. Nogales, S.G. Wolf, and K.H. Downing, "Structure of the αβ tubulin dimer by electron crystallography", *Nature*, Vol. 391, 1998, pp. 199–203.
435. M.A. Ko, *et al.*, "Plk4 Haploinsufficiency Causes Mitotic Infidelity and Carcinogenesis", *Nature Genetics*, Vol. 37, No. 6, 2005, pp. 883–888.
436. F.A. Barr, H.H.W. Silljé, and E.A. Nigg, "Polo-like Kinases and the Orchestration of Cell Division", *Nature Reviews Molecular Cell Biology*, Vol. 5, 2004, pp. 429–439.
437. Z. Zhou, *et al.*, "MRI Detection of Breast Cancer Micrometastases with a Fibronectin-Targeting Contrast Agent", *Nature Communications*, Vol. 6, 2015, p. 7984. doi:10.1038/ncomms8984.
438. C.L. Rieder, S. Faruki, and A. Khodjakov, "The Centrosome in Vertebrates: More Than a Microtubule-Organizing Center", *Trends in Cell Biology*, Vol. 11, No. 10, 2001, pp. 413–419.
439. D. Havelka and M. Cifra, "Calculation of the Electromagnetic Field Around a Microtubule", *Acta Polytechnica*, Vol. 49, 2009, pp. 58–63.
440. S. Doxsey, D. McCollum, and W. Theurkauf, "Centrosomes in Cellular Regulation", *Annual Review of Cell and Developmental Biology*, Vol. 21, 2005, pp. 411–434.
441. C.W. Brownlee, *et al.*, "The Protein Phosphatase 2A Regulatory Submit Twins Stabilizes Plk4 to Induce Centriole Amplification", *Journal of Cell Biology*, Vol. 195, No. 2, 2011, pp. 231–243.
442. M. Bettencourt-Dias, *et al.*, "Genome-Wide Survey of Protein Kinases Required for Cell Cycle Progression", *Nature*, Vol. 432, 2004, pp. 980–987.
443. C.J. Swallow, *et al.*, "Sak/Plk4 and Mitotic Fidelity", *Oncogene*, Vol. 24, 2005, pp. 306–312.
444. S. Duensing, "A Tentative Classification of Centrosome Abnormalities in Cancer", *Cell Biology International*, Vol. 29, 2005, pp. 352–359.
445. E.A. Nigg, "Mitotic Kinases as Regulators of Cell Division and Its Checkpoints", *Nature Reviews Molecular Cell Biology*, Vol. 2, 2001, pp. 21–31.
446. S. Leidel, *et al.*, "SAS-6 Defines a Protein Family Required for Centrosome Duplication in *C. elegans* and in Human Cells", *Nature Cell Biology*, Vol. 7, No. 2, 2005, pp. 115–125.
447. M. Delattre, C. Canard, and P. Gonczy, "Sequential Protein Recruitment in *C. elegans* Centriole Formation", *Current Biology*, Vol. 16, 2006, pp. 1844–1849.

448. B.A.A. Weaver and D.W. Cleveland, "Does Aneuploidy Cause Cancer?" *Current Opinion in Cell Biology*, Vol. 18, 2006, pp. 658–667.
449. A. Dammermann, P.S. Maddox, A. Desai, and K. Oegema, "SAS-4 Is Recruited to a Dynamic Structure in Newly Forming Centrioles that is Stabilized by the ϒ-Tubulin-Mediated Addition of Centriolar Microtubules", *Journal of Cell Biology*, Vol. 180, No. 4, 2008, pp. 771–785.
450. S.K. Dutcher, "Long-Lost Relatives Reappear: Identification of New Members of the Tubulin Superfamily", *Current Opinion in Microbiology*, Vol. 6, 2003, pp. 634–640.
451. M. Moritz, *et al.*, "Structure of the γ-Tubulin Ring Complex: A Template for Microtubule Nucleation", *Nature Cell Biology*, Vol. 2, No. 6, 2000, pp. 365–370.
452. J.M. Kollman, *et al.*, "Microtubule Nucleating γ-TuSC Assembles Structures with 13-Fold Microtubule-like Symmetry", *Nature*, Vol. 466, 2010, pp. 879–883.
453. H. Aldaz, *et al.*, "Insights into Microtubule Nucleation from the Crystal Structure of Human ϒ-Tubulin", *Nature*, Vol. 435, 2005, pp. 523–527.
454. S. Lawo, *et al.*, "Subdiffraction Imaging of Centrosomes Reveals Higher-Order Organizational Features of Pericentriolar Material", *Nature Cell Biology*, Vol. 14, No. 11, 2012, pp. 1148–1158. doi:10.1038/ncb2591
455. J. Fu and D.M. Glover, "Structured Illumination of the Interface between Centriole and Peri-Centriolar Material", *Open Biology*, Vol. 2, 2012, pp. 1–30.
456. V. Archambault and D.M. Glover, "Polo-like Kinases: Conservation and Divergence in their Functions and Regulation", *Nature Reviews: Molecular Cell Biology*, Vol. 10, 2009, pp. 265–275.
457. M. Petronczki, P. Lénárt, and J.-M. Peters, "Polo on the Rise—From Mitotic Entry to Cytokinesis with Plk1", *Developmental Cell*, Vol. 14, 2008, pp. 646–659.
458. N.J. Ganem, Z. Storchova, and D. Pellman, "Tetraploidy, Aneuploidy and Cancer", *Current Opinion in Genetics & Development*, Vol. 17, 2007, pp. 157–162.
459. M.A. van Vugt and R.H. Medema, "Getting In and Out of Mitosis with Polo-like Kinase-1", *Oncogene*, Vol. 24, 2005, pp. 2844–2859.
460. S. Warnke, *et al.*, "Polo-like Kinase-2 Is Required for Centriole Duplication in Mammalian Cells", *Current Biology*, Vol. 14, No. 13, 2004, pp. 1200–1207.
461. S.P. Blagden and D.M. Glover, "Polar Expeditions–Provisioning the Centrosome for Mitosis", *Nature Cell Biology*, Vol. 5, No. 6, 2003, pp. 505–511.

462. D. Job, O. Valiron, and B. Oakley, "Microtubule Nucleation", *Current Opinion in Cell Biology*, Vol. 15, 2003, pp. 111–117.
463. T.J. Keating and G.G. Borisky, "Immunostructural Evidence for the Template Mechanism of Microtubule Nucleation", *Nature Cell Biology*, Vol. 2, 2000, pp. 352–357.
464. R. Leguy, *et al.*, "Monomeric ϒ-Tubulin Nucleates Microtubules", *Journal of Biological Chemistry*, Vol. 275, No. 29, 2000, pp. 21975–21980.
465. A. Khodjakov and C.L. Rieder, "Centrosomes Enhance the Fidelity of Cytokinesis in Vertebrates and Are Required for Cell Cycle Progression", *Journal of Cell Biology*, Vol. 153, No. 1, 2001, pp. 237–242.
466. O. Valiron, N. Caudron, and D. Job, "Microtubule Dynamics", *Cellular and Molecular Life Sciences*, Vol. 58, 2001, pp. 2069–2084.
467. Y.F. Inclán and E. Nogales, "Structural Models for the Self-Assembly and Microtubule Interactions of Gamma-, Delta- and Epsilon-tubulin", *Journal of Cell Science*, Vol. 114, 2000, pp. 413–422.
468. C. Wiese and Y. Zheng, "A New Function for the ϒ-Tubulin Ring Complex as a Microtubule Minus-End Cap", *Nature Cell Biology*, Vol. 2, 2000, pp. 358–364.
469. N. Caudron, *et al.*, "A Reassessment of the Factors Affecting Microtubule Assembly and Disassembly In Vitro", *Journal of Molecular Biology*, Vol. 297, 2000, pp. 211–220.
470. J. Hardin, G. Bertoni, and L.J. Kleinsmith, *Becker's World of the Cell*, 8th Edition, Pearson, Washington, DC, 2012.
471. J.D. Watson and E.H.C. Crick, "Molecular Structure of Nucleic Acids: A Structure for Deoxyribose Nucleic Acid", *Nature*, Vol. 171, 1953, pp. 737–738.
472. A. Skibinski and C. Kuperwasser, "The Origin of Breast Tumor Heterogeneity", *Oncogene*, Vol. 34, 2015, pp. 5309–5316.
473. M. Moritz, *et al.*, "Reconstruction of Centrosome Microtubule Nucleation in Drosophila", *Methods in Cell Biology*, Vol. 67, 1998, pp. 141–148.
474. K. Kamath, E. Oroudjev, and M.A. Jordan, "Determination of Microtubule Dynamic Instability in Living Cells", *Methods in Cell Biology*, Vol. 97, 2010, pp. 1–14.
475. E.L. Grishchuk, *et al.*, "Force Production by Disassembling Microtubules", *Nature*, Vol. 438, No. 17, 2005, pp. 385–388.
476. S.M. Wignall and A.M. Villeneuve, "Lateral Microtubule Bundles Promote Chromosome Alignment during Acentrosomal Oocyte Meiosis", *Nature Cell Biology*, Vol. 11, No. 7, 2009, pp. 839–844.
477. S. Dumont and T.J. Mitchison, "Force and Length in the Mitotic Spindle", *Current Biology*, Vol. 19, 2009, pp. R749–R761.

478. M.E. Janson, M.E. deDood, and M. Dogterom, "Dynamic Instability of Microtubules Is Regulated by Force", *Journal of Cell Biology*, Vol. 161, No. 6, 2003, pp. 1029–1034.
479. S.L. Kline-Smith and C.E. Wolczak, "The Microtubule-Destabilizing Kinesin XKCM1 Regulates Microtubule Dynamic Instability in Cells", *Molecular Biology of the Cell*, Vol. 13, 2002, pp. 2718–2731.
480. C.E. Walczak, "Microtubule Dynamics and Tubulin Interacting Proteins", *Current Opinion in Cell Biology*, Vol. 12, 2000, pp. 52–56.
481. B.T. Castle and D.J. Odde, "Optical Control of Microtubule Dynamics in Time and Space", *Cell*, Vol. 162, 2015, pp. 243–245.
482. M. Dogterom, *et al.*, "Force Generation by Dynamic Microtubules", *Current Opinion in Cell Biology*, Vol. 17, 2005, pp. 67–74.
483. S. Petry and R.D. Vale, "Microtubule Nucleation at the Centrosome and Beyond", *Nature Cell Biology*, Vol. 17, No. 9, 2015, pp. 1089–1093.
484. T. Stearns, "Centrioles, in Absentia", *Science*, Vol. 348, No. 6239, 2015, pp. 1091–1092.
485. V.L. Wong, *et al.*, "Reversible Centriole Depletion with an Inhibitor of Polo-like Kinase 4", *Science*, Vol. 348, No. 6239, 2015, pp. 1155–1160.
486. T.K. Akhshi, D. Wernike, and A. Piekny, "Microtubules and Actin Crosstalk in Cell Migration and Division", *Cytoskeleton*, Vol. 71, 2014, pp. 1–23.
487. A.D. Bicek, *et al.*, "Analysis of Microtubule Curvature", *Methods in Cell Biology*, Vol. 83, 2007, Chapter 10, pp. 237–268.
488. S.L. Prosser and L. Pelletier, "Centrosome Biology: The Ins and Outs of Centrosome Assembly", *Current Biology*, Vol. 25, 2015, pp. R656–R659.
489. L. Zitvogel and G. Kroemer, "A p53 Regulated Immune Checkpoint Relevant to Cancer", *Science*, Vol. 349, No. 6247, 2015, pp. 476–477.
490. R. Yang and J. Feldman, "SPD-2/CEP192 and CDK Are Limiting for Microtubule-Organizing Center Function at the Centrosome", *Current Biology*, Vol. 24, 2015, pp. 1924–1931.
491. J.K. Pagan, *et al.*, "Degradation of Cep68 and PCNT Cleavage Mediate Cep215 Removal from the PCM to Allow Centriole Separation, Disengagement and Licensing", *Nature Cell Biology*, Vol. 17, 2015, pp. 31–43.
492. S. Afreen and D. Varma, "Cell Divisions: Molecular Pathways for KMN Kinetochore Recruitment", *Current Biology*, Vol. 24, 2015, pp. R328–R347.
493. A. Akhmanova and C.C. Hoogenraad, "Microtubule Minus-End-Targeting Proteins", *Current Biology*, Vol. 25, 2015, pp. R162–R171.
494. K. Lauber and M. Herrmann, "Tumor Biology: With a Little Help from my Dying Friends", *Current Biology*, Vol. 25, No. 5, 2015, pp. R198–R201.

495. J. Scrofani, *et al.*, "Microtubule Nucleation in Mitosis by a Ran GTP-Dependent Protein Complex", *Current Biology*, Vol. 25, 2015, pp. 131–140.
496. V. Mennella, *et al.*, "Subdiffraction-Resolution Fluorescence Microscopy Reveals a Domain of the Centrosome Critical for Pericentriolar Material Organization", *Nature Cell Biology*, Vol. 14, No. 11, 2012, pp. 1159–1168.
497. B. Vitre, *et al.*, "Chronic Centrosome Amplification without Tumorigenesis", *Proceedings of the National Academy of Sciences of the United States of America*, Vol. 112, 2015, pp. E6321–E6330. doi:10.1073/pnas.1519388112
498. E. Valsami-Jones and I. Lynch, "How Safe Are Nanomaterials?" *Science*, Vol. 350, No. 6259, 2015, pp. 388–389.
499. F. Zhou and S. Roy, "SnapShot: Motile Cilia", *Cell*, Vol. 162, 2015, pp. 224–224.e1. doi:10.1016/j.cell.2015.06
500. B.G. Lambrus, *et al.*, "p53 Protects Against Genome Instability Following Centriole Duplication Failure", *Journal of Cell Biology*, Vol. 210, No. 1, 2015, pp. 63–77.
501. T.C. Moyer, *et al.*, "Binding of STIL to Plk4 Activates Kinase Activity to Promote Centriole Assembly", *Journal of Cell Biology*, Vol. 209, No. 6, 2015, pp. 863–878.
502. F. Chen, P.W. Tillberg, and E.S. Boyden, "Expansion Microscopy", *Science*, Vol. 347, No. 6221, 2015, pp. 543–548.
503. J.G. DeLuca, "Kinetochore-Microtubule Dynamics and Attachment Stability", *Methods in Cell Biology*, Vol. 97, 2010, Chapter 4, pp. 53–79.
504. P.W. Schafer, "Centrioles: Intercellular Order in Normal and Malignant Cells", *Journal of Thoracic and Cardiovascular Surgery*, Vol. 63, No. 3, 1972, pp. 472–477.
505. C. Wiese and Y. Zheng, "A New Function for the ϒ-Tubulin Ring Complex as a Microtubule Minus-end Cap", *Nature Cell Biology*, Vol. 2, 2000, pp. 358–364.
506. B.R. Oakley, V. Paolillo, and Y. Zheng, "Gamma-Tubulin Complexes in Microtubule Nucleation and Beyond", *Molecular Biology of the Cell*, Vol. 26, 2015, pp. 2957–2962.
507. N. Teixido-Travesa, J. Roig, and J. Luders, "The Where, When and How of Microtubule Nucleation—One Ring to Rule them all", *Journal of Cell Science*, Vol. 125, 2012, pp. 4445–4456.
508. A. Bouissou, *et al.*, "Gamma-Tubulin Ring Complexes Regulate Microtubule Plus End Dynamics", *Journal of Cell Biology*, Vol. 187, No. 3, 2009, pp. 327–334.

509. L. Cassimeris, *et al.*, "XMAP215 Is a Long Thin Molecule that Does Not Increase Microtubule Stiffness", *Journal of Cell Science*, Vol. 114, 2001, pp. 3025–3033.
510. F.P. Beer and E.R. Johnston, Jr., *Mechanics of Materials*, 2nd Edition, McGraw Hill, New York, 1992.
511. R.L. Huston, "A Review of Centriole Activity, and Wrongful Activity, During Cell Division", *Advances in Bioscience and Biotechnology*, Vol. 7, 2016, pp. 169–182.
512. M. Assenheimer, *et al.*, "The T-SCAN Technology: Electrical Impedence as a Diagnostic Tool for Breast Cancer Detection", *Physiological Measurement*, Vol. 22, 2001, pp. 1–8.
513. J. Pokorný, "Endogenous Electromagnetic Forces in Living Cells: Implications for Transfer of Reaction Components", *Electro- and Magnetobiology*, Vol. 20, No. 1, 2001, pp. 59–73.
514. M.S. Markov, "Magnetic Field Theory: A Review", *Electromagnetic Biology and Medicine*, Vol. 26, 2007, pp. 1–23.
515. H. Berg, "Bioelectric and Biomagnetic Methods for Cancer Research and Therapy–A Survey", *Electromagnetic Biology and Medicine*, Vol. 24, 2005, pp. 423–440.
516. K.J. Böhm, *et al.*, "Movement and Alignment of Microtubules in Electric Fields and Electric-Dipole-Moment Estimates", *Electromagnetic Biology and Medicine*, Vol. 24, 2005, pp. 319–330.
517. M. Sataric, B.M. Sataric, and J.A. Tuszynski, "Nonlinear Model of Microtubule Dynamics", *Electromagnetic Biology and Medicine*, Vol. 24, 2005, pp. 255–264.
518. D. Keller, *et al.*, "Mechanisms of HsSAS-6 Assembly Promoting Centriole Formation in Human Cells", *Journal of Cell Biology*, Vol. 204, No. 5, 2014, pp. 697–712.
519. J. Owen, Q. Pankhurst, and E. Stride, "Magnetic Targeting and Ultrasound Mediated Drug Delivery: Benefits, Limitations and Combination", *International Journal of Hyperthermia*, Vol. 28, No. 4, 2012, pp. 362–373.
520. Y. Zhang, *et al.*, "Self-Assembled Coatings on Individual Monodisperse Magnetite Nanoparticles for Efficient Intracellular Uptake", *Biomedical Microdevices*, Vol. 6, 2004, pp. 33–40.
521. A. Meunier and N. Spassky, "Centriole Continuity: Out with the New, In with the Old", *Current Opinion in Cell Biology*, Vol. 38, 2016, pp. 60–67. doi:10.1016/j.ceb.2016.02.007
522. P. Zhang, *et al.*, "Electrical Stimulation Inhibits Neointimal Hyperplasia after Abdominal Aorta Balloon Injury Through the PTEN/p27Kip1 Pathway",

Acta Biochimica et Biophysica Sinica (Shanghai), Vol. 42, No. 11, 2010, pp. 807–815. doi:10.1093/abbs/gmq089

523. P. Guichard, *et al.*, "Procentriole Assembly Revealed by Cryo-Electron Tomography", *The EMBO Journal*, Vol. 29, No. 9, 2010, pp. 1565–1572. doi:10.1038/emboj.2010.45
524. P.W. Schafer, 1980. Personal communication.
525. H. Josephs and R.L. Huston, *Dynamics of Mechanical Systems*, CRC Press, Boca Raton, FL, 2002.
526. R.L. Huston, "Mechanics of Centriole Microtubules", *Advances in Biosciences and Biotechnology*, Vol. 7, 2016, pp. 266–277.
527. L. Lau, *et al.*, "STED Microscopy with Optimized Labeling Density Reveals 9-Fold Arrangement of a Centriole Protein", *Biophysical Journal*, Vol. 120, No. 12, 2012, pp. 2926–2935. doi:10.1016/j.bpj.2012.05.015
528. M. Hilbert, *et al.*, "SAS-6 Engineering Reveals Interdependence between Cartwheel and Microtubules in Determining Centriole Architecture", *Nature, Cell Biology*, Vol. 18, No. 4, 2016, pp. 393–403. doi:10.1038/ncb3329
529. A. Suzuki, *et al.*, "How the Kinetochore Couples Microtubule Force and Centromere Stretch to Move Chromosomes", *Nature, Cell Biology*, Vol. 18, No. 4, 2016, pp. 382–392. doi:10.1038/ncb3323
530. B. Short, "Mapping the Kinetochore Delta", *Journal of Cell Biology*, Vol. 212, 2016, p. 259.
531. T.J. Pereira, *et al.*, "Inducible Fluorescent Speckle Microscopy", *Journal of Cell Biology*, Vol. 212, 2016, pp. 245–255.
532. L. Lau, *et al.*, "STED Microscopy with Optimized Labeling Density Reveals 9-Fold Arrangement of a Centriole Protein", *Biophysical Journal*, Vol. 120, No. 12, 2012, pp. 2926–2935. doi:10.1016/j.bpj.2012.05.015
533. K.F. Sonnen, *et al.*, "3D-Structured Illumination Microscopy Provides Novel Insight into Architecture of Human Centrosomes", *Biology Open*, Vol. 1, 2012, pp. 965–976.
534. P. Gönczy, "Centrosomes and Cancer: Revisiting a Long-Standing Relationship", *Nature Reviews Cancer*, Vol. 15, 2015, pp. 639–642.
535. J.E. Sillibourne, *et al.*, "Assessing the Localization of Centrosomal Proteins by PALM/STORM Nanoscopy", *Cytoskeleton*, Vol. 68, 2011, pp. 619–627.
536. S. Lawo, *et al.*, "Subdiffraction Imaging of Centrosomes Reveals Higher-Order Organizational Features of Pericentriolar Material", *Nature Cell Biology*, Vol. 14, No. 11, 2012, pp. 1148–1158. doi:10.1038/ncb2591
537. L. Schermelleh, R. Heintzmann, and H. Leonhardt, "A Guide to Super-Resolution Fluorescence Microscopy", *Journal of Cell Biology*, Vol. 190, No. 2, 2010, pp. 165–175. doi:10.1083/jcb.201002018

538. F. Lampert and S. Westermann, "A Blueprint for Kinetochores—New Insight into the Molecular Mechanics of Cell Division", *Nature Reviews: Molecular Cell Biology*, Vol. 12, 2011, pp. 407–412.
539. I.M. Cheeseman and A. Desai, "Molecular Architecture of the Kinetochore-Microtubule Interface", *Nature Reviews, Molecular Cell Biology*, Vol. 19, 2008, pp. 33–46.
540. X. Wan, *et al.*, "Protein Architecture of the Human Kinetochore Microtubule Attachment Site", *Cell*, Vol. 137, 2009, pp. 672–684.
541. S. Santaguida and A. Musacchio, "The Life and Miracles of Kinetochores", *The EMBO Journal*, Vol. 28, 2009, pp. 2511–2531.
542. S.M. Moghimi, A.C. Hunter, and J.C. Murray, "Nanomedicine: Current Status and Future Prospects", *FASEB Journal*, Vol. 19, 2005, pp. 311–330.
543. C. Boutin, "To Give Cancer a Deadly Fever, NIST Explores Better Nanoparticle Design."
544. J.M. Rosenholm, *et al.*, "Targeting of Porous Hybrid Silica Nanoparticles to Cancer Cells", *ACS Nano*, Vol. 3, No. 1, pp. 197–206. doi:10.1021/nn800781r
545. C. Wong, *et al.*, "Multistage Nanoparticle Delivery System for Deep Penetration into Tumor Tissue", *Proceedings National Academy of Sciences*, Vol. 108, No. 6, 2011, pp. 2426–2431.
546. B. Haley and E. Frenkel, "Nanoparticles for Drug Delivery in Cancer Treatment", *Urologic Oncology: Seminars and Original Investigations*, Vol. 26, 2008, pp. 57–64.
547. J.D. Byrne, T. Betancourt, and L. Branon-Peppas, "Active Targeting Schemes for Nanoparticle Systems in Cancer Therapeutics", *Advanced Drug Delivery Reviews*, Vol. 60, 2008, pp. 1615–1626.
548. L. Zhang, *et al.*, "Nanoparticles in Medicine: Therapeutic Applications and Developments", *Clinical Pharmacology and Therapeutics*, Vol. 83, No. 5, 2008, pp. 761–769.
549. M.M. Lin, *et al.*, "Development of Superparamagnetic Iron Oxide Nanoparticles (SPIONS) for Translation to Clinical Applications", *IEEE Transactions on Nanobioscience*, Vol. 7, No. 4, 2008, pp. 298–305.
550. Y. Zhang, N. Kohler, and M. Zhang, "Surface Modification of Superparamagnetic Magnetite Nanoparticles and Their Intracellular Uptake", *Biomaterials*, Vol. 23, 2002, pp. 1553–1561.
551. J. Dobson, "Magnetic Nanoparticles for Drug Delivery", *Drug Development Research*, Vol. 67, 2006, pp. 55–60.
552. A. Jordan, *et al.*, "The Effect of Thermotherapy using Magnetic Nanoparticles on Rat Malignant Glioma", *Journal of Neuro-Oncology*, Vol. 78, 2006, 2006, pp. 7–14.

553. J.G. Parvani, *et al.*, "Silencing β3 Integrin by Targeted ECO/siRNA Nanoparticles Inhibits EMT and Metastasis of Triple-Negative Breast Cancer", *Cancer Research*, Vol. 75, 2015, pp. 2316–2325. doi:10.1158/0008-5472
554. E. Russlahti, S.N. Bhatia, and M.J. Sailor, "Targeting of Drugs and Nanoparticles to Tumors", *Journal of Cell Biology*, Vol. 188, 2010, pp. 759–768.
555. K.N. Sugahara, *et al.*, "Tissue-Penetrating Delivery of Compounds and Nanoparticles into Tumors", *Cancer Cell*, Vol. 16, 2009, pp. 510–520.
556. D. Peer, *et al.*, "Nanocarriers as an Emerging Platform for Cancer Therapy", *Nature Nanotechnology*, Vol. 2, 2007, pp. 751–760. doi:10.1038/nnano.2007.387
557. J.-H. Park, *et al.*, "Biodegradable Luminescent Porous Silicon Nanoparticles for In Vivo Applications", *Nature Materials*, Vol. 8, 2009a, pp. 331–336.
558. P.P. Karmali, *et al.*, "Targeting of Albumin-Embedded Paclitaxel Nanoparticles to Tumors", *Nanomedicine, Nanotechnology, Biology, and Medicine*, Vol. 5, 2009, pp. 73–82. doi:10.1016/j.nano.2008.07.007
559. D.T. Butcher, T. Alliston, and V.M. Weaver, "A Tense Situation: Forcing Tumor Progression", *Nature Reviews Cancer*, Vol. 9, 2009, pp. 108–122.
560. P.C. Chen, S.C. Mwakwari, and A.K. Oyelere, "Gold Nanoparticles: From Nanomedicine to Nanosensing", *Nanotechnology Science and Applications*, Vol. 1, 2008, pp. 45–65.
561. W. Cai, *et al.*, "Applications of Gold Nanoparticles in Cancer Nanotechnology", *Nanotechnology Science and Applications*, Vol. 1, 2008, pp. 17–32.
562. B. Bohunicky and S.A. Mousa, "Biosensors: The New Wave in Cancer Diagnosis", *Nanotechnology Science and Applications*, Vol. 4, 2001, pp. 1–10.
563. P.S. Low, W.A. Henne, and D.D. Doomeweerd, "Discovery and Development of Folic-Acid-Based Receptor Targeting for Imaging and Therapy of Cancer and Inflammatory Diseases", *Accounts of Chemical Research*, Vol. 41, No. 1, 2008, pp. 120–129.
564. N. Ramakrishnan, *et al.*, "Biophysically Inspired Model for Functionalized Nanocarrier Adhesion to Cell Surface: Roles of Protein Expression and Mechanical Factors", *Royal Society Open Science*, Vol. 3, June 29, 2016, pp. 1–40.
565. B. Garipcan, *et al.*, "Image Analysis of Endothelial Microstructure and Endothelial Cell Dimensions of Human Arteries–A Preliminary Study", *Advanced Engineering Materials*, Vol. 13, 2011, pp. 854–857.
566. A. Schroeder, *et al.*, "Treating Metastatic Cancer with Nanotechnology", *Nature Reviews Cancer*, Vol. 12, 2012, pp. 39–48.

567. I.J. Majoros, *et al.*, "Poly (amidoamine) Dendrimer-Based Multifunctional Engineered Nanodevice for Cancer Therapy", *Journal of Medicinal Chemistry*, Vol. 48, 2005, pp. 5892–5899.
568. J.F. Kukowska-Latallo, *et al.*, "Nanoparticle Targeting of Anticancer Drug Improves Therapeutic Response in Animal Model of Human Epithelial Cancer", *Cancer Research*, Vol. 65, No. 12, 2005, pp. 5317–5324.
569. M. Liong, *et al.*, "Multifunctional Inorganic Nanoparticles for Imaging, Targeting, and Drug Delivery", *ACS Nano*, Vol. 2, No. 5, 2008, pp. 889–896.
570. H.A. Pohl, "Cellular Spin Resonance", *Journal of Theoretical Biology*, Vol. 93, 1981, pp. 207–213.
571. H.A. Pohl and T. Branden, "Cellular Spin Resonance of Aging Yeast and Mouse Sarcoma Cells", *Journal of Biological Physics*, Vol. 19, 1982, pp. 17–30.
572. I.A. Kent and T.P. Lele, "Microtubule Based Force Generation", *Wiley Interdisciplinary Reviews: Nanomedicine and Nanobiotechnology*, 2016.
573. E.A. Murphy, *et al.*, "Nanoparticle-Mediated Drug Delivery to Tumor Vasculature Suppresses Metastasis", *Proceedings National Academy of Science*, Vol. 105, 2008, pp. 9343–9348.
574. G.G. Ganev, "Pathological Significance of Cilia and Centrioles in Bone and Soft Tissue Tumors", *Journal of Orthopaedic Science*, Vol. 2, 1997, pp. 137–145.
575. J.E. Klebba, *et al.*, "Two Polo-like Kinase 4 Binding Domains in Asterless Perform Distinct Roles in Regulating Kinase Stability", *Journal of Cell Biology*, Vol. 208, No. 4, 2015, pp. 401–414.
576. R.L. Huston, "A Review of Electromagnetic Activity in Cellular Mechanics", *Advances in Bioscience and Biotechnology*, Vol. 7, 2016c, pp. 360–371.
577. N.A. Campbell, *Biology*, Benjamin/Cummings, Redwood City, CA, 1987, p. 231.
578. B.J. Galleta, *et al.*, "Asterless Is Required for Centriole Length Control and Sperm Development", *Journal of Cell Biology*, Vol. 213, No. 4, 2016, pp. 435–474.
579. M.A. Jordan and L. Wilson, "Microtubules as a Target for Anticancer Drugs", *Nature Reviews: Cancer*, Vol. 4, No. 4, 2004, pp. 253–265.
580. R. Nadkarni, S. Berkley, and C. Fradin, "A Comparison of Methods to Measure the Mangetic Moment of Magnetotactic Bacteria through Analysis of Their Trajectories in External Magnetic Fields", *Plos One*, Vol. 8, 2013, p. 1.
581. S. Hashimoto, *et al.*, "The Measurement of Small Magnetic Signals from Magnetic Nanoparticles Attached to the Cell Surface and Surrounding

Living Cells using a General-Purpose SQUID Magnetometer", *Physics in Medicine and Biology*, Vol. 54, 2009, pp. 2571–2583.

582. C.P. Brangwynne, *et al.*, "Microtubules Can Bear Enhanced Compressive Loads in Living Cells Because of Lateral Reinforcement", *Journal of Cell Biology*, Vol. 173, No. 5, 2006, pp. 733–761. doi:10.1083/jcb.200601060
583. C.L. Asbury, J.F. Tien, and T.N. Davis, "Kinetochores' Gripping Feat: Conformational Wave or Biased Diffusion?" *Trends in Cell Biology*, Vol. 21, 2011, pp. 38–46.
584. C. Garzan-Coral, "A Force-Generating Machinery Maintains the Spindle at the Cell Center during Mitosis", *Science*, Vol. 352, 2016, pp. 1124–1127.
585. J.-H. Lee, *et al.*, "Artificially Engineered Magnetic Nanoparticles for Ultra-Sensitive Molecular Imaging", *Nature Medicine*, Vol. 13, 2007, pp. 95–99.
586. A.F. Powers, *et al.*, "The Ndc80 Kinetochore Complex Forms Load Bearing Attachments to Dynamic Microtubule Tips via Biased Diffusion", *Cell*, Vol. 136, 2009, pp. 865–875.
587. V.A. Volkov, *et al.*, "Long Tethers Provide High-Force Coupling of the Dam1 Ring to Shortening Microtubules", *Proceedings of the National Academy of Sciences of the United States of America*, Vol. 110, 2013, pp. 7708–7713.
588. J.F. Tien, *et al.*, "Cooperation of the Dam1 and Ndc80 Kinetochore Complexes Enhances Microtubule Coupling and Is Regulated by Aurora B", *Journal of Cell Biology*, Vol. 189, No. 4, 2010, pp. 713–23. doi:10.1083/jcb.200910142
589. M.E. Presenti, J.R. Weir, and A. Musacchio, "Progress in the Structural and Functional Characterization of Kinetochores", *Current Opinion in Structural Biology*, Vol. 37, 2016, pp. 152–163.
590. M. Sanhaji, *et al.*, "Mitotic Centromere-Associated Kinesin (MCAK): A Potential Cancer Drug Target", *Oncotarget*, Vol. 2, 2011, pp. 935–947. doi:10.18632/oncotarget.416
591. J. Wu, *et al.*, "How Dynein and Microtubules Rotate the Nucleus", *Journal of Cellular Physiology*, Vol. 226, 2016, pp. 2666–2674.
592. F.J. McNally, "Mechanisms of Spindle Positioning", *Journal of Cell Biology*, Vol. 200, 2013, pp. 131–159.
593. W. Zimmerman, *et al.*, "Mitosis-Specific Anchoring of γ-Tubulin Complexes by Pericentrin Controls Spindle Organization and Mitotic Entry", *Molecular Biology of the Cell*, Vol. 15, 2004, pp. 3672–3657.
594. Y. Zheng, *et al.*, "Nucleation of Microtubule Assembly by a γ-Tubulin-Containing Ring Complex", *Nature*, Vol. 378, 1995, pp. 578–583.

595. I.M. Cheeseman and A. Desai, "Molecular Architecture of the Kinetochore-Microtubule Interface", *Nature Reviews, Molecular Cell Biology*, Vol. 19, 2008, pp. 33–46.
596. T. Mitchison and M. Kirschner, "Dynamic Instability of Microtubule Growth", *Nature*, Vol. 312, 1984, pp. 237–242.
597. J.R. Cooper, *et al.*, "Catalysis of the Microtubule on-Rate Is the Major Parameter Regulating the Deploymerase Activity of MCAK", *Nature Structural and Molecular Biology*, Vol. 17, 2010, pp. 77–82.
598. R. Bahtz, *et al.*, "GCP6 Is a Substrate of Plk4 and Required for Centriole Duplication", *Journal of Cell Science*, Vol. 125, 2011, pp. 486–496.
599. C.-J.C. Tang, *et al.*, "The Human Microcephaly Protein STIL Interacts with CPAP and Is Required for Procentriole Formation", *The EMBP Journal*, Vol. 30, 2011, pp. 4790–4804.
600. K. Lee and K. Rhee, "Separase-Dependent Cleavage of Pericentrin B Is Necessary and Sufficient for Centriole Disengagement During Mitosis", *Cell Cycle*, Vol. 11, 2012, pp. 2476–2485.
601. W.-J. Wong, *et al.*, "The Conversion of Centrioles to Centrosomes: Essential Coupling of Duplication with Segregation", *Journal of Cell Biology*, Vol. 183, 2011, pp. 727–739.
602. B.J. Galletta, *et al.*, "Asterless Is Required for Centriole Length Control and Sperm Development", *Journal of Cell Biology*, Vol. 213, No. 4, 2016, pp. 435–474.
603. Y. Matsumoto, *et al.*, "Vascular Bursts Enhance Permeability of Tumour Blood Vessels and Improve Nanoparticle Delivery", *Nature Nanotechnology*, Vol. 11, 2016, pp. 533–539.
604. C.L. Asbury and T.N. Davis, "Insights into the Kinetochore", *Structure*, Vol. 16, 2008, pp. 834–836.
605. T. Mitchinson and M. Kirachner, "Microtubule Assembly Nucleated by Isolated Centrosomes", *Nature*, Vol. 312, 1984b, pp. 232–237.
606. C. Ciferri, *et al.*, "Implications for Kinetochore-Microtubule Attachment from the Structure of an Engineered Ndc80 Complex", *Cell*, Vol. 133, 2008, pp. 427–439.
607. A.P. Joglekar, *et al.*, "Molecular Architecture of a Kinetochore–Microtubule Attachment Site", *Nature Cell Biology*, Vol. 8, 2006, pp. 581–585.
608. A.P. Joglekar, K. Bloom, and E.D. Salmon, "In Vivo Protein Architecture of the Eukaryotic Kinetochore with Nanometer Scale Accuracy", *Current Biology*, Vol. 19, 2009, pp. 694–699.
609. H. Maiato, *et al.*, "The Dynamic Kinetochore-Microtubule Interface", *Journal of Cell Science*, Vol. 117, 2004, pp. 5461–5477.

610. T.U. Tanaka and A. Desai, "Kinetochore-Microtubule Interactions: The Means to the End", *Current Opinion in Cell Biology*, Vol. 20, 2008, pp. 53–63.
611. C.B. O'Connell and A.L. Khodjakov, "Cooperative Mechanisms of Mitotic Spindle Formation", *Journal of Cell Science*, Vol. 120, 2007, pp. 1717–1722.
612. P.T. Conduit, "Microtubule Organization: A Complex Relation", *Journal of Cell Biology (JCB)*, Vol. 213, 2016, pp. 809–812.
613. C. Xu, "Modification of Superparamagnetic Nanoparticles for Biomedical Applications", Ph.D. Dissertation, Department of Chemistry, Brown University, Providence, RI, 2009.
614. B.P. Bouchet and A. Akhmanova, "Microtubules in 3D Cell Motility", *Journal of Cell Science*, Vol. 130, 2017, pp. 39–50.
615. C. Aumeier, *et al.*, "Self-Repair Promotes Microtubule Rescue", *Nature Cell Biology*, Vol. 18, No. 10, 2016, pp. 1054–1064.
616. C.-W. Chang, *et al.*, "CEP295 Interacts with Microtubules and Is Required for Centriole Elongation", *Journal of Cell Science*, Vol. 129, 2016, pp. 2501–2513.
617. K.C. Slep, "A Microtubule Dynamics Reconstitutional Convention", *Journal of Cell Biology*, Vol. 205, No. 3, 2016, pp. 305–315.
618. C.R. Alarcón and S.F. Tavazoie, "Cancer: Endothelial-Cell Killing Promotes Metastasis", *Nature News & Views*, Vol. 536, 2016, pp. 154–155.
619. L. Schaedel, *et al.*, "Microtubules Self-Repair in Response to Mechanical Stress", *Nature Materials*, Vol. 14, 2015, pp. 1156–1165.
620. F. Bordeleau, *et al.*, "Matrix Stiffening Promotes a Tumor Vasculature Phenotype", *Proceedings of the National Academy of Sciences*, Vol. 14, 2017, pp. 492–497.
621. J.W. Raff and R. Basto, "Centrosome Amplification and Cancer: A Question of Sufficiency", *Developmental Cell*, Vol. 40, 2017, pp. 217–218.
622. M.K. Gardner, "Cell Biology: Microtubule Collisions to the Rescue", *Current Biology*, Vol. 26, 2016, pp. R1287–R1289.
623. L.E. Muñoz, *et al.*, "Nanoparticles Size-Dependently Initiate Self-Limiting NETosis-driven Inflammation", *Proceedings National Academy of Sciences*, Vol. 113, 2016, pp. E5856–E5865.
624. A.Z. Wang, R. Langer, and O.C. Farokhzad, "Nanoparticle Delivery of Cancer Drugs", *Annual Review of Medicine*, Vol. 63, 2012, pp. 185–198.
625. Y. Matsumura and H. Maeda, "A New Concept for Macromolecular Therapeutics in Cancer Chemotherapy: Mechanism of Tumoritropic Accumulation of Proteins and the Antitumor Agent Smancs", *Cancer Research*, Vol. 46, 1986, pp. 6382–6392.

626. A.Z. Wang and J.E. Tepper, "Nanotechnology in Radiation Oncology", *Journal of Clinical Oncology*, Vol. 32, No. 26, 2014, pp. 2879–2885.
627. R.F. Service, "Nanotechnology Takes Aim at Cancer", *Science*, Vol. 310, 2005, pp. 1132–1134.
628. Y. Shamay, *et al.*, "P-Selection Is a Nanotherapeutic Delivery Target in the Tumor", *Science Translational Medicine*, Vol. 29, 2016, p. 345ra87.
629. J. Salogiannis and S.L. Reck-Peterson, "Hitchhiking: A Non-Canonical Mode of Microtubule-Based Transport", *Trends in Cell Biology*, Vol. 27, 2017, pp. 141–150.
630. T. Afrin, *et al.*, "Buckling of Microtubules on Elastic Media via Breakable Bonds", *Biochemical and Biophysical Research Communications*, Vol. 480, 2016, pp. 132–138.
631. C. Aumeier, *et al.*, "Self-Repair Promotes Microtubule Rescue", *Nature Cell Biology*, Vol. 18, No. 10, 2016, pp. 1054–1064.
632. M. Falco, *et al.*, "Tumour Biomarkers: Homeostasis as a Novel Prognostic Indicator", *Open Biology*, Vol. 6, 2016, pp. 1–31.
633. S.A. Godinho and B. Pellman, "Causes and Consequences of Centrosome Abnormalities in Cancer", *Philosophical Transactions of the Royal Society of London. Series B, Biological Sciences*, Vol. 369, 2014, 13 pages.
634. J. Reina and C. Gonzalez, "When Fate Follows Age: Unequal Centrosomes in Asymmetric Cell Division", *Philosophical Transactions of the Royal Society of London. Series B, Biological Science*, Vol. 369, 2014, 13 pages.
635. R.M. Rios, "The Centrosome-Golgi Apparatus Nexus", *Philosophical Transactions of the Royal Society of London. Series B, Biological Science*, Vol. 369, 2014, 17 pages.
636. E.N. Firat-Karalar and T. Stearns, "The Centriole Duplication Cycle", *Philosophical Transactions of the Royal Society of London. Series B, Biological Sciences*, Vol. 369, 2014, p. 20130460.
637. C.N. Johnson, *et al.*, "Magnetic Relaxometry with an Atomic Magnetometer and SQUID Sensors on Targeted Cancer Cells", *Journal of Magnetism and Magnetic Materials*, Vol. 324, 2012, pp. 2613–2619.
638. H.J. Hathaway, *et al.*, "Detection of Breast Cancer Cells using Targeted Magnetic Nanoparticles and Ultra-Sensitive Magnetic Field Sensors", *Breast Cancer Research*, Vol. 13, No. 3, 2011, p. R108.
639. M. Hirono, "Cartwheel Assembly", *Philosophical Transactions of the Royal Society of London. Series B, Biological Sciences*, Vol. 369, 2014, 16 pages.
640. M. Winey and E. O'Toole, "Centriole Structure", *Philosophical Transactions of Royal Society of the London. Series B, Biological Sciences*, Vol. 369, 2014, 14 pages.

641. M. Bornens and P. Gonczy, "Centrosomes Back in the Limelight", *Philosophical Transactions Royal Society, London B, Biological Sciences*, Vol. 369, 2014, 15 pages.
642. G. Sluder, "One to Only Two: A Short History of the Centrosome and its Duplication", *Philosophical Transactions of the Royal Society. Series B, London*, Vol. 369, 2014, 8 pages.
643. J.B. Woodruff, O. Wueseke, and A.A. Hyman, "Pericentriolar Material Structure and Dynamics", *Philosophical Transactions Royal Society of London. Series B*, Vol. 369, 2014, 10 pages.
644. F.G. Agrican, E. Schiebel, and B.R. Mardin, "Separate to operate: Control of centrosome positioning and separation", *Philosophical Transactions Royal Society of London. Series B*, Vol. 369, 2014, 10 pages.
645. P.L. Chavali, M. Putz, and F. Gergely, "Small Organelle, Big Responsibility: The Role of Centrosomes in Development and Disease", *Philosophical Transactions Royal Society of London, Series B*, Vol. 369, 2014, 12 pages.
646. A. Dart, "Extra! Extra! Read all about it", *Nature Reviews Cancer*, Vol. 17, 2017, p. 143. doi:10.1038/nrc2017.12
647. M.S. Levine, *et al.*, "Centrosome Amplification Is Sufficient to Promote Spontaneous Tumorigenesis in Mammals, *Developmental Cell*, Vol. 40, 2017, pp. 313–322.
648. J. Xiao and T. Ha, "Flipping Nanoscopy on its Head", *Science*, Vol. 355, 2017, pp. 582–584.
649. F. Balzarotti, *et al.*, "Nanometer Resolution Imaging and Tracking of Fluorescent Molecules with Minimal Photon Fluxes", *Science*, Vol. 355, 2017, pp. 606–612.
650. Y. Zhao, "Centrosome Functions as a Molecular Dynamo in the Living Cell", *Advances in Bioscience and Biotechnology*, Vol. 6, 2015, pp. 452–455.
651. Y. Zhao and Q. Zhan, "Electric Fields Generated by Synchronized, Oscillation of Microtubules, Centrosomes and Chromosomes Regulate the Dynamics of Mitosis and Meiosis", *Theoretical Biology and Medical Modeling*, Vol. 9, 2012a, p. 26.
652. Y. Zhao and Q. Zhan, "Electric Oscillation and Coupling of Chromatin Regulate Chromosome Packaging and Transcription in Eukaryotic cells", *Theoretical Biology and Medical Modelling*, Vol. 9, 2012b, pp. 27–38.
653. K. Matsuo, *et al.*, "Kendrin Is a Novel Substrate for Separase Involved in the Licensing of Centriole Duplication", *Current Biology*, Vol. 22, 2012, pp. 915–921.
654. A.V. Singh, "Biotechnological Applications of Supersonic Cluster Beam-Deposited Nanostructured Thin Films: Bottom-up Engineering to

Optimize Cell-Protein-Surface Interactions", *Journal of Biomedical Materials Research*, Vol. 101A, 2013, pp. 2984–3008.

655. N. Korzeniewski, *et al.*, "Daughter Centriole Elongation Is Controlled by Proteolysis", *Molecular Biology of the Cell*, Vol. 21, 2010, pp. 2942–2951.
656. K.R. Hurley, *et al.*, "Characterization of Magnetic Nanoparticles in Biological Matrices", *Analytical Chemistry*, Vol. 87, 2015, pp. 11611–11619.
657. D. Hanahan and R.A. Weinberg, "The Hallmarks of Cancer", *Cell*, Vol. 100, 2000, pp. 57–70.
658. D. Hanahan and R.A. Weinberg, "Hallmarks of Cancer: The Next Generation", *Cell*, Vol. 144, 2011, pp. 646–674.
659. D. LeSage, *et al.*, "Optical Magnetic Imaging of Living Cells", *Nature*, Vol. 496, 2013, pp. 486–489.
660. D. Sabino, *et al.*, "Moesin Is a Major Regulator of Centrosome Behavior in Epithelial Cells with Extra Centrosomes", *Current Biology*, Vol. 25, No. 7, 2015, pp. 879–889.
661. D.R. Glenn, *et al.*, "Single Cell Magnetic Imaging Using a Quantum Diamond Microscope", *Nature Methods*, Vol. 12, 2015, pp. 736–738.
662. O. Sadik, "The Two Faces of Nanotechnology", *American Scientist*, Vol. 105, 2017, pp. 208–211.
663. G.K. Chan, S.-T. Lin, and T.J. Yen, "Kinetochore Structure and Function", *Trends in Cell Biology*, Vol. 15, 2005, pp. 587–598.
664. M. Kwon, "Using Cell Culture Models of Centrosome Amplification to Study Centrosome Clustering in Cancer", *Methods in Molecular Biology*, Vol. 1413, 2016, pp. 367–392.
665. C.A.K. Borrebaeck, "Precision Diagnostics: Moving Towards Protein Biomarker Signatures of Clinical Utility in Cancer", *Nature Reviews Cancer*, Vol. 17, 2017, pp. 199–204.
666. J. Shi, *et al.*, "Cancer Nanomedicine: Progress, Challenges and Opportunities", *Nature Reviews Cancer*, Vol. 17, 2017, pp. 20–33.
667. J. Xiao and T. Ha, "Flipping Nanoscopy on Its Head", *Science*, Vol. 355, 2017, pp. 582–584.
668. F. Balzarotti, *et al.*, "Nanometer Resolution Imaging and Tracking of Fluorescent Molecules with Minimal Photon Fluxes", *Science*, Vol. 355, 2017, pp. 606–612.
669. S. Seaman, *et al.*, "Eradication of Tumors through Simultaneous Ablation of CD276/B7-H3-Positive Tumor Cells and Tumor Vasculature", *Cancer Cell*, Vol. 31, 2017, pp. 501–515.
670. D.R. Green, "Cancer and Apoptosis: Who Is Built to Last?" *Cancer Cell*, Vol. 31, 2017, pp. 2–4.

671. B.K. Lee, Y. Yun, and K. Park, "PLA Micro- and Nano-Particles", *Advanced Drug Delivery Reviews*, Vol. 107, 2016, pp. 176–191.
672. T. Davoli, *et al.*, "Tumor Aneuploidy Correlates with Markers of Immune Evasion and with Reduced Response to Immunotherapy", *Science*, Vol. 355, 2017, 37 pages.
673. L. Zhang, *et al.*, "27T Ultra-High Static Magnetic Field Changes Orientation and Morphology of Mitotic Spindles in Human Cells", *Elife*, Vol. 6, 2017, p. e22911.
674. J. A. Knoblich, "Asymmetric Cell Division: Recent Developments and their Implication for Tumour Biology", *Nature Reviews/Molecular Cell Biology*, Vol. 11, 2010, pp. 849–860.
675. C. Gonzalez, "Spindle Orientation, Asymmetric Division and Tumour Suppression in Drosophila Stem Cells", *Nature Review*, Vol. 8, 2007, pp. 462–472.
676. D. J. Muller, *et al.*, "Force ·Probing Surfaces of Living Cells to Molecular Resolution", *Nature, Chemical Biology,* Vol. 5, 2009, pp. 383–390.
677. R. Afrin, T. Yamada, and A. Ikai, "Analogies of Force Curves Obtained on the Live Cell Membrane Using Chemically Modified AFM Probes", *Ultramicroscopy*, Vol. 100, 2004, pp 187–195.
678. S. Veno and S. Yamaguchi-Sek, "Magnetic Control of Biological Cell Growth", *Biomagnetics Principles and Applications of Biomagnetic Stimulation and Imaging*, Edited by S. Veno and M. Sekino, CRC Press, 2016, Chapter 7, pp. 186–212.
679. K. Walter, "Intelligent Nanoparticles Could Treat Cancer", *Rand Magazine*, Internet October 24, 2017.
680. S. Sen, S. Subramanial, and D.E. Discher, "Indentation and Adhesive Probing of a Cell Membrane with AFM: Theoretical Model and Experiments", *Biophysical Journal*, Vol. 89, 2005, pp. 3203–3213.
681. F.-D. Naal, *et al.*, "High Hydrostatic Pressure-Induced Cell Death in Human Chondrocytes and Chondrosarcoma Cells", *Anticancer Research*, Vol. 25, 2005, pp. 1977–1982.
682. D.B. Pettigrew, *et al.*, "Co-oriented Asymmetric, Neurtite Outgrowth from Neuronal Explants in Tissue Culture."
683. J.A. Borgia, *et al.*, "*Correlation of Angiogenesis Biomarkers with Early Metastatic Progression in NSCLC as Determined using a Multiplexed Immunoassay Kit*", EMD Millipore, Merck KGGA, Darmstadt, Germany, 2017.
684. S.K. Frandsen, *et al.*, "Normal and Malignant Cells Exhibit Differential Responses to Calcium Electroporation", *Cancer Research, Therapeutics,*

Targets and Chemical Biology, Vol. 77, 2017, pp. 4389–4401. doi:10.1158/0008-5472.CAN-16-1611
685. J. Roostalu and T. Surrey, "Microtubule Nucleation: Beyond the Template", *Nature Reviews Molecular Cell Biology*, Vol. 18, 2017, pp. 702–710.
686. S.J. Sahl, S.W. Hell, and S. Jacobs, "Fluorescence Nanoscopy in Cell Biology", *Nature Reviews Molecular Cell Biology*, Vol. 18, 2017, pp. 685–701.
687. S.M. Hinshaw and S.C. Harrison, "Kinetochore Function from the Bottom Up", *Trends in Cell Biology*, Vol. 28, No. 1, 2018, pp. 22–33.
688. D. Passeri, *et al.*, "Magnetic Force Microscopy", *Biomatter*, Vol. 4, 2014, p. 1. doi:10.4161/biom.29507
689. G. Marteil, *et al*, "Over-Elongation of Centrioles in Cancer Promotes Centriole Amplification and Chromosome Missegregation", *Nature Communications*, Vol. 9, 2018, p. 1258.
690. Z. Kuncic and S. Lacombe, "Nanoparticle Radio-Enhancement Principles, Progress and Application to Cancer Treatment", *Physics in Medicine & Biology*, Vol. 63, 2018, p. 02TR01.
691. S.K. Frandsen, *et al.*, "Difference in Membrane Repair Capacity Between Cancer Cell Lines and a Normal Cell Line", *Journal of Membrane Biology*, Vol. 249, 2016, pp. 569–576.
692. B. Zhivotovsky and S. Orrenius, "Calcium and Cell Death Mechanisms: A Perspective from the Cell Death Community", *Cell Calcium*, Vol. 50, 2011, pp. 211–221.
693. E.L. Hansen, *et al.*, "Dose-Dependent ATP Depletion and Cancer Cell death following Calcium Electroporation, Relative Effect of Calcium Concentration and Electric Field Strength", *Plos One*, Vol. 10, 2015, p. e0122973. doi:10.1371/journal.pone.0122973
694. A. Zielichowska, *et al.*, "Applications of Calcium Electroporation to Effective Apoptosis Induction in Fibrosarcoma Cells and Stimulation of Normal Muscle Cells", *Bioelectrochemistry*, Vol. 109, 2016, pp. 70–78.
695. L. Kurzawa, *et al.*, "Dissipation of Contractile Forces: The Missing Piece in Cell Mechanics", *Molecular Biology of the Cell*, Vol. 28, No. 14, 2017, pp. 1825–1832. doi:10.1091/mbc.e16-09-0672
696. Y. Aylon and M. Oren, "Tumor Suppression by p53: Bring in the Hippo", *Cancer Cell*, Vol. 32, 2017, pp. 397–399.
697. A Jord, *et al.*, "Calibrated Mitotic Oscillator Drives Motile Ciliogenesis", *Science*, Vol. 358, 2017, pp. 803–806.

698. J. Burclaff and J.C. Mills, "Cell Biology: Healthy Skin Rejects Cancer", *Nature*, Vol. 548, August 2017, pp. 289–290.
699. K. Vukušić, *et al.*, "Microtubule Sliding Within the Bridging Fiber Pushes Kinetochore Fibers Apart to Segregate Chromosomes", *Developmental Cell*, Vol. 43, 2017, pp. 11–23.
700. T. Müller-Reichert, R. Kiewisz, and S. Redamann, "Mitotic Spindles Revisited — New Insights from 3D Electron Microscopy", *Journal of Cell Science*, Vol. 131, 2018, p. 211383.
701. S. Werner, A. Pimenta-Marques, and M. Battencourt-Dias, "Maintaining Centrosomes and Cilia", *Journal of Cell Science*, Vol. 130, 2017, pp. 3789–3800. doi:10.1242/jcs.203505
702. E.A. Nigg, L. Cajanek, and C. Arquint, "The Centrosome Duplication Cycle in Health and Disease", *FEBS Letters*, Vol. 588, 2014, pp. 2366–2372.
703. A. Pitaval, *et al.*, "Microtubule Stabilization Drives 3D Centrosome Migration to Initiate Primary Ciliogenesis", *Journal of Cell Biology*, Vol. 216, 2017, p. 3713.
704. T.D. Pollard, "Nine Unanswered Questions about Cytokinesis", *Journal of Cell Biology, ICB Home,* Vol. 916, 2017 Archive, p. 3007 ff.
705. S. Gilberto and M. Peter, "Dynamic Ubiquitin Signaling in Cell Cycle Regulation", *Journal of Cell Biology*, Vol. 216, 2017, p. 2259.
706. J.M. Valles, Jr., "Model of Magnetic Field-Induced Mitotic Apparatus Reorientation in Frog Eggs", *Biophysical Journal*, Vol. 82, 2002, pp. 1260–1265.
707. D.K. Ahirwar, *et al.*, "Non-Contact Method for Directing Electrotaxis", *Scientific Reports*, Vol. 5, 2015, pp. 11005.
708. K. Walter (Digital Reporter), "Gold Nanoparticles could Enable More Targeted Drugs", 2018, from an article published in *Nature Communications*.
709. G. Marteil, *et al*, "Over-Elongation of Centrioles in Cancer Promotes Centriole Amplification and Chromosome Missegregation", *Nature Communications*, Vol. 9, 2018, p. 1258.
710. J. Sun, M. Zborowski, and J. Chalmens, "Quantification of Both the Presence, and Oxidation State, of Mn in Bacillus Atrophaeus Spores and its Imparting of Magnetic Susceptibility to the Spores", *Biotechnology and Bioengineering*, Vol. 108, No. 5, 2011, pp. 1119–1129.
711. X. Jin, J.J. Chalmers, and M. Zborowski, "Iron Transport in Cancer Cell Culture Suspensions Measured by Cell Magnetophoresis", *Analytical Chemistry*, Vol. 84, 2012, pp 4520–4526.

712. J.J. Chalmers, *et al*, "Femtogram Resolution of Iron Content on a Per Cell Basis: Ex Vivo Storage of Human Red Blood Cells Leads to Loss of Hemoglobin", *Analytical Chemistry*, Vol. 14, 2017, pp. 3702–3709.
713. K.-H. Han and A.B. Frazier, "Diamagnetic Capture Mode Magnetophoretic Microseparator for Blood Cells", *Journal of Microelectromechanical Systems*, Vol. 14, No. 6, 2005, pp 1412–1431.
714. L. Wang, P. Park, and C-Y Lin, "Characterization of Stem Cell Attributes in Human Osteosarcoma Cell Lines", *Cancer Biology and Therapy*, Vol. 8, No. 6, 2009, pp. 543–552.
715. C-Y Lin, *et al.*, "Cancer Stem Cell Markers: What Is Their Diagnostic Value?", *Expert Opinion Medical Diagnosis*, Vol. 4, No. 6, pp. 474–481.
716. L. Wong, *et al.*, "Prospective Identification of Tumorigenic Osteosarcoma Cancer Stem Cells in OS99-1 Cells Based on High Aldehyde Dehydrogenase Activity", *International Journal of Cancer*, Vol. 128, 2010, pp. 294–303.
717. S. Beseme, *et al.*, "Transcriptional Changes in Cancer Cells Induced by Exposure to a Healing Method", *Dose Response*, 2018, pp. 1–8.
718. N. Lecland and J. Lüders, "Imaging and Quantifying the Dynamics of γ-Tubulin at Microtubule Minus Ends in Mitotic Spindles", *The Mitotic Spindle—Methods and Protocols*, Edited by P. Chang and R. Ohi, Humana Press, Springer Science and Business Media, New York, 2018, Chapter 5.
719. M.D.D. Miell and A.F. Straight, "In Vitro Kinetochore Assembly", *The Mitotic Spindle—Methods and Protocols*, Edited by P. Chang and R. Ohi, Humana Press, Springer Science and Business Media, New York, 2018, Chapter 8.
720. T.Y. Yoo and D. J. Needleman, "Studying Kinetochores In Vivo using FLIM-FRET", *The Mitotic Spindle—Methods and Protocol*, Edited by P. Chang and R. Ohi, Humana Press, Springer Science and Business Media, 2018, New York, Chapter 11.
721. C. Sacristan, *et al.*, "Dynamic Kinetochore Size Regulation Promotes Microtubule Capture and Chromosome Biorientation in Mitosis", *Nature Cell Biology*, Vol. 20, 2018, pp. 800–810.
722. J. Lüders, "XMAP215 Joins Microtubule Nucleation Team", *Nature Cell Biology*, Vol. 20, 2018, pp. 507–520.
723. M. Zanic, "Measuring the Effects of Microtubule-Associated Proteins on Microtubule Dynamics In Vitro", *The Mitotic Spindle—Methods and Protocols*, Edited by P. Chang and R. Ohi, Humana Press, Springer Science and Business Media, New York, 2018, Chapter 4.
724. N.H. Farina, *et al.*, "Nanoparticle-Based Targeted Cancer Strategies for Non-Invasive Prostate Cancer Intervention", *Journal of Cell Physiology*, Vol. 233, 2018, pp. 6408–6417.

725. J.J. Chen and L. Galluzzi, "Fighting Resistant Cancers with Iron", *Trends in Cell Biology*, Vol. 28, No. 2, 2018, pp. 74–78.
726. M. Mashreghi, *et al.*, "Angiogenesis Biomarkers and their Targeting Ligands as Potential Targets for Tumor Angiogenesis", *Journal of Cell Physiology*, Vol. 233, 2018, pp. 2949–2965.
727. B. Lacroix, *et al.*, "Microtubule Dynamics Scale with Cell Size to Set Spindle Length and Assembly Timing", *Developmental Cell*, Vol. 45, 2018, pp. 496–511.
728. J. Paz and J. Lüders, "Microtubule-Organizing Centers: Towards a Minimal Parts List", *Trends in Cell Biology*, Vol. 28, No. 3, 2018, pp. 176–187.
729. C.A.M. Lopez, *et al.*, "Centrosome Amplification Arises Before Neoplasia and Increases Upon P53 Loss in Tumorigenesis", *Journal of Cell Biology*, Vol. 217, No. 7, 2018 Archive, p. 2353.
730. J. Paz and J. Lüders, "Microtubule-Organizing Centers: Towards a Minimal Parts List", *Trends in Cell Biology*, Vol. 28, No. 3, 2018, pp. 176–186.
731. J.J. Chen and L. Galluzzi, "Fighting Resistant Cancers with Iron", *Trends in Cell Biology*, Vol. 28, No. 2, 2018, pp. 74–78.
732. J. Lončarek and M. Bettencourt-Dias, "Building the Right Centriole for Each Cell Type", *Journal of Cell Biology*, Vol. 217, No. 3, 2018, p. 823.
733. V. Sikirzhytski, *et al.*, "Microtubules Assemble Near Most Kinetochores During Early Prometaphase in Human Cells", *Journal of Cell Biology*, Vol. 217, No. 8, 2018, p. 2641.
734. L.M. Rice, "A New Look for the Growing Microtubule End?", *Journal of Cell Biology*, Vol. 217, No. 8, 2018, p. 2609.
735. H. Bae, *et al.*, "Carbon-Coated Iron Oxide Nanoparticles as Contrast Agents in Magnetic Resonance Imaging", *Springer Open Journal, Nanoscale Research Letters*, Vol. 7, 2012, p. 44.
736. M. Takeda, M. Sami, and Y.-C. Wang, "A Homeostatic Apical Microtubule Network Shortens Cells for Epithelial Folding via a Basal Polarity Shift", *Nature Cell Biology*, Vol. 20, 2018, pp. 36–45.
737. F.J. McNally and A. Roll-Mecak, "Microtubule-Severing Enzymes: From Cellular Functions to Molecular Mechanism", *Journal of Cell Biology*, Vol. 217, No. 12, pp. 4057–4069.
738. A. Vemu, *et al.*, "Severing Enzymes Amplify Microtubule Arrays through Lattice GTP-Tubulin Incorporation", *Science*, Vol. 361, 2018, p. 768.
739. C. Sacristan, *et al.*, "Dynamic Kinetochore Size Regulation Promotes Microtubule Capture and Chromosome Biorientation in Mitosis", *Nature Cell Biology*, Vol. 20, 2018, pp. 800–810.

740. A. Singh, *et al.*, "Polarized Microtubule Dynamics Directs Cell Mechanics and Coordinates Forces During Epithelial Morphogenesis", *Nature Cell Biology*, Vol. 20, 2018, pp. 1126–1133.
741. D.G. Booth and W.C. Earnshaw, "Ki-67 and the Chromosome Periphery Compartment in Mitosis", *Trends in Cell Biology*, Vol. 27, No. 12, 2017, pp. 906–913.
742. F. Bosveld, A. Ainslie, and Y. Bellaïche, "Sequential Activities of Dynein, Mud and Asp in Centrosome-Spindle Coupling Maintain Centrosome Number upon Mitosis", *Journal of Cell Science*, Vol. 130, 2017, pp. 3557–3567.
743. Z. Yue, *et al.*, "Mechanisms Mitigating Problems Associated with Multiple Kinetochores on One Microtubule in Early Mitosis", *Journal of Cell Science*, Vol. 130, 2017, pp. 2266–2276.
744. E.E.M. Furlong and M. Levine, "Developmental Enhancers and Chromosome Topology", *Science*, Vol. 361, 2018, pp. 1341–1345.
745. J.H. Gibcus, *et al.*, "A Pathway for Mitotic Chromosome Formation", *Science*, Vol. 359, No. 652, 2018.
746. S. Watanuki, *et al.*, "Characterization of Centriole Duplication in Human Epidermis, Bowen's Disease, and Squamous Cell Carcinoma", *Journal of Dermatological Science*, Vol. 91, 2018, pp. 9–18.
747. G.D. Gupta and L. Pelletier, "Centrosome Biology: Polymer-Based Centrosome Maturation", *Current Biology*, Vol. 27, 2017, pp. R833–R852.
748. C.A.M. Lopez, *et al.*, "Centrosome Amplification Arises Before Neoplasia and Increases Upon P53 Loss in Tumorigenesis", *Journal of Cell Biology*, Vol. 217, No. 7, 2018, pp. 2353–2363.
749. J. Lončarek and M. Bettencourt-Dias, "Building the Right Centriole for Each Cell Type", *Journal of Cell Biology*, Vol. 217, No. 3, 2018, p. 823–835.
750. N. Barvitenko, *et al.*, "Integration of Intracellular Signaling: Biological Analogues of Wires, Processors and Memories Organized by a Centrosome 3D Reference System", *BioSystems*, Vol. 173, 2018, pp. 191–206.
751. M. Arbi, *et al.*, "Controlling Centriole Numbers: Geminin Family Member as Master Regulators of Centriole Amplification and Multiciliogenesis", *Springer Chromosome*, Vol. 127, 2018, pp. 151–174.
752. A.D. Sanchez and J.L. Feldman, "Microtubule Organizing Centers: From the Centrosome to Non-Centrosomal Sites", *Current Opinion in Cell Biology*, Vol. 44, 2017, pp. 93–101.
753. I.M. Tolić, "Mitotic Spindle: Kinetochore Fibers Hold on Tight to Interpolar Bundles", *European Biophysics Journal*, Vol. 47, 2018, pp. 191–203.

754. J.W. Raff and R. Basto, "Centrosome Amplification and Cancer: A Question of Sufficiency", *Developmental Cell*, Vol. 40, 2017, pp. 217–219.
755. P. Ganesan and D. Narayanasamy, "Lipid Nanoparticles: Different Preparation Techniques, Characterization, Hurdles, and Strategies for the Production of Solid Lipid Nanoparticles and Nanostructured Lipid Carriers for Oral Drug Delivery", *Sustainable Chemistry and Pharmacy*, Vol. 6, 2017, pp. 37–56.
756. D.P. Ignacio, V.C. Cottman, and A.T. Dawes, "Centriole Biogenesis: Symmetry Breaking and Site Selection", *Trends in Cell Biology*, Vol. 29, 2019, pp. 3–5.
757. K. Vukušić, *et al.*, "Microtubule Sliding Within the Bridging Fiber Pushes Kinetochore Fibers Apart to Segregate Chromosomes", *Developmental Cell*, Vol. 43, 2017, pp. 11–23.
758. A.D. Sanchez and J.L. Feldman, "Microtubule Organizing Centers: From the Centrosome to Non-Centrosomal Sites", *Current Opinion in Cell Biology*, Vol. 44, 2017, pp. 93–101.
759. M.P. Gavilan, *et al.*, "The Dual Role of the Centrosome is Organizing the Microtubule Network in Interphase", *EMBO Reports*, Vol. 19, 2018, pp. 1–21.
760. G.D. Gupta and L. Pelletier, "Centrosome Biology: Polymer-Based Centrosome Maturation", *Current Biology*, Vol. 27, 2017, pp. R836–R838.
761. J.K. Mando and I.M. Chassman, "The Kinetochore-Microtubule Interface at a Glance", *Journal of Cell Science*, Vol. 131, 2018, p. 7.
762. M. Karki, N. Keyhaninejad, and C.B. Shuster, "Precocious Centriole Disengagement and Centrosome Fragmentation Induced by Mitotic Delay", *Nature Communications*, Vol. 8, 2017, 15803.
763. P. Guichard, *et al.*, "Cell-Free Reconstitution Reveals Centriole Cartwheel Assembly Mechanisms", *Nature Communications*, Vol. 8, 2017, p. 14813.
764. M.S. Levine, *et al.*, "Centrosome Amplification Is Sufficient to Promote Spontaneous Tumorigenesis in Mammals, *Developmental Cell*, Vol. 40, 2017, pp. 313–322.
765. C. Nuñez, S.V. Estévez, and M. del Pilar Chantada, "Inorganic Nanoparticles in the Diagnosis and Treatment of Breast Cancer", *Journal of Biological Inorganic Chemistry*, Vol. 23, 2018, pp. 332–343.
766. R. Blackwell, *et al.*, "Contributions of Microtubule Dynamic Instability and Rotational Diffusion to Kinetochore Capture", *Biophysical Journal*, Vol. 112, 1917, pp. 552–563.
767. Y. Lin and L. Pelletier, "A Magic Bullet for Targeting Cancers with Supernumerary Centrosomes", *The EMBO Journal*, Vol. 38, 2018, p. e101134.

768. K. Sztandera, M. Gorzkiewicz, and B. Klajnert-Maculewicz, "Gold Nanoparticles in Cancer Treatment", *Molecular Particles in Cancer Treatment*, Vol. 16, 2019, pp. 1–23.
769. I. Gasic and T.J. Mitchison, "Autoregulation and Repair in Microtubule Homeostasis", *Current Opinion in Cell Biology*, Vol. 56, 2019, pp. 80–87.
770. J. Kamenz and S. Hauf, "Time to Split Up: Dynamics of a Chromosome Separation", *Trends in Cell Biology*, Vol. 27, No. 1, 2017, pp. 42–54.

APPENDIX

1. Centriole Notes

**	Review Notes
	They are small orthogonal hollow cylinders (organelles) lying adjacent to the nucleus.
127 124	Size: 400–500 nm long and 200 nm in diameter.
127	The two cylinders are called: (1) "mother" and "daughter" with the daughter being attached to the proximal side of the mother. The axis of the mother intersects the axis of the mother, but not vice versa.
80 121	The daughter is shorter than the mother, about 80% long.
124 131	The mother centriole is more mature than the daughter.
	Each centriole is composed of nine sets (blades) of microtubules, three (3) each.
116 127 133	At the base (proximal ends) of the centriole pair (diplosomes) is a cloud of electron dense material known as the microtubule organizing center (MTOC).
	The MTOC is also known as pericentriolar material.
111 134	The MTOC is made up of hundreds of proteins.
	Centrioles and the MTOC are also known as the centrosome.

(*Continued*)

(*Continued*)

**	Review Notes
124 134	The centrosome is the only organelle in the body without a covering membrane.
47 60 109 111 127 199 208	The centriole plays the central role in mitosis. When mitosis is about to begin, each of the centrioles develops a daughter centriole at its proximal end and on its side. As these, daughter centrioles grow the original pair of centrioles and become four — two pairs — with the mother centriole having a new daughter and the daughter centriole having a daughter of its own.
	Once there are two pairs of centrioles, they begin to separate. The mother–new daughter is then known as the "mature" pair and the daughter–daughter is known as the "immature" pair.
116	Once the centriole pairs are separated, the immature pair departs and migrates to the opposite side of the nucleus.
	The two pairs then pull apart from one another (force at a distance), separating the nucleus and the remainder of the cell (cytoplasm) approximately dividing the organelles.
	As the centrioles are duplicating, the DNA is duplicating.
	If the centriole geometry is disturbed, there may be more than two centrioles in the dividing cell — so-called "supernumerary centrioles."
	When the centrioles are disturbed, the DNA is damaged — known as "aneuploidy."
90	Centriole geometry can be disturbed by more than one daughter growing at the proximal end of the mother centriole. Centriole damage can also be disturbed by loss of perpendicularity of the centrioles.
	Proteins Plk4 and SAS-6 are needed for centriole duplication. But too much can lead to abnormal geometry.
199	Centrioles form the polarity of the cell.

(*Continued*)

(*Continued*)

**	**Review Notes**
	Cancer cells have supernumerary centrioles [1, 4, 6 8, 9, 10, 18 21, 23 90, 95, 96, 100, 103, 104 107, 105, 110, 113, 143, 145, 152, 155, 192, 193, 195, 196, 202].
	Supernumerary centriole is known as "centrosome amplification."
	Centriole existence has been known for at least 50 years and centrosomes for 100 years [6, 54, 55, 56, 58, 60, 198, 200].
10	Aberrant centrosomes might offer a target for therapy.
30	Nanoparticles can easily pass through cell membrane. Nanoparticles are 1–100 nm in diameter.
60	Centrioles are the only organelle besides the nucleus that undergoes exact duplication.
60	The center of a centriole is called the lumen (ϒ-tubulin).
	Ref. 90 shows pictures (Figure 2) of multiple daughter centrioles in a flower-type configuration.
106, 127	The daughter centriole matures during mitosis.
124	Centrioles now of interest (2001).
133	The centriole is the organizer and power unit within a centrosome.
135	Without a membrane about the centrosome proteins and other chemicals, and particles can go in and out without significant resistance.
140 194	Centriole duplication and DNA replication occur at the same time.
194 209	The pericentriolar domain around a centriole pair is a cloud of material ~1 μm in diameter.
194	The centrosome extends to the cytoplasm and the organelles via microtubules [194].
183	Centriole duplication appears to occur simultaneously with DNA replication.

(*Continued*)

(*Continued*)

**	Review Notes
183	Plk1 is required to convert new centrioles to centrosomes as the cells progress from mitosis to interphase.
193	The daughter centriole arising on the side of the mother centriole is called "canonical assembly."
194	Centriole duplication occurs at the onset of DNA replication.
194	The pericentriolar domain is a cloud of 1 μm in diameter extending to the cytoplasm and orgauries via microtubules.
194	There are more than 100 proteins in the pericentriole materials.
197	STIL is recruited to nascent daughter centrioles at the onset of centriole duplication.
197	STIL cooperates with Plk4 and SAS-6 to control the centriole number.
199	Quote: "...the vicinity at the base of each centriole provides an environment that promotes the assembly of a procentriole which is oriented approximately perpendicular to each centriole."
199	SPD-2 and ZYG-1 are recruited before central tube formation, whereas SAS-5 and SAS-6 are recruited slightly later. This suggests that SPD-2 and ZYG-1 are regulatory rather than structured.
199	At the onset of procentriolar assembly, a cortwell due to protein SAS-6 is the first to appear. Then there is the appearance of MTs at the ends of the spokes.
203	Centriole assembly is triggered by an upstream signal mediated by SPD-2 and ZYG-1.
206	Plk4 is needed to initiate centriole duplication, but too much can lead to supernumerary centrioles.
209	Centrosome is ~1 μm.
209	Quote: "Indeed, it would seem that the answers to some of the big questions of tumorigenesis are hiding in small places like the centrosome."

(*Continued*)

(*Continued*)

**	**Review Notes**
221	Multiple centrioles cause cancer.
231	Centrosomes help establish the mitotic spindle, but defects can lead to cancer. These are: 1. Supernumerary centrioles 2. Excess pericentriolar material 3. Disrupted centriole barrel structure 4. Unincorporated microtubule complexes 5. Centriole of unusual length 6. Centriole functioning as auxiliary, basal bodies 7. Mispositioned centrosomes
233	Extra centrosomes may contribute to elevated mitotic index as seen in tumors.
234	Electron dense granules ~70–100 nm in diameter are observed in the centrosome.
236	The earliest event in cartwheel assembly is likely to be the recruitment of Plk4 on the surface of the proximal end of the mother centriole.
236	Centrioles have uniform length within a given cell type.
237	The daughter centriole arises from a "dot" of Plk4 on the proximal end of the mother.
237	It is not clear that intact mechanisms lead to the concentration of Plk4 and whether the choice of the site is at random or regulated by proteins.
237	How does Plk4 attract SAS-6 to form a contracted structure?
240	Pericentriolar material is electron dense.
240	Centrioles have the same size in all mammals.
240	During duplication, the mother centriole keeps the PCM.
240	The block to centriole reduplication is intrinsic to already duplicated centrioles that are engaged in the diplosome.
240	It is not clear why only one daughter forms near the wall of the mother. There is an unknown symmetry breaking mechanism.

(*Continued*)

(*Continued*)

**	Review Notes
240	Adding Plk4 (SAK) can cause flower petal duplication.
240	PCM prevents de nova duplication. Is this an EM effect?
248	Centrosomes are electron dense.
248	In normal cells there is only one point for daughter cells to form.
255	Abnormalities in centrosomes regarding the principal causes of CIN in cancer cells.
260	Engagement of centrioles with each other prevents duplication, but disengagement allows for duplication.
262	A licensing mechanism appears to govern centrosome duplication.
262	Centrosomes regarding the MTOC. Pericentriolar material (PCM) is "electron dense" matter.
262	As mitosis begins, the two centrioles lose their orthogonal arrangement and split slightly maintaining a flexible connection.
262	Daughter centriole always form perpendicular to the mother and at the proximal end.
262	It is unclear why daughter elongation stops when its length approaches that of the mother centriole.
262	It appears that regulatory constraints, as opposed to structural constraints, dictate the length and form of the daughter centrioles.
262	Centrosome duplication is coupled with DNA replication. This may be altered in tumorigenesis.
264	P53 seems to be a regulator of centrosome development.
265	Centriole duplication is controlled by the centrosome and not the nucleus.
265	Loss of p53 results in loss of duplication regulation.
266	P53 mutation can lead to centrosome aberration.
267	Radiation causes cell duplication.

(*Continued*)

(*Continued*)

**	**Review Notes**
269	Centrioles have been studied since the advent of electron microscopy.
270	Centrosome amplification (CA) is a "hallmark" of cancer cells.
275	Centrosome duplication occurs during the S-phase.
276	Quote: "While centrioles are relatively simple in construction, the mechanics of centriole biogenesis remain an important but yet poorly understood process."
276	Centriole duplication is similar to DNA replication.
276	Plk4, SAS-6, Cap64, SAS-5/Ana2/STIL produces flower-like daughter centrioles. Proteins preventing overduplication: (Talch 2, p. 1028): SKP1/SkpA, COL1, Stimb/beta-TrCP, nucleophos-min, SKP2, p53, CDK1, BRCA1, Cdc14B, Geminin, Orel.
277	Plk4 needed for duplication, but too much can lead to multiple daughter centrioles.
279	Centrosomes duplicate during S-phase.
279	Centrosome is MTOC.
279	Centrosome: two centrioles surrounded by electron dense material.
279	Centriole 9 MT triplets ~0.5 μm long and 0.2 μm in diameter.
279	Centrioles are surrounded by ~300 proteins, mostly coiled-coil structures.
282	Multiple centrosomes can occur in four ways: 1. Overduplication of centrioles (i.e. the "flower") 2. Failure to undergo cytokinesis (cell division) 3. Separation of centriole pair 4. Centrosome occurring without centriole
283	Centrosomal defects lead to tumorigenesis and CIN.
283	Centrosomal defects occur early in premalignant tissue.
283	Loss of p53 does not promote centrosome amplification.
283	Centrosome amplification in the absence of cytokinesis is rare.

(*Continued*)

(*Continued*)

**	**Review Notes**
284	Centrosome duplication requires cyclin/cdk2 kinase. Duplication depends upon cyclin E/cdk2 activity.
288	Centrosome abnormalities and chromosome abnormalities go together.
290	Preventing duplication of Mpsl in the centrosome causes centrosome reduplication.
291	Supernumerary centrioles lead to abnormal mitosis and tumorgenesis.
291	Multiple rounds of duplication are called "reduplication." Mechanism for reduplication are poorly understood (zero).
291	If the daughter centriole develops full length before separation, this may lead to reduplication.
291	Delayed disengagement promotes reduplication.
292	Plk1 is an activator of centriole disengagement. Plk1 and separase act at different times during the M-phase to license centrosome duplication.
292	During S-phase a new daughter occurs due to the influence of Cdh2/cyclin E.
293	Physical engagement between mother and daughter centrioles create a block to further centriole assembly.
292	Centriole disengagement is poorly understood.
292	Orthogonal configuration of the centriole plays a controlling role as preventing centriole overduplication.
293	Methods preventing centriole overduplication are poorly understood. Physical removal of the daughter induces reduplication of the mother in the S-phase.
293	Proposal: the size of the PCM cloud associated with the mother centriole restricts the number of daughters that can form simultaneously. If the PCM is too large multiple daughters can form.

(*Continued*)

(*Continued*)

**	**Review Notes**
293	New idea: daughters are formed in the PCM before attachment to the mother.
293	As long as the daughter is attached to the mother, the formation of additional daughters is prevented.
294	SAK aka Plk4 is a necessary regulator of centriole formation but overexpression leads to centriole amplification. SAK may be regulated by the cell itself.
295	Centrioles may assemble de novo after centrosome removal.
298	Centriole disengagement requires the protease separase.
300	C-Napl is necessary for regulated division.
332	The regulators for centriole duplication are the same as those for centrosome separation.
333	Quote: "The mother centriole is not a template but instead a platform for regulating the assembly of the daughter centriole" (really a quote?)
334	The loss of cell polarity is a hallmark of cancer.
334	Regarding cell polarity: the apical and basal sections of a cell are separated by a physical barrier called the "apical junction complex" (AJC).
335	The centrosome is an electron-dense material.
335	The centrosome increases in size at the onset of mitosis. This is called "centrosome maturation."
335	Although some details of the centriole structure are known, relatively little is known about the pericentriolar material (PCM). The reason is the rather homogeneous electron density of the PCM.
336	Centrosome abnormalities are common in cancer.
336	Centrosome clustering is essential for survival of cells with supernumerary centrioles.

(*Continued*)

(*Continued*)

**	**Review Notes**
337	Discusses the PAR complex — molecules which in normal cells regulate cell polarity but in abnormal cells it produces cancer.
337	During mitosis the organelles get arranged to form an axis of symmetry. This organization is known as the apical-basal polarity.
338	Not much new — not previously known. The focus is on the centrosome and PCM.
338	There are as many as 300 proteins in the PCM.
338	Overexpression of SAK/Plk4 leads to centrosome amplification. SAS-6: these proteins could be licensing proteins.
339	The focus of the paper is on CPAP — a protein related to SAS-4. CPAP controls the centriole length. Too little → insufficient daughter development. Too much → long daughter length → supernumerary centrioles → cancer.
339	The amount of PCM is proportional to centriole size.
340	SAS-4 is like tubulin in that it is incorporated into centrioles during their duplication. SAS-4 is necessary for centriole duplication.
340	Quote: "...(centrosome) duplication has long fascinated cell biologists, but the mechanism regulating it is poorly understood."
340	Quote: "Centriolar SAS-4 levels correlate with centrosome size."
340	Partial depletion of SAS-4 leads to defective centrioles.
341	SAS-6 self-assemblies and this forms the structure of the centriole core.
370	Centriole replication may be accelerated by a rectangular pulsed electrical current stimulation (RPECS).
370	At a rest state, the mother centriole has a strong positive charge whereas the daughter centriole is not charged. At an activated state the daughter centriole also has a strong positive charge.

(*Continued*)

(Continued)

**	Review Notes
370	When a pair of mother/daughter centrioles has a significantly large positive charge, the repulsive force overcomes the static force keeping them together.
370	Strong repulsive electric field forces drive centrioles apart.
370	Quote: "A strong externally applied imposed EMF could induce aberrant centrioles and centrosomes that cause life disease." (Known by Paul Schafer 40 years ago)
371	Centriole duplication appears to be driven by elements of the cytoplasm instead of by the nucleus.
372	On plant cells: The authors develop a theoretical model of plant cell mitosis — cells without centrioles. They conclude that mitosis is driven by electromagnetic forces similar to those in eukaryotic cells. (A number of assumptions are made.)
377	Procentrioles appear in male butterflies even with the centrioles still being engaged — but they do not grow and develop. This appears to be an unimportant anomaly.
378	The α,β-tubulin determines curvature at the distal ends of the MTs.
379	Dynein and NuMA exert tension forces to stabilize the minus ends of the MTs.
383	Plk1 is required for maturation of the centrioles. Quote: "This ill-understood centriole maturation process is almost two cycles long, and as a consequence, each cycling cell contains three generations of centrioles. Only one centriole per cell (the oldest one) is fully structurally mature and decorated with subdistal and distal appendages." (Vorobjev and Chentso 1982 and Paintraund *et al.* 1992)
386	2006 Quote: "The induced electric field by externally applied rectangular pulsed electric current stimulation (RPECS) could be mile compared with the spontaneous and natural electric field of the centrioles."

(Continued)

(*Continued*)

**	Review Notes
387	(1982) In interphase the structure of the torso centrioles differ: the mother has appendages but the daughter does not.
388	(1992) Quote: "... a comprehensive description of the overall organization of the centrosome in animal cells is lacking."
393	(2015) Hyperactivity of Plk4 induces centriole amplification.
394	Asterless (Asℓ) is a binding protein of Plk4. Asℓ acts as a scaffolding protein.
394	"...how Asℓ overexpression drives centriole amplification is unknown."
394	"Plk4 is attracted to the centrioles and deposits itself like a spot (asymmetrically) on each centriole as the centrioles are ready for duplication. It then provides a basis for developing new daughter centrioles."
394	Asℓ is thought to target plk4 toward the centriole.
394	Asℓ depleted cells do not duplicate.
395	Plk4 is a regulator of centriole duplication and it depends upon asterless.
395	Cep192 recruits Cep152 and Plk4 (Cep152 is Asterless).
396	Plk4 is a regulator of centriole duplication.
397	Asterless (Cep152) provides a molecular platform for Plk4. It interacts with the polo box of Plk4.
397	Too much or too little Plk4 causes a failure of centriole assembly.
398	Plk4/SAK is a regulator of centriole biogenesis. CEP152 interacts with Plk4 and is required for centriole duplication — an overduplication.
398	Interfering with CEP152 function prevents recruitment of Plk4 to the centrosome.
400	Defects in centrioles are found in cancer.
400	The first sign of a daughter centriole is the electron dense cartwheel (~25 nm in diameter) at the proximal end of the mother centriole.

(*Continued*)

(Continued)

**	Review Notes
401	Not much new here.
402	Proteins required for centriole formation: Plk4/SAK, Asterless/CEP152, SAS-6, SAS-5/Ana2/STIL, CEP135, and SAS-4/CPAP
402	Asterless (Asℓ) is a scaffolding protein which recruits Plk4 to the site on the daughter for assembly.
402	SAS-6 is a ninefold symmetric scaffold for the daughter centriole base.
407	Plk4 needs to be at a certain level before centriole duplication occurs, but then the Plk4 self-destruct before too many centrioles are formed.
409	Plk4 phosphorylates itself to promote its degradation. This guards against duplication instability.
414	Supernumerary centrosomes are a key cause of cancer.
414	Centrioles can occur de novo, but that is unlikely to occur if other centrioles already exist.
415	Protein Bld10p is a tip agent that connects the MTs to the SAS-6 cartwheel.
416	Abnormal centriole numbers are detected in tumor cells.
416	Cullin1 is a protein localized to maternal centrioles and is critical for suppressing centriole overduplication.
416	Multiple daughters can form from a single centriole. These are called "centriole flowers."
416	After Plk4, recruitment of SAS-6 is critical for procentriole assembly.
416	Cullin1 is critical for degradation of Plk4.
417	Abnormal centriole numbers lead to cancer. Overexpression of SAK/Plk4 leads to cancer.
417	There is a picture and drawing of a flowering centriole.
419	Plk1 promotes mitosis entry.
419	Plk1 is involved in chromosome maturation and separation.

(Continued)

(Continued)

**	Review Notes
419	Plk4 is involved in centriole replication.
1	Abnormal centriole numbers are detected in virtually all cancers.
2	Centrioles may be visualized by electron microscopy.
3	Centrioles can recruit pericentriolar material (PCM) to form microtubules from the beginning of the cell cycle.
3	Mother centrioles and daughter centrioles have different abilities to recruit microtubule material. Daughter centrioles have a large portion of their distal ends uncovered and have only their proximal ends embedded in the PCM.
3	Daughter centrioles become mature during mitosis.
3	Plk1 is needed for centriole disengagement.
4	Supernumerary centrioles probably contribute to the development of cancer.
5	Healthy centrioles duplicate only once per cell cycle.
5	Aberrations in centrioles are associated with cancer.
5	Plk4 and SAS-6 are required for centriole duplication. Plk1 is required for disengagement.
6	Aberrant centrioles lead to cancer.
6	Centriole existence has been known for over 50 years but they are just now becoming better understood.
6	There is a connection between extra centrioles and tumor growth.
7	The proximal end is negative and the distal end is positive.
8	Centriole defects promote cancer.
9	Extra centrosomes are suggested to be a common contributor to cancer.
14	Cancer cells have aberrant centrosome numbers.
17	Cancer cells have a wide range of centrosome abnormalities. Centrosome dysfunction results in cancer.
20	Centrioles form centrosome and cilia.

(Continued)

(*Continued*)

**	**Review Notes**
21	Cancer cells have centrosomal abnormalities.
22	G1 → S-phase → G2 → Mitosis. Centrioles are duplicated in S-phase and they mature in G2 phase.
23	Centrosome disruption can lead to disruption in cell tissue. Tumor cells have supernumerary centrioles and multiple centrosomes.
26	To form a new centriole, we need centrin-2, Plk4, and SAS-6.
27	Precise physical properties of the centrosome are yet to be determined.
420	Abnormal centriole numbers produce cancer.
420	Overexpression of SAK/Plk4, DSAS-6, and DSAS-4 leads to de novo centriole-like structures.
421	"Plk4 is the conserved master regulator of centriole assembly." "Taken together, our findings suggest that Plk4 directly generates its own phosphodegron and can do so without the assistance of an additional kinase."
422	In movement about the nucleus, the centriole speed was as fast as 10 μm/s.
423	During mitosis the mother stays near the center while the daughter migrates away.
424	ε-Tubulin is a component of the sub-distal appendages of the centriole. ε-Tubulin is required for centriole duplication and organization of pericentriolar material.
429	There is a direct link between centrosome activity and DNA replication.
430	"Centrosome" is from the Latin meaning "central body" (color drawings).
431	MT nucleation depends upon Υ-tubulin. The Υ-tubulin small complex (Υ-TuSC) then forms a scaffold for the α,β-tubulin dimers to initiate polymerization.

(*Continued*)

(*Continued*)

**	Review Notes
431	The pericentriolar material (PCM) is a highly structured consecutive layers of proteins enabling nucleation.
439	Asterless (*asℓ*) is needed for stable centrosome development.
440	RLH paper.
441	
448	The centrosome, in addition to being the MTOC is also essential for cell division vitality.
450	A review paper: indicates that centrosomes bind more than 100 regulatory proteins. The cellular regulation is the focus of the review.
451	Plk4 initiates centriole duplication and is regulated. The protein phophatase 2A regulatory subunit twins stabilize Plk4.
451	Plk4 localizes to the "parent" centriole as a spot and then it modifies the site.
454	Centrosome anomalies are detected in virtually all human cancers.
454	This is a review paper: It classifies abnormalities into three groups: (1) primary centrosome abnormalities, (2) transient centrosome accumulation, and (3) permanent centrosome accumulation.
454	Centrosome aberration is not necessarily a cause of chromosome instability (CIN).
454	"Centrosome anomalies are an astonishingly frequent finding in malignant tumors."
454	Centrosome abnormalities are likely a result of cellular insults.
455	(A 2001 review by E.A. Nigg). The focus is upon the many kinases needed for fidelity in cell division (M-phase) (mitosis follows after S-phase).
455	"M-phase comprises mitosis and cytokinesis. Mitosis is usually divided into five stages: (1) prophase, (2) prometaphase, (3) metaphase, (4) anaphase, and (5) telophase."

(*Continued*)

(*Continued*)

**	**Review Notes**
455	Before proceeding to M-phase, two events must occur in S-phase: (1) DNA replication and (2) centrosome duplication. (These are believed to occur simultaneously — RH.)
456	SAS-6 (human homologue GFP-HsSAS-6) is recruited to centrioles for daughter formation.
456	SAS-6 is recruited to the mother centriole and then associated with the emerging daughter centriole.
457	This paper presents the now widely accepted procedure of centriole duplication, but with differently named proteins.
457	(2006) "Centriole formation is a poorly understood process."
457	Five proteins have been identified for centriole formation: ZYG-1, SAS-4, SAS-5, SAS-6, and SPD-2. The order is: SPD-2, ZYG-1, SAS-5, SAS-6, and SAS-4 — for C. Elegans.
457	Just before centriole formation, the original pair will split slightly.
457	Excellent pictures and diagrams.
459	C. Elegans embryo: ϒ-tubulin stabilizes the formation of new centrioles.
459	The small size of embrioles makes it difficult to observe the kinematics — only snap shots.
464	The centrosome is not simply a cloud of proteins. Instead it is structured layers of matter.
465	(2012) Behavior of the centrosome in the cell cycle is still poorly understood.
465	The centrosome increases in size upon entry into mitosis.
465	Plk4 is at the head of the pathway for centriole formation. Asterless (*asℓ*) (Cep152 in humans) is a scaffold for Plk4 and SAS-6.
465	SAS-6 assembles the cartwheel structure.
465	Plk1 is required for the dramatic increase in pericentriolar material (PCM).

(*Continued*)

(*Continued*)

**	**Review Notes**
468	(2007) Author (Ganem) suggests there is a strong link between aneuploidy (chromosome instability) and cancer. In a later paper (2009), he suggests there is a link between extra centrosomes and chromosome instability. Papers [9] [195] [235] (duplicate entries).
471	Centrosomes increase dramatically after mitosis entry.
475	Cells without centrioles/centrosomes do not survive or thrive: Quote "These data reveal that the primary role of the centrosome in somatic cells is not to form the spindle but instead to ensure cytokinesis and subsequent cell cycle progression."
494	This is a brief summary and review of a report appearing in the same issue, *Science*, June 5, 2015. It states that irregular centriole function is related to cancer.
495	Cancer cells have extra centrosomes.
495	The authors developed a chemical ("centrinone") which inhibits Plk4. This destroyed (or damaged) the centrosomes. Normal cells died. Cancer cells continued to multiply.
498	Tumor cells have excess centrosomes.
498	This paper is a summary of a paper by Pagan *et al.* (*Nature Cell Biology*, Vol. 17, pp. 31–43). The paper says that the centriole pairs are loosely connected (i.e. disengaged). During mitosis (S-phase) each centriole develops a daughter centriole which is tightly connected to the mother centriole. As S-phase continues the two mother centrioles separate called "centriole disjunction" and then the tight connection between the mother and daughter loosens: called "centriole disengagement."
499	The centrosome increases dramatically in size upon mitosis.
499	The centrosome is not a featureless cloud, but instead an organized structure.
501	The MTOC has other functions when mitosis is not occurring.

(*Continued*)

(*Continued*)

**	Review Notes
502	This is the paper summarized by [498].
508	Under certain stimulated conditions (absence of p53 and overexpression of Plk4) supernumerary centrioles occur without resulting in a tumor.
510	This paper provides a brief description of cilia.
511	P53 is a protector against irregular cell development in the absence of Plk4. Aberrations in centrosome number can lead to cancer.
512	Plk4 primes the binding of STIL to SAS-6. Also, supernumerary centrosomes promote cancer.
513	This paper suggests physically expanding a cell to get better images (could be used to look for centrioles).
516	ϒ-TuRC is a lock washer shaped protein complex which produces MT and thus helps build the centriole.
517	ϒ-TuRC has 13 ϒ-tubulin molecules containing as many as 5 ϒ-TuSC.
512	ϒ-Tubulin provides a template for MT nucleation.
519	Quote: "Biochemical and physiological processes are based on the flow of electrical charges…"
525	There are in vitro studies suggesting significant alterations in cell division for wound healing.
525	EMF are not widely used in therapy due to an absence of a well-established mechanism for EMF bioeffects.
525	MF (magnetic fields) can affect cell proliferations both by increasing a slow rate (as in wound healing) and/or by inhibiting too high of cell proliferation.
529	SAS-6 imparts ninefold symmetry to the centrioles. SAS-6 forms a torus around the base of the parent centriole.
529	Paper discusses the use of STORM technology for imaging (buffer-enhanced 3D STORM).

(*Continued*)

(*Continued*)

**	Review Notes
529	3D STORM helps overcome the minute size of the centriole for imaging.
587	This paper explores how Asterless (*Asℓ*) and Plk4 interact. Overexpression of *Asℓ* recruits Plk4 but then regulates Plk4 for stability in new centriole development.
590	Asterless (*Asℓ*) is an ortholog of Cep 152.
610	This paper focuses on the protein GCP6 and its role in centriole construction. • GCP6 is needed by Plk4 (SAK) but it is not clear how • GCP6 is part of the ϒ-TuRC • GCP6 is localized to the perimeter of the centrosome at a distance away from the centrioles.
611	The paper focuses on STIL and CPAP for procentriole formation. STIL makes an asymmetric location on the daughter cell — required for new centriole formation. "Overexpression of STIL induces a formation of multiple around the parent centriole." Conversely, too little STIL prevents normal centriole development: CPAP induces centriole elongation. Quote: "...our results indicate that two MCPH-associated proteins STIL and CPAP interact with each other and are requested for precentriole formation, implying a central role of centriole biogenesis in MCPH." MCPH is "microcephaly."
612	Protein: pericentrim B is necessary and sufficient for disengagement during mitosis.
613	Centrioles cannot duplicate without pericentriolar material (PCM). They acquire PCM by becoming modified with a Plk1 dependence early in mitosis.
614	Authors show that asterless (*Asℓ*) is required for centriole (length control but not for pericentriolar material (PCM)) recruitment.

(*Continued*)

(*Continued*)

**	Review Notes
624	The author argues that the ϒ-tubulin ring complex (ϒ-TuRC) varies between different species.
628	The protein CEP295 interacts with the microtubules to elongate them. Too much → long centrioles; too little → short centriole.
633	Enlarged centrosomes (supernumerary centrioles) lead to cancer.
646	The paper is focused on centrosome symmetry in cells that divide by "self-renewing asymmetric division (SRAD)."
648	Defects in centriole number are associated with cancer. This is a review paper on how centrioles are assembled and controlled. Centriole proteins at the site of assembly: Plk4, Saas6, and STIL. Plk4 localizes at a dot and then recruits Sass6. It is not clear what targets Plk4 to the site on the mother centriole. It is an asymmetric event. The concentration of Plk4 is critical to limiting duplication to one site. Once the Plk4 is established a cartwheel with ninefold symmetry based on Sass6 is assembled. At the base of the ninefold is a stalk which keeps the centrioles connected. ϒ-Tubulin forms at the ends of the cartwheel to form microtubules. There is a triplet of microtubules A, B, and C (A is near the center and is the longest). The kinetics of the protein reactions and interactions appear to be unknown. Separase is required for separation of the centrioles. The authors present a model for centriole development with Plk4 being a limiting step.
651	The initial stage of centriole assembly is the cartwheel assembly — nine spokes. The central part is SAS-6 — Bld12p. The distal part is Cep135/Bld10p.
652	Centrioles are among the largest protein-based structures found in cell types. They are ~250 nm in diameter and 500 nm long. The paper is a review of ultrastructural observations about centrioles and associated structures. The paper largely repeats what is already known about centrioles from other papers.

(*Continued*)

(*Continued*)

**	Review Notes
	Centrioles are responsible for assembly of microtubules. The centrosome, which surrounds the centrioles is sometimes called the "microtubule organizing center" (MTOC). Here are some definitions: (1) Diplosome — a centriole and a developing centriole (2) Procentriole — a developing centriole (3) Basal body — a mother centriole
653	The terms: "centrosome," "MTOC," and "pericentriole material" are all the same. They are believed to be "electron dense."
653	The microtubule blades: ABC are close to each other. They provide stability to each other. This in turn provides stability for the microtubule.
653	The centriole pair is anchored to the nucleus.
654	This paper reviews theories as to why only one procentriole occurs with a mother centriole. The author states that the answer is still unknown.
655	Pericentriolar material (PCM) concentrates tubulin and serves as the primary organizing center for microtubules. The PCM has no membrane cover. Instead the protein constitutes the separate.
655	PCM appears as an amorphous mass surrounding the centriole. PCM proteins can self-assemble. PCM forms toroids about the centrioles.
656	The authors present a review and discussion on how centriole pairs separate. There is no mention of electromagnetic forces.
657	The author identifies various diseases aside from tumorigenesis that can occur with centriolar disfunction.
658	"Extra centrosomes are sufficient to initiate tumorigenesis in mammals." This quote is based upon findings in the article: M.S. Levin, *et al.*, "Centrosome Amplification Is Sufficient to Promote Spontaneous Tumorigenesis in Mammals", *Developmental Cell*, Vol. 12, 2016, 022, 2017.

(*Continued*)

(*Continued*)

**	**Review Notes**
659	Supernumerary centrioles promote tumorigenesis. Centrosome amplification results in aneuploidy.
659	Overexpression of Plk4 promotes centrosome amplification.
665	Authors state that kendrin regulates separase in centriole disengagement.
667	Centrioles can be elongated by chemicals.
672	Extra centrosomes are found in cancer cells. Sak overexpression leads to centrosome amplification (Sak is Plk4).
676	The presence of extra centrosomes imposes the tendency for catastrophic multipolar divisions.
677	Many efforts have been given to find cancer biomarkers. The paper provides little hope for finding prominent biomarkers (my assessment).
679	This article states that with the naked eye the smallest object we can see is approximately the size of the human hair. With a magnifying glass we can see about 10 times smaller. With light microscopy we can see about 300 times smaller to about 250 nm. With fluorescence microscopy we can see structures 20–30 nm across. (However, most proteins are less than 5 nm across.)
680	The authors introduce "MINFLUX" which localizes photon emitters in space. It minimizes the fluorescence photons needed for precision resolution. MINFLUX attains about 1 nm precision.
692	The paper presents a modelling of cell membrane response to introduction of an atomic force microscopy (AFM) probe.

**Refers to chronological listing of literatures reviewed.

2. Microtubule (MT) Notes

**	Review Notes
	Microtubules are hollow cylinders made up of filaments of α- and β-tubulin.
	OD ~ 25 nm, ID ~ 16 nm.
62 74	13 filaments around the circumference.
	Viewed from the side, the α and β tubules form a helix — they may thus point cw or ccw.
42	MTs are stiff in tension-compression, but weak in shear.
46 87	MTs are electrically polar.
47	MTs growing end is "positive."
59	MT fundamental frequency is ~465 MHz.
63 64 76	Good pictures.
67 68 87	MTs generate electric fields.
	α,β-tubulin together is known as a "dimer."
87	Vibration of MT generates an electric field.
87	The electromagnetic field of a cell can attract dielectric particles to the cell.
87	MT surface is slippery, allowing for longitudinal vibration.
88	Tubulin heterodimers have electric dipole moments.
	There may be a variety of types of MTs.
129	α,β-tubulin defines polarity.

(*Continued*)

(*Continued*)

**	Review Notes
127	ϒ-tubulin is at the negative end of an MT.
134	Negative end of MT in electron dense material.
137	α-tubulin: negative side; β-tubulin: positive side.
182	Outside of MT is smooth, inside is deeply corrugated.
127	In a centriole barrel, the MTs are not the same length: in a blade, they are labeled A, B, C with A > B > C in length and A on the inside, B in the middle, and C on the outside.
198	ϒ-tubulin is the linchpin of microtubule nucleation.
	The centrosome (the centrioles and pericentriole materials) (PCM) is the primary site of microtubule nucleation — aka MTOC (microtubule organizing center).
199	MTs first appear at the ends of the spokes of SAS-6 early on in the centriole duplication process.
240	ϒ-tubulin nucleates the MT.
361	MTs are electrically polar. Vibrations generate an oscillating electric field. α-β dimer produces a moment of: $337D = 1.12 \times 10^{-27}$cm.
362	Longitudinal vibration of MTs generate an endogenous electromagnetic field.
363	α, β-dimer dipole moment ~ 100 Debye (10^{-26} cm). Movement produces ~200–400 Debye.
364	Highest MT vibration frequency ~10^8 Hz, lowest frequency ~10^6 Hz.
364	MTs have 13 or 14 protofilaments.
364	α, β-dimer weighs 55,000 daltons. 1 dalton is mass of a proton ~1.66×10^{-27} kg.
364	Half-life of MT ~ 10 min.
364	Diameter of tubulin ~ 4 nm.
367	MTs are likely to be a source of cellular EMFs.
380	MTs do not generate, but instead maintain cell polarity.

(*Continued*)

(*Continued*)

**	Review Notes
384	MTs look alike in almost all species. How tubulin heterogeneity controls MT functions is still relatively unknown.
390	AFM can image in a liquid environment.
391	AFM studies of MTs are rare (2010).
391	Provides procedures for AFM studies.
411	ϒ-Tubulin plays a key role in MT formation. It is necessary but not sufficient.
411	ϒ-Tubulin was discovered ~ 1989.
412	Claims that centrosome separation and spindle forces are due to MT compressive forces.
13	MTs are ferromagnetic.
13	"It has been demonstrated that it is possible to orient arrays of MTs by the application of an electric field" (see also (38)).
25	MTs grow and shrink. Polarity of the tubulin dimer reflects the polarity of the MT.
26	To form a new centriole, we need centrin-2, Plk4, and SAS 6.
27	(2009) Surface charge density of a centrosome is $\sim e^3$ ($e = 1.6 \times 10^{-19}$ C). This is much smaller than the charge of an MT.
28	Paper presents a model of how alternating electric fields control microtubule-based transport.
29	(2011) A very impressive paper that asserts that electromagnetic fields are developed by oscillating MT and that the higher the frequency the greater is the generalized field. The paper also states that the electromagnetic field has cellular effects that go beyond the chemical effects.
29	MTs can vibrate in kHz to GHz regions.
424	Δ-Tubulin is required for the formation of triplet MTs.
425	ϒ-Tubulin is required for initiating MTs.
426	This paper presents a review of α- and β-tubulin structure. There are detailed color drawings of the tubulin in the figures.

(*Continued*)

(*Continued*)

**	**Review Notes**
426	MTs exist in all eukaryotes.
426	The tubulin dimers are connected head to tail.
426	MTs are highly dynamic and can switch stochastically between growing and shrinking phases.
427	Of the five tubulins α, β, Υ, δ, and ε, α and β make up the MT, Υ at the base for nucleation, δ is associated with the centriole, and ε with the pericentriolar material. (In (424) δ is associated with the triplets.)
428	(2000) MT assembly is initiated by Υ-tubulin. Other proteins (Spc 97p and Spc 98p) form a core of the forming MT.
431	MT nucleation depends upon Υ-tubulin. The Υ-tubulin small complex (Υ-TuSC) then forms a scaffold for the α, β-tubulin dimers to initiate polymerization.
432	(2010) Energy to MTs from: (1) (GTP)/hydrolysis of guanosine triphosphate, (2) energy transfer by motor (moving) proteins, (3) energy released by mitochondria as "wasted" energy in the citric acid cycle.
432	Paper considers axial (longitudinal) vibration of MTs.
432	Hetrodimers have an electric dipole moment of over 1,000 Debye (10^{-26} cm)
432	MT vibration frequencies from kHz to GHz.
433	Paper shows various ways MTs could oscillate.
434	Paper referred by David Lemmon.
435	Presents several modes of vibration of MTs with frequencies from MHz to GHz.
437	Widely referenced paper. Treats entire MT as a thin shell. Uses $E = 0.5 \pm 1$ GPa.
438	Nip is a centrosomal Plk1 substrate which is controlled by Plk1 in the development of MTs.
440 441	RH papers.

(*Continued*)

(*Continued*)

**	Review Notes
442	(2004) The electric dipole moment has only been calculated via computer simulation (not measured).
442	(2004) There exists little experimental about the electrical properties of tubulin and MTs.
442	MTs are not steady-state systems; energy is continually being put in and taken out.
443	Net charge on a tubulin dimer is ~0.19e (lower than calculated earlier).
444	Good pictures of α- and β-tubulin. The two tubulins are almost always nearly the same.
449	"Tubulin dimers have a strong electric dipole moment."
449	Tubulin dimer dipole moment: over 1,000 Debye = 110^{-26} cm (Coulomb moment).
449	MT oscillation frequency: from MHz to GHz.
460	α, β, ϒ-tubulin are found in all eukaryotic organisms and δ and ε-tubulin are found in organisms with triplet MTs.
461	ϒ-tubulin ring complex (ϒ-TuRC) forms a template for MT nucleation.
462	ϒ-tubulin small complex (ϒ-TuSC) helps form the 13-fold MT structure.
462	13-fold structure allows protofilaments to move straight without being twisted by three or four symmetries.
463	New MTs require ϒ-tubulin.
472	MT assembly is usually seeded by ϒ-tubulin ring complexes (ϒ-TuRC) and ϒ-tubulin small complexes (ϒ-TuSC).
473	Authors indicate that MT nucleation follows a pointed template (helix) as opposed to the protofilament model (2000).
474	(2000) ϒ-tubulin is required for MT nucleation but the mechanism is not understood (a near quote).

(*Continued*)

(*Continued*)

**	Review Notes
476	(2001) "The precise ways in which microtubules achieve their cellular function is not fully understood yet."
476	(2001) "MT arrays in cells are dynamic, capable of assembly, disassembly, and rearrangement on a time scale of seconds to minutes."
476	MT dynamics in vitro and in vivo are similar.
477	$\alpha\beta$-tubulin dimers self-assemble to form the MTs.
477	ϒ-Tubulin nucleates the MT. ϒ-tubulin is believed to be capable of longitudinal assembly (2000).
477	δ-Tubulin may seed triplets.
477	δ-Tubulin interacts with α-tubulin. ε-Tubulin interacts with β-tubulin.
478	The focus of the paper is on the behavior of ϒ-tubulin in nucleating MTs.
478	Tubulin dimers can sell-assemble in vitro.
478	ϒ-Tubulin is required for MT nucleation.
478	"ϒ-TuRC resembles a lock washer ..."
478	"ϒ-TuRC is like the first turn of the MT helix."
478	Two models are discussed for MT nucleation: (1) template and (2) protofilament. In template, the ϒ-TuRC is the first turn in an assembled helix. In protofilament, the ϒ-TuRC inserts the dimers longitudinally. The authors state that the in vitro results favor the template model.
479	MT nucleation rate remains constant during polymer assembly.
483	ϒ-Tubulin ring complex (ϒ-TuRC) required for Drosophila centrosomes to nucleate microtubules.
484	MTs are dynamically unstable exhibiting periodic rapid growth and shortening, β-tubulin is at the minus end.

(*Continued*)

(*Continued*)

**	Review Notes
485	(2000) The paper proposes that MT forces or organelles are formed by the curved filaments coming off the positive end of the MT attaching to an organelle.
486	This paper discusses the assembly of MTs about the chromosomes.
487	This is a review paper about the interplay of forces and the assembly of spindle dynamics.
487	Little is known about the underlying mechanical principles governing the assembly and function of the spindle.
487	Most animal spindles can be thought of as the superposition of kinetochore, nonkinetochore, and astral microtubules.
488	MTs alternating between elongating and shortening (apparently at random) is a process called "dynamic instability."
488	The time of elongation is known as "catastrophy time."
488	The catastrophy time is limited by limiting the role of tubulin addition — perhaps by force.
489	My dynamics can be described by four parameters: "(1) growth, (2) shrinkage, (3) catastrophy (transition from growth to shrinkage), and (4) rescue (transition from shrinkage to growth)." (edited quote)
490	Similar to [489] in the four phases.
491	Tubulin is abundant.
492	MTs can push and pull — a closely regulated process. The pushing and pulling seems to be attributed to adding and subtracting tubulin.
492	"Catastrophe" refers to a switch from growing to shrinking.
492	MTs interact with kinetochores and provide for the separation of chromosomes.
493	(2015) This paper presents a review of factors and behavior of ϒ-tubulin — a key MT protein.

(*Continued*)

(*Continued*)

**	Review Notes
493	MTs regarding nucleated from the centrosome (the MTOC).
493	The MTs are nucleated by the ϒ-TuRC (ϒ-tubulin ring complex).
493	The ϒ-TuRC and the ϒ-TuSC form a cone at the negative end of the MT (from papers 461, 473, and 478).
493	MTs can originate outside the MTOC, for example, at the kinetochore.
496	MTs interact with actin to form the shape, migration, and division of the cell. Actin is an abundant protein. (The monomeric form is called globulr or G-actin; the polymeric form is filamentous or F-actin.)
497	$EI = (1 \text{ to } 200) \times 10^{-24}$ Nm^2 (in vitro) $I = 10^{-32}$ m (I calculate 1.668×10^{-32} m^4) $E = 10^9$ Pa, where Pa is N/m^2.
503	Two ways are postulated as the ways the kinetochore attaches to the mitotic spindle.
504	This paper explores MT minus end regulation, but does not seem to come to firm conclusions.
504	This paper states that better understanding of MT dynamics is needed for new cancer strategies.
504	Paper states that the plus end dynamics are better understood than the minus end dynamics.
506	MTs can be developed from chromosomes. The development occurs via ϒTuRC, but the exact mechanism remains unknown. MTs from the chromosomes are essential for the success of the mitotic spindle.
507	The primary role of the pericentriolar material (PCM) is to anchor the MTs through the ϒ-tubulin ring complex (ϒTuRC). The PCM is **not** an amorphous material.
514	Kinetochores are large protein structures built on centromeric DNA. Attachments between MTs and kinetochores must be stable to insure successful mitosis.

(*Continued*)

(*Continued*)

**	**Review Notes**
516	ϒ-TuRC is a lock washer-shaped protein complex.
517	ϒ-TuRC is a 13 ϒ-tubulin molecule containing as many as 5 ϒ-TuSC.
518	(2012) ϒ-tubulin behavior is poorly understood.
519	ϒ-TuRC is formed of ϒ-TuSC.
519	ϒ-TuRC provides a stabilizing effect of the MT dynamics — but not clear how.
520	EI for MT $\approx 1.7 \times 10^{-23}$ Nm^2.
524	The magnitude of the electric field $\|E\|$ near a microtubule can be as large as 10^8 *V/m* or more. The force *F* on a dipole is $F = \nabla (P \cdot E)$ where *P* is the dipole moment and *E* is electric field intensity.
524	MTs are composed of subunits which are strong electric dipoles.
528	Paper proposes ferromagnetic properties for MTs.
591 100	Current MT drugs reduce the MT mass, but they also work at lower concentrations.
594	MTs can withstand higher compressive leads in a cell than in vitro due to surrounding lateral reinforcement.
595	(2011) The paper presents models for MT-kinetochore coupling.
598	Authors present models for the kinetochore/microtubule connection. The focus is on the activity of Ndc80 (not defined) in making the coupling.
599	The authors present some evidence and a theoretical basis for ring attachment between kinetochores and microtubules.
600	J.F. Tein, *et al.* (8) "Cooperation of the Dam1 and Ndc80 Kinetochore Complexes Enhances Microtubule Coupling and is Regulated by Aurora T3", *Journal of Cell Biology,* Vol. 189, 2010, pp. 713–741 (copy by mistake).

(*Continued*)

(*Continued*)

**	**Review Notes**
600	Not clear how microtubules and kinetochores couple Dam1 is from the kinetochores and Nda80 is from budding yeast. The study shows that Dam1 and Ndc80 are compatible.
601	"Kinetochores are macromolecular complexes built on a specialized chromatin domain called the centromere. Kinetochores provide a site of attachment for spindle microtubules during mitosis" (from the abstract). Similarly spelled words: Centrosome — centriole surrounding Centromere — from kinetochores Chromosome — nucleus (DNA).
602	The paper focuses on mitotic centromere associated kinesin (MEAK). Restriction of MEAK can prevent cancer cells from forming and also metastasis. MEAK is found throughout the cell but especially at the centromere and centrosome. MEAK is positively charged.
603	Authors present a model showing that the microtubule rotates the nucleus.
604	The author discusses mitosis section when the nucleus is not centered.
605	The author states that pericentrin provides the anchor for the ϒ-tubulin ring complex (ϒ-TuRC).
606	(1995) ϒ-tubulin is required for microtubule nucleation in vivo. ϒ-tubulin is a complex form of Xenopus with at least seven different proteins. It has an open ring structure.
607	Kinetochores appear to be more complex than microtubules. Kinetochores have more than 80 components. The connection with microtubules is unclear.
608	Paper requested by Roger Adelman. The essence of the paper is that microtubules are always shrinking and elongating, but not at the same time.

(*Continued*)

(*Continued*)

**	Review Notes
609	The focus of the paper is on the role of MCAK in coupling with the microtubule end. MCAK is a kinesin 13 (a catalysis). The acronym is *mitotic centromere associated kinesin*. MCAK is a regulator with a positive charge. The positive charges in the neck of the MCAK depolymerizes microtubules.
609	The mechanism of MCAK is not known. Several models are proposed: (quote) (i) A loose tether that allows the microtubule to diffuse along the microtubule lattice. (ii) An obstructive element that prevents tight bonding to the lattice. (iii) A "pry-bar" that destabilizes lateral interactions at the microtubule end. (iv) A facilitator of cross-linking between protofilament peels at the depolymerizing ends of microtubules.
618	The NDe80 complex is the point of contact between the kinetochores and the microtubules. The NDe80 has four subunits.
619	Authors propose a ring model for kinetochores connection to a microtubule (2006).
620	The authors use two-color fluorescence microscopy to image the proteins for budding yeast kinetochore-microtubule connection. The claim is that budding yeast connection is the same as in eukaryotic cells.
621	(2004) Paper says exact structure of kinetochore is unknown. However, the paper provides a very good picture model of the microtubule-kinetochore connection. The paper says that microtubule ends begin to explore cytoplasm after nucleus softening.

(*Continued*)

(*Continued*)

**	Review Notes
622	Kinetochore components comprise the direct interface with spindle microtubules. Kinetochores initially attach to the surface of a single microtubule. The kinetochore is then transported poleward. Nde80 is the important constituent that interacts with the microtubule.
623	The authors discuss the interactions of proteins to speed the kinetochore–microtubule connection.
624	Authors argue that the ϒ-tubulin ring complex (ϒ-TuRC) varies between different species.
626	Authors discuss the important effects of microtubules on 3D cell functions.
627	Microtubules can obtain rescue by incorporating α, β-dimer from the centrosome.
628	The protein CEP295 interacts with the microtubule to elongate them. Too much $\rightarrow$ long centrioles; too little $\rightarrow$ short centriole.
629	The authors summarize the results of Moriwaki and Goshima: "Flux Factors Can Reconstitute All Three Phases of Microtubule Polymerization Dynamics", *Journal of Cell Biology* (2016).
631	Damaged microtubules are able to incorporate new tubulin dimers and thereby recover their stiffness. The hollow cross-section of microtubules gives them stiffness.
631	Microtubules soften under mechanical stress, but softened microtubules can still self-repair. They have self-healing properties.
631	Figure 6 provides a good graphic model of microtubule damage ("catastrophe") and self-repair ("rescue").
634	Colliding microtubules can preserve their length.

(*Continued*)

(Continued)

**	Review Notes
641	Authors suggest that organelles move within a cell by attaching themselves to microtubules and other recruited molecular motors.
642	Authors use computer simulation to study microtubule buckling behavior.
643	Authors state that microtubule rescue, after catastrophy, can occur by α, β-dimers filling the gaps left by the catastrophy.
647	The author states that microtubule nucleation can occur at either the centrioles or at the Golgi apparatus.
675	(See also (550) and (621)) Kinetochores stochastically encounter microtubules and also possess quality control eliminating errors.

**Refers to chronological listing of literatures reviewed

3. Cancer Notes

**	Review Notes
	Cancer cells have a wide range of centrosome abnormalities [17, 21, 23, 1, 5, 91, 145, 194, 202, 205].
	Centrosome dysfunction causes genomic instability and cancer [17, 105, 107, 114, 146, 195, 202, 205, 210].
	Supernumerary centrioles contribute to cancer [4, 6, 9, 90, 96, 104, 105, 152, 183, 194, 195, 196, 205, 208].
	Centriole aberrations cause cancer. [5, 6, 8, 95, 104, 105, 113, 152, 193, 194, 202, 207].
61	Lack of adhesion leads to detachment of cancer cells — metastasis.
61	EM disturbances cause cancer.
63	Cancer cells are softer than normal cells.
70	Low frequency AC electric fields inhibit growth of tumors.
72	AC frequencies >75 Hz or intensities >8.5 μA causes cell death.
	Plk4 (aka SAK) overexpression induces multiple centrioles [206, 193, 121, 90, 73].
78	Cancer cells are less polarized than normal cells due to ion flow in and out of the cell (2013).
91 119	Cancer can occur when a single maternal centriole assembles multiple daughters concurrently.

(*Continued*)

(*Continued*)

**	Review Notes
	Cancer cells have supernumerary centrioles [96, 103, 143, 200, 202].
96	Cancer cells survive and thrive due to their clustering of supernumerary centrioles.
96 201	Cancer cells may die if the clustering of centrosomes can be prevented.
	Centrosome multiplication is associated with polyploidization [202, 113, 100, 107].
103	Centrosome amplification (CA) increases with tumor development.
139	Cancer cells have greater negative surface charge than normal cells (see [78]). [78] says cancer cells are *less* polarized than normal cells due to ion flow in and out of the cell [142].
139	There is a small steady extra cellular voltage gradient between cancerous tissue and neighboring normal tissue.
130	DC current affects the migration direction of metastatic cells.
142	Hypothesis: Cell membrane potential of cancer cells is lower than that of normal cells.
142	Cells with membrane potential below −36 mV are destined for cell division.
194	CA and DNA damage occur at the same time.
201	Cancer cells have centrosome clustering.
183	It has long been known that there is a strong correlation between aberrant centrosome number and aberrant chromosome numbers are related [73] (Hardy and Zecharias).
186	The steps required for metastasis are similar for all tumor cells.
191	Angiogenesis: Blood vessels form in direct response to tissue demands.
194	Centrosome aberration are common in cancer cells.
194	Errors in centrosome duplication cause aneuploidy and cancer.

(*Continued*)

(*Continued*)

**	**Review Notes**
194	Supernumerary centrosomes occur in almost all human solid tumor cancer cells.
194	Aneuploidy (the gain or loss of a chromosome) is a cause of cancer.
195	Extracentrosomes alone are sufficient to promote centrosome segregation.
195	There is a direct link between extrachromosomes and chromosomal instability (CIN) — which occurs in cancer.
197	STIL cooperates with SAS-6 and Plk4 in the control of centriole number.
201	Ned aka HSET a minus ended kinase is needed for extra chromosome development, but not for regular cell development.
201	Cancer cells have centrosome clustering.
201	Disruption of centrosome clustering can selectively kill cancer cells.
202	Experiments on breast cancer tumors show a 60%–80% amplified centrosome.
202	There is a linear correlation between aneuploidy and centrosome instability. Centrosome instability is an early event in cancer.
202	Centrosome instability may be an early sign of cancer and thus a target for treatment.
205	Centrosomes are abnormal in aggressive carcinomas.
205	Centrosome defects contribute to the earliest stages of cancer development via CIN.
205	Centrosome defect can arise due to dearth of p53.
206	Cancer has extra centrosomes.
206	Overexpression of Plk4 promotes the formation of extra centrosomes.

(*Continued*)

(*Continued*)

**	**Review Notes**
207	Tumors have supernumerary centrosomes.
209	Quote: "Indeed, it would seem that the answers to some of the big questions of tumorigenesis are hiding in small places like the centrosome."
210	Centrosome malfunction causes chromosome damage.
210	P53 is a tumor suppressing gene. Absence of p53 allows for centrosome replication.
218	There is a difference between mature blood vessels and angiogenic blood vessels. The angiogenic blood vessels are weak and they leak.
220	Many cancers arise in epithelia, the tissue that lines all body cavities.
220	Metastatic cells are softer than normal cells.
229	Supernumerary centrosomes lead to multipolar spindles seen in human cancer.
231	Centrosomes help establish the mitotic spindle, but defects in chromosomes can lead to cancer. Here are the ways: (1) Supernumerary centrioles (2) Excess pericentriolar material (3) Disrupted centriole barrel structure (4) Unincorporated microtubule complexes (5) Centrioles of unusual length (6) Centrioles functioning as ciliary basal bodies (7) Mispositioned centrosomes.
233	Extra centrosomes may contribute to an elevated mitotic index as seen in tumors.
235	Extra centrosomes produce CIN.
244	Multiple centrosomes cause cancer cells to divide.
244	Centrosomal clustering leads to division of cancer cells. Centriole clustering leads to CIN.

(*Continued*)

(*Continued*)

**	**Review Notes**
245	Centrosomes in tumors are atypical in shape, size, and composition and in multiplicity.
245	All tumor cells have centrosome defects.
246	Numerous centrosomes are seen in cancer cells. The number of extra centrosomes may be a measure of the degree of the cancer.
248	Overexpression of SAK/Plk4 and SAS-6 causes the "flower" phenomenon (multiple daughters).
249	P53 — regulated centrosome duplication but [250] disputes this.
251	The presence of more than two centrosomes during mitosis is characteristic of cancer.
251	There is a survival of the fittest of cancer cells via clustering of centrosomes.
255	Centrosome abnormality is present in "almost all types of solid tumors."
255	CIN accelerates tumor progression.
255	Centrosome abnormalities are the principal cause of CIN in cancer.
255	Quote: "... it is vital to understand how duplication, numerical integrity and mitotic function of the centrosome are controlled at the molecular level. Understanding this should provide fertile ground for designing more effective cancer diagnosis and intervention protocols."
255	Quote: "The fact that many oncogenes and tumor suppressors control centrosome duplication and function further highlights the suggestion that centrosome cancer targets."
256	Abnormal centrosome numbers produce tumors.
262	Unclear if centrosome abnormalities are a cause or a result of tumorigenesis.
262	Supernumerary centrioles may be due to an uncoupling with DNA duplication.

(*Continued*)

(*Continued*)

**	**Review Notes**
263	Supernumerary centrioles produce aneuploidy.
263	Quote: "... a condition that favors the overproduction of centrosomes could contribute directly to the initiation of chromosome imbalance."
264	CA leads to CIN.
264	All major cancers have CA.
201	Quote: "The existence of mechanisms that suppress multipolar mitosis raises the possibility of a novel therapeutic strategy for cancer: Drugs that interfere with centrosome clustering mechanisms would be lethal to tumor cells containing multiple centrosomes" pc33?
267	Radiation causes centriole duplication.
270	CA is a "hallmark" of cancer cells. It is the major underlying mechanism for multipolar mitosis, CIN, and aneuploidy.
270	Overexpression of Aurora-A kinase is a cause of CA, aneuploidy.
270	CA has been a target for cancer therapy via cherico for 10 years.
573	CA gives rise to CIN; CA is associated with spindle abnormalities in breast cancer.
275	CA occurs in human cancers and is a source of CIN.
277	Plk4 needed for centriole duplication, but too much leads to multiple daughter centrioles.
277	Supernumerary centrioles tend to cluster together.
283	Centrosomal defects lead to tumorigenesis and CIN.
283	Centrosomal defects occur early in premalignant tissue.
283	DNA damage occurs with changes in centrosome number.
286	CA can arise independently of ER or p53.
286	The aggressiveness of breast cancer is proportional to the degree of amplified centrosomes.

(*Continued*)

(*Continued*)

**	**Review Notes**
287	ϒ-Radiation for cancer treatment causes CA which leads to cell death — very interesting in that CA normally leads to cancer.
288	Little new here but centrosome abnormalities and chromosome abnormalities go together.
290	A prolonged S-phase produces extra centrosomes aka centrosome reduplication.
291	Supernumerary centrioles lead to abnormal mitosis and tumorigenesis.
297	*Six hallmarks of cancer:* (1) Sustaining proliferation signaling (2) Evading growth suppressors (3) Resisting cell death (4) Enabling replicative immortality (5) Inducing angiogenesis (6) Activating invasive and metastasis. Plus three more: (1) Repreprograming energy metabolism (2) Evading immune destruction. Acquiring normal cells in the tumor to create a tumor microenvironment.
299	(3) Plk1 is overexpressed in cancer cells.
302	Aurora A is a cancer susceptibility gene. Aurora A is a serinel/threonime kinase required for centrosome and spindle formation during mitosis.
315	Quote: "Differences in the expression of cellular receptors between normal and tumor cells represent a greater opportunity for targeting imaging probes to those cellular surface materials."
313	Tumor cells have heterogeneous expression levels of targeted receptors.

(*Continued*)

(*Continued*)

**	**Review Notes**
320	Angiogenesis studied after but yet poorly understood. Resistance to anti-angiogenesis drugs develops quickly.
320	Angiogenesis usually arises from capillaries rather than from larger vessels of veins and arteries.
320	Tumor blood vessels are leaky, irregular, tortuous and enlarged. They have hypoxia and thus promote VEGFA (VEGF - vascular endothelium growth factor). Endothelium — a simple sheet of flattened cells lining the blood vessel.
322	An optimistic article about the potential of anti-angiogenesis.
324	Anti-angiogenesis therapy works only temporarily. It is not clear why.
336	Centrosome abnormalities are common in cancer.
336	(2011) Quote: "Thus it is tempting to speculate that the chromosome instability generated by extra centrosomes can contribute to tumorigenesis."
336	Centrosome clustering is essential for survival of cells with supernumerary centrosomes.
337	Discusses PAR complex — molecules which in normal cells regulates cell polarity, but in abnormal cells it produces cancer.
342	Hallmarks of cancer: • Limitless replication potential • Insensitivity to regulatory signals • Evasion of apoptosis (cell death) • Sustained proliferation signals • Induction of angiogenesis • Invasion through capillary walls • Metastasis
343	Dying cells recruit new blood vessels out of the capillaries.
342	Leaky vasculature of tumors makes drug delivery more difficult.

(*Continued*)

(*Continued*)

**	**Review Notes**
342	Targeting the endothelial cells of tumor blood vessels is an attractive strategy for inhibition of the growth of the tumor.
342	VEGF.
343	As cells start to die they secrete growth factors that trigger the budding of new blood vessels from the surrounding capillaries (angiogenesis). Rapid angiogenesis promotes irregular vasculature.
352	Fast growing cancer cells recruit new blood vessels and with an imbalance of angiogenic regulation. This leads to highly disorganized and dilated vessels with numerous pores (leaky).
352	Microenvironment around tumor cells is different than that of normal cells. The supply of oxygen and nutrients is not sufficient to maintain growth rate. Therefore, the tumors use glycolysis to get energy and it results in an acidic environment. Glycosis — an enzymatic breakdown of carbohydrate.
353	Magnetitie nanoparticles (40 nm) with positive surface charge were internalized into human breast cancer whereas they were not internalized into normal cells.
362	Malignancy can occur due to magnetic field.
363	Quote: "Physical processes in cancer cell are disturbed due to suppression of oxidative metabolism in mitochondria (Warburg effect)."
363	Suppressive of oxidative metabolism is a general feature of cancer cells.
363	Disturbances of EM activity in cancer cells might be responsible for malignant properties.
365	Low electric fields kill cancer.
366	Low frequency EM fields are successful in treating tumors.

(*Continued*)

(*Continued*)

**	**Review Notes**
368	Not much new here, but there is a potentially useful quote: "Measurement of differences of the electromagnetic field between healthy and cancer cells using microelectronic and nanotechnological systems will crucially contribute to understanding the cancer process."
381	Stroma cells are surrounding environmental cells of cancer cells. They evolve with cancer cells. P53 acts against stroma cells. HSF1 enhances stroma cells (HSF1 is "heat shock factor 1").
382	Tp53 (p53) is the most frequently mutated gene in cancer, being altered in ~50% of human malignancies" p53 is thus a target for therapies. Until recently p53 was regarded as "undruggable," but the authors suggest that that is no longer the case and new drugs are being tested to interact with p53.
389	Chromosome amplification is a factor in cancer.
389	Moesin is a FERM dominant protein. It enhances the formation of CA and clustering.
389	Moesin is highly expressed in basal type breast cancer.
393	(2015) Hyperactivity of Plk4 induces centriole amplification.
393	Plk4 is overexpressed in cancer.
394	"... how *Asℓ* overexpression drives centriole amplification is unknown."
398	There is a correlation between excess centrosomes aneuploidy and cancer.
399	Plk4 is a principal player in the formation of centriole flowers (multiple daughters at a single maternal centriole).
400	Defects in centrioles are found in cancer.
403	CA is a common feature of cancer cells.
404	Too much SAK/Plk4 promotes tumors.
405	Plk1 is highly expressed in tumors.
410	Abnormal centriole dynamics promotes cancer.

(*Continued*)

(*Continued*)

**	**Review Notes**
410	Aneuploidy is a common feature of cancer.
413	Cancer tumors have an abnormal number of centrosomes.
414	Supernumerary centrosomes are a key cause of cancer.
415	Abnormal centriole numbers are detected in tumor cells.
416	Multiple daughters can develop from a single centriole mother. These are called "centriole flowers."
417	Abnormal centriole numbers lead to cancer. Overexpression of SAK/plk4 leads to cancer.
417	There is a picture and image of a flowering centriole.
418	Loss of p53 is not by itself a primary cause of aneuploidy. This is a 2005 review summarizing the then current knowledge of aneuploidy and p53.
418	Aneuploidy — the presence of an abnormal number of chromosomes in the nucleus. This is a common finding in cancer.
1	Abnormal centriole numbers are detected in virtually all cancers.
1	Overexpression of Plk4 stimulates the concurrent formation of multiple daughter centrioles at single maternal centrioles.
4	Supernumerary centrioles probably contribute to the development of cancer.
5	Aberrations with centrioles are associated with cancer.
6	Aberrant centrioles lead to cancer.
6	There is a connection between extra centrioles and tumor growth.
8	Centriole defects promote cancer.
9	There is a direct link between extra chromosomes and CIN.
9	Cancer cells multiply 10–100 times more frequently than normal cells.
9	Extracentrosomes are suggested to be a common contributor to cancer.

(*Continued*)

(*Continued*)

**	**Review Notes**
10	Human tumors have centrosome aberrations.
14	Cancer cells have aberrant centrosome numbers.
16	Electric fields inhibited the growth of dermal tumors in mice.
16	Sequentially programmed magnetic field theory (SPMF) has been shown to arrest brain-tumor growth in a patient with previous poor prognosis.
17	Cancer cells have a wide range of centrosome abnormalities. Centrosome dysfunction results in cancer.
21	Cancer cells have centrosomal abnormalities.
23	Centrosome disruption can lead to disruption in cell tissue. Tumor cells have supernumerary centrioles and multiple.
24	SPMF may be a cancer therapy.
27	Small electric fields can influence the properties of cell division.
30	(2011) Nanoparticles (NPs) easily penetrate cellular membranes. NPs have a range in size from 1 to 100 nm.
419	Plk1 is overexpressed in many cancers.
420	Abnormal centriole numbers produce cancer.
420	Overexpression of SAK/Plk4, SDas6, and DSas4 leads to de novo centriole-like structures.
440	RH papers.
441	
445	Reduced Plk4 leads to cancer.
446	All Plks are similar.
447	Paper referred by M. Schulz. The hypothesis is that a high molecular weight protein: fibronectin in the extracellular matrix of metastatic breast cancer — can be used as a biomarker.
447	(See above) A tiny magnet is used to hold toxins together, then attack the fibronectin biomarker.
451	Supernumerary centrioles can cause aneuploidy and tumorigenesis.

(*Continued*)

(*Continued*)

**	**Review Notes**
453	Cancer can occur with a weakening of the controls of proteins during mitosis.
454	Centrosome anomalies are detected in virtually all human cancers.
454	Centrosome abbreviations is not necessarily a cause of CIN.
454	"Centrosome anomalies are an astonishingly frequent finding in malignant tumors."
458	Paper suggests that aneuploidy promotes tumorigenesis.
466	There is an uptick in Plk1 in human tumors.
467	Plk1 is a potential drug target for cancer therapy.
468	(2007) Author (Ganem) suggests there is a strong link between aneuploidy (CIN) and cancer. In a later paper (2009), Ganem suggests there is a link between extra centrosomes and CIN Papers [9, 195, 235] (duplicate entries).
469	Plk1 is overexpressed in tumors.
482	Paper lists breast cancer types.
491	Small molecular inhibitors of MT development are used in chemotherapy. This paper presents a way to enhance the chemical effect via light.
494	This is a brief summary and review of a report appearing in the same issue: *Science*, June 5, 2015. It states that irregular centriole function is related to cancer.
495	Cancer cells have extra centrosomes.
495	The authors developed a chemical ("centronone") which inhibits Plk4. This destroyed (or damaged) the centrosomes. Normal cells died. Cancer cells continued to multiply.
498	Tumor cells have excess centrosomes.
504	This paper states that a better understanding of MT dynamics is needed for new cancer treatment strategies.
505	Cancer cells use the dying normal cells as fuel for growth.

(*Continued*)

(*Continued*)

**	**Review Notes**
508	Under certain stimulated conditions (absence of p53 and overexpression of Plk4) supernumerary centrioles occur without resulting in a tumor.
511	P53 is a protector against irregular cell development in the absence of Plk4. Aberrations in centrosome number can lead to cancer.
512	Plk4 primes the binding of STIL to SAS-6. Also, supernumerary centrosomes promote cancer.
513	This paper suggests physically expanding a cell to get better images (could be used to look for centrioles).
523	(2001) This paper presents a theoretical model of breast cancer based upon electromagnetic effects. It documents via references that there is a difference in tumor EM than in normal tissue.
525	Little research has been given to exploring the beneficial effects of magnetic and electromagnetic fields for treatment of injury or disease.
525	Magnetic fields (MF) can affect cell proliferation either by increasing a slow rate (as in wound healing) and/or by inhibiting too high of cell proliferation.
529	Depletion of SAS-6 results in a failure of centriole formation. Overexpression of SAS-6 results in supernumerary centrioles.
530	"Magnetic targeting" uses an externally applied magnetic field to move a magnetically charged drug carrier to a tumor.
545	Extra centrosomes promote cancer and CIN.
554	"Iron oxide balls, just a few nanometers in diameter heat up when exposed to powerful magnetic field." Neatly stacked NPs need a stronger field than random NPs.
555	In cancer cells folic acid is 10 times as great as that in normal cells. The author proposes folic acid as a biomarker.

(*Continued*)

(*Continued*)

**	**Review Notes**
555	Dendrimers are nano-sized polymers ("starburst polymers") which can efficiently be targeted for tumors.
556	The author proposes a three-stage delivery of NPs to tumors: (1) blood borne, (2) into large tumor vasculature, and (3) breakdown of 100–100 nm NP and into the tumor core.
557	NPs may be designed for binding to cell membranes or to cell interior biomarkers.
557	A tumor obtains nutrients "by passive diffusion until it reaches a size of ~2 $(mm)^3$. Then to continue growing, new blood vessels must develop ('angiogenesis')."
557	Angiogenesis is *not* orderly and tumors have poor vasculature in some regions and strong vasculature in others.
557	Tumor blood vessels are abnormal and leaky. The leak gap is ~100–780 nm (normal gap size is 5–10 nm).
557	High pressure in tumor blood vessels tends to push out chemo drugs. The pressure is higher in the tumor center and lower in the periphery.
557	*Bio NPs*: (1) *Dendrimers* — These are branching tree-like structures. (2) *Micelles* — These are spherical globular structures with a hydrophobic central core. (3) *Nanospheres* — These are spherical structures composed of a matrix of curved bars. (4) *Nanocapsules* — Fluid filled pouches. (5) *Fullerenes* — Carbon bucky ball. (6) *Nanotube* — Hollow carbon cylinders. (7) *Liposome* — Fluid filled spheres.
558	This is a review paper on current targets for nanoparticles. These are: (1) angiogenesis, (2) uncontrolled cell proliferation, and (3) tumor cell targeting.

(*Continued*)

(*Continued*)

**	Review Notes
558	A $2(mm)^3$ tumor → angiogenesis: Leaker blood vessels have 100 nm + 2 μm dia holes. High internal pressure.
560	Superparamagnetic iron oxide nanoparticles (SPIONS) become magnetic in a magnetic field. They are, however, very dependent on their structure — particularly size and size distribution.
560	Folic acid is a biomarker for SPIONS.
560	SPIONS are applied intravenous and also by direct injection into the tumor.
561	Superparamagnetic nanoparticles were surface-modified with poly(ethylene glycol) (PEG) and folic acid respectively (quote from abstract). The PEG nanoparticles appeared to perform better (i.e., more uptake into the cancer cells) than the folic acid particles, although for breast cancer both performed well.
562	Magnetic NPs can be maneuvered through the body by external magnetic fields. But if the NPs are too deep the procedure becomes less effective.
563	Authors injected SPIONS into rat glioma tumors and then heated them, shrinking the tumor.
565	All tumors have molecular markers that do not occur in normal tissue.
565	This is a review paper focusing on molecular markers created by tumor vasculature.
565	Drugs only penetrate a short distance into the tumor. Tumor blood vessels have large diameters and high blood pressure causing blood outflow. Tumor blood vessels are tortuous, uneven in diameter and leaky.
566	The peptide: iRGD goes not only into the tumor extra vasculature but it also penetrates into the tumor itself.
568	Paper reports on tests of two NPs: nanoshears (Ns) and nanoworms (Nw). These were SPIONS. The Nws were more effective.

(*Continued*)

(*Continued*)

**	**Review Notes**
570	NPs can be targeted to extravascular tumor tissue.
571	As a tumor grows, its size, pushing against normal tissue, creates pressure within the tumor and thus contributes to metastasis.
572	This is a review article. Gold NPs (AuNPs) are discussed for diagnostic imaging. The authors provide a listing of delivery systems.
572	AuNPs are relatively easy to manufacture.
572	Delivery: (1) particle size reduction, (2) peptide platform, (3) small molecule, and (4) proteins.
574	This paper lists primarily blood tests used to indicate the presence of tumors. It is based on a search for biomarkers.
575	This paper presents folic acid as a biomarker for NPs.
578	This is a review paper summarizing the targeting of NPs to specific metastases.
580	The authors use poly amidoamine dendrimer dendrites (tree-like NPs) to deliver toxins to tumors by focusing on folic acid as a biomarker.
580	The folate receptor is over expressed in breast, ovary endometrium, kidney, lung, head and neck, brain, and myeloid cancer.
581	The paper describes the use of silica NPs with a superparamagnetic iron oxide core and attachments. Folic acid is the tumor biomarker.
585	The objective of the paper is to report on NP targeting of blood vessels feeding the tumor. It states that integrin $\alpha\nu\beta3$ is found in angiogenesis. $\alpha\nu\beta3$ is thus a biomarker.
586	There is a significantly higher number of centrioles in malignant tumors compared with the number in benign tumors and normal tissue. There is no significant difference in the number of centrioles between benign tumors and normal tissue.

(*Continued*)

(*Continued*)

**	**Review Notes**
599 100	Current MT cancer drugs reduce the MT mass, but they also work at lower concentrations.
593	This is a disturbing paper: It speaks of measuring small magnetic moments with a general purpose, superconducting quantum interference device magnetometers. It discusses counting SPIONS in and about normal and cancer cells. It says nothing about the magnetism of the cells themselves. It talks about: "... magnetic nanoparticles biologically reacting with living cells."
602	The authors focus on mitotic centromere associated kinesin (MEAK). Restriction of MEAK can prevent cancer cells from forming and also metastasis. MEAK is found throughout the cell but especially at the centromere and centrosome. MEAK is positively charged.
625	In an alternating magnetic field, the SPION changes its magnetic field direction. Therefore, if a SPION can get inside a cell, an externally applied alternating field can cause the SPION to heat the cells to death.
630	Cancer cells kill endothelial cells of blood vessels causing a gap for the cancer cells to get out of the blood vessel and build a tumor in the adjacent tissue.
632	Blue vessels supporting tumors are tortuous, weak, and leaky. Tumors, however, are solid and thus can easily penetrate through the weak vascular and break into it.
633	Enlarged centrosomes (i.e., supernumerary centrioles) lead to cancer.
637	Due to enhanced vascular about a tumor, proteins accumulate about a tumor creating a possible biomarker.
644	Authors provide protein biomarkers for common cancers.

(*Continued*)

(Continued)

**	Review Notes
645	CA is a hallmark of cancer. Researchers believe that centrosome defects cause cancer. Centrosome defects could be biomarkers for advanced disease. The defects could be in the centrioles. Indeed, centrosome defects could be due to supernumerary centrioles. Centrosome clustering is prevalent in cancer. Tumors are heterogeneous having cells with normal and abnormal centrosomes.
659	Supernumerary centrioles promote tumorigenesis. CA results in aneuploidy.
669	This is a lengthy paper describing the various characteristics of cancer cells: (1) Surrounding proliferation signaling (2) Evading cell suppressors (3) Resisting cell death (4) Enabling replicative immorality (5) Inducing angiogenesis (6) Activating invasion and metastasis.
670	Cancer cells also recruit normal cells to acquire these characteristics. They discuss in detail the six characteristics of [669].
672	Extra centrosomes are found in cancer cells. Most cancers are of epithelial origin.
676	The presence of extra centrosomes imposes the tendency for catastrophic multipolar division.
677	Many efforts have been given to find cancer biomarkers. The paper provides little hope for prominent biomarkers. (My assessment)
678	Most nanomedicine is delivered via blood vessels.
678	"EPR effect" is the tendency of cells to allow penetration and retention of nanoparticles. This is a major advantage of nanomedicine. More studies are needed to identify tumors with EPR characteristics.

(Continued)

(*Continued*)

**	Review Notes
678	Normal vasculature evolves in about 1000 days. Tumor vasculature evolves in about 10 days. Tumor vasculature is poorly understood. Tumor vasculature is leaky.
681	(1) CD276/B7-H3 is broadly overexpressed in both cancer cells and tumor vasculature. (2) Both angiogenic and non-angiogenic tumor vasculature express CD276. (3) Anti-CD276-drug conjugates display potent anti-tumor and anti-metastatic activity. (4) Pyrrolobenzodiazepine dimers are ideally suited for targeting tumor vasculature.
682	This is a preview paper with much speculation as to which chemotherapy might be more effective in causing cancer cells to die.
684	Tumors with extensive aneuploidy (high numbers of chromosomes) have fewer markers for immune cell to attack.
686	Asymmetric cell division can lead to tumorigenesis.
687	The focus of the paper is that cancer originates in defective stem cells. Stem cells are unique in that they duplicate asymmetrically producing two unequal daughters — one producing ordinary tissue cells and the other a stem cell itself.
693	Authors apply >350 MPa pressure on bone and other tumor cells. The cells' membranes were "rattled."
694	No new useful information.
696	
697	Mechanisms of microtubule nucleation is still not well understood.
698	This is a review paper on the state-of-the-art of fluorescence nanoscopy — aka "super resolution microscopy."

(*Continued*)

(Continued)

**	Review Notes
698	Resolution of super resolution microscopy is typically 20–50 nm and even down to 1 nm.
698	Dynamic nanoscopy is an undeveloped area.
699	Kinetochores enable the process of connecting chromosomal DNA to microtubules. Cancer cells exhibit defects in this process.
700	"Magnetic force microscopy (MFM) is an atomic force microscopy (AFM) based technique in which an AFM tip with a magnetic coating is used to probe local magnetic fields with the typical AFM spatial resolution."
701	The authors relate centriole elongation to tumorigenesis. The causes are not well understood. Centriole elongation promotes deregulation of events.
702	The paper reviews the use of nanoparticles with photon radiation to enhance the targeted tissue damage. It may be a good paper for studying nano treatment.
703	"Electroporation-based treatments and other therapies that permeabilize the plasma membranes have been shown to be more devastating to malignant cells than to normal cells." "Electroporation is increasingly being used in cancer treatment strategies. It is a method where application of short, high-voltage pulses induces transient permeabilization of the cell membrane and thus allows the passage of ions and molecules into and out of the cell." The authors' experiments show that dye passage through a malignant cell is more rapid than dye passage through a normal cell.
704	The authors discuss the characteristics, inter relationship, and mechanisms in Ca^{24} regulation of cell death (from the abstract).

(Continued)

(*Continued*)

**	**Review Notes**
705	Electroporation is a method for increasing the permeability of cell membranes. "This study aims at understanding the relationship between applied electric field, calcium concentration, ATP depletion and efficacy" (from abstract)
706	"The electroporation (EF) supports various types of anticancer therapies by the selective transport of cytostatics. Increase intracellular calcium level by DP may be a new approach to fibrosarcoma treatment. Calcium is one of the most important factors of cell proliferation, differentiation and cell death (apoptosis or necrosis ...)" (from abstract).
707	The authors focus on traction forces within the cell, insinuating that they are analogous to muscle forces — but not clear how the forces emerge. *No* mention of electromagnetic forces.
708	Authors state that the vibratory process needed to drive cell division forward also performs other functions.
710	Skin cancer cells reject mutant cells. But due to the plasticity of skin cells, the mutant cells can slip between healthy cells.
711	"The forces that drive chromosome segregation in mitosis in human cells remain poorly understood" — from the abstract. Fiber crosslinking is thought to be a principal factor.
712	The mitotic spindle functions to ensure faithful segregation of chromosomes during cell division (from the abstract). This is a review paper on the imaging of mitotic spindles based primarily on advances in 3D electron microscopy.
713	Centrosomes are thought to be stable. This is a review paper which briefly summarizes centrosome assembly.
714	Centrosome aberrations have long been implicated in cancer. The process is not yet fully understood. This is a short review of the gene products involved.

(*Continued*)

(Continued)

**	Review Notes
714	The pericentriolar protein matrix is not an amorphous body but instead a structured system, producing centrioles.
715	The paper title is discussed but no apparent mention of electromagnetic forces.
716	The author lists nine questions related to mitosis which are not yet fully understood: 1. What proteins participate? 2. What specifies the site of the furrow? 3. How does the contractile assemble? 4. How is the ring attached to the plasma membrane?

**Refers to chronological listing of literatures reviewed.

4. Electromagnetic Notes

**	**Review Notes**
19, 126	Electric fields determine the spatial structure of the MT.
7	For an MT the proximal end is negative (–) and the distal end is positive (+).
10	Cells with aberrant centrosomes might offer therapeutic windows for drugs.
41, 34, 32	Electric fields are generated by synchronized oscillations of microtubules and centrosomes. These oscillations facilitate events during mitosis.
33	Electric fields are critical for the proper functioning of biological processes.
33	It is difficult to experimentally access the effect of electric fields on intracellular activity.
34	Oscillations of MTs form the basis for EM fields (2012).
34	Chromosomes orient themselves with their long axis along the field direction (2012).
34	Electric fields not yet well understood (2012).
13, 36	MTs are ferromagnetic (2004).
13, 37, 38	The oriented assemblies of dipoles in the MT wall causes the ferromagnetism.

(Continued)

(*Continued*)

**	**Review Notes**
13, 38	It is possible to orient arrays of MTs by the application of an electric field. It is speculated that intercellular electric fields may induce MT orientation.
40	Dipole forces play a role in MT formation.
46, 59	MTs are electrically polar with vibration modes from kHz to low THz (46) [465 MHz (59)].
47	The fastest growing end of MTs is the distal end and regarded as positive.
49, 199, 207	The centrosomes (the PCM) provide polarity (2012).
51, 144	Biosystems generate electromagnetic fields.
	Paul Schaeffer papers.
59	Tumor cells have greater damping of EM waves (2011).
61	Disturbances of EM fields produce cancer (2008).
67	The peak EM activity occurs during mitosis.
68	Provides calculation of EM fields via axial longitudinal vibration of MTs (2012).
68, 117, 129, 41, 67	The α- and β-tubulin heterodimers have a high electric dipole leading to high electric polarity of MTs.
70, 71	Low frequency AC fields inhibit growth of tumors (2009).
87	The electromagnetic field of a cell can attract small dielectric particles to the cell (2005).
88	The electric dipole moment is ~1000 Debye (10^{-2} cm).
88	The negative charge is associated with the β-monomer.

(*Continued*)

(*Continued*)

**	Review Notes
96	(2009) Suggests that effective cancer therapy may be a mechanism that targets centrosome clusters. Suggests further that cancer cell will die if the centrosome cluster is prevented or destroyed.
99	Centrioles may recruit blood vessels via EM forces.
116	During mitosis the daughter centriole moves around the nucleus to the other side. The mother centriole stays in place.
116	During mitosis the daughter centriole "rocks" (oscillates).
116	The mother centriole is stronger than the daughter centriole.
117	Tubulin is a source of polarity.
127, 130, 134, 194	The proximal end of the centriole is surrounded by electron dense material (pericentriole material).
132	Speculates that centrioles have circumferential polarity.
134	Speculates that the centriole recruits and organizes the proteins for the centrosome.
135	The nucleus determines the polarity of the cell.
135	ϒ-Tubulin occurs at the negative end of the MTs.
139	Although cell electric field is small, it is nevertheless still important (2005).
139, 154, 142	Cancer cells have greater negative surface charge than normal cells [139, 154]. There is a small negative voltage gradient between cancer cells and neighboring normal tissue [139, 142].
142	Cells with membrane potential below −36 mV are destined for cell division.

(*Continued*)

(*Continued*)

**	Review Notes
147	Breast cancer can be detected by surface electropotentials.
149	Kinetochores (protein at the end of the mitotic structure) exert forces on each other at a distance (2013).
151	Electric fields enhance wound healing (2011).
153	Cell polarity is disrupted in cancer cells (2013).
154, 155	Metastatic cells respond to electric fields more than non-metastasis cells [154]. Highly metastatic breast cancer cells move toward the anode [155].
155	Transepithelial potentials range from a few millivolts to 10s of millivolts. In tests they measure 50–500 mV/mm.
156	Static electric and magnetic fields affect in vivo cell behavior.
161, 162	Electric stimulation (0.1 V @ 50 Hz) enhances vascular endothelial growth factor (blood vessels) [161]. Electric potentials exist in region of vascular growth. Electric stimulation induces angiogenesis. Wounds have a potential difference of 40–200 mV/mm [162].
162	Rapid cell growth causes significant changes in cell surface charge.
162	Breast cancer tumors have potentials so large compared with surrounding non-cancerous tissue that the potential can be measured at the skin (2003).
167	How angiogenesis arises is a "largely unresolved matter." Cells divide when subjected to a direct current electric field (EF) and the cleavage plane is perpendicular to the EF vector.
170	The microtubules orient the spindle along the axis of polarity.
181	Nanoparticles can be used to target specific cell receptors.
181	As iron oxide nanoparticles become superparamagnetic below about 30 nm, larger particles must be used for application requiring a stable remnant magnetization.

(*Continued*)

(Continued)

**	Review Notes
186	The steps required for metastasis are similar for all cancer cells.
189	External electromagnetic field is used to activate magnetically charged nanoparticles.
190	This paper supports Roger Adelman idea of using a probe to detect cancer in the colon by measuring the electromagnetic field.
190	The premise of this paper is that malignant colon tissue has different electrical properties than non-malignant tissue.
191	Angiogenesis: blood vessels form in direct response to tissue demands — is this due to EM?
201	Disruption in centrosome clustering can selectively kill cancer cells.
240	Pericentriolar material is electron dense.
240	PCM prevents de nova duplication (Is this an electromagnetic effect?).
248	Centrosomes are electron dense.
254	Quote: Definition of *cell polarity*. "Cell polarity can be defined as a structurally and functionality asymmetric organization in which the non-random positioning of each organelle, the function of which contributes to cell asymmetry, is preserved and transmitted through cell division." — that is, not an electrical phenomenon.
277	Supernumerary centrioles tend to cluster together.
292	Orthogonal configuration of centrioles plays a controlling role in preventing centriole overduplication.
311	Inside surface of blood vessels of tumors is *negatively* charged.
313	Quote: "In general positively charged nontargeted IO nanoparticles bind to cells through electrostatic interaction with the negatively charged cell membranes and are then internalized by cells."

(Continued)

(*Continued*)

**	Review Notes
328	Magnetic effects can be significant (little else in the paper).
343	Negative NP surface charge seems to have no effect but positive NP charge seems to have a negative effect on plasma exposure.
343	The presence of surface charge can alter the way the cells respond to NP.
347	Negatively surface charged NP seems to be attracted to tumors greater than positively charged NP.
350	The inside surface of blood vessels contain many negatively charged components.
353	Magnetite nanoparticles (40 nm) with positive surface charge were internalized into human breast cancer whereas they were not internalized into normal cells.
359	Says cell-to-cell communication may or may not exist.
360	Cancer cells seem to affect other cells at a distance.
361	MTs are electrically polar. Vibrations generate an oscillating electric field. A-β dimer moment is: $337D = 1.12\times10^{-27}$cm, D — Debye.
362	Longitudinal vibration of MTs generates an electromagnetic field.
362	Malignancy can occur due to EM field.
362	Mitochondria generate an electric field.
363	Electromagnetic activity between cancer and healthy cells may differ. Nanomeasurement is needed.
363	Each mitochondrion is surrounded by a static electric field. It diminishes linearly. It is negligible beyond 10 nm.
363	Disturbance in EM activity in cancer cells might be responsible for malignant properties.
365	Low electric fields kill cancer cells.

(*Continued*)

(*Continued*)

**	**Review Notes**
366	Low frequency EM fields are successful in treating malignant tumors.
367	Quote: "A static electric field is generated by a static charged particle. Both an electric field and a magnetic field are generated if a charged particle moves at a constant velocity. The term electromagnetic field is reserved for a condition where a charged particle is accelerated. In most cases the acceleration of a charged particle takes place in the form of an acceleration. Therefore, the electric and magnetic fields also oscillate."
367	MTs are the likely source of cellular EMFs.
368	Potentially useful quote: "Measurement of differences of the electromagnetic field between healthy and cancer cells using microelectronic and nanotechnological systems will crucially contribute to understanding of the cancer process."
369	Discusses technical aspects and problems of EM measurement. Quote: "In conclusion, the precise direct measurement of biological EMFs … has not been accomplished yet."
370	Centriole replication may be accelerated by a rectangular pulsed electrical current stimulation (RPECS).
370	Injury currents play a role in wound healing.
370	Wounds treated with RPECS healed ~15% faster than the control wounds.
370	Quote: "As early as the 1860s the founder of the science of bioelectricity, DuBois-Raymond discovered that an injured bleeding finger is electrically positive compared with an uninjured finger."

(*Continued*)

(*Continued*)

**	Review Notes
370	At a rest state the mother centriole has a strong positive charge whereas the daughter centriole is not charged. At an activated state the daughter centriole also has a strong positive charge.
370	When a pair of mother/daughter centrioles have a significantly large positive charge, the repulsive force overcomes the static force keeping them together.
370	Strong repulsing electric field forces drive centrioles apart.
370	Quote: "A strong externally applied imposed EMF could induce aberrant centrioles and centrosomes that cause life disease" (known by Paul Schafer 40 years ago).
373	Documents experiments that wound generate a positive potential. Reportedly discovered ~1860 by du Bois-Raymond.
374	There exists direct current potential in living tissue. The current is observed in wound healing.
374	In animals with amputated limbs, the stump has a negative polarity. Artificially increasing the polarity enhances new growth.
374	Quote: "Further studies are needed to elucidate the mechanisms accounting for these findings..."
385	(1972) "Electrical parameters are potent stimulating factors for mammalian cells" (weak quote).
386	2006 Quote "The induced electric field by externally applied RPECS (rectangular pulsed electric current stimulation) could be mild compared with the spontaneous and natural electric field of the centrioles."
395	Plk4 attraction to centrioles depends upon electrostatic interaction.
13	MTs are ferromagnetic.

(*Continued*)

(*Continued*)

**	**Review Notes**
13	"It has been demonstrated that it is possible to orient arrays of MTs by the application of an electric field..." [see also 38].
15	Electric fields inhibited the growth of dermal lesions in mice.
16	Sequentially programmed magnetic field theory (SPMF) has been shown to arrest brain tumor growth in a patient with previous poor prognosis.
18	There appears to be an "organizing field" in mitosis, that is not entirely driven by either chemical or electromagnetic effects.
19	(1998) Paper introduces a theory that electric fields determine the spatial structure of the MTs.
24	SPMF may be a cancer therapy.
27	(2009) Surface charge density of a centrosome is $\sim e^3$. This is much smaller than the charge of an MT.
27	Small electric fields can influence the properties of cell division.
27	MT surface charge is 1–2 orders of magnitude greater than that of a centrosome.
28	Paper presents a model of how alternating electric fields control microtubule-based transport
29	(2011) A very impressive paper that asserts that electromagnetic fields are developed by oscillating MT and that the higher the frequency the greater is the generated field. The paper also states that the electromagnetic field has cellular effects that go beyond the chemical effect.
31	(2011) Better understanding of cell signaling is needed.
432	(2010) Energy supply to MTs form: (1) GTP (hydrolysis of guanosine triphosphate), (2) energy transfer by motor (moving) proteins, and (3) energy released by mitochondria as "wasted" energy in the citric acid cycle.

(*Continued*)

(Continued)

**	Review Notes
432	Paper considers axial (longitudinal) vibrations of MTs.
432	Hetrodimers have an electric dipole moment of over 1000 Debye (10^{-26} cm).
432	MT vibration frequencies: kHz–GHz.
434	Paper referred by David Lemmon.
436	Indicates possible DNA harm from low frequency (<300 Hz) electromagnetic fields.
440	R Huston papers.
441	
442	(2004) The electric dipole moment has only been calculated via computer simulations (not measured).
442	(2004) There exists little experimental evidence about the electrical properties of tubulin and MTs.
443	Net charge on a tubulin dimer is 0.19e (lower than that calculated earlier).
449	"Tubulin dimers have a strong electric dipole moment."
449	Tubulin dimer dipole moment: over 1000 Debye = 10^{-26} cm.
523	(2001) This paper presents a theoretical model of breast cancer based of EM effects. It documents via references that there is a difference in tumor EM than in normal tissue EM.
524	MTs are composed of subunits that are strong electric dipoles.
525	Little research beneficial effects of magnetic and electromagnetic fields for treatment of injury or disease.
525	EM has been used for the study of fractures.
525	EM therapeutics may be divided into five fields: (1) Permanent magnets (2) Low frequency sine waves (3) Pulsed electromagnetic fields (4) Pulsed radiofrequency fields (5) Transcranial magnetic/electric stimulation.

(Continued)

(*Continued*)

**	**Review Notes**
525	Quote: "Biochemical and physiological processes are based on the flow of electrical charges..."
525	The limited acceptance of EMF therapy is due to: (1) Physicians not familiar with the techniques (2) Regulation (3) Public concerns about safety.
525	There are in vitro studies suggesting significant alterations in cell division for wound healing.
525	EMF are not widely used in therapy due to the absence of well-established mechanism for EMF benefits.
525	Magnetic fields (MF) can affect cell proliferation either by increasing a slow rate (as in wound healing) or by inhibiting too rapid cell proliferation.
526	Electromagnetic mechanisms are not fully understood.
527	This paper describes experiments attempting to measure the dipole moment on an α/β dimer. Results: ~6000 Debye (others have suggested 1000 Debye).
528	Paper proposes ferromagnetic properties for MTs.
530	"Magnetic targeting" uses an externally applied magnetic field to move a magnetically charged drug carrier to a tumor.
530	Paramagnetic materials magnetize weakly in the direction of the magnetic field. Diamagnetic materials magnetize in the opposite direction.
530	Ferro-magnetic materials tend to aggregate.
530	Superparamagnetic materials provide a compromise between strong magnetization and avoidance of particle aggregation.
530	Superparamagnetic materials have attractive features: (1) Ability to be visualized by MRI (2) Ability to be localized to a specific site (3) Ability to produce localized healing.

(*Continued*)

(*Continued*)

**	Review Notes
592	Magnetostatic bacteria have organelles that confer a magnetic moment on the cell.
593	This is a disturbing paper: It speaks of measuring small magnetic moments with a general purpose superconducting quantum interference device (SQUID) magnetometer. The paper discusses counting super paramagnetic iron oxide particles (SPIONS) in and about normal and cancer cells. It says nothing about the magnetism of the cells themselves.
593	It talks about: "... magnetic nanoparticles biologically reacting with living cells."
596	Authors claim that magnetic forces can affect the positions of the mitotic spindle.
597	The authors discuss the development of magnetic nanoparticles (2007).
625	Electron "spin" is due to a magnetic moment pointing "up" or "down."
625	Types of magnetism: (1) *Paramagnetism*: The magnetic moments tend to be randomly oriented when there is no overall magnetic field. In an applied magnetic field these moments align parallel to the field. Then the magnetism of the material is proportional to the applied field. (2) *Ferromagnetism* (3) *Antiferromagnetism*: Magnetic moments align opposite to each other. Thus there is no overall magnetism. (4) *Farrimagnetism*: The aligned magnetic moments are not of the same size — thus reduced overall magnetism. (5) *Superparamagnetism*: The material is composed of small ferromagnetic clusters — so small that they can easily change magnetic direction. Therefore, the material has no overall magnetic direction. But in an applied magnetic field, the material becomes magnetized and aligned with the magnetic field.

(*Continued*)

(*Continued*)

**	**Review Notes**
625	When SPIONS are placed in a strong magnetic field, their magnetism far exceeds that of biomaterials (i.e., cells).
649	SQUID is useful for detecting cancer cells. It employs a magnetic relaxometry device. The magnetic moment of cancer cells is linear with the number of cells.
649	In magnetic relaxometry, magnetic nanoparticles are briefly magnetized by small magnetic field pulses. Then the decaying magnetic field is measured. Application with biosystems need small particles ~10 nm.
650	In this paper the authors use "antibody-conjugated magnetic nanoparticles" which adhere to the breast cancer cells. They then use SQUID sensors to detect the cancer cells. SQUID is based on "magnetic relaxometry."
649	In magnetic relaxometry magnetic nanoparticles (super paramagnetic nanoparticles) are briefly magnetized by a small magnetic pulse. Then the decaying magnetic fields are detected and measured.
649	With biomaterials, the particles not bound to tissue fall away and decay within microseconds. The remainder are bound particles which can then image the biotissue.
662	The author purposes that the centrosome electrically and magnetically govern cellular mechanics. A figure illustrates the fields.
663	Authors present a theoretical model showing that electric fields govern cellular mechanics. The field is generated by microtubules.
664	Authors propose that oscillation of chromatin produces an electromagnetic field.
671	Procedure: Use bacteria with a nanoscale layer of nitrogen vacancy (NV) centered on the surface of a diamond chip. The bacteria are placed on the surface of the diamond. Then the quantum spin rates are viewed via chains of magnetic nanoparticles ("magnetosomes").

(*Continued*)

(*Continued*)

**	**Review Notes**
673	The paper describes a "quantum diamond microscope": this instrument uses NV centers in diamond for correlated magnetic and fluorescence imaging. The microscope employs a dense layer of fluorescence quantum sensors, based on NV color centers near the surface of a diamond chip on which the sample of interest is placed.
677	Many efforts have been devoted to finding cancer biomarkers. Paper provides little hope for finding prominent biomarkers (my assessment).
678	Finding biomarkers is a major goal, but few studies have been made to find them. Angiogenesis may be a good candidate.
685	In vitro microtubules align themselves along the static magnetic field (SMF). However, in living cells the effects of SMF greater than 20 T have not been reported. The authors have shown that for human cells, the spindle orients at 27 T. The results show that the magnetic torque acts on both microtubules and chromosomes. The preferred direction depends more on the orientation of the chromosomes than on the microtubules. The paper focuses on the effect of magnetism on microtubules and chromosomes rather than on measuring the magnetic field.
688	The paper discusses forces but ignores electromagnetic forces.
690	This is a book chapter on a general discussion of magnetic induced cell growth (e.g., bond healing).

**Refers to chronological listing of literatures reviewed.

5. Nano Notes

**	Review Notes
181	Nanomagnetic action is "action at a distance."
181	Nanoparticles can be used to target specific cell receptors.
181	Nanoparticles need not be magnetically blocked. They may be superparametic. (I believe this refers to nanoparticles <30 nm (less).) P141
181	As iron oxide nanoparticles become superparamagnetic below 30 nm, larger particles must be used for application requiring a stable remnant magnetization.
188	Magnetic nanoparticles (MNPs) were attached to integrin cell receptors for 21 days with no adverse results.
188	Quote: "Magnetic particle technology presents a valuable tool for tissue engineering which permits the delivery of highly localized mechano-inductive forces directly to cells."
189	This paper presents a review of superparamedic nanoparticles (SPION) in medicine (<10 nm).
189	Use of external magnetic fields to activate nanoparticles.
189	Electromagnetic nanoparticles can kill tumors either by chemical release or by vibration (heating) (oscillating external fields).

(*Continued*)

(*Continued*)

**	Review Notes
201	Disruption in centrosome clustering can selectively kill cancer cells.
202	Centrosome instability may be an early target for cancer treatment.
211	Douglas Shi paper: nanocomposite particle targeted to breast cancer in vitro.
212	Carbon nanotubes (CNTs) may be toxic.
212	Quote: "Localization of nanocarriers to tumors can be accomplished by using magnetic fields to concentrate the nanocarrier directly in tumors. While magnetic hyperthermia can be used to directly kill cancer cells or as a trigger for controlled release."
213	A review article with 112 references. Efficient penetya and retention (EPR) — a measure of effectiveness of nanoparticle treatment.
213	Gold nanoparticles can pass the blood brain barrier.
213	Quote: "The use of nanomaterials to treat and image cancer is arguably the most active area of nanomedicine research. Intrinsic properties like surface charge, structure, and biocompatibility can be harnessed for cancer therapy."
213	Quote: "Drug-loaded nanoparticles can be targeted to surface molecules over-expressed in cancer therapy."
214	Concept: magnets can be used to pull magnetically charged nanoparticles through the blood vessels to tumors. However, the problem is the decrease in magnetic field away from the magnets — particularly for targets within deep tissue.
215	Single-walled carbon nanotubes (SWNT) were used to target tumors in mice — reported in 2008. First successful known treatment to date.

(*Continued*)

(Continued)

**	Review Notes
216	Pristine CNTs are highly toxic due to their insolubility. CNT can cross cell membranes.
216	More than one function (targeting, imaging, drug delivery) can be accomplished with a single CNT.
217	"...yet our ability to efficiently have these materials (nanoparticles) to regions of disease *in vivo* remains very limited."
217	Nanoparticles have been developed with unique electromagnetic properties.
218	Quote: "Nanotechnology is still in its infancy."
219	This is a listing of the many ways nanoparticles can deliver therapy.
268	Nanometer described.
285	A primary goal of bionanotechnology is the delivery of therapeutics to malignant tumors. The use of iron oxide is to reduce toxicity.
308	A review. Not much help, except for references. The focus is on targeting proteins.
309	Advocates the use of polyethylene glycol as a cover for nanoparticles.
309	Three requirements for nanoparticles: (1) They need to be stable (2) They need to accumulate in the tumor (3) They need efficient release.
309	Nanoparticle ideal: efficient payload to target: >10% in 4 h, and cleared from circulation in short time: 4–10 h.
310	MNPs have the ability to function at the cellular level.
311	Therapeutic nanoparticles should be 10–100 nm in diameter.
311	A way to get nanoparticles to tumors is by the leaking blood vessels. Particles with diameter of 100 nm can leak through the blood vessels.

(Continued)

(*Continued*)

**	Review Notes
311	Inside surface of blood vessels of tumors is *negatively* charged.
311	Nanoparticles can carry a large payload of drugs.
312	Nanoparticles are larger than molecules but smaller than cells.
312	Quote: "Future directions for this work will be in the field of 'smart' nanoparticles."
312	Quote: "Differences in the expression of cellular receptors between normal and tumor cells represent a great opportunity for targeting imaging probes to those cellular surface molecules."
313	Chemicals attracted by tumors: Her-2/new; $\alpha_v\beta_{31j}$ integrin PMSA and MUC-1; and CREK[4]
313	Quote: "In general, positively charged nontargeted IO nanoparticles bind to cells by electrostatic interaction with the negatively charged cell membranes and are then internalized by the cells…"
313	Tumor cells have heterogeneous expression levels of targeted receptors.
313	There remains a challenge to load up nanoparticles with toxic drugs.
314	Fe_3O_4 nanoparticle effect is enhanced by covering it with a drug: tetraheptylammonium. This causes drug-resistant leukemia cells to uptake the nanoparticle. (This is evidence of the need for more nanoparticle research.)
315	Nanoparticle size $1\ \text{nm} < \text{NP} < 100\ \text{nm}$.
315	NPs differ in properties from bulk materials (magnetically) due to the surface effects.
315	(2007) The most promising application of NPs is the diagnosis and treatment of cancer. Idea is to target the tumor.
315	One idea is to implant a magnetic compound in a tumor and then use magnetically charged NPs for therapy.

(*Continued*)

(*Continued*)

**	**Review Notes**
316	An obstacle for therapeutically charged nanoparticles is the membrane layers of the tumor and cell.
317	Popular chemical used: doxorubicin (aka J&J doxil).
318	Not much new. Main focus appears to be on imaging.
319	Approved use of liposomes and formulations of doxorubicin for Kapasi sarcoma.
319	Quote: "There are still bottlenecks in the existing 'homing' technologies… This emphasizes the need for novel targeting strategies."
321	A review of nanoparticle technology up to 2002.
323	The physical properties of materials radically charge when scaled down to a nanoparticle.
323	Nanoparticles can be targeted to specific cells and locations within the body.
323	Nanoparticles must be the right size to avoid leaking into capillaries, but not too big to get stuck. Optimal range: 20 nm < NP < 100 nm.
323	Nanoparticles can be delivered via leaky blood vessels of tumors, but the distribution is heterogeneous and unpredictable.
323	Antitumor drugs liposomal doxorubicin.
323	CNTs belong to the family of fullerenes and are made of graphene sheets. They have diameters 0.5–3.0 nm and length 20–1000 nm.
323	CNTs can cross cell membranes ("nanoneedles") without disrupting the membrane.
323	Paramagnetic nanoparticles are used to enhance imagine.
323	Nanotechnology is a rapidly growing field.
325	Quote (2006): "Nanoparticle-based drug-delivery works only temporarily." It is not clear why.

(*Continued*)

(*Continued*)

**	Review Notes
325	There are two approaches: active and passive. *Active* involved linking ligands to nanoparticles that are tumor specific. *Passive* takes advantage of the inherent size of nanoparticles and the unique properties of tumor vasculature.
325	Quantum dots are novel semiconducting monocrystals — used for imaging. Nanoshells (~10–300 nm) are also used for imaging.
325	Gold nanoparticles are used for diagnosis and therapy. They are used to enhance sensitivity to external beam radiation.
326	Tumor blood vessels has structural abnormalities ("architectural anarchy" [330]) that hinder therapies. They are large in diameter. They promote high local back pressure. They leak.
327	SWNTs absorb near-infrared light (NIR). These nanoparticles can get inside the cells and kill it by heating via absorbed light.
329	Gold nanoparticles were injected under the skin and then healed by laser. The tumors were ablated. It is not clear if this would work for deep tumors or exactly how the nanoparticles get into the tumors. The main contribution of the paper appears to be the advantage of using NIR nanoparticles.
330	Quote: "Only 1 to 10 parts per 100,000 of intravenously administered monoclonal antibodies reach their parenchycal targets *in vivo*."
330	Fullerenes and CNT were discovered by Richard Smolley.
330	Tumor blood vessels have "architectural anarchy."
331	2008 Quote: "Targeted delivery of drug molecules to organs or specific sites is one of the most challenging research areas in the pharmaceutical sciences."

(*Continued*)

(*Continued*)

**	**Review Notes**
331	The paper focuses on lung disease. It is a review but to the date of the paper (2008) no success.
342	Irregular tumor blood vessels make it difficult to deliver (via blood) to tumors.
342	It is necessary to deliver drugs to intercellular targets.
342	Advantages of nanotechnology: • Increase drug concentration in the tumor • Decrease drug release during transit • Can be organelle specific.
342	Size: 10–500 nm, but particles with 50–100 nm can penetrate large tumors.
342	NPs target endothelial cells in blood vessels (anti-angiogenesis)
342	Quote: "Although organelle-specific delivery is challenging, when achieved, it can dramatically increase the interaction of the bioactive molecules with target organelles." (2010)
342	(2014) There is excitement that NP could contribute to the success of cancer treatment.
342	Goal is enhancing internalization of NP into cells.
342	Animal studies show that 50–100 nm can penetrate large tumors.
342	"ACTIVE TARGETING": Targeting a specific tumor.
342	Targeting the endothelial cells of tumor blood vessels is an attractive strategy for inhibition of the growth of the tumor.
343	The pace of companies doing proprietary research has considerably accelerated.
343	EPR effect.
348	MNPs are used for imaging.
348	This is a review of design parameters about MNPs performance *in vivo*.

(*Continued*)

(*Continued*)

**	Review Notes
348	SPION
348	Quote: "Upon intravascular administration NPs immediately encounter blood, a high ionic strength heterogeneous solution that can induce NP agglomeration, altering their magnetic properties and inducing particle sequestration"... "In each case the NP is in danger of prematurely binding to or being taken up by cells before reaching its target tissue."
348	Chemotherapeutic drugs: doxorubicin, cisplatin, and paclitaxel
348	Problem: With intravenous administration, the drugs are quickly cleared from circulation, and only a small fraction reaches the tumor. Even then the drug may not get into the depths of the tumor long enough to be effective.
349	A brief review paper but nothing new in getting NP to tumors.
349	Only a small fraction of intravenous administrated nanoparticles reach the tumor. Even then the particles may be rapidly cleared from the tumor having little therapeutic effect.
350	Nanoparticles are 10–100 nm. The threshold for first pass elimination by the kidneys is 10 nm. The upper bound size is not well defined.
350	The vasculature of tumors is known to be leaky to nanoparticles.
350	Normal blood vessels keep nanoparticles larger than 1–2 np.
350	Nanoparticles, 50–100 nm, can penetrate through large tumors.
350	The primary purpose of targeting of ligands is to enhance cellular uptake into cancer cells rather than simply accumulate in the tumor.

(*Continued*)

(*Continued*)

**	Review Notes
351	Cell penetration in scaffolds of cells is difficult. The authors develop magnetic NPs coated with chitosan for enhancing cellular invasion under a magnetic field.
352	Paper describes the potential benefits of nanoparticles for cancer treatment.
353	Magnetite NPs (40 nm) with positive surface charge were internalized into human breast cancer whereas they were not internalized into normal cells.
354, 355	Nothing much new. Reasserts the potential of nanotherapy.
356	Nanoworms: (a strong of nanoparticles). Nanoworms have better targeting ability than single nanoparticles.
357	A review article on application of nanoparticles in medicine. Not much new. So far nanotherapy is still relatively ineffective — need for multifunctional nanoparticles.
358	Nanotherapy has great potential, but challenges remain. Need particles that are internalized by the cells. The choice of an appropriate particle is not obvious.
363	Nano measurements are needed to detect EM differences between cancer and normal cells.
12	Nanotechnology needs further development before it can be widespread clinically useful.
30	(2011) Nanoparticles (NPs) easily penetrate cellular membranes; NPs have a range in size from 1 to 100 nm.
440, 441	RH papers.
447	Paper found by M. Schulz. The hypothesis is that a high molecular weight protein: fibronectin in the extra cellular matrix of metastatic breast cancer. It can be used as a biomarker.

(*Continued*)

(*Continued*)

**	Review Notes
447	(see above) A tiny magnet is used to hold toxins together; then attacks the fibronectin biomarker.
509	"There is still no consensus on the toxicity of nanomaterials." There are numerous open questions related to type of nanoparticles, age, and injestion potential.
530	Paramagnetic materials magnetize weakly in the direction of the magnetic field. Dimagnetic materials magnetize in the opposite direction.
530	Ferromagnetic materials tend to aggregate.
530	Superparametic materials have attractive features: (1) Ability to be visualized by MRI (2) Ability to be localized to a specific site (3) Ability to produce localized heating.
531	MNPs cluster together and become ineffective for bioapplications.
531	Clusters of magnetic nanoparticles differ magnetically from the individual nanoparticles.
531	MNPs have a large surface area/volume ratio and tend to agglomerate to reduce surface energy.
531	An effective approach to preventing particle agglomeration is to coat the particles with biocompatible polymers.
546	Discusses PALM/STORM nanotechnology. Our results demonstrate that both PALM and STORM imaging have great potential as alternatives to immune-EM."
553	Paper provides a listing of common NPs in use.
553	Small NPs (1–20 nm): colloidal gold iron oxide crystals and quantum dots.
553	For cancer treatment, an NP must be stable until it reaches the tumor and be able to carry drugs.

(*Continued*)

(*Continued*)

**	Review Notes
553	NPs with medical applications: (1) *Superparamagnetic iron oxide crystals*. These have an iron core ~4–5 nm in diameter surrounded by a polymer. These particles become magnetic when brought into a magnetic field. (2) *Quantum dots*. Developed in 1980s. They absorb light and then emit it in a different color. Useful in diagnostics. (3) *Dendrimers*. These can be of various shapes and carry drugs on their surface. (4) *Polymeric micelles*. These are less than 50 nm in diameter. They are used to deliver water insoluable drugs. (5) *Liposomes*. These are large NPs 100–1000s nm in size. Used for drug delivery. (6) *Nanospheres*. Spherical 100–1000 nm encapsulating drugs. (7) *Aquasomes*. These are carbohydrate-ceramic NPs. They are spherical, 60–300 nm in diameter, and used for drug delivery. (8) *Polyplexes/lipopolyplexes*. These are assemblies of nanomaterials.
554	"Iron oxide balls, just a few tens of nanometers in diameter heat up when exposed to a powerful magnetic field." Neatly stacked NPs need a stronger field than more random NPs.
555	In cancer cells folic acid is 10 times as great as that in normal cells. The author proposes folic acid as a biomarker.
555	Dendrimers are nano-sized polymers ("Starburst polymers") which can be efficiently targeted for tumors.
556	Author proposes a three-stage delivery of NPs to tumors. (1) Blood borne, (2) into large tumor vasculature, and (3) break down of 100–100 nm NP and into the tumor core.
557	NPs may be designed for binding to a cell membrane or to a cell interior biomarker.

(*Continued*)

(*Continued*)

**	**Review Notes**
557	Tumor blood vessels are abnormal and leaky. The leak gap is 100–780 nm. (Normal leak gap is 5–10 nm.)
557	*Bio NPs* (1) *Dendrimers* — Branching tree-like structures. (2) *Micelles* — Spherical globular structures with a hydrophobic sphere-like core. (3) *Nanospheres* — Spherical structure composed of a matrix of curved bars. (4) *Nanocapsules* — Fluid filled pouches. (5) *Fullerenes* — Carbon "bucky ball." (6) *Nanotube* — Hollow carbon structure. (7) *Liposome* — Closed fluid filled sphere with a number of cover layers.
558	This is a review paper on current targets for nanoparticles. These are: (1) angiogenesis, (2) uncontrolled cell proliferation, and (3) tumor cell targeting.
558	Nanoparticles: (1) dendrimers, (2) liposomes, (3) polymeric nano-articles, (4) micelles, (5) protein nanoparticles, (6) ceramic nanoparticles, (7) viral nanoparticles, (8) metallic nanoparticles, and (9) CNTs.
559	Paper lists five NPs: (1) liposomes, (2) polymer drugs, (3) polymeric, (4) dendrites, and (5) iron oxide.
560	Superparamagnetic iron oxide nanoparticles (SPIONS) become magnetic in a magnetic field. They are however very dependent on their structure — particularly the size and size distribution.
560	Folic acid is a biomarker for SPIONS.
560	SPIONS are applied intravenous and also by direct injection into the tumor.
560	SPIONS have great potential but more research is needed for efficacy.

(*Continued*)

(*Continued*)

**	**Review Notes**
561	Superparamagnetic nanoparticles are surface-modified with poly (ethylene glycol) (PEG) and folic acid respectively (quote from abstract). The PEG nanoparticles appeared to perform better (i.e., more uptake into the cancer cells) than the folic acid particles, although for breast cancer both performed well.
562	Magnetic NPs can be maneuvered through the body by external magnetic fields. But if the NPs are too deep, the procedure becomes less effective.
563	Authors injected SPIONS into rat glioma tumors and then heated them, shrinking the tumor.
565	All tumors have molecular markers that do not occur in normal tissue.
565	This is a review paper focusing on molecular markers created by tumor vasculature.
566	The peptide: iRGD goes not only into the tumor extra vasculature but it also penetrates into the tumor itself.
567	This is a review article on NPs (2007). The leaky vasculature allows NPs to escape from the blood stream.
567	Desired properties of drug carrying nanoparticles: (1) Protect the therapeutic drug from degradation (2) Prevent drugs from prematurely interacting with nearby tissue (3) Enhance drug absorption into target tissue (4) Improve intercellular penetration of the drug (5) Exhibit high differential uptake into the target cells compared with normal cells (6) Be made of biocompatible material (7) Have extended circulating half-life.
567	*Nanoparticles* (1) Liposomes, (2) biodegradable polymers, (3) chemo-therapeutic, (4) surface functionality, (5) spacer group, (6) targeting molecule, (7) inorganic particle: metallic shell, (8) amphipathnic molecules, (9) dendrimer, and (10) CNT.

(*Continued*)

(*Continued*)

**	Review Notes
568	Paper reports on tests of two NPs: nanospheres (NS) and nanoworms (Nw). These were SPIONS. The Nw had better results.
569	The objective of the authors is to develop non-toxic NPs. They offer luminescent porous silicon nanoparticles (LPS_iNPs) which self-destruct in a mouse model (renally). Dextran coated LPS_iNPs (D-LIP_iNPs) are good for tumor imaging.
570	NPs can be targeted to extravascular tumor tissue.
572	This is a review article. Gold NPs (AuNPs) are discussed for diagnostic imaging. The authors also provide a listing of delivery systems.
572	AuNPs are relatively easy to manufacture.
572	Deliver: (1) particle size reduction, (2) peptide platform, (3) small molecule, and (4) proteins.
573	This is a review paper on gold NPs. It lists these: nanospheres, nanorods, nanoshells, and nanocages.
575	This paper presents folic acid as a biomarker for NPs.
577	Artery endothelial cells are 13 ± 4.1 μm wide and 25.8 ± 8.5 μm long.
578	This is a review paper summarizing the targeting of NPs to specific metastases.
579	The authors discuss the use of poly (amidoamine) dendrimer as an NP. It is non-toxic
580	The authors use poly amidoamine dendrimer dendrites (tree-like NPs) to deliver toxins to tumors by focusing on folic acid as a biomarker.
581	The paper describes the use of silica NPs with superparamagnetic iron oxide core and attachments. Folic acid is the tumor biomarker.

(*Continued*)

(Continued)

**	Review Notes
585	The objective of the paper is to report on NP targeting of blood vessels feeding the tumor. It states that integrin $\alpha v\beta 3$ is found in angiogenesis. $\alpha v\beta 3$ is thus a biomarker.
585	The authors use RGD as their NP. The chemo is doxorubicin.
593	This is a disturbing paper: It speaks of measuring small magnetic moments with a general purpose superconducting quantum interference device (SQUID) magnetometer. It discusses everything SPIONS in and about normal and cancer cells. It says nothing about the magnetism of cells themselves. It talks about: "...magnetic nanoparticles biologically reacting with living cells."
597	The authors discuss the development of magnetic nanoparticles (2007).
615	Authors state that NPs are too large to penetrate tumors and suggest using vascular outbursts as a way to get in (2016).
616	Kinetochores form a bridge between the chromosomes and the microtubules. The paper implies that *all* kinetochores need to be attached. The kinetochore is a collection of 60 proteins. The core component of a kinetochore is the Nde80 complex.
625	Visualizes NPs put into the blood stream (like chemo). NPs may have globular, plate-like, rod like, or complex geometries.
625	When SPIONS are placed in a strong magnetic field, their magnetism far exceeds that of biomaterials (i.e., cells). SPIONS: 50 nm with a 20-nm core. In an alternating magnetic field, the SPION changes its magnetic field direction. Therefore, if a SPION can get inside a cell, an externally applied alternating field can cause the SPION to heat cells to death. Also SPION effectiveness could be increased by reducing its size.

(Continued)

(*Continued*)

**	**Review Notes**
635	Nanoparticles 10 nm or less in diameter are not captured by tissue, but nanoparticles 10–40 nm can cause inflammation, capturing, and neutrophil extracellular traps.
636	This is a review paper on nanoparticles used up to 2012 for cancer treatment. Here are kinds of nanoparticles: (1) Liposomes (2) Nanoparticle albumin-bound (NAB) technology (3) Polymeric nanoparticles (4) Dendrimers (5) Metal nanoparticles (6) Molecular targeted nanoparticles
636	(1) *Liposomes* — Developed in 1965 were among the first developed. Now (2012) at least 11 varied formulations. They are biocompatible and biodegradable. They can also be coated with polymers, for example, PEG (2) *NAB* — Uses albumin as a therapeutic carrier. That is, a nanoparticle composed of albumin with hydrophobic interior. (3) *Polymeric nanoparticles* — Made of biocompatible and biodegradable polymers. They form a shell for therapeutic compounds. (4) *Dendrimers* — Macromolecules made of natural elements such as amino acids, sugar, and nucleotides. They are "dendrimer-drug" composites. (5) *Metal nanoparticles* — Made of inert metals such as gold or titanium and used for controlled release of therapeutic drugs. But, although biocompatible, they remain in the body and can accumulate to a toxic level. (6) *Molecular targeted nanoparticles* — Best hope to date. They are intended to be attracted to tumor markers.
636	Provides a table of various nanoparticles.
638	Nanoparticles can aid in imaging for radiation therapy.

(*Continued*)

(Continued)

**	**Review Notes**
639	An historical article.
640	P-selectin is a biomarker (found in tumor blood vessels). Authors suggest using fucoid, a polysaccharide that binds to P-selection to deliver therapeutic drugs to tumors.
649	SQUID is useful for detecting cancer cells. It employs a magnetic relaxometry method.
649	The magnetic moment of cancer cells is linear with the number of cells.
649	In magnetic relaxometry, MNPs are briefly magnetized by small magnetic pulses. Then the decaying magnetic field is measured. Application with biosystems need small particles ~10 nm.
649	In magnetic relaxometry nanoparticles (superparametic nano particles) are briefly magnetized by a small magnetic pulse. Then the decaying magnetic fields are detected and measured. With biomaterials, the particles not bound to the tissue fall away and decay within microseconds. The remainder are the bound particles which can then image the biotissue.
650	In this paper the authors use "multibody-conjugated magnetic nanoparticles" which adhere to the breast cancer cells. They then use SQUID sensors to detect the cancer cells. SQUID is based on "magnetic relaxometry."
660	Naked eye can see hairs; a magnifying glass can enlarge it ~10x; light microscopy can enlarge it to 300x. Fluorescence "nanoscopy" can image objects 20–30 nm in diameter. Most proteins are 5 nm across. A new concept in nanoscopy MINFLUX goes down to 2–3 nm.
660	A common procedure in nanoscopy is to localize a molecule where the signal is strongest.

(Continued)

(*Continued*)

**	Review Notes
666	Nanoparticles have been used for 1000 years in aerosol sprays. Suitable metals for nanoparticle production: gold, silver, titanium, and zinconia. Iron is excluded since it is incompatible with cells.
666	Aside from metals, peptides, proteins, carbohydrates, and lipid moieties are suitable for nanoparticle materials (peptides are components of protein). Also suitable are polymers.
666	A supersonic beam colliding with another beam produces frozen nanoparticles. The author presents a table of nanostructured beams suitable for biomedicine.
668	MNPs have been found in teeth and brain matter. A characteristic technique for magnetic field detection is to use SQUID.
668	In summary, the focus of the paper is to discuss ways of determining the effects of nanoparticles on biological materials.
671	Procedure: Uses bacteria with a nanoscale layer of NV centered on the surface of a diamond chip. The bacteria are placed on the surface of the diamond. Then the quantum spin rates are viewed via chains of MNPs ("magnetosomes").
674	Nanotechnology can have toxicity, but it can also be useful in mechanical applications.
678	Nanoparticle treatment is spurred on by the limits of conventional cancer treatments. There is an incomplete understanding of the interaction of nanoparticles and cell proteins.
678	The number of nanomedicine articles appearing is roughly doubling every year.
678	Most nanomedicine is delivered via blood vessels.
678	The "EPR effect" is the tendency of cells to allow penetration and retention of nanoparticles. EPR is a major advantage of nanomedicine. It is the foundation of nanomedicine. More studies are needed to identify tumors with EPR characteristics.

(*Continued*)

(Continued)

**	Review Notes
678	Gold, silica, and polystyrenes are good for protein nanoparticle interaction.
678	Nanomedicine is likely to develop its own toxic chemicals.
683	Polylactic acid in micro and nanoparticle sizes is biodegradable, biocompatible, and non-toxic. Their disadvantages are: (1) Low drug loading capacity (2) Low encapsulation efficiency (3) Low terminal sterilization.
691	Tumor cells can be weakened or destroyed at temperatures between 107° and 113° Fahrenheit.

**Refers to chronological listing of literatures reviewed.

6. Protein Notes

**	Review Notes
	Abhorrent centriole duplication requires cyclin D/CDK2 and Plk4 [1, 6, 90, 109, 121, 148, 164, 174, 193, 206].
5,	Plk4 and SAS-6 are required for centriole duplication.
164	Disengagement requires Plk1.
	It is believed that excessive Plk4 or SAS-6 may contribute to extra daughter centrioles (mainly Plk4). [1, 6, 73, 109, 121, 148, 164, 174, 193, 206].
8	K-ras activation could lead to cancer.
	Plk4 is aka SAK.
73	Plk4 is an unstable protein.
	E cyclin-dependent kinase is aka CDK2.
91	Centriole duplication arises from Plk1 and separase.
91	A strict regulation of Plk4 is required for healthy cell mitosis (2012).
91	CANDI is a centrosomal protein that controls the stability of Plk4.
91	CANDI synergizes with Plk1 to stimulate overduplication of centrioles.
102	KSP — a kinesin motor protein needed to establish the mitotic spindle bipolarity. There is no other known role for KSP.
109	Plk4 excess causes simultaneous multiple procentrioles in each parent centrite.
111	The protease separase is a mechanism for restricting chromosome duplication.

(*Continued*)

(Continued)

**	Review Notes
111	Separase promotes centrosome separation; Plk4 (aka SAK) regulates centriole formation.
111	The centrosome contains 100 different proteins.
121	Plk4 (SAK) is essential for centriole duplication.
127	P53 is a tumor suppressor. Its exact role is unknown [180, 207, 210]
127	The centriole barrels contain only a few proteins: α, B, ϒ-tubulins.
127	ϒ-Tubulin occurs within the proximal end of the centriole. Some believe ϒ-tubulin is responsible for centriole duplication and even the geometric structure of the MTs.
131	The protein calnexin makes a daughter centriole mature during mitosis.
157 159 199	Sas-6, a conserved procentriole protein provides the ninefold symmetry to the centriole scaffold.
158	Sas-6 is self-assembling. It concentrates at the core of the cartwheel.
164	Plk4 is the most structurally divergent of the polo-kinase family.
165	Sas-6 is required for centriole formation.
174	Cep152 acts with Plk4 to initiate centriole formation (2010).
180	There are five core centriolar components: Cep192, Plk4, STIL, Sas-6, and Sas4.
183	Plk1 is required to convert new centrioles to centrosomes as the cells progress from mitosis to interphase.
193	Centriole biogenesis involved SAK (Plk4), SAS-6 organizes the nine procentriolar units.
193	Plk4 (SAK) is an upstream regulator of canonical formation of the centrioles.
	STIL is a centrosomal protein whose overexpression results in excess centriole formation [196] (2012). STIL is recruited by nascent daughter centrioles at the onset of centriole duplication [197]. STIL cooperates with SAS-6 and Plk4 in control of centriole number.

(Continued)

(Continued)

**	**Review Notes**
198	ϒ-Tubulin is the inch-pin of MT nucleation.
199	Principal centriole proteins: ZYG-1 (Plk4/SAS), SPD-2, SAS-4, SAS-5, and SAS-6 (2008). SPD-2 and ZYG-1 are recruited before central tube formation whereas SAS-5 and SAS-6 are recruited slightly later. This suggests that SPD-2 and ZYG-1 are regulatory rather than structural.
201	Ned aka HSET is a minus end kinase needed for extra centrosome development.
206	Plk4 stability is regulated by its own kinase.
210	Aurora excess produces cancer. Aka AurA (Aurora A kinase).
193	Centriole biogenesis involves SAK, Plk4, and SAS-6.
193	SAK and Plk4 overexpression leads to multiple centrosomes.
193	SAK/Plk4 is an upstream regulator of canonical centriole formation.
194	There are ~100 proteins in the centrosome.
196 197	STIL is a centrosomal protein whose overexpression results in excess centriole formation. But too little leads to a loss of centriole formation.
196	STIL recruits SAS-6.
197	STIL cooperates with SAS-6 and Plk4 in controlling the centriole number.
198	Significant proteins ZYG-1 (Plk4, SAS), SPD-2, SAS-4, SAS-6, and SAS-6.
199	SPD-2 and ZYG-1 are recruited before the central tube formation whereas SAS-5 and SAS-6 are recruited slightly later. This suggests that SPD-2 and ZYG-1 are regularly rather than structural.
199	SAS-6 appears early at the side of the mother centriole and establishes the ninefold symmetry spokes. MTs then appear at the ends of the spokes.
201	Nod aka HSET is a minus ended kinase needed for extra centrosome development, but not for regular cell development.

(Continued)

(*Continued*)

**	**Review Notes**
203	Centriole assembly begins with the formation of a central tube followed by the peripheral assembly of nine single microtubules. Tube formation and assembly is dependent upon SAS-5 and SAS-6 proteins whereas the assembly of singlet microtubules onto the central tube depends upon SAS-4.
203	Centriole assembly is triggered by an upstream signal mediated by SPD-2 and ZYG-1.
205	Dearth of p53 leads to centrosome defects.
206	Centrosome duplication arises due to centriole duplication. Plk4 plays a key role in initiating centriole duplication. But overexpression of Plk4 promotes centriole overduplication.
206	Plk4 stability is regulated by its own kinase activity.
207	P53 is a tumor suppressor.
210	Aurora excess produces cancer.
226	Plk4 is the "central player" in cell duplication — but its activity is still not well understood.
230	States that ninein (a protein), and not ϒ-tubulin, forms both the anchoring and the placement of the minus end of microtubules.
231	Electron dense granules ~70–100 nm in diameter are observed in the centrosome.
237	The daughter centriole arises from a "dot" of Plk4 on the surface of the proximal end of the mother.
242	SAS-6 is required for centrosome duplication. Also to make a centriole Aurora-4 (SPD-5) and ϒ-tubulin is required (3-proteins). SAS-6 is required for centriole assembly.
243	Explains how the ninefold spiral occurs.
247	SAK/Plk4 is needed for centriole development.
248	Overexpression of SAK/Plk4 and SAS-6 causes the "flower" to form (multiple daughters).
249	P53 regulates centrosome duplication but [250] disputes this.

(*Continued*)

(*Continued*)

**	**Review Notes**
258	Studies the role of p53 in centrosome duplication regulation.
259	
261	Proteins needed for cell duplication: kinase ZYG-1; coiled-coil proteins: SAS-4, SAS-5, SAS-6, and SPD-2
262	SAS-4 and ZYG-1 govern daughter formation. A licensing mechanism appears to govern centrosome duplication.
264	P53 seems to be a regulator of centrosome development.
265	Loss of p53 results in loss of duplication regulation.
266	P53 mutation can lead to centrosome aberration.
270	Overexpression of Aurora-A kinase is a cause of CA, aneuploidy, etc.
275	Loss of p53 induces CIN.
276	Plk4, SAS-6, Cap63, SAB-5/Ana2/STIL *procued* like daughter centrioles. Proteins *preventing* over duplication: (Tak62 p1028) SKP1/SKpA, CUL1, Stimb/beta–TrCP, Nucleophosmin, SKP2, p53, CDK1, BRCA1, Cdc1rB, Geminin, Orel
277	Plk4 is the main regulator of centriole duplication. Necessary for duplication, but too much leads to multiple daughter cells.
280	Cells lacking p53 result in chromosome transmission errors. P53 is directly involved in centrosome duplication regulation.
280	Advanced human tumors have deleted p53.
280	CDK2/CyclinE decouples the centrosome.
281	Removing p53 does not result in aneuploidy (seemingly a surprise in view of Fukasawa).
283	Loss of p53 does not promote centrosome amplification.
284	Centrosome duplication requires cyclin/cdk2 kinase. Duplication depends upon cyclinD/cdk2 activity.
289	Overexpression of Aurora-A kinase causes centrosome amplification. Overexpression of Plk1 and Aurora B also increases centrosome number. Absence of p53 exacerbates this situation.

(*Continued*)

(*Continued*)

**	Review Notes
290	Preventing degradation of Mps1 in the centrosome causes centrosome reduplication.
291	Reduplication requires Plk1. Absence of Plk1 prevents reduplication.
292	Plk1 is an activator of centriole disengagement. Plk1 and separase act at different times during M-phase to license centrosome duplication.
292	The role of separase in centriole disengagement is not clear.
294	SAK aka Plk4 is a necessary regulator of centriole formation but overexpression leads to centriole amplification. SAK may be regulated by the cell itself.
298	Centriole disengagement requires the protease separase.
299	Plk1 is overexpressed in cancer cells.
300	C-Napl is necessary for regulated division.
301	This is a review article on Plks. Plks are increasingly seen as key regulators of mitosis, meiosis, and cytokinesis.
302	Kinases cooperate with Cdks in the orchestration of cell division.
302	Plk1 excess and/or deficiency can cause cancer.
302	Quote: "With Cdks on the conductor, the Plks have the role of first violin."
302	Aurora A is a cancer susceptibility gene. Definition: Aurora A: A serine/threonine kinase required for centrosome and spindle formation during mitosis.
302	Cdk1 — Cyclin B — the principal mitotic serine/threosine kinase which regulates entry into mitosis. Cdk1 is a catalytic subunit. Cyclin B is a regulatory sub-unit.
303	A review: says that Cdk1-Cyclin B is *not* a regulator of mitosis (opposite of [302]). But [302] is in 2004, [303] is in 2001.

(*Continued*)

(*Continued*)

**	**Review Notes**
304	Mitotic progression and chromosome duplication is regulated by a group of protein kinase families including: Cyclin dependent kinase (Cdk) Polo Aurora NIMA Bub Msp1.
306	E2F3 inactivation results in centrosome amplification.
307	Mpsl is required for spindle assembly but not for centrosome duplication.
338	There are as many as 300 proteins in the PCM.
338	Overexpression of SAK/Plk4 leads to centrosome amplification. Also SAS-6. These proteins could be thought of as licensing proteins.
339	Focus of the paper is on CPAP — a protein related to SAS-4. CPAP controls the centriole length. Too little → insufficient daughter development. Too much → long daughter length → supernumerary centrioles → cancer.
339	The amount of PCM is proportional to centriole size.
339	CPAP overexpression results in supernumerary centrioles.
340	SAS-4 is like tubulin in that it is incorporated into centrioles during their duplication. SAS-4 is necessary for centriole duplication.
340	SAS-4 is a stably associated centriole component that is incorporated during centriole duplication.
341	SAS-6 self-assemblies and this forms the structure of the centriole core.
342	P53 is a tumor suppressor protein — a central regulator of stress response.

(*Continued*)

(Continued)

**	Review Notes
376	Overexpression of cyclin B_2 and p53 antagonistically regulates centrosome separation.
376	Cyclins B_1 and B_2 both drive tumorgenesis when they are overexpressed. P53 mediates this by negatively regulating aurora A kinase activity.
378	The α, β-tubulin determines curvature at the distal ends of MTs.
379	Dynein and NuMA exert tension forces to stabilize the minus ends of the MTs.
380	MTs do not generate but instead maintain cell polarity.
378	The α, β-tubulin determines curvature at the distal ends of MTs.
379	Dynein and NuMA exert tension forces to stabilize the minus ends of MTs.
381	Stroma cells are surrounding environmental cells of cancer cells. They evolve with cancer cells. P53 acts against stroma cells. HSF1 enhances stroma cells. (HSF1 is "heat shock factor 1.")
382	"Tp53 (p53) is the most frequently mutated gene in cancer, being altered in ~ 50% of human malignancies." P53 is a target for therapies. Until recently, p53 was regarded as "undruggable," but the authors suggest that is no longer the case and new drugs are being tested to interact with p53.
389	Moesin is a FERM domain protein. It enhances the formation of centrosome amplification and clustering.
389	Moesin is highly expressed in basal type breast cancer.
393	Plk4 is an enzyme — a master regulator of cell division; its hyperactivity induces centriole amplification (2015).
393	Plk4 undergoes autoinhibition and is thus self-regulating.
393	Plk4 is overexpressed cancer.
393	Plk4 has an extra polo box, different from other Plk family members.
394	Asterless (*Asℓ*) is a binding partner of Plk4. *Asℓ* acts like a scaffolding protein. *Asℓ* is Cep152 in humans.

(Continued)

(Continued)

**	Review Notes
394	"...how *Asℓ* overexpression drives centriole amplification is unknown."
394	Plk4 is attracted to the centrioles and deposits itself like a spot (asymmetrically) on each centriole as they are ready to duplicate. It then provides a basis for developing new daughter centrioles. (my paraphrase)
394	*Asℓ* is thought to target Plk4 toward the centriole.
394	*Asℓ* depleted cells do not duplicate.
395	Plk4 is a regulator of centriole duplication and it depends upon asterless.
395	Cep192 recruits Cep152 and Plk4 (Cep152 is asterless).
396	Plk4 is a regulator of centriole duplication.
397	Asterless (*Ceo152) provides a molecular platform for Plk4. It interacts with the polo box of Plk4.
397	Too much or too little Plk4 causes failure of centriole assembly.
398	Plk4/SAK is a regulator of centriole biogenesis CEP152 (*Asℓ*) which interacts with Plk4 and is required for centriole duplication
398	Interfering with CEP152 function prevents the recruitment of Plk4 to the centrosome.
398	Plk4 appears to regulate itself.
399	Plk4 is a principal player in the formation of centriole flowers (multiple daughters at a single maternal centriole).
400	Cartwheel proteins: Bld10 and SAS4/CPAP, SAS-6. SAS-6 is a coiled-coil protein?
401	Not much helpful here.
402	Proteins required for centriole formation: Plk4/SAK Asterless/CEP152, SAS-6, SAS-5/Ana/STIL, CEP135, and SAS-4/CPAP
402	Asterless (*Asℓ*) is a scaffolding protein which recruits Plk4 to the site on the daughter for assembly.
402	SAS-6 is a ninefold symmetric scaffold for the daughter centriole base.

(Continued)

(*Continued*)

**	Review Notes
403	*Asℓ* is required for initiation of centriole duplication.
404	SAK/Plks is the most divergent of all the Plks.
404	Figure 2 shows a drawing of a speculative model of Plk4 behavior, including the polo box and autophosphorylation.
405	Figure 3 shows a model of polo box action and phosphorylation.
405	Plk1 is highly expressed in tumors.
406	Not much new here, but this is a highly referenced paper.
407	Plk4 needs to be at a certain level before centriole duplication occurs, but then the Plk4 self-destructs before too many centrioles are formed.
408	Plk4 autophosphorylates.
409	Plk4 phosphorylates itself to promote its degradation. This guards against duplication instability.
415	Protein Bld10p is a tip agent that connects the MTs to the SAS-6 spike (cartwheel).
416	Cullin1 is a protein which localizes to maternal centrioles and is critical for suppressing centriole overduplication.
416	Cullin1 is critical for the degradation of Plk4.
416	After Plk4, recruitment of SAS-6 is critical for procentriole assembly.
418	Loss of p53 is not by itself a primary cause of aneuploidy. This paper is a review summarizing the current (2005) knowledge between aneuploidy and p53.
419	This is a review focusing on Plk1 and family members plk2–4 and their roles in mitosis.
419	Plk1 promotes mitosis entry.
419	Plk1 is involved in a chromosome maturation and separation.
419	Plk4 is involved in centriole replication.
419	Plk1 does not only phosphorylate other proteins, it can also phosphorylate itself.

(*Continued*)

(Continued)

**	**Review Notes**
1	Overexpression of Plk4 stimulates the concurrent formation of multiple daughter centrioles at single maternal centrioles.
1	The harmful effects of Plk4 are enhanced by E/CDK2.
3	Plk1 is needed for centriole disengagement.
5	Plk4 and SAS-6 are required for centriole duplication. Disengagement requires Plk1.
420	Overexpression of SAK/Plk4, DSas6, and DSas4 leads to de novo centriole-like structures.
421	"Plk4 is the conserved master regulator of centriole assembly." "Taken together, our findings suggest that Plk4 directly generates its own phosphodegron and can do so without the assistance of additional kinase."
424	ε-Tubulin is a component of the sub-distal appendages of the centriole. ε-Tubulin is required for centriole duplication and organization of pericentriolar material.
424	δ-Tubulin is required for the formation of triplet MTs.
424	ϒ, δ, and ε-tubulin localize in the centrosome.
425	ϒ-Tubulin is required for initiating MTs.
427	Of the five tubulins (α, β, ϒ, δ, and ε), α and β make up the MT. ϒ is at the base for nucleation, δ is associated with the centriole, and ε with the pericentriolar material. (In [424] δ is associated with the triplets.)
428	(2000) MT assembly is initiated by ϒ-tubulin. ϒ-Tubulin and other proteins (Spc 97p and Spc 98p) form the core of the forming MT.
438	Nip is a centrosomal substrate which is controlled by Plk1 in the development of MTs.
439	Asterless (*asℓ*) is needed for stable centrosomal development.
440	RLH papers.
441	
445	Reduced Plk4 leads to cancer.
446	Discusses various proteins associated with the Plks.

(Continued)

(*Continued*)

**	Review Notes
446	All Plks are similar.
451	Plk4 initiates centriole duplication and is regulated. The protein Phosphatase 2A regulatory submit twins stabilizes Plk4.
451	The mechanism stabilizing Plk4 is still unknown.
451	Plk4 localizes to the "parent" centriole as a spot and then it modifies the site.
452	As many as 80 proteins, and possibly more, are needed for cell cycle stability. This is a review-survey paper on the protein.
453	SAK/Plk4 is different than other polo-like kinases in having only a single polo box.
453	SAK/Plk4 is destroyed by APC/C-dependent proteolysis.
456	SAS-6 is recruited to the mother centriole and then associated with the emerging daughter centriole.
457	Five proteins have been identified for centriole formation: ZYG-1, SAS-4, SAS-5, SAS-6, and SPD-2. The order is SPD-2, ZYG-1, SAS-5, SAS-6, and SAS-4 — for *C. elegans*.
459	*C. elegans* embryo: ϒ-tubulin stabilizes the formation of new centrioles.
459	*C. elegans* requires SAS-4, SAS-5, SAS-6, and ZYG-1 to form new centrioles.
460	A, β, ϒ-Tubulin are found in all eukaryotic organisms and δ and ε-tubulin are found in organisms with triplet MTs.
461	ϒ-Tubulin ring complex (ϒ-TuRC) forms a template for MT nucleation.
462	ϒ-Tubulin small complex (ϒ-TuSC) helps form the 13-fold MT structure.
463	New MTs require ϒ-tubulin.
464	The centrosome is not simply a cloud of proteins. Instead it is structured layers of matter.
465	Plk4 is at the head of the pathway for centriole formation. Asterless (*asℓ*) (Cep 152 in humans) is a scaffold for Plk4 and SAS-6.
465	ZYG-1 is Plk4, SAS-6 is Ana2 in Drosophila.

(*Continued*)

(*Continued*)

**	**Review Notes**
465	SAS-4 provides for protein–protein interaction.
465	Plk1 is required for the dramatic increase in PCM.
465	Plk1 recruits ϒ-tubulin.
466	Plks are characterized by their polo-box domain.
466	Plks are controlled by phosphorylation, proteolysis, and transcription.
466	There is an uptick in Plk1 in human tumors.
467	Plk1 is a regulator of cell division. Plk1 is recruited to centrosomes, kinetochores, and the spindle.
467	Plk1 is a potential drug target for cancer therapy.
469	Plk1 is overexpressed in tumors.
469	Plk1 regulates many processes. Plk1 and Plk4 (SAK) localize to many structures via their polo box.
469	Plk1 has two polo boxes, but Plk4 has only one polo box.
470	Plk2 is required for centriole duplication.
475	MT assembly is usually seeded by ϒ-tubulin ring complexes (ϒ-TuRC) and ϒ-tubulin small complex (ϒ-TuSC).
474	(2000) ϒ-tubulin is required for MT nucleation, but the mechanism is not understood (a near quote).
477	αβ-Tubulin heterodimers self-assemble to form the MTs.
477	ϒ-Tubulin nucleates the MT. ϒ-Tubulin is believed to be capable of longitudinal assembly.
477	Δ-Tubulin may seed triplets.
477	ϒ-TuRC is called "gammasome" (gamma tubulin ring complex); ϒ-TuSC (gamma tubulin small complex).
477	δ-Tubulin interacts with α-tubulin. ε-Tubulin interacts with β-tubulin (2000).
478	The focus of this paper is on the behavior of ϒ-TuRC in nucleating MTs.
478	Tubulin dimers self-assemble in vitro.

(*Continued*)

(Continued)

**	Review Notes
478	ϒ-Tubulin is required for MT assembly.
478	"ϒ-TuRC resembles a lock washer...," like the first turns of the MT helix.
496	MTs interact with actin to form the shape, migration, and division of the cell. Actin is an abundant protein. The monomeric forms is call globular of G-actin; the polymeric form is filamentous for F-actin.
500	P53 is a tumor suppressor.
508	Under certain stimulated conditions (absence of p53 and overexpression of Plk4) supernumerary centrioles occur without resulting in a tumor.
511	P53 is a protector against irregular cell development in the absence of Plk4. Aberrations in centrosome number can lead to cancer.
512	Plk4 primes the binding of STIL to SAS-6. Also, supernumerary centrosomes promote cancer.
516	ϒ-TuRC is a lock washer shaped protein complex.
517	ϒ-TuRC is a 13-ϒ-tubulin molecule — containing as many as 5 ϒ-TuSC.
518	(2012) ϒ-tubulin behavior is poorly understood.
518	ϒ-Tubulin provides a template for MT nucleation (good pictures).
519	ϒ-TuRC is formed by ϒ-TuSC.
519	ϒ-TuRC provides a stabilizing effect of the MT dynamics — but how is not clear.
529	SAS-6 imparts ninefold symmetry of the centrioles. SAS-6 forms a torus around the base of the parent centriole.
529	Depletion of SAS-6 results in failure of centriole formation. Overexpression of SAS-6 results in supernumerary centrioles.

(Continued)

(*Continued*)

**	**Review Notes**
610	This paper focuses on the protein GCP6 and its role in centriole construction. • GCP6 is needed by Plk4 (SAK) but it is not clear how. • GCP6 is localized to the perimeter of the centrosome — at a distance from the centrioles. • GCP6 is part of the ϒ-TuRC.
611	This paper focuses on STIL and CPAP for procentriole formation. STIL makes an asymmetric location on the daughter cell — required for new centriole formation. "Overexpression of STIL induces a formation of the multiple procentrioles around the parent centriole." Conversely too little STIL presents normal development. CPAP induces centriole elongation. Quote: "our results indicate that the two MCPH associated proteins STIL and CPAP interact with each other and are required for procentriole formation, implying a central role of centriole biogenesis in MCPH." MCPH is microcephaly.
648	In the formation of the microtubules, the kinetics of the protein action and interaction appears to be unknown.
659	Overexpression of Plk4 promotes centrosome amplification.
660	Most proteins are 5 nm across.
678	There is an incomplete understanding of the interaction of nanoparticles and cell proteins.
689	The paper describes force curves on the cell membrane.

**Refers to chronological listing of literatures reviewed.

7. Chronological Literature References Review Notes

Category	**	Review Notes
Centriole cancer	1	Abnormal centriole members are detected in virtually all cancers.
Protein cancer	1	Overexpression of Plk4 stimulates the concurrent formation of multiple daughter centrioles at single maternal centrioles.
Protein	1	The harmful effects of Plk4 are enhanced by E/CDK2.
Centrioles	2	Centrioles are visualized by electron microscopy.
Centrioles	3	Centrioles can recruit pericentriolar material (PCM) to form microtubules from the beginning of the cell cycle.
Centrioles	3	Mother centrioles and daughter centrioles have different abilities to recruit microtubule material. Daughter centrioles have a large portion of their distal ends uncovered and have only the proximal ends imbedded in the PCM.
Centrioles	3	Daughter centrioles become mature during mitosis.
Centriole protein	3	Plk1 is needed for centriole disengagement.
Centrioles cancer	4	Supernumerary centrioles probably contribute to the development of cancer.
Centrioles	5	Healthy centrioles duplicate only once per cell cycle.
Centrioles cancer	5	Aberrations in centrioles are associated with cancer.
Cent DNA	5	Centriole duplication is like DNA replication.
Cent	5	Excess PLK4 triggers the formation of supernumerary centrioles.
	5	A color model of centriole duplication.

(*Continued*)

(*Continued*)

Category	**	Review Notes
Cent	5	Speculation that mother/daughter separation occurs before centriole duplication.
	5	Speculation: "... proteins required for centrosome clustering might be attractive targets for selective elimination of tumor cells..."
Nigg 2011	5	In mitosis, the cell receiving the older mother centriole is able to form primary cilium earlier than the other cell.
	5	PLK1 needed for separation.
Centrioles	6	Centriole existence has been known for over 50 years but they are just now becoming better understood.
Centrioles cancer	6	Aberrant centrioles lead to cancer.
Centrioles cancer	6	There is a connection between extra centrioles and tumor growth.
Centrioles	7	The proximal end is negative and the distal end is positive.
Centrioles cancer	8	Centriole defects promote cancer.
Cancer	9	C1N is a conspicuous feature of human cancers.
Cancer	9	Cancer cells multiply 10–100 times more frequently than normal cells.
Centrioles cancer	9	Extra centrosomes are suggested to be a common contributor to cancer.
Cancer	10	Human tumors have centrosome aberrations.
Nano	11	A paper on nanotechnology and DNA.
Nano	12	Nanotechnology needs further development before it can be widespread clinically useful.
MT, EM	13	Microtubules are ferromagnetic.
MT, EM	13	"It has been demonstrated that it is possible to orient arrays of MTs by the application of an electric field ..." (see also 38).
Centriole Cancer	14	Cancer cells have aberrant centrosome numbers.
Cancer EM	15	Electric fields inhibited the growth of dermal tumors in mice.
EM cancer	16	Sequentially programmed magnetic field (SPMF) therapy. SPMF has been shown to arrest brain tumor growth in a patient with previous poor prognosis.

(*Continued*)

(Continued)

Category	**	Review Notes
Centriole cancer	17	Cancer cells have a wide range of centrosome abnormalities. Centrosome dysfunction results in cancer.
EM	18	There appears to be an "organizing field" in mitosis that is not entirely driven by either chemical or electromagnetic effects.
EM	19	(1998) Introduces a theory that electric fields determine the spatial structures of the MTs.
Centrioles	20	Centrioles form centrosomes and cilia.
Centriole cancer	21	Cancer cells have centrosomal abnormalities.
Centrioles	22	G1 $\rightarrow$ S-phase $\rightarrow$ G2 $\rightarrow$ Mitosis. Centrioles are duplicated in S-phase and they mature in G2 phase.
Centriole cancer	23	Centrosome disruption can lead to disruption in cell tissue. Tumor cells have supernumerary centrioles and multiple centrosomes.
Cancer EM	24	SPMF Therapy may be a cancer therapy.
MT	25	MTs grow and shrink. Polarity of tubulin dimer reflects the polarity of the MT. In mitosis, the "pushing" from the two ends of the spindle centers the nucleus.
MT	26	To form a new centriole we need centrin-3, Plk4, and SAS-6.
EM, MT	27	(2009) Surface charge density of a centrosome is $\sim e^3$ $e = 1.6 \times 10^{-19}$ C. This is much smaller than the charge of an MT.
Centriole	27	Precise physical properties of the centrosome are yet to be determined.
Cancer EM	27	Small electric fields can influence the properties of cell division.
EM	27	MT surface charge is 1–2 orders of magnitude greater than that of a centrosome.
EM, MT	28	Paper presents a model of how alternating electric fields control microtubule-based transport.
EM, MT	29	(2011) A very impressive paper that asserts that electromagnetic fields are developed by oscillating MT and that the higher the frequency the greater is the generated field. The paper also states that the electromagnetic field has cellular effects that go beyond the chemical effects.
MT	29	Microtubules can vibrate in kHz to GHz regions.

(Continued)

(*Continued*)

Category	**	Review Notes
Cancer Nano	30	(2011) Nanoparticles (NPs) easily penetrate cellular membranes. NPs have a range in size from 1 to 100 nm.
EM	31	(2011) Better understanding of cell signaling is needed.
EM	31	Signals by cells can be transmitted through a polystyrene wall — even with cells growing in different petri dishes (2011).
	32	Electric fields generated by synchronized oscillations of microtubules, centrosomes, and chromatin fibers facilitate events during mitosis and meiosis (2012).
	32	Cancer cells have greater entopic states than normal cells. Microtubules of cancer cells generate an electromagnetic field with various frequencies (2012).
	32	Above a certain field intensity, chromosomes orient themselves with their long axis in the field direction.
	32	There is a pressing need for more detailed mathematical models of chromosomes and centrioles to predict cellular events (2012).
EM	33	Electric fields are found in and around every living cell and are critical for the proper functioning of biological processes.
EM	33	It is difficult to experimentally assess the effect of electrical fields on intracellular activity.
EM Vibs	34	Oscillations of microtubules form the basis for the electromagnetic fields (2012).
EM	34	Chromosomes orient themselves with their long axis along the field direction.
	34	Physical mechanisms regulating a chromosome packaging has not been fully characterized.
EM Res Keck	34	Electric fields not yet well understood.
EM Res Expl	35	Magnetic force microscopy is an application of atomic force microscopy (AFM) where a probe coated with a magnetic film is vibrated slightly above a surface.
EM Expl	35	Electric force microscopy (EFM)
EM Expl	35	EFM and MFM can characterize the electromagnetic field above a nanosurface.
EM	36	Microtubules are ferromagnetic [13 — 2004] — proposed by Hol in 1985 and modeled by M.V. Sataric in 1993 (1985).

(*Continued*)

(*Continued*)

Category	**	Review Notes
EM	37 38	Oriented assemblies of dipoles in the MT wall is responsible for the ferromagnetism [13 — 2004]/from [37 — 1993 and 38 — 1999].
EM	38	It is possible to orient arrays of MTs by the application of an electric field. It is speculated that inter-cellular electric fields may induce MT orientation [38 — 1999] (from [13 — 1985]).
1998 protein	39	Each α and β-tubulin has a "relative molecular mass of about 55,000 Daltons" (?)
	40	MT size 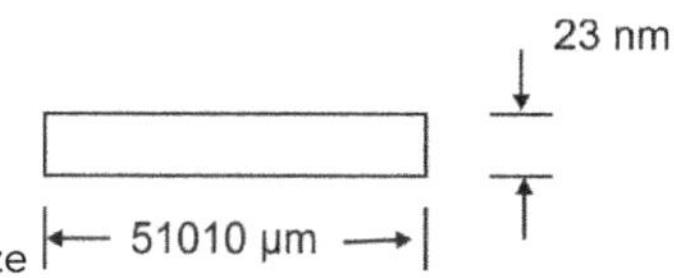 (2005) Is analytical/computation — *not* expt. Microtubules form the mitotic spindles [40] that segregate the chromosomes. Dipole — dipole forces parallel to MT surfaces are negligible. But not so for perpendicular to surface! Dipole forces play significant role in MT formation.
P. Schaeffer Keck EM	41	MT can generate electrodynamic fields. Cancer can occur from disturbed electrodynamic fields. Nanotechnology is promising for investigating electromagnetic fields. "Long time — cancer — dependent on chemical and genetic changes."
MT EM	42	Microtubules made of -tubulin and β-tubulin. Microtubules help define cell morphology. They are building blocks for cilia and flagella. They are stiff in length but work in shear. 100G < E.
	43	1985 big paper.
MT EM	44	MT still not clearly understood. "Travelling kink" discussed. Experimental work called for.
	45	Confirms earlier findings.
	46	MTs are electrically polar and have vibration modes from kHz to low THz. MTs are highly deformable hollow tubes. MTs undergo growth and shrinkage, unless the positive end is stabilized by a capping structure. MTs are near mitochondrion. MTs are orthotropic structures.

(*Continued*)

(Continued)

Category	**	Review Notes
Keck	47	"many new facts may change over view of cells very little, whereas a few can transform the whole picture" p. vii (1994).
MT EM	47	For MT: the fastest growing and is defined as "positive," pp. 805–806.
MT	47	Due to their spiral geometry, an MT may point either CW or CCW, p. 805.
Cent	47	Centrioles arise by the duplication of existing centrioles, p. 819.
MT	47	Picture of growing/shrinking MT, p. 810.
	47	Actin filaments form a contractile ring to separate dividing cells, p. 912.
Cent	47	Daughter centrioles begin to grow near the base of each parent centriole and at a right angle to it.
Model c	48	16 nm ~ 2 dimer lengths.
MT	48	MTs have variable growth rates, and sometimes shorten as much *as five tubulin* layers.
EM Keck	49	Centrosomes (PCM) provide polarity (charge).
	50	GTP cap exists (an MT tip tubulin).
EM	51	Biosystems generate electromagnetic fields.
	52	Confirms earlier findings.
	53	Understanding cell mechanics at micro/nano-level needed AFM good tool. Malignant tumors are stiffer (softer?) than surrounding tissue.
History	54,55,56	Paul W. Schafer papers.
	57	Confirms earlier findings.
History	58	Paul W. Schafer paper (1972).
EM	59	J. Pokorny, *et al.*: Tumor cells have greater damping in electromagnetic waves on the MT.
MT	59	MT shear modulus: $G = 5$ N/m^2, that is, $G = 5$ Pa.
History	60	W.F. Marshall and J.L. Rosenbaum Centrioles seen over a century ago. Centrioles needed to form centrosome.
Cent	60	Pre-existing centrioles are the templates for making new centrioles.
Cent	60	The center of a centriole is called the lumen: γ-tubulin.

(Continued)

(*Continued*)

Category	**	Review Notes
Cent Keck	60	~2000 Long way to go before understanding centriole functions and duplication.
Cancer	61	Loss of adhesiveness plays a major role in the detachment of cancer cells (metastasis).
EM/Cancer	61	Disturbances in electromagnetic fields produce cancer.
EM	61	Cells have electromagnetic fields.
MT	62	MTs are orthotropic. This anisotropy lowers torsion and radial wave velocity, but it does not affect longitudinal velocities. This would not occur if MTs were isotropic.
MT	62	MTs are the most rigid of structures.
MT	62	MT: 25 nm OD; 15 nm ID. 13 parallel protofilaments bonded laterally. Helical structure.
MT	62	Lots of recent studies of MT.
MT	62	Shear modulus and elastic modulus in circumferential direction are lower than in the longitudinal direction.
MT	62	Authors neglect helix effect.
MT	62	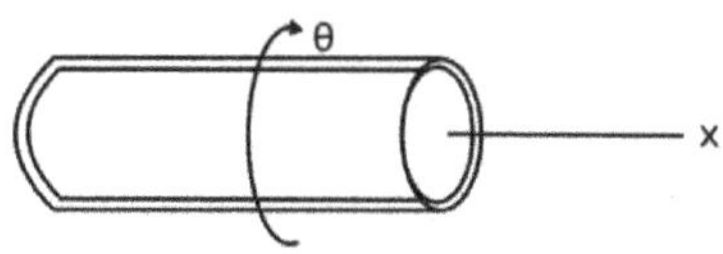 $E_x = 0.5$ to 2 Gpa $\upsilon = 0.3$ $E_\theta = 1$ to 4 Mpa $\gamma = 1.47$ $G = 1$ MPa
MT	62	Wave velocities given but no frequencies. Torsional wave → 26.1 m/s; longitudinal wave → 824.8 m/s.
MT	63	Young's modulus $E = 1.96$ Pa Bending stiffness EI = 2.6×10^{-23} Nm^2
MT	63	Good pictures
Cancer	63	Cancer cells are softer (less stiff) than normal cells (Table 5), p. 4002.
Cancer	63	Disruptions of tubulin reduced adhesive of cancer cells to blood vessels during metastasis.
MT	64	"Circumferential" modulus ~0.1 "longitudinal module" ⇒ orthotropic shell model — (same as 62).
MT	64	Isotropic model overestimates frequencies except in longitudinal and bending modes. (Specifically, in radial modes is the major difference.)

(*Continued*)

(*Continued*)

Category	**	Review Notes
	64	Good picture
	64	$E = 0.56$ GPa $\upsilon = 0.3$ isotropic model $E_x = 0.16$ GPa $\upsilon_x = 0.3$ orthotropic model
MT	65	MT can withstand compressive loads due to adjacent and surrounding tissue.
MT	66	Temperature may affect MT properties. Testing may lead to misleading results.
MT	66	Claims to report elastic moduli for the first time.
MT	66	MTs are anisotropic and inhomogeneous.
MT	66	MTs are similar to carbon nanotube ropes.
	66	External MT dia. $1/E_{bending} = 1/E + (10/3G)(D^2_{ext} + D^2_{int})/L^2$ – MT length shear modulus (G); Internal MT dia. (D_{int})
EM MT	67	MTs can generate electric fields.
EM Cancer	67	The peak of electromagnetic activity occurs during mitosis.
MT EM	68	Calculation of electric field generation via axial longitudinal vibration of MTs.
EM	68	α, β tubulin heterodimers have high electric dipole leading to high electrical polarity of MTs.
MT	68	There are several hundred MTs in a cell.
MT	68	2010 Says Kis [66 — 2002] finds $E > 100$ MPa and $G = 1.4 \pm 0.4$ MPa. (orthotropic)
Vibs	69	AFM (2004) measures cell vibs of 0.8–1.6 kHz.
EM Cancer	70	Low frequency AC electric fields inhibit growth of tumors (2004).
	71	Follow-up of [70] 2007.
Cancer EM	72	AC frequencies >75 Hz or intensities (>8.5 μA) causes cell death.
Protein cancer	73	Plk4 (aka SAK) overexpression induces formation of multiple centrioles.
MT	74	Protofilament — a column of an MT (as opposed to a helix of dimers). Typically, 13 protofilaments in an MT.
Vibs 2005	74	Vibs go faster along protofilaments than along the helix.

(*Continued*)

(*Continued*)

Category	**	Review Notes
MT	74	Uses a model of a lattice of dimers.
MT	74	Rupture strength of MT ~0.4–0.5 Pa.
MT	74	MT: $E = 10^6–10^9$ Pa.
	74	"Phonon" elastic vibration of subunits about equilibrium position.
	74	Frequency of phonon vibration $10^7–10^{10}$ Hz (Pokorny) But here: $\omega_x = 3.15 \times 10^{11}$ Hz $\omega_y = 4.8 \times 10^{10}$ Hz
MT 2004	75	MT buckle at ~600 Pa ↕ Wow!
	75	E ~ 1.6 GPa, MT buckle at 14.8 MPa.
MT pictures	76	Good drawings of tubulin.
MT 2005	76	Table of elastic properties $E = 1.2$ to 2.7×10^9 Pa [8–11] $E = 1$ to 5×10^8 Pa [5, 12] $G = 1.4 \times 10^6$ Pa $\nu = 0.3$ to 0.4 $EI = (16$ to $45) \times 10^{-24}$ Nm^2 [see 77]
MT 1993	77	$EI = 2.2 \times 10^{-23}$ Nm^2 [same as 76].
Cancer EM	78	Cancer cells are *less*? Polarized than normal cells (due to ion flow into and out of a cell).
2001	79	Talks about a "matrix" and protein "skeleton" which may or may not play a role in mitosis.
EM	80	The centrosome transmits polarity to daughter centrioles.
Centriole	80	The daughter centriole is only ~80% as long as the more mature mother centriole.
Centriole	81	Speculation about nine-ness: the number 9 arises due to the inner diameter of the centriole being fixed and thus 9 blades of 3 are needed along the circumference. (Really?)
2011	82	A review.
2007	83	PAR3 — a polarizing protein. PAR proteins are still not well understood.
MT	84	Computer modeling of tubulin → MT MTs appear to be zipped up lattices of filament. Each tubulin monomer has more than 400 amino acids.

(*Continued*)

(*Continued*)

Category	**	Review Notes
		Centriole MT probably differs from mitotic spindle MTα and β tubulin are very similar. MT dissection 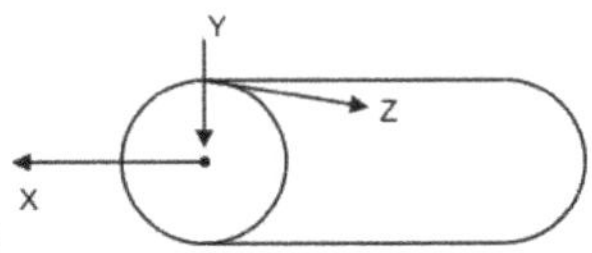
Centriole	85	Flies can live without centrioles but they are not normal and soon die.
EM MT	86	Difficult to read. The conclusion seems to be that MT carry current from polarity in the centrosome (2007).
	87	Seemingly important paper: (1) MTs form electrically polar structures (2) MTs generate endogenous electromagnetic field (3) The electromagnetic field can generate a driving force for transport (4) *Vibrating MTs generate the field* (5) The majority of proteins are polar (6) MTs are the main organizers of the cytoskeleton (7) The electromagnetic field of the cell can attract small dielectric particles to the cell (8) Tubulin heterodimers are strong electric dipoles with dipole moment greater than 1000 Debye is 10^{-26} cm (check). (9) An ionic charge layer is formed around MT. A similar charge layer may be formed inside the MT in the cylindrical cavity also (10) The static electric potential around an MT is effectively screened within two Debye lengths (a Debye length is ~0.8 nm) (11) Debye length ~0.8 nm (12) MTs are very elastic. They have large deformation at low stress. They do not become plastic until strain is ~50–60%. (13) "Any analysis that neglects interfacial boundary conditions cannot yield correct results." (14) *MTs are treadmills, tubulin dimers ad to the proximal and leave at the distal end* (15) MT surface is *slippery* allowing for vibration.
	88	*Cytoskeleton* — a protein network giving the shape to the cell.
	88	The structures of the cytoskeleton with electric dipole moments could generate oscillating electric fields and interact with external electromagnetic fields.

(*Continued*)

(*Continued*)

Category	**	Review Notes
MT EM	88	The subunits of microtubules: tubulin heterodimers — have electric dipole moments.
2003	88	Microtubules satisfy the basic requirements for excitation of vibrations and the generation of endogenous oscillating electric field.
Pokorny	88	Damping of vibrations in biological structure is an important question — particularly viscous damping by surrounding structures.
Vib	88	Vibrations of MT filaments are thought to be longitudinal (see Figure 1).
	88	*Cytoplasm* — The contents of a cell outside the nucleus.
	88	*Cytosol* — Contents of the cytoplasm except for the organelles.
* explore this	88	The electric dipole moment is ~1000 Debye (10^{-26} cm) for a heterodimer. Motion of electrons or protons can induce ~200–400 Debye.
EM	88	The negative charge is associated with the β-monomer.
MT	88	"The charges in the microtubule are at most 2 nm from the surface with ~ exponential dependence of the electric potential in the charge layer may be assumed." (See Figure 1.) 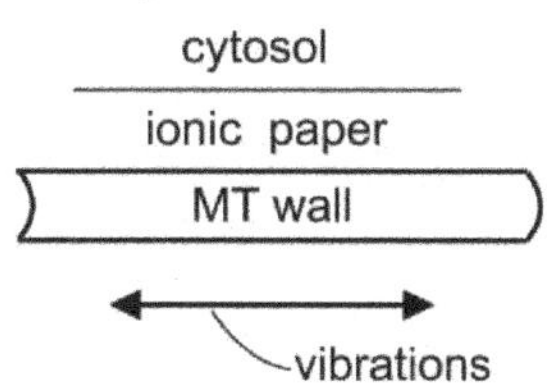
MT 1998	89	αβ dimer tubulin and connections between dimers form the protofilament — has a drawing (Figure 1).
Cent Can	90	"Abnormal centrosome members are detected in virtually all cancers".
2007	90	A single maternal centriole can generate multiple daughter centrioles.
Protein	90	Multiple daughter centrioles arise from a presence of Plk4 and E/cyclin-dependent kinase2 (CDK2).
Cent CC	90	Figure 2 shows a picture of multiple daughter centrioles in a flower-type configuration.
Cancer	91	Cancer cells have chromosome aberrancies.

(*Continued*)

(*Continued*)

Category	**	Review Notes
Cent CC protein 2012	91	Centriole duplication arises from Plk1 and separase (Polo-like kinase 1). The right regulation of Plk4 is needed for healthy cell mitosis.
Cancer 2012	91	A single maternal centriole assembles multiple daughters concurrently.
CC protein	91	"..CAND1 is a centrosomal protein that controls the stability of Plk4, a master regulator of centriole biogenesis. CAND1 was found to synergize with Plk4 to stimulate centriole overduplication..."
EM 2012	92	How polarity is initiated is not well understood.
EM 2004	93	Similar to 92 polarity starts in centrosome. The focus, however, is on sperm/embryos.
Cent	94	"The mechanisms directing centriole formation are incompletely understood and represent a fundamental open question in biology" (2011).
Cent	94	The ninefold symmetry probably comes from a geometric characteristic of a protein.
Cancer CC 2006	95	Centrosome aberrations lead to cancers. (This is a review by Nigg.)
EM Keck	96	Very interesting paper — suggests targeting cancer cells due to their supernumerary centrosomes.
Cancer	96	Cancer cells have supernumerary centrosomes.
Keck Cc 2009	96	Cancer cells survive (and thrive) due to their clustering of the supernumerary centrosomes.(This may be a plus in electromagnetic detection and as an attraction for charged nanoparticles.)
MT	96	Mother centrioles contain a set of distal appendages for anchoring MTs — daughter centrioles do not have this appendage.
Centriole cluster in cancer cells	96	Suggests that effective cancer therapy may be a mechanism that targets the *centrosome cluster*.
CC 2009	96	Suggests that cancer cells will die if the clustering of centrosomes can be prevented.
Keck 2010	97 98	A good reference on nanotechnology application.
EM 2010	99	The centrioles may recruit blood vessels.

(*Continued*)

(Continued)

Category	**	Review Notes
Can 1999	100	Centrosome multiplication is associated with polyploidization. Ploid: Number of chromosomes, that is, multiple centrosomes may cause, or be caused by multiple centrosomes. Centrosome — DNA associated duplication in the nucleus. This occurs with non-malignant cells as in wound healing.
2004	101	Focuses upon drugs to destroy MT. Such drugs may also be useful in destroying MT in blood vessels which support tumors.
MT	101	There may be a variety of types of MTs.
2004	101	Some MTs predominate in cancer cells. They also develop drug resistance.
Keck	101	"it has been argued that microtubules represent the best cancer target to be identified so far, ..."
	101	MT exchange tubulin during mitosis in 10–30 s.
2007	102	Discussion about drugs to stop mitosis.
Protein	102	KSP is a kinesin motor protein needed to establish the mitotic spindle bipolarity. There is no other known role for KSP.
Protein	102	Anti-mitotic drugs are used to inhibit the development of KSP.
Cancer	103	Centrosome amplification (more than two centrosomes) is characteristic of cancers.
Cancer	103	Multipolar spindles unequally distribute chromosomes to daughter cells.
MT 2004	103	"When the cell enters mitosis, the daughter centrosomes nucleate the astral arrays that contribute most of the microtubules to the formation of the spindle.
Cancer	103	Tumor cell populations proliferate and the extent of the centrosome amplification appears to increase progressively.
Cancer 2001	104	Aneuploid tumor cells are caused by supernumerary centrosomes. Aneuploidy: the gain or loss of one or more chromosomes from a diploid genome.
	104	This is a very insightful paper: it basically asserts that multiple centrosomes lead to defective daughter centrioles.

(Continued)

(Continued)

Category	**	Review Notes
	104	Defective daughter centrioles can lead to cells that die (apoptosis) but also to a few that survive and thrive and multiply repeatedly and quickly and then form tumors.
	104	Multiple centrioles lead to a chaos of daughter cells — Most of these die but a few thrive and produce tumors.
	105	Centrosome amplification is correlated with chromosomal instability and tumor development.
2002	105	Removal of centrioles leads to failure to duplicate.
	105	DNA copying and centriole duplication are related.
2000	106	Study of yeast cells (who have no centrioles) and attempts to draw conclusions about eukaryotic cells.
	106	Daughter centrioles mature during mitosis.
Cancer 2002	107	Centrosome amplification (too many centrioles) leads to genomic instability, abnormal mitosis, and aneuploidy.
Protein 2002	108	Plk regulates centrosome duplication.
Protein Plk4 excess	109	This paper focuses on how excessive Plk4 leads to multiple centrioles.
2007 protein	109	Plk4 excess causes *simultaneous* multiple procentrioles in each parent centriole.
Keck	109	"Centrioles and basal bodies are assembled and how their numbers are controlled within cells constitute long-standing unresolved questions."
Centrioles	109	"...it now appears that pre-existing centrioles act primarily as solid-state platforms to accelerate the assembly process."
Centriole	109	... ninefold symmetry (termed "cast wheel") has been proposed to serve as a scaffold for the assembly of centrioles.
2007	109	"pre-existing centrioles represent 'solid-state platforms' to facilitate centriole formulation rather than general templates" (Nigg 2007).
Picture	109	Mitosis diagram in Figure 7 [109].
Cancer 2008	110	Centrosome amplification initiates tumors in flies.
2007	111	Viruses often assemble in proximity to centrosomes.

(Continued)

(Continued)

Category	**	Review Notes
Keck	111	"The question of how cells keep centriole numbers constant over successive cell divisions continues to represent one of the most mysterious problems in contemporary cell biology."
Keck	111	Another mystery is: What determines the length of a centriole MT?
Centriole 2002	112	This paper discusses how the centrioles develop the MTOC and thus the centrosome.
Picture	112	A few good pictures.
Cancer 2003	113	Centrosome abnormalities promote chromosome instability and pre-invasive cancer.
Cancer	114	Supernumerary centrioles cause multipolar mitosis and genomic instability.
2008 cancer EM	114	Supernumerary centrioles occur by a centriole producing more than a single daughter centriole.
Keck	114	Mechanism preventing centriole "overdevelopment" in the wall of a mother centriole is not well understood.
Keck	115	An older paper (1990). (i) Centriole pairs are ⊥ only up to a point in the mitosis. (ii) Centriole doubling and DNA replication are similar but DNA replication is much better understood.
Centriole	116	A diplosome is a centriole pair.
2000	116	Centrioles maintain their perpendicularity through mitosis.
Keck	116	Centriole function within the centrosome is still unclear.
	116	Centrioles are likely to play an important role in centrosome reproduction.
	116	Centrin is a small protein concentrating within the distal end of the centriole.
	116	Mother and daughter centrioles differ in their behavior and contributions.
EM	116	During mitosis the daughter centriole moves away and around the nucleus to the opposite side.
Centriole EM	116	During mitosis the mother centriole stays in place near the *centriole*. Thus in mitosis there are centriole pairs on opposite sides of the nucleus.
EM	116	During mitosis the daughter centriole rocks (or oscillates).

(Continued)

(Continued)

Category	**	Review Notes
EM	116	The mother centriole is stronger than the daughter centriole.
EM MT	117	Tubulin is apparently a source of polarity from embryo testing.
	118	See 112.
2007 Keck paper intro	119	As of 2007, very little known about centriole biogenesis.
Keck 2010	120	It is unclear how the cartwheel of a centriole is formed.
	120	SAS-6 self-assembles in stable tetramers.
	120	Tetramers (fourfold molecules forming a polymer).
Centriole	120	Tetramers are thought to be building blocks of a central microtubule of a centriole [the first time of reference to a central MT in a centriole (I thought there were two of these)].
	120	Centrioles recruit a pericentriolar matrix forming the centrosome.
Keck	120	When a centriole duplicates, it is unknown how the daughter develops in that it does not use the existing centriole as a template (i.e., where does its cartwheel come from?)
	120	Centriole biogenesis uses a procentriole.
	120	It has been proposed that the central tubule promotes the ninefold symmetry.
Keck	120	The composition and biogenesis of the central centriole are unknown.
Centriole growth	121	Daughter centriole is 80% of the length of the mother centriole.
2007 Centriole	121	Mother centrioles are distinguished by two sets of nine appendages at their distal ends — which are thought to be required for anchoring microtubules at the centriole.
	121	Tubulin levels increase dramatically prior to mitosis concomitant with the MT-associated problem required for spindle formation.
Protein	121	Plk4/SAK is essential for centriole duplication.
Protein cancer	121	Overexpression of Plk4/SAK leads to centrosome amplification.
2007	121	Centriole elongation reaches full length during cell cycle.

(Continued)

(Continued)

Category	**	Review Notes
Centriole	121	Centrosome separation is believed to require disassembly of the linkage connecting the two centriole pairs, but not the centrioles within a pair.
2007	122	Micrograph images of centriole ends show two (centrioles? MTs?) in the center.
2006	123	Centrioles position and reorient themselves as need in migrating cells as in wound healing.
Paper	124	(2001) Centrioles now of interest.
Centriole	124	Centrioles and chromosomes (DNA) are the only organelles able to self-duplicate.
	124	Centrioles are barred-shaped 0.5 m (500 nm) long, 0.2 m (200 nm) in diameter.
	124	Mother centriole has extra appendages.
Paper Keck	124	"The exact function of centrioles is still unclear."
	124	Centrioles elongate at the distal end.
	124	2001 + distal, − proximal.
	124	Global influence — still unclear.
Picture	125	(2011) Interesting pictures.
EM	126	(1988) Electric fields determine spatial orientation of microtubules and actin filaments — proposed theory.
Paper	127	The triplets of MTs are not equal in length: outer is largest, inner is shortest. A — inner, B — mid, and C — outer.
1996	127	Good picture (good for making a drawing).
	127	Centrosome: centrioles plus surrounding glob of pericentriole material.
	127	Size: 0.2 m dia and 0.4 m long.
	127	Mother centrioles have distal appendage.
	127	Mature centrioles have a simple connector of the three MT sets at the proximal end. Immature centrioles have a more complex connecting set at the proximal end (cartwheel structure). Both mature and immature centrioles have a dense ring at the distal end. (This suggests that the distal ends are complete and that growing occurs at the proximal end — coming out of the paracentriolar material.)
Paper	127	Right hand rule: Long axis of daughter centriole (immature centriole) intersects the mother (mature) centriole — but not vice versa. That is the axis of the mother does not intersect the axis of the daughter.

(Continued)

(*Continued*)

Category	**	Review Notes
1996	127	Maturation occurs during mitosis. This includes the growing of appendages near the distal end.
Mitosis	127	Cenexin (?) is acquired during mitosis.
	127	States that orthogonal orientation is lost, centrioles separate, and then acquire cenexin to mature.
EM	127	The proximal end of a mother centriole is surrounded by *electron-dense* pericentriole material.
Cancer protein	127	"P53" is a tumor suppressor.
MTs protein	127	Centriole barrels contain only a few proteins namely: α and β-tubulin. Inside the cylinder there is also γ-tubulin. γ-tubulin also occurs within the proximal end. (Some believe γ-tubulin is responsible for centriole duplication and/or growth of MTs.)
	127	DNA is probably not present in centrioles.
	127	Centriole geometry known for over 50 years, but not studied thoroughly. Instead the focus is on centrosomes. (But the centriole is at the cone of the centrosome!)
2003	128	Centriole dimensions 100–250 m.
MT EM	129	αβ dimer defines polarity: negative (proximal); positive (distal).
Picture	129	P 2725 Figure 2. Possible guide of αβ drawing (2001).
	129	P 2727 Table listing roles for αβ, γ, δ, … tubulin. α, β, γ seem to be most impacted with γ forming the basis for αβ.
Paper 2001	130	γ-tubulin is necessary to initiate the assembly of αβ dimers.
	130	P51 Figure 2: Cross-section drawing of a basal body (e.g., centriole barrel). It looks like two MTs also in the center.
EM 2001	130	At the proximal end of the basal body is an amorphous "electron dense" disk.
Proteins	130	Tubulins are proteins.
1995 Protein cenexin Centrioles	131	Paper provides some good descriptions of centrioles. (1) They say that the acquisition of cenexin makes a daughter centriole mature during mitosis. (2) It is one of a few organelles which exist as a single copy.

(*Continued*)

(*Continued*)

Category	**	Review Notes
		(3) Each daughter cell receives an old centriole generated previously and a new one from the current cycle. (4) The "old" (mature) centriole is capable of nucleating a primary cilium at its distal end. (5) The mature centriole has fibrous appendages. (6) It is not known if there are any unique proteins in the centrioles.
EM 1999	132	Investigates basal bodies which are claimed to be structurally similar to centrioles.
	132	Speculates that centrioles have circumferential polarity.
	132	Mother and daughter centrioles are orthogonal. The axis of the daughter intersects the axis of the mother but not the reverse.
Centriole	133	Centriole is the origin, organizer, and power within the centrosome.
1999	133	Proximal end of centrioles is negative; distal end is positive.
MT	133	MTs within the centriole are highly stable, but this exceptional stability is "yet to be elucidated".
1999	133	The two centrioles are not equivalent either structurally or biochemically.
Centriole	133	Minus ends of centriole triplets bind and concentrate γ-tubulin.
EM 2012	134	Somewhat speculative paper on how centrosomes are formed. Current opinion: the centriole recruits and organizes the proteins for the centrosome.
	134	The centrosome consists of two centrioles surrounded by a cloud of proteins forming the PCM.
	134	At the onset of mitosis, the PCM increases in size.
EM MT	134	Microtubules (at the base) are surrounded by a region of dense proteins (electron-dense material?)
Centriole	134	No membrane about the centrosome.
2012	134	PCM is a fibrous matrix of coiled-coil proteins.
	134	MT nucleation comes from γ-tubulin.
	134	Suggestion: Dynamically diffusing pools of protein stochastically came into contact with the centrosome and then stay and go (speculation). What about electromagnetic effects?

(*Continued*)

(Continued)

Category	**	Review Notes
Centrosomes	134	The number of proteins in the centrosome may be >1000.
2012	134	Chromosome is template based or self-organizing.
Membrane absence	135	Without a membrane proteins can come into and out of the centrosome without difficulty.
EM	135	This is simply an editor's overview of what's in the specific issue: suggests that the nucleus determined the polarity of a cell.
EM	135	γ-tubulin cap the minus ends of MTs.
2011	136	Minor additional details about αβ-dimers in MTs.
MT 2007	137	Interesting paper (somewhat speculative) about the base structure of an MT. The focus is on γ-tubulin: Two types γ-tubulin small complex, γ-TuSC and γ-tubulin ring complex, γ-TuRC.
	137	γ-TuRC forms the MT base and it only contacts the α-tubulin.
MT	137	The β-tubulin is on the positive side of the dimer and the α-tubulin on the negative side.
	137	γ-TuRC determines the number of filaments in an MT.
1985	138	Talks a bit about wound healing.
EM	139	Quote: "... the existence of an EF and the flow of current are interdependent and inseparable events. The EF is a spatial organizer, because it can impose directional movement on chemicals in extracellular environments."
EM	139	EF in cells is small. Therefore, some have suggested that is unimportant, or "nonphysiological." The authors say this is a "misconception."
EM cancer	139	Cancer cells have greater negative surface charge than normal cells.
EM 2005	139	There is a small steady extracellular voltage gradient between cancerous tissue and neighboring normal tissue.
Cancer 2005	139	DC current affects the migration direction of metastatic cancer cells (migration direction of metasticizing cells).
	139	It is not clear how or why this happens.
2012	140	New technology enables research.

(Continued)

(Continued)

Category	**	Review Notes
Centrioles DNA	140	Centriole assembly is "tightly coordinated" with DNA duplication.
Picture	140	Good centriole picture.
2009	141	A fluids paper.
	141	Charge of MT lower than tubulin monomers α and β: [4.2 compared to 5.55 (α) and 5.37 (β)].
	141	Centrosome charge orients the tubulin dimer.
EM MT	141	Centrosome and MTs have same polarity.
Cancer EM	142	Hypothesis: The cell membrane potential of cancer cells is lower than that of normal cells.
1986 EM	142	Cells with membrane potential below −35 mV are destined for cell division.
Paper Keck	143	Centriole development and disengagement is not well understood (2006).
	143	Three essential events for centrosome duplication: (1) centriole disengagement (2) centriole growth (3) centrosome separation.
Cancer centriole	143	Human cancers exhibit centrosome amplification.
EM 2007	144	Electric fields are found around all living cells.
Pictures 2009	145	Beautiful pictures and summary of centriole behavior.
Cancer	145	Abnormalities in the centrosome occur in cancer cells.
2008	146	Supernumerary centrioles cause genomic instability.
Cancer centriole	146	The attachment of the daughter centriole to the wall of the mother inhibits the formation of additional daughters (? What about the flower pedal analogy?).
Centriole	146	As long as the daughter is attached to the mother, formation of additional daughters is not possible.
	146	"A single mother centriole is capable of repeatedly duplicating at least four times in the same cell cycle. However, continuous associations of the daughter centriole with its mother prevents the mother centriole from reduplicating."
EM 1998	147	Regarding breast cancer: the authors found high electric charges in breast cancer biopsies. "Thus, altered skin-surface electropotentials may indicate the presence of underlying abnormal proliferation that is indicative of cancer."

(Continued)

(*Continued*)

Category	**	Review Notes
Protein 2007	148	The authors find that Plk4 induces multiple centriole development.
EM 2013	149	Kinetochores exert forces upon each other at a distance (kinetochore: protein structure at the ends of the mitotic spindle).
2013	150	Multicilliated cells (e.g., mucous membranes) have small cilia (hair-like rods) each attached to a centriole. Such cells may have 100 or more centrioles grown anew from a structure known as the "deuterosome." (This is different from mitosis.)
EM 2011	151	Electric fields (EF) enhance wound healing, but the mechanism is unknown.
	151	Sound healing: It is believed that in a wound there is a break in the polarity (like a broken circuit). This causes cells to migrate to close the circuit. Once this happens the "healing" stops. EFs are believed to amplify the effect of the wound promoting the healing.
Cancer	152	Centrosome amplification promotes malignancies.
2013	152	Centrosome amplification/DNA damage seems to occur simultaneously. Which is the cause and which is the effect? Answer is unknown.
EM	153	Cell polarity is disrupted in cancer (2013).
EM 2001	154	Metastatic cells respond to an electric field more than non-metastatic cells.
EM	154	Cancer cells are known to have negative surface polarity.
EM	155	Galvanotactic mechanisms may be involved in metastatic disease.
2004 EM	155	Transepithelial potentials range from a few millivolts to 10s of millivolts.
EM	155	In tests they use 40–500 mV/mm.
EM	155	In most cases cells move toward the cathodes but highly metastatic breast cancer cells move toward the anode.
Keck	156	"There is an almost universal need in disease treatment to increase the efficacy and delivery of bioactive agents against target cells."
2000 EM	156	Static fields impact cell metabolism.

(*Continued*)

(Continued)

Category	**	Review Notes
EM	156	"...static electric and static magnetic fields can significantly increase the in vivo reaction of mammalian cells..."
2013 paper	157	A molecular understanding of how centrioles form is just beginning to emerge.
Protein 2013	157	SAS-6, a conserved procentriole protein, self-assembles to provide ninefold symmetry to the centriole scaffold.
Protein paper	157	SAS-6 polymers display the ninefold symmetry (quotes on References 16 and 17) (good references if needed in a paper).
Protein 2010	158	"It remains unclear how the cartwheel is formed" SAS-6 is self-assembling.
Protein	158	SAS-6 concentrates at the core of the cartwheel.
2011 protein	159	According to 157: "...the authors of this paper solved the structure of SAS-6 domains and predicted that it organize to a central tubule with i-fold symmetry."
2011	160	According to 157: "Using structural biology, the authors of this paper solved the structure of SAS-6 domains and predicted that it organize to a central tubule with 9-fold symmetry."
EM	161	Electrical stimulation (0.1 V @ 50 Hz) enhances vascular endothelial growth factor (VEGF). Blood vessels.
EM 2003	162	Endogenous electric potentials exist in regions of vasculature growth.
EM	162	Electrical stimulation induces significant angiogenesis and VEGF.
EM	162	Wounds have a potential difference of 40–200 mV/mm.
EM Keck	162	"Rapid uncontrolled proliferation of cells also causes significant changes in cell surface charge."
2003 EM Keck	162	Breast cancer tumors have potential so large compared with surrounding non-cancerous tissue that the potential can be measured at the skin.
2002	163	Tumor cells induce the formation of blood supply from preexisting vasculature.
Angiogenic	163	Fluid shear induces blood cells to divide.
2002	163	High blood flow induces angiogenesis.
Protein 2005	164	Plk4 is the most structurally divergent of the Polo-kinase family.

(Continued)

(Continued)

Category	**	Review Notes
Protein	164	Plk4 is a key regulator in centriole duplication.
2007	165	Mechanism ensuring cell duplication is not well understood.
Protein	165	SAS-6 is identified as a protein required for centriole formation.
2011	166	Individual multiciliated cells have dozens of cilia that beat in a directed manner termed rotational polarity.
2011	166	"... polarized microtubules are an integral component of early steps of cell polarity."
	167	Angiogenesis: Precancerous tissue acquires angiogenic capacity.
2000	167	VEGF
EM Keck	167	How angiogenesis arises in a "largely unresolved matter."
2005	168	Centrosomes control cell cycle progression.
	168	Quote: "One particularly enticing idea is that the centrosome might anchor signal-transduction pathways and serve as a central site that receives and integrates signals from outside the cell and facilitates the conversion of these signals into cellular functions in the cell interior."
EM 1999	169	Cells divide when subjected to a direct current electric field (EF) and the cleavage plane is perpendicular to the EF vector.
EM 2012 Keck, paper	170	Quote: "The generation of cell polarity is one of the most intriguing symmetry-breaking events in biology."
EM	170	The microtubules orient the spindle along the axis of polarity.
Keck EM	171	Little is known about how cell polarity influences cell division.
2011	171	"A challenge for the future is to identify the molecular mechanisms behind the coordination between cell shape/polarity and cell division."
1996	172	"...practically any event occurring at the membrane level is brought about within an electrostatic context, which may modulate, or even replace, the biochemical interactions of the involved macromolecules."

(Continued)

(*Continued*)

Category	**	Review Notes
EM	172	"In short, the chemistry of protein has an electrostatic dimension that cannot be ignored where considering their signaling properties."
1991	173	Angiogenesis occurs infrequently in adults.
Angiogenesis	173	There are many stimulators of angiogenesis.
	173	~1991 no yet good way to block angiogenesis.
Protein 2010	174	Cep152 acts with Plk4 to initiate centriole formation.
	174	Centrioles organize two organelles: the centrosomes and the cilium.
Protein	174	Plk4 overexpression causes multiple centrioles to form adjacent to the two existing centrioles.
Paper protein	174	Little is known about how Plk4 initiates centriole assembly.
2010	175	(Article from M. Schulz) In vitro cancer cells killed by externally energized nanoparticles.
Paper Boveni	176	"As early as one century ago, Theodor Boveri proposed that cancer commonly arises through aneuploidization linked to abnormal mitoses, and postulated the underlying cause of this phenotype to be aberrant centrosomes" [3]. [3] T. Boveri, "Zur Frage der Entstehung malignerr Tumoren," *Gustav Fischer Verlag*, Jena, 1914.
DNA cancer	176	The theme of this paper is that DNA damage promotes centrosome misapplication.
2004	177	"...centrosomes might have an important function in regulating cell cycle progression..."
EM 1994	178	Cancer is injected and then an electric field is applied. This increases the uptake of the chemo by the tumor.
1999	179	Conclusion: "Strong evidence supports the hypothesis that VEGF is a key mediator of angiogenesis associated with various disorders".
Protein 2012	180	Five core centriolar components: CEP192, Plk4, STIL (SAS, SAS-6, and SAS4).
Protein	180	P53 opposes cell cycle progression in cells with extra centrosomes.
Protein	180	The exact role of p53 is unknown.
Cancer	180	Aneuploidy is a common feature of cancer cells.

(*Continued*)

(*Continued*)

Category	**	Review Notes
	180	"Although the mechanisms linking centrosome amplification to chromosome instability are well understood, it is uncertain if centrosome amplification by itself can be sufficient to drive tumor formation."
EM nano	181	Nanomagnetic action is "action at a distance."
Keck 2008	181	Nanoparticles can be used to target specific cell receptors.
Nano	181	The techniques have evolved to target specific cell receptors.
2008	181	More recently, Wang and Ingher [7–9] used magnetically blocked nanoparticles binding to integrin receptors.
	181	The particles need not be magnetically blocked — they may be "superparamagnetic." (I believe this refers to nanoparticles $\leq$ 30 nm.)
Nano	181	"As iron oxide nanoparticles become superparamagnetic below about 30 nm, larger particles must be used for applications requiring a stable remnant magnetization..."
MT 1999	182	The inside and outside of the MT surfaces are different: The outside is smoother with shallow undulation and is rather flat. Alternatively, the inside surface is deeply corrugated.
	183	There are "older" and "younger" centriole pairs after mitosis.
	183	The older pair is more mature and can initiate ciliary axoneme — a hair-like projection.
EM Keck paper	183	Centrosomes influence cell shape, polarity, and motility.
Centriole	183	"In cycling cells, exactly one new centriole forms adjacent to each preexisting centriole, reminiscent of the replication of DNA."
Cancer	183	It has long been known that there is a strong correlation between aberrant centrosome number and aberrant chromosome number [73] (Hardy and Zacharias).
Cancer	183	Extracentrosomes create serious mitosis problems.
Protein	183	Plk1 is required to convert new centrioles to centrosomes as the cells progress from mitosis to interphase.
2000	184	Mechanical stresses applied to a cell surface alters the adenosine 5-monophosphate (part of an RNA molecule) signaling.

(*Continued*)

(Continued)

Category	**	Review Notes
Paper 2005	185	Good reference for historical account of cancer analysis.
EM 1997 centrioles	186	The steps required for metastasis are similar for all tumor cells.
	186	Steps for metastasis: (1) vessel formation (angiogenesis), (2) cell attachment, (3) invasion (matrix degradation), and (4) cell proliferation.
1997	186	Adhesion molecules play a central role in cellular attachment.
	187	Superparamagnetic (30 nm) particles are used to attempt to control cells in mechanical actuation.
Nano 2008	188	Magnetic nanoparticles were attached to integrin cell receptors for 21 days (as opposed to 2 h) with no adverse results.
EM	188	"Magnetic particle technology presents a valuable tool for tissue engineering which permits the delivery of highly localized mechano-inductive forces directly to cells."
Nano 2005	189	A review of using superparamagnetic nanoparticles (SPION) in medicine (<10 nm).
EM	189	Uses external magnetic field to activate the SPION.
2005	189	SPION can kill tumors either by chemo drug delivery or by vibration (heating 1 by oscillating external field).
	189	Discusses ways to manufacture SPION.
	189	Smallest diameter of capillaries in the body is 4 m.
R. Adelman angiogenesis 1989	190	A paper supporting Roger Adelman idea to use a probe to detect colon polyps/tumors in their beginning stages by measuring magnetic fields among rapidly dividing cells.
EM	190	The premise of this paper is that malignant colon tissue has different electrical properties than non-malignant tissue.
Angiogenesis 1997	191	Angiogenesis paper: The blood vessels form in direct response to tissue demands.
2002	192	Enlarged centrosome disrupts chromosomes in breast cancer development.
Cancer 2008	193	Cancers are associated with abnormalities in centrosome structure.

(Continued)

(*Continued*)

Category	**	**Review Notes**
Protein	193	Centriole biogenesis involves SAK/PLK4, and SAS-6 organizes the nine precentriolar units.
Centriole 2008	193	The daughter centriole arising on the side of the mother centriole is called "canonical assembly."
Protein	193	SAK/PLK4 is an upstream regulator of canonical centriole formation.
Protein cancer	193	An overexpression of SAK/PLK4 leads to formation of multiple centrosomes.
Paper 2001	194	Paper refers to Boveri: Boveri, T. *Zur Frage der Entstehung amligner Tumoren*, Fischer-Verlag: Jena, 1914.
Cancer	194	Centrosome aberrations are common in cancer cells.
Cancer DNA	194	Centrosome duplication corresponds with DNA replication.
Cancer	194	Errors in centrosome duplication cause aneuploidy and cancer.
Centriole DNA	194	Centriole duplication occurs at the onset of DNA replication.
Cancer DNA chromosomes	194	Aneuploidy — numeric chromosomal imbalance.
Cancer 2002	194	Suggests that aneuploidy is a cause of cancer.
Centriole	194	The pericentriolar domain around the centriole pair is a cloud of pericentriolar material ~1 m in diameter. The centrosome extends to the surrounding cytoplasm and organelles via microtubules.
EM	194	The outer surface of the centriole is associated with electron dense satellite structures along its length.
Protein	194	There are ~100 proteins in the centrosome, but these are believed to simply associate with the centrosome to ensure proper segregation distribution during mitosis.
Cancer 2009	195	Extra centrosomes alone are sufficient to promote chromosome missegregation.
Cancer	195	There is a direct mechanistic link between extracentrosomes and chromosomal instability (CIN) — which occurs in human cancer.
Protein 2012	196	STIL is a centrosomal protein whose overexpression results in excess centriole formation.
Cancer	196	Extra centrosome number leads to cancer.

(*Continued*)

(Continued)

Category	**	Review Notes
2012	196	Too little STIL leads to a loss of centriole formation.
Protein 2012 centriole	196	STIL recruits SAS-6.
Cancer protein 2012	197	STIL is required for centriole duplication, but an overexpression produces multiple daughter centrioles around each mother.
Centriole protein	197	STIL is recruited to nascent daughter centrioles at the onset of centriole duplication.
Cancer	197	STIL cooperates with SAS-6 and PLK4 in the control of centriole number.
2007	198	Presents a new idea about centriole function: "vital function of centrioles lies in their ability to become … basal bodies which produce flagella or cilia" (quote from a reference).
Paper	198	States that Theodor Boveri described a centrosome over 100 years ago.
2007	198	Centrioles are still intriguing to cell biologists.
Protein	198	γ-tubulin is the linch-pin of microtubule nucleation and therefore a prime player in microtubule organization.
EM	199	Centrosome determines cell polarity.
2008 MT	199	Centrosome: Centrioles and PCM. This is the primary site of microtubule nucleation — aka microtubule organizing center (MTOC).
Centriole	199	"…the vicinity of the base of each centriole provides an environment that promotes assembly of a procentriole which is oriented approximately perpendicular to each centriole".
Protein	199	Proteins: ZYG-1 (Plk4, SAS), SPD-2, SAS-4, SAS-5, and SAS-6
Protein 2008	199	SPD-2 and ZYG-1 are recruited before central tube formation whereas SAS-5 and SAS-6 are recruited slightly later. This suggests that SPD-2 and ZYG-1 are regulatory rather than structural.
Picture	199	Some good pictures.
Centriole	199	At the onset of procentriolar assembly, the cartwheel is the first ninefold structure to appear. Then there is the appearance of microtubules at the tips of the spokes.
Centriole 2008	199	The cartwheel is the structural scaffold onto which the centriolar microtubules assemble.

(Continued)

(Continued)

Category	**	Review Notes
Protein	199	It is speculated that SAS-6 dictates the ninefold symmetry of the centriole.
Paper 2004	200	References Boveri
Aneuploidy ø supernumer-ary centrioles	200	Aneuploidy is defined as an alteration in the number of chromosomes — not the number of centrioles. That is, aneuploidy and supernumerary centrioles are not the same. However, when aneuploidy occurs so also are supernumerary centrioles.
Cancer	200	A small fraction of cancers are not aneuploidic but they still have defective DNA repair mechanisms.
2004	200	CIN — chromosome instability — not centrosome instability
Paper	200	Several routes lead to chromosome instability.
Protein cancer	201	Ned aka HSET a minus ended kinase is needed for extra centrosome development, but not for regular cell development.
Cancer EM 2008	201	Extra centrosomes can lead to aneuploidy (defective chromosomes) and rapid cell divisions, and change in cell polarity.
Cancer	201	Cancer cells have centrosome clustering.
	201	Disruption in centrosome clustering can selectively kill cancer cells.
Paper	201	Refers to Boveri (1929): T. Boveri, *The Origin of Malignant Tumors*, Williams and Wilkins, Baltimore, MD, 1929.
Quote (Keck) see C-64	201	"The existence of mechanisms that suppress multipolar mitoses raises the possibility of a novel therapeutic strategy for cancer: Drugs that interfere with centrosome clustering mechanisms could be lethal to tumor cells containing multiple centrosomes."
Copy by error	202	W.L. Lingle, *et al.* (8), "Centrosome amplification drives chromosomal instability in breast tumor development", *Proceedings of the National Academy of Science*; Vol. 99, 2002, pp. 1978–1983.
Cancer 2002	202	Experiments on breast cancer tumors show 60–80% aneuploidy and 80% with amplified centrosomes. "Amplified centrosomes" refers to amplified size, centrosome number, and centrosome microtubule nucleation.
Cancer	202	There is a linear correlation between aneuploidy and centrosome instability.

(Continued)

(Continued)

Category	**	Review Notes
Cancer 192	202	Centrosome instability is an early event in cancer.
Cancer nano	202	Centrosome instability may be an early sign of cancer (diagnosis) and a target for treatment.
Centriole	203	Mechanisms underlying daughter centriole formation remain unclear.
Protein 2006	203	"Centriole assembly begins with the formation of a central tube followed by the peripheral assembly of nine single + microtubules. Tube formation and assembly is dependent upon SAS-5 and SAS-6 proteins whereas the assembly of singlet microtubules onto the central tube depends upon SAS-4."
	203	"Centriole assembly is triggered by an upstream signal mediated by SPD-2 and ZYG-1.
Pictures paper 2007	204	See 145. Good pictures. Mentions Boveri.
Cancer 2003	205	Centrosomes are abnormal in aggressive carcinomas.
Cancer	205	Centrosome defects are implicated in chromosome instability and loss of cell control.
Cancer	205	Centrosome defects contribute to earliest stages of cancer development via chromosome instability.
Cancer	205	Speculation: Centrosome defects are universal indicator of most, if not all, carcinomas.
2003 protein	205	Centrosome defects can arise from many factors: protein changes, p53 mutation...
	205	It is unlikely that p53 mutations can account for all the findings.
Cancer	205	Speculation: Centrosome instability is the source of cancer.
Cancer	206	Extra centrosomes are found in many cancers.
Centriole DNA protein	206	Centrosome duplication is controlled by centriole replication. Plk4 plays a key role in initiating centriole duplication.
Protein cancer 2010	206	Overexpression of Plk4 promotes centriole overduplication and for the formation of extra centrosomes.
Protein	206	Plk4 stability is regulated by its own kinase activity.
Cancer	207	Centrosome defects lead to aneuploidy and then to tumors.

(Continued)

(*Continued*)

Category	**	Review Notes
1999 EM	207	Centrosomes may be involved in cell polarity.
Protein	207	P53 is a tumor suppressor.
Cancer	208	Tumors have supernumerary centrosomes.
Cancer	208	"Centrosome amplification" is supernumerary centrosomes.
Cancer	208	Centrosome amplification correlates with tumorigenesis in rat mammary glands.
Paper	209	(1998) Refers to Boveri "...over 100 years ago."
Centriole	209	Centrosome ~1 m in diameter.
Centriole paper Keck	209	"Indeed, it would seem that the answers to some of the big questions of tumorigenesis are hiding in small places like the centrosome".
Cancer	210	Centrosome malfunction causes chromosome damage.
Paper 2001	210	References Theodor Boveri. Says Boveri's idea was more or less forgotten until about 5 years ago (written in 2001).
Protein	210	P53 is a tumor suppressor gene.
Protein	210	Absence of p53 allows centrosome replication.
Protein	210	Aurora excess produces cancer.
	211	Donglu Shi paper: Nanocomposite particle targeted to breast cancer (in vitro).
	212	D. Shi, N.M. Bedford, and H-S Cho, "Engineered Multifunctional Nanocarriers for Cancer Diagnosis and Therapeutics," *Small*, Vol X, 2011, pp. 1–19 (copied by mistake).
	212	CNTs may be toxic.
	212	"Localization of nanocarriers to tumors can be accomplished by using magnetic fields to concentrate the nanocarrier directly in tumors, while magnetic hyperthermia can be used to directly kill cancer cells or as a trigger for controlled release."
	213	(2013) — a review article on nanomedicine. Objective of nanomedicine: enhanced specificity, efficacy, and safety. It has 112 references.
	213	Nanomedicine may be targeted for specific application — for example, breast cancer.
	213	EPR — efficient penetration and retention.

(*Continued*)

(*Continued*)

Category	**	Review Notes
	213	Nanomedicine (to date) is good for magnetic resonance and improved imaging.
	213	Currently the focus is on assisting drug delivery. (Gold nanoparticles can pass the blood-brain barrier.)
	213	Quote (on conquering cancer) "The use of nanomaterials to treat and image cancer is arguably the most active area of nanomedicine research. Intrinsic properties like surface charge, structure, and biocompatibility can be harnessed for cancer therapy."
	213	"Drug-loaded nanoparticles can be targeting to surface molecules over-expressed in cancer cells or tumor vasculature."
2011	214	Concept: Magnets can be used to pull magnetically charged nanoparticles through the blood vessels to tumors. However, the problem is the decrease in magnetic field away from the magnets — particularly for targets within deep tissue.
	214	MRN — magnetic resonance navigation.
2008	215	About carbon nanotubes (CNT).
	215	SWNT — single-walled carbon nanotubes.
	215	SWNT show promise in tumor-targeted accumulation in mice. (First successful treatment known to date.)
2005	216	About CNT "... (CNT) have emerged as a new alternative and efficient tool for transporting and translocating therapeutic molecules."
	216	Pristine CNT are highly toxic due to their insolubility. CNT can cross cell membranes.
	216	More than one function (targeting, imaging, drug delivery) can be accomplished with a single CNT.
2011	217	"... yet our ability to efficiently home these materials (nanoparticles) to regions of disease *in vivo* remains very limited."
	217	Nanoparticles have been developed with unique electromagnetic properties.
Cancer nanotubes	218	Nanotechnology — delivery of drugs as opposed to finding the tumor.
Nano EPR	218	A measure of nanotechnology/nanoparticles: EPR — enhanced permeability and retention

(*Continued*)

(*Continued*)

Category	**	Review Notes
Angiogenic	218	There is a difference between mature blood vessels and angiogenic blood vessels. The angiogenic are weak and they peak.
	218	"Nanotechnology is still in its infancy."
Nano	219	Listing of the many ways nanoparticles can deliver therapy.
Cancer	220	Many cancers arising in epithelia, the tissue that lines all body cavities.
EM	220	Polarity allows transport across the epithelia layer.
EM cancer	220	Polarity charges (negative) in cancer "... whether loss of cell polarity is causal to human cancer are questions still under investigation."
	220	Metastatic cells are softer.
Centriole	221	Multiple centrioles cause cancer.
Cancer EM	222	"Loss of polarity is considered both a hallmark and precondition for human cancer."
	222	Suggests cell polarity is an electrical event (at least for the proteins) but no $\pm$ signs, no voltage given.
Polarity	222	"Cell polarity — is defined as an asymmetrical cell shape."
Polarity	223	Abstract beginning: "Cell polarity is defined as asymmetry of cell shape, protein distributions and cell functions."
	224	Confirms previous information.
Protein	225	Pars are cell polarity proteins.
	225	Tumor cells have loss of polarity.
Protein	226	Plk4 is the "central player" in cell duplication — but its activity is still not well understood.
Protein cancer	227	Loss of polarity (shape) is a possible source of tumor growth.
Polarity	228	Loss of polarity (shape) ("tissue architecture") are characteristics of cancer.
Cancer	229	Supernumerary centrosomes lead to multipolar spindles seen in human cancer.
	229	This is a review with many useful references.
Keck	229	Multipolar spindles as a feature is starting to be explored therapeutically.

(*Continued*)

(*Continued*)

Category	**	Review Notes
Protein	230	States that ninein (a protein) (and not γ-tubulin) forms both the anchoring and the placement of the minus end of microtubules.
Cancer centrosome defects Centriole duplication	231	Centrosomes help establish the mitotic spindle but defects in chromosomes can lead to cancer. Alternatives include: (1) Supernumerary centrioles (2) Excess pericentriolar material (3) Disrupted centriole barrel structure (4) Unincorporated microtubule complexes (5) Centrioles of unusual length (6) Centriole functioning as ciliary basal bodies (7) 7) Mispositioned centrosomes.
MTOC centriole	232	Two views for the MTOC: (1) a tubulin producer (like a spider web) and (2) a capston (a reeling in/out) of the MTs. (Author says both.)
Cancer	233	Extra centrosomes may contribute to elevated mitotic index as seen in tumors.
Proteins	234	Electron dense granules ~70–100 nm in diameter are observed in the centrosome.
Cancer	235	Extra centrosomes produce centrosome instabilities (CIN).
Centriole	236	The earliest event in cartwheel assembly is likely to be the recruitment of Plk4 on the surface of the proximal end of the mother centriole.
Centriole Good pictures	236	"A mystery surrounding centriole growth is their ability through several cell divisions and constancy of their length within a cell type."
Centriole protein	237	An interesting paper: States that the daughter centriole arises from a "dot" of Plk4 on the proximal end of the mother.
Centriole	237	"It is not clear what mechanisms lead to the concentration of Plk4 ... and whether the choice of the site is stochastic — or regulation by other proteins."
Centriole	237	How does Plk4 attract SAS-6 to form a cartwheel structure?
	238	A good source of references.
Keck	238	"...we are only starting to unveil the role of centrosomes and cilia in human diseases and many exciting questions still lie ahead."

(*Continued*)

(*Continued*)

Category	**	Review Notes
Centriole geometry	239	Details of appendages of centrioles and a model.
Centriole duplication	240	PCM is electron dense.
	240	Centrioles have the same size/shape in all mammals.
	240	Centrioles duplicate from preexisting centrioles.
	240	During duplication, the mother centriole maintains the PCM.
	240	Centrioles can be made to occur via injecting Plk4 but this is *rare* and not naturally occurring. Indeed, when a centriole pair exists, new centrioles will not develop ("lock and license").
	240	Once duplication begins the cartwheel protein occurs (SAS-6?).
	240	The mother–daughter pair when duplicated became a "diplosome". Early in mitosis the diplosome breaks down.
	240	γ-tubulin is important in centriole formation because it nucleates the microtubules.
	240	A mechanism exists that prevents formation of additional procentrioles. Like DNA duplication is blocked.
	240	"The block to centriole reduplication is intrinsic to already-duplicated centrioles that are engaged in the diplosome."
	240	It is not clear why only one daughter forms near the wall of the mother. There is a symmetry-breaking mechanism — not clear what it is.
	240	Cells have intrinsic asymmetry.
	240	Adding Plk4 (Sak) can cause flower-pedal duplication.
	240	PCM presents de novo duplication. (Is this an electromagnetic effect?)
Centriole	241	General description of centrioles. Long list of references.
Pictures	241	Good pictures.
Protein	242	SAS-6 is required for centrosome duplication. Also, to make a centriole Aurora-A (SPD-5) and γ-tubulin are needed (three proteins).
	242	SAS-6 is required for centriole assembly.

(*Continued*)

(*Continued*)

Category	**	Review Notes
Protein	243	Explains how the ninefold spiral occurs.
Protein	244	Multiple centrosomes cause cancer cells to divide.
Cancer centrioles	244	Centrosomal clustering leads to successful division of cancer cells.
Keck	244	Mechanisms for centriole clustering are not completely understood.
Centrioles	245	Excellent 1998 paper.
Cancer	245	Centrosomes in tumors were atypical in shape, size, and composition and were present in multiple copies.
	245	All tumor cell lines had centrosome defects. Centrosome defects are found in the most common human malignancies.
	245	Speculates that centrosome defects lead to chromosome instability (CIN) — probable.
Cancer centrioles	246	Numerous centrioles seen in cancer tissues.
	246	The number of extra centrosomes may be a measure of the degree of the cancer and also a possible means of treatment.
Protein	247	SAK/Plk4 needed for centriole development.
Centriole cancer proteins	248	This is a review paper with 225 references.
	248	Centrosome is electron dense.
Good pictures	248	In normal cells there is only one time point for daughter cells to form.
	248	Overexpression of SAK/Plk4 and SAS-6 causes the formation of the "flower" (multidaughter) phenomena.
Protein	249	Asserts that p53 regulates centrosome duplication.
Protein	250	Disputes 249 "… we demonstrate here that p53 status has no direct influence on centrosome duplication in each of several mammalian cell lines."
Cancer centriole	251	The presence of more than two centrosomes during mitosis is characteristic of cancer.
Cancer centriole	251	There is a survival of the fittest of cancer cells (via *clustering* of centrosomes?).
Cancer	251	Clustering
Protein	252	Protein

(*Continued*)

(*Continued*)

Category	**	Review Notes
Protein	253	Protein
Definition polarity	254	Definition of cell polarity: "Cell polarity can be defined as a structurally and functionally asymmetric organization in which the non-random positioning of each organelle, the function of which contributes to cell asymmetry, is preserved and transmitted through cell division."
Centriole	254	The centriole has defined polarity that can be transmitted to the cell.
	254	Polarity is transmitted throughout the cell during division.
Cancer	255	Centrosome abnormality is present in "almost all types of solid tumours".
Cancer	255	Chromosome instability (CIN) accelerates tumor progression.
Centrioles cancer CIN	255	Abnormalities in centrosomes are the principal causes of chromosome instability (CIN) in cancer.
Cancer Keck	255	"... it is vital to understand how duplication, numerical integrity and mitotic function of the centrosome are controlled at the molecular level. Understanding this should provide fertile ground for designing more effective cancer diagnosis and intervention protocols."
Cancer Keck	255	"The fact that many oncogenes and tumour suppressors control centrosome duplication and function further highlights the suggestion that centrosomes might prove to be effective cancer targets."
Keck	255	"...controlling centrosome number and function should be the focus of further study."
Cancer	256	Centrosomal clustering is a kind of preventative for multiple spindles.
	256	Supernumerary centrioles do not always produce cancer [but cancer needs supernumerary centrioles — my hypothesis].
Cancer	257	Abnormal centrosome numbers produce tumors.
	257	Idea: abnormal centrosome numbers might create an elongation of mitosis time enabling damaged DNA to be discarded.
Protein	258	Studies the role of p53 in centrosome duplication regulation.

(*Continued*)

(Continued)

Category	**	Review Notes
Protein	259	Similar to 258.
Cancer centrioles	260	Engagement of centrioles with each other prevents duplication, but disengagement allows for duplication.
Protein	261	Identifies proteins needed for cell duplication: kinase ZYG-1, coiled-coil protein: SAS-4, SAS-5, SAS-6, and SPD-2.
Cell mitosis	262	SAS-4 and ZYG-1 govern daughter formation.
Protein	262	A licensing mechanism appears to govern centrosome duplication.
Centriole geometry	262	Centrosomes are the MTOC. PCM is "electron dense" matter.
Centriole duplicate and DNA	262	As mitosis begins: the two centrioles lose their orthogonal arrangement and split slightly while maintaining a flexible connection.
Centriole	263	Daughter centrioles always form perpendicular to the mother and at the proximal end.
	263	It is unclear why daughter elongation stops when its length approaches that of the mother centriole.
Centriole control	263	It appears that regulatory constraints, as opposed to structure constraints, dictate the length and form of the daughter centrioles.
	263	Centrosome duplication is coupled with DNA replication — this may be altered in tumorigenesis.
	263	Unclear if centrosome abnormalities are a cause or a result of tumorigenesis.
	263	Supernumerary centrioles may be due to an uncoupling with DNA duplication.
Cancer	263	Definition: "Aneuploidy" — the gain or loss of one or more chromosomes from a diploid genome.
	263	Quote: "...a condition that favors the overproduction of centrosomes could contribute directly to the initiation of chromosome imbalance."
Cancer	264	Centrosome amplification leads to chromosome instability.
Protein	264	Centrosome amplification leads to loss of p53.
Protein	264	P53 appears to be a regulator of centrosome development.

(Continued)

(*Continued*)

Category	**	Review Notes
Cancer	264	Listing cancers with centrosome amplification breast, bladder, brain, bone, liver, lung, colon, prostate, pancreas, ovary, testicle, cervix, gallbladder, bile duct, adrenal, head and neck squamous cell, leukemia, and lymphoma.
Cancer	265	Centriole duplication is controlled by the centrosome and not the nucleus.
Protein	265	Loss of p53 results in a loss of duplication regulation.
Proteins	266	P53 mutation can lead to centrosome aberration.
Radiation → centriole duplication	267	Radiation causes centriole duplication.
Nano	268	Nanomotor
Centrioles paper Keck	269	Centrioles have been studied since the advent of electron microscopy.
History	270	Twenty-first century has seen a surge of interest in centrosome research.
Cancer centrioles	270	Centrosome amplification (CA) is a "hallmark" of cancer cells.
Cancer centriole	270	CA is the major underlying mechanism for multipolar mitoses, CIN, and aneuploidy.
	270	Centrosome description
Protein	270	Overexpression of Aurora-A kinase is a cause of CA, aneuploidy, etc.
Cancer	270	CA has been a target for cancer therapy via chem for 10 years.
Cancer	271	Chromosomal abnormalities (CIN).
Cancer	272	Centrosome abnormality → bladder cancer
Cancer	273	Centrosome amplification gives rise to CIN for bladder cancer.
Cancer	274	Centrosome amplification is associated with spindle abnormalities in breast cancers.
Cancer	275	Centrosome amplification occurs in human cancers and a source of chromosomal instability (CIN).
Protein	275	Loss of p53 induces centrosome instability (CIN).
Centriole	275	Centrosomes duplicate during S-phase.

(*Continued*)

(*Continued*)

Category	**	Review Notes
Centrioles paper Keck	276	"While centrioles are relatively simple in construction, the mechanics of centriole biogenesis remain an important yet poorly understood process."
Centriole	276	"Centriole duplication is remarkably similar to DNA replication." The older is a foundation for the new.
Centriole cancer protein	276	Plk4, SAS-6, Cep63, SAS-5/Ana2/STIL Asterless/Cep152, Twins/B55 produce flower-like daughter centrioles — that is, multiple daughters (Table 1, p. 1027). Proteins *preventing* overduplication: Table 2, p. 1029. DKP1/SkpA, CUL1, Stimb/beta-TrCP, Nucleophosmin, SKP2, p53, CDK1, BRCA1, Cdc14B, Geminin, Orel.
Proteins	277	Discusses proteins involved in centriole duplication and centrosome amplification.
Protein cancer	277	Plk4 is the main regulator of centriole duplication; overexpression of Plk4 leads to the formation of multiple daughter centrioles.
Listing of proteins Cancer	277	Supernumerary centriole tends to cluster together enabling division.
	278	Aneuploidy — when daughter cells inherit too many or too few chromosomes.
Centriole	279	Centrosomes duplicate during S-phase.
	279	(2007) Little is known about the mechanisms of centriole duplication or of their control.
	279	Centrosome is the MTOC.
	279	Centrosome: two centrioles surrounded by electron dense material.
	279	Centriole: 9 MT triplets ~0.5 m long and ~0.2 m in dia.
	279	Centrioles are surrounded by ~300 proteins, mostly with coiled-coil structure.
Protein	280	Cells lacking p53 result in chromosome transmission errors. P53 is directly involved in centrosome duplication regulation.
Protein	280	Advanced human tumors have a deleted p53 allele.
Protein	280	CDK2/cyclin E decouples the centrosome.
Protein	281	Removing p53 does not result in aneuploidy — (seemingly a surprise in view of Fukasama).

(*Continued*)

(Continued)

Category	**	Review Notes
Centrioles	282	This is an introduction paper to a series of reviews. What is new is the idea that multiple centrosomes can occur in four ways: (1) Over duplication of centrioles (the "flower") (2) Failure to undergo cytokinesis (i.e., cell dividing) (3) Separation of the centriole pair (4) 4) Centrosome occurring without centriole.
Centriole cancer	283	Centrosomal defects lead to tumorigenesis. The paper reviews how centrosomal defects lead to chromosomal instability (CIN)
Centriole cancer	283	Centrosomal defects occur early in premalignant tissue.
Protein centriole	283	Loss of p53 does not promote centrosome amplification.
	283	DNA damage occurs with changes in centrosome number.
Centriole	283	Centrosome amplification in the absence of cytokinesis is rare.
Protein	283	The loss of p53 is not a promoter of centrosome amplification.
Protein	284	Centrosome duplication requires cyclin/cdk2 kinase. Duplication depends upon cyclin E/cdk2 activity.
Nano	285	A primary goal of bionanotechnology is delivery of therapeutics to malignant tumors. The use of iron oxide is to reduce toxicity.
Cancer	286	Centrosome amplification can arise independently of ER or p53 status.
Cancer	286	The aggressiveness of breast cancer is proportional to the degree of amplified centrosomes.
Cancer	287	γ-radiation for cancer treatment causes centrosome amplification which leads to cell death — very interesting in that centrosome amplification normally leads to cancer.
Centriole cancer	288	Confirms previous findings.
Protein	289	Overexpression of Aurora-A kinase causes centrosome amplification in cultured cells. Overexpression of Plk1 and Aurora B also increases centrosome numbers. Absence of p53 exacerbates this behavior.
Protein	290	Supernumerary centrosomes promote the assembly of abnormal mitotic spindles. A prolonged S-phase produces extra centrosomes ("centrosome reduplication").

(Continued)

(Continued)

Category	**	Review Notes
Protein centrioles	290	Preventing degradation of Mps1 at the centrosome causes centrosome reduplication.
Cancer	291	Supernumerary centrioles lead to abnormal mitosis and tumorigenesis.
Centriole	291	Multiple rounds of duplication are called "reduplication". Mechanisms for reduplication are poorly understood (2010).
Protein	291	Reduplication requires Plk1. Absence of Plk1 prevents reduplication.
Centriole	291	If the daughter centriole develops full length before separation, this may lead to reduplication.
Centriole	291	Delayed disengagement promotes reduplication.
Protein	292	Plk1 is an activator of centriole disengagement Plk1 and separase act at different times during M-phase to license centrosome duplication.
Centriole	292	During S-phase a new daughter occurs due to influence of Cdk2/cyclin E.
	292	Physical engagement between mother and daughter centrioles create a block to further centriole assembly.
	292	Centriole disengagement is poorly understood.
	292	The role of separase in centriole disengagement is not clear.
	292	Orthogonal configuration of centrioles plays a central role in preventing centriole overduplication.
	293	Methods preventing centriole overduplication are poorly understood. Physical removal of the daughter induces a reduplication of the mother in the S-phase.
	293	Proposal: The size of the PCM cloud associated with the mother centriole restricts the number of daughters that can form simultaneously: if it is too large multiple daughters can form.
	293	New idea: Daughters are formed in the PCM before attachment to the mother.
Centriole	293	As long as the daughter is attached to the mother formation of additional daughters is prevented.
Centriole protein	294	SAK a.k.a. Plk4 is a necessary regulator of centriole formation but overexpression leads to centriole amplification. SAK itself may be regulated by the cell itself.

(Continued)

(*Continued*)

Category	**	Review Notes
Centriole	295	Centrioles can assemble de novo after centrosome removal.
	296	Confirming past results.
Cancer	297	Six hallmarks of cancer: (1) Sustaining proliferative signaling (2) Evading growth suppressors (3) Resisting cell death (4) Enabling replicative immortality (5) Inducing angiogenesis (6) Activating invasion and metastasis. Plus three more: (1) Reprogramming energy metabolism (2) Evading immune destruction (3) 3) Acquiring normal cells in tumors to create a tumor microenvironment.
	298	Centriole disengagement requires the protease separase.
Protein	299	Plk1 is overexpressed in cancer cells.
Protein	300	C-Napl is necessary for regulated division.
Protein	301	Plk1 helps initiate cytokinesis by recruiting RhoGet Ect2.
Protein	302	This is a review article on Plks. Plks are increasingly recognized as key regulators of mitosis, meiosis, and cytokinesis.
	302	Kinases cooperate with Cdks in the orchestration of cell division.
	302	Plk1 excess and/or deficiency can cause cancer. Therefore cell division is sensitive to the level of Plk1.
	302	Quote: With Cdks as the conductors, the Plks have the role of first violin.
	302	Aurora A is a cancer susceptibility gene. Definition: Aurora A: A serine/threonine kinase required for centrosome and spindle formation during mitosis.
Protein	302	Cdk1 — Cyclin B — the principal mitotic serine/threonine kinase which regulates entry into mitosis. Cdk1 is a catalytic subunit. Cyclin B is a regulatory subunit.
Protein	303	A review: Says that Cdk1-Cyclin B is not a regulator of mitosis (opposite of 302). But 302 is in 2004, 303 is in 2001.

(*Continued*)

(*Continued*)

Category	**	Review Notes
Protein	304	Centrosome duplication is regulated by proteins: cycle kinase CDK2 and Cyclin E or A.
Protein	305	Mitotic progression and chromosome duplication is regulated by a group of protein kinase families, including: Ckd (cyclin dependent kinase) Polo Aurora NIMA Bub Mps1.
Protein	306	E2F3 inactivation results in centrosome amplification.
	307	Mps1 is required for spindle assembly but not for centrosome duplication.
Nano	308	A review. Good reference list. Mostly about targeting proteins.
Nano	309	Advocates use of poly ethylene glycol as a cover for nanoparticles.
Nano	309	Lists three requirements for nanoparticles to deliver the payload to the tumor: (1) They need to be stable in the blood circulation (2) The nanoparticles need to accumulate in the tumor (3) The drug needs to be released efficiently in the tumor.
Nano	309	Ideal: (1) Payload to target efficiently (>10% in 4 h) (potent) (2) Cleared from circulation in a short time (4–10 h) solo.
Nano	310	Magnetic nanoparticles (MNPs) have the ability to function at the cellular level.
Nano	311	Therapeutic nanoparticles should be 10–100 nm in diameter.
Nano	311	A way to get nanoparticles to tumors is by the leaking blood vessels. Particles with diameter of 100s of nm can leak through the blood vessels.
Nano EM	311	Inside surface of blood vessels of tumors is *negatively charged*.
Nano	311	Nanoparticles can carry a large payload of drugs.
Nano	312	Nanomaterials are longer than molecules but smaller than cells.

(*Continued*)

(*Continued*)

Category	**	**Review Notes**
Nano	312	Focus of paper is to getting nanoparticles past the blood-brain barrier.
Nano	312	Quote: "Future directions for this work will be in the field of "smart" nanoparticles..."
EM nano cancer paper	313	Quote: "Differences in the expression of cellular receptors between normal and tumor cells represent a great opportunity for targeting imaging probes to those cellular surface molecules."
Nano	313	Chemicals attracted by tumors: Her-2/neu, β_3 integrin, PMSA and MUC-1; and CREKA.
Nano	313	Quote: "In general, positively charged nontargeting IO nanoparticles bind to cells through electrostatic interactin with the negatively charged cell membranes and are then internalized by cells..." (IO: iron oxide).
Cancer	313	Tumor cells have heterogeneous expression levels of targeting receptors.
Nano	313	There is still a challenge to load up nanoparticles with toxic drugs.
Nano	314	Fe_3O_4 nanoparticle effect is enhanced by covering it with a drug: tetraheptylammonium. This caused drug resistant leukemia cells to uptake the nanoparticle. (This suggests that more research is needed on nanoparticles.)
Nano	315	Nanoparticles (np) 1 np < NP < 100 np.
Nano	315	NPs differ from bulk material due to magnetic and electronic properties and surface phenomena.
Nano	315	(2007) The most promising application of NPs is the diagnosis and treatment of cancer. Idea is to target tumor.
Nano	315	An idea is to implant a magnetic compound in the tumor and then use magnetically charged NPs for therapy.
Nano	316	Obstacle for therapeutic nanoparticles is the membrane layers of the tumor and cell.
Nano	317	Chemical used: doxorubicin (J&J Doxil).
Nano	317	Need to deliver material to subcellular location within the cancer cell and bind to the unique *moieties* of a cancer cell.

(*Continued*)

(*Continued*)

Category	**	Review Notes
Nano	318	Main four appears to be on imaging.
Nano	319	Nanotechnology used for early detection, imaging or drug delivery.
Nano	319	Approved: Use of liposomes and formulations of doxorubicin for Kapasi sarcoma.
Nano	319	Quote: "There are still bottlenecks in the existing 'homing' technologies, ... This emphasizes the need for novel targeting strategies."
Cancer	320	Angiogenesis studied after but yet poorly understood. Resistance to anti-angiogenesis drugs develops quickly — presumably through signaling mechanisms.
Cancer	320	Angiogenesis usually arises from capillaries rather than from large vessels (arteries, veins).
Cancer	320	Tumor blood vessels are irregular, tortuous, enlarged, and leaky. They have hypoxia and thus promote VEGFA. (endothelium — a single sheet of flattened cells lining a blood vessel.)
Nano	321	A review of nanoparticles in use up to 2002.
Cancer	322	An optimistic article about the potential of anti-angiogenesis (2000).
Nano	323	The physical and chemical properties of materials can radically change when scaled down to a nanoparticle.
Nano	323	Nanoparticles can be targeting to specific cells and locations within the body.
Nano	323	Nanoparticles must be of the right size to avoid leaking into capillaries but not too large to get stuck — for interstitial injection. Optimum range 20–100 nm.
Nano	323	Nanoparticles can be delivered via the leaky blood vessels of tumors. But the distribution is heterogeneous and unpredictable.
	323	Antitumor drugs: cioplatin and doxorubicin. The most promising is liposomal doxorubicin (lipid delivery of doxorubicin).
	323	CNT belongs to the family of fullerenes and are made up of graphene sheets.
	323	CNT diameters 0.5–3.0 nm and length 20–1000 nm.
	323	CNT can cross the cell membrane ("nanoneedles") without disrupting the membrane.

(*Continued*)

(Continued)

Category	**	Review Notes
	323	Listing of nanoparticles (1) *Superparamagnetic iron oxide crystals* 4–5 nm in dia They possess large magnetic moments when brought into a magnetic field, but the magnetic effect disappears when the magnetic field is removed. (2) *Quantum dots* Nano-scale crystalline of cadmium selenide that can transform light. (3) *Dendrimers* Polymers (4) *Polymeric micelles* Spherical with dia less than 50 nm. (5) *Liposomes* Closed vesicles — heterogeneous in size ranging from 100 nm to 1000+ nm aka "small unilamellar vesicles," "large unilamellar vesicles. (6) *Nanospheres* Spherical objects 10s–100s nm. (7) *Aquasomes* (carbohydrate-ceramic nanoparticles). (8) *Polyplexes/Lipopolyplexes* Assemblies which form between nucleic acids and polycations or cationic liposomes.
Cancer	324	Antiangiogenic therapy works only temporarily. It is not clear why.
Nano	325	Quote: "Nanoparticle-based drug-delivery systems continue to evolve and are the subject of ongoing trials in clinical oncology."
Nano	325	Two approaches: active and passive. *Active* involves linking ligands to nanoparticles that are tumor specific. *Passive* takes advantage of the inherent size of nanoparticles and unique properties of tumor vasculature.
Nano	325	Quantum dots are novel semiconductor nanocrystals — and for imaging. Nanoshells (approximately 10–300 nm) are also used for imaging.
Nano	325	Gold nanoparticles re used for diagnosis and therapy. They are used to enhance sensitivity to external beam radiation.
Nano	325	Paramagnetic nanoparticles — used to enhance imaging.

(Continued)

(*Continued*)

Category	**	Review Notes
Nano	325	Nanotechnology is a rapidly growing field.
Cancer	326	Tumor blood vessels have structural abnormalities that hinder therapies. They are large in diameter, they promote high back pressure. They leak.
Nano	327	SWNTs absorb near-infrared light (NIR). These nanotubes can get inside the cell and kill it by heating via the absorbed light.
EM	328	Magnetic effects can be significant (little else).
Nano	329	Gold nanoparticles were injected into tumors under the skin and then heated by laser. The tumors were ablated. It is not clear if this would work for deep tumors or how the nanoparticles were gotten into the tumor. The main contribution appears to be the advantage of using NIR-absorbing nanoparticles.
Nano	330	Quote: "... only between 1 and 10 parts per 100,000 of intravenously administered monoclonal antibodies reach their parenchymal targets in vivo.
	330	Fullerenes and CNT discovered by Richard Smalley.
	330	Tumor blood vessels have "architectural anarchy."
Nano	331	2008 Quote: "Targeted delivery of drug molecules to organs or special sites is one of the most challenging research areas in pharmaceutical sciences."
Nano	331	Paper focuses on lung disease. It is a review but to the date of the paper no significant successes.
Centriole	332	States that the regulators for centriole duplication are the same as those for chromosome separation.
Centriole	333	"The mother centriole is not a template but a platform for regulating the assembly of the daughter centriole."
Centriole	334	Regarding cell polarity: The apical and basal sections of a cell are separated by a physical barrier called the "apical junction complex."
Centriole	335	Centrosome is electron dense material.
Centriole	335	Centrosome increases in size at the onset of mitosis. This is called "centrosome maturation."
Centriole	335	Although important details of the centriole structure are known, relatively little is known about the PCM. The reason is the rather homogeneous electron density of the PCM.

(*Continued*)

(Continued)

Category	**	Review Notes
Centriole cancer	336	Centrosome abnormalities are common in cancer.
Cancer	336	(2011) Quote: "Thus, it is tempting to speculate that the chromosome instability generated by extra centrosomes can contribute to tumorigenesis."
Cancer	336	Centrosome clustering has been observed in tumors.
Cancer	336	Centrosome clustering is essential for survival of cells with supernumerary centrosomes.
Centriole cancer	337	Discusses the Par complex-molecules which in normal cells regulate cell polarity, but in abnormal cells it produces cancer.
Centriole	337	During mitosis the organelles get arranged to form an axis of symmetry. This organization is known as apical-basal polarity.
Centriole	338	The focus is on the centrosome and PCM — as many as 300 proteins in the PCM.
Centriole protein	338	Overexpression of SAK/Plk4 leads to centrosome amplification, also SAS-6. These proteins could be licensing proteins.
Protein centriole	339	Focus of paper is on CPAP — a protein related to SAS-4. CPAP controls the centriole length. Too little → insufficient daughter development. Too much → long daughter length → supernumerary centrioles → cancer.
Centriole protein	339	The amount of PCM is proportional to centriole size.
Protein	339	CPAP overexpression results in supernumerary centrioles.
Protein centrioles	340	SAS-4 is like tubulin in that it is incorporated into centrioles during their duplication. SAS-4 is necessary for centriole duplication.
Centriole	340	(2003) Quote: "... (centrosome) duplication has long fascinated cell biologists, but the mechanism regulating it is poorly understood."
Protein	340	Quote: "SAS-4 is a stably associated centriole component that is incorporated during centriole duplication."
Centriole	340	Quote: "Centriolar SAS-4 levels correlate with centrosome size."
Centriole	340	Partial depletion of SAS-4 leads to defective centrioles.

(Continued)

(Continued)

Category	**	Review Notes
Protein centriole	341	SAS-6 self-assembles and this forms the structure of the centriole core.
Nano	342	A review paper. It states that there is a necessity to deliver nanoparticles to the cell interior.
Cancer	342	Leaky vasculature of tumors makes drug delivery difficult.
Cancer	342	Six hallmarks of cancer: (1) limitless replication potential (2) insensitivity to growth inhibition (3) evasion of apoptosis (4) sustained proliferation signals (5) induction of angiogenesis (6) invasion and metastasis.
Nano	342	Advantages of nano therapy (1) increase drug concentrations (2) improve the pharmakinetics (3) decrease drug release during transit (4) improve drug stability (5) improve cellular internalization and organelle-specific delivery of the drug due to surface characteristics.
Nano	342	Animal studies show that slightly positively charged nanoparticles 50–100 nm can penetrate large tumors.
Definition	342	"Active targeting": Targeting a specific tumor.
Cancer	342	Targeting the endothelial cells of tumor blood vessels is an attractive strategy for inhibition of the growth of the tumor.
Cancer	342	VEGF
Nano	343	The pace of companies doing proprietary research has considerably accelerated.
Nano	343	EPR effect.
Cancer	343	As cells start to die they secrete growth factors that trigger the budding of new blood vessels from the surrounding capillaries (angiogenesis). Rapid angiogenesis promotes irregular vasculature.
Glossary	343	Extravasation → drug out of blood vessels.
Electro magnetics	343	The presence of surface charge can alter the way the cells respond to the NP.
Glossary	343	Active targeting is also called "lizard-mediated targeting."

(Continued)

(*Continued*)

Category	**	Review Notes
	344	Relates to patents.
	345	Omit
Electro magnetism	346	Difficult to read. Suggests NPs have slight negative charge and particle size of 150 nm.
Electro magnetism	347	Negatively surface charged NPs seem to be attracted to tumors greater than positively charged NPs.
Nano	348	MNPs are used for imaging.
	348	This is a review of design parameters about MNP performance in vivo
	348	SPION
	348	Quote: "Upon intravascular administration, NPs immediately encounter blood, a high ionic strength, heterogeneous solution, that can induce NP agglomeration, altering these magnetic properties and inducing partial sequestration." "…In each case the NP is in danger of prematurely binding to or being taken up by cells before reaching its target tissue."
Quote for proposal	348	Quote: "As we move forward, better characterization tools are needed to both evaluate new MNPs and better understand their behavior in the body. Controlled studies of individual physicochemical parameters will offer NP engineers the basic understanding to successfully build more elaborate and functional MNPs."
Nano	348	Chemotherapeutic drugs: doxorubicin, cisplatin, and paclitaxel.
	348	Problem: With intravenous administration, the drugs are quickly cleared from circulation, and only a very small fraction reaches the tumor. Even then the drug may not get into the depths of the tumor long enough to be effective.
Nano	349	A brief review paper on nanoparticles, but nothing new in getting the nanoparticle to the tumors.
Nano	349	Only a small fraction of intravenous administered nanoparticles reach the tumor. Even then the particles may be rapidly cleared from the tumor having little therapeutic effect.

(*Continued*)

(*Continued*)

Category	**	Review Notes
Nano	350	Nanoparticles are 10–100 nm. The threshold for first pass elimination by the kidneys is 10 nm. The upper bound size is not well defined.
	350	The vasculature of tumor is known to be leaky to nanoparticles.
	350	Nanoparticles 50–100 nm can penetrate through large tumors.
	350	Normal blood vessels keep nanoparticles larger than 1–2 nm.
EM	350	The inside surface of blood vessels contain many negatively charged components.
Nano	350	The primary role of the targeting of ligands is to enhance cellular uptake into cancer cells rather than simply accumulate in the tumor.
Nano	351	Cell penetration in scaffolds of cells is difficult. The authors develop magnetic NPs coated with chitosan for enhancing cellular invasion under a magnetic field.
Nano	352	Paper describes the potential benefits of nanoparticles for cancer treatment.
Cancer	352	Fast growing cancer cells recruit new blood vessels — and with an imbalance of angiogenic regulators. This leads to highly disorganized and dilated vessels with numerous pores (leaky).
Cancer	352	Microenvironment around tumor cells is different than that around normal cells. The supply of oxygen and nutrients is not sufficient to maintain the growth rate. Therefore the tumors use glycolysis to get energy and it results in an acidic environment.
EM, Nano Cancer	353	Magnetite nanoparticles (40 nm) with positive surface charge were internalized into human breast cancer whereas they were not internalized into normal cells.
Nano	354	Confirms past research. Reasserts the potential of nanotherapy.
Nano	355	Confirms past research. Reasserts the potential of nanotherapy.
Nano	356	Discusses the use of nanoworms — a string of nanoparticles. The tumor targeting ability of nanoworms is found to be better than single nanoparticles.

(*Continued*)

(*Continued*)

Category	**	Review Notes
Nano	356	The search is for ligands for attracting the nanoparticles.
Nano	357	A review article on current application of nanoparticles in medicine.
Nano Keck	357	So far nanotherapy is relatively ineffective — need multifunctional nanoparticles.
Nano Keck	358	Nanotherapy has great potential but challenges remain. Need for particles that are internalized by the cells. The choice of an appropriate particle is not obvious.
EM	359	Says cell-to-cell communication may or may not exist.
EM	360	Cancer cells seem to affect other cells at a distance.
MT, EM	361	MTs are electrically polar. Vibrations generate an oscillating electric field.
	361	Calculates intensity of the electric field based upon the dipole of the tubulin -β dimer. The moment is $P = 337\text{D} = 1.12 \times 10^{-27}$ cm.
MT EM	362	Longitudinal vibration of MTs generate an endogenome electromagnetic field.
Cancer	362	Malignancy can occur due to magnetic field.
MT	362	MT vibs ~10–30 MHz and 100–200 MHz.
EM	362	Mitochondria generate an electric field.
Cancer	363	Quote: "Physical processes in cancer cell are disturbed due to suppression of oxidative metabolism in mitochondria (Warburg effect)."
EM Nano	363	Electromagnetic activity between cancer and healthy cells may differ. Nano measurements are needed.
EM	363	Each mitochondrion is surrounded by a layer of strong static electric field. It diminishes linearly with distance from the mitochrondrion.
EM	363	Static electric field of mitochondrion is negligible beyond 10 nm.
Cell	363	Mitochondria make up ~ 22% of cell volume. There are ~1000 per cell. Spherical diameter of ~0.6 m with separation distance of ~0.4 m.
MT	363	Dipole moment of α,β-tubulin ~1000 Debye (10^{-26} cm). Induced moment from movement is ~200–400 Debye. This also produces polarization changes.
Cancer	363	Suppressive of oxidative metabolism is a general feature of cancer cells.

(*Continued*)

(Continued)

Category	**	Review Notes
EM, Cancer	363	Disturbances of the electromagnetic activity in cancer cells might be responsible for malignant properties.
Keck	363	Need nano measurements of electromagnetic properties of MTs.
MT Vibs	364	Highest MT vib frequency ~10^8 Hz. MTs have 13–14 protofilaments α,β tubulin each have molecular weight of 55,000 Daltons. 1 Dalton is the mass of a proton ~1.66×10^{-27} kg — lowest vib frequency ~10^6 Hz.
MT	364	Half-life of MT ~10 min.
MT	364	Diameter of tubulin ~4 nm.
EM	365	Low electric fields kill cancer cells.
EM	366	Low frequency EM fields are successful in treating malignant tumors.
EM	367	Quote: "A static electric field is generated by a static charged particle. Both an electric field and a magnetic field are generated if a charged particle moves at a constant velocity. The term electromagnetic field is reserved for a condition where a charged particle is accelerated. In most cases the acceleration of a charged particle takes place in the form of an oscillation. Therefore, the electric and magnetic fields also oscillate."
EM	367	MTs are likely to be a source of cellular EMFs, but four other models are cited.
EM	367	Discusses biological effects of low-frequency EMFs.
	367	Suggests that cells can interact via EMFs.
Cancer	368	A potentially useful quote: "Measurement of differences of the electromagnetic field between healthy and cancer cells using microelectronic and nanotechnological systems will crucially contribute to understanding of the cancer process."
Keck EM	369	Discusses technical aspects and problems of EM measurement. Quote: "In conclusion the precise direct measurement of biological EMFs — has not been accomplished yet." "... we may draw the conclusions that the nanoelectronics should be used."

(Continued)

(Continued)

Category	**	Review Notes
EM centriole	370	Centriole replication may be accelerated by a rectangle pulsed electrical current stimulation (RPECS).
EM	370	Injury current play sa role in wound healing.
EM	370	Wounds treated with RPECS healed ~15% faster than the control wounds.
EM	370	Quote: "As early as the 1860s the fastener of the science of bioelectricity, DuBois-Reymond, discovered that an injured bleeding finger is electrically positive compared with an uninjured finger."
EM centriole	370	At a rest state, the mother centriole has a strong positive charge whereas the daughter centriole is not charged. At an activated state the daughter centriole also has a strong positive charge.
EM centriole	370	When a pair of mother/daughter centrioles have a significantly large positive charge the repulsive force overcomes the static force keeping them together.
	370	Strong repulsing electric field forces drive centrioles apart.
EM centriole	370	Quote: "A strong externally applied imposed EMF could induce aberrant centrioles and centrosomes that cause life desire" (Known by Paul Schafer 40 years ago).
Centriole	371	Centriole duplication appears to be driven by elements of the cytoplasm instead of by the nucleus.
Centriole	372	On plant cells. Authors develop a theoretical model of plant cell mitosis — cells without centrioles. They conclude that mitosis is driven by electromagnetic forces similar to those in eukaryotic cells. A number of assumptions are made.
EM	373	Documents experiments showing that wounds generate a positive potential. Reportedly discovered circa 1860 by duBois-Reymond.
EM	374	There exists direct current potential in living tissue. This current is observed in wound healing.
EM	374	In animals with amputated limbs the stump has a negative polarity. Artificially increasing this polarity enhances the new growth.
EM	374	Quote: "Further studies are needed to elucidate the mechanisms accounting for these findings..."

(Continued)

(*Continued*)

Category	**	Review Notes
Protein	376	Overexpression of cyclin B_2 leads to an excess of Plk1 and deregulates mitosis.
Protein	376	Cyclin B_2 and p53 antagonistically regulate centrosome separation.
Protein	376	Cyclins B_1 and B_2 both drive tumorigenesis when they are overexpressed. p53 mediates this by negatively regulating aurora A kinase activity.
Centrioles	377	Procentrioles appear in male butterflies even with the centrioles still being engaged. But they do not grow and develop. This appears to be an unimportant anomaly.
Protein MT centrioles	378	The α,β tubulin determines curvature at the distal ends of MTs.
MT proteins Centrioles	379	Dynein and NuMA exert tension forces to stabilize the minus ends of MTs.
MT	380	MTs do not generate, but instead maintain cell polarity.
Cancer protein	381	Stroma cells are surrounding environmental cells to cancer cells. They evolve with cancer cells. P53 acts against stoma cells. HSF1 enhances stoma cells (HSF1 "heat shock factor 1").
Protein cancer	382	"TP53 (p53) is the most frequently mutated gene in cancer, being altered in approximately 50% of human malignancies." p53 is thus a target for therapies. Until recently p53 was regarded as "undruggable," but the authors suggest that is no longer the case and new drugs are being tested to interact with p53.
Centrioles	383	Plk1 is required for maturation of the centrioles. Quote: "This ill-understood centriole maturation process is almost two cycles long, and as a consequence, each cycling cell contains three generations of centrioles. Only one centriole per cell (the oldest one) is fully structurally mature and decorated with subdistal and distal appendages" (Vorobjev and Centsov, 1982 and Paintrand *et al.*, 1992).
MT	384	Microtubules look alike in almost all species. How tubulin heterogeneity controls MT functions is still relatively unknown.
EM	385	(1972) "electrical parameters … are potent, stimulatory factors for mammalian cells."

(*Continued*)

(Continued)

Category	**	Review Notes
Centriole	386	2006 Quote: "The induced electric field by externally applied RPECS (rectangular pulsed electric current stimulation) could be mild compared with the spontaneous and natural electric field of the centrioles."
Centrioles	387	(1982) In interphase the structures of the two centrioles differ: the mother has appendages, the daughter does not.
Centrioles	387	After cells have left the mitotic cycle replication of centrioles does not take place.
Centrioles	388	(1992) Quote: "...a comprehensive description of the overall organization of the centrosome in animal cells is lacking."
Cancer	389	Centrosome amplification is a factor in cancer.
Cancer protein	389	Moesin is an FERM domain protein. It enhances the formation of centrosome amplification and clustering.
Cancer protein	389	Moesin is highly expressed in basal type breast cancers.
MT	390	AFM can image in a liquid environment.
MT	391	2010 AFM studies of MTs are rare.
MT	391	Provides procedures for AFM studies.
	392	Biology Smart Book
Centriole protein	393	(2015) Plk4 is an enzyme — a master regulator of cell division.
Protein centriole	393	Plk4 is the master regulator of cell division. Its hyperactivity induces centriole amplification.
Protein	393	Plk4 undergoes autoinhibition and is thus self-regulating.
Cancer	393	Plk4 is overexpressed in cancer.
Protein	393	Plk4 has a third polo box, different from other Plk family members.
Protein centriole	394	Asterless is a binding partner of Plk4. Asterless (Asl) acts as a scaffolding protein.
Protein centriole cancer	394	"...how Asl overexpression drives centriole amplification is unknown."
Protein	394	"Plk4 is primarily regulated by protein turnover and efficiently promotes its own destruction to suppress centriole overduplication."
Protein	394	"...Plk4 forms a homodimer between its first two boxes — unlike other Plks."

(Continued)

(Continued)

Category	**	Review Notes
Protein centriole	394	"Plk4 targets mitotic centrioles, appearing as a single asymmetric spot on each centriole. Plk4 within each spot modifies the site on the centriole enabling the spawning of a single daughter centriole during the S phase."
Protein	394	Asl is Cep152 in humans.
Centriole	394	Asl is thought to target Plk4 toward the centriole.
Centriole	394	Asl-depleted cells do not duplicate.
Centriole protein	395	Plk4 is a regulator of centriole duplication and it depends upon asterless.
Protein centriole	395	Cep192 recruits Cep152 and Plk4 (Cep 152 is Asterless).
EM	395	Plk4 attraction to centrioles depends upon electrostatic interaction.
Protein centriole	396	Plk4 is a regulator of centriole duplication.
Protein	396	Plk4 has a single polo box.
Protein centriole	397	Asterless (Asl) (human ortholog CEP152) provides a molecular platform for Plk4. The amino terminus interacts with the polo box of Plk4.
Protein centriole	397	Too much or too little Plk4 causes a failure of centriole assembly.
Protein	397	Asterless is an ortholog of CEP152.
Protein centriole	398	Plk4/SAK is a regulator of centriole biogenesis.
Protein, centriole cancer	398	CEP152 interacts with Plk4 and is required for centriole overduplication.
Cancer	398	There is a correlation between excess centrosomes, aneuploidy, and cancer.
Protein	398	Interfering with CEP152 function prevents recruitment of Plk4 to the centrosome.
Protein	398	Plk4 appears to regulate itself.
Protein cancer	399	Plk4 is a principal player in the formation of centriole flowers (multiple daughters at a single maternal centriole).
Centriole cancer	400	Defects in centrioles are found in cancer.
Centriole	400	First sign of daughter centriole is electron dense cartwheel (~25 nm in dia.) at the proximal end of the mother centriole?
Protein	400	Cartwheel proteins: Bld10 and SAS-4/CPAP, SAS-6? SAS-6 is a coiled-coil protein.

(Continued)

(Continued)

Category	**	Review Notes
Centriole proteins	401	Confirms previous work: focus is on evolution and centriole difference in species.
Centriole	401	Centrioles are found in eukaryotic cells.
Protein centriole	402	Proteins required for centriole formation: Plk4/SAK, Asterless/Cep152, SAS-6, SAS-6/Ana2/STIL, CEP135, and SAS-9/CPAP.
Centriole protein	402	Asterless (Asl) is a scaffolding protein which recruits Plk4 to the site on the daughter for assembly.
Centriole protein	402	SAS-6 ninefold symmetric cartwheel scaffold for the daughter centriole base.
Protein	403	CEP152 is the ortholog of asterless (Asl).
Cancer	403	Centrosome amplification is a common feature of cancer cells.
Centriole protein	403	Asl is required for initiation of centriole duplication.
Protein	404	SAK/Plk4 is the most divergent of all the Plks.
Cancer	404	Too much SAK/Plk promotes tumors.
Protein	404	Figure 2 shows a drawing of a speculative model of Plk4 behavior, including polo box and autophosphorylation.
Protein	405	Polo — originally named after the phenotype in flies.
Protein	405	PBD — Polo box domain.
Protein	405	Figure 3 — A model of PB actions and phosphorylation.
Protein cancer	405	Plk1 is highly expressed in tumors.
Protein	406	Confirms previous studies, but the paper is frequently references.
Protein centriole	407	Plk4 needs to be at a certain level before centriole duplication occurs, but then Plk4 self-destructs before too many centrioles are formed.
Protein	408	Confirms earlier studies. Plk4 autophosphorylates.
Protein centriole	409	Plk4 phosphorylates itself to promote its degradation. Plk autoregulation guards against duplication instability.
Cancer	410	Abnormal centrosome dynamics promotes cancer.
Cancer	410	Aneuploidy is a common feature of cancer.
MT	411	γ-tubulin plays key role in MT formation. Necessary but not sufficient.
MT	411	γ-tubulin discovered ~1989.

(Continued)

(Continued)

Category	**	Review Notes
MT	412	Claims that centrosome separation and spindle forces are due to MT compressive forces.
Cancer	413	Cancer tumors have an abnormal number of chromosomes.
Centriole cancer	414	Supernumerary centrosomes are a key cause of cancer.
Centriole	414	Centrioles can occur de novo, but this may not occur if other centrioles already exist.
Centriole protein	415	Protein Bld10p is a tip agent that connects the MTs to the cartwheel.
Cancer	416	Abnormal centriole numbers are detected in tumor cells.
Centrioles protein	416	Cullin1, a protein localizes to maternal centrioles and is critical for suppressing centriole overduplication.
Centrioles cancer	416	Multiple daughters can form from a single centriole. These are called "centriole flowers."
Protein	416	Cullin1 is critical for the degradation of Plk4.
Centriole protein	416	After Plk4 recruitment of SAS-6 is critical for procentriole assembly.
Cancer centrioles	417	Abnormal centrioles lead to cancer. Overexpression of SAK/Plk4 leads to cancer.
Cancer centriole	417	There is a picture of a lowering centriole.
Protein	418	Loss of p53 is not by itself a primary cause of aneuploidy. The paper is a review summarizing the current knowledge between aneuploidy and p53.
Cancer	418	Aneuploidy — the presence of abnormal chromosome numbers. This is a common finding human cancer.
Protein	419	A review focusing on Plk1 and family members Plk2-4, and their roles in mitosis.
Protein centriole	419	Plk1 promotes mitotic entry.
Protein	419	Plk1 is involved in chromosome maturation and separation.
Protein centriole	419	Plk4 is involved in centriole replication.
Protein	419	Plk1 does not only phosphorylate other proteins, it can also phosphorylate itself.
Cancer	419	Plk1 is overexpressed in many cancers.
Centrioles cancer	420	Abnormal centriole numbers produce cancer.

(Continued)

(Continued)

Category	**	Review Notes
Proteins centrioles cancer	420	Overexpression of SAK (Plk4), DSas6, and DSas4 leads to de novo centriole-like structures.
Centriole protein	421	"Plk4 is the conserved master regulator of centriole assembly." "Taken together, our findings suggest that Plk4 directly generates its own phosphodegron and can do so without the assistance of an additional kinase(s)."
Centriole	422	In movement about the nucleus the centriole speed was as fast as 10 m/min.
Centriole	423	During mitosis the mother stays near the center while the daughter migrates away.
Protein centrioles	424	γ-Tubulin is a component of the sub-distal appendages of the centriole. γ-Tubulin is required for centriole duplication and organization of pericentriolar material.
Proteins centrioles	424	δ-Tubulin is required for the formation of triplet microtubules.
Protein	424	γ, δ, and γ-tubulin localize in the centrosome.
Protein MT	425	γ-Tubulin is required for initiating MTs.
MTs protein	425	γ-Tubulin is a protein required for initiating the polymerization of MTs.
MT	426	This paper is in a review journal, but it is not a traditional literature review paper. Instead it is more of a review of the α and β-tubulin structure. There are detailed color drawings of the tubulin in the figures.
MT	426	MTs exist in all eukaryotes.
MT	426	The tubulin dimers are connected head to tail.
MT	426	MTs are highly dynamic and can switch stochastically between growing and shrinking phases.
Protein MT	427	Of the five tubulins α, β, γ, δ, and ε, α and β make up the MT, γ is at the base for nucleation, δ is associated with the centriole, and γ with the pericentriolar material. (In 424 δ is associated with the triplets.)
Protein MT	428	(2000) MT assembly is initiated by γ-tubulin. γ-Tubulin and other proteins (Spc97p and Spc98p) form a core of the forming MT.
Centriole	429	There is a direct link between centrosome activity and DNA replication.

(Continued)

(Continued)

Category	**	Review Notes
Centriole	430	"Centrosome" from Latin meaning "central body." Color drawings.
MT, centriole	431	MT nucleation depends upon γ-tubulin. The γ-tubulin small complex (γ-TuSC) then forms a scaffold for the α/β tubulin dimers to initiate polymerization.
Centriole	431	The PCM is a highly structured consecutive layers of proteins enabling nucleation.
EM, MT	432	(2010) Energy supply to MTs from: (1) GTP (hydrolysis of guanosine triphosphate), (2) energy transfer by motor (moving) proteins, and (3) energy released from mitochondria as "wasted" energy in the citric acid cycle.
EM, MT	432	Paper considers axial longitudinal vibration of MT.
EM, MT	432	Hetrodimers have an electric dipole moment of over 1000 Debye (10^{-26} cm).
EM MT	432	Frequencies of MT vibrations span from kHz to GHz.
MT	433	Shows various ways an MT could oscillate.
EM, MT	434	Paper referred from David Lemmon.
MT	435	Presents several modes of vibration of MTs, with frequencies from MHz to GHz.
EM	436	Indicates possible DNA harm from low frequency (<300 Hz) electromagnetic fields.
MT	437	Widely referenced paper. Treats entire MT as a thin shell. Uses $E = 0.5 \pm 1$ GPa.
Protein, MT	438	Nip is a centrosomal Plk1 substrate which is controlled by Plk1 in the development of MTs.
Protein centriole	439	Asterless (Asl) is needed for stable centrosome development.
	440	RLH paper.
	441	RLH paper.
MT, EM	442	(2004) The electric dipole moment has only been calculated via computer simulations (not measured).
MT, EM	442	(2004) There exists little experimental evidence about the electrical properties of tubulin and MTs.
MT	442	MTs are not steady-state systems; energy is continually being put in and taken out.
EM, MT	443	Net charge on a tubulin dimer is 0.19e (lower than that calculated earlier).

(Continued)

(*Continued*)

Category	**	Review Notes
MT	444	Good pictures of α and β-tubulin. The tubulins are almost the same.
Protein	445	Reduced Plk4 leads to cancer.
Protein	446	Discusses various proteins associated with the Plks. All Plks are similar.
Cancer nano	447	Paper found by M. Schulz. The hypothesis of the paper is that a high molecular weight protein: fibronectin in the extracellular matrix (ECM) of metastatic breast cancers can be used as a biomarker.
Cancer nano	447	"A tiny magnet" is used to hold toxins together then attack the fibronectin biomarker.
Centriole	448	The centrosome, in addition to being an MTOC is also essential for cell division vitality.
MT EM	449	"Tubulin dimers have a strong electric dipole moment."
MT EM	449	Tubulin dimer dipole moment: over 1000 Debye = 10^{-26} cm.
MT	449	MT oscillation frequency: from MHz to GHz.
Centriole	450	A review paper: Indicates that centrosomes bind more than 100 regulatory proteins. The cellular regulation is the focus of the review.
Protein centrioles	451	Plk4 initiates centriole duplication and is regulated. The protein phosphatase 2A regulatory submit twins stabilizes Plk4.
Cancer	451	Supernumerary centrioles can cause aneuploidy and tumorigenesis.
Protein	451	The mechanism stabilizing Plk4 is still unknown.
Centriole protein	451	qPlk4 localizes to the "parent" centriole as a spot and modifies the site.
Protein	452	As many as 80 proteins, and probably more, are needed for cell cycle stability. This is a review-survey paper on the proteins.
Protein	453	SAK/Plk4 is different than other polo-like kinases in having only a single polo box.
Protein	453	SAK/Plk4 is destroyed by APC/C dependent proteolysis.
Cancer	453	Cancer can occur with a weakening of the controls of proteins during mitosis.
Centrioles cancer	454	"Centrosome anomalies are detected in virtually all human cancers."

(*Continued*)

(Continued)

Category	**	Review Notes
Centrioles	454	This is a review paper: It classifies abnormalities into three groups: (1) primary centrosome abnormalities, (2) transient centrosome accumulation, and (3) permanent centrosome accumulation.
Centrioles cancer	454	Centrosome aberrations is not necessarily a cause of chromosome instability (CIN).
Centrioles cancer	454	"Centrosome anomalies are an astonishingly frequent finding in malignant tumors."
Centrioles	454	Centrosome abnormalities are likely a result of cellular insults.
Centrioles mitosis	455	A 2001 review by E.A. Nigg. The focus is upon the many kinases needed for fidelity in cell division (M-phase). (M-phase follows after S-phase.)
Centriole mitosis	455	M-phase comprises mitosis and cytokinesis. Mitosis is usually divided into five stages: (1) prophase, (2) prometaphase, (3) metaphase, (4) anaphase, and (5) telophase.
Centriole mitosis	455	Before proceeding to M-phase, two events must occur in S-phase: (1) DNA replication and (2) centrosome duplication. (These are believed to occur simultaneously — RH.)
Protein centriole	456	SAS-6 (human homologue GFP-Hs SAS-6) is recruited to centrioles for daughter formation.
Protein centriole	456	SAS-6 is recruited to the mother centriole and then associated with the emerging daughter centriole.
Centriole	457	This paper presents the now widely accepted procedure of centriole duplication, but with differently named proteins.
Centriole	457	(2006) Centriole formation is a poorly understood process.
Protein centriole	457	Five proteins have been identified for centriole formation: ZYG-1, SAS-4, SAS-5, SAS-6, and SPD-2. The order is SPD-2, ZYG-1, SAS-5, SAS-6, and SAS-4 — for *C. elegans*.
Centrioles	457	Just before centriole formation, the original pair splits slightly.
Centriole (papers)	457	Excellent pictures and diagrams.
Cancer	458	Suggests that aneuploidy promotes tumorigenesis.
Centrioles protein	459	*C. elegans* embryo: γ-tubulin stabilizes the formation of new centrioles.

(Continued)

(*Continued*)

Category	**	Review Notes
Centrioles protein	459	*C. elegans* requires SAS-4, SAS-5, SAS-6, and ZYG-1 to form new centrioles.
Centrioles	459	The small size of centrioles makes it difficult to observe the kinematics — only snapshots.
Protein	460	α, β, γ-tubulin are found in all eukaryotic organisms and δ and ε-tubulin are found in organisms with triplet MTs.
Protein	461	γ-tubulin ring complex (γ-TuRC) forms a template for MT nucleation.
Protein	462	γ-tubulin small complex (γ-TuSC) helps form the 13-fold MT structure.
MT	462	13-fold structure allows protofilaments to move straight without being twisted by 3 or 4 symmetries.
Protein MT	463	New MTs require γ-tubulin.
Centrioles protein	464	The centrosome is not simply a cloud of proteins. Instead it is structured layers of matter.
Centriole	465	Behavior of centrosome in the cell cycle is still poorly understood (2012).
Centriole	465	The centrosome increases in size upon entry into mitosis.
Centriole protein	465	Plk4 is at the head of the pathway for centriole formation Asterless (Asl) (Cep152 in humans) is a scaffold for Plk4 and SAS-6.
Protein	465	Zyg-1 is Plk4, SAS-6 is Ana2 in Drosophila.
Centriole	465	SAS-6 assembles the cartwheel structure.
Protein	465	SAS-4 provides for protein–protein interaction.
Protein centriole	465	Plk1 is required for the dramatic increase in PCM.
Protein	465	Plk1 recruits γ-tubulin.
Protein	466	Plks are characterized by their polo box domain.
Protein	466	Plks are controlled by phosphorylation, proteolysis, and transcription.
Protein cancer	466	There is an uptick in Plk1 in human tumors.
Protein	467	Plk1 is a regulator of cell division. Plk1 is recruited to centrosomes, kinetochores, and the spindle.
Cancer	467	Plk1 is a potential drug target for cancer therapy.

(*Continued*)

(Continued)

Category	**	Review Notes
Cancer centriole	468	(2007) Author Ganem suggests there is a strong link between aneuploidy (chromosome instability) and cancer. In a later paper (2009), he suggests there is a link between extra centrosomes and chromosome instability. Papers 9, 195, 235 (duplicate entries).
Protein cancer	469	Plk1 is overexpressed in tumors.
Protein	469	Plk1 regulates many processes. Plk1 and Plk4 (SAK) localizes to many structures via their polo box.
Protein	469	Plk2 is required for centriole duplication.
Protein	470	Plk2 is required for centriole duplication.
Centrioles mitosis	471	Centrosomes enlarge dramatically after mitotic entry.
Protein MT	472	MT assembly is usually seeded by γ-tubulin ring complexes γ-TuRC: γ-tubulin ring complex; γ-TuSC: γ-tubulin small complex.
MT	473	Authors indicate that MT nucleation follows a pointed template (helix) as opposed to a protofilament model (2000).
MT protein	474	(2000) γ-tubulin is required for MT nucleation but the mechanism is not understood (a near quote).
Centriole	475	Cells without centrioles/centrosomes do not survive or thrive: Quote: "these data reveal that the primary role of the centrosome in somatic cells is not to form the spindle but instead to ensure cytokinesis and subsequent cell cycle progression."
MT	476	(2001) "The precise ways in which microtubules achieve their cellular functions is not fully understood yet." "MT arrays in cells are dynamic, capable of assembly, disassembly, and rearrangement on a time scale of seconds to minutes."
MT	476	Dynamics in vitro and in vivo are similar.
Protein MT	477	αβ-tubulin heterodimers self-assemble to form the MTs.
Protein MT	477	γ-Tubulin nucleates the MT. γ-Tubulin is believed to be capable of longitudinal assembly.
Protein MT	477	δ-Tubulin may seed the triplets.
Protein	477	γ-TuRC is called "gammasome" (gamma tubulin ring complex). γ-TuSC (gamma tubulin small complex).
Protein	477	δ-Tubulin interacts with α-tubulin. γ-Tubulin interacts with β-tubulin.

(Continued)

(*Continued*)

Category	**	Review Notes
Protein MT	478	The focus of the paper is on the behavior of the γ-TuRC in nucleating MTs.
Protein	478	Tubulin dimers can self-assemble in vitro.
Protein MT	478	γ-Tubulin is required for MT nucleation.
Protein MT	478	"γ-TuRC resembles a lock washer..." "like the first turn of the MT helix"
MT	478	Two models are discussed for MT nucleation: (1) "template" and (2) "protofilament." In template the γ-TuRC is the first turn in an assembled helix. In protofilament, the γ-TuRC uncurls the dimers longitudinally. The authors state that the in vitro results favor the template model.
MT	479	MT nucleation rate remains constant during polymer assembly.
	480	*The World of the Cell* book.
	481	Watson and Crick (DNA article).
Cancer	482	Breast cancer and types.
MT	483	γ-Tubulin ring complex is required for Drosophila centrosomes to nucleate microtubules.
MT	484	MTs are dynamically unstable exhibiting periods of rapid growth and shortening. β-tubulin is at the + end.
MT	485	(2005) The paper proposes that MT forces on organelles are formed by the curved filaments coming off the positive end of the MT attaching to an organelle.
MT mitosis	486	This paper discusses the assembly of MTs about the chromosomes.
MT mitosis	487	Little is known about the underlying mechanical principles that govern the assembly and function of the spindle.
MT mitosis	487	Most animal spindles can be thought of as the superposition of kinetochore, nonkinetochore, and astral microtubules.
MT mitosis	487	This is a review paper about the interplay of forces and assembly of spindle dynamics.
Glossary MT	488	MTs alternating between elongating and shortening (apparently at random) is a process called "dynamic instability."
Glossary MT	488	The time of elongation is known as "catastrophe time."

(*Continued*)

(*Continued*)

Category	**	Review Notes
MT	488	Force reduces the catastrophe time by limiting the rate of tubulin addition.
MT Glossary	489	MT dynamics can be described by four parameters: (1) growth, (2) shrinkage, (3) catastrophe (transition from growth to shrinkage), and (4) rescue (transition from shrinkage to growth (edited quote).
MT	490	Similar to *489* in the four phases.
MT	491	Small molecular inhibitors of MT development are used in chemotherapy.
Cancer	491	Small molecular inhibitors of MT development are used in chemotherapy. This paper presents a way to enhance the chemical effect via light.
MT	492	MTs can push and pull — a closely regulated process. The pushing and pulling seem to be attributed to adding and subtracting tubulin.
MT	492	"Catastrophe" refers to a switch from growing to shrinking.
MT	492	MTs interact with kinetochores and provide for the separation of the chromosomes.
MT	493	(2015) This is a review of factors and behavior of γ-tubulin — a key MT protein.
MT	493	MTs are nucleated from the centrosome (the MTOC).
MT	493	The MTs are nucleated by the γ-TuRC (γ-tubulin ring complex).
MT	493	The γ-TuRC and the γ-TuSC form a cone at the negative end of MT (from papers *461, 473, 478*).
MT	493	MTs can originate outside of the MTOC, for example, at the kinetochore.
Centriole cancer	494	This is a brief summary and review of a report appearing in the same issue: *Science*, June 5, 2015. It states that irregular centriole function is related to cancer.
Centriole cancer	495	Cancer cells have extra centrosomes.
Centriole cancer	495	The authors developed a chemical ("centrinone") which inhibits Plk4. This destroyed (or damaged) the centrosomes. Normal cells died. Cancer cells continued to multiply.
MT protein	496	MTs interact with actin to form the shape, migration, and division of the cell. Actin is an abundant protein. The monomeric form is called globular or G-actin; the polymeric form is filamentous or F-actin.

(*Continued*)

(*Continued*)

Category	**	Review Notes
MT	497	The paper proposes ways to measure curvature of MTs.
MT	497	EI = (1 to 200) × 10^{-24} Nm^2 (in vitro) $I = 10^{-32}\ m^4$ (I calculate $1.668 \times 10^{-32}\ m^4$) E ≈ 10^9Pa, where Pa is N/m^2.
Centrioles cancer	498	Tumor cells have excess centrosomes.
Centrioles	498	This paper is a summary of a paper by Pagan *et al.* (Nat. Cell. Biol. Vol. 17, pp. 31–43). The paper says that the centriole pair are loosely connected (i.e., "disengaged"). During mitosis (S-phase), each centriole develops a daughter centriole which is tightly connected to the mother centriole. As S-phase continues the two mother centrioles separate called "centriole disjunction" and then the tight connection between the mother and daughter centrioles loosens called "centriole disengagement."
Centriole	499	The centrosome increases dramatically in size upon mitosis.
Centriole	499	The centrosome is not a featureless cloud, but instead an organized structure.
Protein	500	p53 is a tumor suppressor.
Centriole	501	The MTOC has other functions when mitosis is not occurring.
Centrioles	502	This is the paper summarized by *498*.
MT	503	Two ways are postulated as the ways the kinetochore attaches to the mitotic spindle.
MT	504	Paper explores MT minus end regulation, but does not seem to come to firm conclusions.
MT cancer	504	Paper states the better understanding of MT dynamics is needed for new cancer treatment strategies.
MT	504	Paper states that plus end dynamics are better understood than minus end dynamics.
Cancer	505	Cancer cells use the dying normal cells as fuel for growth.
MT	506	MTs can be developed from chromosomes. The development occurs via γ-TuRC, but the exact mechanism remains unknown. MTs from the chromosomes are essential for the success of the mitotic spindle.

(*Continued*)

(*Continued*)

Category	**	Review Notes
MT	507	The primary role of the PCM is to anchor the microtubules through the γ-tubulin ring complex, γ-TuRC. The PCM is *not* an amorphous material.
Centriole cancer protein	508	Under certain simulated conditions (absence of p53 and overexpression of Plk4) supernumerary centrioles occur without resulting in a tumor.
Nano	509	"There is still no consensus on the toxicity of nanomaterials." There are numerous open questions related to type of nanoparticles, age, and ingestion potential.
Centriole	510	A brief description of cilia.
Centriole protein cancer	511	p53 is a protector against irregular cell development in the absence of Plk4. Aberrations in centrosome number can lead to cancer.
Centriole protein cancer	512	Plk4 primes the binding of STIL to SAS-6. Also, supernumerary centrosomes promote cancer.
Cancer centrioles	513	This paper suggests physically expanding a cell to get better images. (Could be used to look for centrioles.)
MT	514	Kinetochores are large protein structures built on centrosomeric DNA. Attachments between MTs and kinetochores must be stable to ensure successful mitosis.
Centrioles cancer	515	This is another Paul Schafer paper describing centriole irregularities in cancer cells.
MT protein centriole	516	γ-TuRC is a lock washer-shaped protein complex.
MT protein	517	γ-TuRC is a 13 γ-tubulin molecule — containing as many as five γ-TuSC.
MT protein	518	(2012) γ-tubulin behavior is poorly understood.
MT protein	518	γ-Tubulin provides a *template* for MT nucleation (good pictures).
MT protein	519	γ-TuRC is formed of γ-TuSC.
Protein vibrations	519	γ-TuRC provides a stabilizing effect of the MT dynamic but how this is done is unclear.
MT	520	EI for an MT ≈ 1.7×10^{-23} Nm^2.
	521	Mechanics of materials text.
	522	First review paper.

(*Continued*)

(*Continued*)

Category	**	Review Notes
EM, cancer	523	(2001) This paper presents a theoretical model of breast cancer based upon electromagnetic effects. It documents via references that there is a difference in tumor EM than in normal tissue.
EM, MT	524	Electric field near a microtubule can be 10^8 v/m or more. Force F on a dipole is: $F = \nabla(p \cdot E)$ where p is the dipole moment and E is the electric field intensity.
EM, MT	524	Microtubules are composed of subunits that are strong electric dipoles.
EM, cancer	525	Little research has been given to exploring beneficial effects of magnetic and electromagnetic fields for treatment of injury or disease.
EM	525	EM has been used for study of fractures.
EM	525	EMF therapeutics may be divided into five fields: (1) Permanent magnetics (2) Low-frequency sine waves (3) Pulsed electromagnetic fields (4) Pulsed radiofrequency fields (5) Transcranial magnetic/electric stimulation.
EM	525	The limited acceptance/use of EMF therapy is due to: (1) Physicians not familiar with the technique (2) Regulation (3) Public concern about safety.
EM, centriole	525	Quote: "Biochemical and physiological processes are based on the flow of electrical charges…"
EM, centrioles	525	There are in vitro studies suggesting significant alterations in cell division for wound healing.
EM, centriole	525	EMF are not widely used "due to the absence of a well-established mechanism for EMF bioeffects."
EM centriole cancer	525	Magnetic fields (MF) can affect cell proliferation both by increasing a slow rate (as in slow would healing) and by inhibiting too high of proliferation.
EM	526	Electromagnetic "mechanisms are not fully understood."
EM	527	This paper describes experiments attempting to measure the dipole moment on a α/β dimer. Results ~6000 Debye. (Others have suggested 1000 Debye.)
EM MT	528	Proposes ferromagnetic properties for MTs.

(*Continued*)

(Continued)

Category	**	Review Notes
EM	528	The electrical effect can cross the entire cell.
Centriole protein	529	SAS-6 imparts ninefold symmetry of the centrioles. SAS-6 forms a torus around the base of the parent centriole.
Protein cancer	529	(6) Depletion of SAS-6 results in a failure of centriole formation. Overexpression of SAS-6 results in supernumerary centrioles.
Centriole	529	(4) Use of STORM microscopy for imaging (sulfur-enhanced 3D STORM)
Centriole	529	ED STORM helps overcome the minute size of the centriole for imaging.
Cancer EM	530	"Magnetic targeting" uses an externally applied magnetic field to move a magnetically charged drug carrier to a tumor.
EM Nano	530	Paramagnetic materials magnetize weakly in the direction of the magnetic field. Diamagnetic materials magnetize in the opposite direction.
EM Nano	530	Ferro-magnetic materials tend to aggregate.
EM Nano	530	Superparamagnetic materials provide a compromise between strong magnetization and avoidance of particle aggregation.
EM Nano	530	Superparamagnetic materials have attractive features: (1) Ability to be visualized via MRI (2) Ability to be localized to a specific site (3) Ability to produce localized heating
Nano	531	Magnetic nanoparticles cluster together and become ineffective for bio-applications
Nano	531	Clusters of magnetic nanoparticles differ magnetically from the individual nanoparticles.
Nano	531	Magnetic nanoparticles have a large surface area/volume ratio and tend to agglomerate to reduce surface energy.
Nano	531	An effective approach to preventing particle agglomeration is to coat the particles with biocompatible polymers.
Centrioles	532	Centrioles are necessary for eukaryotic cell reproductions.
	533	Electric stimulation has an effect on cell behavior.

(Continued)

(Continued)

Category	**	Review Notes
Centrioles	534	Paper asserts that in the MT triplet blades only the A MT is nucleated through the γ-TuRC and that A forms a template for the B and C MTs.
EM	535	Paul Schafer personal communication. High magnetic field can disrupt centriole geometry.
	536	Dynamics book.
Centrioles	537	MT review paper on microtubules
Microscopy	538	STED: Stimulated emission depletion.
Protein centriole	539	Paper discusses SAS-6 activity.
Centrioles	540	Kinetochores pull the chromosomes apart.
MT	541	Kinetochores are not thin flat plates. Measurements have been based on single-molecule, high-resolution colocalization. Researchers believe that is not accurate enough.
Microscopy	542	Theoretical paper: Predicts better results when laser imaging and photoblocking related by the golden ratio.
Microscopy	543	Stimulated emission microscopy (STED). This is a fluorescence microscopy.
Centriole protein	544	At the proximal end of the centriole SAS-6, Plk4, and STIL localize. Plk4 forms a dot before SAS-6.
Centriole protein		Plk4 is an early marker for new centriole formation.
Centriole		"...definitive information about the spatial organization of various components remains scarce."
Microscopy		Recently super resolution microscopy techniques have become available (2012).
Microscopy		Structural illumination microscopy (SIM) Stimulated emission depletion (STED).
MT		At the beginning of mitosis MT nucleation is stimulated by recruitment and exchange of PCM components.
Centriole MT		PCM undergoes a drastic expansion at the beginning of mitosis.
Centriole MT protein		Plk4 is an early marker (the first) for new centrioles. It falls onto a ring stained by Cep152. Also Plk4 colocalizes with SAS-6.
Centriole MT		The site for MT nucleation is determined on the mother centriole.

(Continued)

(Continued)

Category	**	Review Notes
Centriole MT		It is not clear whether the nucleation site is chosen at random or is somehow predetermined.
Centrioles	545	Centrioles recruit PCM. This material forms concentric layers around the centriole.
Cancer		Extra centrosomes promote cancer and chromosomal instability.
Centrioles cancer		Extra centrosomes cluster together.
Nano imaging	546	Discusses PALM/STORM nanotechnology. "Our results demonstrate that both PALM and STORM imaging have great potential as alternatives to immune-EM."
Centrioles	547	The exact architecture of PCM is not known but it appears to have an intricate mode of assembly. The paper purports to present the higher-order PCM organization in regulating centrosome assembly.
Centrioles		Kinetochores and centrioles have no membrane cover. The centrosome is like a vapor cover over the centriole. That is, the centrosome takes the place of the membrane.
Centriole		The PCM is largely featureless in EM imaging.
Imaging	548	The rapidly developing field of fluorescence is leading to new insights into cell biology.
Centrioles	549	Kinetochore assembly is restricted to centrosomes which are chromosomal domains.
Centrioles	550	Kinetochores have more than 80 components (proteins). These proteins are organized. Minor errors in the structure can lead to aneuploidy and tumorigenesis. Major errors cause cell death.
Centriole MTs	551	"Robust spindle microtubule (MT) plus-end attachment is 'end-on', and MTs bound to kinetochores are known as kinetochore microtubules (kMTs)."
Centriole MT	552	Kinetochores have four functions: (1) Connect to centromeric chromatin (2) Connect to MTs (3) Monitor the state of the kinetochores/MT attachment (4) Correct improper attachments.
MT		"The primary function of kinetochores is to create load-bearing attachments between chromosomes and microtubules in a dividing mother cell."
Nano	553	Paper provides a listing of common NPs in use.

(Continued)

(*Continued*)

Category	**	Review Notes
Nano		Small NPs (1–20 nm): colloidal gold iron oxide crystals and quantum dots.
Nano		For cancer treatment an NP must be stable until it reaches the tumor and be able to carry drugs.
Nano		Nanoparticles (NPs) with medical application: (1) *Superparamagnetic iron oxide crystals* These have an iron core ~4–5 np in diameter surrounded by a polymer. These particles become magnetic when brought into a magnetic field. (2) *Quantum dots* Developed in 1980s. They absorb light and then emit it in a different color. Used in diagnostics. (3) *Dendrimers* These can be of various shape and carry drugs on their surface. (4) *Polymeric micelles* These are less than 50 nm in diameter. They are used to deliver water insoluble drugs. (5) *Liposomes* These are large NPs 100s–1000s nm in size. Used for drug delivery. (6) *Nanospheres* Spherical 10s–100s of nm, encapsulating drugs. (7) *Aquasomes* These are carbohydrate-ceramic NPs. They are spherical 60–300 nm in diameter and used for drug delivery. (8) *Polyplexes/lipopolyplexes* These are assemblies of nano material.
Nano cancer	554	"Iron oxide balls, just a few tens of nanometers in diameter heat up when exposed to powerful magnetic field." Neatly stacked NPs needed a stronger field than the more random.
Nano cancer	555	In cancer cells, folic acid is 10 times as great as that in normal cells. Author proposes folic acid as a biomarker.
Nano cancer	555	Dendrimers are nano-sized polymers ("starburst polymers which can be efficiently targeted for tumors").
Nano cancer	556	Author proposes a three-stage delivery of NP to tumor: (1) blood borne, (2) into large tumor vasculature, and (3) break down 100–100 nm NP and into tumor core.

(*Continued*)

(*Continued*)

Category	**	Review Notes
Nano cancer	557	NP may be designed for binding to cell membrane or to cell interior biomarkers.
Cancer		A tumor obtains nutrients "by passive diffusion until it reaches a size of ~2 mm^3." Then to continue growing, new blood vessels must develop ("angiogenesis").
Cancer		Angiogenesis is not orderly and tumors have poor vasculature in some regions and strong vasculature in others.
Cancer nano		Tumor blood vessels are abnormal and are leaky. The leak gap is ~100–780 nm (normal cell gap is 5–10 nm).
Cancer		High pressure in tumor blood vessels tends to push out chemo drugs. The pressure is higher in the tumor center and lower on the periphery.
Nano/cancer		*Bio NPs:* (1) *Dendrimers* — These are branching, tree-like structures. (2) *Micelles* — Spherical, globular structures, with a hydrophobic central core. (3) *Nanospheres* — A spherical structure composed of a matrix of curved bars. (4) *Nanocapsules* — Fluid filled pouch. (5) *Fullerenes* — Carbon "bucky ball." (6) *Nanotube* — Hollow carbon cylinder. (7) *Liposome* — Closed fluid filled sphere with a number of cover layers.
Nano cancer	558	This is a review paper on current targets for nanoparticles. These are: (1) antiogenesis, (2) uncontrolled cell proliferation, and (3) tumor cell targeting.
Nano		Nanoparticles: (1) dendrimers, (2) liposomes, (3) polymeric nanoparticles, (4) micelles, (5) protein nanoparticles, (6) ceramic nanoparticles, (7) viral nanoparticles, (8) metallic nanoparticles, and (9) carbon nanotubes.
Cancer		2 mm^3 tumor — angiogenesis. Leaky blood vessels 100 nm to 2 m holes. High internal pressure.
Nano cancer	559	Application of nanotechnology in medicine is called "nanomedicine."
Nano cancer		Without targeting nanoparticle half-life is short and often with side effects.

(*Continued*)

(*Continued*)

Category	**	Review Notes
Nano		Paper lists five nanoparticles: (1) liposomes, (2) polymer drugs, (3) polymeric, (4) dendrites, and (5) iron oxide.
Nano/cancer	560	Folic acid is a target for SPIONs.
Nano cancer		Application: intravenous and local injection to the tumor site
Nano		SPIONs have great potential but more research needed for efficacy.
Nano cancer	561	Superparamagnetic nanoparticles were surface-modified with poly (ethylene glycol) (PEG) and folic acid, respectively (quote from abstract). The PEG nanoparticles appeared to perform better (i.e., more uptake into the cancer cells) than the folic acid particles, although for breast cancer both performed well.
Nano cancer	562	Magnetic NPs can be maneuvered through the body by external magnetic fields. But if the NPs are too deep, the procedure becomes less effective.
Nano cancer	563	Authors injected superparametic iron oxide nanoparticles into rate glioma tumors and then heated them, shrinking the tumor.
Nanocancer	564	Discusses potential treatment of triple-negative breast cancer with nanoparticles: ECO/siRNA.
Nano cancer	565	All tumors have molecular markers that do not occur in normal tissue.
Nano cancer		This is a review paper focusing on molecular marker created by tumor vasculature.
Cancer		Drugs only penetrate a short distance into the tumor. Tumor blood vessels have large diameters and high pressure causing blood outflow. Tumor blood vessels are tortuous, uneven in diameter, and leaky.
Nano cancer	566	Use of a tumor homing peptide that goes not only into the tumor extravasculature but also penetrates into the tumor itself. The peptide is "iRGD."
Nano cancer	567	This is a review article on nanoparticles (2007). The leaky vasculature allows nanoparticles to escape from the flood stream into the tumor.
Nano		Desired properties of nanoparticles: (1) protect drug from degradation (2) prevent drugs from prematurely interacting with nearby tissue (3) enhance drug absorption into target tissue

(*Continued*)

(Continued)

Category	**	Review Notes
		(4) improve intercellular penetration of the drug (5) exhibit high differential uptake into target cells compared with normal cells (6) be made of biocompatible material (7) have extended circulating half-life.
Nano		*Nanoparticles*: (1) liposomes, (2) biodegradable polymer, (3) chemo-therapeutic, (4) surface functionality, (5) space group, (6) targeting molecule, (7) inorganic particle metallic shell, (8) amphipatnic molecule, (9) dendrimer, and (10) carbon nanotube.
Nano cancer	568	Paper reports on tests of two NPs: nanospheres (NS) and nanoworms (NW). The NPs used superparamagnetic iron oxide. The NW had better results than the NS.
EM		Authors used 3D SQUID magnetometry to measure the electromagnetism.
Nano	569	The objective of the authors is to develop non-toxic nanoparticles. They offer luminescent porous silicon nanoparticles (LPSiNPs) which self-destruct in a mouse model (renally). Dextran coated LPSiNPs (D-LPSiNPs) are good for tumor imaging.
Nano cancer	570	Nanoparticles can be targeted to extravascular tumor tissue.
Cancer	571	As a tumor grows, its size, pushing against other organs creates pressure within the tumor and thus contributes to metastasis.
Nano cancer	572	This is a review article. Gold nanoparticles (AuNPs) are discussed for diagnostic imaging. The authors also provide a listing of delivery systems.
Nano cancer		AuNPs are relatively easy to manufacture.
Nano cancer		Delivery: (1) particle size reduction, (2) peptide platform, (3) small molecule, and (4) proteins.
Nano	573	This is a review paper on gold nanoparticles. It lists these: nanospheres, nanorods, nanoshells, and nanocages.
Cancer	574	This paper lists primarily blood tests used to indicate the presence of tumors. It is based on a search for biomarkers.
Nano cancer	575	This paper presents folic acid as a biomarker for nanoparticles.

(Continued)

(*Continued*)

Category	**	Review Notes
Nano	576	The paper presents a computational model for the adhesion of nanoparticles to cell membranes.
Nano	577	Artery endothelial cells are 13 α 4.1 m wide and 25.8 α8.5 m long.
Nano cancer	578	This is a review paper summarizing the targeting of nanoparticles to specific metastases.
Nano	579	The authors discuss the use of poly (amidoamine) (PAMAM) dendrimer as a nanoparticle. It is non-toxic.
Nano cancer	580	The authors use PAMAM [see 479] dendritic (tree-line nanoparticles) to deliver toxins to tumors by focusing on folic acid as a targeting agent.
Cancer		The folate receptor is over-expressed in breast, ovary, endometrium, kidney, lung, head and neck, brain, and myeloid cancers.
Nano cancer	581	The paper describes the use of silica nanoparticles with superparametic iron oxide core and attachments. Folic acid is the tumor biomarker.
EM	582	Cells were observed to spin at differing rates depending upon an applied ac field frequency.
EM	583	[See 582].
MT	584	This is a review paper on how MTs and kinetochore link together. MTs in dynamic instability undergo cycles of polymerization and depolymerization controlled by GTP-processes GTP ("rescue" and "catastrophe").
MT		GTP must be bound to β-tubulin for individual dimers to be added.
MT		There are different views of how MTs and kinetochore connect to each other: one is via a ring from the kinetochore encircling the MT; the other is to have MT-like strings immersing with the catastrophe dimers of an MT.
Cancer nano	585	The objective of the paper is to report on NP targeting of blood vessels feeding the tumor. It states that integrin αvβ3 is found in angiogenesis. αvβ3 is thus a biomarker.
Nano	585	The authors use RGD as their NP. The chemo is doxorubicin.
Cancer	586	There is a significantly higher number of centrioles in malignant tumors compared with the number in benign tumors and normal tissue. There is no significant difference in the number of centrioles between benign tumors and normal tissue.

(*Continued*)

(Continued)

Category	**	Review Notes
Centrioles	587	This paper explores how Asterless (Asl) and Plk4 interact. Overexpression of Asl recruits Plk4, but then regulates Plk4 for stability, in new centriole development.
RLH Paper	588	RLH paper.
Biology book	589	Biology book.
Centriole	590	Asterless (As ℓ) is an ortholog of Cep 152. As ℓ is the Drosophila melanogaster orthologue of Cep 152.
MT Cancer	591 (101)	Current MT drugs reduce the MT mass. But the drugs also work at lower concentrations.
EM	592	To me this is disturbing. It is disturbing because it speaks of measuring small magnetic moments with a general purpose superconducting quantum interference device (SQUID) magnetometer. It discusses counting SPIONS (copyrighted as "Resovist") in and about normal and cancer cells. It says nothing (that I have seen) about the magnetism of the cells themselves.
NP Cancer	593	(see above) It talks about ... "magnetic nanoparticles biologically reacting with living cells."
MT	594	MTs can withstand higher compressive loads in a cell than in vitro due to surrounding lateral reinforcement.
MT Kinetochores	595	(2011) The authors present two models for microtubule-kinetochore coupling: wave and biased diffusion.
EM	596	Authors claim that magnetic forces can affect the positions of the mitotic spindle.
Nano/EM	597	The paper discusses the development of magnetic nanoparticles (2007).
MT	598	Authors present models for the kinetochore/microtubule connection. The focus on the activity of Ndc80 (not defined) in making the coupling.
MT	599	Provides some evidence and a theoretical basis for ring attachment between kinetochores and microtubules.
MT	600	Not clear how microtubules and kinetochores couple. Dam1 is from the kinetochores and Ndc80 is from budding yeast. The study shows that Dam1 and Ndc80 are compatible.

(Continued)

(Continued)

Category	**	Review Notes
MT	601	"Kinetochores are macromolecular complexes built on a specialized chromatin domain called the centromere. Kinetochores provide a site of attachment for spindle microtubules during mitosis." (from the abstract) Similar spelled words: centromere — kinetochores; chromosome — nucleosis (DNA); centrosome — centriole surrounding.
MT Cancer	602	The paper focuses on mitotic centromere-associated kinesin (MCAK). Restriction of MCAK can present cancer cells from forming and mutating. MCAK is found throughout the cell but especially at the centromere and centrosome. MCAK is positively charged.
MT	603	Authors present a model showing the microtubule rotate the nucleus.
MT	604	The author discusses mitosis when the nucleus is not centered.
MT	605	Author states the pericentrin provides the anchor for the γ-tubulin ring complex, γ -TuRC.
	606	(1995) γ -tubulin is required for microtubulin nucleation in vivo. γ -Tubulin is a complex form of Xenopus with at least seven different proteins. It has an open ring structure.
MT	607	Kinetochores appear to be more complex than microtubules. Kinetochores have more than 80 compounds. The connection however is not clear.
MT	608	Paper requested by Roger Adelman. The essence of the paper suggests that microtubules are frequently shrinking and elongating, but rarely both at the same time.
MT	609	The focus of the paper is on the role of MCAK in coupling with the microtubule ends. MCAK is a kinesin 13 (a catalysis). The acronym is *mitotic-centromere-associated-kinesin*. MCAK is a regulator with a positive charge. The positive charges in the mech of MCAK depolymerizes microtubules.
MT	609	The mechanism of MCAK is not known. Several models are proposed: (a quote) (1) A loose tether that allows the molecule to diffuse along the microtubule lattice. (2) An obstructive element that prevents tight binding to the lattice.

(Continued)

(Continued)

Category	**	Review Notes
		(3) A "pry-bar" that destabilizes lateral interactions at the microtubule end. (4) A facilitator of cross-linking between protofilament peels at the depolymerizing ends of microtubules.
Centrioles Protein	610	This paper focuses on the protein GCP6 and its role in centriole construction. – GCP6 is needed by Plk4 (SAK) but it is not clear how. – GCP6 is part of the γ-TuRC. – GCP6 is localized to the perimeter of the centrosome — at a distance from the centrioles.
Centriole Protein	611	This paper focuses on STIL and CPAP for procentriole formation. STIL makes an asymmetric location on the daughter cell — required for new centriole formation. "Overexpression of STIL induces a formation of multiple procentrioles around the parent centriole." Conversely, too little STIL prevents normal centriole development. CPAP induces centriole elongation. Quote: "...our results indicate that the two MCPH-associated proteins STIL and CPAP interact with each other and are required for procentriole formation, implying a central role of centriole biogenesis in MCPH." MCPH is "microcephaly."
Centriole	612	Protein: Pericentrin B is necessary and sufficient for disengagement during mitosis.
Centriole	613	Centrioles cannot duplicate without PCM. "They acquire PCM by becoming modified with a Plk1 dependence early in mitosis."
Centriole	614	Authors show that Asterless (As ℓ) is required for centriole length control but not for PCM recruitment.
NP	615	Authors state that NPs are too large to penetrate tumors and suggest using vascular outbursts as a way to get in (2016).
NP	616	Kinetochores form a bridge between the chromosomes and the microtubules. The paper implies that *all* kinetochores need to be attached.

(Continued)

(*Continued*)

Category	**	Review Notes
NP	616	The kinetochore is a collection of 60 proteins. The core component of the kinetochore is the Ndc80 complex.
MT	617	(5)This is a companion paper to paper 608.
MT	618	— The NDc80 has four subunits.
MT	619	Proposes a ring model for kinetochore connection to a microtubule (2006).
MT	620	The authors use two-color fluorescence microscopy to image the proteins for budding yeast kinetochore-microtubule connection. The claim is that budding yeast connection is the same as in eukaryotic cells.
MT	621	(2005) Paper says exact structure of kinetochore is unknown.
	621	Very good picture model of MT kinetochore connection.
MT	621	Paper says that MT ends begin to explore cytoplasm after nucleus softening.
MT	622	Kinetochore components comprise the direct interface with spindle microtubules.
MT	622	Kinetochores initially attach to the surface of a single microtubule. The kinetochore is then "transported" poleward.
MT	622	Ndc80 is the important constituent that interacts with the MT.
MT	623	Paper discusses interaction of proteins to speed the kinetochore-microtubule connection.
Centriole MT	624	Paper argues that the γ-tubulin ring complex (g -TuRC) varies between different species.
Nano	625	Visualizes NPs put into the blood stream (like chemo).
Nano	625	NPs may have globular, plate-like, rod-like, or complex geometries.
EM	625	Electron "spin" is due to a magnetic moment pointing "up" or "down."
EM	625	Types of magnetism: (1) *Paramagnetism* The magnetic moments tend to be randomly oriented when there is no overall magnetic field. In an applied magnetic field these moments align parallel to the field. Then the magnetism of the material is proportional to the applied field. (2) *Ferromagnetism*

(*Continued*)

(Continued)

Category	**	Review Notes
		(3) *Antiferro-magnetism* Magnetic moments align opposite to each other. Thus there is no overall magnetism. (4) *Farrimagnetism* The aligned magnetic moments are not of the same size — thus reduced overall magnetism. (5) *Superparamagnetism* The material is composed of small ferromagnetic clusters — so small that they can easily change magnetic direction. Therefore, the material has no overall magnetic direction. But in an applied magnetic field, the material becomes magnetized and aligned with the applied field.
Nano EM	625	When SPIONS are placed in a strong magnetic field, their magnetism far exceeds that of biomaterials (i.e., cells).
Nano	625	SPIONS: 50 nm with a 20-nM core.
Nano Cancer	625	In an alternating magnetic field, the SPION changes its magnetic field direction. Therefore if a SPION can get inside a cell, an externally applied alternating field can cause the SPION to heat the cell to death.
Nano	625	SPION effectiveness could be increased by reducing its size.
MT	626	Authors discuss the important effects of microtubules on 3D cell functions.
MT	627	Microtubules can obtain rescue by incorporating α, β – dimers from the centrosome.
Centrioles MT	628	The protein CEP 295 interacts with the microtubules to elongate them. Too much $\rightarrow$ long centrioles; too little $\rightarrow$ short centrioles.
MT	629	Paper summarizes results of Moriwaki and Goshima, "Five factors can reconstitute all three phases of microtubule polymerizative dynamics," *The Journal of Cell Biology* (2016).
Cancer	630	Cancer cells kill endothelial cells of blood vessels causing a gap for the cancer cells to get out of the blood vessel and build a tumor in the adjacent tissue.
MT	631	This paper is a reference in paper 627.
MT	631	Damaged microtubules are able to incorporate new tubulin dimers and thereby recover their stiffness.
MT	631	The hollow cross-section of microtubules gives them stiffness.

(Continued)

(*Continued*)

Category	**	Review Notes
MT	631	Microtubules soften under mechanical stress, but softened microtubules can self-repair. They have self-heating properties.
MT	631	Figure 6 provides a good graphic model of microtubule damage ("catastrophe") and self-repair ("rescue").
Cancer	632	Blood vessels supporting tumors or tortuous, weak, and leaky. Tumors, however, are solid, and thus can easily punch through the weak vasculature and break into it.
Centrioles Cancer	633	Enlarged centrosome (i.e., supernumerary centrioles) leads to cancer.
MT	634	Colliding microtubules can pressure their lengths.
NP	635	NPs 10 nm or less are not captured by tissue, but NPs 10–40 nm can cause inflammation, capturing and neutrophil extracellular traps.
NP	636	This is a review paper on nanoparticles use up to 2012 for cancer treatment.
NP	636	These are the following kinds of nanoparticle considered: (1) liposomes, (2) nanoparticle albumin-bound (NAB) technology, (3) polymeric nanoparticles, (4) dendrimers, (5) metal nanoparticles, and (6) molecular targeted nanoparticles.
NP	636	*Liposomes* — Developed in 1965, among the first developed. Now (2012) at least 11 varied formulations. They are biocompatible and biodegradable. They can also be coated with polymers, for example, PEG. *NAB* — Uses albumin as a therapeutic carrier. That is a nanoparticle composed of albumin with hydrophobic interior. *Polymeric nanoparticles* — Made of biocompatible and biodegradable polymers. They form a shell for therapeutic compounds. *Dendrimers* — Macromolecules made of natural elements such as amino acids, sugars, and nucleotides. They are "dendrimer-drug" composites. *Metal nanoparticles* — Made of inert metals such as gold or titanium and used for controlled release of therapeutic drugs. But, although biocompatible, they remain in the body and can accumulate to a toxic level. *Molecular targeted nanoparticles* — Best hope to date. It tries to use tumor biomarkers.

(*Continued*)

(*Continued*)

Category	**	Review Notes
NP	636	Provides a table of various nanoparticles.
Cancer	637	Due to enhanced vascular about tumor proteins accumulate around the tumor creating a possible biomarker.
NP	638	Nanoparticles can aid in imaging for radiation therapy.
NP	639	A historic article.
NP	640	P-selectin is a biomarker (found in tumor blood vessels). Authors suggest using fucoidan (tucoidan?), a polysaccharide that binds to P-selectin to deliver therapeutic drugs to tumors.
MT	641	Authors suggest that organelles move within a cell by attaching themselves to microtubules and other recruited molecular motors.
MT	642	Authors use computer simulation to study microtubule buckling behavior.
MT	643	Authors state that microtubule rescue, after catastrophe, can occur by α,β-dimers filling the gaps left by the catastrophe.
Cancer	644	Authors provide protein biomarkers for common cancers.
Cancer	645	Centrosome amplification is a hallmark of cancer. Researchers believe that centrosome defects cause cancer.
Cancer	645	Centrosome defects could be biomarkers for advanced disease. The defects could be in the centriole.
Cancer	645	Centrosome amplification could be due to supernumerary centrioles.
Cancer	645	Centrosome clustering is prevalent in cancer.
Cancer	645	Tumors are heterogeneous having cells with normal and abnormal centrosomes.
Centrioles	646	The paper is focused on centrosome asymmetry in cells that divide by "self-renewing asymmetric division."
MT	647	The author states that microtubule nucleation can occur at either the centrioles or at the golgi apparatus.
Centriole	648	Defects in centriole number are associated with cancer.
Centriole	648	This is a review paper on how centrioles are assembled and controlled.

(*Continued*)

(*Continued*)

Category	**	Review Notes
Centriole	648	Centriole proteins at the site of assembly: Plk4, SASS-6, and STIL. Plk4 localizes at a dog and then recruits SASS-6. It is not clear what targets Plk4 to the site on the mother centriole. It is an asymmetric event.
Centriole	648	The concentration of Plk4 is critical to limiting duplication to one site.
Centriole	648	Once the Plk4 is established a cartwheel with ninefold symmetry based on SASS-6 is assembled. At the base of the ninefold is a stalk which keeps the centrioles connected.
Centriole	648	γ-Tubulin forms on the ends of the cartwheel to form microtubules. There is a triplet of microtubules A, B, and C (A near center and longest).
Centriole protein	648	The kinetics of the protein reactions and interactions appear to be unknown.
Centriole	648	Separase is required for disengagement.
Centriole	648	Authors present a model for centriole development. One of the steps is that Plk4 is limiting.
EM, NP	649	SQUID is useful for detecting cancer cells. It employs a magnetic relaxometry method.
EM, NP	649	The magnetic moments of cancer cells are linear with the number of cells.
EM, NP	649	In magnetics relaxometry, magnetic nanoparticles are briefly magnetized by small magnetic field pulse. Then the decaying magnetic fields is measured. Application with biosystems need small particles ~10 nm.
EM, NP	649	In magnetic relaxometry magnetic nanoparticles (superparamagnetic nanoparticles) are briefly magnetized by a small magnetic pulse. Then the decaying magnetic fields are detected and measured. With biomaterials the particles not bound to the tissue fall away and decay within microseconds. The remainder are the bound particles which can then image the biotissue.
NP, EM	650	In this paper, the authors use "antibody-[conjugated magnetic nanoparticles" which adhere to the breast cancer cells. They then use SQUID sensors to detect the cancer cells. SQUID is based on "magnetic relaxometry."

(*Continued*)

(Continued)

Category	**	Review Notes
Centriole	651	The initial stage of centriole assembly is the continued assembly — nine spokes. The central part is SAS6-Bld12p. The distal part is Cep135/Bld10p.
Centriole	652	Centrioles are among the largest protein-based structures found in cell types. They are ~ 250 nm in diameter and 500 nm long.
Centriole	652	Paper is a review of ultrastructural observations about centrioles and associated structures. The paper largely repeats what is already known about centrioles from earlier papers.
Centrioles	653	Centrioles are responsible for the assembly of microtubules.
Centriole	653	Centrosome is sometimes called the MTOC.
Centriole	653	"Diplosome" — a centriole and a developing centriole.
Centriole	653	"Procentriole" — a developing centriole.
Centriole	653	"Basal body" — mother centriole.
Centriole	653	Centrosome, MTOC, and "PCM" are all the same — "electron dense material."
Centriole	653	In the microtubule blades: ABC they are close to each other and thus provides stability to one another. This helps to establish the stability of the microtubules.
Centriole	653	The centriole pair is anchored to the nucleus.
Centriole	654	Paper reviews various theories as to why only one procentriole occurs with a mother centriole. The author states that the answer is still unknown.
Centrioles	655	PCM concentrates tubulin and serves as the primary organizing center for microtubules. The PCM has no membrane cover. Instead the proteins contribute the separation.
Centrioles	655	PCM appears as an amorphous mass surrounding the centrioles.
Centrioles	655	PCM proteins can self-assemble
Centrioles	655	PCM forms toroids about the centrioles
Centriole	656	The authors present a review and discussion as how centriole pairs separate. No mention of electromagnetic forces.
Centriole	657	Identifies various diseases, aside from tumorigenesis, that can occur with centrosome disfunction.

(Continued)

(*Continued*)

Category	**	Review Notes
Centriole cancer	658	"extra centrosomes were sufficient to initiate tumorigenesis in mammals." This quote is based upon findings in the article: M.S. Levine, *et al.*, "Centrosome amplification is sufficient to promote spontaneous tumorigenesis in mammals," *Developmental Cell*, 2016, 12, 022, 2017.
Centriole cancer	659	Supernumerary centrioles promote tumorigenesis.
Centriole cancer	659	Centrosome amplification results in aneuploidy.
Centrioles protein	659	Overexpression of Plk4 promotes centrosome amplification.
Nano	660	Naked eye can see hair; a magnifying glass can enlarge it ~10X; light microscopy can enlarge the hair 500X. Fluorescense "nanoscopy" can image objects 20–30 nm in diameter.
	661	
EM	662	Author proposes that the centrosome electrically and magnetically govern cellular mechanics. A figure illustrates the fields.
EM	663	Authors present a theoretical model showing that electric fields governing cellular mechanics. The field is generated by microtubules.
EM	664	Authors propose that oscillation of chromatin produces an electromagnetic field.
Centriole	665	Authors state that kendrin regulates separase in centriole disengagement
NP	666	Nanoparticles have been used for 1800 years in aerosol sprays.
NP	666	Suitable metals for nanoparticle production: gold, silver, titania, and zirconia. Iron is excluded since it is incompatible with cells. Aside from metals, peptides, proteins, carbohydrates, and lipid moieties are suitable for nanoparticle materials (peptides are components of protein). Also suitable: polymers.
NP	666	Manufacture by supersonic beams is gaining attention.
NP	666	Supersonic Chister Bean Deposits (SCBD).
NP	666	A supersonic beam colliding with another beam produces frozen nanoparticles.
NP	666	Author presents a table of nanostructured beam suitable for biomedicine.

(*Continued*)

(Continued)

Category	**	Review Notes
Centriole	667	Centrioles can be elongated by chemicals.
NP	668	Magnetic nanoparticles have been found in teeth and brain matter.
NP	668	Characterization techniques for pristine magnetic NPs — SQUID.
	668	The focus of the paper is to discuss ways of determining the effects of nanoparticles on biological materials.
	669	This is a lengthy paper describing the various characteristics of cancer cells. They list six characteristics: (1) Sustaining proliferative signaling (2) Evading growth suppressors (3) Resisting cell death (4) Enabling replicative immortality (5) Inducing angiogenesis (6) Activating invasion and metastasis.
	670	Cancer cells also recruit normal cells to acquire these six characteristics. The authors discuss in detail the six characteristics listed in 669 (above).
	671	Procedure: Uses bacteria with a monoscale layer of nitrogen vacancy (NV) centered on the surface of a diamond chip. The bacteria are placed on the surface of the diamond. Then the quantum spin rates are viewed via chains of magnetic nanoparticles ("magnetosomes").
	672	Extra centrosomes are found in cancer cells.
	672	Most cancers are of epithelial origin.
	672	Sak overexpression leads to centrosome amplification (Sak is Plk4).
	672	Uses Drosophlia wings in their studies.
	673	Quantum diamond microscope: "This instrument uses nitrogen-vacancy (NV) centers in diamond for correlated magnetic and fluorescence imaging."
	673	"The quantum diamond microscope employs a dense layer of fluorescent quantum sensors, based on nitrogen-vacancy (NV) color centers, near the surface of a diamond chip on which the sample of interest is placed."
	674	Nanotechnology can have toxicity but can also be useful in medical applications.

(Continued)

(Continued)

Category	**	Review Notes
	675	[See also 550 and 621] Kinetochores stochastically encounter microtubules and also possess quality control eliminating errors.
Centrioles cancer	676	Full article not found. The presence of extra centrosomes imposes the tendency for catastrophic multipolar division.
Cancer EM	677	Many efforts given to find cancer biomarkers. Paper provides little hope for prominent biomarker (my assessment).
NP	678	NP treatment is spurred on by limits of conventional cancer treatments.
NP Protein	678	There is incomplete understanding of the interaction of NPs and cell proteins.
NP	678	Number of nanomedicine articles appearing is roughly doubling every year.
EM	678	Finding biomarkers is a major goal.
Cancer NP	678	Most nanomedicine is delivered via blood vessels.
NP Cancer	678	EPR effect (tendency of cells to allow permutation and retention of NPs) is a major advantage of nanomedicine. It is the foundation of nanomedicine. More studies are needed to identify tumors with EPR characteristics.
EM	678	Few studies have been made on finding biomarkers. Surgiogenesis may be a good candidate.
NP	678	Gold, silica, and polystyrenes are good protein/NP interactors.
Cancer	678	Normal vasculature evolves ~1000 days. Tumor vasculature evolves ~10 days.
Cancer	678	Tumor vasculature is poorly understood.
Cancer	678	Tumor vasculature is leaky.
NP Cancer	678	Nanomedicine is likely to develop its own toxic chemicals.
Centriole	679	This is a preview article for the article by Balzarotti, *et al.* and other articles [See 680].
Centriole	679	The article states that with the naked eye the smallest object we can see is approximately the size of the human hair. With a magnifying glass we can see 100 times smaller; with light microscopy we can see ~300 times smaller (~250 nm). With fluorescence nanoscopy we can see structures 20–30 nm across. However, most proteins are less than 5 nm across. MINFLUX.

(Continued)

(Continued)

Category	**	Review Notes
Centriole	680	The authors introduce "MINFLUX" which localized photon emitters in space. It minimizes the fluorescence photons needed for precision resolution.
Centriole	680	MINFLUX attains ~ 1 nm precision.
Cancer	681	From "Highlights": (1) CD276/B7-H3 is broadly overexpressed in both cancer cells and tumor vasculature. (2) Both angiogenic and non-angiogenic tumor vasculature express CD276. (3) Anti-CD276-drug conjugates display potent anti-tumor and anti-metastatic activities. (4) Pyrrolobenzodiazepine dimers are ideally suited for targeting tumor vasculature.
Cancer	682	This is a preview paper with much speculation as to which chemotherapy might be more effective in causing cancer cells to die.
NP	683	Polylactic acid in micro and nanoparticle sizes is biodegradable, biocompatible, and non-toxic. Their disadvantages are: (1) low drug loading capacity, (2) low encapsulation efficiency, and (3) terminal sterilization.
Cancer	684	Tumors with extensive aneuploidy (high number of chromosomes) have fewer markers for immune cells to attack
EM	685	In vitro microtubules align themselves along the static magnetic field (SMF). However, in living cells the effects of SMF > 20T have not been reported. The authors have shown that for human cells the spindle orients in 27T SMF.
EM	685	The results show that the magnetic torque acts on both microtubules and chromosomes. The preferred direction depends more on the orientation of the chromosomes than on the microtubules.
EM	685	Paper focuses on effects of magnetism on microtubules and chromosomes rather than on measuring the magnetic fields.
Cancer	686	Authors provide model.
Cancer	687	The focus of the paper is that cancer originates in defective stem cells.
Cancer	687	Stem cells are unique in that they duplicate asymmetrically producing two unequal daughters — one to produce ordinary tissue cells, and the other, a stem cell itself.

(Continued)

(*Continued*)

Category	**	Review Notes
EM	688	Discusses forces but ignores EM forces.
Protein	689	Paper describes for curves on cell membrane.
EM	690	Book chapter of a general discussion of magnetic induced cell growth (e.g., bone healing).
NP	691	"Tumor cells can be weakened or destroyed without affecting normal tissue but only under temperatures between 107 and 113 degrees Fahrenheit."
Centriole	692	The paper presents a modeling of cell membrane response to induction by an AFM probe.
Cancer	693	Authors apply > 350 MPa pressure on bone and other tumor cells. The cell membranes are "ruffled."
Cancer	694	No new useful information.
Cancer	695	Paper describes the development of tumor blood vessels.
Cancer	696	Blank
Cancer	697	Mechanisms of microtubules nucleation is still not well understood.
Microscopy	698	This is a review paper on the state of the art of fluorescence microscopy — aka "super resolution microscopy."
Microscopy	698	Resolution of super resolution microscopy is typically 20–50 mm and even down to 1 mm.
Microscopy	698	Dynamic nanoscopy is an undeveloped area.
MT	699	Kinetochores enable the process of connecting chromosomal DNA to microtubules. Cancer cells exhibit defects in this process.
MT	700	"Magnetic force microscopy (MFM) is an atomic force microscopy (AFM) based technique in which an AFM tip with a magnetic coating is used to probe local magnetic fields with the typical AFM spatial resolution."
MT	701	The authors relate centriole elongation to tumorigenesis. The causes are not well understood. Centriole elongation promotes deregulation of events.
Nano	702	The paper reviews the use of nanoparticles with photon radiation to enhance the targeted tissue damage. It may be a good paper for studying nano treatment.

(*Continued*)

(*Continued*)

Category	**	Review Notes
Cancer	703	"Electroporation-based treatments and other therapies that permeabilize the plasma membranes have been shown to be more devastating to malignant cells than to normal cells."
Cancer	703	"Electroporation is increasingly being used in cancer treatment strategies. It is a method where application of short, high-voltage pulses induces transient permeabilization of the cell membrane and thus allows the passage of ions and molecules into and out of the cell."
Cancer	703	The authors' experiments show that dye passage through a malignant cell is more rapid than dye passage through a normal cell.
Cancer	704	The authors discuss the characteristics, interrelationship, and mechanisms in Ca^{2+} regulation of cell death (from the abstract).
Cancer	705	Electroporation is a method for increasing the permeability of cell membrane. "This study aims at understanding the relationship between applied electric field, calcium concentration, ATP depletion and efficacy" (from abstract).
Cancer	706	"The electroporation (EP) supports various types of anticancer therapies by the selective transport of cytostatics. Increase intracellular calcium level by EP may be a new approach to fibrosarcoma treatment. Calcium is one of the most important factors of cell proliferation, differentiation and cell death (apoptosis or necrosis)..." — from the abstract.
EM	707	The authors focus on traction forces within the cell, insinuating that they are analogous to muscle forces — but not clear how the forces emerge. *No* mention of electromagnetic forces.
Cancer protein	708	Authors summarize the role of p53 in tumor activation and suppression.
Centriole	709	Authors state that the vibratory process needed to drive cell division forward also performs other functions.
Cancer	710	Skin cancer cells reject mutant cells. But due to the plasticity of skin cells, the mutant cells can slide between healthy cells.

(*Continued*)

(Continued)

Category	**	Review Notes
Microtubule	711	"The forces that drive chromosome segregation in mitosis in human cells remain poorly understood" — from the abstract. Fiber crosslinking is thought to be a principal factor.
Microtubules	712	"The mitosis spindle functions to ensure faithful segregation of chromosomes during cell division." — from the abstract. This is a review paper on the imaging of mitotic spindles based primarily on advances in 3D electron microscopy.
Microtubules	713	Centrosomes are thought to be stable. This is a review paper which briefly summarizes centrosome assembly.
Centrioles Microtubules Cancer Protein	714	Centrosome aberrations have long been implicated in cancer. The process is not yet fully understood. This is a short review of the gene products involved.
Cell	714	The pericentriolar protein matrix (PCMI) is not an amorphous body but instead a structured system, producing centrioles.
Microtubules	715	The paper title is discussed but no apparent mention of electromagnetic forces.
Mitosis microtubules	716	The author lists nine questions related to mitosis which are not yet fully understood: (1) What proteins participate? (2) What specifies the site of the furrow? (3) How does the contractile assemble? (4) How is the ring attached to the plasma membrane? (5) What triggers constriction of the contractile ring? (6) How do rings produce force to form the furrow? (7) How does the ring disassemble as it contracts? (8) How does the cell membrane expand in the furrow? (9) How do daughter cells separate?
Cell	717	Paper discusses cell cycle regulation. Quote: "Ubiquitylation is a cardinal cellular modification and is long known for driving cell cycle transition."
EM	718	"...static magnetic fields can alter the geometry in early cell cleavages..."
EM	719	This is a preliminary paper out of Ohio State. The claim is that movement of breast cancer cells is slowed by an electric field — smaller than that had previously been thought. Underlying mechanics are not well-understood.

(Continued)

(*Continued*)

Category	**	Review Notes
NP	720	Gold nanoparticles can bind with biological protein.
Centriole cancer	721	Title problem is discussed. Long centrioles can lead to centriole fragmentation and tumorigenesis.
	722	Manganese in spores (clusters) produce a magnetic effect.
	723	Authors state that magnetic effects on a cell indicate the presence of either iron or manganese.
	724	Authors discuss a method for detecting hemoglobin changes in red blood cells.
	725	The paper discusses a method for separating cells due to their magnetic properties.
	726	Children's Hospital Biology paper
	727	Children's Hospital Biology paper
	728	Children's Hospital Biology paper
	729	Brian Thompson's "Healer's paper"
	730	This is a biology methodology paper. It presents a detailed description of the imaging methodology for the minus end of the microtubules. Minus ends can attach to each other.
	731	As with 730, this is a biology methodology paper. Kinetochores are made up of 100 proteins.
	732	As with 730 and 731, this is a biology methodology chapter. FLIM — fluorescence lifetime imaging FRET — Forster resonance energy transfer.
	733	Kinetochores are the open-ended round structures which expand in diameter during mitosis to capture the ends of the microtubules. If they fail to make an attachment, problems can occur leading to errors.
	734	The principal contribution is the assertion that the tubulin polymerase XMAP215 is "required for efficient nucleation in cooperation with γ-tubulin."
	735	As with the other papers in The Mitotic Spindle book, this paper is a biology paper describing measurements. Here the author is measuring the effects of microtubule-associated proteins on microtubule dynamics (in vitro). It is stated that growth and shrinkage of microtubules is essential for the development of the mitotic spindle.

(*Continued*)

(Continued)

Category	**	Review Notes
	736	This is a mini review presenting strategies for nano-treatment of nano-invasive prostate cancer treatment relative advantages and disadvantages of various treatment methods. MicroRNAs — a clan of small non-coding RNA that control biological processes — can aid in cancer diagnosis. There are different kinds of MicroRNAs. The paper provides a table of MicroRNAs.
Nano	737	The main contribution appears to be simply that iron is a good potential substance for cancer treatment.
MT	738	It now is known that kinetochores transmit chromosomal DNA to microtubules.
Cancer	739	Tumors need new blood vessels to grow. MicroRNAs are candidates for biomarkers. Paper summarizes various biomarkers.
MT	740	Paper states what might be expected: That is, the microtubule growth rate scales with the cell size.
MT	741	There are hundreds of proteins in the centrosome. These proteins have other functions besides microtubule development.
Cancer	742	Paper on people who are healers. Recommended by Brian Thompson.
Cancer Centrioles	743	Centrosome abnormalities are a hallmark of human cancers. Centrosome amplification is a pre-malignant condition. Centrosome amplification also correlates well with the suppression of P53.
		Centrosome amplification can lead to aberrant mitotic spindle.
MT	744	The authors try to identify the proteins essential for the centrosome, the MTOC. The microtubule originates from the γ-TuRC. Approximately 13 molecules of γ-tubulin form a helix which becomes the template for microtubule assembly.
Centriole	745	This is a review of various features of kinetochores. Their principal mechanical function is to connect to the microtubules. Incorrect attachments can reconnect.
Cancer	746	Iron may be a form of chemotherapy.
Centriole	747	This is a description and general review focusing on proteins. Centrioles form the centrosome.

(Continued)

(Continued)

Category	**	Review Notes
MT	748	Kinetochores are formed by lateral microtubules which assemble quickly into end-on shape.
MT	749	The αβ-tubulin constructs are different outside the microtubule within. "If we understood the biochemistry of the αβ-tubulin conformational cycle we could predict the end structure. Conversely if we knew the end structure we could infer the biochemistry. The end structure and the biochemistry are both unknown.
NP	750	Ferrite nanoparticles were created with carbon hydrothermally. They are spherical with diameter 90 nm. The carbon strongly cooled the iron.
EM	751	Cell shortening is a result of negative-end effects.
MT	752	There are microtubule enzymes that cause internal breaks in the MT. This can lead to disorders. The mechanism is not well understood.
MT	753	The breaking up of microtubules is a poorly understood phenomenon.
MT	754	The kinetochore expands to capture a microtubule.
MT	755	In early cell development in Drosophila wing epithelium development, each cell has its own microtubule that can bear compressive forces.
MT	756	Chromosome periphery is a complex network of proteins. Mitotic chromosomes have five major regions, one of which is the kinetochore.
Centrioles	757	Dynein, Mud, and Asp control form and regulate the centrosome.
MT	758	The kinetochore interacts with the lateral side of the microtubule and then is subsequently tethered at the microtubule end. Multiple kinetochores could occasionally interact with a single microtubule. Although multiple kinetochores can interact with the lateral side of microtubule only one kinetochore can form a sustained attachment. Kinetochores can slide along a microtubule. This sliding enables only one kinetochore to be attached to the end of the microtubule.
EM	759	It is unknown how information is transmitted from one organization to the other.
NP	760	Discusses the use of iron oxide nanoparticles in MRI imaging.

(Continued)

(Continued)

Category	**	Review Notes
Centrioles	761	In mitosis the cells compact their chromosomes into dense rod-like structures to ensure reliable transmission to the daughter cell.
Centriole	762	Multiple centrioles are produced.
Centriole	763	Centrosome increases in size during mitosis.
Cancer Centriole	764	Authors state that the most early sign of cancer may be centrosome amplification and that this corresponds with a loss of p53.
Centrioles	765	Authors state that there are variations in centrioles depending upon the type of cell they are in.
EM	766	Information within a cell is transmitted on though there were connecting wires.
NP iron	767	Carbon coated iron spheres.
Centriole	768	The chromosomes and centrosomes must duplicate and separate at the same time and rate.
MT	769	Generally, microtubules are developed in the centrosome also known as (aka) MTOC but on occasion they can be developed elsewhere.
MT	770	Pulling microtubules separate duplicated chromosomes. The microtubules are attached to the chromosomes via kinetochores. Kinetochores appear to be attached laterally to the microtubules along the length of the microtubules.
Centrioles	771	Enlarged centrosomes and cancer are common. Although the author believes that the enlarged centrosome is the cause of cancer, others believe cancer causes the enlarged centrosome.
MT	772	The objective is to find the minimal number of proteins needed to make a microtubule.
Nano	773	Lipid nanoparticles enhance drug absorption.
Centriole	774	For duplication centrioles it is unclear where the site of duplication is selected. It is suggested that "two separate but interesting" feedback mechanisms (autocatalysis activation and substrate depletion) are capable of selecting a single site for centriole biogenesis.
MT	775	Bridging fibers are required for prober kinetochore separation.
MT	776	Microtubules can assemble outside the centrosome (the MTOC).

(Continued)

(*Continued*)

Category	**	Review Notes
MT	777	The centrosome is the principal effective site for MT development.t
MT	778	Still elusive how centrosomes are developed. The PCM contains hundreds of proteins.
MT	779	The kinetochore serves as a link between the microtubules and the DNA.
Centriole	780	When cells experience mitotic delay, there is an increase in centriole disengagement.
Cancer	781	Authors state that centrosome amplification is sufficient for tumorigenesis.
Centriole	782	There appear to still be many questions about centriole assembly via cartwheel mechanisms.
Centriole	783	It is still not clear how centrosomes increase in size and in microtubule nucleation during mitosis.
Centriole	784	Centrosome amplification is sufficient to promote tumorigenesis.
Nano	785	Nanoparticles can be useful in breast cancer diagnosis and treatment. Both organic and inorganic can be used.
MT	786	Microtubule dynamic instability allows for the connection with kinetochores.
MT	787	The authors propose preventing cancers by targeting centrosome development.
Nano	788	Gold nanoparticles provide good biocompatibility.
MT	789	Perpetual growth and shrinkage provide for microtubule vitality.
MT	790	Separation of chromosome in cell division is sensitive and any errors can lead to cell death.
Cancer	791	Extra chromosomes are frequent in human cancers.

** Refers to chronological listing of literatures reviewed.

Index

Lightning Source UK Ltd.
Milton Keynes UK
UKHW050002200220
358970UK00006B/62

9 789811 208942

Islam and Healing

Loss and Recovery of an Indo-Muslim Medical Tradition, 1600–1900

Seema Alavi

palgrave
macmillan

South Asian edition first published 2007 by
PERMANENT BLACK
'Himalayana', Mall Road, Ranikhet Cantt
Ranikhet 263645
perblack@gmail.com

This edition published 2008 by
PALGRAVE MACMILLAN

Palgrave Macmillan in the UK is an imprint of Macmillan Publishers Limited, registered in England, company number 785998, Houndmills, Basinsgtoke, Hampshire RG21 6XS.

Palgrave Macmillan in the US is a division of St Martin's Press LLC, 175 Fifth Avenue, New York, NY 10010.

Palgrave Macmillan is the global academic imprint of the companies and has companies and representatives throughout the world.

Palgrave® and Macmillan® are registered trademarks in the United States, the United Kingdom, Europe and other countries.

ISBN-13: 978–0–230–55438–2 hardback
ISBN-10: 0–230–55438–5 hardback

This book is printed on paper suitable for recycling and made from fully managed and sustained forest sources. Logging, pulping and manufacturing processes are expected to conform to the environmental regulations of the country of origin.

A catalogue record for this book is available from the British Library.

Library of Congress Cataloging-in-Publication Data

Alavi, Seema.

Islam and healing: loss and recovery of an Indo-Muslim medical tradition, 1600-1900/Seema Alavi.

p.cm.
Includes bibliographical references and index.
ISBN-13: 078-0-239-55438-2 (hardback: alk paper)
ISBN-10: 0-230-55438-5 (hardback: alk paper)
1. Medicine, Arab—India—History. 2. Medicine, Greek and Roman—India—History. 3. Medicine — India—History I. Title. [DNLM: 1. Medicine, Unani—history—India. 2. History, Modern 1601—- India. 3. Islam—history—India. . 4 Religion and Medicine — India. WZ 80.5.A8 A472i 2008]
R605.A583 2008
610.938—dc22 2008014350

10 9 8 7 6 5 4 3 2 1
17 16 15 14 13 12 11 10 09 08

Printed and bound in Great Britain by
CPI Antony Rowe, Chippenham and Eastbourne

For

my parents

Roshan and Shariq Alavi

Contents

Preface and Acknowledgements

My maternal grandmother Begum Ejaz Jahan (1915–2000) introduced me to the world of Tibb-i-Unani. She was the granddaughter of Hakim Abd al Aziz, the founder of the Azizi family of Lucknow hakims. She made me realize that Unani was not just a system of medicine but a form of healing—a way of life. The stories of intimate friendships that her family shared with the British civil surgeons of Lucknow encouraged me to think afresh about our colonial experience and its impact on our everyday lives. My grandmother's maternal home, the Takmil-ut-Tibb College in Lucknow, epitomized for me the entanglement of the local medical culture with the global contours of medical science. At the same time her pride in her family's exclusive status as *the* scholarly hakims of the city, different from *neem hakims* (spurious hakims), urged me to also explore the story of Tibb-i-Unani from within a very stratified tradition. This book is the result of my endeavours to understand the social history of North India via a documentation and analysis of the history and transformation of the Unani healing tradition. It studies Unani texts and its practitioners from the seventeenth to the twentieth centuries so as to unravel the complex social history of the period.

Claudia Liebeskind's essays on the Azizi family encouraged me to take the first step into the history of my grandmother's family. Margrit Pernau's boundless energy and enthusiasm provided the final push that converted those initial hesitant steps into a full-time obsession. Muzaffar Alam patiently heard the nuances of the argument and helped strengthen the pre-modern aspects of my work. Sumit Sarkar and Javed Majeed offered valuable advice at the initial stages of research.

The research for this book was carried out in Lucknow, Aligarh, New Delhi, and London. In Lucknow I was lucky to get valuable documents and help from members of the Azizi family: my mother's maternal uncle Abdul Rahim (grandson of Hakim Abd al Aziz), my maternal uncle Khwaja Shakir Husain (great-grandson of Hakim Abd al Aziz), and Syed Imtiaz Ali (scion of the Azizi family and former Secretary, Takmil-ut-Tibb College, Lucknow). I am grateful to the staff of the Shibli Library in Nadwat-ul-Ulama, the Takmil-ut-Tibb College Library, and the Amir-ud-Daula Public Library in Lucknow. I owe a special word of thanks to Obaidur Rahman Nadwi for his valuable assistance in the city. In New Delhi, Sajjad Rizvi provided valuable editorial assistance in preparing the glossary. In Aligarh, Hakim Syed Zillur Rahman of Tibbia College was generous with his time and documents. In London, the British Library and the Wellcome Trust provided a mine of rich material in Persian, Urdu, and English. My friend Katherine Prior offered both intellectual support and the comfort of her home to enable me to access these libraries. I am grateful to my friend Guy Attewell of the Wellcome Institute, London, for sharing his time and ideas with me. His newly published book on Tibb-i-Unani in Hyderabad will add valuable weight to some of my own findings on North India.

The award of a Smuts Fellowship in 2002–3 enabled me to research this book full-time in the intellectually stimulating environment of Cambridge University. Here, the intellectual support and friendship of Chris and Susan Bayly, Gordon Johnson, the late Raj Chandavarkar, Tim Harper, Richard Drayton, and Francesca Orsini shaped my work in no small measure. I benefited also from the rich collections of the Cambridge University Library and the Centre for South Asian Studies. My loving friends of my student days in Britain—Katherine Prior, Mike Hirst, Simon Dunkley, Elke Nachtigall, and Colm O'Higgins—always made me feel at home.

A year-long Harvard-Yenching Fellowship at Harvard University in 2005 gave me the peace and quiet to devote myself to writing. I am grateful to the staff of the Widener Library at Harvard for readily making available all that I needed. Uncountable dinners and animated intellectual discussions at the kitchen table of Ayesha Jalal and Sugata Bose made the process of writing exhilarating and stimulating. This work would never have seen the light of day without their warmth, friendship, and boundless hospitality. I was lucky that my year in Harvard overlapped with the short teaching stint there of Chris Bayly who was, as always, generous with his time and ideas. His presence enlivened my stay in Harvard both intellectually and socially. Others who made my stay in the US memorable by their friendship and intellectual inputs include Upinder Singh, Sunil Sharma, Robert Traverse, Durba Ghosh, Maya Jasanoff, Sana Aiyar, Neeti Nair, Elaine Witham, Seung Mi Han, Zhou Xiang, Li Kang, Engseng Ho, Karim and Leila Fawaz, Shruti Kapila, Syed Akbar Hyder, Anand A. Yang, Vijay Pinch, Michael H. Fisher, Kenton and Marlie Clymer, Cheryl and Charles Martin, Indrani Chatterji, and Sumit Guha.

Over the last four years I have benefited immensely by presenting aspects of this work in numerous conferences and talks that I delivered in India, Europe, and North America. I am grateful in particular to other friends and colleagues at Delhi University, Jamia Millia Islamia, Max Müller Bhawan, and Calcutta University; at the universities of Cambridge, Oxford, and London (SOAS) in the UK; at the Social Science Research Centre, Berlin, Germany; at Harvard University and the University of Texas, Austin; and at Illinois University, George Washington University in Washington DC, University of Washington in Seattle, and Duke University.

Finally, my friends and family in India have been a bedrock of emotional support, intellectual sustenance, and warm companionship. I owe very special gratitude to my scholarly editor Rukun Advani, who combines high standards of professionalism with endearing qualities of friendship. I thank also my wonderful friends Mukul Kesavan, Shohini Ghosh, Sabeena Gadihhoke, Uma Singh, Farida Khan, Meena Bhargava, Radhika Singha, Ravi Vasudevan, Dilip Menon, Ahmed Zaheer, and Zakia Zaheer for being with me through thick and thin and accepting all my eccentricities with a smile.

My brother Nasir Alavi and sister-in-law Farah, and their two beautiful daughters Maryam and Ayesha, know very well how much I owe them. And of course my parents Roshan and Shariq Alavi have always been my source of inspiration and strength. Their confidence in my abilities has given me that extra stamina to move seamlessly in a journey of intellectual pursuit. To them I dedicate this book with love.

To make this book accessible, I have not used diacritical marks but spelt 'native' terms in accordance with current North Indian pronunciation. A glossary at the end of the book explains all such terms.

Introduction

The Argument

INDO-MUSLIM MEDICINE, OR THE UNANI TRADITION AS IT IS generally known in the subcontinent, came and developed in India along with Mughal political culture. While it healed the body, this tradition also had a profound bearing on the fabric of society. This book tries to show the nature and extent of the interaction of this medical tradition with Indian society and politics as they evolved and flourished over a period of three hundred years: 1600–1900.

I draw on a wide variety of sources: some critical Persian texts for the pre-colonial phase, the colonial English archive, Urdu texts and pamphlets, local newspapers, and private family records. This variety of material enables a revision of the historiography of India's encounter with Western medicine. I believe this book also represents the first major effort at telling the story of the Indo-Muslim medical tradition and its subsequent transformation by locating it in both the pre-colonial and colonial time frame. Straddling the centuries, it pioneers the idea of using the Mughal legacy—rather than the British colonial frame—to properly comprehend India's medical encounter with the West.

I show how, in the period of high colonialism, established medical practitioners kept their tradition intellectually alive. They struggled to preserve and recast the Mughal legacy, control knowledge, and consolidate doctrinaire languages of power. They felt threatened by the newcomers who had risen to prominence by taking advantage of opportunities provided by the spread of print culture and education. This created in-house differences and distinctions within Unani that, by contrast, makes the threat from 'colonial medicine' seem pale.

My book does not undermine the role of the state, but it does focus more on how the in-house struggle, rather than the threat from colonial power, defined Indo-Muslim medicine's take on 'modernity' and coloured its imagining of the 'nation' as both territorially rooted and universally envisaged.

Historiographical Location of the Argument

Historians of the Mughal period rejected the orientalist literature that saw Indian science and medicine as having stagnated and withered with the consolidation of Indo-Muslim rule.[1] These historians also offered insights into advances in science, technology, and medicine in pre-modern India. However, their focus remained state-centric. They saw medicine mainly as a part of imperial welfare measures and demonstrated the control of state over society through imperial interventions in the realm of medical care. Thus, studying state-run hospitals interested historians of medieval India.[2] Medical technology too was analysed primarily to understand the state: details of medical technological progress were spelt out to indicate the level of state patronage of technology.[3] Only recently, the significance of Mughal medical texts and their practitioners to empire-building is being recognized.[4] Appreciable as these efforts are, they are unable to break out of the confines of the 'powerful state syndrome'. They lay out a simple correspondence between practitioners and the Mughal bureaucracy, making the former appear as yet another arm of governance.[5] Even the cursory studies of medical texts that there are, are fitted into the imperial

[1] For such orientalist views on Indian science and medicine, see Kopf 1969.

[2] Askari 1957: 7–21; Subba Reddy 1957: 102–5.

[3] As with most features relating to Mughal history, the focus remained on Akbar and his interventions in the field of science. Habib 1997: 129–48. For discussions on the Mughal rulers in general, see Habib 1980: 1–34; see also Khan 2001: 26–39; Qaiser 1982. Qaiser showed the larger European impact on Mughal technology and culture in the fifteenth–sixteenth centuries. He was the first to shift the discussion away from the state to larger interconnections between the Mughals and the world beyond.

[4] For one of the early sketchy attempt in this direction, see Verma & Keswani 1974: 127–42.

[5] Rezavi 2001: 40–65.

edifice of the state.[6] Such researches view medicine as a science that is seen as an administrative arm of government. Since they locate the state as being external to society, medicine too gets situated in these as being outside society and culture. It follows that the medical tradition perishes with the decline of its patron state.

This book does not disprivilege the state. But it sees the state as a fluid entity located within society, constantly piecing together a fractured political culture.[7] It shows that, in the Mughal empire, medicine was a form of healing central to the building of an imperial political culture.[8] Indeed, the healing tradition expanded and was consolidated in tandem with the entrenchment of the Indo-Muslim fabric of the region's society. Thus, my argument is that the Mughal medical tradition has a far greater bearing on Indian society and politics than has been indicated in earlier works. And the entanglement of this medical tradition with Indian society continues much after the decline of the Mughal empire.[9]

The opening chapter shows how, in Mughal India, the medical tradition lay at the centre of society and politics. Scholar physicians (*hakims*) of Unani medicine emphasized outward bodily deportment as critical to health. Very like the intellectual upper classes of the French absolutist state, they identified with the court and produced Persian medical encyclopaedic texts that codified elite social manners as health regimes. The proximity of hakims to the court ensured that

[6] Moosvi 2001: 66–70.

[7] For this view of the Mughal state, see Alam & Subrahmanyam 1998; Hasan 2006.

[8] I borrow the idea of medicine as healing rather than as science from the Harvard medical anthropologists Arthur Kleinman, Mary Good, and Byron Good. See their introduction in Good, Brodwin, Good, & Kleinman 1994: 1–20. In contrast, there is a view that sees Unani as a system of science. See Burgel 1998; Ahmed &Qadeer n.d.

[9] Arnold 2000. David Arnold has sensitized us to the spilling over of the Mughal medical and scientific culture into the eighteenth century, and its bearing on the European impact on India in the period of colonialism. Yet, in the ultimate analysis, he sees the introduction of 'colonial medicine' as overpowering enough to cause a definite disjunction in local pre-modern medical culture. My book disagrees with the overemphasis on 'colonial medicine' and its role in Arnold's work.

health became part of aristocratic virtue. In Mughal India, medical texts carved out the contours of socially acceptable behaviour and civility. And, not surprisingly, they were produced and circulated along with Mughal etiquette literature, namely the *akhlaaq*. Medical texts and authors may have been patronized by the state, and thus contributed to the building of political sovereignty: but they impacted society and culture in longer-term ways. They laid down norms of civility that continued to be negotiated even after the collapse of the empire.

The location of this tradition in society enables us to take the story of its transformation into the eighteenth century. This was the period when the formal institutional edifice of the Mughal state collapsed. I show that the exclusive control of elite intellectuals and the upper classes—predominantly Muslim—on Mughal norms of Persianate civility was threatened in the eighteenth century. In this period of Mughal decline, the threat came from regions where, increasingly, Persian was getting vernacularized and becoming popular, making Persian medical and etiquette literature more easily accessible to a wider section of society. It also made access to gentlemanly status relatively easy. In this period of transition to linguistically diverse regional state forms, the intellectual class and medical families tried to protect and preserve their knowledge by detaching it from court culture and—its now relatively popular—Persianate norms of civility. They resorted increasingly to Arabic as the language in which medical knowledge was to be preserved and disseminated. Arabic medical literature, unlike the Persian texts, did not simultaneously comprise the author's expression of social reality. Instead, these Arabic medical texts were translations or compilations of assorted fragments collected from the Graeco-Arabic works of Avicenna, Hippocrates, and Galen. This transition from Persian to Arabic texts had consequences for both late Mughal notions of civility as well as for the idea of health as aristocratic virtue. Arabic texts made Unani scientistic and dissociated it from its Persianate Mughal encasement within ideas of elite bodily deportment. The transition thus created a new medical community based on Arabic learning.

My book discusses the impact of Western medicine, which came to India through the British conquest starting in 1757, in the backdrop of this robust Indo-Muslim medical tradition. Recent studies on the impact of Western medicine on traditional Indian healing have seen this intervention as ideologically motivated and all-powerful. They have argued that, by the 1830s, colonial interference stifled home-grown healing practices and marginalized them completely.[10] I counter this assertion and problematize the idea of colonial medicine as the site of culture and power. I argue that the Persianate and Arabicized pre-modern tradition had already influenced cultural norms and society in pervasive ways which blunted early colonial administrative intrusions. Thus the shift in Unani learning from Persianate aristocratic virtue to Arabic science in the period of transition to colonialism was shaped both by the new learning that came in from the Arab lands as well as the Western context. Indeed, the English East India Company's intrusion, far from being all-powerful and pervasive, was mediated through the indigenous linguistic cultures and communities within which Unani was sustained. Thus, at the turn of the nineteenth century, the Company perforce leaned on a new community of Urdu medical literature to interlocute the Unani tradition. New literatures introduced new agents of knowledge and new ideas of health in society; they created new 'communities of literature'.[11] These intensified Unani's shift away from aristocratic virtue towards the upholding of medicine as science.

The discussion on Unani's early encounter with the English Company's doctors shows that the competition between the predominantly Muslim communities of Arabic and Persian medical learning, and Englishmen, generated multiple ways of articulating medical authority. The encounter popularized new ideas of well being while keeping the flame of old learning burning. This period saw medical ideas ranging from health as aristocratic virtue to healing as scientistic medical wisdom. It also saw a shift in focus from health as individual well being to wider debates about the well being of society as a whole. These several ideas about health survived the reforms of the 1830s.[12] They continued to simmer in families, in the hands of individual scribes,

[10] Arnold 1993; Pati & Harrison 2001: 41; Gupta 1976: 368–78; Palit 2006: 123–62.

[11] Pollock 2003: 1–36.

[12] The 1830s are viewed as a watershed in conventional scholarship. Scholars such as D. Arnold, M. Harrison, G. Prakash, and J. Langford have hitherto argued that the 1830s signalled a turn towards coercion on the part of the Company

and in the courts of the regional powers, even as, officially, the colonial state embarked on a more stringent policy as far as traditional learning was concerned.

Historians of empire have been interested in Indian and African experiences with Western medicine inasmuch as they draw connections between imperialism, the development of British medicine, and the medical profession.[13] In the context of South Asian studies, the political economy of the colonial state, and its agendas of domination, control, and surveillance have overshadowed studies of India's medical encounter with the West. Deepak Kumar for instance views colonial science as an instrument of state power that was perceived in society as part of the cultural categories of racial arrogance.[14] Bernard Cohn argued long ago for the centrality of colonial knowledge production as the primary agency of power in British India.[15] And following Edward Said's influential *Orientalism*, colonialism increasingly came to be studied as a cultural project by a range of historians.[16] In the context of colonial medicine, David Arnold elaborated on the cultural project of 'colonizing the body' and the 'civilizing mission' of European science. Others, such as Gyan Prakash, underlined the 'cultural authority' of Western science in colonial India.[17] It was also argued that

as it abandoned orientalism, 'colonised the body', and introduced new-style Western knowledge with the help of the English language. See Arnold 1993; Prakash 1999; Langford 2002; Pati & Harrison 2001. A notable exception is the recent work of Kavita Sivaramakrishnan, which shows the continuation in Punjab of indigenous medical learning under the aegis of Sikh ascetics, a tradition that continued until the 1850s. See Sivaramakrishnan 2006: 14–31.

[13] Johnson 1973. John Iliffe has shown how European practitioners in East Africa racialized the authority of the Western medical professional by subordinating and marginalizing African practitioners in the government medical service. Iliffe 1998. For similar trends in West Africa, see Patton 1996. See also Vaughn 1991: 1–28. For how the imperatives of empire changed Britain's medical profession from a national to imperial service, see Haynes 2006: 130–56.

[14] Kumar 1995.

[15] Cohn 1996.

[16] Dirks 1992.

[17] Arnold 1993: 1–11; see also Arnold 2000: 1–8; Prakash 1999: 3–14, 156–8; Catanach 1986: 216–43; Panikkar 1995: 145–75.

instances of epidemics and the consequent needs of public welfare were deployed to extend the colonial state's influence over society.[18] Yet others, while accepting the state-centric paradigm of India's medical experience, indicated the limits and limitations of the state's intrusion in the realm of public health.[19]

Such researches view the introduction of Western doctors in Asia and Africa as central to understanding British medical tradition as it developed both in the metropolis as well as the colony: they have enriched studies of both empire and colonialism. Yet, remarkable as these studies are, they are invariably state-centric: their narrative remains insular. I turn the spotlight away from the agendas of empire and the colonial state to the Indian players who contributed to the agility of the Indo-Muslim tradition over the course of colonial rule in India. I question the idea of Western models of civil society being imported into India and used as tools of colonial administration and domination. Instead, I elaborate on the fact that supposedly colonial concepts like 'public health' had indigenous subcontinental origins. For instance, I show how the pre-colonial 'scientistic' notion of medicine, which was divorced from Persianate civility and which reached out to larger society, became in fact the base of the early British idea of public health. It is undeniable that British intervention introduced a new medical ethos, in terms of a professional medical public servant with new referents of status and authority. But early public health concerns in India were not entirely modelled on borrowed Western models of civil society. On the contrary, far from being a Western import, they encapsulated ideas of public welfare and social well being that emerged out of the West's contestation with indigenous patrons of such spaces. Local medical patrons, along with Urdu print culture, played a critical role in sketching out the public welfare concerns of the British in

[18] Arnold 1986: 145–51; Ramasubban 1988: 40–43.

[19] Harrison 1994: 228–34, who shows how intervention was shaped by the concern with public reaction. Also Ramanna 2001: 233–48, who shows how medical attendance in dispensaries and hospitals in Bombay and Madras after the plague epidemic showed little increase. See also Kumar 1998, who shows how, after the 1857 mutiny, discrimination in government service checked the growth of Western medicine for much of the remaining century.

nineteenth-century India; indeed, they helped create a public health manifesto. My book explains how this India-specific manifesto was exported to England to handle the cholera epidemic of the 1830s.

The book then shifts focus to the making of a social consensus over early-nineteenth-century colonial 'human welfare' schemes—the dispensary, the sanitation plans to prevent cholera, and the smallpox vaccination drives. It argues that public health policies emerged as a consequence of fierce contest between indigenous elites and colonial masters over control of the medical public sphere. It comes as no surprise that medical interventions are not resisted outright; and they do not represent the success story of a strong colonial state either. Indeed, both within these institutions and outside them, older healing communities and patrons of health compete with the state to maintain their hold on individual well being. This wider dynamics of early-nineteenth-century society, rather than an aggressive colonial state with an ideological interest in Indian healing, encases the nature of change in early colonial India. Unani's encounter with colonialism shows that, in the realm of medicine for instance, a range of pre-colonial medical authorities and ways of being a medic persist alongside the universalizing language of Western medical science. The Anglicist reforms of the 1830s were unable to extinguish this syncretic medical culture. The plurality of the culture limited the ability of the state to establish a pervasive and hegemonizing medical discourse that 'colonizes the body' and extinguishes local medical tradition. Colonial medical knowledge remains contingent on the historical situation and is itself socially constructed more than being constructionist.

At one level, the discussion reveals how a relatively weak colonial state negotiated with, and in the process sustained, powerful 'communities of literature'. Their importance became particularly significant as the state battled unsuccessfully to control the epidemics of smallpox and cholera in the early nineteenth century. Indeed, until the mid nineteenth century an epidemic-ravaged state tried to harness its larger project of 'human welfare' to their existing knowledge of materia medica. It competed with local patrons to insinuate itself as their patron. So valuable was local knowledge to the state that it used print not to 'revolutionize' their intellectual base, but rather to establish its authorship over their medical wisdom. In turn, local medical communities realized their indispensability to the state and used every opportunity to contest colonial medical authority and patronage, even as they cooperated with it to contain epidemics. In this contest, local drugs entered the colonial lexicon. But now Persian-driven ideas of health as aristocratic virtue and individual well being were being pushed to the fringe, even while the more scientistic ideas of Arabic-driven 'communities of literature' survived. Their medical wisdom was now increasingly used not just for individual well being but also in the wider context of social well being.

At another level, I elaborate on the entrenchment of 'communities of literature' in British institutions such as the dispensary. I show how these communities were incorporated into the colonial discursive framework, as laid out in its dispensaries. This process intensified as the colonial medical infrastructure expanded. In the 1840s, the dispensary employed native doctors schooled in Urdu medical literature produced by the Company. It also employed men from the Arabic and Persian medical communities: the Company and village communities jointly financed the dispensary. This provided endorsement to the intermingling of the different medical communities that the institution of the dispensary upheld. The incorporation of these different communities of literature in the dispensary meant that the colonial framework offered a space to continue the intellectual interactions that had been a feature of Indian society.

But this had its own risks: it could backfire against the state. Often, the state's efforts to monopolize medical knowledge was resisted by its own native doctors. They used ideas in the Persian and Arabic medical traditions which the dispensary itself kept alive to challenge the imposition of Western ideas of medicine as science, reason, and rationality. Outside the dispensary, longer-term contestations with colonial authority went on within the institution of the family. Here, Unani consolidated its learning, relied on Arabic, and confined itself within a narrow clientele of families and close associates.

My book also thus ties in with scholarship that questions the centrality of the Foucauldian coupling of knowledge and power in colonial studies. Scholars working in both colonial and non-colonial societies have indicated the limitations of invented cultures, of discourse theory, of the vulnerability of 'tradition' seen as an entity, and of the ability of colonial regimes to carry on their 'revolutionizing' projects through the use of print capitalism. In the context of Islamic societies, the works of cultural anthropologists such as Clifford Geertz and historians such as Juan Cole have cautioned against accepting the overriding status of any single political or religious culture. Their studies indicate that a culture elaborates its symbols of authority over time. Thus, Islamic culture, they argue, remained a universalistic umbrella under which a variety of localisms thrived.[20] In the context of India, C.A. Bayly suggests a similar elaboration of local identities and structures, and the survival of 'traditional patriotism' under the umbrella of British colonial culture. Knowledge, according to Bayly, does not flow in a capillary way generated out of conditions created by colonialism alone. Rather, it throbs in 'traditional patriotisms'.[21] The Unani story, too, shows that colonialism provided an umbrella under which this medical tradition reconfigured itself. It neither reified nor Westernized itself as it took on the colonial challenge; rather, it called the shots and chose its terms of engagement with the state with remarkable selectivity.

Historians of medicine have, again, studied the late-nineteenth-century reforms initiated by Indian healers within the colonial frame. David Arnold sees the Indian response to epidemics as driven by a set of cultural values that were triggered by colonial intrusion in the medical public sphere.[22] Others take the argument further, seeing figures involved in the 'cultural revival' of indigenous medicine as symbols for national regeneration.[23] More recently, Claudia Liebeskind, N. Qaiser, and Guy Attewell reveal the plurality of the Unani defence that lay embedded in its overt anti-colonial shrill. They argue that medical reformers projected Unani as a particularistic 'scientific-rational' system that stood in contrast to the value-neutrality and universalism of Western medicine.[24] Raj Chandavarkar in the context of Indian response to the plague epidemic, and Kavita Sivaramakrishnan in the light of

[20] Cole 1992: 1–30. Geertz 1968.

[21] Bayly 1996. For the limitations of discourse theories in colonial Africa, see Vaughn 1991: 1–28.

[22] Arnold 1987: 55–90.

[23] Kumar 1992: 172–90.

[24] Liebeskind 2002: 58–75; Qaiser 2001: 317–55; Attewell 2004.

Ayurveda's reaction to Western medicine, stress instead the need to see in-house churnings as a political process. They argue that, rather than a simple anti-colonial stance, a local tradition's clamour for change represented particularistic negotiations and the accommodation of varied interests during a period of adjustment to colonial rule.[25]

In this book, discussions on the Unani reforms of the late nineteenth century follow Sivaramakrishnan's line of argument and move the discussion out of the 'colonial state *vs* indigenous assertion of authority' frame. Like her, I discuss reforms against the backdrop of the plural medical culture that colonialism itself helped sustain. I argue that Unani healing underwent tremendous internal change at a time when Western notions of medicine as a universal and professionalized science competed with particularistic local ideas of well being. This historic conjecture created a new Unani. New men used the Urdu vernacular press to reach out to new kinds of medical knowledges that lay both within and beyond the purview of older medical communities.

With this longer rope, the new Unani broke loose of the control of both the older linguistically particularistic medical communities as well as the colonial medical apparatus while maintaining its links with plural medical traditions. It borrowed the idea of vernacularizing and popularizing medical knowledge from the community of Urdu medical literature. It used the Persian idea of health as individual comportment with Arabic scientistic learning as its intellectual core. To these it added other knowledges culled from the religious and cultural cosmologies of North Indian society. These included healing through the traditions of the Prophet (*Tibb-i-Nabawi*), and the use of charms produced by fakirs, and talismans, and other such folk remedies. The new Unani also used the vernacular printing press to generate a different kind of medical literature, a new community of Urdu medical literature. In the end, this community presented an uncontrollable realm of medical knowledge. Both the colonial state and the linguistically particularistic medical communities of literature—traditional medical families—were equally helpless in their attempts to control this new Unani.

My book locates the late-nineteenth and early-twentieth century Western-style reforms in Unani healing to this wider social phenomenon of mobility and change. It argues that the social complexities

[25] Chandavarkar 1992: 203–40; Sivaramakrishnan 2006.

of the Unani tradition, rather than European science, provide the discursive framework for its self-definition as modern medicine. On the surface, Muslim medical reformists' claim that Unani was modern medicine, and their demand for its professionalization, seem a survival technique against colonialism, a strategy to realign with colonial scientific instrumentation. But colonialism was only one aspect of their problem. The reforms, in effect, reveal the social cry of the old order of Unani which is being overwhelmed by the popularity of the new Unani. The reforms reflect the effort of the old guard to regain lost prestige and reassert control over their medicine. Reformists adopt the vocabulary of colonial medicine with its emphasis on science, reason, and rationality. But these are not deployed only to align Unani to Western science.

The creation of the new Unani of the late nineteenth century and the era of in-house reforms poses several questions about the ways conventional historiography has viewed science, medicine, and its connections with Western modernity. How do we understand the Unani medical tradition in this period of high nationalism? Was it reified and preservationist,[26] invented,[27] a subverted rendering of a colonial construct,[28] derived and emulative of 'colonial medicine'.[29] Or was it culturally constituted so as to defy these binaries?

Ever since the influential work of Benedict Anderson, the binaries of tradition and modernity in colonial encounters and histories of nationalism have been questioned. This is particularly true for scholars

[26] A Foucauldian rendering of the colonial encounter as elaborated by Partha Chatterjee 1993.

[27] Hobsbawm & Ranger 1983. They show the role of the indigenous elite in inventing a nationalist tradition. They point to the reality and fiction divide in the history of nation.

[28] Langford 2002. She shows how the Ayurveda medical tradition in the same period was seen as a spiritual cultural system of practices by its practitioners. They subverted this colonial construct to their political advantage.

[29] In the manner in which all kinds of scientific and medical traditions in this period derived from Western Enlightenment notions of science, reason, and rationality. Chakrabarty 2000; Prakash 1999; Arnold 2000. In any case, all these historians talk only about the 'Hindu' or the Ayurveda medical tradition and its practitioners when they discuss the construction of alternative modernities in the early twentieth century.

who give a cultural turn to research on nationalism.[30] Historians like Manu Goswami have indicated the simultaneous rendering of global universal and particularistic strands that culturally constituted the nation: the imagined national space as culturally particularistic was constituted from within a universal colonial frame, making the binaries of 'modern' and 'traditional' superfluous.[31] My analysis of the new Unani shows that it too defied such binaries in the period of high colonialism. Unani diversified to embrace new knowledges that lay both within and outside the colonial and literary communities.[32]

My discussion of late-nineteenth-century Unani reforms shows that the tide of global capitalism, which brought universal ideas of medicine and health management into India, did not sever Unani from its cultural core. Thus, in the late nineteenth century, resistance to colonial medical authority derives not by emulation or subversion of Western-style modernity, nor by an invented tradition narrowly culled from new influences by elites. It draws on a variety of linguistically defined strands of medical knowledges and authorities that persisted through colonialism: Arabic medical learning with its Greek cosmological imprints that stressed natural law, ideas of comportment and ethics as disseminated through Persianate book production, and Urdu print culture with its Western leanings in relation to health and the body. To these were added the reformist and revivalist religious ideas of late-nineteenth-century India, and the ideas of Western science.

Studies by Deepak Kumar, Jean Langford, and others have shown how the early-twentieth-century 'nationalist' contours of indigenous medicine derived from a direct anti-colonial stance.[33] Sivraramakrishnan again, in the case of Ayurveda, shifts the focus inwards, to the

[30] Metcalf 2004: 173–90.

[31] Goswami 2004.

[32] I follow the concept of tradition as laid out by Hans G. Gadamer and Martin Gwyer. They see tradition as 'embracive' and as a carrier of modernity. They argue against the setting up of tradition as an analytical category separate from the global and the modern. And they make a case for the interplay of the global and the local that invigorates tradition. Instead of modernity being opposed to tradition, modernity is manufactured by an invigorated tradition. Thus, tradition carries modernity rather than opposes it. See Gadamer 2004; Geyer & Paulmann 2001.

[33] Kumar 1992; Langford 2002.

internal rooting of the nationalist profile. My book, similarly, locates the origins of Unani's march towards a 'national' medicine within its own tradition. It explains its nationalist stance in terms of particularistic negotiations and accommodations that its traditional bearers struck with in-house contenders. It understands the efforts of Muslim families of high learning (Arabic and Persian) to carve out a 'national' space for Unani in the backdrop of internal churnings that brought new men into prominence who threatened the earlier social standing. Traditional hakims leaned on the vernacular press which was itself controlled by families similarly constituted by high learning traditions. Print capitalists helped traditional Muslim families of hakims to carve out a 'national' space for their medicine so as to differentiate it from the new Unani. As these older communities of Arabic and Persian literature geared up to fight the new hakims, colonialism paled into the background as a lesser evil. Indeed, Western medicine's infrastructural frame and vocabulary of reason, rationality, and professionalization provided the grid around which the old families battled the new Unani.

Even though such communities used the colonial discursive framework, they did not align Unani to Western science. Instead, colonial ideas of surgery and professionalization were cannibalized so as to embellish Unani's own robust learning traditions. Even then, the distance from colonialism was always maintained by giving a specific history to Unani high learning in Hindustan—a history that went back to antiquity and located its core in Hindustan, and which was constituted as a distinct cultural space-locale. As the older families mapped Unani on to the imagined cultural space of the nation, they projected it as a part of Indian tradition. This made Unani modernity and scientificity unique. It stood in stark contrast to Western 'modernity', which had no history or territoriality in Hindustan.

Finally, my book looks at the formation of Muslim elite institutions of Unani learning in 1900. These institutions, on the surface, marked the culmination of Western-style professionalization of tradition.[34] But I caution against such simplistic understandings and urge the need to see their emergence as a product of the power struggle that riveted the tradition from within. Here I discuss the Takmil-ut-Tibb College in Lucknow, established by the Azizi family in 1902, to show

[34] Liebeskind 2002; Attewell 2004.

how it epitomized the exclusive monopoly of an individual family on a freshly constructed 'national' Unani. The discussion here shows how families consolidated their hold over Unani high knowledge by protecting it from the religious communitarian politics of the time, which threatened to bring new contenders into the domain of medical knowledge. Such families zealously argued that Unani had a professional community identity rather than a Muslim one, and that it should therefore remain out of the purview of revivalist and reformists ulema who interfered increasingly in the educational frameworks of society. This was one way to consolidate exclusive family hold on medical knowledge and shield it against the new contenders who emerged in the form of religious leaders and Urdu-read new hakims. Raising the professional identity issue of Unani thus became imperative in the preservation of their familial and professional status as traditional custodians of unalloyed Unani knowledge.

The Unani institutions of the early twentieth century thus put ideas of professionalization to new uses. In the process, they were distanced even more from the colonial intellectual grid where they had originated. And yet the link with colonial knowledge was never severed, for Unani was being mapped on the territorial borders of Hindustan as sketched out by the colonial medical infrastructure. At the same time, Unani's unique history in Hindustan was being written to highlight its difference from colonial medicine, which could make no such territorial or historical claims: Unani was in this sense different from colonial medical modernity, even if its 'national' space was culled out of that larger framework. This balancing act worked with some degree of success until the 1920s. However, the balancing between the national territorial, the Muslim communitarian, and the Islamic universal was more difficult. Fissures appeared in Unani ranks over this strategy once the communal politics of the region gained momentum, prevailing on Unani and colouring it with a Muslim hue.

In this sense, my study of Unani institutions adds new dimensions to our understanding of Muslim politics in this period. Muslim engagement with Western learning and national politics has been mainly studied from the viewpoint of religious seminaries such as Deoband and Farangi Mahal.[35] Or else, the highlights have seemed the efforts

[35] Robinson 2001; Metcalf 1982.

of Sir Syed Ahmad Khan to establish an Oxford-style school for Muslims that made Western education compatible with Islam.[36] In earlier works on such themes, the religiously determined communitarian discourse on Muslim identity and politics was located as antagonistically opposed to the more enlightened discourse represented by people like Syed Ahmad.[37] Barbara Metcalf's work on Deoband questioned such dichotomies and showed how the traditionally educated were not necessarily sunk in 'stagnation' and 'rigidity'. She convincingly argued that Muslim reform and renewal were structured on Islamic thought and institutions even as they assimilated the modern organizational style of new educational institutions.[38] Christian Troll, complementing this analysis, has brought out the dichotomies within the 'progressive' movement of Syed Ahmad Khan in particular.[39]

Such studies have blurred the borderlines of reformists and revivalists and indicated the compatibility of new Western-style learning with traditional religious knowledge. However, the view from the seminaries of Muslim learning—of both the religious and 'secular' kind—continues to perpetuate the idea that Islamicate elites created an exclusive Muslim cultural space in this period; and that this space, irrespective of the motivations behind it, was their nation.[40] This cultural space has then been seen as fanning the communal politics of the region, and creating ideas of separatism and conflict with the 'nationalist' politics represented by Congress and its Muslim members.[41]

My study of Unani families and their politics indicates that there existed a range of Muslim professionals whose identity as Muslims remained subordinated to their profile as a community of professionals. Motivated by professional considerations of preserving their status as hakims, they constructed a 'national' space for their medical tradition.

[36] Lelyveld 1991.

[37] Faruqi 1963.

[38] Metcalf 1982.

[39] Troll 1978.

[40] For a difference of opinion on why the Muslim cultural space was carved out by elites, see Robinson 2001: 177–209. He argues against the materialist argument of Paul Brass: see Brass 1974.

[41] Robinson 2001: 177–209. For a discussion on nationalist Mulsims, see Hasan 1979; also Hasan 1987.

This space was culled out of the colonial medical discursive framework and, while remaining closely connected with colonialism's intellectual grid, was distinct from it in having a history and cultural roots in Hindustan that were inclusive—and not religiously or exclusively Muslim. This process of carving out an inclusive cultural space was not easy. Social pressures from all corners caused deep fissures within the Unani profession. Yet on the whole it withstood these pressures. The frenzied religious politics of the 1920s created problems, and yet the effort to balance the professional with the religious communitarian was never given up. Members of such professional families did not directly participate in nationalist or Muslim politics of the time. They were keener to safeguard their professional space.

This adds a new dimension to our understanding of the Muslim politics of the time. It is not insignificant that, even though not actively involved in the Congress or nationalist politics of the period, in 1947 not a single member from the Lucknow family of hakims opted to migrate to Pakistan. Their story shows that there was no simple march to either a 'secular' territorial nation state or a Muslim separatist one. Islamic universalism, Muslim communitarianism, and professionalism all combined in such people to offer us a different glimpse of Muslim politics and society in pre-independent India. Indeed, they also compellingly show that throughout the nineteenth century Muslim healers had to balance the territorial national with the Muslim communitarian and Islamic universal. It was for this historical reason and basis that their politics in the period of high nationalism was far from being a simple case of anti-colonialism. It was, rather, foremost about handling in-house contestations over control of Unani knowledge and the multi-directional pulls on its profile.

My study of Unani healing, in short, provides fresh insights into the role of languages, medical literature, education, and the links of these arenas with social mobility, identity, status, and the politics of Muslims in colonial North India.

1

Indo-Muslim Medicine: Unani in Pre-modern India

Unani and the Dar al-Islam: Eighth–Fifteenth Centuries

IN THE PERIOD OF THE CLASSICAL EMPIRES OF ISLAM (EIGHTH to thirteenth centuries), an integrated Islamic world system flourished that was knitted together by trade, politics, and cultural contact.[1] Scholars and traders moved with ease along trading and cultural contours that spanned the Middle East, Central Asia and China, and included the Indian Ocean area. India was thus very much part of this world. It did not fall under the political hegemony of the Islamic empires, but it was part of the trading and cultural ambit of what Muslims called the Dar al-Islam.

Dar al-Islam was important not only for trade of commodities but also for the trade in ideas. Knowledge of medicine, in particular, was of immense interest to the caliphs of the Islamic empires. Medical ideas were appropriated from the Graeco-Roman philosophical tradition represented by Hippocrates, Aristotle, and Galen. During their early contact with the Christian Byzantine empire and the Iranian Sasanian empire, Muslims found this intellectual system flourishing in conquered lands. They appropriated it and made it part of their intellectual legacy. In the famous medical institution at Jundi Shapur, in south-west Persia,

[1] Voll 1994: 13–26.

for example, Christian and Jewish physicians upheld the humoral theory of health derived from Hippocrates and Aristotle. They also acknowledged their intellectual debt to Galen, the physician who practised in Rome in the first century AD.

This Graeco-Roman medicine that politically expanding Islam adopted was represented largely as a form of Galenism. This meant a synthesis of the various Greek philosophical influences that had shaped Galen's notion of human procreation and the body. Galen's ideas derived from Greek culture and cosmology, which itself was rooted in the maintenance of a primarily agrarian political economy and territorial control by the ruler.[2] Thus, Galen appropriated the idea of the world as divided between the 'spheres' or planets. These had perfect form and movement, but neither generation nor decay. According to Galen there was a sublunar half to the world in which form and matter were continuously in transformation. Like Hippocrates, he saw the earth as located in the sublunar sphere, where nature conditioned generation. He explained the functioning of nature's laws using Aristotle's understanding, which was structured around a series of oppositions and correspondences: hot and cold, wet and dry. In this Galen was influenced by the analogical reasoning of Greek cosmology which treated the universe as a macrocosm perfectly reflected in the human microcosm. Thus he, like Aristotle and Hippocrates, believed that the human body itself had a hierarchy of oppositions and correspondences—the humours—that had to be kept in balance with the elements of nature so as to maintain good health.[3] And for him the body was itself created out of the mixing of two opposites: hot semen and cold ovum. Health, as framed by Greek cosmology, was very much part of social well being, the human body being dependent on the maintenance of a balance within nature.

So, the Greek medical theory that Muslims came in contact with was structured on a philosophical understanding of the workings of nature as propounded by and associated with Aristotle. Primarily a political theorist, Aristotle's medical ideas mirrored the concerns of his

[2] Galdston 1969: 20–2.

[3] For a lucid description of Greek cosmology and its organizing episteme for Graeco-Roman medicine, see Good & Good 1992: 259–65.

political theory of governance. His spotlight on nature brought into salience the role of the king, who was meant to manage and administer the bounties of nature so as to maintain social equilibrium. The kingdom was like the body and the king its main physician. Just as the humoral equilibrium of the body had to be maintained for good health, so social equilibrium in society was to be engineered by a just king to maintain a stable polity. As the propagator of this theory Aristotle, the foremost political theorist, was revered in medical circles as well.[4]

The Umayyads (661–750) and the Abbasids (750–1258) showed immense interest in Greek medical and political philosophical literature. It became part of the core of their intellectual heritage. The caliphs Al Mansur, Harun Rashid, and Al Mamun were instrumental in translating the corpus of Greek and Syriac medical literature into Arabic and integrating this Graeco-Roman intellectual legacy of philosophy and medicine into the Arab sciences. Special departments, the Bayt-al Hikmat, were set up in Baghdad to oversee the translation and compilation of Graeco-Roman texts.[5] Here Arab Christian physicians like Hunayn ibn Ishaq (808–73), and his son Ishaq and nephew Hubaysh, translated from Greek into Arabic texts such as Hippocrates' *Aphorisms* and Galen's *De Differentiis Febrium* and *De Typis Febrium*, among others.[6]

In this period, medical knowledge came also to Damascus and Baghdad from India. Trade in Indian medicines between the Middle East and Hindustan had been in the hands of Arab traders much before the consolidation of the caliphate.[7] But Indian medical texts also attracted the Persians and Arabs. In fact the physician Perzoes from the Sassanian empire had been dispatched to India to procure medical literature which was then translated into Persian.[8] But it was in the period of the Umayyad and the Abbasid caliphates that intellectual exchange of ideas in medicine and science between the two regions began in some earnest. Yahya bin Khalid, the Arab general who laid the

[4] For a discussion on the idea of medical and political theories upholding the power of the king, see Galdston 1969: 20–2.

[5] For an account of translations and scribes involved, see Elgood 1951: 58–96. See also Siddiqi 1959: 17–29.

[6] Elgood 1951: 106–20; Siddiqi 1959: 22. Hubaysh translated Galen's books on anatomy.

[7] See Nagrami 1989: 197–8 for details of this trade.

[8] Elgood 1951: 51–2.

foundation of the caliphate, not only invited Hindustani vaids (medicine men) to Baghdad, but also sent a man there to collect information about Indian plants and herbs. Such emissaries interacted with certain Brahmins, who were the repositories of medicinal knowledge, which remained confined in family records.[9] In the Abbasid caliphate, as early as in the reign of Caliph Mansur, an important intellectual from Hindustan is reported to have been in Baghdad and presented his medical book, *Siddhanta*, to the caliph.[10] This kind of exchange increased in the time of his successor, Harun Rashid, who invited several vaids from Hindustan to cure him of his illness. Many of them were associated with the department he set up to promote the collection of scholarly manuscripts of all disciplines—the Bayt-al Hikmat. These included the famous Vaid ibn Saleh, Manka, Bazigar, Qalbarqal, Bahlah, Ibn Dhan, and Khatif Hindi.[11] Two of these, Manka and Ibn Dhan, settled in Baghdad. Ibn Dhan was made the chief officer of the biggest hospital, or *shifakhaana*, in Baghdad, and he translated into Arabic the famous Sanskrit books on Ayurveda—*Sushruta* on surgery and *Sameeka* on the cures for poisons.[12] Large-scale translations of Sanskrit medical literature into Arabic went on under the aegis of the Bayt-al Hikmat in Baghdad.[13] Manka translated into Persian a Sanskrit book on poisons—*Sarat*,[14] and Sanjahl, another Indian, translated the work of Charaka into Persian: the latter had already been translated into Arabic.

The Abbasids also dispatched men of learning to Hindustan to explore both texts and materia medica. Harun Rashid's prime minister sent an agent to Hindustan to collect and bring back medicines in use

[9] Ibid.: 198.

[10] Latif 1949: 5. The book was translated into Arabic by Ibrahim Fazari.

[11] Nagrami 1989: 198; Siddiqi 1959: 30–45.

[12] Ibid.: 201; Latif 1949: 5.

[13] Nagrami 1989: 199–201. See the list of Sanskrit books translated into Arabic.

[14] Elgood 1951: 101. Of particular significance to Unani is the translation of the Sanskrit book on surgery, *Sushruta*, which is the first lesson (*qaidah*) on the art of surgery (*jarrahat*). It is the first book of its kind, including comprehensive discussions on diseases requiring surgery (*amraaz i-jarrahiyat*), surgical tools (*allat-i-jarrahiyat*), and cures through surgery (*ilaj i-jarrahiyat*). It discusses at least a hundred surgical implements. It also includes discussions on the basic principles of dayaagiri (midwifery) that are considered seminal to the profession even now.

there.[15] Many such agents learnt Sanskrit, which they saw as the key to Indian medical knowledge. One such scholar was Muhammad bin Ismael Tanoji, an expert on the ninth-century physician who lived in Baghdad, al Razi. He lived in Hindustan for years, learnt Sanskrit, and gathered expertise in psychology (*himyaat*). And he was not alone. Since Indian courts in this period were also keen to hold discussions (*munaazra*) on culture and religion with foreigners, many scholars travelled to India with enthusiasm. They often ended up living in India and translating Sanskrit texts into Arabic and vice versa.[16] Al Beruni's was an important case in point.[17] Via such travellers a rich literature was generated in Arabic, Persian, and Sanskrit, often under state patronage, which reflected this dialogue.

By the thirteenth century Unani medicine had acquired a truly global character that corresponded to the cultural, political, and trading contours of Dar al Islam. And because the contours of Dar al Islam embraced Christian, Jewish, Hindu, and a range of other civilizations, it could never be imprinted only with the seal of Islam. Indeed it took on the flavour of the different civilizations over which it expanded. In the Indian context, therefore, it can be safely called Indo-Muslim medicine. However, it is more generally known as Unani in the subcontinent. This is because the Muslims acknowledged its cultural heterogeneity and christened their medical tradition on the name of the country they had adopted it from—Greece. And hence the name 'Yunani' or Unani, i.e. Greek. The truly global and trans-religious character of Unani was best reflected in the names of medicines that Unani offered: *Qist-i-Hindi* (qist from Hindustan), *Saazakh-i-Hindi*, *Aaloo Bukhaara* (fruit from Bukhara), *Ajwain-i-Khurasani* (ajwain from Khurasan), *Salih-i-Misri* (salih from Egypt), *Namak-i-Lahori* (salt from Lahore), *Dar Cheeni* (herb from China), *Gul-i-Armani* (flower from Armenia), *Gul-i-Multani* (flower from Multan), *Gul-i-Dakhghastani*, *Afsanteen-i-Rumi* (afsanteen from Rome), *Gulnaar-i-Farsee* (fruit from Persia), *Asgand-i-Nagori* (asgond from Nagor), etc.

[15] Latif 1949: 11.

[16] Ibid.: 13.

[17] Ibid.: 7, 13. He translated at least twenty-two books in Sanskrit into Arabic.

This variety of medical influences stimulated Islamic authors to classify the borrowed Greek medical knowledge in large encyclopaedic medical works. This huge compendium of medical literature in Arabic included not just translations of Hippocrates and Galen but also many lesser works written later and attributed to them. The encyclopaedic compendiums borrowed freely from earlier works and did not always acknowledge their sources, a classic example being the enormously influential *Canon of Medicine* (*al Qanun fi tibb*) by Ibn Sina (Avicenna). The text stands at the end of a long tradition of Galenic scholarship that had formalized and idealized Galen's work, as Galen himself had done with Hippocrates.[18]

Texts like the *Canon* that were compiled as anthologies of information culled from a variety of sources, and only occasionally containing references, became a characteristic feature of medieval Islamic manuscript production. They were comparable to the florilegia (literally 'nosegays of readings') genre of anthological literature in medieval Europe.[19] And, as in medieval Europe, in the Islamic florilegia literature reverence for ancient authority coincided with extraordinary indifference to textual authenticity. Thus, the famous Abbasid translator Hunain Ibn Ishaq is said to have translated or supervised the translation of more than 129 works of Galen into Arabic and Syriac. Many of these then appeared as fragments in other medical encyclopaedic texts of Arab authors like al Jurjani and Avicenna (Ibn Sina). But many texts attributed to Galen in this list were not authentic, and were in fact spurious.[20] Medieval medical encyclopaedic compendiums, often 'de-authorized' or attributed spurious authorship, 'decanonized', and 'detheorized' ancient authorities of knowledge even as they certified their own knowledge via the status of the knower or the source.[21] Such

[18] Dols 1984: 8–10.

[19] Wallis 1995: 105. Some florilegia circulate as canonized anthologies, others are assembled ad hoc by the compiler of the volume. The choice of selections is not random and always means something.

[20] Max 1984: 701.

[21] This kind of knowledge which does not rely on the methodology of the known but more on the source or status of the knower is referred to as gnostic knowledge by Don Bates. He contrasts this with epistemic knowledge, which is more methodology-oriented. See Bates 1995: 1–22.

texts in Arabic were stylistically austere. They offered nuggets of medical wisdom. And they delinked medicine from ideas about individual comportment.

Unani's medical rationality thus derived both from Graeco-Roman philosophical traditions as well as Ayurvedic ideas. This wide range of influences resulted in the development of a medical writing style that was encyclopedic and which did not necessarily acknowledge or accurately record its sources. This was particularly true in the borrowings from Ayurveda, which was very much a part of Unani's intellectual legacy but remained unacknowledged.[22] In fact after al Razi Arab medical writers seldom made any reference to Indian medical works excepting *Charaka* and *Sushruta*, to which some of them referred in connection with drugs and medicaments. For instance Ibn Sina, in his encyclopaedic *Canon*, quotes Indian opinions only on the toxicity of leeches and only mentions the name of Charaka in connection with therapeutics. Al Razi is equally silent. Later medical encyclopaedias, like the twelfth-century *Zakhirah-i-Khwarzmshahi* of al Jurjani, are equally silent on Ayurvedic influences, even though they are so clearly visible, specially in their sections on surgery.

Of course, not acknowledging sources can be explained as being part of the medieval Islamic literary style; yet it also had to do with the fact that Unani's core, which lay entrenched in pre-Islamic philosophical traditions, had made it evolve as a system of medicine that saw itself separate from any distinct religio-cultural belief systems. In contrast, Ayurveda was entrenched in the Hindu idea that saw the physician—the vaid—as the incarnation of God. Thus, the most famous vaid, the physician Charaka, was seen as the incarnation of the Hindu serpent god with a hundred heads; and the surgeon Sushruta was seen to have divine powers as well because he was a student of the divine physician Dhanvantari who had been recovered from the ocean along with thirteen gems.[23] Ayurveda had mixed notions of healing based on magic and

[22] Elgood 1951: 371–2. He makes the point that in the period of the Abbasids, Unani borrowed extensively from Ayurveda both its pharmacopoeia as well as medical theory, but soon forgot its debt to India. The notable exception is *Firdaus-ul-Hikmat* of Ali ibn Rabban al-Tabari, who composed a general work on medicine for his patron Caliph al Mutawakkil. This text had thirty-six chapters devoted to Indian medicine.

[23] Sinh Jee 1927: 32–3. Also see Dunn 1976: 133–58.

superstition with psychosomatic cures of the Graeco-Roman type; but its links with Hindu spirituality remained. Unani therefore steered clear of the diagnostic theories of Ayurveda. It confined itself to benefiting from Ayurveda's surgical and pharmaceutical resources.

Unani medicine derived its cultural authority by combining ideas of Islamic cosmology and procreation with the Graeco-Roman and Sanskritic cosmological frameworks that framed the health of society. Thus the Islamic formulation about the essential unity of being (*tauhid*), the Quranic idea of seven earths and heavens, the Divine pedestal, and the Throne, and the sacred history and sacred geography that the Quran generated were integrated to Greek concepts of the universe, the latter being divided into a non-generative sphere and a sublunary sphere of generation and transformation (which included the mineral, plant, and animal kingdoms). The essential unity of being (*tauhid*) so central to Islamic philosophy made it accept without problem the Aristotelian idea of the human body as a microcosm of the macrocosm represented by universe–nature. In brief, Unani medicine took over Greek ideas of nature's functioning to understand the health of the human body, and, like Greek medicine, saw health as part of both nature and social well being.[24] In this manner, the sacred cosmology of Islam built itself on Greek cosmology and physiology and earned for itself a 'cosmopolitan' cask.

In the period of classical Islam, Unani remained very protective about its Graeco-Roman intellectual legacy and the humoral theory central to health. The four humours— red and black bile, blood, and phlegm—had to be balanced for good health. These had also to be in harmony with the four elements: earth, air, fire, and water. The idea of a porous body conditioned by its environment was the key to Unani diagnostics and therapeutics. To this basic idea was added the knowledge of Aristotle's rationality, which focused on the functioning of organs, and Galen's contribution to human anatomy.

The development of Unani as a mechanistic medicine separated from religion and deriving from pre-Islamic philosophical traditions caused a clash with theologians and those who saw medicine as a cultural system of beliefs embedded in the Quran and the Hadith (traditions of the Prophet). And theologians were not the only opponents.

[24] For a discussion of this, see Good 1992; and Khan 1986: 21–9.

As Islamic mysticism—Sufism—grew in popularity and respectability in the later Abbasid caliphate, its thinkers attributed healing powers to saints and shrines and questioned the claims of medicine. For instance, the prominent Sufi thinker al Ghazali of the twelfth century was a staunch critic of medicine.[25]

The rejection of Unani by the theocracy traditionalists was particularly strong towards the end of the Abbasid caliphate. This period saw in the hakim a serious lack of true faith. A religion *vs* medicine debate was triggered at the end of the period of the caliphates. As the debate picked up, a defence of medicine began by people who claimed religion for medicine. A certain Abd-al-Wadud son of Abd-al-Malik, who lived in this period, wrote an essay translatable as 'the blamelessness of making a living from the craft of medicine' (*Fi Dhamm at-takassub bi-sina at at tibb*). Here, he argues that medicine cannot be seen as anti-religion since its subject (*maudu*) is man—the noblest of God's creation. This, he argues, not only makes medicine fall within the framework of religion, but also renders it the noblest of crafts. And there were many others who defended medicine in this manner.[26]

The most strident resistance to Unani came from a competitive movement based on indigenous Arab medical traditions. These traditions were sanctified by being attributed to the Holy Prophet—who may have given medical advice occasionally. This medical tradition referred to the Quran and the Hadith as the highest referents of legitimacy and transformed tribal Bedouin medical lore into an intrinsic part of the holy legacy of the Prophet.[27] This Islamically inspired medical system is referred to by anthropologists and historians as Prophetic medicine, and it competed with the Arabic-Persian or Galenic-driven Unani.[28]

By the fourteenth century, exclusive texts on Prophetic medicine began to be written as a genre of medical literature, separate from

[25] Dols 1984: 40.

[26] For an interesting discussion on the medicine *vs* religion debate in the twelfth–thirteenth centuries see Rosenthal 1990: 519–32.

[27] For a discussion on whether the Prophet's medicine was of divine origin, or a codified tribal practice, see Burgel 1992: 44–62. Also see Elgood 1951: 63–7; Rahman 1987: 41.

[28] Burgel 1992; Dunn 1992: 133–58.

Unani. These derived from the Hadith compendium. As Prophetic medicine came into its own, it expanded on the Hadith works and ran into innumerable volumes, each sometimes 400 pages. Its texts aimed to counter the authority of Galenic medicine via its appropriation of the Prophet's authority and to thus weaken the hold of physicians. This literature was written by traditionalists who were attempting to spiritualize and Islamicize medicine, deploying the Quran and the Hadith. They steered clear of medical theory and instead offered advice and prescriptions on the prevention of disease. They also tried to make medicine comprehensible and accessible to the layman.[29] They defined health in the Islamic tradition to mean spiritual, psychological, physical, and moral well being. Medication was not rejected, but was seen as a part of this Islamic notion of well being.[30] Thus the two fourteenth-century authors, Ibn Qayyim al Jauziya and al Surramarri, were among the early contributors to the genre of Prophetic medicine. These authors in fact combined divinely ordained medicines associated with the Prophet with those offered by medical pharmacopoeias.[31]

Finally, the Greek and Indian imprint, rather than the Islamic influence, was most evident in the Unani theory of governance corresponding with a similarly structured political theory of governance. Such correspondence had been a feature of both Greek and Indian political culture. The correspondence of medical and political theories of governance had, as we saw, been a running theme in Greek philosophy whereby Aristotle made the king alert to the maintenance of the physical health of both his physical body and that of his body politic or kingdom. Unani found this idea replicated in Ayurveda, where, as we saw, medical texts and texts on politics—the *Dharmashastras*—were put together by the same Brahmanical castes to consolidate the position of high-caste kings.[32] This idea of medical theory aiding a political theory of governance found echoes in Islamic theories of political governance as well, penned by Muslim philosophers such as Nasir al-Din Tusi in the thirteenth century. Muzaffar Alam has shown that the *akhlaaq* genre of

[29] Rahman 1987: 42.

[30] Ibid.: 45.

[31] Ibid.: 42–3.

[32] For an excellent discussion of this, see Trawick 1995: 279–96.

Persian literature, which discussed theories of governance in Islamic lands, discussed shariat in its non-juridical sense, and defined justice (*adl*) using the terminology of medical rationality derived from Greek philosophical traditions. Thus, a just society was a healthy society; justice was defined as a state of social harmony, co-ordination, and balance between different classes; as the maintenance of the equilibrium of a healthy body was the job of the physician, the role of the king lay in ensuring social balance.[33]

This notion of medical theory underlying political governance was, as we will see below, used more and more in Mughal India, as Unani medicine tried to wedge itself in society under the court patronage of a minority Muslim ruling house within a predominantly Hindu country. Indeed, the role of the king for social balance between conflicting interests gathered urgency, for the Mughals had to strive to maintain power over a religiously diverse society. In fact, in Mughal India, the court's decision to offer patronage to Unani learning and practice extended the role of the king as one who was not just metaphorically responsible for the health of society, but as the personage who must quite literally ensure the physical health of his subjects.

Unani in India

The weakening of the Abbasid caliphate from the thirteenth century ended the political and trading hegemony of imperialist Islam. But it did not affect the real and imagined contours of Dar al-Islam, along which ideas and men continued to move freely. The dialogue between Unani and Ayurveda, for instance, that had been initiated in the earlier years, continued unaffected by political change. Indeed, it only increased when Hindustan came under the political ambit of Muslims with the establishment of the Turkish Delhi Sultanate and later the Mughal empire. Both Turkish sultans and Mughal emperors were great patrons of medical learning. In the courts of the Delhi sultans, patronage was provided to translate Sanskrit texts on medicine into Persian, and Muslim physicians learnt Sanskrit to access medical manuscripts. Many of their books written in Hindustan were modelled on the Persianate

[33] Alam 2004: 55, 59, 65.

encyclopaedic style that borrowed freely from many sources. For instance, Zia Muhammad Masood Rasheed Zangi, a noted physician of Muhammad bin Tughlaq's period, learnt Sanskrit. He wrote a famous text on medicine, *Majmuah Ziai*, which reflected his use of important Persian manuscripts as well as Sanskrit texts.[34]

The appropriation of Indian healing techniques soon began to change the content of the medical literature, even if stylistically it continued to be written in the Persianate encyclopaedic form. For instance, the *Tibb-i-Firoz Shahi* of the Delhi sultan Firuz Tughlaq was stylistically modelled on the *Canon* of Avicenna and the *Zakhirah Khwarzmshahi* of al Jurjani, even as it included some description and treatments not mentioned in those texts.[35] Some of these treatments included amulets with Quranic verses, charms, talismans, and religious charts and drawings that had never been part of the Unani canon. These forms of treatment were more clearly listed in other medical texts of the period, such as the *Rahat al-Insan* of Ilias bin Shahab (1385).[36] This trend continued in the period of the Lodis, where physicians like Bahwa bin Khwas Khan learnt Sanskrit and studied its literature to enhance their medical knowledge.[37] And the well-known physician from Herat and Khurasan, Hakim Yusuf bin Muhammad bin Yusuf al-Tabib Khurasani, wrote with equal confidence on both Unani and Ayurveda in the reign of the first Mughal emperor Babur.[38] He dedicated his masterpiece, *Qasidahdar hifz i-sihhat*, to Babur: this was a metrical work on health preservation in 42 verses, and had an introductory short preface in prose.[39] Indeed, in the Mughal empire Indian vaids and hakims worked together as they shared therapies in the hospitals of Delhi and surrounding areas.[40]

[34] Nagrami 1989: 393.

[35] *Physicians-authors of Graeco-Arab Medicine in India*, Institute of History of Medicine and Medical Research (New Delhi: n.d.), p. 8.

[36] Ibid.: 12–13.

[37] Ibid.: 394.

[38] Ibid.: 408–9. See his list of books. Of particular importance was his *Tibbi Lughat*, a medical dictionary in Arabic.

[39] WMS. Per. 199 (D), 3 folios, Wellcome Library, London.

[40] Siddiqi 1959: xxxvi. In the period of Aurangzeb a hospital at Etawah founded by the local nawab, Khayr Andish Khan Kamboh, exemplified such co-operation

The Mughal Empire: The Medical Labour Market

The courts of virtually all the Mughal emperors attracted physicians from Central Asia and Iran. The exodus from there turned into a flow to India in the seventeenth century, when intellectuals preferred to leave a Safavid Iran rocked by internal strife. The Iranian hakims that flocked to India attained high status at the Mughal court; they constituted in fact the highest echelons within the medical hierarchy. Their major contribution was the production of a rich medical literature in India.

Medical texts written in India contributed also in the making of Mughal political culture. Following their precursors, they built upon the Graeco-Arab medical tradition both in their conception of the body and in their engagement with political sovereignty. Like the Graeco-Arab literature, they upheld the humoral theory of health and used it as a model to reinforce the theory of kingship. But what made Mughal medical texts different from the Arab ones was their elite authorship and textuality in the powerful court language of Persian. In the Mughal empire, Persian replaced Arabic as the language of science and medicine. And since Persian was not just the language of power and administration but also represented a distinct cultural lifestyle characterized by literary finesse and etiquette, it lent its aesthetics to medicine as well. Unani medicine in Mughal India in fact acquired a distinct stature and dignity because it was composed in Persian. And, not surprisingly, these Persian medical texts defined the ideal healthy body as one that was accomplished also in social etiquette—for example the proper norms of conduct and correspondence.

Because Unani thrived in the Mughal empire under court patronage, its association with royalty meant that medical knowledge became embodied among structures of high aristocracy, the courts, high-profile teachers, and a textuality tied to the elite language of the empire. In

---

as both vaids and hakims worked there together. In this hospital hakims like Abdu'r Razzaq Naysaburi, Abdul Majid Isphahani, Mirza Muhammad Ali Bukhari, Muhammad Adil, and Muhammad Azam worked with Indian vaids like Kanwal Nayn, Sukanand, and Naynsukh.

Mughal India there was always a tension between Unani as an aristocratic virtue tightly controlled by scribes, noblemen, and elite families, and knowledge as medical wisdom. The former represented high knowledge and was shielded in ornate Persian texts; it was epitomized by the well-heeled Persianate hakim. The latter was available in relatively simpler Persian and user-friendly Arabic manuals that were taught at the madrasas. Running across these two arenas was the emperor, the fountainhead of political power, and agents who controlled knowledge as it lay dispersed between these two ideal types, and who also wielded immense power.

Medical services were provided to society by a range of other professionals as well: ordinary hakims who had trained in the Arabic texts of the madrasas, vaids, Sufis, and faqirs. Together, they constituted a thriving medical labour market. Persianate court physicians were at the top of this market, and through their writings they helped legitimize royal authority. But the other contenders who provided services to ordinary people also reinforced the authority of the king as the welfare manager of society, for their services too ensured physical and moral health within society. Patronage to and control of this market were a critical part of Mughal governance.[41]

The Persianate Hakim

With the expansion and consolidation of Mughal political culture, the Unani tradition evolved. Its 'high knowledge' had a considerable impact on Indo-Muslim society, even as it continued to be shaped by the Persianate veneer of the time. 'High knowledge' in Mughal India was produced and controlled by Iranian physician families of hakims. These families constituted the medical community of Persian literature. As noted earlier, the antecedents of these hakims lay in Central Asia

[41] I borrow the idea of the medical labour market from Dirk Kolff's usage of the military labour market in pre-colonial India. Kolff sees this market as critical to an understanding of the transition to British rule in the eighteenth century. See Kolff 1990: 2–3. See also Siddiqi 1959: xxxvi. For instance, as noted above, in the hospital founded during the reign of Aurangzeb, by Nawab Khayr Andish Khan Kamboh at Etawah, Hindu vaids and Muslim hakims looked after the patients.

and Iran. Both Babur and Humayun had personal debts to repay to the Safavid court.[42] The Mughals were happy to extend shelter and patronage to Persian physicians (and litterateurs) who had fled to the Mughal court to avoid the distractions and dangers caused by internal strife in Safavid Iran.[43] Among the more famous of such migrant physicians in Delhi were Abul Fath Gilani in the sixteenth century, and Nur-ul-Din Muhammad Abd-ullah bin hakim Ayn-ul-Mulk of Shiraz in the seventeenth century. Another notable migrant from the same city was Muhammad Akbar Shah Arzani.

All these hakims wrote medical texts in Persian while living in Delhi. In Mughal India court patronage to Unani physicians, including to those mentioned above, ran parallel to the high status accorded to intellectuals like Abul Fazl—who wrote tracts on theories of governance for the empire. Indeed, political theory at the time normally projected the king as the sole manager of the kingdom, which he maintained by a judicious management of nature (irrigation, agricultural production, distribution, and the maintenance of harmonious relations between people), mirrored the medical theory of Unani, which too viewed, as we saw, the proper administration of nature and social conduct as intrinsic to the maintenance of the human body. For instance, both the physician Shirazi and the political writer Abul Fazl share the legacy of Greece and Rome. Abul Fazl likens the body politic to animal constitution, arguing that just as the equilibrium of animal constitution is dependent upon the equilibrium of elements, so political constitution is well tempered by a judicious division of ranks. Each rank contributes to the prosperity of the 'ideal city' with goods and virtues.[44] Abul Fazl, in fact, identifies various ranks in the body politic and their correspondence with the elements of nature. Thus, the warrior rank corresponds to fire, merchants and artificers to air, intellectuals (jurists, physicians, mathematicians, astronomers) to water, and husbandmen to earth.[45]

[42] Elgood 1951: 372–3. Shah Tahmasp of Iran had extended protection to Babur's sister. Later, Emperor Humayun had sought refuge in the Safavid court after he was defeated by the Afghans in India.

[43] For a list of Iranian physicians in Mughal India, see *Physicians-authors of Graeco-Arab Medicine in India.*

[44] Rizvi 1975: 366.

[45] Ibid.: 367–8.

Abul Fazl sees the emperor very much as the great physician and describes his role as preserver of the health of empire in his Introduction to the Persian translation of the *Mahabharat*:

> The generous heart [of His Excellency] is temperamentally inclined towards the well being of all classes of people, friend and foe, relations and strangers are all equal in his farsighted view. This [consideration for all] is the best method for the physicians of bodies, and should be highly appropriate for the physician of the soul [as well]. Why should this beneficence then not be the [distinctive] feature of [His Excellency], the chief physician of the chronic ailments of the human soul? He has noticed the increasing conflict between the different sects of Muslims on the one hand and the Jews and the Hindus on the other—the sagacious mind [of His Excellency] then decided to arrange the translations of the sacred books of both the communities, so that with the blessing of the most revered and perfect soul [the Emperor] of the age, they both refrain from indulging in hostility and disputes . . .[46]

Such conflation of medical theory and political theory was not surprising, given that Mughal medics and political theorists derived alike from the Graeco-Arabic intellectual legacy. Shirazi borrowed directly from the Greek humoral theory of health; Fazl built on the ideas of Aristotle and Plato; Greek ideas filtered into Mughal political literature through the thirteenth century Islamic political theorist Nasir al-Din Tusi, who too was influenced by the philosophical traditions of Plato and Aristotle.[47] Tusi projected the king as the sole repository of intellect who was needed for the social equilibrium essential for a just and humane society, and Fazl, influenced by the writings of Tusi, saw the king in very similar terms.

This correspondence in theories of health and politics made Unani of special interest to the Mughal emperors. Most Iranian hakims dedicated their medical works to the emperor and extolled his power. And, not surprisingly, they came to constitute the inner core of court society. Many, like the hakim Gilani, held important administrative and political positions at the court of Akbar and were envied by other courtiers.

[46] Alam 2004: 65–6.
[47] Ibid.

It was said of Gilani that 'he enjoyed much proximity to the King, and had so much persuasive influence over the King's temperament that even Jafer Barmaki did not have on [the Abbasid caliph] Harun Rashid'.[48] Gilani was appointed *sadr* and *amin* of the province of Bengal in 1579; in 1580 the *sadarat* of the capital was entrusted to him; the following year he was made its amin and diwan; he was duly honoured with a rank of 800.[49] And this trend of hakims in political positions continued in the period of Shah Jahan, who bestowed on a hakim the title of Ayn al-Mulk. During the period of Aurangzeb a hakim held the office of *Diwan-i-Buyutat*—accounts officer for the royal household—in Akbarabad and received robes of honour and presents at various occasions.[50]

The Mughals invested in medical learning just as they encouraged the production of a political literature that upheld the power of the king. Thus it seemed as politically imperative for the Mughal emperors to translate Sanskrit medical manuscripts and incorporate them into Unani texts as to translate Sanskritic Hindu scriptures and political tracts and accommodate them into the freshly evolved political definition of shariat that implied justice or *adl*.[51]

The larger political imperative that framed this Mughal interest in Unani meant that encyclopaedic medical manuscripts written in India were not practical or prescriptive. Neither had they much utilitarian value beyond the professional community of Persian-knowing medical scholars and bibliophiles. But they did bestow authority and dignity to the scribe because of the respect for the written word, particularly in Persian, within Mughal society.[52] Written generally as commissioned texts for the king, the medical literature glorified him: such considerations outweighed the usefulness of the information they contained. Many Mughal manuscripts, like the *Tibb-i-Darashikohi* of Shirazi,

[48] Ibid.

[49] Azmi 2004: 68.

[50] Ibid.: 202.

[51] For a discussion of the non-juristic definition of shariat, see Alam 2004: 26–80.

[52] For an excellent account of the history of medieval European book production and the power of the written word in the writing of medical manuscripts in Italy over the tenth to thirteenth centuries, see Jones 1990: 1–29.

were ornate and catered to bibliophiles amongst the royalty and high aristocracy.[53] Indeed, reading them and possessing them were an essential part of gentlemanly accomplishment and signified status and noble upbringing. Most Mughal emperors and nobles maintained libraries where such texts were housed for consultation by learned men.[54]

In Mughal India, to be cultured and urbane in the Persianate style was to be cosmopolitan and in tune with the historical, literary, and medical cultures of the world. Medics, litterateurs, and historians strove to integrate narratives of selfhood with those of the larger world. Thus, Mughal historians wrote world histories into which they located their own history of empire, and physicians wrote global medical encyclopaedias to situate Unani medicine in a global context. It followed that a good hakim was also a litterateur, a cultured man who had a smattering of useful knowledge about the world. He displayed this knowledge in his encyclopaedic writings and laid out the norms of proper conduct and correspondence so that people might relate to the wider world of aesthetics and culture. Not surprisingly, he defined a healthy body as a cultured body—one that had mastered the etiquettes and correspondence skills laid out in the literary genres of Persian texts.

Mughal Medical Literature: The Tibb-i-Darashikohi and the Mizan-i-Tibb

Among the encyclopaedic medical texts that borrowed freely from Indian medical practice as well as Graeco-Arabic literature was Gilani's compendium *Fattahi*, which is a commentary on Chaghmini's *Qanunchah*. As in pre-Mughal Delhi, stylistically the text followed the pattern of the Persianate medical literature of Galen, Avicenna, and al Jurjani: it was a wide canvas medical encyclopaedia that synthesized information freely from a variety of influences, certified knowledge by deploying ancient authority without necessarily taking care over sources,

[53] Ibid. This was very similar to the illuminated ornate medical manuscripts of medieval Italy.

[54] The famous Delhi poet Abdur Rahim Khan Khanan, the son of Akbar's regent Bairam Khan, had a library in Delhi that housed many Persian manuscripts. Many learned men visited the library for study and self-improvement. Law 1916: 170.

information, and internal consistency.[55] The same was the case with Nur-ul-Din Muhammad Shirazi, who wrote the *Alfaz-ul-Adwiyah*, 'a vocabulary of drugs', in 1628–9 for his patron Shah Jahan. He based his text on the *Sihah al-adwiyah* of Husayn al-Ansari. The 1850 copy of the text, compiled by the copyist Ghulam Mahomed, has 138 folios and is divided into one preface (*muqaddimah*), one body (*natijah*) and one conclusion (*khatimah*).[56] The text reveals that Shirazi collected drugs from local Ayurveda pharmacopoeias and combined these with those from Persia. He listed some 1441 items in the Arabic, Persian, and Hindavi languages.[57] Shirazi's medica caters to the humoral concept of health, qualified by natures and elements, that was central both to Unani and Ayurveda. But it nowhere acknowledges its debts to Ayurveda, or any other texts or authors. Yet wide-ranging texts, written with court patronage, endowed the scribe with aristocratic virtue all the same.

Persian medical texts were unique in underscoring the idea of health as individual well being and aristocratic virtue. They emphasized the salience of individual comportment, proper conduct, and correspondence as central to personal well being. *Tibb-i-Darashikohi* by Shirazi, written in 1645–6 and dedicated to Shah Jahan's son Dara Shikoh (1615–59), is an important case in point.[58] The original copy of the text no longer exists. We have no idea how many copies existed in the Mughal empire—probably very few, as not a single copy has survived. But we do have a 1780s copy, prepared in Surat for a Parsi hakim called Bizhan. This is a huge encyclopedic and ornate text in three volumes that cover approximately 3000 pages. It follows the Islamic encyclopedic tradition of Avicenna's *Canon*. And, like the *Canon*, it is said to have borrowed extensively from a variety of healing traditions and pharmacopoeias without always acknowledging them.[59] It is unique because of its ornate Persian style, its obeisance to the court, and the wide historical canvas that it sketches for Unani. It offers a definitive history of Unani spanning the pre-Islamic Graeco-Arab world, the period of

[55] Azmi 2004: 69–70.

[56] WMS. Per. 580, Wellcome Library, London.

[57] Ibid.

[58] Elgood 1951: 374. A copy of the text is in Bibliotheque Nationale Paris, no. 342, suppl. The MS contains 1711 folios.

[59] Ibid.

the caliphates, and then down to the pre-Mughal Indian medical tradition. The scribe underlines his aristocratic virtue and those of his readers through this wide canvas. He upholds the king's authority as manager of the health of the people, thus obtaining state sanction for his exclusive control of medical knowledge. State approval was forthcoming because he argues the idea of health being about individual well being as well as aristocratic virtue; his medical ideas reinforce the social hierarchies of society. Shirazi thus delves into the philosophy of health even as he seeks to administer the body through the external agencies of proper conduct, comportment, and diet. For him, health is about individual well being, and an ideal state that can be achieved by a select few through proper comportment.

The first volume of the *Tibb-i-Darashikohi*, copied by the scribe Mohan Rai Kayasth, is on *ilmi-tibb* (theoretical medicine). It focuses on the making of the individual (*insaan*) and explains the meaning of reality (*haqeeqat*). It uses the philosophical wisdom of Hippocrates, Aristotle, Galen, and Avicenna. It does not always show due reverence to its sources. It incorporates their idea of the healthy individual as being in harmony with nature: thus, it gives details of botany, minerals, and zoological descriptions to emphasize the idea of well being structured in the relations between the individual and his cultural context .

The second volume, copied in 1193 hijri, concentrates on *amali tibb* (practical medicine). It follows Hippocrates' dietetics as a guarantee of health. It offers copious details of comportment, conduct, diet, and deportment that are critical for harmony between individual and society. It argues that proper individual comportment prevents disease and keeps the body healthy. Here health is very clearly about social well being and gentlemanly conduct, in the Persianate style.[60]

The third volume discusses surgery, bloodletting, and the diseases of women and children.[61] It details the aspects of medical science that must be resorted to in cases of ill health.

The tradition of writing medical encyclopaedias in India corresponded to a similar trend in the penning of world histories by Mughal scholars. As Muzaffar Alam has shown, one such important text of

[60] *Ilajat-i-Dara Shikuhi*, 342A, vol. 2, Paris.

[61] *Ilajat-i-Dara Shikuhi*, 857–9. Supplement Persan 342, 342A, 342B.

global history was the *Rauzat al-Tahirin*. Tahir Muhammad Imad al-Din Hasan Sabzawari compiled the book in 1603. His work integrated the history of his master, Akbar, into a larger history of the world, once again spanning the pre-Islamic Greek philosophers, the Pious Caliphs, and India's pre-Islamic kingdoms. All the people Sabzawari included in his history were, as the title suggests, among the *tahirin*—namely, the pure, clean, and holy. Sabzawari's book was a history of mankind and the author saw himself as an inheritor of the heritage of universal humankind.[62]

This tradition of writing medical encyclopaedias continued in the period of Aurangzeb, who is said to have weathered the worst political and economic crisis in the empire and overseen its demise. In this period of imperial crisis (1700), Muhammad Akbar Shah Arzani produced several works. The more famous ones are the *Tibb-i-Akbari* or Akbar's medicine, and the *Mizan-i-Tibb* or the scales of medicine. Both follow the tradition of Shirazi in borrowing freely from local influences in their choice of drugs and treatments. The *Tibb-i-Akbari* is the Persian translation of the Arabic text *Sharh al-Asbab wa al-Alamat* of Nafis bin Aiwaz Kirmani (d. 1449). But in 1700 Arzani constructed and added his own commentary to it as well.[63] The *Tibb-i-Akbari* tries to emulate the *Tibb-i-Darashikohi* both stylistically and in its content. It was composed in 1700, after the Mughal conquest of the Deccan. It covers a wide canvas in its twenty-seven chapters, which deal with the history of Unani, symptomatic treatments of local and general diseases, and the properties of medicines and comportment regimens for individual good health.[64]

But the *Mizan-i-Tibb* is different from the average Mughal text. In its concise and user-friendly style it bends more towards providing

[62] Alam 2004: 67.

[63] Azmi 2004: 227.

[64] For the British Library copy of the *Mujarrabat-i-Akbari* manuscript, see Add Ms. 17954. This has 114 folios. The different copies of the *Tibb-i-Akbari* at the Wellcome Institute Library include those copied in the years 1790, 1813, 1818, 1826. WMS. Per.172, 374 folios, 1790, copied in Peshawar; WMS. Per. 564, 185 folios, 1813 AD copyist Haji Muhammad; WMS. Per. 566, 171 folios, 1818; WMS. Per. 563, copied for Sultan Mahmud, 1826; WMS. Per. 165, 525 folios, copied in the eighteenth century by Abd Allah for Sikandar Ali and Chiragh Ali. See Keshavarz 1986: 159.

useful medical knowledge as a guarantee of individual well being. Hakim Arzani makes his intention to write a popular text of Unani clear on the opening page: 'My children and relatives were too busy to devote time to the study of the science of medicine. So I wrote this brief text to make simple the teaching of medicine. Even though it is a brief treatise it has many benefits. I named it *Mizan-i-Tibb*. I hope that God, the biggest hakim, should make it successful.'[65] The *Mizan-i-Tibb* claims to be a handbook of medicine for beginners.[66] So, stylistically, it stands in sharp contrast to the voluminous and ornate Persian medical texts of the period. It is a short work of 48 folios, written in ordinary nastaliq style. The 1742 copy of the text seems to have been written by more than one scribe. On folio 14 the scribe appears to have changed. However, the nastaliq calligraphy is maintained; and the original scribe resumes writing after this one folio. More noteworthy is the effort of a third scribe, who makes corrections in the spellings and contents of the text, and also offers supplementary clarifications and explanations in the margin. These marginal notes are in Persian. At times, the scribe translates the Persian into Hindi, using Devanagari to make it easy for the lay reader to recognize the medicines recommended.[67] And at times the equivalent Hindustani names of Persian medicines are listed in the margins to make it easy for the lay reader to follow prescriptions. Thus, the Hindustani word *kewra* (a sweet fragrant edible potion for digestion) is inscribed in the margin to explain its Persian equivalent, listed in the text as *sharbat-i-kadar*.[68]

The *Mizan* is divided into three sections (*maqalahs*), each of which lay out medical wisdom rather than aristocratic virtue as the key to good health. The text makes clean break from the Mughal practice of offering doses of aristocratic virtue to maintain health, laying out

[65] Add Mss. 17949, 48 folios. British Library, London. This is a 1742 copy. See folio 1, preface.

[66] The different copies of the *Mizan-i-Tibb* at the Wellcome Institute Library include those copied in the years 1811, 1818, and 1844. See WMS. Per. 179 (A), 112 folios dated 1811; WMS. Per. 515 (C), 95 folios, 1818, copied by Mirza Awlad Ali; WMS. Per. 514, 111 folios, 1839; WMS. Per. 126, 1844. Keshavarz 1986: 84.

[67] Add Mss. 17949, 48 folios. British Library, London. This is a 1742 copy, f. 15. There is a word in Devanagari inscribed in the margin.

[68] Ibid.: f. 1.

instead fresh norms for the making of a gentleman, norms that have more to do with medical knowledge than comportment; the Mughal-style dictums on individual comportment are extremely conspicuous by their absence. The first section is on *kaifiyats* (temperaments), the second on the properties of simple and compound drugs, and the last on the symptoms and remedies for local diseases.[69]

Arzani explains to the lay reader the four *kaifiyats* of the human body. These are: hot (*hararat*), cold (*sardi*), moist, and dry. The symptoms of the first are excessive thirst, burning, and paleness of the body. They can be caused by excessive bile or blood secretion. The former cause leads to the body turning yellow, the tongue becoming hard, and the mouth tasting bitter. This is followed by weakness and a sensation of coldness. The latter makes the body lethargic, red in colour, and causes bleeding from the gums and nose. Arzani explains that, in contrast, the cold condition of the body (*sardi*) leads to lack of thirst and a burning sensation. In the case of excessive *sardi*, there is phlegm. The body becomes white, lethargic, and soft. Indigestion and a running nose take over. This condition is due to an excessive secretion of yellow or black bile. Man becomes thin and weak, and loses his appetite. The body becomes black or pale, depending on the colour of the bile that causes this condition. Arzani clarifies to his reader that dryness and moistness of the body are related to heat and cold conditions. And these are in turn connected to the secretion of blood and bile.[70]

The *Mizan* does not stop at spelling out clearly the four conditions of the body and their symptoms. It also makes intelligible to the lay reader the physiological changes that cause these conditions. And here a narrow medical definition of health in terms of body physiology rather than comportment is highlighted. Arzani—in sharp contrast to the *Tibb-i-Darashikohi*, which emphasizes comportment as critical to physical fitness—maintains that a body that secretes its four humours (blood, phlegm, yellow bile, and black bile) in the correct proportion is a healthy body.[71] In the last section, on medicines, Arzani lists simple and compound medicines rather than deportment norms through

[69] Ibid.
[70] Ibid.: f. 1.
[71] Ibid.

which the balance of the four humours can be maintained.[72] Indeed, this section extends all through the text whenever diseases relating to the gall bladder, urinary bladder, illness specific to males, and females, colic pains (*quolaj*), and pleurisy (*zaat-ul-janb*) are discussed along with their symptoms and cures. Emphasis on the knowledge of medical physiology and medicines that offered correctives made such wisdom, rather than aristocratic virtue, the cornerstone of health and 'civility'. It redfined a learned gentleman.

Both these texts by Arzani come nowhere close to the encyclopaedic intellectual range of Shirazi's *Tibb-i-Darashikohi*. They are not even comparable stylistically or in their notions of health. They reflect the drift from health as aristocratic virtue to health as medical wisdom—science. To some extent, they reflect turn-of-the-century changes in late Mughal society. Written at the beginning of the eighteenth century the *Mizan*'s easy-to-read style and accessibility reflect the 'vernacularization' of Persian and the loosening of the monopoly of scribes, families, and court over the medical knowledge that it embodied.[73] The simplicity of the text reflected also the austerity of Arzani's patron Aurangzeb and his redefinition of gentlemanly conduct in the language of doctrinal Islam. The texts radiate the austerity associated with doctrinal languages like Arabic with its emphasis on the utilitarian character of science and theology.[74]

[72] Ibid.: ff. 1–7. The medicines that are listed include: *kasni, kismis, gul-i-surkh, aab-i-limu, shikanji, sharbat-i-unnab, sharbat-i-sandal, sharbat kewra kadr*. These folios have listing of the disease, its symptoms and the medicines recommended.

[73] Alam 1998: 317–49. Alam shows the attention the Mughals were forced to give to Hindavi/Braj in the early eighteenth century as a result of regional assertion. He argues that this affected the dominance of Persian. Mughal attempts to purge Persian of Hindavi influences followed. So did the move to Persianize the vernaculars.

[74] It is not surprising that the use of the *Mizan-i-Tibb* and the *Tibb-i-Akbari* outlived the Mughal empire. They were seen not just as library items in the collections of the elites, but used as teaching primers until the early twentieth century. They were recopied by different copyists throughout the late eighteenth and nineteenth centuries and used as texts in the syllabus of the Persian schools set up by zamindars and rajas in Bengal province. They were in use also at the Calcutta Madrasa. See British Library catalogue, Egerton 1006 & 1007, being

Indeed, the influence of Arabic religious literature was very evident in the reading patterns and writing styles of late Mughal society. We noticed above that Arzani translated the *Tibb-i-Akbari* directly from an Arabic text which circulated in Indian society. He complemented this with the translation into Persian of Jalal-ul-Din al Suyuti's Arabic text *Tibb-i-Nabawi* or the medicine of the Prophet.[75] This text defined comportment along the Arab tradition of the Prophet's life, in contrast with the more elaborate Persian forms of comportment popularized by the *Tibb-i-Darashikohi*. The translation of such Arabic texts was new to Mughal medical collections. Their inclusion in Mughal compilations, along with the circulation of easy-to-read Persian medical manuals, signalled the weakening control of Persian-knowing families over medical knowledge. It also reflected the vernacularization of Persian while revealing the increasing influence of new forms of knowledge streaming in from the Arabic-speaking world in the late eighteenth century.

The tension between a Persian-derived Unani grounded in aristocratic virtue, and an Arabic-oriented revealed science had always existed in Mughal society, for hakims of smaller stature who served in Mughal hospitals and interacted with the public were conditioned in the more scientistic Arabic-style healing. This rung of practitioners acquired its rigorous professional training and the specialized knowledge of medicine in Mughal madrasas which combined Persian learning about polite conduct and literary flair with the rigour of Arabic texts on theology. Medicine, science, and natural philosophy continued to be taught in Persian. Medical texts were taught in the madrasa curriculum as part of natural philosophy, along with theology, logic, and rhetoric.[76] However, Persian texts of 'high knowledge' rarely figured in the madrasa curriculum. Instead, texts of medical wisdom, such as Persian translations of Avicenna's Arabic *Canon*, were prescribed. Medical learning was given a Persianate cultural polish in the family clinics of the leading hakims of the city. Because students from madrasas trained as apprentices to Iranian hakims of the court, many gained access to ideas within the

---

vols 1 & 2 of *Tibb-i-Akbari*. For the medical syllabus of the Calcutta Madrasa, see Leitner 1882: 74. See also India Office Collections (IOC)/v/24/942, 6, 39.

[75] Elgood 1966: 8; Elgood 1951: 63.

[76] Alam 1998: 326.

ornate Persian medical manuscripts that their masters had authored or collected—this crowned their scholastic training in the works of Avicenna in the madrasa. The tension between the two traditions intensified in the mid-eighteenth century, when Persian-knowing families experienced a weakening of their hold over medical knowledge.

New Medical Learning in Arabic: Unani in Eighteenth-century India

In the eighteenth century, the collapse of the Mughal empire alongside the popularization of Persian made scribes and families recreate their monopoly over medical knowledge by hooking it on to the doctrinal language, Arabic. The cloud of Arabic had loomed over medical knowledge from the days of Aurangzeb, but in the eighteenth century it became the language used by scribes and families to store medical knowledge away from the reach of society at large. It was used to restore the prestige of older families and scribes who were worried about the spilling out of their medical knowledge by the increasing accessibility of Persian.[77]

The increased use of Arabic in medical circles initiated a new kind of medical learning. Medical texts in Arabic, like the religious knowledge of the period, were influenced by the austere doctrinal trends flowing in from the Arab world—the Wahabi-style return to the scriptures and the person of the Prophet.[78] These trimmed medicine of its comportment frills and projected it as science, aristocratic virtue going largely out of the window. But unlike religious knowledge, the austere medical literature was never popularized. It remained under the control of scribes and families who taught select students; the control of medical wisdom gave them power. Away from the gaze of regional courts, they consolidated their hold over the new knowledge. Persian medical learning did not die out entirely, for regional courts continued to patronize its production and maintain those who still sought to preserve it. This

[77] The shift to Arabic began to give medical knowledge an Islamic slant. And this may have intersected with the other forms of Islamization that were under way at the regional level in this period. See Eaton 2001: 189-202; Eaton 1993.

[78] For a discussion on the new learning in religious knowledge, see Robinson 2000: 44–65.

kept alive the idea of health as individual aristocratic virtue. Simultaneously, new ideas of health as medical wisdom took root in society.

Arabic Medical Learning

The collapse of the Mughal empire in the early eighteenth century caused a devolution in the patronage of medicine to new regional courts, local rajas, and zamindars. In these regional states, the dignity of medical science in the Persian language was preserved and its centrality in the making of an accomplished gentleman continued. Regional courts like those of Oudh in the north and Tipu Sultan in the south maintained huge libraries well equipped both with copies of Mughal Persian medical texts as well as fresh texts composed in Mughal style.

In the early nineteenth century Alloy Sprenger, assistant to the British Resident at Lucknow, surveyed the 'Muhammadan libraries' of Lucknow and reported that they had thousands of Persian manuscripts. These ranged from history, philosophy, and literature, to science, astronomy, medicine, archery, hunting, and sports. A small collection of Arabic manuscripts also existed. In the kings' libraries alone—the Topkhana and the Farhad Baksh—there were at least 6000 Persian and 310 Arabic manuscripts.[79] Besides these royal repositories, there existed in the city private libraries, e.g. in the Moti Mahal, maintained by rich bibliophiles. These too were plush with Persian manuscripts.

A big component of these collections was medical manuscripts: for instance, the Topkhana library of the king had forty-eight manuscripts on Unani medicine. The texts included the oldest Persian medical books, like the *Al-a-gharaz al Tabiah wa almbahat al almeiah* of Ismael al Hasan Jorjany, who dedicated it to Abu al Muzaffar, son of the Khwarzm shah. Books dedicated to the Mughal emperors Shah Jahan and Aurangzeb, such as the *Ganj Baadadard Sahib Qarani* by Imanallah, and Arzani's *Qarabadeen Qadiri*, also figured in the collection. Sprenger noted a range of Arabic medical authors as well in these collections.[80]

[79] A. Sprenger, assistant to Resident in Lucknow, to H.M. Elliott, Foreign Sec. to Govt of India, 18 March 1849, Selections from the records of the Govt of India, Foreign Dept, no. cccxxxiv, serial no. 82. Report of the researches into the Muhammadan libraries of Lucknow, by Sprenger 1896: 18–19. He catalogued only 11,453 manuscripts in the Topkhana.

[80] Ibid.: 19, 29–30.

In the southern kingdom of Mysore, Tipu Sultan too had a huge collection of Persian medical manuscripts in his library. These included a copy of the *Zakhirah-i-Khwarzmshahi*, the *Dastur al ilaj* of the pre-Mughal period, and the *Tibb-i-Akbari*, the *Qarabadeen-i-Qadiri*, and the *Tejurribat-i-Akbari* of Muhammad Arzani. Some home-grown Persian manuscripts, like the *Bihr-al Munafi* or the 'Sea of Profit' (a treatise on midwifery) of Mulud Muhammad, were dedicated to Tipu.[81] The collection included a few Arabic medical texts, e.g. Avicenna's *Canon*, and the commentary on the *Sharh-i-Mujiz* by Ala-ud-din Ali Bin Abu al Hazim al Qureishi.[82]

Beyond such courts and capital cities, the rajas, zamindars, and nobles, who had become entrenched within local society, looked for new means to legitimate their authority. Consequently both the authoriality and textuality of Persianate medical knowledge began to be realigned. At one level, it was critical to break out of the Mughal norms of gentlemanly culture. But, more importantly, the vernacularization and popularization of Persian weakened the court language's efficacy as a tool through which families could control medical knowledge; families looked towards Arabic to reconfigure their monopoly. Thus, outside the capital cities of post-Mughal society, the dignity of medical science shifted from the Persian in which it had rested since the time of the Mughals, to Arabic, which was now projected as the custodian of the Arabic sciences, and in the late eighteenth century one could be a cultured gentleman without having read a single Persian text of medicine.

The drift towards a doctrinal language, Arabic, to control tightly medical knowledge stood out in a period marked by regional assertion. It also fitted in with the general orientation of society towards the new religious learning that was streaming in from the Arab lands. In the late eighteenth century the Wahabi-inspired ideas of Arabic scriptures and the salience of the Prophet gave religious knowledge a new, austere, and Arab orientation. Medical knowledge succumbed to this trend. In fact this trend helped it become the exclusive preserve of medical individuals, families, and their clientele. In short, Islamization through Arab learning

[81] For Persian medical manuscripts in Tipu Sultan's collection, see Stewart 1809: 108–13.

[82] Ibid.: pp. 114–16.

helped communities of Perso-Arabic medical literature retain their status and added new dimensions to the formation of regional identities.[83]

But the Arabization came with a price. Persian had accommodated into itself the Hindu and Muslim scientific traditions, in keeping with an encompassing and inclusive Mughal political culture. The preservation of medical learning in Arabic now fractured this legacy along religious lines, for Arabic was widely perceived as the universal language of the medical science of *Muslims*. And, at the fag end of the Mughal innings, when regional identities were being sharply articulated in local languages, it was Muslims who established pan-regional links in science, medicine, astronomy, and astrology in Arabic. Hindus meanwhile increasingly consolidated their scientific legacy in Sanskrit. Therefore, the distancing of Unani from Persian meant a moving away also from the composite imperial medical culture that it represented. The break was, however, gradual and never complete, for Persian language and literature continued to be taught in village and family schools widely attended by both Hindus and Muslims. All the same, the science and medicine that were taught in Arabic schools predominantly attracted Muslim students.

As Unani withdrew into the Arabic fold, the religious divide became noticeable. Of the 62 Arabic students in a village school in south Bihar, only two were Hindus and the rest Muslims. In contrast, in the Persian school there were 865 Hindus and 559 Muslims.[84] The reason for a more mixed population in the Persian schools was that they taught more elementary and grammatical works, forms of correspondence, popular poems and tales, and almost no sciences: these were relegated to the Arabic schools. Similarly, in the district of Burdwan in Bengal, there were 93 Persian schools with 899 Hindu and Muslim students, and only 8 Arabic schools with 51 students—out of these, only 4 were Hindus.[85] In this manner the preservation of medical knowledge in

[83] Asim Roy talks about the Perso-Arabic reading elites isolating themselves from the evolving syncretistic regional culture of pre-colonial west Bengal. His focus is more on their preservation of religious knowledge through the resort to Arabic. Roy 1983.

[84] Adam 1941: 28–83.

[85] Ibid.

Arabic and its communitarian slant added a new element to the vernacularization of new religious knowledge, and regional identity formation in eighteenth-century India.

As Arabic came to be perceived as the language of high knowledge, the rajas, rich families, and notables too separated Persian and Arabic learning. They made the latter exclusively the language of scientistic medicine and the former the sole language of polite culture and Persianate accoutrement. As Persian slipped beyond their control, Muslim elites monitored medical knowledge more and more tightly by relegating it to a few Arabic family schools. As medicine and theology became the preserve of these Arabic schools, and literary norms and etiquette the hallmark of Persian schools, an exclusivist hakim was gradually created. Unlike the Mughal hakim, he was not necessarily *au fait* with the literature of polite conduct. Health for him was about medical wisdom and useful knowledge, not aristocratic virtue.

Tighter control was most evident via the small number of Arabic learning institutes set up by medical families. These contrasted with the proliferation of Persian language and literature schools. Persian schools made Persianate Mughal cultural norms of conduct and deportment more easily accessible to a greater range of people, and this popularization also redefined what it meant to be a cultured individual in the late-eighteenth-century countryside. The markers of a well-heeled individual were no more measured by his knowledge of medical learning, logic, and natural philosophy. Instead, Persianate literary knowledge and familiarity with the literature on polite conduct or *akhlaaq*, and correspondence or *insha*, was enough to confer status. This new cultured man, deficient in medical knowledge, emerged alongside the medical professional who specialized narrowly in Arabic medical texts and who was trained by scholars in Arabic schools.

From the late 1820s to the mid-1830s the English East India Company conducted many surveys on the state of native education in North India. Almost all reporters commented on the proliferation of family-run Persian and Arabic schools in the towns and countryside of North India. These were said to be of recent origin and exemplified the change in post-Mughal society. In 1829 an education committee reported that Bareilly district topped the list in instruction. More than 300 seminaries existed in the district, which included Persian schools '*maktabkhaanas*', Hindi schools '*chutsals*', Sanskrit schools '*putsals*', and 17 teachers who hawked their services teaching the Arabic language and sciences.[86]

The Arabic schools were few and with limited students, reflecting an exclusive clientele and hakim families. I.M. Boulderson, the Collector of Bareilly, said the 17 persons who hawked their services as Arabic teachers collected students in their houses, which were then called madrasas. On the whole, 114 students in the district obtained instruction from them in the language and sciences.

Such family schools received no endowments from regional rulers. They relied on individual or family investments. This ensured their continuing hold over a limited clientele. The houses of notables and families doubled up as schools and were designated madrasas or seminaries. In many such madrasas, Persian and, occasionally, English were also taught. But the sciences were taught exclusively through Arabic texts. I. Davidson, Sub-secretary to the Board of Revenue, reported to the government various details about Arabic teachers and their instruction styles. Most, he said, came from reputed Arabic-knowing families. Many came from theological families with long traditions of mastery over Arabic religious knowledge. They were all keen to control medical knowledge and restrict it to a chosen few.

His list of medical instructors using Arabic texts ran as follows: Maulvi Niaz Ahmad, a 'learned man' and an 'excellent astrologer' who instructed 12 youths in the various sciences. Hafiz Ghulam Ahmad Qazi and Maulvi Ghulam Hazrat were 'good' Arabic scholars who instructed people in sciences and 'whatever they wanted'. Maulvi Ghod Muhammad—he had 10 pupils for Arabic instruction. Maulvi Syed Mahomed, Maulvi Suttullah, Maulvi Muhiuddin and Maulvi Yaqub Ali had pupils ranging from 4–15. Maulvi Mullook Shah had 10 pupils and instructed them in the sciences. Akbar Abdul Rahim had 2 pupils whom he educated in the medical arts exclusively. Muzzur-ud-din Husain Sheikh had 15 students whom he instructed in Arabic, Urdu, and Persian languages. He was supposed to be a 'good' mathematician. Boulderson pointed out one Kunwar Rutten Singh as an Arabic

[86] L no. 463, Education Committee to Vice President in Council, 9 July 1827, Extract, Bengal Pol. Consult, 5 June 1829, Boards Collection, F/4/1170, file no. 30640, 511.

teacher in the city who also gave instruction in English. His son Kunwar Dowlat Singh was also an Arabic teacher of the area.[87]

In contrast, the Persian schools were numerous and had far greater numbers of students. There were reported to be 1485 students in the district studying in such schools in 1827. Persian literary learning was of course not monopolized by individuals and families: ordinary people made their houses sites of Persian instruction, hiring teachers to instruct students. Even though their salary was as low as Rs 3–4 per month to Rs 8–10 per month, such teachers were much respected. They obtained gifts (*nazar*), food, and clothes from their students and families. Boulderson gave a detailed break-up of schools in every pargana of the district. He concluded that the major change that had set in from Mughal times to the early nineteenth century was that both Persian language and learning and the knowledge of sciences in Arabic had become more accessible to ordinary people. This was particularly true of Persian learning: 'we gather that under the former governments none but ushrafs viz Brahmins, Rajputs, Bukkals, Kaits and Khattri among Hindus and Sheikhs, Syuds, Moguls and Pathans of the Muhammadans were permitted to study sciences or even learn Persian language. But now all learn Persian.'[88]

In the 1820s the education committee set up by the Company to survey native education lamented that the texts taught in Persian schools were of little relevance to present-day requirements since they were 'chiefly epistolary compositions or compilations'. Their object was the 'acquirement of skill in correspondence, or rather in the use of approved modes of phraseology and address, an accomplishment of some value under the Muhomeddan administration, but little calculated to exercise and invigorate the intellectual faculties'.[89] This was corroborated by Boulderson, who noted that Persian schools taught no medical texts. They concentrated on literary texts of high culture associated with the

[87] I.M. Boulderson, Collector, Bareilly, to Holt Mackenzie, Sec. to Gov. Gen., 29 January 1827, Extract, Bengal Pol. Consult, 5 June 1829, Boards Collection, F/4/1170, file no. 30640, 633–6.

[88] Ibid., I.M. Boulderson, Collector Bareilly, to Holt Mackenzie, Sec. to Gov. Gen., 29 January 1827, 617, 582–7.

[89] Ibid., L no. 463, Education Committee to Vice President in Council, 9 July 1827, p. 515.

Mughals. These included Mughal favourites, such as the '*Boostan*' and '*Goshtan*' (*Gulistan*), romance texts read and enjoyed by the Mughal emperors and nobles. They also had texts like *Zuleikha*, *Neel Daman*, *Laila Majnu*, and *Bahar Danish*. Also listed in the curriculum were texts by court historians like Abul Fazl, and the Mughal *insha* literature, e.g. *Insha Khalifa*. The masnavis of important poets, the divan of the Persian poet Hafiz, *Divan Nasir Allee*, *Divan Shumsher Khan*, and the *Rookat Alumgiri* were also important readings.[90]

Indeed, in the Persian schools of the North West Provinces and Agra, so great was the emphasis on 'polite education' and the neglect of arithmetic and the sciences that a course of study was considered incomplete if it did not include the usual forms of 'address and correspondence'. In the 1830s J. Kerr, who reviewed indigenous schools in the area, lamented that 'large fields of knowledge which would open and improve the minds of the pupils are left uncultivated, such as to go no further in history, geography and natural philosophy'.[91]

A similar trend was evident in the Punjab as well. At the time of annexation, British commentators reported that medicine was taught only in the Arabic schools along with Arabic law and literature. The better Arabic schools were housed in exclusive buildings. Here, teachers taught Aristotelian natural philosophy, logic, rhetoric, and philosophy through Arabic texts like Aristotle's *Shifaa*. But medicine was taught by Arabic-knowing individuals in exclusive enclaves as well. These ranged from family homes to school sheds. Indeed, in the Punjab too Arabic was also used by influential families to tighten their control over the fast popularization of high etiquette Persian culture. Families tried here to monopolize their control over comportment as well through the use of Arabic. Many of the seminaries taught Arabic texts on polite conduct and correspondence rather than leave the students to Persian literary influence. Thus, the polite Arabic letter writer, the *Ajab-ul-Ajaib*, was taught in schools along with the Arabic medical texts of Avicenna.[92]

[90] Ibid., I.M. Boulderson, Collector, Bareilly, to Holt Mackenzie, Sec. to Gov. Gen., 29 January 1827, p. 587.

[91] Kerr 1852: 155.

[92] Leitner 1882: 74.

In the 1830s William Adam, who reported on the state of native education in Bengal and Bihar, noted a similar trend in the area. That medical knowledge was narrowly controlled was evident from the fewer Arabic schools in which it was taught. For instance, in the south Bihar district there were 291 schools, of which 279 were Persian and only 12 Arabic. In the former there were 1424 Persian students and the latter had only 62.[93] And again, as in Bareilly, many Persian schools had endowments from the local raja. In contrast, Arabic schools were established by families, individuals, and notables. The endowments for Arabic schools ranged from modest contributions by an individual teacher of Rs 250–Rs 200, to large ones that ranged from Rs 15,000 to Rs 50,000 by local Muslim notables.[94]

There was a near-absence of medical texts in the curriculum of Persian schools. They focused on Persian literary texts, e.g. *Insha-i-Herkern*, on forms of correspondence, verses of Persian poets like Urfi, Hafiz, Ghani, and Khakani, and the works of Khusro.[95] Only in rare instances were the austere-style Persian medical texts of the later Mughal period listed. These included the more simplistic Persian translations of Arabic texts like the *Tibb-i-Akbar* of Aurangzeb's Hakim Arzani.[96] The *Mizan-i-Tibb* of the same author, also austere in style, and imparting medical wisdom, was sometimes used.

Indeed the predominance of Arabic as the vehicle of science and medicine in the region prompted British observers to suggest that medical books in the Arabic language taught at the Calcutta Madrasa be introduced in government-run Punjab schools. This they felt would only reinforce and systematize their medical instruction.[97] In 1829 the education committee similarly suggested that the medical instruction going on in the Arabic family schools of Bareilly could also be improved and systematized by the introduction of the Arabic scientific texts used in the Calcutta Madrasa.[98]

[93] Adam 1941: 286–7.

[94] Ibid.: 281–3.

[95] Ibid.: 284.

[96] Adam 1838: 70–3.

[97] Leitner 1882: 74.

[98] L no. 463, Education Committee to Vice President in Council, 9 July

For Bareilly, Boulderson gave a list of Arabic texts that the students read in the family schools. Significantly, medicine was taught in conjunction with other doctrinaire disciplines like law, logic, and theology. These included texts on law and jurisprudence like the *Ilm Fiqh*, commentaries on the Hidaya—*Shareh Hidayah*; logic—*Ilm Mantaq*; and Arabic-language primers, *qaidah.* Such texts rubbed shoulders with medical texts like the *Sharh-i-Mujiz* and the *Mizan-i-Tibb*. All these were oriented towards the revealed Arabic sciences and were influenced by the new learning from Arab lands. The only Mughal medical text included here was the early-eighteenth-century text *Mizan-i-Tibb* of Arzani. The *Mizan*, as we saw above, was brief, austere in style, and scientistic.

Adam reported a similar trend in the Arabic schools of Bengal and Bihar. These schools too concentrated on texts on astronomy and natural philosophy that included medicine. Often, the same families and individuals which preserved medical knowledge in Arabic continued to earn their living by popularizing literary knowledge in Persian. Indeed, they used their knowledge of Arabic to write different kinds of popular literary texts in Persian. For instance, in the district of south Bihar, two maulvis who were highly rated in Arabic learning had, Adam reported, authored many texts. Maulvi Ghulam Hussein of Sahibgunj wrote the Persian text *Jam-i-Bahadur Khani*, which he culled from a range of Arabic works on arithmetic, astronomy, algebra, and the natural sciences, using his extensive knowledge of Arabic. He also compiled the astronomical tables called the *Zij Bahadur Khani*. Another Arabic scholar, Maulvi Mohiyuddin of Erki in the *thana* of Jehanabad, composed a 288-page-long Persian text, *Sharh-i-Abdul Rasul*, on Arabic syntax. He also wrote in Arabic a 32-page-long manuscript called *Majmua Taqrir Mantiq Amani*, which was a text on logic.[99]

Medical knowledge taught in family schools thus articulated the new idea of health as medical wisdom—as science rather than aristocratic virtue. As medicine withdrew into the Arabic fold, its popularization

---

1827, Extract, Bengal Pol. Consult, 5 June 1829, Boards Collection, F/4/1170, file no. 30640, 515.

[99] Adam 1941: 286.

was checked. Families disseminated knowledge in Arabic to a few. As Boulderson said of the Arabic-driven learning, 'both the masters and pupils are of a higher grade and have given their attention to scientific pursuits'.[100]

[100] L no. 463, Education Committee to Vice President in Council, 9 July 1827, Extract, Bengal Pol. Consult, 5 June 1829, Boards Collection, F/4/1170, file no. 30640, 639.

2

Encounter with the West: The English East India Company

Introduction

IN THE LATE EIGHTEENTH CENTURY, THE ENGLISH EAST INDIA Company inherited the fractured medical legacy of the Mughal empire. More than by the orientalist orientations or the anglicist urges of its mixed bag of officers, the Company was swept by the huge tide of change in post-Mughal society that had already begun to disembody medical knowledge from the clutches of the royal court, single-copy exclusive texts, the Persian language, the nobility, and elite family backgrounds. The shift from Persian to Arabic as the language of medical science and the disembodied and communitarian profile of medical knowledge that it generated were taken to their culmination by the Company.

Unani's association with Arabic ensured that the English too treated it as a medical system.[1] The Company sharpened internal demarcations in the realms of knowledge that had been created through the setting up of Persian and Arabic schools by local elites. It allowed regional courts to patronize Persian medical literature. It shifted Arabic medical learning from family schools and village teachers to an impersonal

[1] This obviously contradicts the position held by David Arnold, Jean Langford, and others who argue that the Company's motives were more ideological than material in the encounter with Unani. See Arnold 1993; Langford 2002. The notable exception is Sivaramakrishnan 2006: 14–31, who documents the preservation of indigenous medical learning by Sikh ascetics in pre-annexation Punjab.

setting—the Calcutta Madrasa—which was for Muslim students only. Simultaneously, a Sanskrit College with a medical class was set up for Hindu learning.

However, the Company's intrusion also introduced critical changes. Change came much before the anglicist reforms of the 1830s which introduced English-language instruction and Western knowledge into India. And the change was rooted in and shaped by the Company's use of the indigenous linguistic cultures and medical knowledges which it sustained. Thus, from the outset, the Company preserved a disembodied form of Arabic medical learning in the new institutions of learning that it set up. This kept alive the production of Arabic texts and the new learning they embodied. To this the Company added Arabic translations of European texts on anatomy. These strengthened already existing scientistic ideas of Arabic healing. More importantly, the Company used Urdu and the printing press to popularize these new medical ideas.[2] The Company's intervention also decisively shifted health to the wider domain of social well being, and linked medical status to public service. In the process, new referents of medical authority were devised for 'older' medical knowledges. This process of change was remarkably rapid and occurred long before the period of the 1830s—the period normally defined as a watershed in the conventional scholarship.[3]

[2] Pernau 2006: 1–34. The use of Urdu translations of European works of science is generally attributed to the 1840s, when institutions like Delhi College, with their thrust on translations came up for major educational reforms as a consequence of the anglicist pressure on the Company. Pernau argues that the 1840s marks the beginning of the process of 'cultural translation' and dissemination of British rule to Indians. This chapter suggests an earlier drive towards cultural translation. The establishment of the Native Medical Institution (henceforth NMI) and the lithographic press in 1823 initiated large-scale Urdu translations of medical manuscripts—more than two decades before Delhi College took up the task in earnest.

[3] Scholars such as Arnold, Prakash, and Langford have hitherto argued that the 1830s signalled a turn towards coercion on the part of the Company as it abandoned orientalism, 'colonised the body', and introduced new-style Western knowledge with the help of the English language. For a different view, see Sivaramakrishnan 2006: 14–31. She shows how in the Punjab 'indigenous' learning was kept alive by Sikh ascetic sects until the 1850s, when the British land reforms caused a major disjunction. These put an end to court grants that had sustained learning and had a detrimental effect on 'native' education.

The present chapter shows that the anglicist reforms of the 1830s were in fact limited in their impact. Without a doubt, institutions of Arabic and Urdu Unani learning were abolished. New English-language institutions of Western medical knowledge and instruction were established. However, neither of these changes extinguished the medical ethos already established via the earlier interaction between the Company and indigenous medical knowledges and linguistic cultures. A variety of ways of articulating medical authority, created in the period of transition to colonial rule, survived the reforms of the 1830s. These articulations were used by hakims in the late nineteenth century to organize their resistance to increasing colonial intrusion.

The Calcutta Madrasa

In 1781 Warren Hastings extended patronage to Arabic learning of all kinds by setting up the Calcutta Madrasa. Hastings had endowed a madrasa in Calcutta to foster learning in Arabic and Persian and train a class of maulvis or Muslim jurists. They were trained to assist in the interpretation of Islamic law.[4] Hastings was convinced that the madrasa would provide the Company the Muslim public servants it needed to administer its freshly acquired territory. He argued that since the Company had decided to operate within the legal framework it inherited, it had to depend on the experience of Muslim officers. However, he found that Muslims were not well versed in Arabic and Persian languages and legal texts, and so established the madrasa to educate Muslims, believing that the transition to Company rule had deprived them of resources to maintain their own educational institutions for Arabic and Persian learning.[5] In 1781, he erected a building for the madrasa at his own expense. He obtained as government endowment lands with a revenue yield which was estimated at about Rs 29,000 per annum. This took care of the expense of maintaining teachers and scholars and keeping the college in repair. It was decided that

[4] In 1791 an equivalent school, Sanskrit College, was opened in Benares for Hindus. This was a brainchild of the British Resident Jonathan Duncan. For the early history of the madrasa and the Sanskrit school, see Kerr 1853 II: 66–94, 134–71.

[5] L no. 12, Warren Hastings to Board of Directors, n.d., Extract Rev. Letter to Bengal, 28 October 1814, Boards Collection, F/4/708, file no. 19201, pp. 73–9.

vacancies in the *faujdari* (military) courts would be filled by students of the madrasa.[6]

The madrasa divided students into five classes with separate teachers. It instructed one set of students, belonging to the learned and highest classes of Muslims, in Arabic language and sciences, these being taught from Arabic books. The sciences taught in the madrasa were in fact entirely through the medium of Arabic texts and included natural philosophy—including medicine (*hikmat*)—theology, law, astronomy, arithmetic, geometry, logic, and rhetoric. Students were expected to finish the course in seven years.[7] In the early nineteenth century a separate medical class was added. This not only continued Arabic medical learning but also familiarized students with facts culled from Western medical tracts.[8] Indeed, these hybrid intellectuals were expected to be the bridge to cultivate a taste for European medical knowledge in society.

In 1829, Surgeon John Tytler, orientalist scholar and presidency surgeon, declared that the madrasa produced scholar-physicians. In response to the government's anxiety about the usefulness of colleges like the madrasa, he made it clear that the madrasa did not merely produce public servants. Their job was, he declared, to teach Arabic medical texts and 'to diffuse among them [students] a taste of European

[6] Ibid.: 77. For details of the regulations of the madrasa, see L no. 14, regulations of the madrasa, ibid.: 91–7.

[7] L no. 12, Warren Hastings to Board of Directors, n.d., Extract Rev. Letter to Bengal, 28 October 1814, Boards Collection, F/4/708, file no. 19201, pp. 91–7. The madrasa had a rich library for which in 1820 Rs 6818.3.7 was sanctioned by the government. See H. Mackenzie, Sec. to Govt, to Madrasa Committee, 21 July 1820, ibid.: 167.

[8] See for medical class, report of the general committee for the diffusion of instruction among the natives of India for the year 1829, Bengal Public Collection, political letter transferred to public dept, 27 August 1830, Pub. Dept, 29 September 1830, Boards Collection, F/4/1255, file no. 5050 (1), p. 55. Later, in 1829, the madrasa introduced an English department as well which catered to a lower strata of Muslim society who, for a small fee, learnt only the English language. In 1849, at a high fee of Rs 100 per month, an Anglo-Arabic class was started to instruct in the English language such Arabic students as wanted to learn English. See, recommendations on madrasa to the government by the council of education, *Selections from the Records of the Bengal Government, no. XIV, Papers Relating to the Establishment of the Presidency College of Bengal*, Calcutta, 1854, MF 1/871, p. 7.

literature and science [in Arabic]. If the latter is accomplished colleges would have said to have done their duty.'[9] Alloy Sprenger, the principal of the madrasa, shared this view. In 1854 he had convinced the council for education that medical knowledge in Arabic as well as Persian was so central to the 'character of a [Muslim] man of liberal education' that Muslims could not be educated in Western medicine 'without a knowledge of their own literature'.[10] The madrasa thus had the arduous task of creating a Muslim scholar-hakim who had acquired both Arabic and European knowledges.

The medical class of the madrasa was conducted primarily by John Tytler. He took full advantage of the shift to Arabic learning that had been initiated by individuals and families locally. He observed 'that the orientals are always disposed to receive a work written in it [Arabic] with respect which they might reject or despise in a vernacular tongue'.[11] Like the notables of the countryside, he was convinced about the significant role Arabic would play as a language that would pull medicine out of the shackles of Persian culture. Like them, he saw Arabic as the universal language of science and medicine and contrasted its wider appeal with the narrow geographical confines of Persian, Urdu, and other vernaculars. He clubbed Persian with the vernaculars and contrasted it with Arabic, which he said was understood 'from Malacca to Morocco', whereas the vernaculars were understood only in a few provinces of India.[12]

[9] J. Tytler to P. Breton, H.H. Wilson and J. Adam, Members Medical Board, 19 August 1829, Bengal Pub. Collection, 27 August 1830, Boards Collection, F/4/1255, file no. 5050 (1), pp. 726–7.

[10] The council said that this question of language and identity was not true of the Hindus as far as their relation with Sanskrit was concerned. The Hindu was content to leave his learned language to the exclusive enjoyment of the pandit and not carry its burden. He was therefore more accepting of the English language than was the Muslim. The Council saw the failure of the English class at the madrasa as a result of the Muslim attitude to English. See *Selections from the records of the Bengal Government, no. XIV, Papers Relating to the Establishment of the Presidency College of Bengal*, pp. 10–11.

[11] Tytler 1828: v.

[12] Hooper 1830: 8. Tytler reduced his tirade against the vernaculars in 1834 when the introduction of English as the language of all instruction seemed imminent. He preferred the vernaculars, e.g. Urdu, to English as the language of medical science and utility. He cited the success of the NMI in providing public

Realizing the immense potential of Arabic as a conduit through which medical knowledge could be gradually extracted from its earlier encasements, Tytler not only learnt the language himself but set about the task of studying and editing Unani medical texts in Arabic.[13] This was required not only because Arabic texts were taught at the madrasa, but also because Tytler felt that European medical tracts could be understood best by the students if translated into Arabic. As he undertook the arduous task of translation, he saw that he needed to study the Arabic medical texts most carefully. This was essential so as to find in them words, technical terms, and explanation techniques for the Greek and Latin technical lexicon of European literature. Tytler was rarely disappointed in his search for such equivalences because, he noted, the Arabs had not retained any Greek word when they translated from the Greek: 'since such is the genius of the language and of the people it is hopeless and unprofitable to strive against it. We only thereby run the risk of increasing their dislike to our sciences when they find them associated with words which to their ears must sound unintelligible and barbarous'.[14]

Tytler collected, through gifts and purchase, Arabic medical texts and edited them for use by students. The idea was to make medical texts as user-friendly as possible. In cases where more than one copy of the manuscript existed, he compared them and produced a single comprehensive edited version. In 1832 he edited two manuscripts of the Arabic translation of Hoanin Ben Motawukkue's *Aphorisms of Hippocrates*, called *Fusool-i-Abkrat*. One of these copies he procured from a native physician in Patna; the other was loaned to him by his trusted hakim assistant, Abdul Majeed.[15] Other Arabic texts used by him for teaching included the *Sharh-i-Mujiz* by Muhammad Kazim.

---

servants as a case in point to prove the efficacy of medical instruction in the Urdu language. See Tytler 1834: 1–4.

[13] He says that he learnt Arabic in the fourteen years after his arrival in India in 1813, when he was posted in the upper provinces. He learnt it because the government needed oriental scholars. And he says he used his own resources to learn the language and did not take any special favours from the government. See TR-15, 'Memorial of Surgeon John Tytler Addressed to the Court of Directors of the East India Company', pp. 35–6.

[14] Tytler 1830: 12.

[15] Tytler 1832.

Arabic Unani texts taught in the madrasa were of the same type as were taught at family schools: they lacked the Persian comportment frills and were prescriptive and comprehensive. They made Unani user-friendly by their concentration on disease, symptoms, and medication. They gave advice on disease and medication rather than the physiognomic details and their philosophical relations with nature. For instance, the *Sharh* was divided into four sections that dealt with the general principles of medicine, compound medicines, treatment of local diseases, and treatment of general diseases. It focused on medicine as science rather than comportment.

In the Calcutta Madrasa the scientistic literature was complemented by translations into Arabic of European literature on body anatomy. Anatomy only reinforced the new idea of health as science, and distanced this notion from individual bodily deportment. Such translated literature introduced the importance of anatomy into the diagnosis and treatment of disease. Since body anatomy had universal referents, medical knowledge that was based on it saw all individuals as equals. Health was thus in a sense about levelling social differences between people, for it questioned the idea of hierarchies between those with similar body anatomies. In this sense the new Arabic literature based on European texts gave the idea of aristocratic virtue and individual comportment a new irrelevance. Tytler hoped that the inclusion of body anatomy in medical knowledge would make people question the social hierarchies created by Persian medical texts.

Most of the European texts translated were on anatomy and chemistry. These included a brief 22-page text, 'a short anatomical description of the heart', as well as the larger work of Robert Hooper called *The Anatomist's Vade Mecum*. Using these, Tytler offered a scathing critique of the Mughal idea of health as aristocratic virtue. He explicitly justified the translations of anatomy texts by arguing that the study of body anatomy enabled recognition of certain essential similarities among all castes, religions, and communities in India. The discipline, he argued, was a social leveller, underlining the fact that all human beings had the same anatomy. Thus, healing was about offering correct inputs to make all individuals perform similar body functions. Tytler argued that medicine which depended on anatomy was different from the individual-comportment-driven ideas of healing that perpetuated social hierarchies.

Contrasting Mughal Persian texts with the anatomical literature, he said: 'anatomy has a most powerful influence in counteracting prejudices that arise from birth or station or caste, by demonstrating that however mankind may differ in their externals, their internal organization is the same. Before the knife of the anatomist every artifical distinction of society disappears—.'[16]

He went on to emphasize the change in the medical culture that the study of anatomy would introduce, for, once on the dissection table 'the prince will not be found to possess a single muscle or a single artery beyond the meanest of his subjects; and the Christian, the Muhammadan and the Hindoo, the Brahmin and the Sudra have precisely the same organs and same faculties'.[17] He was convinced that such reflections would be obvious to all who study anatomy and it would be impressed on all that 'the different denominations are the offspring of the same benevolent Being who hath made of one blood all nations of men'.[18] In sum, Tytler's anatomical texts reinforced the idea of medicine as science that had been initiated by Arabic texts.

He used the printing press to popularize this idea. He hoped the press would help break the monopoly of families and individuals on such medical knowledge. The government lithographic press now churned out multiple copies of his Arabic translations of European texts. It demystified medical knowledge and moved it out of its single copy confines. This happened with Unani manuscripts in the Arabic as well as with the Arabic translations of European texts.

Tytler translated Hooper's text on anatomy, *The Anatomist's Vade Mecum*, into Arabic. The Arabic version, called *Anis ul Musharrahin*, was produced in several copies by the Lithographic Education Press. Tytler emphasized the great advance that this style of Arabic book production represented over the earlier forms of knowledge dissimination:

> The teacher can instruct only a limited number of scholars; his influence is confined both in space and time, and when period of instruction is over, effect of it can be lost—the influence of the book is very different. They travel through all parts of the country and can be studied at all

[16] Tytler 1830: 14.
[17] Ibid.: 15.
[18] Ibid.

times, by all classes of people, old as well as young, and when their contents are forgotten they may be reexamined—.[19]

He reiterated the relatively expansive reach of a book as compared to that of a teacher who taught from a single-copy manuscript: 'I may also add as perhaps the most advantageous circumstance of all, that thousands are able to purchase a few books who might be utterly unable to afford the expense of the teacher.'[20] So great was the reliance on the printed book in the madrasa that its committee regularly lamented that texts were in short supply; each student could not have his own copy.[21]

The press thus weakened the monopoly of families over Arabic knowledge. Yet the madrasa never severed its links with individuals and families who possessed Arabic learning. D. Ruddell, secretary of the madrasa committee, appointed the 'learned' Hakim Zulfiqar Ali to teach students not only Arabic medical texts like the *Sharh*, but also anatomical and medical facts from books composed at the NMI by both surgeons Peter Breton and Tytler.[22] In 1829 the offer to depute a qualified Muslim student from the NMI to assist Zulfiqar Ali was vetoed on grounds of the 'unpropriety of attaching to the college any other natives than those who were well acquainted with the Arabic language'.[23] Indeed, Ruddel, on the death of Zulfiqar Ali, thought of discontinuing the medical class because it was getting difficult to find a competent Arabic-knowing substitute. Eventually, the dead hakim's son was appointed as a teacher of medicine. He earned great repute

[19] Ibid.: 3.

[20] Ibid.

[21] L no. 978, H. Shakespeare and Colleagues, Madrasa Committee, to W.C. Bentinck, Gov. Gen., 28 May 1830, Bengal Pub. Consult., 27 August 1830, Boards Collection, F/4/1255, file no. 5050 (1), p. 100.

[22] P. Breton, Supdt Native Medical Institution (henceforth NMI), to H.H. Wilson, Sec. Gen. Committee of Public Instruction, 5 August 1829, Boards Collection, Bengal Pub. Consult, 27 August 1830, F/4/1255, file no. 5050 (1), pp. 710–11.

[23] P. Breton, Supdt NMI, to H.H. Wilson, Sec. Gen. Committee of Public Instruction, 5 August 1829, Boards Collection, Bengal Pub. Consult, 27 August 1830, F/4/1255, file no. 5050 (1), pp. 710–1.

amongst the students and staff on account of his acquaintance with Arabic medical texts.[24]

Tytler's favourite assistant, Hakim Abdul Majeed, taught Unani medicine at the Calcutta Madrasa for several years. His career reveals wonderfully how the Company rekindled the fire of Arabic learning, using people from the same stock that had ignited it in the late eighteenth century. The hakim was accomplished in Arabic medicine and theology. The Company was first introduced to him in Mysore, where he was attached as a teacher of over twelve years' experience to the children of Tipu Sultan. He was particularly close to the family of Prince Moizuddin. British officers in Mysore such as Lt. Col. U. Wright, superintendent of Tipu Sultan's departments, kept him in employment. They depended on him as a 'learned and respectable' man who helped them in 'investigating disputes on the establishment' as a result of his linguistic skills and network of contacts.[25] He remained a great asset to British officers like T.A. Cobbe, superintendent of the affairs of the Mysore princes, as a translator who helped in the setting up of the new administration in Mysore.[26] His command over Arabic medical and theological texts brought him to the notice of Calcutta orientalist scholars H.H. Wilson, A. Galloway, Peter Breton, and Tytler. The last employed him as lecturer of medicine at the Calcutta Madrasa, and also as translator in the Committee of Public Instruction and assistant in the NMI.

In all these positions Hakim Majeed was actively engaged for more than fifteen years in either teaching Arabic medical texts or composing and translating medical texts into Arabic.[27] He earned a decent living through these various services and got a salary of Rs 60 per month just from the NMI.[28] His knowledge of Arabic legal texts got him, in 1831,

[24] Ibid.: 712.

[25] L no. 16, Hakim Abdul Majeed to William C. Bentinck, Gov. Gen., 16 February 1835, Consult, 25 February 1835, Bengal Pub. Consult, 5 January–6 May 1835, P/13/12.

[26] Ibid. See letters of reference attached: H.H. Wilson, n.d.; A. Galloway, 16 March 1832; J. Tytler, 16 March 1834.

[27] Ibid.

[28] L no. 26, Hakim Majeed to H.T. Prinsep, Sec. to Govt in the Gen. Dept, n.d., Consult, 1 April 1835, Bengal Pub. Consult, 5 January–6 May 1835, P/13/

appointment as acting Qazi-ul-Quzat to the Sadr Diwani and Nizamat Adalat.[29] Commenting on his role in the madrasa and his scholarly works in the NMI publications section, Tytler said he was 'fully competent were he willing both to render intelligible to students the Arabian medical writers and also to explain to them almost any European medical tract which I have seen published.'[30]

By the 1830s the Company had in short rekindled and reinvigorated the Arabic-driven medical culture it found in India. It disembodied Arabic medical learning by locating it in the madrasa. It moved it out of its manuscript base and into a wider sphere of circulation made possible by the printing press. It had introduced anatomy texts into the corpus of medical literature in Arabic. These were projected as social levellers that had the potential to break hierarchies sustained in particular by Persian texts.

The 1830s madrasa reforms only underlined the importance of this new invigorated medical knowledge now easily available in Arabic. In this period, Ruddell, secretary of the madrasa committee, made the

---

12. In 1835, after the abolition of the NMI and the shift away from Arabic learning, men like Hakim Abdul Majeed were left jobless and in dire financial crisis. The hakim petitioned the government to be sanctioned an allowance of Rs 60 per month, which was his last drawn salary at the NMI, so as to maintain his large family. But the government refused this on the grounds of there being no precedent for such allowances to men not in their service. In 1835, the government also refused to appoint Hakim Majeed to the newly founded Calcutta Medical College on the grounds that he did not know the English language. Pandits of the NMI, like Madhusudan Gupt and Prasad Pandit, were adjusted at the college because they knew English. See L no. 128, Medical Board to W.C. Bentinck, 17 February 1835, and L no. 133, J.C.C. Sutherland, Sec. General Committee of Public Instruction, to Lt. Col. W. Casement, Sec. to Govt, 30 April 1835, Consult, 4 May 1835, India Military Consult, 4–18 May 1835, P/35/16.

[29] L no. 16, Hakim Abdul Majeed to William C. Bentinck, Gov. Gen., 16 February 1835, Consult, 25 February 1835, Bengal Pub. Consult, 5 January–6 May 1835, P/13/12.

[30] J. Tytler, to P. Breton, H.H. Wilson and J. Adam, Members Medical Board, 19 August 1829, Bengal Pub. Collection, 27 August 1830, Boards Collection, F/4/1255, file no. 5050 (1), p. 726.

Arabic *Sharh-i-Mujiz* the core text for the student entrance exam.[31] Simultaneously, the reforms strengthened the stress on anatomical studies. They introduced a regimen of practical training in body anatomy.[32] Students of the madrasa, even though not trained as practitioners, were expected to enhance their medical knowledge by attending practical training at the native hospital in Calcutta.[33] Later, when this was abolished, the feasibility of sending them for practical instruction to the military-run NMI was contemplated. This never materialized on account of the interference that it would cause with the military rules that monitored students there.[34]

The linkages of medicine with science and anatomy that were seen as social levellers introduced the idea of the larger role that medicine could play in society. It made the issue of health central to the well being of society. The pressures to vernacularize medical knowledge followed. So did the move to link medical status with public service. Neither of these demands were easy in an age marked by intense debate over the issue of the language of education in British India.

The Language of Medical Education

It was not a smooth ride for the Calcutta medical class and its Arabic orientation. There was a lobby that wanted to throw away the indigenous

[31] Madrasa Exam Report 1829/30, Enclosure no. 2, D. Ruddell, Sec. Madrasa Committee, to H. Shakespeare and Colleagues, Madrasa Committee, 22 February 1830, Bengal Pub. Consult, 27 August 1830, Boards Collection, F/4/1255, file no. 5050 (1), pp. 151–2. He suggested that the reading of the Arabic grammar text *Muqamat* be reduced to the first twenty-five stories only. In its place the *Mujiz* be substituted as the qualifying text for admission and scholarship.

[32] P. Breton, Supdt, NMI, J. Tytler, Presidency Surgeon, J. Adam, Sec. Medical Board, to H.H. Wilson, Sec. to Gen. Committee of Public Instruction, 24 August 1829, Bengal Pub. Collection, 27 August 1830, Boards Collection, F/4/1255, file no. 5050 (1), p. 702. The students of the madrasa, even though not trained as practitioners, were expected to enhance their medical knowledge by attending practical training at the native hospital in Calcutta.

[33] Ibid.

[34] See also L no. 84, J. Tytler, Presidency Surgeon, to J. Hutchinson, Secretary Medical Board, 11 April 1832, Consult, 7 May 1832, Bengal Military Consultation (henceforth BMC), 23 April–14 May 1832, P/34/21.

framework altogether and replace it with one dominated by the English language. And in the 1820s critiques of the madrasa came also from those who questioned the relatively superior scientificity that it allocated to Arabic over Sanskrit and Persian.

Throughout the late 1820s, under the governor generalship of Bentinck, a larger debate went on between the general committee of public instruction headed by the orientalist scholar H.H. Wilson, and the Delhi Committee with members like C.E. Colebrooke, T.C. Metcalfe, and Charles Trevelyan, on the feasibility of creating a proper intellectual environment for the introduction of Western science and literature. Trevelyan and his colleagues were critical of the madrasa-style experiment which supported oriental learning and enriched it by adding to it the translations of European medical literature. They felt that this policy could never lay the ground for the eventual transition to European learning. What was needed was a clean break from the Arabic and Persian linguistic, literary, and scientific tradition. They argued that this was particularly important because the Muslim scientific literature in Arabic was a critical part of the intellectual's social and religious identity. It was rendered 'inveterate and unchangeable by being identified with their religion'. And thus translations of European texts in Arabic could never wean Muslims away from their intellectual legacy.[35] They would, on the contrary, entrench them further in their intellectual past. Trevelyan and his team wanted to bypass the minority Muslim intellectual class and its knowledge base, and target instead Muslims who 'are entirely uneducated and attached to no previous system' to create a new class of Muslims educated only in European sciences in the English language. This class could then carry forward European knowledge in the vernacular to the masses. They argued that this would save the expense on translations and, more importantly, enable them to 'cease to carry on with such imperfect weapons an open warfare against the old system entrenched as it is behind religious feelings and pre-conceived opinion.' They hoped that by creating a new class of educated Muslims the 'system of the Arabians—will by degree cease

[35] C.E.Colebrooke, T.C. Metcalfe, C.E. Trevelyan, and J. Ranken, Delhi Committee for Public Instruction, to H. H.Wilson, Sec. to Committee of Public Instruction, Calcutta, 12 February 1829, Boards Collection, file no. 30639, F/4/1170, p. 324.

from neglect and want of notice without being at all brought into collision with the new literature.'[36]

The Delhi committee did not have full support from H.H. Wilson, secretary of the general committee, on their proposal. But the governor general, Bentinck, was with them in doubting the wisdom of the large expense entailed in translations. He too wondered about the feasibility of translations of European works into Arabic and Persian considering that 'very few of the Muhammedans, only the learned ones know it'.[37]

The differences of opinion over the Arabic *vs* Western medical and scientific literature was only one aspect of the education debate. Parallel to it was a discussion which approved working within the received linguistic framework. But the protagonists of this position wanted the existing hierarchies realigned. And here, the personal career interests of Company surgeon bureaucrats and their professional rivalries played no small role in the advocacies that they made about privileging Arabic over Sanskrit, or vice versa. For instance, the supremacy of Arabic as the language of medicine and science was disputed by medical officers like Peter Breton, who argued that Sanskrit should be given top status instead of Arabic. Breton taught anatomy and surgery at the NMI of which he was also the superintendent. He interacted with pandits, and with their help published innumerable tracts on body anatomy and physiology using Devanagari, with an additional Arabic script serving as appendage. But he did not make as much profit out of this as did his colleague Tytler, who translated many medical texts into the Arabic language and script. Tytler was in charge of the government lithographic press and its publications. He hogged most translation contracts for himself and minted a fortune during his career in India.[38] It was no

[36] Ibid.: 326.

[37] A. Sterling, Deputy Sec. to Govt, to the Gen. Committee of Public Instruction, 26 December 1828, Extract Bengal Pub. Consult, 5 June 1829, Boards Collection, file no. 30639, F/4/1170, p. 422.

[38] In 1833 Charles Trevelyan revealed that John Tytler had commissioned all the recent Arabic translations for himself. Between September 1829 and March 1833 Tytler, who already had a salary as a result of being the superintendent of the NMI, received an extra Rs 17,000 for such translations. He further calculated that, in spite of spending over Rs 100,000 on printing since 1824, the General Committee of Public Instruction had not sponsored a single publication in

surprise, then, that Breton was keen on a switch away from Arabic translations: it would have pulled the rug from under Tytler's feet and presumably helped position it nearer his own.

Breton doubted the wisdom of teaching Arabic medical literature and translations of European texts in the madrasa. Arabic, according to him, was not a language suitable for the teaching of medical science to beginners, for these texts 'not being elementary are far too abstruse for the comprehension of novices.'[39] He believed Arabic medical instruction in the madrasa had no future. This was particularly true after the death of Hakim Zulfiqar Ali, for no adequate substitute seemed available of the same calibre as the 'old hakim'.[40] Indeed Breton went as far as recommending discontinuation of the Arabic medical class of the madrasa and amalgamation of its students with those of the Hindu and Sanskrit colleges. He argued that here all students would be trained in the European medical texts in native languages and be better prepared towards becoming private practitioners.[41]

Tytler, of course, was quick to spring to the defence of Arabic medicine. He read into Breton's critique of the limitations of Arabic knowledge a disparaging sentiment that doubted the ability of Muslims to carry forward Western medical science, and he contradicted this by speaking on their behalf. He charted out a history of Muslim intellectual accretion and a legacy which showed they had always been happy to include Western learning in their linguistic fold. In view of this, he was sure that 'were classes for medicine and natural philosophy (which have always been favourite studies among them) established at the madrasa in European principles, they would be extremely popular.'[42]

Bengali or Urdu; and out of a total of 23,190 volumes, only 2030 had been in Hindi, none of which was a translation of a European work of science. Cited in K.H. Prior, Introduction to C.E. Trevelyan's 'On the Education of the People of India (1838)', unpublished Ms., p. 13.

[39] P. Breton, Supdt, NMI, to H.H. Wilson, Sec. Gen. Committee of Public Instruction, 5 August 1829, Bengal Pub. Collection, 27 August 1830, Boards Collection, F/4/1255, file no. 5050 (1), p. 709.

[40] Ibid.

[41] Ibid.: 711.

[42] J. Tytler, Lecturer Hindu College, to H.H. Wilson, Sec. Gen. Committee of Public Instruction, 19 August 1829, Bengal Pub. Collection, 27 August 1830, Boards Collection, F/4/1255, file no. 5050 (1), p. 723.

Tytler compared Arabic to Latin and argued that students at the madrasa were best equipped to be educated in medical science, for, like students of Latin in Europe, they had mastered the 'learned languages' which are the 'instruments by which knowledge in the sciences is to be acquired.'[43]

Tytler's passionate defence of Arabic learning sprang no doubt from vested career concerns. Yet, more than the local Muslim elites who were convinced of the superiority of Arabic medical texts, he was actually in a position to privilege them in the hierarchy of knowledge that the Company was working with. He refuted Breton's charge by elaborating on the contents of the popularly used Arabic medical text in the madrasa, the *Mujiz*, arguing that it was a simple and comprehensive book divided into a theory and a practical part. The first had a Galenic account of healthy and morbid physiology, the second was a catalogue of diseases and their remedies. According to Tytler, this was so elementary that any writer 'professedly proposing to be elementary could hardly adopt any arrangement greatly preferable to this'.[44] He felt that Galen was the base of most Arabic medical authors, including Avicenna: most modern Arabic medical texts were abridgements of his work. This made the task of teaching Arabic medicine simple rather than abstruse—as suggested by Breton. All that was needed in obtaining a key to the Arabian authors was to 'be familiar with Galen and the humoral pathology', and this was not a difficulty of very great magnitude.[45] However, his vociferous defence of the consolidation of Unani learning in Arabic had the unintended effect of only sharpening its specifically Muslim profile. This process had already begun within village Arabic schools.

The Native Medical Institution and the Medical Community of Urdu Literature

At the turn of the eighteenth century the Company was under pressure from the anglicists. Yet it was no less influenced by the orientalists, who now had personal investments and a stake in the maintenance of the old order. In the 1820s the Company took a middle path: it continued with the madrasa instruction and its scholar-physicians, but it did not

[43] Ibid.: 724.
[44] Ibid.: 725.
[45] Ibid.

ignore Trevelyan's pleas entirely. As the ground was prepared for English instruction of a standard sufficient to impart Western medical knowledge, it looked to the vernaculars for help. It chose a locally spoken vernacular—Urdu—to carry the knowledge of European science to natives, and to fashion a new-style indigenous practitioner different both from the scholar physician educated at the Calcutta Madrasa and the Mughal gentleman hakim. This was a practitioner who epitomized the commodified nature of medical knowledge. Thus, much before the anglicist-driven educational reforms of the late 1830s, the Company had introduced a new medical ethos in India through use of the Urdu language.

Urdu was a pidgin tongue. It borrowed from Arabic, Persian, and Hindi. It was linked to no particular community, profession, royal court, or scientific tradition. It was the language of ordinary people and of soldiers in camp. In the early nineteenth century, the Company's interest in Urdu exalted this pidgin's status. Urdu became the language of Western-style medical education in India. In the 1820s the Company-sponsored NMI became the place for medical learning in Urdu, and the government lithographic press became the site for Urdu book production. Urdu print culture, with its thrust on popularizing medical knowledge and producing mass-scale medical texts for sale to public servants, reinforced the new ideas about health and body anatomy that the madrasa epitomized. But, more importantly, it pushed medical knowledge from its concern with individual well being to a larger anxiety about social well being. It also created a native doctor to carry forward this new role that medicine was meant to play, a professional medical man who had easy access to useful scientific medical knowledge in the vernaculars, and the status of a public servant. His medical status saddled him with the responsibility to ensure social well being and made him the harbinger of a new and enlarged concept of 'public'. This new public excluded those who refused to come under the purview of the new men of medical science. Thus, much before the anglicists pushed for reforms in the late 1830s, the Company introduced a new Urdu-driven medical ethos into India.

Background of the Native Medical Institution

At the General Hospital in Calcutta for European soldiers the British assistant surgeon was assisted by a compounder, dresser, and assistant.

These jobs were provided by locals, people of ordinary background called 'native doctors'. There was a growing demand for such men by the Bengal army. The need only increased when the army expanded its regiments by the recruitment of high-caste Hindu and Muslim sepoys.

Native doctors were employees of the army. Some had quite basic literacy and skills acquired at seminaries and native establishments. However, such training centres were few in number. Generally, they learnt to read and write Nagari while in the service of the army. L. Nicolson, the presidency surgeon, reported that the early batches of native doctors had no medical educational background. They were trained while on the job in basic medical skills by the surgeon to whom they were attached. Most of this training took place not in a classroom but at the bedside of the patient. It was here that British doctors instructed native doctors on matters of medical practice. Many times, passages from medical journals were read out to them. The native doctors noted this medicinal knowledge with a piece of chalk on the floor, at the foot of the patient's bed. Later, they memorized it. This helped them administer the correct dosages to the patient. Native doctors always carried a medical box around their necks. Medicines not already part of such box collections were prepared by European apothecaries at the hospital dispensary, under the supervision of the British assistant surgeon. The native doctor attached to the surgeon was expected to learn such methods of preparation.[46]

Recruitment was not restricted to any particular caste. In fact any coolie attached to the army, once he became well versed in the Nagari script and qualified in basic hospital skills, could rise to become a native doctor. Over the course of time, many such trained men passed on their skills to their family and close associates. Such trained associates were always welcome as recruits in army hospitals. Very much like the recruitment of sepoys, native doctors, in the early phase of Company rule, brought their trained sons or family and friends as potential recruits. And as in the case of sepoys, these men were invariably the preferred candidates for medical service in the army.[47] They started their careers as dressers in the medical establishment and rose to become

[46] L no. 95, Enclosure, L. Nicolson, Presidency Surgeon, to W.H. Bannerman, Officiating Sec. Finance Committee, 11 August 1833, Consult, 3 October 1833, BMC, 5 September–3 October 1833, P/34/47.

[47] L no. 95, Enclosure, Surgeon, J. McDowell, to R.B.C. Bannervium,

compounders and finally native doctors, the last being the highest position in the hierarchy of native subordinate medical staff.[48]

Thus, the large-scale general hospitals attached to the headquarters of every division of the army acted as a nursery for churning out native doctors. Their status was defined not by their family, caste, or scholastic achievement but by their being public servants and employees of the army. Indeed, their status was similar to that of the Company's sepoy. And, like sepoys, many native doctors created a sphere of influence for themselves in local society. Nicolson commented on one such successful native doctor who served him. He was conversant in English and could read and write the language with ease. Indeed, he could read some Latin as well. And, once the divisional general hospitals were abolished in 1807–8 and the native doctors of his generation retrenched, he became one of the eminent practitioners of Calcutta. He was so successful that, some years before his death, he drove his own carriage.'[49]

By 1812 the closure of the general hospitals led to a shortage of native doctors. The regimental hospitals which replaced the general hospitals had to recruit 'orphan boys' as hospital apprentices, to be trained on the job so as to replace the dying creed of native doctors. Even this did not help satisfy the increasing demand for trained native doctors in the army. Their presence was missed even more as the army expanded in the first three decades of the nineteenth century. In 1822, James Jameson, secretary to the medical board, wrote in exasperation to the government that even as the training centre for native doctors had considerably diminished, the demand for them had increased as the military expanded its establishment hugely. He strongly recommended the formation of an institute to produce native doctors who would be better trained and equipped to cater to the medical needs and social sensitivities of the Company's high-caste Hindu as well as Muslim sepoys.[50]

Officiating Sec. to Finance Committee, 13 August 1830, Consult, 3 October 1833, BMC, 5 September–3 October 1833, P/34/47.

[48] L no. 95, Enclosure, L. Nicolson, Presidency Surgeon, to W.H. Bannerman, Officiating Sec. Finance Committee, 11 August 1833, Consult, 3 October 1833, BMC, 5 September–3 October 1833, P/34/47.

[49] Ibid.

[50] L no. 100. Memorandum by J. Jameson, Sec. Medical Board, Extract BMC, 24 May 1822, Boards Collection, F/4/737, file no. 20085, p. 17.

By 1822 Jameson had convinced the government of the absolute necessity of establishing an institution to train natives in basic medical knowledge and skills. These trained men would provide a ready supply of public servants to the army and also to jails and municipalities. The government accepted the recommendations of the medical board for the establishment of such an institution in Calcutta and passed a General Order towards this end in 1822.[51] The NMI was to be part of the Company's military establishment.

Social Profile of the Native Medical Institution

The NMI began opeations in 1823. It was located in a rented house at Park Street, Calcutta, on the estate of its superintendent Peter Breton. The superintendent lived on the estate and the teaching and residential areas were all on one campus. On his death in 1830, a rent of Rs 230 was paid to the estate's executors.[52] The compact nature of the NMI's layout cemented a camaraderie structured on the learning of new medical knowledge in the vernaculars, and on public service.

The NMI avoided confrontation with the established medical men of pre-colonial India. As we saw, such medics studied Unani medicine in the Arabic schools of rural India that were run by local charismatic teachers, or else in the Calcutta Madrasa. The NMI targeted students who did not have the necessary background to access medical knowledge via families, courts, charismatic teachers, and doctrinaire languages like Arabic. Students who were not more than twenty years of age and of 'respectable family background' were the preferred recruits.[53] In 1835 the absence of boys from established families of hakims and vaids in the NMI was evident when, out of the 67 students, only 1 was recorded as the son of a hakim; 6 of them were sons of native doctors,

[51] For GO on NMI see James Hutchinson, Code of Regulations for Medical Dept of the Bengal establishment compiled by order of government under Supdt of Medical Board, Calcutta, 1838, L/Mil/17/2/450, p. 265.

[52] L no. 149, J. Hutchinson, Sec. Med. Board, to Lt. Col. W. Casement, Sec. to Govt in Mil. Dept, 6 January 1831, & L no. 150, Lt. Col. W. Casement, Sec. to Govt, to J. Hutchinson, Sec. to Med. Board, 14 January 1831, Consult, 14 January 1831, BMC, Consult, 7–21 January 1831, P/33/66.

[53] L no. 100, J. Jameson, Sec. Med. Board, Memorandum on Institution for Education of Native Doctors, 1 May 1822, Extract BMC, 24 May 1822, Boards Collection, F/4/737, file no. 20085, p. 17.

and the bulk of them from military families of non-commissioned native officers: 5 sons of subedars, 16 sons of jemadars, 15 sons of havaldars, 3 sons of naiks, 3 sons of sepoys, 3 sons of pandits, 1 son of a 'Tisser'/'Tifser', 1 son of a 'Moolsoodee', 1 son of a thanadar, and 1 son of a munshi.[54]

At the NMI such men could start their medical studies in Urdu on a clean slate. A basic knowledge of Urdu, rather than considerations of high status, caste, and religion, was the criterion for selection. This basic educational qualification was the ability to read and write Hindustani in Nagari and Persian characters. Later, Governor General Bentinck insisted that a minimal knowledge of arithmetic be included as an added qualification, this being essential to their training in the compounding of medicines. The recruits were to be neither too old nor very young and in good health at the time of admission.[55]

The 1831 list of students appointed to the service of government showed a nice mix of Hindus and Muslims who came from upper India and Bengal. As native doctors they were scattered all over the Bengal Presidency in both civil and military postings. The list included Pursad Singh, Shiekh Peer Bux—posted in Sagor; Gangaprasad Tiwari, Mir Riaz Ali, Sheikh Tegh Ali—all posted in Barrackpore; Sheikh Muskin—Meerut; Guneas Prasad—Bhagalpur; Sheikh Inayat Hosain and Runjwonym Dass—governor general's household; Mir Qasim Ali and Shaikh Shahed Ali—posted in Ramri; Kashi Prasad—assistant surgeon to commander-in-chief; Sheikh Hidayat Ali—posted to lunatic asylum; Buldoo Tiwari—posted to Balasore; and Rung Lal—posted to Etawah.[56] In 1833 there were 38 Muslim students in the NMI and the rest of the 60-plus students were Hindus.[57]

[54] L no. 179, Members Native Medical Education Committee to W.C. Bentinck, n.d., Consult, 28 January 1835, India Mil. Consult, 23–28 January 1835, P/35/9.

[55] L no 104, Minute of Gov. Gen. Bentinck, 29 August 1833, Consult, 3 October 1833, BMC, 5 September–3 October 1833, P/34/47.

[56] L no. 125 Enclosure, J Tytler, Supdt, NMI, to J. Hutchinson, Sec. to Med Board, 28 March 1831, list of pupils appointed to service of government from the NMI in 1831, Consult, 15 April 1831, BMC, 15 April–6 May 1831, P/34/3.

[57] L no. 166, T. Wood, Engineering President and Member Special Committee,

The NMI systematized medical instruction and laid out strict codes of medical apprenticeship and training. Indeed, it linked medical instruction with professional public service, and this under the superintendence of the military. The superintending surgeons of military divisons were to recommend potential students of 'unexceptionable caste, character, steady habits and good capacity' to the medical board. Each recommendation was accompanied with a descriptive roll that notified the name, caste if Hindu and 'tribe' if Muslim, and the age of the applicant. The students presented themselves to the secretary of the medical board. Once enrolled, they were dispatched to the NMI and placed with its superintendent. Initially, not more than twenty students were admitted to the NMI at a time. If more were found fit, they were put on a waitlist and inducted when vacancies occurred.[58] Later, this number was increased to fifty.[59] In 1835 the institution had sixty-seven students.[60] Each student was financially supported at public charge and a sum of Rs 8/month was sanctioned for him.[61] In 1825 this amount was revised and increased to Rs 10/month. At the end of two years of study it was raised to Rs 12/month.[62]

The recruits were soon put through a rigorous schedule of lectures and practical training. This training from scratch was justified by many surgeon-scholars on the grounds that the students were lacking in any kind of medical education. They justified the creation of a fresh class of native doctors by painting a sorry picture of Unani Arabic

---

to Lt. Col. W. Casement, Sec. to Govt, 22 August 1833, Consult, 30 August 1833, BMC, 23–30 August 1833, P/34/46.

[58] L no. 133, J. Crawford, Sec. Med. Board, to Lt. Col. Casement, Sec. to Govt in Mil. Dept, 10 February 1823, BMC, 7–14 February 1823, Consult, Fort William (henceforth FW), 14 February 1823, P/30/10.

[59] W. Casement, Sec. Govt, to Sec. Med. Board, 30 December 1825, Consult, 30 December 1825, BMC, 30 December 1825, P/31/41.

[60] L no. 179, Members Native Medical Education Committee to W.C. Bentinck, n.d., Consult, 28 January 1835, India Mil. Consult, 23–28 January 1835, P/35/9.

[61] L no. 108, GO, 21 June 1822 for the NMI, Extract BMC, 24 May 1822, Boards Collection, F/4/737, file no. 20085, pp. 62, 68.

[62] W. Casement, Sec. Govt, to Sec. Med. Board, 30 December 1825, Consult, 30 December 1825, BMC, 30 December 1825, P/31/41.

medical learning, arguing it had failed to reach out to the masses: it had, they said, been reduced to the hands of 'conjurers, jugglers and cheats'. Surgeon H.H. Wilson was convinced that the NMI was a necessity because 'the natives of India—have neither oral nor written tuition in the very rudiments of the [medical] art, anatomy has never been studied by them in the human subject, pathology, surgery therefore is flawed because it's based on anatomy. Mixtures of medicine are compounded in defiance of chemistry, pharmacy and common sense.'[63]

Vernacularizing Medical Knowledge: The Native Medical Institution and the Urdu Lithographic Press

The native doctor as a trained public servant projected the new medical concern for general well being. The press popularized this idea. The government printing press played a critical role in both training the native doctor and educating society on medicine's new role. And it was therefore not of little significance that the government-owned lithographic press that specialized in Urdu tracts and the NMI both got a kickstart in the same year—1823.[64] Rind was made the first superintendent of the government lithographic press in the same year that NMI got Peter Breton—the superintendent who took over after the untimely death of the founder, James Jameson, in less than a year of his superintendentship.[65] And both men were trained 'in the study and practice of surgery' and qualified to serve as assistant surgeons in the

[63] L no. 99, Enclosure, Surgeon H.H. Wilson, to R. Bannerman, Officiating Sec. to Finance Committee, 20 August 1830, Consult, 3 October 1833, BMC, 5 September–3 October 1833, P/34/47.

[64] L no. 38, C. Lushington, Sec. to Govt, to Committee for Reporting on Mr Rind's plan for the formation of a government lithographic establishment, 29 March 1823, Consult, 29 March 1823, Bengal Pub. Consult, 20–29 March 1823, P/11/1. In 1823 the government agreed to buy the private Lithographic Press run by Mr Rind for Rs 1650 and ordered four additional ones from Messrs. Gessop and Co. at Rs 1200.

[65] Ibid. See for details of the setting up of the government lithographic press L no. 40, J.N. Rind, Supdt, Lithographic Press, to Lithographic Committee, 25 October 1828, Extract Bengal Pub. Consult, 10 December 1828, Boards Collection, F/4/1290, file 51650, pp. 62–71.

service of the Company. Dr Rind, a native of Stirling, Scotland, obtained his nomination for a posting in Bengal via William Elphinstone in 1814.[66] And once he headed the press he obtained an allowance of Rs 400/month and house rent plus an office rent of Rs 200.[67] In October 1824 Breton, who came from an established Norman family of merchants in Southampton, became the first superintendent to be confirmed to the office of NMI. After having cleared the language test of Fort William College, he resigned from the Ramghur Corps, where he was surgeon, and joined as superintendent on a salary of Rs 1600 per month.[68] Breton, along with his sons Thomas and Frederick, had been serving in India for some time and, like Rind, he created a rather cosy financial and social niche for himself in society.

Both the superintendent of the NMI and that of the lithographic press were expected to possess a good knowledge of Urdu.[69] Indeed, the NMI superintendent was the main lecturer for the students and was expected to author medical manuals for his classroom instructions. The superintendent of the lithographic press helped him discharge medical instructions indirectly by printing the Urdu literature, for orders as high as 800 copies of a single tract. For instance, between January and December 1826 the lithographic press had done 150 impressions of whole-sheet drawings only for the NMI.[70]

The lithographic press took orders to print routine government

[66] Petition of James Rind to the Court of Directors of EIC L/Mil/9/366, pp. 90–4.

[67] L no. 40, J.N. Rind, Supdt, Lithographic Press, to Lithographic Committee, 25 October 1828, Extract Bengal Pub. Consult, 10 December 1828, Boards Collection, F/4/1290, file 51650, pp. 62–71.

[68] L no. 138, P. Breton, Supdt, NMI, to Lord Amherst, Gov. Gen. in Council, n.d., Consult, 19 April 1824, BMC, 8–19 April 1824, P/30/47. See also L nos. 139 and 140 for the confirmation of Breton's salary at Rs 1600/month. For family details of Breton, see V.C.P. Hodson, *List of the Officers of the Bengal Army 1758–1834*, London, 1927, part 1, pp. 200–1.

[69] G. Proctor, Sec. Med. Board, to C. Lushington, Sec. to Govt in Gen. Dept, 20 August 1823, Consult, 21 August 1823, Bengal Pub. Consult, 14 August–4 September 1823, P/11/8.

[70] L no. 44, Govt Lithographic Committee to Sec. Prinsep, 31 March 1829, Extract Bengal Pub. Consult, 5 May 1829, Boards Collection, F/4/1290, file no. 51650, p. 49.

correspondence and stationery for the marine board, medical board, judge and magistrate's offices, military board, etc. But it was the specialized texts and contracts of the NMI that earned it fame. In 1828 the government lithographic committee justified the greater usefulness of the press as compared to private ones by pointing out its great efficiency in producing NMI publications. It noted that Breton had 'already derived the most essential aid from the use of the establishment under our superintendence'.[71] Indeed Breton had free access to the press. Until his death in 1830 the NMI contract for printing at the lithographic press needed no formal government approval.[72] Breton worked hand in hand with the lithographic press, which printed his manuals—both for students as well as for circulation amongst native doctors and surgeons posted in far-flung regions of India.

The Urdu books disseminated existing Western medical knowledge that had already been echoed in the Arabic literature of the Calcutta Madrasa and family schools. A notable addition now was popularization of body anatomy. Breton was one of the main authors of this new Urdu literature. His long superintendentship over six years, 1824–30, revealed the role of the NMI as the nodal point of Urdu medical book production and new medical knowledge. The ideas that Breton popularized were similar to those embodied in Arabic texts used by hakim families and the madrasa, but differed from those that controlled Arabic learning in their desire to vernacularize and popularize this medical knowledge.

In less than a year of his taking over the NMI, Breton requested free access to the government lithographic press for the printing of his medical texts in Urdu. These included his works on body anatomy, pharmacology, orthopaedics, and surgery. By October 1824 he had produced before the medical board sixteen copies of his vocabulary of medical terms in the Urdu and English language published by the lithographic press.[73] The following year he submitted to them

[71] L no. 39, Govt Lithographic Committee to Sec. H.T. Prinsep, 19 November 1828, Extract Bengal Pub. Consult, 10 December 1828, Boards Collection, F/4/1290, file no. 51650, p. 59.

[72] For Breton's easy relationship with the press, see L no. 73, A. Russell, Member Medical Board, to W. Casement, Sec. Govt, 20 March 1824, Consult, 1 April 1824, BMC, 1–8 April 1824, P/30/46.

[73] L no. 180, Enclosure, P. Breton, Supdt, NMI, to G. Proctor, Sec. to Med.

twenty-four copies of his Urdu version of a London Pharmacopoeia, also produced at the same press, while he retained another 100 copies at the NMI awaiting government instructions for their distribution.[74] By 1829 his list of tracts composed for the NMI included anatomy texts on osteology, a description of thoracic and abdominal viscera, of the brain and appendages, of the eye and appendages, of the ear and appendages, of the urinary organs, and of the male generative organs. His texts on pharmaceutical issues included treatises on materia medica, a translation of the London Pharmacopoeia, treatises on the component parts of the air and the posological table. The medical treatises included those on vaccination, cholera morbus, vegetable poison, mineral poison and the venom of serpents, rheumatism, intermittent fever, and the medical topography of the Ceded Provinces. The surgical treatises included those on suspended animation, cataract, hydrocele, and dislocation.[75]

Breton said these tracts were so simply written that they would be of use to students. They would also help medical officers, who could, even 'with a slight acquaintance of the Hindustani language', understand and instruct native doctors via them. Breton's request was supported by the medical board and got a sympathetic hearing from the government, which allowed him free and easy access to the lithographic press for such Urdu publications.[76]

Even while the printed medical book was in the making, the dependence on the old-style scribe and painter continued. In a political culture where the written word commanded authority, the transition from manuscript to the lithographic print carried along with it the scribes and artists who commanded respect in society. This was both

Board, 14 October 1824, Consult, 18 March 1825, BMC, 11–25 March 1825, P/31/16.

[74] L no. 181, Enclosure, P. Breton, Supdt, NMI, to G. Proctor, Sec. to Med. Board, 7 March 1825, Consult, 18 March 1825, BMC, 11–25 March 1825, P/31/16.

[75] L no. 125, Enclosure List of Treatises prepared by Dr Breton for use of NMI, J. Tytler, Supdt, NMI, to J. Hutchinson, Sec. to Med. Board, 28 March 1831, Consult, 15 April 1831, BMC, 15 April–6 May 1831, P/34/3.

[76] L no. 73, P. Breton, Supdt, NMI, to G. Proctor, Sec. to Med. Board, 13 March 1824, Consult, 1 April 1824, BMC, 1–8 April 1824, P/30/46.

of strategic importance as well as a practical necessity. Urdu print culture did not erase the earlier notion of 'public' where the calligrapher, artist, scribe, and elite readers enjoyed status. This limited public, with its elitist contours of 'civility', continued. The early print literature only expanded the contours of this sphere slightly to include people who now had easy access to books in a language familiar to them. This increase happened because the new print culture meant a proliferation of Urdu book production. The larger ambit created then resulted in new norms of what ought to constitute a 'gentleman'.

Breton remained dependent on older communities of medical learning when producing his texts, even as he broke their stranglehold over medical knowledge. The best ever published anatomical engravings, known as the Lizars anatomical plates, were bought by him at a cost of Rs 130 to aid his staff in the publication of Urdu texts on anatomy.[77] He argued that he needed both maulvis and pandits because the tracts needed to be in Urdu or Hindustani, both in the Devanagari as well as Persian characters. This was essential because Hindu and Muslim students could read only one kind of script, respectively. Ideally, a person knowing both scripts would have been most useful to him; but he said such a person was almost impossible to find.

He entertained in his private service Maulvi Asadullah, a former teacher of the Fort William College, to asssit him with Persian characters, and a pandit on Rs 25/month. Later, Asadullah was appointed as Munshi to the NMI. His request for regularizing the services of his private pandit on Rs 40/month were also acceded to by the government.[78] An artist or draughtsman, Rahim Buksh, was hired by him at Rs 16/month to paint blank anatomical drawings of arteries that came out of the lithographic press for his various tracts on the abdominal and thoracic viscera. Rahim Buksh later demanded and obtained a

[77] L no. 136, J. Adam, Sec. Med. Board to Lt. Col. W. Casement, 1 August 1828, & reply L no. 137, Lt. Col W. Casement, to J. Adam, Sec. Med. Board, 1 August 1828, Consult, 16 August 1828, BMC, 16–29 August 1828, P/33/6.

[78] Ibid. Breton was also allowed on his request access to the library of Fort William. See for government's sanction of all his requests. L no. 75, Lt. Col. W. Casement, Sec. to Govt, to Board, 1 April 1824, Consult, 1 April 1824, BMC, 1–8 April 1824, P/30/46.

hike in his salary to Rs 20/month—this was when he not only had to paint but also draw some of the anatomical plates for the tracts.[79] Breton also had in his service native surgeons like the Calcutta Muslim named Saut Cauree, who had been operating people for cataract using a couching needle, and performing operations for dropsy, hydrocele and spleen for twenty-five years. He trained students especially in cataract surgery, within the NMI. After some initial practice on the eyes of goats and sheep, students operated on human subjects. One of the students, Pursun Singh, was particularly praised for his newly acquired surgical skills. Many of Breton's tracts on cataract and the eye were influenced by the couching technique which, he felt, suited the native practitioner better than the Western mode of using a crystalline lens.[80]

And the NMI itself provided a nodal point of employment. Many, like Persaud Pandit who helped with the Devanagari script, worked both for the lithographic press as well as the NMI.[81] The services of this entourage became critical because it was made a requirement that one of the qualifications for the person chosen to superintend the NMI be the capability 'of reading the native system of medicine and of discussing and conversing with his pupils on all subjects of medical science in intelligible if not accurate terms.'[82] Jameson, secretary to the medical board, was of the view that this level of interaction was possible only

[79] L no 100, J. Taylor, Supdt, NMI, to Sec. Med. Board, 15 February 1833, Consult, 27 February 1833, BMC, 19 February–5 March 1833, P/34/38; see also in same Consult, L no. 101, Lt. Col. Casement, Sec. to Govt, to Med. Board, 27 February 1833.

[80] L no. 180, Enclosure, P. Breton, Supdt, NMI, to G. Proctor, Sec. to Med. Board, 14 October 1824, Consult, 18 March 1825, BMC, 11–25 March 1825, P/31/16.

[81] L no. 128, Medical Board to W.C. Bentinck, Gov. Gen India, 17 February 1835, Consult, 4 May 1835, India Mil. Consult, 4–18 May 1835, P/35/16.

[82] L no. 100, J.Jameson, Sec. Med. Board, Memorandum on Institution for Education of Native Doctors, 1 May 1822, Extract, BMC, 24 May 1822, Boards Collection, F/4/737, file no. 20085, p. 24. Jameson's successor, J. Crawford, was of similar view. See L no. 117, J. Crawford, Sec. Med. Board, to Lt. Col. Casement, Sec. to Govt, 7 March 1823, Consult, 29 March 1823, BMC, 29 March–11 April 1823, P/30/13.

with the assistance of a 'learned native physician' whose services would be 'indispensible to discharge Western medical science.'[83] Both he and his successor J. Crawford, convinced the government about the important role of Urdu and its practitioners in the dissemination of Western medical knowledge. But for this the superintendent himself had to know the Urdu language and the Persian script. Unlike the British medical teachers of the Calcutta Madrasa, who had to be well versed in Arabic, he was to possess a 'well grounded knowledge of the Persian and Hindustani both written and colloquial.'[84] In 1823 a government order made it compulsory that the person selected to superintend the NMI clear an exam, in the Persian script and the Hindustani language, conducted by Fort William College, before he was confirmed to his post. A certificate from the college had to certify that he not only knew the written and colloquial forms of these languages but also that he was capable of reading native treatises on medicine and discussing them with students.[85]

Thus, both the NMI and the press remained dependent on local men of learning and scribes. The lithographic press hired pandits, maulvis, munshis, hakims, and local draughtsmen and painters to assist in technical terminology and prepare anatomical plates for its productions. Until 1829 it had on its staff a local Persian and Nagari writer at a salary each of Rs 35/month and a painter at a salary of Rs 16/month who were exclusively employed in work connected with the NMI. In 1830, when financial concerns trimmed the lithographic establishment, these three staffers were transferred to the NMI and their salaries charged to its account.[86] Beside these staffers was an entire entourage of local printers, spongers, paper cutters, stone polishers,

[83] Ibid.

[84] Ibid.

[85] L no. 118, GO by Gov. Gen. in Council, 29 March 1823, Consult, 29 March 1823, BMC, 29 March–11 April 1823, P/30/13.

[86] Extract Public Letter from Bengal, 27 July 1830, Boards Collection, F/4/1243, file no. 40981, p. 17; see L no. 38, Govt Lithographic Committee to H. Prinsep, Sec. to Govt, 30 May 1830, ibid.: 207. See also L no. 38, Lithographic Committee to H. Prinsep, Sec. to Govt, 30 May 1829, and reply being L no. 4, H. Prinsep, Sec. to Govt, to Lithographic Committee, 9 June 1829, Consult, 9 June 1829, Bengal Pub. Consult, P/12/44.

ink grinders, and book binders salaried with the press.[87] Alongside these native staffers, the NMI had European draughtsmen and painters to assist locals in making 'charts, maps and anatomical plates and such other drawings that need more than ordinary skills'.[88] An English artist, Mr Pearson, was one such European assistant employed for making anatomical plates. He was on the personal salary of Superintendent Rind at Rs 150/month. In 1825 Rind complained that he had spent from his private funds Rs 4000 to maintain Pearson and his predecessor. He wanted the government to take Pearson into its service.[89]

The large local staff was essential to meet the high demand on the press. By 1829 so great was the pressure on the press by the NMI that Rind reported it required the constant use of a third of his establishment, i.e. two of the six presses. And while Breton complained that his orders had to wait because of the rush at the press, John Adam, secretary of the medical board, suggested the setting aside of at least one press at the government lithographic establishment to meet the NMI orders.[90]

[87] See for details L no. 26, J.N. Rind, Supdt, Govt Lithographic Press, to Lithographic Committee, 30 April 1825, Consult, 2 June 1825, Bengal Pub. Consult, 19 May–9 June 1825, P/11/37. See also minutes of the Govt Lithographic Committee, 14 September 1829, Boards Collection, F/4/1243, pp. 275–6. In 1829 the government's lithographic establishment consisted of the following strength: 4 printers at Rs 7/month, 4 spongers at Rs 7/month, 8 levermen at Rs 6/month, 1 ink grinder at Rs 6/month, 1 paper cutter with security at Rs 10/month, 1 paper cutter without security at Rs 6/month, 2 stone polishers at Rs 6/month, 2 inkers at Rs 6 /month, 4 supernumerarists at Rs 6/month, 2 coolies at Rs 4/month, 1 duftery at Rs 7/month, 2 watchmen at Rs 5/8 per month. The total cost of the establishement was Rs 200.

[88] L no. 27, J.N. Rind, Supdt, Govt Lithographic Press, to Govt Lithographic Committee, 15 January 1829, Extract Bengal Pub. Consult, 30 January 1829, Boards Collection, F/4/1290, file no. 51650, p. 103.

[89] L no. 26, J.N. Rind, Supdt, Govt Lithographic Press, to Lithographic Committee, 30 April 1825, Consult, 2 June 1825, Bengal Pub. Consult, 19 May–9 June 1825, P/11/37. Four other European draughtsmen were employed: Mr Fitzpatrick @ Rs 200/month. He was the mathematical and geography draftsman; Mr Schorn @ Rs 150/month. He was the lithographic copyist; and Mr Archer @ Rs 100/month. He arranged work for the printing superintendent; Mr Black @ Rs 60 /month. He also assisted the superintendent, and helped in the English writing as well as in ruling forms on lithographic stone.

[90] L no. 143, J. Adam, Sec. Med. Board, to Lt. Col. Casement, Sec. to Govt,

Breton's successor John Tytler, who took over the superintendentship of the NMI in 1830, continued with the tradition of composing anatomical texts. Tytler was a Scotsman who became an assistant surgeon in Bengal in 1813. This was after he was found qualified in 'physic' to serve in the Company's presidencies in India.[91] In 1830, upon the death of Peter Breton, he gave up his position of presidency surgeon and took charge of the NMI. He made some changes to Breton's staff of writers, whom he thought inefficient. He employed his trusted informer and translator, 'the most learned and intelligent native physician in Calcutta' in the Committee of Public Instruction, Hakim Abdul Majeed. Maulvi Abdul Rahim who was an able English translator of the government, was also employed by him. And a pandit of the Sanskrit College was taken on his rolls as well.[92] He did not enjoy the same free access to the press as Breton, and had to have his texts cleared by the medical board before he could approach the press. Many times, his texts—like one on gunshot wounds—were rejected by the board on the grounds that they did not include adequate references and extracts from the Western surgical texts like those by Cooper.[93] Tytler's uneasy relationship with the board was also because he overdid his publishing, causing suspicion in some members about the personal profits accruing to him and the possible neglect of his duties as superintendent.[94]

---

12 November 1829, and reply dated 11 December 1829, Consult, 11 December 1829, BMC, 11–26 December 1829, P/33/42.

[91] L/Mil/9/365. Papers of assistant surgeons, baptism and service entry for John Tytler, pp. 255–9.

[92] L no. 84, J. Tytler, Supdt, NMI, to J. Hutchinson, Sec. to Med. Board, 11 April 1832, Consult, 7 May 1832, BMC, 23 April–14 May 1832, P/34/21.

[93] Enclosure L no. 179, Members of Native Medical Education Committee to W.C. Bentinck, n.d., in L no. 178, M.J. Bramley, Asstt. Surgeon & J. Grant, President and Member Native Medical Education Committee, 20 October 1834, Consult, 28 January 1835, India Mil. Consult, 23 –28 January 1835, P/35/9. Also see L no. 84, J. Tytler, Supdt, NMI, to J. Hutchins, Sec. to Med. Board, 11 April 1832, Consult, 7 May 1832, BMC, 23 April–14 May 1832, P/34/21. This has Tytler's resentment on these rejections and his reasons for not including all the concerned Western tracts on grounds of making his text bulky. This he said would make the production costs high, and cause inconvenience to the native officers and men who lugged these texts as they marched with the regiments. Bulky texts would also be difficult to read.

[94] See Prior, 'Trevelyan', pp. 13–14.

However, misgivings notwithstanding, he put together as impressive a list of anatomical publications as Breton. In 1832, on approval by the medical board, the government sanctioned his text on vascular and absorbent systems and ordered 400 copies of it each in Urdu and Persian, and an equal number in the Devanagari script, to be produced at the lithographic press for the use of the NMI.[95]

By 1831, before this large order was sanctioned and despite the slowing down of his publications because of interference by the board, Tytler had translated or composed for students of the NMI the following tracts to be published at the lithographic press: Urdu translation of the 1824 edition of the *London Pharmacopoeia* in Persian and Nagari characters; 100 copies each in Urdu translation—in Persian and Nagari characters—of J. Hutchinson's Bengali tracts on fevers, a five-page Urdu tract on the foetus composed by him, and a tract on cataract in Urdu.

These were in addition to the tracts, chiefly on anatomy and surgery, that he composed and circulated to students in manuscript form. These texts had been returned by the press because of objections raised by the medical board on matters of content. These included: Introduction to the materia medica—37 pages; account of the most useful articles of the materia medica—178 pages; a system of osteology that had descriptions of all the bones in the skeleton, attachment of muscles and ligaments, passage of blood vessels and nerves—69 pages; account of the abdominal and respiratory muscles—9 pages; account of the vascular system—41 pages; tract on gunshot wounds that had extracts in translation from Hooper's chapter on gunshot wounds and Tytler's comments on how to use them in practice—6 pages; tracts on injuries of the head—14 pages; tract on hernia–36 pages.[96]

The NMI received a range of anatomical and pharmacy literature from Europe. Tytler was always enthusiastic about translating these into Urdu. In 1835 a list of books sent for the NMI from England included Claquet's anatomy descriptions, Fisher on smallpox, Cooper on hernia, Tuain's *Anatomy*, Thomson's *Dispensatory*, Thomas's *Practice*

[95] L no. 122, Members Med. Board to C.T. Metcalfe, VP in Council, 19 June 1832, and L no. 123, Lt. Col. Casement, Sec. to Govt, to Med. Board, 25 June 1832, Consult, 25 June 1832, BMC, 18 June–2 July 1832, P/34/23.

[96] L no. 84, J. Tytler, Supdt, Med. Board, to J. Hutchins, Sec. to Med. Board, 11 April 1832, Consult, 7 May 1832, BMC, 23 April–14 May 1832, P/34/21.

of Physic, Cooper's *Surgical Dictionary*, Paris's *Pharmacoloquia*, Alison's *Pathology*, *Smith's Botany* by Hooker, and the *Medical Gazette*, volumes 1–14.[97] Tytler argued that translating these was a sure way to diffuse new ideas of medical knowledge. So frustrated was he at the lack of free access to the government lithographic press that he suggested the use of typography for his compositions over lithography. Indeed, he made an appeal for the large-scale publication of the texts of the NMI, specially since they were in great demand 'in every institution of education in India'. He wanted every student to be given a collection of these books, to carry with him for revision of his knowledge when he left the NMI.[98]

Such Urdu texts on anatomy and pharmacy popularized the new idea of medical knowledge being about medical wisdom and science. They contrasted with Mughal encyclopaedic medical texts like the *Darashikohi* in their production style, the relatively small range they covered, the ordinariness of their authors, and the pedestrian rather than regal status they commanded because of being in the military camp language—Urdu. They differed also from the less ornate but dense and difficult-to-access tracts in the Arabic language, like the *Mujiz*, used in the Calcutta Madrasa as well as in family medical schools. In the Urdu medical literature produced by the NMI, specialized knowledge of anatomy, surgical skills, and chemistry compensated for the absence of an encyclopedic range, the physical ornateness of earlier book production, and the weight and dignity of the doctrinaire Arabic used in the past. Urdu as the language of the new medical literature reached those hitherto excluded from its ambit.

Demonstrating the New Knowledge: From Scholar to Practitioner

In the NMI, the knowledge of anatomy and surgery gained from texts was elaborated in lectures and practicals. This helped translate medical wisdom into medical practice. It welded a practitioner with a scholar.

[97] L no. 84, Enclosure, Members Med. Board to C.T. Metcalfe, Gov. Gen. India, 1 September 1835, Consult, 14 September 1835, BMC, 31 August–21 September 1835, P/35/22.

[98] L no. 84, J. Tytler, Supdt, Med. Board, to J. Hutchins, Sec. to Med. Board, 11 April 1832, Consult, 7 May 1832, BMC, 23 April–14 May 1832, P/34/21.

Every Tuesday and Thursday, pupils attended lectures given in Hindustani on anatomy. They also participated in experiments in chemistry and preparing materials used in medicine and surgery: neutral salts, acids, spirits, camphor, benzoin, sulphur, etc. On Friday night they were made to read from 8 to 10 p.m. Urdu texts prepared for them at the NMI: the pharmacopoeias, materia medicas, and tracts on anatomy. Each student was provided a copy of every text produced by the NMI. At this session, senior students were expected to memorize the names and features of the bones of the human body, the component parts of the brain, the thoracic and abdominal viscera, the structure of the eye and its appendages, as well as answer questions on the subjects. They were also expected to monitor the readings of junior students on these themes.[99]

They received practical knowledge of anatomy at the General Hospital and Company dispensaries. Here they observed British surgeons dissect the human body. In 1825 an assistant surgeon, William Twining, posted at the General Hospital in Calcutta, regularly demonstrated to them the anatomical details of bodies he dissected. And the apothecary, Mr Reid, at the Calingah dispensary, located close to the NMI, trained students in chemistry. Students also got clinical experience in their interactions with patients at these institutes.[100] This was followed up in the classroom by anatomical wax models of the arm, leg, and other body parts that came from London.[101]

In 1830, when John Tytler took over as superintendent, he retained this thrust on anatomy lessons. He systematized its instruction and made it more practical and rigorous. Students were divided into four classes: the lowest class started with anatomy and the highest ended with surgery. In between they studied materia medica and physic. The teaching of anatomy ran through all the four classes of the NMI. Tytler himself introduced the discipline in an introductory lecture. He followed this up with a demonstration on 'real bones': he made students aware of their different forms, the attachment of muscles, and the passage of

[99] L no. 7, enclosure, P. Breton, Supdt, NMI, to J. Adam, Sec. to Med. Board, 25 March 1829, Consult, 15 May 1829, BMC, 8–15 May 1829, P/33/26.

[100] Ibid.

[101] L no. 13, Lt Col. W. Casement, Sec. to Govt, to President and Members of Med. Board, 11 September 1829, Consult, 11 September 1829, BMC, 4–11 September 1829, P/33/34.

vessels and nerves. After this demonstration he explained the soft parts—the viscera of thorax, abdomen, pelvis, brain, and sense organs—on the bodies of sheep and goats. He compared the physiology of the animal body to that of the human. And through this comparison he highlighted their difference. This training was complemented at the general hospitals and dispensaries which the students attended to observe human dissections.[102]

Tytler showed due sensitivity to the religious beliefs of his students when dissecting. He used anatomical models shipped for him from England.[103] He also obtained from upcountry a 'very complete set of bones' and took much pains, he said, 'cleaning them so that they presented nothing repulsive to the fastidious'. Men of the 'Dome' caste (low caste) were encouraged by him to collect human bones from river banks for lecture demonstrations.[104] He also took care of superstitions regarding dingy spaces and foul smells, providing them a 'commodious' dry and well lighted lecture room' in place of the old one. He believed that his efforts had a positive effect in softening students towards accepting ideas about anatomy and dissections. Good students were employed by him as dissectors in his anatomy classes, and he wished this practice to be formalized. He pointed out that the mellowing down of students was evident when they asked him to use the animal body to explain anatomical facts. Indeed, one of them—a Hindu called Durohun Lall—brought to him a human skull which his friend had picked up from the banks of the river so as to be used in classroom demonstrations. Encouraged by this incident, Tytler asked the board to send him, for the purpose of teaching, 'a well prepared human skeleton, a set of separate bones, an injected head, trunk, preparation of lymphates, a pelvis with ligaments, a foetal head, a heart and lungs.'[105]

[102] L no. 84, J. Tytler, Supdt, Med. Board, to J. Hutchins, Sec. to Med. Board, 11 April 1832, Consult, 7 May 1832, BMC, 23 April–14 May 1832, P/34/21.

[103] L no. 6, Lt. Col. Casement, to Med. Board, 27 December 1833, Consult, 27 December 1833, BMC, 5–27 December 1833, P/34/50.

[104] L no. 91, J. Tytler, Supdt, NMI, to Sec. Med. Board, 23 December 1833, Consult, 16 January 1834, BMC, 16–29 January 1834, P/34/56. He said he needed to pay *bakshish* (tip) to the 'Dome' and listed this as one of the expenses. To this he added the money spent on buying charcoal for the experiments and oil for lamps so that the students could carry on their night study.

[105] L no. 84, J. Tytler, Supdt, Med. Board, to J. Hutchins, Sec. to Med.

Texts on anatomy and chemistry and practical knowledge acquired in the NMI had a life beyond the classroom as well. Awareness of the new medical ethos slowly spread through society via the wide range of service gentry attracted to the press for employment from all over northern India. Such knowledge was disseminated through the person of the native doctor as well, and texts literally moved around with the marching regiments, who had their native doctors. Almost all regimental doctors carried Urdu translations of the pharmacopoeia. Many stitched copies of the translated *London Pharmacopoeia* were sent to general and native hospitals, dispensaries, marine and police departments, and the garrison surgeons of Fort William for consultation. Whenever the demand for any Urdu medical text came from surgeons of the native corps scattered all over India, Breton was always happy to assist the lithographic press with his editorial expertise and staff so that they could meet the demand. The government was invariably supportive of the endeavour.[106] In 1829 Breton reported to the medical board the wide circulation of his Urdu medical literature, which reached students, European and native assistants of the NMI, and government dispensaries and hospitals. He indicated that it had a readership outside the Bengal Presidency. Copies of the texts were dispatched to the Madras and Bombay presidencies, the Hindu and Muhammedan colleges in Calcutta, Fort William College, the Medical and Physical Society of Calcutta, and the Asiatic Society of Bengal. They had a clientele beyond these English enclaves as well. Breton reported that they were sent on demand to the hakims of the Raja of Lahore, to the French settlement at Pondicherry, and to those native doctors as well who had never been trained at the NMI.[107] A happy Breton never failed to underline the great difference the lithographic press had made to the diffusion of the

---

Board, 11 April 1832, Consult, 7 May 1832, BMC, 23 April–14 May 1832, P/34/21.

[106] In 1828 the government gave the sanction to the Lithographic Press for one such big contract for Urdu texts. The request was made by surgeons of native corps and civil stations. See L no. 149, J. Adam, Sec. Med. Board to Lt. Col. W. Casement, 7 February 1828 & L no. 150, Lt. Col. W Casement, to J. Adam, Sec. to Med. Board, 15 February 1828, Consult, 15 February 1828, BMC, 15–22 February 1828, P/32/55.

[107] L no. 7, enclosure , P. Breton, Supdt, NMI, report on the NMI, to J. Adam, Sec. to Med. Board, 25 March 1829, Consult, 15 May 1829, BMC, 8–15 May

new medical knowledge at minimum financial strain to the government. This was particularly true of the production of anatomical plates which had, so far, remained a major lacuna in medical education. With the press in place, he said, all that was required was 'paper and ink'. He always had words of appreciation for Surgeon Rind, the superintendent of the press, whom he commended for his 'exertions and solicitude— to render me every assistance in his power in printing material as speedily as I can prepare them for the use of students.'[108]

New Medical Knowledge and Public Service: Caring for Society

Authority to the native doctor came from the new knowledge in Urdu. But it also came from his being a public servant in the coveted service of the Company's army. A government order of 21 June 1822 clarified that students admitted to the NMI would be enlisted as soldiers. And, like all other soldiers, they were to be fully supported financially throughout their stint at the NMI by government. The medical board examined them at the end of their three years and issued them a certificate which enabled them to work as native doctors in the military.

Most sepoy regiments of the Company had a Hindu and a Muslim native doctor. These took care of the sepoys, whose caste identities the Company zealously guarded.[109] In the past, the pampered sepoys had refused to be attended by native doctors who were of low caste or not of respectable status. This had prompted the military authorities to ensure that native doctors came from a respectable family background.[110] Most came from lower-level service-gentry families. They were acculturated in the tradition of 'public service' from their contact

---

1829, P/33/26. See also for dispatch and receipt of Breton's texts to Bombay and Madras, L no. 155, J Adam, Sec. to Med. Board, to Lt. Col. Casement, Sec. to Govt, 13 April 1829 and his reply L no. 156, Consult, 18 April 1829, BMC, 10–18 April 1829, P/33/24.

[108] L no. 181, enclosure, P. Breton, Supdt, NMI, to G. Proctor, Sec. to Med. Board, 7 March 1825, Consult, 18 March 1825, BMC, 11–25 March 1825, P/31/16.

[109] J. Hutchinson, Code of regulations for medical department of the Bengal establishment, compiled by order of government under superintendence of the Medical Board, Calcutta, 1838, L/Mil/17/2/450, Article 4, p. 207.

[110] L no. 95, enclosure in L no. 93, Civil Finance Committee to W.C. Bentinck,

with older regimes. But as Company servants they had more clout and authority. They were primarily military doctors, though they could be sent on civilian postings as well.[111] Many of them functioned as government vaccinators in civilian stations, others worked in local dispensaries.[112] Some of them were employed as assistant teachers at the NMI.[113] As public servants they were responsible for the health of society as a whole.

The harnessing of the new medical knowledge to general well being and public service was evident in the special and separate treatment given to graduates from the NMI. Additional allowances were sanctioned to those with certificates from that institution. Thus, a certified native doctor from the NMI received a monthly salary of Rs 25 when posted in the field, instead of the Rs 20 allocated for those who did not have any NMI certificate. And when on garrison duty in civil areas an NMI doctor received Rs 20 instead of Rs 15 that his untrained colleagues received. The additional Rs 5 was viewed as a special military allowance called *batta*. It was witheld when the doctor went on long leave.[114] Again, as a further incentive, the native doctor got an additional increase

---

9 September 1830, Consult, 3 October 1833, BMC, 5 September–3 October 1833, P/34/47.

[111] Extract from GO relating to NMI, 21 June 1822, Hutchinson 1838: 265. In 1834, native doctors were posted in civil stations in southern central Assam, in Gauhati and in Patna. See L no. 113, Members Medical Board to C.T. Metcalfe, 26 July 1834, Consult, 7 August 1834, BMC, 31 July–21 August 1834, P/34/66; and L no. 55, Members Medical Board to C.T. Metcalfe, 17 October 1834, Consult, 23 October 1834, BMC, 16 October–6 November 1834, P/34/69.

[112] For a native doctor appointed as vaccinator, see L no. 17, Col. J.A. Paul Macgregar, Mil. & Auditor Gen. to W.C. Bentinck, 16 April 1833, Consult, 19 April 1833, BMC, 3–19 April 1833, P/34/40.

[113] Native doctors Jashoda Misser and Ram Ishwar Awasthi were employed at Rs 25/month to be assistant teachers at NMI. See L no. 6, Col. W. Casement, Sec. to Govt of India Mil. Dept, to Med. Board, 26 September 1836, Consult, 26 September 1836, India Mil. Consult, 19–26 September 1836, P/35/45; and L no. 18, Col. W. Casement to Col. J.A. Paul Macgregar, Mil. & Auditor Gen., 19 April 1833, Consult, 19 April 1833, BMC, 3–19 April 1833, P/34/40.

[114] Extract from GO relating to NMI, 21 June 1822, Hutchinson: 1838: 265. See also L no. 135, Maj. J. Stuart, Deputy Sec. to Govt in Mil. Dept, to Med. Board, 27 February 1834, Consult, 27 February 1834, P/34/59. The government discouraged long leave and left it to the discretion of the Commander-in-Chief.

of Rs 5/month after completing seven years of service if he was favourably recommended by his medical officer.[115]

The military also took care of the pension benefits of the doctor in the same way as it did for sepoys and other native officers serving in its regiments. Every native doctor was entitled to an invalid's pension in case he could no longer serve because of wounds received in the service or disorders contracted while on duty. Those who had served for less than seven years with the army received a monthly pension of Rs 7/month, whereas a service period of 7–15 years entitled him to a third of the salary of his field pay if posted with the army. The same percentage of his garrison pay was given to him if on civil duty. All native doctors who had served for more than fifteen years received an invalid's pension of Rs 10/month. This was raised to half their salary after a service of twenty-two years.[116]

The native doctors were governed by military law. This made them rather exclusive, and, very much like the sepoys, they enjoyed special status in village society. Their medical authority derived largely from this military backing. Thus, no native doctor could be summarily dismissed without a court martial. His respectability and authority in local society were protected zealously by the army. The army ordered that under no circumstances were penalties like corporal punishment to be exercised against its doctors.[117] In an 1832 government order, the commander-in-chief took serious note of a case of corporal punishment involving a native doctor. Noting that, 'the commander-in-chief desires it may be understood that he does not consider corporal punishment as a fit sentence for a class of persons whose respectability His Excellency is desirous to maintain.' The order emphasized that the commander-in-chief authorized the court martial to dismiss the native doctor as the most extreme punishment, rather than publicly humiliate him.[118]

Indeed so zealously was the special status of native doctors maintained and protected by the military that, many times, the medical

[115] The medical officer had to certify that the general character and professional conduct of the individual was worthy of this hike in salary. See extract from GO relating to NMI, 21 June 1822, Hutchinson 1838: 265.

[116] Ibid.

[117] Ibid.

[118] Ibid.: 207.

board objected to the excessive 'control of the military over the [NMI] alumni'.[119] The military always reminded the board of the doctor's special status as a military man from the time of his admission in the NMI.[120] Indeed, the two establishments quibbled most over the issue of protecting the religious sensitivities of students. The military, as in the case of its sepoys, took excessive care of students' religious and caste sensitivities. Special dietary provisions, in keeping with their preferences, and wood and water were supplied on the ships they travelled to foreign assignments.[121] This only sharpened the religious profiles among physicians that village Arabic schools had initiated. The medical board did not always approve of this excessive pandering to religious sensibilities. A major conflict ensued in 1833 over the issue of accommodation for the students.

That year, Tytler shifted the NMI to a larger building at the top of Free School Street, on its east side.[122] The medical board complained that no permission was sought by him to make this shift, and only when they went for an inspection was the matter brought to their notice.[123] The rent of this building was Rs 250/month.[124] It belonged to Raja Buddinauth Roy.[125] The medical board was particularly agitated because Tytler not only shifted the building but separated the teaching and residential premises and also moved the superintendent's house

[119] Maj. J. Stuart, Deputy Sec. to Govt Mil. Dept, to Med. Board, 22 May 1834, BMC, 8–22 May 1834, Consult, 22 May 1834, P/34/62.

[120] Ibid.

[121] L no. 43, J. Stuart, Deputy Sec. to Govt Mil. Dept, to Mil. Board, 11 February 1834, Consult, 16 February 1834, BMC, 6–20 February 1834, P/34/58. Native doctors on duty to Arracan were supplied with special provisions on the Company's ship *Ganges*.

[122] L no. 103, Members Med. Board to C. Bentinck, 22 November 1833, Consult, 28 November 1833, BMC, 7–8 November 1833, P/34/49.

[123] L no. 126, Members Med. Board to W.C. Bentinck, 8 August 1833, Consult, 16 August 1833, BMC Consult, 25 July–16 August 1833, P/34/45. They reported that Tytler was given an ejection notice because, due to certain legal wrangles, he had suspended payment of the monthly rent.

[124] L no. 91, Med. Board to C. Bentinck, 4 January 1834, Consult, 16 January 1834, BMC, 16–29 January 1834, P/34/56.

[125] L no. 91, Enclosure, J. Tytler, Supdt, NMI, to Sec. Med. Board, 18 December 1833, BMC, 16–29 January 1834, P/34/56

away form its earlier compact layout. Tytler resisted the board on the grounds of protecting the religious sensitivities of medical students. One of his reasons for the shift to the larger building was to provide space for Muslim students to erect their own temporary out-offices, to be used as cook rooms.[126] Hindu students were provided the same provision of separate cook houses.[127]

Again, he lodged Hindu and Muslim students within separate quarters. Both lived on Short Bazaar Street. The Muslim students had separate lodgings close to Tytler's house, at a safe distance from the Hindu accommodation on the same street. Here, in any year, as many as 35 Muslim students lived in 2 large rooms and 3 out-offices attached to it. The headmaster had a private apartment in the same area, which the students had access to for purposes of private study and classroom instruction.[128] The Hindu residences were reported to be on premises in the opposite direction from those formerly occupied by the NMI on Park Street. The medical board was incensed by the fact that two kinds of students, who sat in the same classrooms and read the same texts, were segregated in this way. The dampness and overcrowded nature of the Muslim quarters was particularly distressing, which they said 'are unfit for habitation particularly to youngmen who are natives of the western provinces to whose constitution the climate of Bengal is sufficiently trying under the most favourable circumstances.'[129] They set up a committee to look into the matter.[130]

[126] L no. 103, Members Med. Board to C. Bentinck, 22 November 1833, Consult, 28 November 1833, BMC, 7–8 November 1833, P/34/49.

[127] L no. 72, Col. J.A.P. Macgregor, to C. Bentinck, 11 December 1833, Consult, 12 December 1833, BMC, 5–27 December 1833, P/34/50.He reported that Rs 40 had been drawn by Tytler for expenses incurred in removing the old cook house for both Hindu and Muslim students of the NMI. He had erected a new building to be used as a cook house.

[128] L no. 166, T. Wood, Engineering president and member special committee, to Lt. Col. W. Casement, 22 August 1833, & L no. 167, Lt. Col. W. Casement to Medical Board, 30 September 1833, Consult, 30 August 1833, BMC, 23–30 August 1833, P/34/46.

[129] L no. 126, Members Med. Board to W.C. Bentinck, 8 August 1833, Consult, 16 August 1833, BMC, 25 July–16 August 1833, P/34/45.

[130] L no. 128, Lt. Col. Casement to T. Wood & H.S. Mercer, 16 August 1833, Consult, 16 August 1833, BMC, 25 July–16 August 1833, P/34/45. These two men were appointed members of the committee.

In 1833 the committee appointed to look into the Muslim lodgings expressed its concern over the state of overcrowding in the building. They declared it almost unfit for living because of its location in a low-lying area which made it damp and ill ventilated. Bentinck took serious note of this report, calling the deplorable conditions 'unnecessary and objectionable'. He ordered the shifting of the NMI to other rented premises, for which he sanctioned as much as Rs 250/month as rent. He argued that this new space be used as a compact campus for both the superintendent as well as all students.[131]

Abolition of the Medical Classes in the Calcutta Madrasa and Native Medical Institute

In the 1830s age of reforms and financial crisis, the medical class of the Calcutta Madrasa and the NMI both came under severe attack from the Civil Finance Committee. The committee found the NMI an uneconomical experiment and recommended its abolition.[132] More importantly, these institutions came under fire from the anglicist lobby. They attacked oreintalist scholars like John Tytler, and saw their institutions as an impediment to the introduction of Western medicine and English instruction in India. With the backing of the new governor general, Bentinck, they supported the establishment of a medical institution that would open up Western medical learning in English to a 'respectable class' of Indians.[133] The future of the medical class at the madrasa and the NMI thus got tied up with the education debate between orientalist scholars and anglicists over the appropriate timing of introducing English instruction in India.[134] And in 1835 Bentinck, at the recommendation of the native medical education committee,

[131] L no. 167, Lt. Col. W. Casement to Medical Board, 30 September 1833, Consult, 30 August 1833, BMC, 23–30 August 1833, P/34/46.

[132] L no. 93, Civil Finance Committee to W.C. Bentinck, 9 September 1830, Consult, 3 October 1833, BMC, 5 September–3 October 1833, P/34/47.

[133] See Dr H.H. Goodeve's paper on the NMI, which makes a case for its abolition. It suggests the setting up of an alternative institute that would impart Western medical training in English to Indians from respectable family backgrounds. This, he argues, would create a class of medical professionals called sub-assistant surgeons. L no. 103, H.H. Goodeve's paper, Consult, 3 October 1833, BMC, 5 September–3 October 1833, P/34/47.

[134] See for the details of this debate Arnold 1993.

abolished the institution and medical class of the madrasa. He recommended the establishment of a new medical college in Calcutta.[135]

The Company, of course, viewed the college as representing a new era in its functioning. From the viewpoint of London, the abolition was viewed as a change of official policy, a victory for the anglicist lobby. Locally, the change was less perceptible. Many of the NMI staffers, from its superintendent John Tytler, to the native hakims, maulvis, pandits, and native doctors like Ram Ishwar Awasthi, made pleas to join the new college. They did not see it as representing any major change as far as the content of medical instruction was concerned.[136] In 1834 Tytler made it clear to the Company that he did not see any contradiction between the old and the new order. He was neither against the introduction of the English language nor Western medical education. Indeed, he said he was 'clearly of the opinion that European literature should be made a permanent object of study; but I do not think the English language should be made an exclusive medium of

[135] For the detailed report of the native medical education committee see L no. 179, Members Native Medical Education committee to W.C. Bentinck, n.d., Consult, 28 January 1835, India Military Consult, 23–28 January 1835, P/35/9. Their objections to the NMI were on grounds of economy, unsatisfactory administrative command structure, and lack of regulations. According to them, knowledge imparted in Urdu was inadequate as it did not teach Western medical science in its entirety. This would be achieved only through English-language texts. For Bentinck's minute abolishing the NMI, see L no. 183, Minute by the Gov. Gen of India, 26 January 1835, Consult, 28 January 1835, India Military Consult, 23–28 January 1835, P/35/9.

[136] In 1835 the Medical Board forwarded the applications of NMI staffers Hakim Abdul Majeed, Madhusudan Gupt, and Prasad Pandit. They wanted appointment at the new college or at the education department. They also requested for a pension since the closure of the NMI had left them jobless. Hakim Majeed could not be adjusted since he did not know English, but the other two were taken into the service of the General Committee of Public Instruction that produced books for the new college. See L no. 128, Medical Board to W.C. Bentinck, 17 February 1835, Consult, 4 May 1835, Indian Mil. Consult, 4–18 May 1835, P/35/16; also see L no. 133, J.C.C. Sutherland, Sec. to General Committee of Public Instruction, to Col. W. Casement, 30 April 1835, Consult, 4 May 1835, Indian Mil. Consult, 4–18 May 1835, P/35/16.

instruction.'[137] The Company thought otherwise. It therefore turned down Tytler's application to be considered for the superintendentship of the new college, despite his continued pleadings for a reconsideration of his case.[138]

British officers in India cautioned against making too much of the change. As late as the 1850s, the recommendations of Alloy Sprenger, principal of the Calcutta Madrasa, urging the Company to show only a philological interest in Arabic learning, was opposed locally. The council reporting on the madrasa advised the government against such an ideological shift. It argued that elite Muslim families continued to pursue a scholarly interest in Arabic sciences. The Company needed to be careful, because for many families medical education still was 'more than just employment'.[139] The note of caution from officers in the field reflected that the medical culture of the period outlived these reforms.

At one level, the Calcutta Medical College continued with the NMI agenda to invigorate the Arabic sciences through the introduction of Western-style anatomy and surgery. The similarity of the agendas of the two institutions was reflected in the transfer of medical items of the NMI to the new college for its use. These included anatomical models, surgical instruments, books like Lizar's *Anatomical Plates*, Tuson's *Mythology*, Dewhurst's *Anatomy of Muscles*, Annesley's *Diseases of India*, and accessories like tables, chairs, thermometers, electric machines, and other professional articles.[140] But the NMI had grafted the new

[137] Tracts 15, Memorial of Surgeon John Tytler of the Bengal Establishment, Addressed to the Court of Directors of the EIC. L. no. ix, J Tytler's reply to the committee of native medical education as to the best language to be used as a medium of instruction, p. 26.

[138] Ibid.: p. 6. See also L no. xi, J. Tytler, to J. Cosmo Melvill, Sec. to EIC, n.d., pp. 31–40. He made a strong case for getting the job of superintendent of the new college.

[139] *Selections from the Records of the Bengal Govt, no. XIV*, pp. 4, 10–11.

[140] L no. 129, Col. W. Casement, Sec. to Govt of India, to Med. Board, 19 February 1835, Consult, 19 February 1835, India Military Consult, 19–24 February 1835, P/35/11; and L no. 9, List of medicines and other articles from NMI to M.J. Bramley, Supdt, of Calcutta Medical College, 19 May 1835, Consult, 1 June 1835, India Military Consult, 25 May–8 June 1835, P/35/17.

knowledge on existing medical communities of literature and Urdu linguistic culture. In contrast, the new college carved out a less compromised way of operation. Bentinck was of the view that the new institution would impart Western medical learning in its *entirety*, including human dissections, to Indians. He was convinced that this would be of great political and economic benefit to the Company. It would be more economical since it would teach Western medicine in the English language and cut out expenses on translation. It would churn out better-qualified Indians. This would reduce the cost of maintaining the huge establishment of Western practitioners in India. And costs would also go down when the learning of oriental languages proved no longer necessary for Britishers posted in India.[141] Bentinck appointed M.J. Bramley as the superintendent of the new institution and dispensed with the services of Tytler. The new man was sanctioned two years on an allowance of Rs 1790.[142]

However, in March 1835 Bentinck perhaps did not realize the limitations of his reforms when the medical board reported to him that, upon the closure of the NMI, they proposed, 'to furnish each of the students with a complete set of all the tracts in the native language and to lodge the remainder in our office to meet future demands.'[143] In May 1835 with their Urdu medical texts firmly in place, they 'left the presidency in boats hired at their own expense under the care of the native assistant teacher Shaikh Waris Ali and Hira Lall.'[144] As the students of the NMI and its staffers dispersed into the qasbas and towns of the North Indian countryside, so did their new ideas and texts. These now became diffused in local society, alongside the flame of

[141] For Bentinck's minute abolishing the NMI see L no. 183, Minute by the Gov. Gen of India, 26 January 1835, Consult, 28 January 1835, India Military Consult, 23–28 January 1835, P/35/9.

[142] L nos 184, 186–190. These are copies of the Government Order of 26 January 1835, abolishing the NMI, that were sent to the Adjutant General, the Medical Board, John Tytler, etc. Consult, 7 May 1832, BMC, 23 April–14 May 1832, P/34/21.

[143] L no. 184, Members Medical Board to W.C. Bentinck, 4 March 1835, Consult, 13 March 1835, India Military Consult, 13–18 March 1835, P/35/13.

[144] J. Hutchinson, Sec. Med. Board, to Members of Medical Board, 16 May 1835, Consult, 1 June 1835, India Military Consult, 25 May–8 June 1835, P/35/17.

Arabic medical learning ignited by the Calcutta Madrasa (even though its medical class had been abolished). The patronage to Persian medical learning, at the level of regional courts, also continued its flicker.

The agility of this plural medical culture was more than evident in the late nineteenth century, as we shall see in chapters 5 and 7, when hakims used Persian, Arabic, and Urdu strands of this culture to contest colonial medical drives in the period of high nationalism.

3

The Practice of Medicine: Public Welfare

Introduction

THIS CHAPTER ELABORATES ON HOW THE PRE-COLONIAL 'scientistic' notion of medicine, divorced from Persianate civility, and the native doctor as public servant set the ground for the early British idea of public health. It shows how these early health concerns were not entirely modelled on borrowed Western models of civil society. Rather, they encapsulated ideas of public welfare and social well being that operated as domains fiercely contested by local patrons and the Company. Indigenous medical patrons—notables, rajas, and elites—and the knowledge base of hakims and other medical communities—played a critical role in the public welfare concerns of the British in nineteenth-century India.

I show here how the Company remained dependent on indigenous medical practitioners and their patrons, who operated as the welfare managers of society. This dependence became particularly significant as the state tried in vain to control the early-nineteenth-century epidemics of smallpox and cholera. It tried to harness its 'public welfare' on to their materia medica. Like them, the Company projected itself as the benevolent master and tried to insinuate itself as the new medical patron; indeed, it competed with older patrons of knowledge to emerge as the new welfare manager. Public welfare became the new site where Unani local knowledge was both shared and contested, its elite patrons both used and dislocated. A rudimentary notion of public health emerged out of this complex relationship between the colonial doctor and Indian society.[1]

The rudimentary public health manifesto written at the time of the cholera epidemic of 1817–18 was a consequence of this interaction. Print lent this shared knowledge a new textuality and authorial—and thus authoritative—status. It facilitated its wider circulation. Significantly, this India-specific manifesto was exported to England to handle the cholera epidemic of the 1830s. The narrative of such a medical culture of early modern India, outlined here, therefore questions the idea of colonialism as the defining site of culture and power.

Health and Public Welfare

From 1802 the fear of spread of smallpox triggered the health concerns of the Company in the direction of disease prevention. Vaccine inoculation against smallpox was one of the early 'welfare' schemes related to the health of the subjects that the Company undertook. The inoculation drives of the Company began to loosely define its commitment not just to therapeutics but also to the prevention of disease, and thereby demonstrated its interest in the 'well being' of the people. Even though the direct connection of health to physical and economic well being emerged only during the course of the 1818 cholera epidemic, the management of the inoculation drives across North India and the Maratha territory sent a positive message regarding the Company's intent in matters of general well being.

The inoculation drives paved the ground for local rulers and the Company to come together to introduce schemes that would prevent disease, and thus guarantee the well being of people. Inoculation had the active support of local physicians, notables, and ruling houses, who

[1] This was different from Britain, where a movement for public health pressurized the state into action in the 1840s and 1850s. This was also in contrast to Qajar Iran, where ideas of public health emerged only in the late nineteenth century as a part of the concern for military health in the wake of epidemics. The sanitary committees and town planning that followed directly borrowed from European models in the late nineteenth century. See Bynum 1994; Ebrahimnejad 2004: 31–49.

co-operated even as they competed with the Company's effort to project itself as the welfare patron of their subjects. This competition for patronage only invigorated the space for 'public welfare'. It created a rudimentary format of what would later be projected as 'public health' concerns.[2]

Regional Rulers and the Company: The Vaccination Drive 1802–1823

The practice of variolation was not entirely unknown in India. The Brahmins of Hindustan practised it in some form as a vocation. From 1767, Company officers like J.Z. Holwell reported with appreciation such Brahmins' success stories. Holwell observed that they arrived in Bengal in teams from Brindaban, Benares, and Allahabad every disease season. They inoculated people using a thorn-like implement to pierce the skin, into which they inserted the cow variole with cotton wool purified with Ganges water. Holwell described the process in detail and found it a successful practice, based on 'rational principles and experiments'. He agreed with the medical 'rationality' the Brahmins' offered of smallpox being caused by 'noxious animalculae' in the air that adhered to certain food items in large numbers and caused disease. Fish, milk, and purified butter (ghee) were such items singled out by the Brahmins. Holwell said that he was persuaded by the cool regimen and free admission of air that they prescribed to bring out the infection from the body of the diseased. Doctors who did not adopt the same mode, he argued, 'making necessary distinction and allowance between the constitutions of the natives and Europeans, have lost many patients, which might otherwise have been saved.'[3] In fact so influential was Holwell's essay and his endorsement of the 'scientificity' of the Brahmins

[2] Vaccination may have been opposed, but the contestations opened a critical public space of enquiry and debate between British doctors and patrons of Indian medical practice. This goes against the views of David Arnold, who argues that the Company's expectations—that vaccination would project its humane and benevolent intentions and would be readily accepted by people—were dashed from the very beginning. Arnold says this was because of cultural and religious antipathy towards the programme; but it was also because of the technical problems of getting cowpox readily into India for vaccination. Arnold 1998: 45–65.

[3] Holwell 1767: 21.

that he was quoted in many important histories of smallpox written in the early nineteenth century. For instance, in an 1815 tract by James Moore, the rationality of the Brahmins in Bengal rubbed shoulders with the successful treatments of William Cullen who, in the nineteenth century, reduced the entire management of both the inoculated and the causal disease to one single plan.[4]

As the Company extended its vaccination drive in the early nineteenth century, it was no surprise to find it receiving minimal co-operation from Brahmin vaccinators in Bengal. In 1802, John Shoolbred, the superintendent general of vaccine inoculation in Bengal, reported that Brahmins had been carrying out a crude form of vaccine inoculation in the region. Arriving in small group, of three or four from as far as Brindaban, Benares, and Allahabad, they would fan out in the region, vaccinating children and getting remunerated. He reported that they saw the Company as a rival; not surprisingly, they were 'determined enemies' of the new practice. To contest the Company's drives, they claimed they knew of the practice much before the Company acquired any knowledge of it. They indicated a passage in a copy of the Sanskrit medical manuscript called *Sudhasan Graha* by the physician Mahadeva to authenticate their claim. The Company doubted the veracity of the passage. Shoolbred said it looked suspicious because it explained the Brahminical practice as identical to that of the British: inoculation being performed from matter originating in the cow for the purpose of destroying susceptibility to smallpox. Sanskritists in the Company administration compared the manuscript with other copies available and found that the passage did not exist in any other text. They concluded that it had been fraudulently inserted in the copy that had been brought to their notice.[5]

The opposition of the Brahmins of Bengal stood in contrast to the enthusiasm of regional rulers. One of the most detailed accounts of the history of inoculation is provided by one Mr Thomas Coats, surgeon at Poona, involved with the Company's inoculation establishment.[6] He reports that the political benefits of the inoculation drive drew the

[4] Moore 1815: 26–32, 275.

[5] Shoolbred 1804: 17, 73–9.

[6] Thomas Coats wrote a pamphlet in Marathi aimed at propagating vaccination in Poona. See a translation of this in Guha 2001: 140–55.

Peshwa into the picture. The Peshwa played his cards carefully as the Company attempted, through its inoculation drives, to insinuate itself as the welfare manager of his subjects. He tried to share the laurels by extending royal hospitality and infrastructural support to Coats. Through his help he tried to assert his rightful role as the welfare patron in what was a Company-sponsored health initiative. The Peshwa set apart a house at Etal Pet for the reception of Mr Coats, which he visited every Sunday. More importantly, he offered his consort as the 'subject' of vaccination, and thus not only put the seal of royal legitimacy on inoculation but also salvaged his role as welfare patron of his people. The involvement of the Peshwa produced the desired effect. By 1812 Mr Coats had inoculated 17,056 people in Poona alone. Altogether, more than 30,000 people had been successfully inoculated by him.[7]

Health guaranteed by inoculation became the medium through which the Company's welfare concerns began to be marked out in Peshwa territory. Coats showed due regard to the people's religious sensitivity. He allowed those successfully inoculated to perform the rituals that signified their having bypassed the disease. As Sumit Guha argues, Coats sacralized the process of vaccination so as to take the sting out of the 'raw secularity' of the innovation.[8] He encouraged people to pray to their local goddess, Bhawani, for the success of cowpox, just as they prayed to her for smallpox.[9] The Company, encouraged by the enormous success of Coats, relieved him from his regular duties and asked him to remain at Poona with a suitable salary for the purpose of conferring 'the blessing of vaccination and affording gratuitous medical aid to persons of all descriptions who apply to him.'[10]

[7] Bengal Public Letter 239, 21 November 1812, Mr Coats, Surgeon, Poona, to M. Elphinstone, Resident Poona, enclosed copy of a short account of the cow-pox as translated into the Mahrattah language by T. Coats, Boards Collection, F/4/427, file no. 10455, pp. 23–46.

[8] Guha 2001: 144.

[9] Bengal Public Letter 239, 21 November 1812, Mr Coats, Surgeon, Poona, to M. Elphinstone, Resident Poona, enclosed copy of a short account of the cow-pox as translated into the Mahrattah language by T. Coats, Boards Collection, F/4/427, file no. 10455, pp. 44–5.

[10] Ibid.: 35. He made his inspections on the eighth day of inoculation and provided certificates in Marathi to those who had taken the infection successfully.

Again, in 1804, from the Maratha territory of Mahadaji Shinde, Mr Ingledew, a vaccine inoculator, reported the full co-operation of the ruler who was willing to offer his only daughter Bala Bhye as a subject of inoculation. Shinde regretted that he had to delay this plan as she was taken ill with a fever.[11] In 1805, from the northern part of India, a similar story was reported by A. Seton, agent of the governor general at Bareilly. He reported of the help he received from the Rohilla ruling houses of Rohilkhand. The government quickly capitalized on this good will. It directed that a depot of vaccine inoculation be established at Bareilly and appointed Mr John Gilman, surgeon at Bareilly, as the subordinate superintendent of vaccine inoculation.[12] The importance the Company gave to this project was reflected in the high salary of Rs 200/month that it paid Gilman. He also had a staff comprising four Brahmin inoculators, a Muslim inoculator, a peon to call in patients, and a writer to assist in keeping the register and making reports. These were allotted a salary of Rs 8/month for each inoculator, and Rs 5/month and Rs 15/month for the peon and writer respectively.[13] Seton also reported on the support he received from local practitioners, who were eager to learn the skill and assist in this disease-prevention programme. According to him, a European surgeon successfully instructed two hakims of Bareilly, Ghalib Ali Khan and Muhammad Aiwaz, and they were then dispatched to Rampur to carry on the inoculation drive.[14]

[11] Shoolbred 1804: 119.

[12] Extract Political Letter from Bengal, 3 January 1806, Consult, 4 October 1805, Boards Collection, F/4/198, file no. 4452, pp. 7–8.

[13] Undated order appended to L no. 29, A. Seton, Agent to Gov. Gen., to Thomas Thornhill, Judge and Magistrate Bareilly, 7 June 1805; and the latter's reply dated 8 June 1806, Extract Bengal Pub. Consult., 4 October 1805, Boards Collection, F/4/198, file no. 4452, pp. 14–18. It is interesting that the Board made a distinction between Muslim and Hindu inoculators, in keeping with the distinction it had been laying out in the realm of 'codifying law'. This distinction in the medical realm continued to be used, as we will see, in the British administration of dispensaries as well.

[14] L no. 29, A. Seton, Agent to Gov. Gen., to Thomas Thornhill, Judge and Magistrate Bareilly, 7 June 1805, Extract Bengal Pub. Consult., 4 October 1805, Boards Collection, F/4/198, file no. 4452, pp. 14–18.

In Rampur he reported that their success was primarily due to the full support of many Rohilla chiefs of 'rank and status'. Added to this was the seal of royal legitimacy the project received with the involvement of the nawab regent Naseemullah Khan.[15] Very much like the Peshwa in Poona, and probably for similar reasons, the nawab not only gave one of his own houses for the accommodation of the surgeons but, more importantly, offered some of his close relatives as medical 'subjects' to the Company's doctors.[16] This benefited the Company in the short term as the seal of royal legitimacy made the inoculation drive immensely popular.

The same was the case in Bhopal, where the seal of legitimacy came from the royal family of the nawab of Bhopal. Assistant Surgeon Spry, who was posted in the city, consolidated this advantage by requesting that the Company set up an inoculation establishment for the ceded and conquered territories of Central India.[17] The Company was convinced that the co-operation of regional rulers in its public health initiative was critical to the success of the inoculation experiment. In 1813, W. Russell, superintendent general of vaccine inoculation at Calcutta, was concerned that Mirza Muhammad Husain Khan, the 8-year-old son of the nawab wazir at Lucknow, got smallpox even after he had been inoculated. This he said has happened 'most unfortunately in a person of his exalted rank, as it is likely from that influence to produce a stronger impression of distrust in the minds of natives'.[18]

The short-term benefits of royalty participation were immense. But the Company was aware that this co-operation generated competition.

[15] A. Seton, Agent of Gov. Gen. at Bareilly, to N.B. Edmonstone, Sec. to Govt in the Sec. and Pol. Dept, 18 March 1804, Extract Bengal Pub. Consult., 4 October 1805, Boards Collection, F/4/198, file no. 4452, pp. 20–3.

[16] A. Seton, Agent of Gov. Gen. at Bareilly, to N.B. Edmonstone, Sec. to the Govt in the Secret & Political Dept, 9 June 1805, Extract Bengal Pub. Consult., 4 October 1805, Boards Collection, F/4/198, file no. 4452, pp. 9–14.

[17] Asstt. Surgeon Spry, to H.I. Prinsep, Sec. to Gov. Gen., 15 July 1832; Asstt. Surgeon Spry to A. Moloney, Agent to Gov. General at Sagor & Narmada territory, 18 June 1821, Extract Fort William Political Consultations, 13 August 1832, Boards Collection, F/4/1486, file no. 58647, pp. 7–8 & 11.

[18] L no. 35, W. Russell, Superintendent General vaccine inoculation, Calcutta, to D. Munro, President and Members of Medical Board, 4 August 1813, Bengal Public Letter, 154–8, 2 October 1813, Boards Collection, F/4/446, file no. 10749, p. 40.

And this had its pitfalls. The offer of family members by Indian rulers to be used as medical subjects was a clever move in a society where royal personhood embodied power and legitimacy. It dented the Company's efforts to insinuate itself as the exclusive patron of subjects' welfare. And the Company was alert to this challenge. Thus Seton, aware of the competition between him and the nawab of Rampur for welfare patron status, recommended the setting up of a permanent and regular establishment at Rampur for the inoculation of children. This, he said, would be seen also as an 'additional mark of the fostering care of the British government'.[19] In 1805 Seton sent a newly vaccinated local child from Bareilly as a 'specimen' to Rampur, along with hakims Ghalib Ali Khan and Muhammad Aiwaz. The scheme had the desired results. The number of chidren vaccinated at Rampur increased considerably.[20]

And yet, despite the risk of making regional rulers score political advantage in what the Company saw as its exclusive 'public health' concern, the advantages of their involvement were far too many. Indeed, in 1808 A. Seton, then Resident at Delhi, was appreciative of the surgeon, Reid, the doctor attached to the Mughal emperor Shah Alam of Delhi, for having involved the latter in the inoculation process. Seton reported that Mr Reid not only theoretically explained the inoculation process to the emperor but also presented him with some lancets. The emperor used these to vaccinate with his own hands many of his royal grandchildren. And the involvement of the emperor and his family gave widespread publicity to the drive as it was reported in the Persian newspapers. Seton thought this was one of the very clever moves of Mr Reid, and, as he said, 'a circumstance which has excited no small degree of attention and is likely to contribute more purpose than anything else to the diffusion of this salutary practice amongst the natives of every rank and description however distant from Delhi since the circumstance has found its way into all the Persian newspapers.'[21]

[19] A. Seton, Agent of Gov. Gen. at Bareilly, to N.B. Edmonstone, Sec. to the Govt in the Sec. & Pol. Dept, 9 June 1805, Extract Bengal Pub. Consult., 4 October 1805, Boards Collection, F/4/198, file no. 4452, pp. 9–14.

[20] L no. 29, A. Seton, Agent of Gov. Gen. in Bareilly, to Thomas Thornhill, Judge and Magistrate Bareilly, 7 June 1805, Extract Bengal Pub. Consult., 4 October 1805, Boards Collection, F/4/198, file n. 4452, pp. 14–18.

[21] L no. 18, A. Seton, Resident at Delhi, to N. B. Edmonstone, Sec. to Govt

If in the core of the erstwhile Mughal empire the now insignificant emperor could be both friend and competitor, in the fringes of the Rajmahal hills in the Bhagalpur district, the hill chiefs who wielded authority were similarly ambivalent partners. The inoculation drives enlisted and relied on their support. This was very much like the military establishment where officers like Augustus Cleveland had recruited the Hill Rangers, through the clan chiefs in the Rajmahal hills, and thus 'tamed' the fringes of empire.[22] Indeed, in 1822 Captain I. Graham, the adjutant to the Hill Rangers, wrote to Captain Caldwell, ADC to the governor general, of the enthusiastic support of the hill chiefs in the inoculation drive. He suggested that the hill chiefs, who comprised non-commissioned officers of the Hill Rangers, could be effectively trained by the surgeon at Bhagalpur. With a small additional allowance to their pay, they could be sent to the hills for three or four months every dry season to vaccinate their people.[23] In 1823 the medical board decided to set up an inoculation establishment at Bhagalpur under a superintendent. The government approved the project and appointed Mr James Thomson as the superintendent.[24] His main task was to move into the hills and select hill chiefs and men who could be trained as 'native assistants'. The help of the more 'intelligent' and 'respectable' inhabitants was solicited by distributing amongst them short accounts in the vernacular tongue on the benefits of inoculation.[25]

By the time the Company was confronted with the fresh problem of managing the 1817 cholera epidemic, it had its vaccination department firmly in place under the charge of the medical board, and with support of local vaccinators and physicians. Each establishment had a superintendent with the following locally recruited staff: a Brahmin

---

in the Pol. Dept, 28 January 1808, Extract Bengal Pol. Consult., 22 February 1808, Boards Collection, F/4/251, file no. 5625, pp. 3–5.

[22] See for the Hill Rangers recruitment, Alavi 1995: 155–93.

[23] L no. 45, Captain I. Graham, Adjutant to Hill Rangers, to Captain Caldwell, ADC to Gov. Gen, 30 August 1822, Extract Bengal Pub. Consult., 3 October 1822, Boards Collection, F/4/761, file no. 20686, pp. 15–19.

[24] C. Lushington, Sec. to Govt, government's resolution, 18 December 1823, Boards Collection, F/4/776, file no. 20984, pp. 17–18.

[25] Medical Board to C. Lushington, Acting Chief Secretary, 20 January 1823, Extract Bengal Pub. Consult., 30 Jan. 1823, Boards Collection, F/4/761, file no. 20686, pp. 43–7.

attending patients at Rs 8 per month, posted across the country; a peon for collecting patients on the days appointed for inoculation and inspection at Rs 5 per month; three inoculating Brahmins at Rs 8 per month; a Muslim inoculator at Rs 8 per month; a writer to maintain records at Rs 15 per month. Each establishment was to maintain a full record of the people, listing in particular their names and religion in its record.[26]

Indigenous Medical Communities, Doctors, and the 1817–1818 Cholera Epidemic: Health as Social Well Being

The inoculation drives of the early nineteenth century brought Company and regional rulers together in patronage competition over rudimentary public health. They both agreed that prevention of disease was as important an investment as therapeutics or cure. This commitment was based on the shared assumption that disease related not just to individual body physiology and its well being but to the larger well being of society. This larger role of medicine had, as we saw, been popularized by Arabic and Urdu print literature of the late eighteenth and early nineteenth centuries. It had been reinforced by the introduction of a medical public service. However, in the realm of medical practice the issue was who would monopolize the new welfare role in society. The competition between regional rulers and the Company was intense. The outbreak of the cholera epidemic in 1817–18 increased their contest over the welfare role, but also strengthened their shared commitment to what later in the century was called 'public health'. A positive fallout of this dynamics was that the knowledge of Indian medical communities was harnessed to the Company's medical understanding of epidemics. But more importantly, the idea of health being integral to social well being also gained popular consensus.

In 1817–18 a fatal epidemic of cholera raged through the town and suburbs of Calcutta and several districts of the Lower Provinces, and threatened to move westwards to the Upper Provinces as well. Medical

[26] L no. 32, circular to superintendents of vaccination, enclosed in L no. 29, R. Leny, Sec. to Medical Board, to C.M.Rickett, Chief Sec. to Govt, 24 June 1816, Consult, Fort William, 6 July 1816, Bengal Pub. Consult, 6–13 July 1816, P/9/13.

opinion was divided on whether cholera was caused by dislocations in the nervous and circulatory system, or an imbalance in body humuors. There was certainly a fast-growing consensus that pollutants in the atmosphere triggered physiological changes. And, as medical opinion narrowed down to the theory of atmospheric pollutants, the state's role in maintaining a clean environment became critical. The state was expected to maintain the health of towns and cities by improving the living conditions of subject populations. In the cholera reports of medical practitioners collected by the medical board, the idea of health being related to the physical, economic, and social well being of subjects was clearly articulated.

R. Tytler, assistant surgeon to Charles Chapman, judge and magistrate of the district of Jessore in Bengal, believed that cholera was caused by the 'vitiated state of the bile'. Very much in the tradition of Hippocrates, he was of the view that the physical circumstances of squalor and congestion in which people lived, combined with the heat and moisture in the weather, had polluted their atmosphere. This adversely affected body humours and caused disease. He was particularly perturbed by the 'confined nature of natives dwellings in the bazaar which are filled with a deleterious atmosphere, consisting of marshy vapours unaffected by a free and salutary circulation of air.'[27] To confirm his view the magistrate of the suburbs of Calcutta, John Eliot, explained the very limited occurrence of the disease among convicts of the Alipur jail because of the 'cleanliness, airiness, and dryness of their habitations'. He felt that the disease had a limited occurrence in the jail also because the inmates were not being employed on very 'fatiguing or unwholesome' tasks. He added that convicts of the same jail got the disease when they were employed in outposts and trusted with unsavoury jobs like cleaning drains, and made to sleep in kutcha buildings after slogging in the sun.[28] Like many others of his time, he was of the view that cholera was linked to pollution in the atmosphere. So, people could be protected if the physical and economic circumstances of living were

[27] R. Tytler, Assistant Surgeon, to Charles Chapman, Judge and Magistrate of Zillah Jessore, n.d., Boards Collection, F/4/610, file no. 15058, pp. 18–19.

[28] John Eliot, Magistrate, Suburbs of Calcutta, to W.B. Bayley, Sec. to Govt in Judicial Dept, 13 September 1817, Extract, Bengal Judicial Consultations, 30 September 1817, Boards Collection, F/4/610, file 15058, pp. 37–9.

improved. Eliot was convinced that health was related to standards of living.

The idea of health and its connection with economic well being was articulated most clearly by H. Young, a surgeon posted in the suburbs, who said that his observation was that even though all types of people were affected by cholera, among them the 'poor and labouring classes' were the most vulnerable and worst affected. And, like John Eliot, he said that this was because the lower classes were unable to protect themselves from the vagaries of weather, temperature, and atmospheric pollutants that carried the disease. He focused on the lack of economic and social well being in this class in his descriptions of their 'huts being crowded together in unwholesome situation, [and] the want of proper ventilation in their buildings.' They followed customs like sleeping on damp floors, with barely a thin mat, and also of starving for long hours and then eating rice and vegetables in a 'cold and aqueaous state'.[29]

Local practitioners had no problems with the Company idea that health was related to social well being and could be maintained through public welfare schemes: by the early nineteenth century this was not a novel concept. William Moorcroft, the British botanist and physician who travelled through the Himalayan area and Punjab in the 1820s, reported that people viewed cholera as a 'public' rather than an individual issue. He said that in areas around the village of Srinagar they practised a ritual called 'burut' to appease their deity, Mahadeo. This was essential to avert cholera, which they saw as a 'public calamity'. They attributed the protection of their village from the 1818 cholera epidemic to this ritual.[30] In 1825 W. Ainslie made similar observations, noting that the medical texts of both Persian origin and those that leaned on Sanskrit astronomical literature shared his view that health was related to atmospheric pollutants. He also reported that local practitioners saw health as part of public cleanliness and personal hygiene. Medication alone was of no use. At best it could be useful as part of this larger issue of society's physical and environmental well being.[31]

[29] L no. 11, H. Young, Surgeon to Suburbs, to C.R. Barwell, Acting Magistrate of Suburbs of Calcutta, 31 August 1818, Extract, Bengal Judicial Consultation, 18 September 1818, Boards Collection, F/4/617, file no. 15372, pp. 5–18.

[30] Moorcroft Papers, MSs Eur.D.236, ff. 28–30.

[31] Ainslie 1825.

As British doctors articulated their community-centred view of disease, they enlisted the support of local practitioners with no difficulty. In the 1817 cholera epidemic in Calcutta, forty to fifty native physicians were temporarily enlisted and supplied with the necessary remedies and detailed instructions in writing. R. Leny, secretary to the medical board, reported that at no point did they show any reluctance to enlist.[32] The number was later increased, given the co-operation both on the medical as well as the administrative side that they offered. They were paid an average salary that ranged between Rs 8 and Rs 10 a month, and which was at par with that of the dispensary-trained 'native doctors' of the administration.[33] These men were supplied with remedies and detailed instructions on their usage. Police posts in their areas were their coordinating point and offered them help.[34]

Medical communities were not just enlisted as additional hands, their knowledge was also harnessed to the Company's welfare schemes. In the case of Calcutta, the cholera remedies recommended showed a combination of local and British cures: opium, Bengal rum diluted in two *chataank* of warm water, and a *chataank* of pepper or any warm aromatic spice boiled in a quart of spirit. The memorandum on remedies recommended that heat could be restored in the patient if warm bricks, rolled up in cloth, were applied to the pit of the stomach, the armpits, and soles of the feet. Dry ginger powder or a beer were recommended as well. Many of the mild purgatives used by native physicians, like rhubarb, were listed as cleaning the bowels and easing pain.[35]

[32] L no. 4, R. Leny, Sec. to Medical Board, to W.B. Bayley, Sec. to Govt in the Judicial Dept, 22 September 1817, Extract, Bengal Judicial Consult, 30 September 1817, Boards Collection, F/4/610, file no. 15058, pp. 48–50.

[33] David Todd, Surgeon at Dacca, to East India Company, 21 October 1818, Extract Bengal Judicial Consultation, 18 September 1818, Boards Collection, file no. 15372, p. 43.

[34] Extract Judicial Letter from Bengal, 29 October 1817, Boards Collection, F/4/610, file no. 15058, pp. 6–8.

[35] L no. 4, R. Leny, Sec. Medical Board, to W.B. Bayley, Sec. to Govt in the Judicial Dept, 22 September 1817, appended memorandum and instructions prepared by the directions of Government-memorandum cholera morbus, Extract, Bengal Judicial Consultation, 30 September 1817, Boards Collection, file no. 15058, pp. 60–73.

The same was the case in Jessore and the Lower Provinces, where native physicians were given a remuneration to obtain 'Bengallee' medicines for treating cholera.[36] The same trend continued outside Bengal. In the Bundelkhand area, both hakims and native physicians freely enlisted to help the administration tackle cholera. Indeed, government restrictions on the sale and limited usage of opium by hakims were often cited as a reason for the increased number of cholera deaths.[37] The importance the medical board gave to native remedies was evident when, in the 22-point questionnaire that it circulated to its practitioners to help them compile their reports on cholera, a specific column listed the remedies resorted to by native practitioners.[38]

Officers and medics posted in the districts were of the view that a proactive state that cared for the well being of its subjects was always appreciated by people; it would in the long term benefit the prospects of British medicine in India. In 1818 the acting magistrate of Bundelkhand, F.C.Smith, optimistically reported that the Company's efforts to combat cholera through spreading its doctors and trained hakims far and wide had already begun to show positive results. A Brahmin family in Bundelkhand who lost a child due to the negligence of a hakim complained to Smith that 'their child had been killed by the hakim from his neglecting to administer English/Wellayut medicines.'[39]

As cholera management intensified, so did the idea of health as an index of well being. On the basis of district-level cholera management reports, the medical board suggested that if the government was interested in issues of health, it had to improve the physical and financial well being of its subjects. The Company needed to invest and involve itself in improving the physical environment of villages, towns, and cities. In 1817, R. Leny, secretary to the medical board, wrote to W.B. Bayley, secretary to government, about the urgent need to invest in the

[36] See Boards Collection, F/4/819, file no. 21848, pp. 66–7.

[37] L no. 25, F.C. Smith, Acting Magistrate of Bundelkhand, to W.B. Bayley, Sec. to Govt in Judicial Dept, 27 July 1818, Extract, Bengal Judicial Consultation, 4 September 1818, Boards Collection, F/4/642, file no. 17708, p. 26.

[38] L no. 101, J. Jameson, Sec. Medical Board, to Native Doctors, 9 October 1818, Extract, BMC, 26 September 1818, Boards Collection, F/4/622, file no. p. 51.

[39] Ibid.

'effectual draining of the town and suburbs [of Calcutta] as well as to better ventilation which at present is everywhere obstructed by trees and rank vegetation of every description'.[40] These he thought were 'powerful auxiliaries' that could be removed to better standards of living and prevent pollution of the atmosphere. The Company, he argued, should also consider improvement in the economic standards of its subjects to improve their 'insufficient diets' and the 'miserable accommodation afforded by the low damp huts of the lower more indigent orders of the natives.'[41]

In 1820 James Jameson, secretary to the medical board, endorsed this view. His lengthy report on cholera, based on an analysis of about a hundred reports from his colleagues and inputs from hakims and other local physicians, concluded that the disease was linked to well being. He attributed the disease to the poor sanitation, dense foliage, high population, and congested dwellings of the poor in Calcutta and its suburbs. The poor and lower classes had suffered hardship from the rapid urbanization of Calcutta. This, coupled with their poor vegetarian diet, had made them more vulnerable to the disease. According to him, a damp climate and population congestion were the main causes of cholera. Thus the sanitized Lucknow cantonment located on a dry sandy soil, escaped the cholera, while the city, located at a low altitude and crowded, was affected by the disease. Similarly, 'the thickly populated and close lanes of Delhi suffered more than the rectangular and spacious streets of Jaipur. And again the low and nasty town of Saharanpur more than the city and cantonment of Meerut.'[42]

Jameson concluded that cholera could be averted if the government invested in public welfare schemes to clean up cities, remapped them to make them less congested, and improved the social well being of its subjects. He made a strong case for improvement in the native portion of Calcutta and the following suggestions to improve the sanitation, salubriousness and beauty of the city: open broad avenues from the circular road or outskirts of the town to some distance into the country; open wide streets with occasional squares and tanks; break up the

[40] L no. 4, R. Leny, Sec. Medical Board, to W.B. Bayley, Sec. to Govt in the Judicial Dept, 22 September 1817, Extract, Bengal Judicial Consultation, 30 September 1817, Boards Collection, F/4/610, file no., 15058, pp. 37–9.

[41] Ibid.

[42] Jameson 1820: 116.

numerous nests of huts in the European part of the city by purchasing ground on which they were built and allotting new ground in its stead in the suburbs; fill up all unwholesome tanks and foul puddles and ditches; improve the drainage; cease using European and Muslim burial grounds; and cut down jungles near towns and suburbs.[43] Jameson had the endorsement of district magistrates who were willing to put up pressure on the government to improve the living conditions of people and provide cities with adequate healthcare services.

W. Lock, magistrate of Murshidabad, pointed to the lack of dispensaries in Bengal. He recommended the opening up of dispensaries in the area superintended by people of a particular caste. Through the persuasion of such staffers, their good example, and good medicines, he hoped the people would be encouraged to visit them at the time of epidemics and be admitted there if required. He hoped the dispensary would also provide employment to young men in Calcutta who were training in Western medicine at the NMI, the Sanskrit College and the Calcutta Madrasa.[44]

The medical board emphasized to the governor general that the idea of public health be considered seriously so as to control and prevent future cholera epidemics. Jameson's report, based as it was on the input of more than a hundred medical practitioners and their local medical contacts, powerfully made the link between cholera and the well being of subjects. It became the dominant working plan of cholera management in Company administration. By 1820 the military secretary had printed more than 300 copies of Jameson's report for circulation to all medical officers. At least 126 of these were sent to Fort St George, Bombay, Prince of Wales, Fort Marlborough, and Singapore; 24 copies were dispatched to London.[45]

So influential was the report that the governor general refused to extend his patronage to any alternative views on the subject. In 1819 R. Tytler, assistant surgeon at Allahabad, requested the governor general's

[43] Ibid.: 111–12.

[44] L no. 3, W. Lock, Magistrate of the City of Murshidabad to W.B. Bayley, Sec. to Govt, 13 October 1817; & Extract from Lock's minute 19 November 1816 [*sic*] 1817, Extract, Bengal Judicial Consult, 25 November 1817, Boards Collection, F/4/610, file, 15058, pp. 198–9, pp. 203–4.

[45] Extract, Military Letter from Bengal, 20 June 1820, Boards Collection, F/4/638, file no. 17663, pp. 1–2.

patronage to publish his manuscript on cholera, this being based on his experiences of handling the disease in Jessore and Allahabad. Patronage was refused because he attributed the disease not to the physical environment of people but to inclement weather which infected the crops and made them so unwholesome that they caused disease if consumed. The secretary to the government wrote to Tytler that the governor general would go by the opinion of the medical board and not extend patronage to his work because 'there would be an obvious incongruity in extending the patronage of government to a publication which may possibly contain opinion at variance with those entertained by the authorities on which his Lordship in council must necessarily rely on questions so strictly of a professional nature.'[46]

The government's positive response to the demand for maintaining well being to ensure public health was particularly significant because a movement with similar demands in London would emerge only in the 1830s, when cholera ravaged the city. In a sense, then, the 1817 memorandum of the medical board, which had the consent of 'native practitioners' and British medics, can be seen as laying the basis of such demands in mid-nineteenth-century Britain.[47]

It is not without significance that at the time of the cholera epidemic in Britain, surgeons with Indian experience were in demand. Men like the surgeon James Adair Lawrie, Professor of Surgery at Andersonian University, Glasgow, who had had postings in India at the time of the 1817 epidemic, were asked to travel to affected areas and, in light of their India experiences, offer help. Lawrie toured cholera-affected Newcastle and wrote tracts that suggested remedies deriving from his Indian experience. And, as in India, here too he stressed that the state provide adequate investment to public health. Very much as in the Indian case, he identified public health with cleaning the environment,

[46] L no. 43, C. Lushington, Sec. to Govt, to R. Tytler, Assistant Surgeon, 30 April 1819, Consult, 30 April 1819, Bengal Pub. Consultation 22 April–21 May 1819, P/9/59. Also see ibid., L no. 42, R. Tytler, Assistant Surgeon, to Charles T. Metcalfe, Chief Secretary to Government, 29 March 1819.

[47] There was therefore a connection between the pre-1830s cholera management discussions in India and the movement for sanitary reforms that picked up under Chadwick in the 1840s in Britain. For a contrary argument, see Singh 2001: 144–65.

dealing with urban congestion, hospitals and the civil police management of people, and concentrating on the class profile of the disease.[48] Lawrie was not the only author who was of this view. Others with Indian experience were also involved, and they too sent their reports to the board. In 1832 Charles Turner Thackrah wrote remarks on the comparative study of the two kinds of choleras in Scotland and India. He too urged the government to reduce overcrowding and control the environment, as was done in India.[49]

Doctors in India wrote cholera management texts so as to provide reference material for their colleagues in Britain. In 1832 Frederick Corbyn's text from Calcutta claimed to have a comprehensive account of cholera management: he had referred to all the literature produced in India and hoped it would 'promote the interest of medical science and welfare of humanity' as well as remove the cholera 'sufferings all over Europe'.[50] Indeed texts and reports from India circulated in periodicals all over Europe. Often, in rival countries like France, they were used to denigrate the claims of Britain as a welfare state. In 1825, the surgeon W.E.E. Conwell of Madras had to submit a rejoinder to the French minister of marines and colonies because British doctors had been given unfair coverage in French periodicals.[51]

And yet in the 1830s the movement for public health in London, and the pressure on the government to evolve into a 'welfare state', so divided opinion that even in India the rumblings of such differences could be heard. Indeed, James Hutchinson, surgeon of the Bengal establishment and secretary to the medical board of Bengal, recommended quarantine, cordons, separation of the sick, evacuation of buildings,

[48] James Adair Lawrie, 'Essay on cholera founded on observations of the disease in various parts of India and in Sunderland, Newcastle and Gateshead, with notices of the treatment and of the civil and hospital police adopted in the towns', T.923, Wellcome Library.

[49] Thackrah 1832.

[50] Corbyn thought the disease was caused by weather changes that caused the skin to humidify. This caused congestion and disease. Physical living conditions that could prevent moisture and a good diet could prevent the disease. See Corbyn 1832: vii–ix.

[51] L no. 12, Medical Board to Maj. Gen. Thomas Munro, Governor in Council, Fort St George, n.d., Boards Collection, F/4/949, file no. 26649, p. 6.

and the fumigation of chambers, apparel, and furniture with acid gases to kill the contagion.[52] This was in sharp contrast to the 1817 recommendations of the medical board. It reflected the pre-1830 era in Britain, when health was only about therapeutics and state policy did not view it as related to a community's well being. Such views were now in a minority, however. In the same period, there were surgeons like James Ronald Martin, the Bengal Presidency surgeon attached to the native hospital, who linked disease to lifestyle and diet, and argued that improvements in these rather than medical advancements were the key to ensure good health in society. The state had to play an active role in this arena. He attributed the disappearance of the malignant fevers of the 1710s not only to medical management based on strides in pathology and physiology, but to 'public health [which] has been amended not through advantages of modern improvements in treatment of disease but through the great measures of prevention of disease, consequent on the progress of the public mind and of governments, in general knowledge leading directly to improved habits of life in communities, improved localities, institutions of the police etc. It is to these institutions that public health is indebted.'[53]

Managing of the cholera epidemic and the inoculation drive created a shared but also often contested public health space in which British doctors and Hindustani healers came together to share their medical practice, and the Company and regional rulers expressed their commitment to public welfare. Though each desired to monopolize this arena, the contest remained inconclusive as it worked out, being framed within cooperative welfare schemes. Thus in 1828, when the king of Oudh reiterated his commitment to the 'welfare and prosperity' of his subjects and the 'comfort of a kingdom committed to his charge', he was clear that this meant keeping his people healthy; but this was a task he could not accomplish alone. As he said, 'he needed the services of British doctors.'[54] Apart from Dr G. Baillie, the surgeon posted in Lucknow, the king requested British doctors—W. Stevenson, a junior surgeon posted with the 14th regiment at the cantonment in Lucknow,

[52] Hutchinson 1832.

[53] Martin 1837: 121–2.

[54] L no. 30, transl of a letter from the King of Oudh to W. Ricketts, Resident at Lucknow, 28 March 1828, Bengal Pol. Consult, 11 April 1828, P/125/47.

Monsieur Arthur Du Bois of Najafgarh; and Mr Lock of Hyderabad—to be posted in his service so as to enable him to carry on his job of a 'welfare patron'.[55] He reinforced his demand by reiterating the joint commitment of his nawabi and the Company towards the 'prosperity and safety' of the people.[56] The Company was quick to affirm its commitment and acceded to his requests.[57]

Significantly, once the British doctors became part of the king's service, they were assimilated to his court society through rituals that conferred on them the insignia of royalty—the *khilat*. Their appropriation into court-sponsored welfare schemes kept the fires of contested patronage over medical welfare burning. The customary practice of the recipient of the *khilat* recognizing his appropriation into the court by a return token gift—the *nazar*—was forthcoming from the British doctors.[58] However, the doctors assessed these transactions for their financial value, less in terms of their cultural significance. This was evident when one of the recipients, Dr Baillie, remarked that the *khilat* was 'a mark of his favour not the less gratifying that the khelaut itself scarcely equals in value the customary nuzzurs'.[59] But there were other advantages in being close to the court. British medics found in the king a suitable counter to disseminate their ideas about Western medicine. The court was an important forum from where their medical tracts could be lithographed and distributed amongst the public for greater medical awareness. P. Breton, superintendent of the NMI, presented his tract, written for 'native' students at his institution, to the king of Oudh for its larger dissemination in the area. Like many others, he was never disappointed in the royal support he received.[60]

[55] Ibid.

[56] Ibid.

[57] A. Sterling, Officiating Sec. to Govt, to Resident at Lucknow, 11 April 1828, Bengal Pol. Consult, 11 April 1828, P/125/47.

[58] L no. 34, G. Baillie, Surgeon at Lucknow, to W. Rickett, Resident Lucknow, 25 June 1828; and W. Stevenson, Asst Surgeon in service of King of Oudh, to W. Rickett, Resident at Lucknow, 24 June 1828, Bengal Pol. Consult, 18 July 1828, P/125/54. Both these doctors, along with Drs Arthur De Bois and I.D. Lock, were honoured with *khilats* and they returned *nazars* to the king.

[59] Ibid.

[60] L no. 53, P. Breton, Superintendent of the NMI, to Persian Secretary, n.d., Calcutta, Bengal Pol. Consult, 18 July 1828, P/125/54.

Co-operation in the public arena of health also led to individual rulers developing trust and confidence in personal physicians, giving a morale boost to such individuals. In 1815 the nawab wazir of Oudh requested that Dr Wilson, the surgeon posted at the Lucknow Residency and attending the nawab wazir as well, be allowed to continue as his personal physician even though he had been promoted to a higher rank and thus was due for a transfer out of Lucknow. The nawab wazir said Wilson should be retained since he was 'thoroughly acquainted with my constitution, and possesses my entire confidence'. He was even willing to pay the additional allowance that was due to Wilson at his new station. The nawab hoped that since the Company was concerned about his welfare, and as health was crucial to his happiness, they would grant his request.[61] The governor general granted the request at minimum losses to Dr Wilson, who was given his new rank and all pension benefits and future promotions on that rank. He was to continue on his old salary until he remained at Lucknow, but the nawab wazir compensated financially for the loss to him entailed in staying put with the nawab.[62] The following year the Company had to appoint an assistant surgeon, Dr B. McLeod, to the Residency so as to relieve the heavy workload of Dr Wilson. And later in the year McLeod took over as the surgeon of Lucknow and as personal physician to the nawab wazir. He was paid Rs 1500 by the nawab for services to his court.[63]

The Mughal emperor Shah Alam also always had a British surgeon posted at the court. In 1808 Dr Reid, the surgeon posted to the Delhi court, complained of being overburdened with patients from the royal household and the city, and inadequate remuneration. Like the surgeons at Lucknow, he was integrated with court society and given honorifics—which he too valued in financial terms. He wanted the Company to give him an additional allowance of Rs 800 since the heavy 'welfare'

[61] L no. 42, Nawab Vizir, to G. Swinton, Persian Secretary to Gov. Gen., received 19 April 1815, Bengal Pol. Consult, 15 Aug. 1815, Boards Collection, F/4/509, file no. 12265A, pp. 11–12.

[62] Extract Political Letter from the Governor General, 15 August 1815, Bengal Political Letter no. 6, Boards Collection, F/4/509, file no. 12265A, pp. 1–6.

[63] See Extract, Political Letter from Bengal, 3 January 1817, Boards Collection, file no. 14453, pp. 1–3; Nawab Vizir to F.V. Raper, Acting Resident Lucknow, 2 October 1817, Extract, Bengal Pol. Consult., 17 October 1817, ibid., p. 15.

responsibilities that the emperor had heaped on him were not adequately compensated by mere honorifics.[64] The Company yielded his request.[65] The practice of having a British doctor at the court of Delhi continued even after Dr Reid was dead. He was replaced by a Dr McWherter.[66]

Beyond the Epidemics: Local Healers and British Doctors Contest Medical Authority

We have seen thus far that early concerns about human welfare and health created a medical public sphere in which men of Western training and traditional healing came together. This created a popular mindset which saw no contradiction in the parallel existence of Western, Unani, and other remedies. In 1813 T.D. Broughton, the commander of the Resident's Escort at the court of the Maratha leader Mahadaji Shinde, reported that the raja of Krishangarh, near Ajmer, regularly sent his vakil to the Resident's camp to procure medicines for himself. He said while he tried English medicines, other vakils of the raja tried both Unani and Ayurvedic remedies, which he described to Broughton in great detail.[67]

Moorcroft's travelogue further corroborates this free-choice society and gives us an excellent picture of the relaxed nature of medical exchange that went on between him and local hakims. Moorcroft reports that, in the Hoshiarpur area of Punjab, cataract operation was an eye surgery carried on by local hakims who doubled as 'barber surgeons' (*jarrahs*) using a 'spear pointed lancet of which a quarter of an inch is bone and the rest of the blade tied with a thread.' The eye was perforated and a copper needle introduced through the wound, and then pushed towards the top of the cataract. It was then brought close to the iris and

[64] L no. 19, Dr Reid to A Seton, Resident at Delhi, 12 January 1808, Extract Political Letter from Bengal, 29 March 1808, Consult, 22 February 1808, Boards Collection, file no. 5625, F/4/251, pp. 5–7.

[65] L no. 38, N.B. Edmonstone, Sec. to Govt to Resident at Delhi, 7 March 1808, Extract Bengal Pol. Consult., 7 March 1808, Boards Collection, F/4/251, file no. 5625; and Extract Pol. Letter from Bengal, 29 March 1808, Consult, 22 February 1808, ibid., pp 1–2.

[66] L no. 197, N.B. Edmonstone, Sec. to Govt, to A. Seton, Resident at Delhi, n.d., Boards Collection, F/4/335. pp. 7–8.

[67] Broughton 1813: 185.

the lens was lodged, at the first attempt, in the vitreous humour. Moorcroft observed all this carefully, with a view to learn. He found such surgery unsuccessful in eight out of ten cases, and noticed that people were averse to getting it performed. He found the European technique that he himself used more successful, and less prone to infection. When he showed the local practitioners his instruments and method of operation, they were quick to acknowledge its superiority. They expressed a desire to be trained in this art of cataract surgery. Moorcroft, eager to impart knowledge, engaged one of the hakims of Hoshiarpur to accompany him in his travels so that he could observe the surgery and serve as apprentice. He promised to 'instruct him on how to perform the operation for the cataract, to bear his expenses and to set him up with instruments—hydrocephatus.'[68]

Moorcroft trained hakims by demonstrating his skills in every village he went to. He was a careful and diligent surgeon. He kept his first patient, a 'Muslim of property', in one of the apartments of an alms house and monitored him from the roof of the house. At every large town, he followed this practice of monitoring his patients as they recuperated: he did not want the operation discredited on account of negligence by his patients. His personalized care paid off. Soon, Moorcroft's practice became so popular in the area that patients flocked to him in large numbers. He had to refuse services to many and ask them to wait until he returned to their region.[69] As he travelled around, hakims of the villages brought their patients to him. Many, like the 'Lama Yoarou', in a hill village in Leh, wanted him to train the hakims in his area. The abba, or head physician of the 'raja of Puskyoom', accompanied him so as to learn this surgery.[70]

Indeed, so highly was Moorcroft regarded—not only by patients but by hakims as well—that on his way from Hemis to Snewrla he reports that the physician of a village who was present at one of his cataract surgeries offered him a beautiful copper teapot worked over with brass 'in fiel' (filigree). This was in recognition of the high medical

[68] Moorcroft Papers, MSS Eur. D. 238. Third Fascicilus of Journal from Joalamookhee to Lahore the Capital of the Sikh Dominion, from 28 March–13 May, ff. 59–60.

[69] Ibid.: f. 60.

[70] Moorcroft Papers, MSs Eur. D 246, ff. 86–7, 116.

status he accorded to Moorcroft. As he said, 'it was the duty of every hakim in the country when the sirdar of hakims visited their residence to show his respect by offering to his acceptance such refreshment as was at his command.' Moorcroft returned the acknowledgement by presenting the physician with a 'gun lancet', an instrument used for opening veins in the forearm. The hakim was familiar with the instrument, but the one presented was superior to anything that he had ever seen, and he was therefore thrilled.[71]

However, in all this co-operation there was an element of contestation over issues of medical authority. This was most evident in discussions over the cure of rajas and members of ruling dynasties. In a society where royal personhood defined power and authority, the cure of the body of the king always brought status to the practitioner, and authority and credibility to his medical tradition. The royal body was thus a critical medical subject over which local and British medical practitioners both co-operated and competed—its cure was an instant step forward to claims of medical authority and legitimacy. As we saw, ruling houses deployed the bodies of their household members as medical subjects in public health programmes such as inoculation. This was a tactical move to steal for themselves the authority that such welfare schemes lent to the Company. A similar contestation between local hakims and Moorcroft went on as they cooperated to cure Ranjit Singh, the maharaja of Punjab.

Maharaja Ranjit Singh was a hypochondriac. He continuously had British doctors and local hakims attending him for all manner of ailments. Indeed, hakims were part of his royal entourage, often acting as political advisers who gave valuable advice on matters relating to war and relations with the British.[72] The maharaja also welcomed medical advice from British dignitaries who visited him, and from European and British physicians who passed through his kingdom. His body, and those of his son and brothers, often became the subject of a medical

[71] Ibid.: ff. 34–5.

[72] Ibid.: f. 113. One such hakim, Azizuddin, was raised to the status of 'wizarat' in addition to his duties as a physician because his advice on not going to war with the British proved useful to the Maharaja. His brother, Nuruddin, was given the charge of the city of Lahore, the care of the artillery, and some other important duties. The third brother, Imamuddin, was made commander of Gobindgurh.

dialogue between Unani practitioners and British doctors. This was a difficult contestation, because the maharaja astutely had every British medicine first tested and experimented on the body of his hakims, thereby underlining both their medical authority as well as implicating them in the success or failure of Western medicine. Drugs, rather than diagnostics and disease management, was in fact the arena where a difference of opinion existed between the hakims and British doctors. Until the early nineteenth century, British cures, very much like Unani, remained largely Hippocrates-oriented. Both diagnosed and managed disease with reference to the balance of humours in relation to the elements.

Thus, in 1820 Moorcroft's recommendations regarding certain complaints of the maharaja had a heavy Hippocrates orientation and caused no stir at all amongst the hakims to whom they were conveyed. Moorcroft says that he was in fact 'flattered by being told that they had highly approved both of the matter of plan and of the manner in which it was arranged.'[73] Very much in the manner of the Unani medical tradition, Moorcroft had made dietary and lifestyle modifications in the maharaja's routine, rather than concentrated on therapeutics. Of course, the only difference was that British diets and choice of clothing were recommended instead of Hindustani ones. Moorcroft recommended soap pills and aloes for the maharaja's constipation. Fine texture clothing with silk stretchings for the summer, and lambswool stockings and Welsh flannel underwear were recommended for the winter. He replaced country rum and spirits with European wines and liquors. The quality of brandy and Madeira port that the maharaja was consuming was found suspect. General Ochterlony's gift of a consignment of wines were thought to be best for him. Ranjit Singh was asked to preserve his health and focus on preventives via physical exercise, and abstaining from certain types of food, rather than look to therapeutics and cures. Moorcroft said that if it was difficult for the maharaja to take long rides in the morning, these could be substituted by his performing the 'gymnastic exercises of the natives'.[74]

Yet, even though the hakims agreed with Moorcroft on the management of disease, neither the maharaja nor they had any trust in his

[73] Ibid.: f. 116.
[74] Ibid.: ff. 114–16.

medicines. It was here that the two systems had to prove their relative efficacy. This was an unequal playing field, for Western medicines had to be first certified by the hakims before being consumed by the maharaja. This blunted the competition for medical authority, because the hakims, by being made certifying authorities, already had the upper hand and were in a position to deny certification to British doctors. Moorcroft reported with surprise that Ranjit Singh 'uses English things in military matters, but wants to enquire ill effect of English medicines before practice under his own eyes—tries them out on somebody else under his own eyes.' Either his hakims tried the medicines in front of him, examined them, and certified them fit for consumption, or, many times, Moorcroft himself consumed them in front of the maharaja to prove they were fit for royalty.[75]

On the issue of medicines for royalty, medical upmanship could be worked out; and here crucial exchanges were also possible. If local substitutes could be found for Western drugs, the economics of health management could benefit the Company in a major way. So would the process of disseminating Western prescriptions, for they would be more acceptable if packaged in familiar materia medica. The maharaja's continuous desire to have his physicians prepare Moorcroft's prescriptions with local drugs was proof of the feasability of this scheme.[76] And by acting as testing laboratories and certifying authorities for Western drugs, hakims not only learnt of these drugs but did so from the position of superior medical authority that the maharaja himself vested in them.

This was very evident during interactions between the hakims and Moorcroft during the fever and illness of Ranjit Singh's son, Kharak Singh. Moorcroft was asked for his advice and help. Once again, the hakims agreed with his diagnosis: the fever was caused by 'bad air' and 'offensive vapour that emanated out of a gutter in the middle of the bazaar'. They agreed also to a shift in the prince's environment as a way to manage his fever—even if they suggested Shalimar instead of Adeenagar, which Moorcroft recommended. But the hakims remained sceptical of the medicines Moorcroft recommended and called them '*sakht*' (violent); they offered instead local substitutes. One of the main

[75] Moorcroft Papers, MSS Eur D 239, ff. 10–12.
[76] Ibid.: f. 11.

hakims, Nooruddin, examined Moorcroft's drugs and offered to test them on women of the zenana. Finally, despite detailed consultations amongst them, and between them and Moorcroft, they were not administered on the prince. Instead, local substitutes were found and the son was cured.[77]

The same sorts of consultations between the hakims and Moorcroft went on in the case of the treatment of Ranjit Singh's brother, Fateh Chand, who was suspected of having cholera. It was with great reluctance that the hakims allowed bloodletting to bring down his fever—their practice of inducing vomiting had not had the desired effect. Moorcroft also injected turpentine and sesame oil into the patient's rectum to ease his bowel movements through a crudely-put-together syringe made of goat's intestine tied to a bamboo. The surgical intrusions worked and the patient recovered.[78] In this instance, Moorcroft's authority as a competent surgeon was established, and the hakims who were steadfast about their own drugs were eager nonetheless to learn his surgical skills.

The steadfastness with which local practitioners held on to their drugs and materia medica and argued their therapeutic effectiveness made such items attractive substitutes for the expensive English drugs that had to be imported into the country. The shared patronage that the Company officers and regional courts extended to medical explorations made the further exploration of Indian materia medica possible. Thus in 1825 M. Rickett, the Resident at Lucknow, gave standing instructions to all officers who passed through the province to send reports of their observations regarding flora and fauna. The botanist Dr Wallich, superintendent general of plantations in the botanical gardens at Calcutta, was one such officer who sent a long report as he travelled through timber forests in the eastern region of the nawab of Oudh's territory. He claimed that in the cinchona bark that he discovered in the forest, and in a species of gentian or 'cherayta' that he grew there, there could well lie the remedies for Tarai fever.[79] Wallich noted that

77 MSs Eur. D 240, ff. 28–33.

78 Ibid.: ff. 59–63.

79 L no. 66, Dr Wallich, Supdt Gen., Plantation, to M. Rickett, Resident at Lucknow, n.d., Bengal Pol. Consult., 15 April 1825, P/124/14.

the nawab encouraged his explorations, provided they remained confined to explorations of a botanist in search of new plant species and did not 'entail any loss of his majesty's government'.[80]

Indeed, the British quest for Indian *nuskhas* or prescriptions, and the desire to learn the local brand of alchemy followed in drug preparation, prompted a British officer, Edward Henry Cradock Monckton, to maintain an entire collection of chemical, alchemical, medical, and miscellaneous recipes that were Indian in origin.[81] This compendium also had a *nuskha* called 'bismillah' (beginning). It comprised a Quranic verse in praise of God that said, '*ya allah ya rahmanoo ya wahato, ya woodoodo*'. This had to be repeated a hundred times every night for forty nights to cure illnesses.[82]

This was a minor part of the vast collection. Monckton's manuscript is in 16 parts; each part reads like a diary. Some parts include a compendium of two or three notebooks. The contents of this compendium are prescriptions, for which the Persian word *nuskha* is used. Most *nuskhas* are alchemical—for making zinc, gold, and silver, and tin locally for domestic and medicinal use. Stylistically, the *nuskha* compendium is no match for the sophisticated ones maintained by Unani physicians—who did not reveal all the details of their components or the sources of their remedies. In contrast, here, the names of the providers of remedies, and at times their places of origin, as also the places from where the ingredients could be obtained, are listed. The providers include a range of non-medical professionals, including fakirs, traders, local officials like qazis, ordinary women, and nawabs and rajas. Some parts of Monckton's collection are appropriately titled 'Receipt Book', indicating the range of providers that made the compilation possible.[83]

For instance the ' alchemical recipe' entitled 'silver' not only details its chemical proportions and the method of preparation, but also

[80] L no. 88, Dr Wallich, Supdt Gen., Plantation, to M. Rickett, Resident at Lucknow, 23 June 1825, Bengal Pol. Consult., 15 July 1825, P/124/22.

[81] MS 3596–3611. Edward Henry Cradock Monckton, Collections of chemical, alchemical, medical and miscellaneous recipes, Indian in origin 1842–53. Wellcome Institute Library, London.

[82] Monckton Collection, MS 3598, no. 5. Pages unnumbered.

[83] Ibid., MS 3596, dated 1842, 'Shumsh and Nukra Receipt Book (1842). No folio or page nos.

indicates that it was provided by Hakim Wohab Ali of Arrah in Bihar. It also indicates one Lalleh Monorut Lal of Mahadewa mohalla in Arrah, through whom the said hakim may be contacted. Similar alchemical recipes and details of providers are listed for *nuskhas* 'to make bheem-sinee caphoor' (camphor), 'jora of gold' (pair of gold), 'lead', etc.[84] Some *nuskhas* give the recipient's name very professionally at the top of the page. For instance, the *nuskha* entitled 'Kushta of copper' has the name of one Khoajeh Ali Shah of Hindawun on its head. The one called 'khak of chandi' (dust of silver) has the name of Nawab Wulleedad Khan of Bulandshahr as the provider. Lithographer of Agra, Wajid Ali, has his name on top of the *nuskha* titled 'to colour silver', and on the one called 'fixed sulphur'. Qazis like Syed Mardan Ali and Syed Aman Ali Khan of Khurjah in the district of Bulandshahr are also prominently identified with many alchemical *nuskhas*.[85] Non-alchemical and purely medical remedies for common ailments like itch, piles, jaundice, etc. also give not just prescriptive detail but also information on the source.[86] In fact, at the end of one such *nuskha* compilation, there is a memo that gives full details of the providers' profession, family background, and very often address as well.[87] Minute details of where the raw materials for these *nuskhas* can be obtained are also scribbled on the sides of many *nuskhas*. For instance, the popularly used ingredient 'kala bhun-gra' is listed as an item to be found in the Kadir or the 'turree of the Ganges'.[88]

Monckton's collection of *nuskhas* mirrored the archetype British focus on Indian alchemy traditions, drugs, and their sources. The diagnostics of Unani and other indigenous systems of medicines *per se* were not a focus of their study. But their pharmaceutical 'recipes' were. Hakims had an inkling of this and were not always forthcoming with

[84] Ibid.

[85] Monckton Collection, MS 3597, no page nos.

[86] Monckton Collection, MS 3598, no. 5. The remedy for itch is listed as follows: put in oil extracted from black seeds of the plant rateela. Remedy for piles provided by Daroga Nasir Ali Khan: a quarter of cow's milk cold with coarse brown sugar dissolved in it. Drink this twice a day for three days or more.

[87] Monckton Collection, MS3598, MS no. 5, no page nos.

[88] Monckton Collection, MS no. 1, no page nos.

information about their *nuskhas*. This is why it is not without significance that most of Monckton's providers were ordinary, non-medical people.

It was with the help of such informants that Moorcroft began his hunt for the sources of the purgative drug, made from the root of the rhubarb plant, sold by druggists in most cities of Hindustan. He discovered its source in the Himalayan ranges of Ladakh and the hills of Himachal Pradesh. He observed that the drugs made from the roots of the Indian rhubarb were inferior to those obtained from Russia and Turkey. Moorcroft's detailed study of the roots of the rhubarb enabled him to ascertain its qualities as compared to those that grew in Russia, Turkey, and China.[89] His interest in local flora and their medicinal values led him to collect botanical specimens from traders, 'nautch' women, and whoever offered him any novelty. He crosschecked the attributes of his myriad collections with local men as he travelled around north India.[90]

The Arrival of Print: Medical Patronage, Textuality, and 'Authoriality'

By the late 1820s the expansion of lithograph printing in India added new dimensions to this contested and yet shared pool of medical knowledges. In this period the manuscript copies of Arabic and Persian developed a new face after their brush with translators and the printing press. Alongside hitherto unpublished knowledge, the repositories of the communities of healers began to be printed. Many such publications became the first-time books of British doctor-authors. Careers were

[89] MSS Eur D244. Ninth Fascicilus of a journal from 16 September to 21 October 1828 in the country of Ladakh. 'Memorandum Respecting Rhubarb', ff. 684–92. Moorcroft's study helped dispel notions that rhubarb came to India from China.

[90] MSS Eur. D 243. Eighth Fascicilus of a journal from the 17 August to September 16 and Ritanka ke joth in Koolloo to Toong-loong ghath in Tartary. No folio numbers. He cross-compared and enquired about certain roots supplied to him by one Mr Turnbull with those given to him by a trader in Joshimath and by a nautch woman in Lahoul.

made as Indian patients became the subjects of a growing genre of anatomical literature. The arrival of print did not dramatically change mentalities. But it did influence the ongoing contest over medical patronage and authority. For instance, the figure of the royal patron had loomed large in pre-print medical manuscripts; in printed texts it was the English Company which was extolled.[91] As financier and sponsor of the new author and text, the Company won its battle of patronage against regional rulers. Similarly, the spotlight shifted from the Indian author to the British translator as Arabic and Persian medical literature began to be translated.

In this period of heightened interest in printing materia medicas, Mughal and late-eighteenth-century medical texts attracted the attention of British doctors. Some of these were translated into English. However, if translations done at the end of the eighteenth century expunged their patrons from the text, those of the nineteenth century went a step further and omitted the author's name altogether. They projected the translator as the author. Print thereby gradually converted Mughal manuscripts into works by British doctors.

In 1793 Francis Gladwin translated into English *Ulfaz Udwiyah*, the materia medica in the Arabic, Persian, and 'Hindevy' languages. Its author was Nur-ul-Din Muhammad Abdullah Shirazi, the Iranian physician to the emperor Shah Jahan.[92] The opening page of Gladwin's text made it clear that the new patron of the materia medica was the British government. Addressing its preface to John Laird, the president and members of the hospital board, he wrote: 'The following work being published at your recommendation for the use of the Honourable Company, I beg leave to place it under your immediate protection'.[93] The materia medica had 1441 items. Even though most entries were in Persian, Arabic, and Hindavi, many were listed in only one of these languages. Each had an English translation and transliteration. Even though Gladwin's translation establishes new patrons, it still maintains a link with the author and closely follows his text without expunging

[91] For an excellent discussion on the role of print in the fading of the patron figure and the arrival of the selfconscious author as the centre of knowledge in late medieval Paris, see Brown 1991: 103–42.

[92] Gladwin 1793.

[93] Ibid. opening page. No page number.

parts or adding a new commentary to drown the voice of the author. Gladwin was not a doctor, and therefore, even though he used his philological knowledge to translate the text, he did not use print or the process of translation to compete for medical authority in the Indian medical market.

This race for medical authority was most glaring in 1833, when George Playfair, superintendent surgeon in the Bengal Service, translated the late-eighteenth-century materia medica, *Taleef Shareef*.[94] The text was originally authored by the Hakim M. Sharif Khan, who died in 1816. One part of it was dedicated to the Mughal emperor Shah Alam (1759–1806). It derived from Sanskrit texts and from the medical experiences of the author, and that of his ancestors. It was meant to fill gaps in the earlier encyclopaedic Mughal medical texts, such as the *Tibb-i-Darashikohi* and the *Dastur at Amal*.[95]

The translated version of the text had 1053 entries on medicinal plants and herbs, transliterated in the Roman script from their Arabic and Persian forms. However, Playfair interfered hugely with the text and overlaid it with a new veneer. The entries were chosen selectively, and he included only those which were chemically tested and approved by him. The translation also includes observations by the author on various medicines. The text has an interesting narrative style as it includes a dialogue between hakims and the author, interlaced with commentaries by the translator. Since the identity of the original author and the Mughal patron is never revealed, the voices of author and translator mix and are difficult to discern. For instance, the entry *neeb* (*Melia azedirachta*) has the following commentary attached:

> It is good for digestion, leprosy, scabies, etc. It is also said by hakim Ali to be excellent in curing obstinate ulcers. I have often used the leaves heated as an application to swellings and boils, it in some cases resolved them. I have added salt to bruised leaves for some purpose, and found that it added to its efficacy. Neem in opinion of Indian physicians is cool and Hakim Urzanee has described its qualities as equal—good for hair, scabies and ring worm, etc.[96]

[94] Playfair 1833.
[95] Storey 1971: 283.
[96] Playfair 1833: 170.

In this entry the reference to 'Indian physicians' makes the identity of the writer closer to that of the British translator doctor. At the same time the identity remains ambiguous since the rest of the entry shows a familiarity with usage of the materia medica that one would associate with hakim Sharif Khan, the author.

The Company printed fresh texts as well. The governor general, specially after the cholera epidemic of 1817–18, was keen that medical men be encouraged to print their Indian experiences for the 'cultivation of science' in Britain. He also reminded them that print could add to their 'reputation and distinction'. Their status would increase if they communicated with the public at home as authors. As the governor general said, 'it will afford for the development of talent—cannot fail to act as a stimulus to laudable ambition which besides giving birth to frequent and copious communication of interesting facts and discoveries to the public may be expected to effect a very beneficent improvement in the general character and habits of the medical officers of this establishment.'[97]

After the cholera epidemic of 1817–18, the medical board circulated a questionnaire to its doctors with twenty-two queries relating to the management of cholera. This was to serve as the format for their detailed reports for publication.[98] The board was keen that the reports be standardized and printed so that they could be used as texts both for public consumption and future medical research in Britain. Indeed, J. Jameson, secretary to the medical board, was of the view that the Company, by taking on the expense of printing the medical texts of its doctors in India, would at long last put an end to their exclusion from their larger professional community. He attributed their inability to 'undertake the superintendence of printing a large work as the cause of their marginality.'[99] He noted that, deprived as they were, 'in this remote part they have not, as in London, and other great cities periodical

[97] L no. 118, I. Young, Sec. to Govt, to J. Jameson, Sec. Medical Board, 26 September 1818, Extract, BMC, 26 September 1818, Boards Collection, file no. 15654, F/4/622, p. 45.

[98] Ibid., L no. 101, J. Jameson, Sec. Medical Board, to all officers, 9 October 1818, pp. 51–60.

[99] Ibid., L no. 123, J. Jameson, Sec. Medical Board, to all medical practitioners, 20 October 1818, p. 71.

journals exclusively devoted to professional subjects, and affording facilities for publishing without expense or trouble such short essays or individual facts, as they might deem worthy of general notice.'[100] He was of the view that these circumstances had adversely affected both the reputation of the individual doctor and the character of the profession at large. The Company endorsed Jameson. It could only benefit from the proposal. As a sponsor and financier of the texts, it was able to claim exclusive patronage of medical learning in the contested medical sphere of India. And printing the Indian medical experience helped not just medical research in Britain but also brought good publicity to the Company at home.

Alongside epidemic management, the Company was interested in receiving also for publication tracts on the materia medica of India. Authors were given the option of remaining anonymous or having their names in print. In 1818 J. Jameson, secretary to the medical board, stressed the need for such publications when he wrote to the government that

> the various species of cutaneous diseases, elephantiasis and other maladies almost peculiar to warm countries, present a wide and nearly untrodden field to the curiosity of future investigators. Whilst much room for experimental research still remains in ascertaining the real power of such articles of the materia medica of the natives, as seem to profess virtues peculiar to themselves, or as might be occasionally substituted with advantage for medicines imported from Europe. So soon as a body of manuscript sufficient to form a sizable volume shall have been collected, the Board will proceed to digest and prepare it for the press.—It will be optional to authors if they wish their names to be made known to the public.[101]

The board lamented that assistant surgeons posted with the army were less than forthcoming in sending their reports and observations. The reports it did receive were of poor quality and not printable. The

[100] Ibid.: 71.

[101] J. Jameson, Sec. Medical Board, to Govt, 20 October 1818, BMC no. 51, arrangement for collecting facts and observations elucidatory of prevailing diseases in India, Military Secretary Office, 1820, Boards Collection, file no. 15654, F/4/622, pp. 77–8.

government continued to support the scheme, even as it regretted the 'absence of habits of literary composition among the medical staff of this army.'[102]

If army doctors disappointed the Company, those posted in its dispensaries were more forthcoming. From the 1830s superintendents of dispensaries were encouraged to put into print their discoveries of local materia medica and cures. The knowledge gained in human anatomy during the course of postmortems and other surgeries was of equal interest to the Company. Both the Company and dispensary surgeons benefited from the print culture. While the Company gained recognition as the patron of medical learning, the authors used the printed reports to both strike roots in India and further their cause as apothecaries and surgeons at home. Indeed, as authors of Indian materia medica, British doctors conferred dignity upon the apothecary. And their printed tracts on anatomy highlighted their superior skills in the field of surgery, helping the cause of both surgeon and apothecary and providing a context to blunt the monopoly of the physician at home.[103] In India, print helped bring the 'hand and the brain' together and thereby helped the cause of the 'literate surgeon'. This was very much like sixteenth-century France and early modern Europe where, as Natalie Davis[104] and Elizabeth Eisenstein[105] have shown, the impact of print

[102] L no. 118, I. Young, Sec. to Govt to Sec. Medical Board, 26 September 1818, Extract BMC, 26 September 1818, Boards Collection, file no. 15654, F/4/622, pp. 44–6.

[103] For the role of print in the fading of the patron figure, see Brown 1991: 103–42 and fn 91 above. The new 'authorial self consciousness' that print created was used by British assistant surgeons in India to fight their professional battles in England. In Britain surgeons, along with the apothecary, functioned outside the learned academies dominated by physicians. In early-nineteenth-century Britain the study of human anatomy was still debated, and it kept surgeons outside medical academies. From the late 1820s surgeons posted in India used print to publish tracts on anatomy. As authors of anatomy texts, they dignified their profession by making it available for study in academies of learning. For details on professional tensions and medical hierarchies in eighteenth-nineteenth-centuries Britain, see Lane 1985; Bynum 1994.

[104] Davis 1975: 189–226.

[105] Eisenstein 1979: 520–74.

blurred the distinctions between physician, surgeon, and apothecary by creating new authors via textuality.

W.H. Ainslie, who received the post of assistant surgeon in the Company's presidencies in January 1805, on being found qualified in the study and practice of surgery, published one of the earliest materia medicas in 1813.[106] This 2-volume text described drugs that were used in India and Eastern countries. It gave an influential commentary on items that were commonly used both in Europe as well as in India so as to establish a link between the materia medica of the two continents. Ainslie freely used Mughal medical texts like the *Ulfaz Udwiyah* and Avicenna's *Qanun-fi-tibb*, used by hakims in India as reference books. For instance, alum and oil of almonds were two of the many items that were stated to be recommended as astringents and for pulmonary complaints by respectively, both hakims and doctors in Britain. However, the title page of the text gave this shared knowledge a definitive author and a new patron: the text was dedicated to the King of England.[107] The author and the patron of the 1813 materia medica shadowed both the reference texts and the native practitioners whose remedies it sought to compile. In 1826 Ainslie published a second edition of his text from London in order to bring his knowledge to the notice of the British public.

William B. O'Shaughnessy was another important case in point. Born into an Irish family O' Shaughnessy, better known as the inventor of the telegraph in India, was an assistant surgeon posted to India in 1833.[108] In Britain his Irish background and training as a surgeon pushed him to the fringes of the well heeled medical society of the academies.[109] In India he made his professional mark as he entered the contested medical market at a time when the arrival of print offered possibilities to both the Company as well as its surgeons to carve out

[106] See his papers of nomination and his petition to the Company asking for service as assistant surgeon, L/Mil/9/358, pp. 5–7.

[107] Ainslie 1826: x, 11–12.

[108] For an entry on his India posting, see L/Mil/9/383, pp. 124–30, British Library, London.

[109] William Hunter, a wealthy surgeon of London, was out of the circuit of London physicians. See Porter: 1985: 7–34.

their exclusive patronage of medical knowledge and medical authority, respectively.

He rose to the position of professor of chemistry and materia medica at the Calcutta Medical College in the 1830s. As a lecturer of materia medica he was enthusiastic in exploring and collecting information on the drugs and herbs of India. He was thus a contestant in the fluid medical market where, as we saw, hakims and doctors co-operated to cure people and ensure public health even as they contested medical authority, chiefly over the efficacy of drugs. However, the use of print in the 1840s gave a breakthrough to O'Shaughnessy. He had always shown an interest in the use of the print for codifying knowledge concerning the apothecary and anatomy, with which he was associated and which had remained an 'undignified' profession in eighteenth-century British society. For instance, in 1839, during a temporary charge of the Asiatic Society at Calcutta, he arranged to get with great difficulty wooden casts from England for the anatomical drawings in Dr Hooper's text on anatomy that was being translated into Sanskrit.[110]

In the 1840s he used print to publish the information he had collected in his interaction with hakims and colleagues. The information was printed in two volumes as the *Bengal Dispensatory* and the *Bengal Pharmacopoeia*. Even though he had the support of the Company, O'Shaughnessy started the project of writing on his personal initiative, spending his own private funds to buy books to consult and self-financing his correspondence and travels for collecting material.[111] In order to experiment with his drugs on patients and ascertain their efficacy, he made a special request to be allowed to carry on private practice alongside his teaching. Since the Company did not allow this to its assistant surgeons posted in medical colleges, O'Shaughnessy made it

[110] W.B. O'Shaughnessy, to H. Wilson, Calcutta, 27 August 1839, MSS Eur. E 301/4, ff. 77–8. He toyed with various options to get the casts, and then requestd Wilson to order from Messrs. Taylor Owalton 500 copies of the woodcuts and steel plaits painted in Royal Octavo size, if all options failed.

[111] He had bought books worth Rs 500 for the preparation of the *Bengal Pharmacopoeia* out of his own money. Even though he did want the government to buy these books from him at cost, he was happy to pay for them in case of any objection. L no. 31, W.B. O'Shaughnessy, Assistant Surgeon and Professor, Calcutta Medical College, to H. Prinsep, Sec. to Govt, 14 August 1839, India Pub. Consult, 3 July–26 December 1839, Consult, 21 August 1839,P/186/90.

clear that his motives were other than financial profit. As he said: 'My avocations are too peculiar and my circle of acquaintance too limited to warrant my ever aspiring to extensive practice. But I wish to be permitted to avail myself of the few opportunities which present themselves to me for improving my knowledge of the practical part of my profession.'[112]

The government agreed to his request, but put brakes on his professional ambitions when it directed him to shelve his experimental researches on the local materia medica and concentrate instead on compiling its different varieties in the style of earlier works by Dr Ainslie, Dr Wallich, Dr Roxburgh, and others. Reluctantly, O'Shaughnessy agreed. Thus, even though he became famous as the 'author' of the *Bengal Dispensatory*, a status that brought him much fame and medical authority, he had basically only compiled existing medical knowledge, using print to codify and stamp as his own the loose spread of local medical knowledge. His *Dispensatory* was a compilation of local knowledge about plants, herbs, and materia medica screened, edited, and codified by O'Shaughnessy. The appropriated local knowledge, when packaged as a printed book with a medical author, lent both the author as well as the knowledge a new sanctity. And the use of the book as a 'class book of materia medica' for the instruction of 'native pupils' only reinforced the authorial self-consciousness of O'Shaughnessy and made the knowledge embodied in the text distinct.[113]

This authoriality was embellished when O'Shaughnessy improved its usefulness as a book of instruction by including in it a section contextualizing the local knowledge he had codified. Thus the *Dispensatory* included instructions on pharmaceutical manipulations; tips on the mode of taking specific gravities and making meteorological observations; an outline of chemistry for the guidance of teachers of native apothecaries; a grammar of botany; vegetable materia medica arranged

[112] L no. 42, W.B O'Shaughnessy, Assistant Surgeon and Professor, Calcutta Medical College, to H. Prinsep, Sec. to Govt in the General Dept, 11 December 1837, India Pub. Consult, 20 October–27 December 1837, Consult, 20 December 1837, P/186/80; L no. 43, H.T. Prinsep, Sec. to Govt, to W.B O'Shaughnessy, Assistant Surgeon and Professor of Chemistry in Calcutta Medical College, 20 December 1837, ibid.

[113] O'Shaughnessy, MDCCCXLII: xvii.

in the natural system; and a brief account of the mode of action of the several therapeutical classes of remedial agents.[114]

And yet this was only in part fulfilment of O'Shaughnessy's professional ambitions as a medical author. He also wanted to chemically test the materia medica so as to make it useful for the general instruction of druggists and practitioners. He regretted that the *Dispensatory* did not deal with the practical side of chemically and clinically experimented materia medica, and offered only pharmacy rules and remedies. The Company needed a text—a pharmacopoeia—of more practical value so as to reduce the cost of importing drugs to India, and also to use chemically tested Indian drugs in Britian. O'Shaughnessy of course craved to be allowed to prepare such a text, because that would help him wrap his authoriality and textuality within a new scientificity. It would help him dig his roots deeper into the medical market of India by emerging as an author, and as one who had transformed local knowledge to an almost unrecognizable form. He hoped that 'the resulting work would at all events prove a useful class book of materia medica for the instruction of his native pupils, and serve as the outline of one which in a subsequent edition more leisurely researches and opportunity might complete to his own satisfaction'.[115]

O'Shaughnessy had to wait for the fruition of such a moment a few years. This was partly because the Company was reluctant to relieve him from his teaching responsibilities at Calcutta and allow him to devote his time to experimental researches for the writing of his *Pharmacopoeia*. But the project was delayed also because in 1841 he had to return to Dublin because of ill health.[116] He resumed the job only on his return to India in 1844. Now, in his second stint, he complained of having lost much of the facilities he formerly enjoyed as 'chemical

[114] Ibid.: xix.

[115] Ibid.: xvii.

[116] See his letters for furlough. Letter of W.B. O'Shaughnessy to F.A.Wise, Sec. to General Committee of Public Instruction, 13 October 1841, enclosed in L no. 43, F.A. Wise to G.A. Bushby, Sec. to Govt in Gen. Dept, 18 October 1841; and L no. 44, Extract from Proceedings of Gov. Gen. of India in Council in Military Dept, 20 October 1841, GO granting furlough to W.B O'Shaughnessy, n.d., Lt. Col. I. Stuart, Sec. to Govt of India, Mil. Dept, India Pub. Consult, 18 August–29 December 1841, Consult, 27 October 1841, P/186/99 .

analyst' for pharmaceutical and clinical experimentation. He was thus unable to test many substances that local physicians brought to his notice, and reserved such tests for the future.[117] All the same, he did publish the *Pharmacopoeia* in 1844.

If print finally launched him as the author of a new text, it was the laboratory which coloured its textuality with scientificity. From the late 1830s the laboratory at the Calcutta Medical College, where O'Shaughnessy carried on chemical trials of many native drugs, played a crucial role in creating the medical authority that the *Pharmacopoeia* lent him. Indeed, the laboratory rooted his new textuality and authoriality in scientific experimentation. It moved the knowledge embodied in his text away from its original habitat and weaned him away from the hakims—the custodians of this knowledge. In this new role, O'Shaughnessy became a very important and busy man. He carried on chemical testing of a range of drugs of Indian origin that came to him via the dispensary. In 1840 he tested the extract of the bark of 'nux vomica', issued by the Company's dispensary to the Calcutta General Hospital and Serampore hospital. His experiments covered a wide span: he experimented the possibility of introducing '*kaladaana*' seeds as a substitute for jalup; the introduction of essential oil of gurgum for copaiboc; the introduction of *Crienum asiaticum* as an emetic; the experimental examination of the cost at which alcohol and numerous country remedies could be manufactured for public use; the examination of the medical uses and chemical properties of the berberry extract of the Himalayas and the 'safsafios' of Assam; experiments on the manufacture of pyroligneous acid from Indian woods; the introduction and the successful use of '*gunjah*', or Indian hemp, in tetanus cholina and many other diseases. He experimented on many Indian drugs, as well as on Indian clays in relation to the manufacture of pottery.[118]

It was this laboratory experience that was reflected in his *Bengal Pharmacopoeia*. Here, the local materia medica, drugs, and remedies were subjected to chemical analysis, which made the *Pharmacopoeia* different from the *Dispensatory*. It not only subjected local materia

[117] O'Shaughnessy MDCCCXLIV: 6.

[118] L no. 1, Notification FW, Gen. Dept, 23 April 1840. Minute by the Gov. Gen. Auckland, n.d.; no. 1B Govt's Resolution attached. India Pub. Consult, 1 April–24 June 1840, Consult, 29 April 1840, P/186/93.

medica to chemical analysis but also scientifically sanitized what O'Shaughnessy called the 'rude chemistry of their [local physicians] forefathers'.[119] He recommended through his committee—which made suggestions on the *Pharmacopoeia*—that each native plant be identified botanically and subjected to adequate chemical analysis. The preparation of its pharmaceutical products also needed to be put to clinical research. It had to be tried in hospital practice as well. He was of the view that the clinical and chemical analysis reports needed to be referred to a committee which would ratify its usage and lay out its chemical formula for prescriptions. Foreign products that were essential to the formula also needed to be spelt out.[120] Unlike the *Dispensatory*, here O'Shaughnessy was more than an editor and codifier of local knowledge. Through his team comprising a botanist, a chemist, an experienced European doctor, and a trustworthy local practitioner, he transformed local knowledge through the laboratory to a form distinct from that of the habitat where it had been collected.

As his reputation as a chemical analyst spread, he was requested by the military to carry on tests on aspects of military technology. A series of experiments on the explosion of gunpowder under water, and the examination of lighting conductors to powder magazines were some of the jobs he did for the army. Encouraged by his performance, the police department too involved him in chemical testing of many of the items that had been used in dangerous accidents. An important case he handled was the analysis of a case of poisoning during the incumbency of Dr Bain as police surgeon. And of course it was his skills as a chemical analyst that led him to carry on experiments on improvements to the galvanic battery for the communication of the telegraphic signal, which revolutionized the telegraph in India.[121] The governor general, in recognition of his vast contribution, designated him 'chemical examiner' and gave him some relief from the medical college, and a salary of Rs 400.[122]

Despite its shortcomings, the *Bengal Pharmacopoeia* was closer in its layout and contents to what its author had envisaged. It was different

[119] O'Shaughnessy MDCCCXLII: xv.
[120] Ibid.: xiv.
[121] Ibid.
[122] Ibid.

from the materia medicas of the past and from the *Dispensatory* because it transformed local medical knowledge into completely new forms by subjecting it to the rigours of clinical and chemical testing. This added a fresh layer of medical authority to O'Shaughnessy, and increased his hold over the fluid medical market and its 'native' contestants. Print codified and widely disseminated this knowledge to 'instruct' local practitioners and druggists on the art of making and prescribing their drugs.

The printed *Pharmacopoeia*, he hoped, would play a role in 'instructing' native druggists and practitioners; it would, he argued, 'educate any intelligent native, speaking English, Hindustani or Bengali, so as to make him perfect master of its contents.'[123] The tables, which listed local substitutes for clinically tested drugs like quinine, was meant as a 'useful guide' for the instruction of 'native medical students and practitioners, subordinate medical establishment of the army hospitals, and in case of emergency of the exhaustion of the supply of a particular article of medical stores, it will enable them in most cases to avail themselves of a good or tolerable substitute from the resources of the bazaar.'[124]

The *Pharmacopoeia* also helped O'Shaughnessy win points against British colleagues who were equally important contestants in the race for medical authority. He emphasized that the *Pharmacopoeia* was different from the materia medica of Dr Ainslie, which he described as a 'useful catalogue of several hundred plants, the products of which are used by native physicians.'[125] He underlined that the difference lay primarily in the fact that in Ainslie's text the plants 'have not been subjected to chemical analysis and clinical investigation.'[126]His *Pharmacopoeia* had tables that located medicinal plants in their natural botanical system, which were then listed in terms of their therapeutical value. His object was to 'facilitate research for new remedies in India, where a vast and rich field is open to the careful experimentalist.'[127] O'Shaughnessy hoped the tables would help officers posted in different parts of India to easily identify the therapeutic value of flora in their regions.

[123] Ibid.: xv.
[124] O'Shaughnessy MDCCCXLIV: 3.
[125] Ibid.: xiv.
[126] Ibid.: xiv.
[127] Ibid.: v.

Via his *Pharmacopoeia* he also wanted to instruct colleagues who he hoped 'will pursue experiments and researches and increase the existing catalogue of medicinal agents and resources.'[128]

His colleagues did not let him down. Through the 1840s, assistant surgeons printed regional materia medicas that emphasized chemical testing of their collections and highlighted the difference of their apothecary from that of the 'natives' because of their knowledge and reliance on chemistry. In 1848 R.H. Irvine, civil surgeon at Patna, wrote a materia medica for the city. The text had 480 entries in English, with country names appended, plus botanical names in Latin and comments on local usage. He collected information from local druggists and apothecaries and critiqued both the corruption and adulterated nature of their establishments, as well as their lack of 'scientific views' about materia medica. He was of the view that his materia medica would be useful only in the hands of 'superior hakims' because it had qualities 'nearly in accordance with modern scientific views.'[129] He complained that 'the natives unacquainted with chemistry never consider the medicinal powers of natural bodies as connected with that of science. From this they lose the great advantage of being able to extract the peculiar principles constituting the efficiency of many bulky and inconvenient substances.'[130] He was of the view that the same ignorance of chemistry was reflected in their prescriptions.

Alongside this authoriality in pharmacy, the authorship of books on anatomy represented another marker of status and dignity for the British doctor. It marked his superiority to the hakim. It also enhanced his status in the circles of British learned physicians, who traditionally looked down on surgeons. Print helped surgeons posted in India to emerge as authors and thus gain status in medical circles at home. In 1839 S. Davies, superintendent of the government dispensary in Patna, appended a list of materia medica collected from the region, but critiqued the humoral theory of its practitioners. He contrasted their lack of surgical skills to the British surgeons' immense knowledge of 'human anatomy, circulation of blood and the mechanical structure of

[128] Ibid.: vi.
[129] Irvine 1848: 2.
[130] Ibid.: 1.

the human body.'[131] However, all these attributes were controversial advances in Britain, where surgery and anatomy were still suspect. Dispensary reports from India, while underlining the superiority of such skills among British doctors, served the cause of the literate surgeon in Britain. Superintendents of dispensaries like those in Kanpur, Bareilly, and Delhi also furthered the cause of the apothecary and the literate surgeon by highlighting their skills and indispensability in printed reports. They indicated the need for clinical and chemical trials to make the materia medica usable, more sophisticated, and reliable.[132] They also put in print their observation that the natives appreciated most the skill of the British surgeon. In contrast, he had no faith in the average British physician. James McRae, superintendent of the Kanpur dispensary, could not have stated his case more clearly when he said in print: 'natives are well aware of our surgical skill, dexterity and judgement and readily have recourse to the dispensary hospital in all accidents and diseases requiring surgical operations. But I fear they have not the same faith in our medical treatment of diseases in general—we can never expect to acquire their confidence as physicians.'[133]

District magistrates posted in India backed the cause of the colonial surgeon. They urged him to author books on anatomy and apothecary, and critique the limitations of the physician. In 1838 H.C. Tucker, the officiating magistrate of Azamgarh, urged the School Book Society and the Committee of Public Instruction to print not just reports but translate into the vernacular and print monographs on medicine written by medical officers in India. He suggested a volume of about 150–200 pages with an illustrated section containing anatomical plates of the human frame, and a section on materia medica, and diseases and their symptoms, lithographed in the Persian and 'Nagri' script. He was of the view that, priced at about Rs 4–5, this would be most suited for the home market as a textbook for schools, while the English version

[131] Report of S. Davies, Superintendent of Govt Dispensary, Patna. Prepared with inputs sent by Ram Eshur Awushtee, Incharge Govt Dispensary, 30 January 1840, pp. 56–7. V/24/732.

[132] J. Duncan, Superintendent Bareilly dispensary, to G. Playfair, Superintending Surgeon, Meerut, 1 February 1841, p. 168, V/24/732.

[133] James Mc Rae, Report on Cawnpur Dispensary, 1 February 1841, p. 131, V/24/32.

would be useful for furthering the career of new authors in Britian as well. The format suggested by Tucker emphasized the importance of putting in print the knowledge gained by officers in India in anatomy and apothecary.[134]

Tucker's enthusiasm for the value of print was probably stoked by the success stories of surgeons like James Annesley, who had printed tracts on anatomy and made a success of their Indian experiences in the late 1820s. Men like him with literary talent used print to mark their authority in India and give dignity to their profession in Britain, where they were traditionally regarded as lower in competence and status than physicians of the academies. Annesley, surgeon to the Madras General Hospital, published a series of illustrated texts over 1825–8. He was with the Madras medical establishment and served in various hospitals of India for twenty-five years. Over 1811–24 he took down notes on the 'symptoms, progress and treatments' of diseases that he handled. He performed many postmortems in order to relate physical symptoms to body anatomy. In 1825 he published his illustrated treatise on the cholera epidemic. And in 1828, once back in England, he published his other carefully indexed notes along with sketches of body anatomy that he drew during the course of performing postmortems in India.[135]

Annesley's text underlined the superior knowledge of anatomy to mark the difference both with local Indian knowledge and that of the physician in Britain. It thus critiqued the universalizing claims of Western physicians and displayed a strong Hippocratic orientation to disease. According to Annesley, climate, soil, vicissitudes of temperature and the seasons, as well as 'modes of living, diet and regimen' of the individual are critical to his health. Disorders of the body emerge, he

[134] L no. 62, H.C. Tucker, Officiating Magistrate, Azamgarh, to E.P. Smith, Officiating Commissioner of circuit for 5th Benares Div., 2 March 1838, Proceedings of Gov. Gen., n.d., in the NWP May 1838, NWP Gov. Gens. Procds. in the Gen. Dept, January–June 1838, P/214/34. The format suggested by Tucker included a section of plates on illustrations of human anatomy gathered from postmortems, followed by chapters of causes and treatments of disease, as well as materia medica confined largely to common bazaar medicines and easily procured European drugs.

[135] Annesley 1828; and Annesley 1825.

argued, when man 'migrates from the climate which contributed to generate the peculiarities of his frame, to one which is remarkably different from that to which he is assimmilted.'[136] However, Annesley felt that the major contribution of his text were the illustrations of body anatomy that it carried, which he regarded very useful for both physicians and surgeons as they related physical symptoms to specific organs of the body. Through his text he not only claimed to have aided in the physician's diagnosis and cause of disease but also advanced the cause of the developing literature on anatomy. He argued that his anatomical diagrams were more accurate and therefore superior because he had drawn them on the basis of observations during postmortems on bodies soon after death. He was of the view that, in warm tropical climates like India, postmortems had to be performed a few hours after death, when a better sense of anatomy was possible because the blood was still in the arteries and had not returned to the veins.[137]

It is not without significance that Annesley published his 2-volume researches with richly illustrated anatomical drawings once he returned to London. He used print as a vehicle to harness his medical experiences in India, carving a niche for himself in British medical academies. These had hitherto been virtually closed to surgeons. The dissection of Indian bodies, cut open soon after death, provided greater accuracy to his knowledge of anatomy. More importantly, by printing the sketches of these more accurately dissected organs, he brought anatomy into the medical public sphere and thus dignified his talent as a 'literate surgeon' of India. Through his publication he hoped to 'apply morbid anatomy to the treatment of internal disease, as it has been so successfully applied to the improvement of surgery.'[138] His illustrated text on cholera was one of the early publications relating the symptoms of the disease to the workings of the nervous system. The latter was, of course, said to be adversely affected by the 'electrical conditions of the air, arising from or accompanied by terrestrial exhalations of a kind unfavourable to animal life.'[139]

[136] Annesley 1828: 7.
[137] Ibid., preface, ix.
[138] Ibid., preface, ix.
[139] Annesley 1825: 138, 147.

The 1820s was a time when many other surgeons in India also used print to underline the value of their profession, both in India and in Britain. In 1825 Peter Breton, surgeon in the service of the East India Company and superintendent of the NMI, published a text that consisted of a vocabulary of the names of the various parts of the human body and their medical and technical terms in English, Arabic, Persian, Hindi, and Sanskrit. Breton contrasted the superior knowledge of anatomy that British surgeons possessed, to the vague and loose sense of anatomy that native physicians possessed. The latter's lack of knowledge of anatomy was reflected in the absence of technical terms for anatomical expressions in Oriental languages: often, a single term served for a range of body parts. He said that this lack of an adequate vocabulary led to diagnostic problems. Breton thus justified the printing of his vocabulary of anatomy as a text that was vital for locals in this critical aspect of medicine. The codification of the knowledge of body anatomy in the form of an instruction manual naturally added to Breton's authority in the medical market.

The Company, eager to earn good publicity in Britain, encouraged Breton to publish his anatomy text. It stood behind him also because he indicated the advantage of such instruction to local surgeons who, even with their inadequate knowledge of medical science and anatomy, performed so many cataract operations: 'With the advantages of instruction in the science of medicine and anatomy much useful assistance may be reasonably expected from the native; a point of no small importance with reference to the civil and military station under the presidency and to the native subjects of the British government in India.'[140] The long list of dictionaries and local works in Arabic, Persian, Hindi, and Sanskrit that Breton appended to his text as reference works that he had consulted enhanced the status of the learned surgeon that print was bringing into sharper focus.

Again in 1828, John Tytler, lecturer of anatomy at the government sponsored Sanskrit College in Calcutta, and superintendent of the NMI whom we met earlier, used his India posting to make a strong case for the study and publication of anatomy texts. He justified its inclusion in the Sanskrit College on the grounds that it was a science that 'has

[140] Breton 1825.

a perpetual and irresistible appeal from the errors of theory to the facts of nature, from falsehood to truth.'[141] In teaching anatomy by real dissection, he argued, he was teaching truth by 'the plainest most direct and most incontrovertible means'.[142] He privileged anatomy over all other branches of science: 'No other science comes close to it. Its truths are obvious and uncontrovertible to everyone who takes the trouble to examine them, but at same time they are discoverable only by actual examination, and it is therefore the very best imaginable means of leading the mind from the study of words to that of things.'[143] Tytler's anatomy classes at the Sanskrit College were a great success and the governor general appreciated his efforts to initiate high-caste students towards knowing body anatomy and, in some cases, performing dissections on animals. He hoped Tytler's students would finalize their training in anatomy at the NMI of Surgeon Breton.[144] The success of Tytler in furthering the cause of anatomy lead to the General Committee of Public Instruction give him huge contracts to translate many Western scientific and medical texts into Arabic and Persian, including, as we saw, Hooper's tract on anatomy.[145]

Print and the Transformation of Medical Knowledge

Print created a new British author for what was hitherto shared medical knowledge. It also codified the new knowledge about body anatomy. Materia medicas, pharmacopoeias, tracts on anatomy, and medical dictionaries had been written in the Mughal period as well; what was

[141] John Tytler, Supdt, of Press and Anatomy Lecturer Sanskrit College, to H.H. Wilson, Sec. to the Committee of Public Instruction, 28 January 1828, Extract, Bengal Pol. Consult., 5 June 1829, Boards Collection F/4/1170, file 30639, p. 404.

[142] Ibid.

[143] Ibid.

[144] A Sterling, Sec. to Govt, to General Committee of Public Instruction, 26 June 1829, ibid., p. 377.

[145] Tytler's greed for appropriating large sums of money for translation works came up as a charge against the publication wing of the general committee for Public Instruction by Charles Trevelyan in 1833 when he launched his pro-English campaign.

new was the dilution of the earlier natural philosophy content and the standardization of new knowledge. An additional feature was the introduction of the vernacular—Hindustani, in the Arabic and Devanagari scripts—as the language of the new printed medical literature.

The materia medicas that the British compiled in India were about Indian medicinal plants and herbs, their botanical names, and their medicinal properties. Published largely for circulation amongst British medical staff and native doctors posted in far-flung districts of India, they were meant to eventually substitute Indian drugs for expensive European drugs. By classifying the diverse Indian materia medica in standard botanical families, local medicinal knowledge was now hooked on to the larger concerns of the evolving British medical sciences.

In 1815 F. Ramsay, assistant surgeon in the Bengal medical service, presented a manuscript for materia medica that included items used both in British pharmacopoeias and those used by Indians. He was of the view that his book would be the 'means of enlightening and improving [the] mass of them [natives].'[146] He wanted the science and knowledge of medicine to be spread amongst those who had no opportunity for instruction in medical schools. But his immediate target were the native doctors who, he thought, were often ignorant of drugs and unable to discharge their responsibilities at their postings. Ramsay's materia medica predictably had a Hindustani version, necessary if its knowledge was to be disseminated in society. But its reliance on the vernacular does not alone make it so different from that of Shirazi's materia medica. Here, instead of listing of medicines in terms of hot, cold, dry, and moist, their universally recognized chemical compounds and botanical names are also noted. Ramsay also used the standard weights and measures introduced by the medical board rather than the range of local ones used in various districts. Native doctors were also provided with a table to convert local weights to these standard ones.[147]

Now, most British materia medicas 'corrected' this lack of a universal scientific language in Indian medical knowledge. In the 1840s H.R. Irvine corrected this anomaly and introduced scientific order by adding

[146] L no. 163, A.F Ramsay, to R. Leny, Sec. to Medical Board, Calcutta, n.d., Boards Collection, F/4/611, file no. 15266, pp. 6–12.

[147] Ibid.

a column on botanical names in the Patna materia medica. While he indicated the many common items it shared with British materia medicas, he was quick to add that it was lacking in scientificity and therefore structured on empiricism, but 'superstitious and licentious'.[148]

The trend to give global standardization to local knowledge can also be seen in later English translations of Indian materia medicas. Translations invariably included additions, such as the botanical names of plants and plant families. These were often parallel to the publication of local prescriptions. For instance, George Playfair, a superintendent surgeon in the Bengal medical service, who as we saw earlier translated the *Taleef Shareef*, in effect created a new style text. In the process of translating this Indian materia medica he reorganized information in universally acceptable scientific categories of botanical names and chemical compositions. In a curious combination of the universal and the local, he appended to his text medical prescriptions (*nuskhas*) collected from local physicians. *Nuskhas* were rarely compiled in the Unani medical tradition: they were part of family knowledge and not to be made public. Playfair's appending of *nuskhas* to his translation thus completely changed the Indo-Persian writing style of the *Taleef Shareef*. But it also jarred the drive towards global standardization that the scientific reclassification of the items of the text reflected. This was truly a new-style medical text.

The impact of the discovery of chemistry in the early nineteenth century had its impact on the British medical literature generated in India. The most obvious shift was from a focus on writing materia medica to compilations of pharmacopoeias. The pharmacopoeia was a truly British medical introduction with no precedents in pre-colonial medical literature. It was, as we saw in the case of the *Bengal Pharmacopoeia*, more advanced than materia medica as it subjected local drugs to scientific tests. It gave details of preparing compound drugs, ethers, and mixtures. These texts were windows through which the revolution in the pharmaceutical industry, inaugurated by the arrival of chemistry, could be gleaned.

Soon, knowledge of the local materia medica was harnessed to the pharmacopoeia. As we saw earlier, O'Shaughnessy, professor of chem-

[148] Irvine 1848: 1.

istry and materia medica at the Calcutta Medical College, compiled the first 'regional pharmacopoeia for Bengal.[149] O'Shaughnessy elaborated its need when he writes of 'tracing the broad distinctions established by remedies on different races of mankind and including in our glance the remarkable varieties of the material from which in different countries one and the same medicinal agent has to be prepared, we cannot admit the wisdom of preparing pharmacopoeia to meet local circumstances and wants by sources of local supply.'[150]

If print changed the contours of medical knowledge by giving it a standardized textuality, the choice of the vernacular Hindustani as the language of these texts changed the consumers of this knowledge. Medical knowledge became accessible to a new range of readers as it moved out of the Arabic–Persian confines in which it had nestled until the 1820s. In this decade, government-sponsored colleges patronized such learning and sponsored the translation of even European medical tracts into these languages.[151] The vernacularized texts helped create a new class of medically aware communities.

Indeed the emergence of this genre of vernacular medical literature and its consumers reflected the fulfilment of the plan envisaged by anglicists like Charles Trevelyan, member of the Delhi Committee of Public Instruction, and his colleagues who wanted to break away from the old order and introduce Western science in Indian society to a fresh set of people. Such men were different from and not entrenched in older modes of Arabic medical and scientific learning. They were critical of the General Committee of Public Instruction which encouraged Arabic and Persian medical learning and languages, and which perpetuated the old order that had developed under the Mughals. Trevelyan, in particular, hoped to create a new class of medically and scientifically aware people by teaching them English as a prelude to the

[149] O'Shaughnessy: MDCCCXLIV; O'Shaughnessy: MDCCCXLII.

[150] O'Shaughnessy: MDCCCXLII: 6–7.

[151] Until the 1830s government-sponsored institutions like the Calcutta Madrasa and the Sanskrit College in Calcutta encouraged Arabic and Persian languages and scientific learning associated with their literature. See for details L no. 12, Gov. Gen. W. Hastings to Board, n.d., pp. 73–9.; L no. 14, 'regulations for the madrasa', pp. 91–7, in Extract Revenue Letter to Bengal, 28 October 1814, Boards Collection, F/4/708, file no. 19201.

development of the vernaculars, in which they would eventually disseminate their newly acquired medical and scientific learning. In 1829 he stated his case clearly: 'we need to target the bulk of people (Hindus and Muslims) who are entirely uneducated and attached to no previous system, while they are ready to adopt our own literature—.'[152] He wanted to sidetrack the 'the small proportion of our subjects (the literary Muhammadans) [who] are attached to a system of their own, by the strongest ties of habit and religious prejudice so as to render it hopeless even to effect any essential improvement in it by translation.'[153] Through this plan he hoped to give a new direction to scientific learning in India by creating a 'fresh class of people who were not weighed down in mind by their own literature deemed learned, venerable and sacred.' He was convinced that by targeting a new class of people 'we shall cease to carry on with such imperfect weapons an open warfare against the old system entrenched as it is behind religious feelings and preconceived opinion and make an immediate acquisition of a large class of people predisposed in favour of the new literature under which they will be entirely educated.'[154]

However, even though Trevelyan's dream of creating a new class of consumers of medical knowledge was fulfilled, the texts—neither in their content nor in the way they had been produced—represented unalloyed European medical science. In fact, as we saw above, they were the product of a truly shared medical knowledge that was as essential for circulation in India as it was critical for the growth of the profession in Britain. Such vernacular texts therefore reflected the dilemmas created in the minds of British medics when balancing the administrative expediencies of empire with the larger needs of a fast-evolving Western medical science tradition. The larger interests of science demanded

[152] C.E. Trevelyan, E. Colebrooke. T.C. Metcalfe and I. Ranken, Delhi Committee of Public Instruction, to H. Wilson, Sec. to Committee of Public Instruction, Calcutta, 12 February 1829, Boards Collection, F/4/1170, file no. 30639, p. 325. For the education debate of the 1830s that culminated in the famous minute of Macaulay, see Moir & Zastoupil 1999; Greaves 1967.

[153] C.E. Trevelyan, E. Colebrooke, T.C. Metcalfe and I. Ranken, Delhi Committee of Public Instruction, to H. Wilson, Sec. to Committee of Public Instruction, Calcutta, 12 February 1829, Boards Collection, F/4/1170, file no. 30639, p. 325.

[154] Ibid.: 327.

that local knowledge not only be connected to Western repositories but located in a language comprehensible to the 'scientific men of every country'. At the same time, the politics of governance required that these locally compiled texts be unesoteric, easily accessible, and comprehensible to the local population. This dilemma was resolved by O'Shaughnessy by debunking Arabic and Persian and relying on the vernaculars for easy dissemination locally. At the same time, English and Latin were used for its circulation in Europe. O'Shaughnessy arranged the descriptive part of the *Bengal Pharmacopoeia* in the following order: Bengali or Hindustani, followed by English, and lastly by Latin. Ramsay translated his materia medica into Hindustani to ensure that its codified and standardized knowledge did not remain locked in esoteric languages. Through the use of the vernacular he hoped it would penetrate levels of society where the Persian-Arabic Unani medical knowledge had never tried to percolate. Ramsay contrasts his unesoteric text to those of pre-colonial India: 'I may say that there must be considerable difficulty in explaining the nature of chemical substances particularly and their action upon the system to those who have never studied the one, nor have even a name for the others.'[155]

Through the 1830s and 1840s the British gave up their policy of translating British medical texts into Arabic and Persian, and resorted to the vernacular as the language to reach out to a new class of people. And, of course, the first such vernacularized text that circulated in Indian society was the Hindustani translation of the *London Pharmacopoeia*.[156] In 1824 Peter Breton translated the 1809 edition of the *Pharmacopoeia* into Hindustani. It was lithographed in both Nagari and Persian characters and circulated. Its great demand and usefulness led to further editions in 1841 by Dr G.G. Spilsbury, surgeon, and his assistant surgeon, Samchurn Dutt. These were only in the Nagari script. Since many Muslim students from the North West Provinces who had were enrolled at the Calcutta Medical College expressed their inability to read the Nagari script, Fred. J. Mouat, secretary to the council of education, with the help of munshi Nasiruddin Ahmad, produced a version in the Persian script in 1845.[157]

[155] L no. 163, A.F. Ramsay, to R. Leny, Sec. Medical Board, Calcutta, n.d., Boards Collection, F/4/611, file no. 5266, p. 11.

[156] Spilsbury & Dutt 1845.

[157] Ibid.

The readership of this translated medical literature shows the creation of medical or medically aware communities. One such important community was that of the service gentry of the qasbas and towns of Oudh and contiguous areas. Comprising Hindus, Muslim scribes, accountants, physicians, and a range of literate people, this class derived status from service in regional courts and looked to the British as their new employers. Tuned to the emerging new referents of service—Urdu and English—they felt left out of the family- and Arabic-defined religious and secular learning routes to status. They obtained these new texts form moving regiments, native doctors, and dispensaries. They took to the vernacular literature as a way of mobility in a society in which they felt rudderless.

In an era of grave social, economic, and ecological crisis that had created a huge clientele of healers and alternative treatments, the vernacularization of medical knowledge provided this service gentry with a welcome avenue towards the pursuit of lucrative careers. Not only did it make the medical profession easier to access, it made people aware of the immense potential of combining Unani medical knowledge with British medicine. A new community of aspiring medics could now enthusiastically participate in the medical sphere as collaborators and competitors of British doctors. From the 1830s it was with their active participation and enthusiasm in the medical public sphere that the institutionalization of medicine began. This happened with the large-scale establishment of dispensaries in India.

4

Dispensaries and Shifakhaanas in Early-Nineteenth-Century India

Introduction

BY THE 1840S THE LANDSCAPE OF NORTH INDIA WAS DOTTED with dispensaries that came up with contributions from the government and village society. They employed both doctors and hakims. And they exemplified yet another facet of the collaborative yet contested space of public welfare.

The dispensary served many purposes. It provided medicines to the sick. But it also doubled as a clinical laboratory where the texts of the new British author were tested. It lent scientificity to the doctor's authority and distanced his medicine further from healing. More importantly, it was the site where Indian communities of medical literature worked and interacted with British doctors. The incorporation of different communities of literature in the dispensary meant that the colonial framework offered a space to continue the intellectual dialogue between hakims and doctors, this having become a feature of Indian society. Here I show how the dispensary worked as a collaborative institution where the strained relations between the two systems of medicine were deftly negotiated.

Outside the dispensary, Unani combated the colonial onslaught by consolidating its learning within the structures of the family. It borrowed the discursive framework of seminary learning to protect its knowledge and disseminate it within a narrow clientele of family and close associates. Thus, both within the dispensary and outside it, Unani healing reconfigured itself and was not overwhelmed by colonialism.

Administration of Dispensaries

In 1837 the medical board wrote to the President of the Council of India about the pressing necessity of establishing dispensaries in the towns of India. The outlay of these establishments provided for six beds for emergency and accident victims.[1] They proposed the following medical establishment and salaries for each dispensary:

Surgeon—monthly salary—Rs 100
Hakim—average salary—Rs 75
3 native doctors and vaccinators—average salary—Rs 15 to Rs 20 each—Rs 51
Bihishti and sweeper—salary—Rs 10
One apprentice—salary—Rs 5
Total: Rs 241

The rules of the dispensary were that the surgeon located himself in the hospital and did not visit the sick at their residence. He was strictly forbidden home visits as these interfered with his duties at the hospital. Native doctors, who also doubled as vaccinators, were requested to live close to the dispensary and remain in constant attendance. Hakims were required in the mornings and evenings, when most patients came to them for prescriptions. However, since emergencies could occur at any time, it was expected that the hakim would not absent himself from his place of duty for long hours.[2]

[1] The medical board made a distinction between the dispensary and the native hospital. It did not encourage patients to be admitted for treatment for any period of time in the dispensary. They argued that resident patients required arrangements for special diets and cooks, keeping their religious sensibilities in mind. This work the native hospitals performed efficiently. See, for details, Proceedings, May 1838, NWP, Gov. General's Proceedings, January–June 1838, P/214/34, Extract of a Letter from the Medical Board to the address of A. Rofs, President of the Council of India, 31 October 1837.

[2] Ibid.

In January 1838 the governor general approved the proposal with some minor modifications. He sanctioned the establishment of at least twenty dispensaries but gave the financial go-ahead for only six, initially. They were to be established under the charge of an assistant surgeon who was granted an allowance of Rs 50 per month for this additional responsibility. A surgeon from the Calcutta Medical College was appointed on a pay scale that ranged from Rs 40 to Rs 100 per month. The governor general ordered that the dispensary was to be supplied with medicines and surgical instruments from the government store upon indent forwarded by the medical board. Significantly, he announced that the dispensary would be a clinical testing site for materia medica compiled by doctors and a place of medical instruction for hakims and recruits.[3]

In 1842 R. Clarke, the magistrate of Bareilly district, expanded the proposal and recommended the setting up of 'branch dispensaries'. This was suggested because the government could not afford to set up government dispensaries in all the parganas. The scheme was to encourage government dispensaries in towns and nodal areas to supply medicines and medical care to the surrounding areas in districts which did not have a dispensary. Eventually, they were to use their surplus funds to lay the foundation of dispensaries in these areas. Clarke called these 'branch dispensaries' and hoped they would create the 'first link of interest between the government and the sick in the distant parganas'.[4]

Clarke took the lead in Pilibhit, where he set up one such branch dispensary, at his own expense, to attend to the sick in the police force. The surplus funds of the Bareilly dispensary, amounting to Rs 464.5.8, were utilized. Later, the government sanctioned Rs 400 for this commendable project. And by 1845 he had whipped up enough enthusiasm in the area to ask for 'native' subscriptions. These poured in from all quarters, to the tune of Rs 3934. By 1846 the project picked up momen-

[3] Ibid., Extract from a Letter from H. Prinsep, Sec., to W.H. Macnaghten, Sec. to Govt of India with the Gov. Gen., 24 January 1838, para 2.

[4] L no. 121, Enclosure in a Letter from Assistant Surgeon Bareilly, 31 December 1847, 'Note on branch dispensaries recently opened in Bareilly district', by J. Balfour, assistant surgeon, Bareilly, 28 December 1847, Proceedings NWP, January 1848, NWP, Lt Governor's Proceedings, General, January–March 1848, P/214/79.

tum in the area . That year, another branch dispensary was established at Behara, near the Tarai area, with a local subscription of Rs 2300. Here, Mr Browne, the tehsildar at Basulpoor, convinced the wealthy zamindars to contribute towards a dispensary patterned on the ones at Pilibhit and Behara. With their contribution and government help, a dispensary came up there in 1847.[5]

Like all government dispensaries, here the overall superintendence was in the hands of a British assistant surgeon. A native doctor was stationed at a salary of Rs 25 per month. He was provided with an assistant with some knowledge of medicine at a salary of Rs 7 per month. The establishment also included a sweeper, *kahar*, *bihishti*, and cook. Native medicines were used in the prescriptions. The expense of the total establishment was valued at Rs 52/month.[6] The government was normally willing to support such dispensaries conditional on the availability of sub-assistant surgeons and proper European superintendence.[7]

Donors and Patients: Creating a Community Consensus for British Medicine

The key features of the dispensary were that it mobilized collective and not individual charity, and encouraged donations rather than bequest.[8] The local population and elites contributed financially towards its establishment and the British added a fare sum to its running cost. By donating money the community purchased a role in the management and working of the dispensary. Each dispensary hired hakims and

[5] Ibid.

[6] Ibid. The dispensary mainly entertained patients requiring surgery: broken bones, etc.

[7] L no. 34, I.H. Battee, Commissioner Kumaon, Nainital, to J. Thornton, Sec. to Government NWP, 15 June 1844, Proceedings, July, NWP, Lt Governor's Proceedings, General, July–September 1849, P/214/86 . This was a condition repeated by the government in the case of most requests for assistance. See for example the case of the Almora dispensary.

[8] Porter 1989: 149–78. The form of the early-nineteenth-century dispensary was very similar to the eighteenth-century infirmary of Georgian England, where collective charity by the living rather than individual charity by bequest created a stage-managed noblesse oblige that helped the community paper over the different levels of cracks in its ranks.

menial staff. The active participation of the hakims and the village community in setting up the dispensary meant that it was part of the consensus that was being forged for the medical authority that it represented.

A general order of 8 May 1825 underlined the fact that local contribution and initiative was a must before government pitched in with its financial share to the establishment of dispensaries. The government followed this rule steadfastly. While it was forthcoming where it saw popular initiative, it was unwilling to contribute when there was no local funding. Thus it turned down the demand for a branch dispensary at Nainital when no local funds were available.[9] However, not everyone approved of the government's steadfastness in demanding that local society contribute to the dispensary. J. Thornton, secretary to government in the NWP and a firm supporter of the dispensary experiment, was of the view that rejections of popular demand for the dispensary were 'discouraging to the proposers and deprives people of much valuable aid which it is desirable to place within their reach and to lead them to appreciate'.[10] He even offered a plan for systematizing the government's contribution. He suggested that the government contribution to the dispensary be frozen at Rs 220 per month. All other charges, excluding European medicines, should be met by local subscription. He requested the government to sanction on these conditions the three dispensaries of Mirzapur, Ghazipur, and Saharanpur.[11]

[9] Ibid., L no. 161, Enclosure in a Letter from the Commissioner of Kumaon, 10 July 1850, being a letter from the Chairman of Municipal Committee, Nainital, to Commissioner of Kumaon, district Nainital, 10 July 1850. N. Richards, Chairman of the Nainital Municipal Committee, wrote to the Commissioner of Kumaon, in the district of Nainital, about the need for state patronage for a branch dispensary of the one at Almora. This was to be established at Nainital. He was of the view that, once the state patronage came, local contributions would pour in. However, his request was turned down on the grounds that local contributions were a prerequisite before the government added its share. See ibid., L no. 162, J. Thornton, Sec. to Govt NWP, to I.H. Batten, Commissioner of Kumaon, 28 August 1850.

[10] L no. 98, J. Thornton, Sec. to Govt, NWP, to Sec. to Govt of India, Home Dept, 8 May 1845, Proceedings, May 1845, NWP, Lt Governor's Proceedings, General, April–July 1845, P/214/68.

[11] Ibid.

Most donors at the local level were elites: rajas, nawabs, jagirdars, zamindars, and merchants. The dispensary initiative was partly philanthropy and charity for the rich, driven by their pressing political need to play the 'welfare manager' role in society. The investment in the dispensary, which was in the end a Company-backed institution, was a way to co-share patronage to medical learning and social welfare. The Company only encouraged initiatives because they were the best way to obtain community consensus for its medical authority.

One of the earliest acts of such philanthropy was evident in Benares, where the city hospital and its four branch dispensaries were set up in 1811. They owed entirely to the philanthropy of local elites and the support of the British administration. In 1787 Jonathan Duncan, the British Resident at Benares, granted a jagir for the foundation of the Benares city hospital and its dispensaries. In 1811 the principal residents of the town took the lead to get a medical complex ready on the jagir because the hakims who were the trustees of the jagir proved inept. Subscriptions soon poured in from the locality and the construction was soon complete. Part of the money collected was used to purchase the residence of the civil surgeon on a rent of Rs 60/month. The balance collections of Rs 24,435 were put in government securities at the disposal of a committee appointed by the government. This committee, also the overseer of the medical complex, comprised the judge, magistrate, governor general's agent, superintending surgeon, chaplain, civil surgeon, commissioner, and 'native gentlemen'. Eventually, the jagir that Duncan had given was resumed. But an annual payment of Rs 3081 in support of the hospital and its branch dispensaries was sanctioned by the government in lieu of the jagir.[12]

Once the hospital came up, the rajas and rich zamindars of Benares contributed money and property to establish dispensaries attached to the hospital. As a result of this help the Chauk dispensary was shifted from its modest location in the gateway area of the city kotwali to a 'large and commodious' house that was obtained on lease after protracted negotiations with the raja of Benares.[13] Similarly, the Ousangunge

[12] L no. 250, E.A. Reade, Commissioner 5th Div. Benares, to William Muir, Sec. to Govt of NWP, Benares, 16 March 1852, Proceedings NWP March 1852, NWP, Lt Governor's Proceedings, General, March–April 1852, P/215/16.

[13] Ibid.

dispensary, initially called the Rajghat dispensary, was shifted to north-east of the city. This was with local zamindari support. The dispensary had had a native doctor occasionally visit it, with 'medicines in his girdle'. Its new site was in the mohulla of Ousangunge. Here, at the site of an old police station, a big building was erected for the dispensary with the assistance of zamindar Babu Deo Narayan Singh. A row of shops was established flanking the dispensary and their rent was used for its maintenance.[14] Singh had on more than one occasion impressed the British administration with his charitable drives. He was rewarded by being appointed a member of the committee of the native hospital.[15] Again, the Beylopoor dispensary in the complex was maintained by the benevolence of the Maharani of Vizianagram. She had made Benares her home after the death of her husband in 1845. The maharani planned to invest a sum that would be sufficient for its permanent maintenance.[16]

But philanthropy, patronage to medical care, and welfare were not the only motives behind the enthusiasm of the Company and regional elites. By the 1850s the British used the dispensary also as a classroom from where they disseminated their vernacularized medical literature. This enabled them to counter the relatively heightened awareness about the theories of Unani healing that circulated in society. The circulation of British-sponsored medical literature on anatomy and materia medica from the dispensary began to increase the awareness in literate elites about European medicine. As familiarity increased, so did the popularity of European medicine—because it not only provided self-help cure manuals but also the possibility of employment in the dispensary. Many Indian elites were convinced that the dispensary provided not just medical care but training in a 'medical practice superior to that which consists in the exercise of some knowledge of simples and belief in the efficacy of charms.'[17]

Such were the views of Babu Devi Dayal Singh, the zamindar of Buthura, who proposed to set up both a school and dispensary on his

[14] Ibid.

[15] Ibid., L no. not cited, W. Muir, Sec. to Govt, NWP, Agra, to Babu Deo Narayan Singh, 31 March 1852.

[16] Ibid., L no. 250, E.A. Reade, Commissioner 5th Div. Benares, to William Muir, Sec. to Govt, NWP, Benares, 16 March 1852.

[17] Ibid.

estate. He wished to make a change so that qualified and licensed medical practitioners found a conducive environment for livelihood and thereby aid the sick. His proposal was to place at the disposal of the civil surgeon the sum of Rs 25/month or Rs 300 annually for a term of five years for maintaining a dispensary and medical establishment near Chobeypur on the Ghazipur road, about ten miles from Benares. He offered to provide a house sufficient for the establishment; he also added a fee of Rs 50 to the medical officer for an occasional home visit.[18] The government appreciated the initiative of the zamindar and sanctioned the dispensary. It ordered a supply of European medicines and equipment for its functioning.[19] The dispensary soon became operative. However, the proposal of the medical school was vetoed by Dr Stokes of the dispensary administration on the grounds that it was located close to the dispensary and he feared the spread of infections to students.

Babu Devi Dayal Singh's view of the dispensary providing knowledge superior to that defined by 'simple knowledge and charms' may have not been widespread. Yet it was ideas like his that the British administration exploited: Babu Devi Dayal had actually summed up for them what the dispensary represented. And thus they encouraged local initiatives for the dispensary; these brought community consensus for their medical system. And government encouragement acted as a catalyst that prompted a wide cross-section of locals to invest in dispensary establishments.

In an age of widespread epidemics and disease, the dispensary evoked popular interest and enthusiasm. In 1850 the sahukars and jagirdars of Ajmer urged Thornton, secretary of the government, to 'vouchsafe to them the same favourable consideration which has been so liberally accorded to the principal towns in the provinces in the establishment of a dispensary for use of towns people and adjacent villages'.[20] They

[18] Ibid.

[19] Ibid., W Muir, Sec. Govt, NWP, to E.A. Reade, Commissioner 5th Division, Benares, 31 March 1852.

[20] L no. 125, Lt. Col. C.G. Discon, Supdt, of Ajmer and Mairwarra to J. Thornton, Sec. to Govt, NWP, Agra, Ajmer, 8 April 1850, Proceedings, NWP April 1850, NWP, Lt Governor's Proceedings, General, April–May 1850, P/215/4.

collected Rs 5400 by voluntary subscriptions from the 'higher classes' for the construction of the dispensary building. The demand was triggered by the severe sickness that affected more than a third of the population of the area in the months from October to December, and the inability of the two hakims maintained by the dargah there to cope with the problem.[21] Lt. Col. C.G. Discon, the superintendent of Ajmer, supported the local initiative and identified government land near the new bazaar, and close to the new gate of the town as the site for the dispensary. However, like Devi Dayal, he was clear that more than to combat disease, the dispensary was needed for the 'scientific treatment of diseases and for the conduct of all operations falling within the province of surgery'.[22] In May 1850 the government approved the establishment of the dispensary.[23] Soon Thornton lent it a political role as well. He felt that the dispensary served also the important political objective of placating Rajput chiefs by underlining British benevolence to their people.[24]

Encouraged by the British response to the Ajmer dispensary, the people of the close-by town of Naya Nagar made a request for a similar 'indulgence'. The traders of the town promised a voluntary collection of Rs 1000 needed for the construction of the dispensary building and hoped for British financial assistance to meet the cost of the establishment.[25] The government sanctioned this branch dispensary.[26]

Prior to Ajmer and Benares, such collaborative dispensary experiments had succeeded in the Rohilkhand province. Here, in the six suddar stations of Pilibhit, Bijnor, Moradabad, Badaun, Bareilly, and

[21] Ibid. There were no aspersions cast on their professional knowledge of disease and medicines, even though C.G. Discon, the superintendent of Ajmer and Mairwarra, questioned their medical expertise .

[22] Ibid.

[23] Ibid., Proceedings, NWP, May 1850, L no. 162, J. Thornton, Sec. to Govt, NWP, to Lt. Col. C.G. Discon, Supdt, Ajmer and Mairwarra, 25 May 1850.

[24] Ibid., Proceedings, NWP April 1850, L no. 125, Lt. Col. C.G. Discon, Supdt, Ajmer and Mairwarra to J.Thornton, Sec. to Govt, NWP, Agra, Ajmer, 8 April 1850. And L no. 126 J. Thornton, Sec. to Govt, NWP, to H.M.Elliot, Sec. to Govt of India, 22 April 1850.

[25] Ibid.

[26] Ibid., Proceedings, NWP, May 1850, L no. 162, J. Thornton, Sec. to Govt, NWP, to Lt. Col. C.G. Discon, Supdt, Ajmer and Mairwarra, 25 May 1850.

Shahjahanpur, dispensaries were set up and sub-assistant surgeons or hakims, in the case of unavailability of a surgeon, appointed as early as 1844. Initially, there were some financial hiccups in their running.[27] But these were tided over in different ways. In the Shahjahanpur dispensary this was done by the use of *nuzul* revenue.[28] In the Saharanpur dispensary, keeping in mind the enthusiastic collections that people had made for the dispensary, the government decided to bend rules to provide medicines from the civil surgeon's store.[29] And in 1845, when a dispensary with a modest local contribution of Rs 42 to Rs 52 a month came up in Ghazipur, the government agreed to contribute on condition that the sub-assistant surgeon, Babu Raj Kisto Chatterji, stayed their permanently. In the event of this not being possible,

[27] L no. 73, W.I. Conolly, Commissioner Rohilkhand, to Sadr Board of Revenue, NWP Agra, 7 February 1844, Proceedings, April 1844, Lt Governor's Proceedings, General, P/214/63. For details of Badaun dispensary setup, see also L no. 79, I Conolly, Commissioner of Rohilkhand to Sadr Board of Revenue NWP, Agra, 8 March 1844. The sub-assistant surgeon at Badaun, Gobind Chunder Das complained of financial stringency, unpaid salaries, and inability to buy bazaar medicines from banias due to lack of funds. See also L no. 70, W.I. Conolly, Commissioner Rohilkhand Division Bareilly, to Sec. to Govt, NWP, 25 June 1844, and the latter's reply, L no. 73, J. Thornton, Sec. to Govt, NWP, to W. Conolly, Commissioner of Rohilkhand, 10 July 1844, Proceedings, July 1844, NWP, Lt Governor's Proceedings, General, July–September 1844, P/214/64. This dispensary even though locally supported by funds was not sanctioned by government because it did not have a sub-assistant surgeon to spare. And unless that and independent European superintendence were ensured the government did not lend governmental support to local initiative. However, in this case the commissioner, I. Conolly, had erroneously given it a stamp of government sanction, not realizing the conditions for such approval. Hence the dispensary ran into financial problems since no aid from government was forthcoming. The medical board had to take decisions on this and also on the Bijnor dispensary, which had a similar history.

[28] L no. 73, W.I. Conolly, Commissioner Rohilkhand to Sadr Board of Revenue, NWP Agra, 7 February 1844; also see L no. 81, J. Thornton, Sec. to Govt, NWP, to M. Elliott, Sec. Board of Revenue, 17 April 1844, Proceedings, April 1844, NWP, Proceedings, General, P/214/63. For Shahjahanpur the buildings not exceeding Rs 2500 had been granted.

[29] Ibid., Proceedings, April 1844, L no. 117, J. Thornton, Sec. to Govt, NWP, to G.F. Francis, Commissioner, Meerut Division, 20 May 1844.

it authorized the committee to draw a sum of Rs 300 from the nuzul funds for its maintenance.[30] And in 1846 on the demand of E. Peploe, Commissioner, 5th Division, the government agreed to designate the Ghazipur dispensary a full-fledged government one. People reiterated the pressing need and popular demand for medical aid in the area and the paucity of funds for maintaining the existing dispensary. Rs 200 per month was to be spent on its maintenance and its affairs were run by a standing committee comprising the commissioner and local agents. The civil surgeon was deputed as its ex officio secretary and any other native or European resident of the place could be included as its member.[31] In the same year a similar collaborative initiative in Mirzapur saw the emergence of yet another dispensary. The native merchants and other wealthy people residing in or connected with the city of Mirzapur contributed to the establishment of this dispensary building with financial support from government.[32]

The wide-scale popular demand for dispensaries is evident also from the list of private subscribers for such institutions at Dehra Dun, Aligarh, Banda, and Saharanpur.[33] The social base of the volunteers who contributed towards the dispensary was mixed and included all classes, religions, and racial types. It included lower classes like grass

[30] P. Smith, Commissioner 5th Div. Ghazipur, to J. Thornton, Sec. to Lt. Gov. NWP, Ghazipur, 31 May 1845, Proceedings, June 1844, NWP, Lt Governor's Proceedings, General, April–July 1845, P/214/68.

[31] L no. 165, Commissioner 5th Div. to J. Thornton, Sec. to Lt. Gov., NWP, Agra, 5 May 1846; and L no. 168, J. Thornton, Sec. to Gov., NWP to Commissioner of Benares Div. Agra, 29 May 1846, Proceedings, May 1846, NWP, Lt Governor's Proceedings, General, Jan–May 1846, P/214/71.

[32] Ibid., Proceedings, May 1845, L no. 99, J. Thornton, Sec. to Govt, NWP, to E.P. Smith, Commissioner 5th Division Benares, Agra, 14 May 1845.

[33] For a list of financial subscribers of these dispensaries see: NWP, Lt Governor's Proceedings, General, 15 August–8 September 1854. Procds. NWP, August, P/215/36. For Dehra Dun, L no. 172 A, Enclosure in letter from superintendent Dehra Dun to the Commissioner 1st Div. Meerut of 9 June; For Aligarh, L no. 177, Enclosure statement of sums received on account of private subscriptions or contributions for the Aligarh charitable dispensary upto 31 December 1851. And for Banda and Saharanpur dispensary see L no. 183A, Enclosure in a letter from Banda dispensary, 15 June by Ramsford, civil surgeon, Banda, 15 June 1854.

cutters, artisans, small-time traders, big zamindars, and leading merchants.[34] Indeed, donations bridged gaps within elites and also between them and the poor. For instance, in the case of the Dehra Dun dispensary the donors ranged from well-known Britishers like the lieutenant governor of the North West Provinces, the former Residency surgeon of Lucknow Dr Logan, the famous surveyor Mr J. Rennell, the assistant surgeon Major Ramsay, Sergeant Major Couch, Mr Guthrie, Captain James, and other such, including one Mrs Dawe and one Mrs Newhouse. Indian contributors included notables and service gentry like a tehsildar named Ahmadullah, Babu Kalicharan, Babu Ram Kumar, Girdhari Lal, the jail daroga, Munshi Muddar Bux, zamindars of villages in the Dhoon, Hardial, the zamindar of Kullaur, Mohan Singh, Dhoom Singh, Dwarka Das, and Goomanie Buzazes. The dispensary had ordinary donors also, who contributed small amounts like Rs 2. Individual contributions from the lieutenant governor to the ordinary man ranged from Rs 2 to Rs 100. These modest contributions from ordinary people were symbolic endorsement for the dispensary.[35]

The total contributions in the more affluent areas of Bareilly and Pilibhit in the Rohilkhand area were much more than in the Dehra

[34] Ordinary individuals listed as Muhammad Husein, and traders like Lalji Mull etc. See L no. 172A, Enclosure in letter from Supdt, Dehra Dun to the commissioner 1st Div. Meerut of 9 June 1854, Proceedings, August 1854, NWP, Lt Governor's Proceedings, General, 15 August–8 September 1854, P/215/36.

[35] L no. 172A, Enclosure in letter from Supdt, Dehra Dun to the commissioner 1st Div. Meerut of 9th June, Proceedings, August, statement of the amount of private subscriptions and contributions on account of the dispensary from its commencement, NWP, Lt Governor's Proceedings, General, 15 August–8 September 1854, P/215/36. The list of contributors and their contributions is as follows: From Landour as private subscription—Rs 516; Mrs New House–Rs 3; Mrs Dawe–Rs 5; Babu Kalicharan–Rs 5; Sergeant Major Couch—Rs 2; Major Ramsay–Rs 25; Mr Guthrie—Rs 16; Hundi from Mumorie—Rs 10; Captain James—Rs 10; Munshi Muddar Bux—Rs 4; Mr Rennell—Rs 3; Babu Ram Kumar—Rs 2; Dr Logan—Rs 100; Girdhari Lall, jail daroga—Rs 11; Lt Governor NWP—Rs 100; Mohan Singh, Dhoom Singh, Dwarka Das, Goomanie Lall Buzazes—Rs 305; Jogedhan Bazaz—Rs 5; Kedarnarain Pandit—Rs 5; Muhammad Husein—Rs 2; Lallji Mull—Rs 2; Ihmadullah tehsildar, contributions from Deyrah bazaar and from zamindars in villages in Dhoon—Rs 660; Hurdial, Zamindar of Kullaur—Rs 2; Total—Rs 2839. Sd/- by R.H. Dunlop, Superentendent Dehra Dun, 9 June 1854.

Dun dispensary. But here also the contributors came from all walks of life. In this region of the erstwhile Rohilkhand state, people contributed liberally, cutting across religion, class, and caste. The dispensary brought them together: scions of the Muslim Rohilla Pathan families, Muslim nawabs, the rich Rajput zamindars, ordinary peasants and the trading community—all contributed to the best of their ability. The individual contributions ranged from Rs 8 to Rs 500.[36] A similar trend that cut across religion, caste, and class was evident in the list of contributors of the Almora dispensary. These included the hill aristocracy and the rajas, rich zamindars, traders, shopkeepers, and ordinary people. The subscriptions ranged from Rs 10 to Rs 400.[37]

The joining of ranks of the community to support the dispensary provided a wide consensus for colonial medical authority. But this solidarity of the village community had the potential to be used against the state as well. Thus, the Company was quick to create a wedge in the

[36] L no. 48, Fleetwood William, collector of Bareilly, to H. Bedcock, commissioner of Rohilkhand, 11 November 1846, Proceedings, December 1846, Lt Governor's Proceedings, General Dept, June–December 1846, P/214/72. He listed the liberal donations from people. He argued that this would encourage others to contribute. His appended a list of contributors: Kalka Parshad—Rs 500, Kalka Dass & Jwalanath—Rs 200, Delaram Makoondram—Rs 300, Sheuber Dass—Rs 300, Nawab Ali Muhammad Khan—Rs 400, Golam Mohamed—Rs 500, Hoseinie Begum—Rs 200, Ashruff Khab—Rs 100, Kuruntin Singh—Rs 100, Lallchand—Rs 100, Cheetram—Rs 100. See also ibid., L no. 49, J. Balfour, Asstt. Surgeon Sec. Dispensary Committee, to F. William, Magistrate Bareilly, 20 August 1846. His list also included a range of ordinary contributors and a total contribution of Rs 3935. The highest donor gave Rs 500 and the lowest was that of Rs 8.

[37] The list includes: Raja Seedut Shah—Rs 400; Rani of Badrinath—Rs 300; Rani of Kedarnath—Rs 200; Rambuksh—Rs 150; Lallah Nuthsunt—Rs 100; Lallah Sri Ram—Rs 100; Bal Bahadur—Rs 50; Captain H. Ramsay—Rs 500; Tulsee—Rs 25; Moolchund—Rs 25; Ram Singh—Rs 15; Sewanund—Rs 10; Ramanund—Rs 10; Balo Dull—Rs 10; Ieyanund—Rs 10; Godha Dhur—Rs 10; Sohanund—Rs 10; Jaykishun—Rs 10; Rampershad—Rs 10; Kurrekh Pullee—Rs 10; 109 subscribers together—Rs 115.7; I. Shackey—Rs 500. Total Rs 2,580.7. L no. 37, Captain H. Ramsay, assistant commissioner Garhwal to I.H. Battee, commissioner in Kumaon, 1 June 1849, enclosed list of subscribers, Proceedings, July 1849, NWP, Lt Governor's Proceedings, General, July–September 1849, P/214/86.

ranks of contributors: it promptly distinguished contributors by rank. It conveyed its gratitude and underlined the rank of the contributors by issuing individual *parwanas* of thanks to people of rank and a general parwana of thank you to the rest. It drew two lists of subscribers: Part One consisted of people of rank who were issued individual parwanas; Part Two included those who were not of rank and hence not worthy of individual parwanas.[38] Thornton finally sent individual parwanas to anyone who had paid more than Rs 100.[39]

The first list included the following people of rank:

1. Ali Muhammad Khan. Rank: Descendant of Rohilla Chief Hafiz Rahmat Khan. Title: Nawab. Amount: Rs 400.
2. Ashraf Khan. Rank: Elakadar and mafeedar. Title: Elakadar. Amount: Rs 100.
3. Lalchand. Rank: Zamindar. Title: Zamindar. Amount:Rs 100.
4. Shoom Singh. Rank: Zamindar. Title: Zamindar. Amount: Rs 100.
5. Gholam Muhammad Elahi Buksh. Rank: Zamindar. Title: Zamindar. Amount: Rs 500.
6. Kalka Dass. Rank: Sahukar. Title: Sahukar. Amount: Rs 500.
7. Kalka Dass Joalanath. Title: Sahukar. Rank: Sahukar. Amount: Rs 200.
8. Della Ram Makoond Ram. Title: Sahukar. Rank: Sahukar. Amount: Rs 300.
9. Shankar Dass. Title: Sahukar. Rank: Sahukar. Amount: Rs 300.
10. Jevan Ram. Title: Sahukar. Rank: Sahukar. Amount: Rs 50.
11. Karut Singh. Title: Zamindar. Rank: Zamindar. Amount: Rs 100.
12. Hoosania Begum. Rank: Begum. Title: Zamindar. Amount: Rs 200.
13. Cheit Ram. Rank: Malguzar. Title: Malguzar. Amount: Rs 100.
14. Bunseedhur. Rank: Khursalee. Title: Khursalee. Amount: Rs 100.

[38] L no. 58, T. William, Collector Bareilly, to H. Pidcock, Commissioner of Rohilkhand, 3 February 1847, Enclosure in Henry Pidcock, Commissioner of Rohilkhand letter to J. Thornton, Sec. to Govt, NWP, 9 February 1847, Proceedings March 1847, NWP, Lt Governor's Proceedings, General, January–March 1847, P/214/74.

[39] Ibid., L no. 59, J. Thornton, Sec. to Govt, NWP, to H. Piddock, Commissioner of Rohilkhand, 10 March 1847.

15. Khoshalee Kulloo. Rank: Arutheeas. Title: Arutheeas. Amount: Rs 50.
16. Bhowanee Dass. Title: Arutheeas. Rank: Arutheeas. Amount: Rs 50.
17. Deedar Bux Elahi Bux. Rank: Boat Builders. Title: Boat Builders. Amount: Rs 100.

Total: Rs 3,250.

The second list comprised ordinary contributors who did not belong to status and rank. This included:

1. Muttra Dass Narain Dass. Title: Buneeas. Rank: —. Amount: Rs 15.
2. Dabee Dass. Rank: —. Title: Mahajan. Amount: Rs 10.
3. Mahubbut Khan and others. Rank: —. Title: Buzzajis. Amount: Rs 125.
4. Gulzarimul. Rank: —. Title: Malguzar. Amount: Rs 25.
5. Ram Sukh Chudumee Soll. Rank: —. Title: Traders. Amount: Rs 20.
6. Manjee Ram and others. Rank: —. Title: Shopkeepers. Amount: Rs 412.
7. Goordial Bhujan Lal. Rank: —. Title: Khursahees. Amount: Rs 25.
8. Hafiz Ghulam Muhammad. Rank: —. Title: Khursahee. Amount: Rs 20.
9. Sheikh Hulleru. Amount: Rs 25.
10. Mullhoo. Amount: Rs 8.

Total: Rs 685.[40]

The Dispensary as a Provider of Medical Care

The dispensary offered medical services to a wide range in society. People were not averse to a visit to the dispensary even if it represented an alien medical system, and though its doctors were of different skin colour, caste, and religion. Rather, faith in individual doctors and class considerations defined choice. The rich rarely visited the dispensary.

[40] Ibid.

But this did not mean that they were averse to the use of Western medicine or medical examination by Western doctors. We saw earlier that most regional rajas, like Ranjit Singh of Punjab, and the nawabs of Oudh, had favourite British doctors whom they trusted and consulted.[41] We also saw that when Dr Wilson was transferred out of Lucknow on promotion, as superintending surgeon to Kanpur, Nawab Ghazi-ud-din Hyder sent an urgent request to the governor general: 'he [Wilson] is thoroughly acquainted with my constitution and possesses my entire confidence.'[42] The governor general heeded this request.[43] And we saw that Shah Alam in Delhi had Dr Reid, the surgeon, posted as his personal physician.[44]

[41] Indeed, more than the system of medicine it was the confidence in the individual doctor that inspired the nawabs to try British medicine. Dr Wilson, in particular, was the hot favourite of the court at Lucknow. On more than one occasion he was suitably rewarded for his many cures. In 1808, Nawab Saadat Ali Khan hosted a grand banquet to celebrate the recovery of John Bailley, the British Resident, from a protracted illness. Dr Wilson was gifted with Rs 5000 as reward for 'his professional labour and skill, which by removing Dr Bailley's distemper restored His Excellency to health'. See L no. 37, J. Bailley, Resident Lucknow, to N.B. Edmonstone, Sec. to Govt Secretariat and Pol. Dept FW, 15 January 1808, Consult, 25 January 1808, Bengal Pol. Consult, P/118/22. The faith in Dr Wilson was particularly noticeable in Ghazi-ud-din Hyder. The nawab trusted Dr Wilson more than the personal physician of his father, Dr Law. The latter was sent by his father to attend to his ailments. He avoided Dr Law: 'I never took his medicines, but privately consulted Dr Wilson whose prescription I always followed'. See L no. 42, Nawab Vizir to Lord Moira, Gov. Gen., 19 April 1815, Extract Bengal Pol. Consult, Boards Collection, F/4/509, file no. 12265A, pp. 11–12.

[42] Ibid.

[43] Ibid., L no. 6, Extract Pol. Letter from Gov. Gen., 15 August 1815, pp. 1–6. Wilson performed the double duty of nawab's physician and Residency surgeon throughout his stay in Lucknow. In 1817, at his request Mr B. McLeod was posted there as assistant surgeon to help him. And later, on the request of the nawab, he was made his personal physician on a salary that the nawab fixed at Rs 1500 a month. See, L. nos 14–15, Extract Political Letter from Bengal, 3 January 1817, Consult, 17 August 1816, Boards Collection, F/4/601, file no. 14453, pp. 1–3; and also letter undated of Nawab Vizir to F.V. Rapier, Acting Resident, Lucknow, 2 October 1817, p. 15.

[44] In Delhi, the Mughal emperor had his entire extended family examined and treated by the British surgeon posted in Delhi—Dr Reid. He piled him with

However, such inclination towards individual British doctors notwithstanding, the elites did not come to the dispensary, which was largely attended by the poor. And here too James McRae, superintendent of the Kanpur dispensary, was pessimistic because of the general apathy of Indians towards the British system of medicine, despite noting their regard and respect for individual 'European talents and acquirements'.[45] Such appreciation for individual European practitioners transcended sensibilities of race and colour. And, on most occasions, Indians flocked to dispensaries by choice when these included their favourite physicians. At other times, necessity, curiosity, and the novelty of the institution brought them there. There were of course always those who came because hakims had failed to cure them.[46]

G. Playfair, superintending surgeon, Delhi, and A. Ross, civil surgeon, Delhi, endorsed the idea that individual European talent, and that of Western-trained Indians, was popularly appreciated. In the case of the Delhi dispensary, he cited the case of the appeal of British-trained sub-assistant surgeon Baboo Rajakishna Dey who had earned such a good reputation that his dispensary was always hugely attended. After his death, nobody came to the dispensary. The few people who visited came exclusively to consult Dr Ross, whose talent too they appreciated.[47] Ross and Playfair also provided a class-based profile of patients who attended the dispensary, pointing out that people who came were generally poor and of the low class. These included artisans, servants of the more wealthy inhabitants, ryots from the neighbouring country, banias, and many indigent people who found it difficult to pay the high fees charged by hakims. They reported that the upper

---

honorifics and praises, but no extra remuneration. This urged the surgeon to obtain an extra allowance of Rs 800 from the British administration for the additional work he performed for the royal family. So dependent was the emperor on British medics that when Reid died he wanted a substitute very urgently. The governor general posted Dr McWherter, assistant surgeon, as surgeon at Delhi and also physician to the emperor. See Boards Collection, F/4/251, file no. 5625 & F/4/335, file no. 7658.

[45] V/24/732, James McRae, Report on Cawnpore Dispensary, 1 February, p. 131.

[46] Ibid.: 131.

[47] V/24/732, Report on Delhi City Dispensary for 1840 by A. Ross and G. Playfair, pp. 156–7.

classes were often dissuaded from visiting the dispensary by hakims who spread canards about British medicine using body parts as ingredients.[48] Doctors like J. Duncan, superintendent of the Bareilly dispensary, also reported that, often, the rich did not attend the dispensary because they did not wish to rub shoulders with the poor. It was, he said, not the considerations of having faith in his medical system, or matters of caste, colour, race, or religion, but the issue of class that deterred them from visiting the dispensary. The very patients who refused to come to the dispensary were happy to visit him in his house for consultation.[49]

Caste-related notions of purity and pollution were shelved when making the choice to go to a Western doctor. This was particularly true at the time of epidemics. Indeed, epidemics like the 1818 cholera one were also moments where British medical practitioners struck a chord with the sick and gained legitimacy over indigenous practitioners such as the hakims. In Bundelkhand district, for instance, successful efforts to reduce cholera-related mortality were appreciated by the people. F.C. Smith, the magistrate of Bundelkhand, reported that the 'sick eagerly crowded for English, rejecting with contempt the native medicines.'[50] And when a Brahmin's child died because the native doctor who did not know the opium-calomel dosage for children gave him 'native medicine', the parents complained bitterly to Smith. They were convinced that 'their child had been killed by the hakim from his neglecting to administer English/wellayut/medicines.'[51]

The Dispensary as a Site for Clinical Trial of Materia Medica

The dispensary provided an avenue for doctors to become new authors. As a laboratory for the clinical testing of the local materia medica, the dispensary encouraged doctors to publish their results and emerge as

[48] Ibid.

[49] Ibid., J. Duncan, Supdt, Bareilly to G.Playfair, Supdt, Surgeon Meerut, 1 February 1841, returns of the Bareilly Dispensary, p. 168.

[50] Ibid., Extract Bengal Judicial Consultation, 4 September 1818, Acting Magistrate Bundelcund to W.B. Bayley, Sec. to Govt in Judicial Dept, FW, 27 July 1818, p. 26.

[51] Ibid.: 27.

such authors. Indeed, the dispensary was a site for experimentation with the local materia medica before such medicines could meander into medical texts and then into the pharmacies of Europe. In 1846 Raj Kristo Chatterji, the sub-assistant surgeon of Ghazipur dispensary, wrote in his report that he was giving a fair trial to all such native medicines 'as are said to possess any peculiar medicinal virtues.'[52] Hinting at the pharmacological laboratory that the dispensary could become, he said that we 'are about to ascertain how far they [materia medica] are capable of superceding the use of some of the most costly medicines of Europe.'[53]

In 1839 the in-charge of the Patna dispensary, Ram Ishwar Awasthi, in his annual report to his dispensary superintendent S. Davies, detailed the native treatment of diseases and catalogued both indigenous and imported medicines sold in the bazaar. He listed their qualities, uses, and doses as prescribed by hakims. Davies tested the Unani drugs on his patients at the dispensary and, once satisfied, listed the materia medica in his statistical report. Even though he was happy to codify Unani drugs, he was critical of Unani as a system of medicine: 'the state of popular medicine differs little from that which has existed for many centuries. The catalog of materia medica has been greatly added to, but no knowledge of the structure of the human body exists among the best informed of native physicians.'[54] He lashed out at its practitioners, saying that 'humoral pathology is the favourite study of the more learned, behind which they can hide their real ignorance and substitute a few technical phrases for real knowledge, and if their practice of medicine is so deplorable, their knowledge of the principles of surgery is still more so, being totally unacquainted with anatomy, having no knowledge of the circulation of blood and being perfectly unacquainted with the mechanical structure of the human body—.'[55] But this critique

[52] L no. 167, Report of Ghazipur Dispensary for last 8 months ending 31 December 1845, by Raj Kristo Chatterji, Sub Assistant Surgeon Ghazipur, 30 April 1846, Proceedings, May 1846, NWP, Lt Governor's Proceedings, General, January–May 1846, P/214/71.

[53] Ibid.

[54] V/24/732, Ram Eshur Awushtee, in charge Govt dispensary, to S. Davies, Supdt, of Patna Dispensary, 30 January 1840. Annual return of Patna Govt dispensary for the year 1839, p. 56.

[55] Ibid.: 57.

of Unani did not stop him from compiling detailed statistical tables of indigenous medicines, and the local Unani regimens or prescriptions (*nuskhas*) for diseases like leprosy, elephantiasis, etc.[56] Moreover, for their circulation in the larger scientific community the names of drugs were translated also into their Latin and English botanical equivalents. In fact, the tables gave the names of local drugs in three columns: English and Latin, Hindi in Roman characters and Hindi and Persian. This was a way to codify Unani knowledge and integrate it to British medicine for use both in India and England.

Awasthi continued to send Davies details of indigenous medicines which he had successfully used on his patients for the treatment of diarrhoea, dysentery, asthma, and coughs. Some of his medicines included the extract of Indian colocynth, the bark of the 'phaulash' tree, the bark of the mango tree, and the juice of 'bakaend' leaves.[57] And in 1856, encouraged by the efficacy of Davies' clinical tests, James Irving, superintendent of the government dispensary at Allahabad, wrote to several dispensary superintendents asking for any Unani prescriptions they had used for indigenous drugs: he was happy to receive any cheap and valuable local remedies they followed at their dispensaries, being keen to test them at his dispensary and then publish his own materia medica.[58]

Given this interest in Indian medicinal plants and drugs, it was not surprising that in most dispensaries country drugs were in use. For instance, in the Bareilly dispensary country drugs were bought from the local bazaar, and comprised *attars* (perfumes) etc. that were in use with local practitioners. However, apprehensions about their purity persisted. Superintendents like Duncan doubted if their chemical composition had been tested by principles of chemistry or even clinically examined.[59] Such fears only intensified the urge to gradually disseminate

[56] Ibid.: 73–7 for Unani *nuskhas* appended to the report.

[57] V/24/732, Report from Ram Eshur Awushtee, in charge of the Patna Dispensary, p. 211.

[58] L no. 446, James Irving, Supdt, Govt dispensary, Allahabad, to Sec. to Govt, NWP, Allahabad, 13 February 1856, Proceeding, February 1856, NWP, Lt Governor's Proceedings, General, 25 February–31 March 1856, P/215/53.

[59] V/24/732, J. Duncan, Supdt, Bareilly, to T.G. Playfair, Supdt, Surgeon, Meerut, 1 February 1841, p. 169.

Western medicines in Indian markets. However, practical exigencies, such as cartage from Europe and the expense involved in shipping consignments, remained major impediments.

If local materia medica was tested and codified, the already printed materia medicas, pharmacopoeias, and tracts on anatomy were freely consulted and put to clinical verification. The dispensary also acted as a clinical and chemical laboratory where the hitherto shared medical knowledge of printed texts was scientifically tested. Dispensary doctors also tested the freshly compiled tracts on anatomy during the physical examination of patients, and over postmortems. Breton's tract on anatomical vocabulary, Monat's text on anatomy, O'Shaughnessy's *Bengal Pharmacopoeia*, and Spilsbury's Hindustani translation of the *London Pharmacopoeia* were all supplied to dispensaries and used extensively there. The superintendents reported back on their shortcomings. In 1841 Duncan at the Bareilly dispensary found the medica of Dr Ainslie 'but an initiatory jejune and hearsay compilation'. He strongly recommended the chemical testing of many of its items before use in the dispensary. He warned that if this was not done, then, 'instead of rendering the dispensary more popular and attractive [it] would I apprehend be productive of much and serious evil'.[60]

Popularizing Western Medicine: The Dispensary as Training School

British doctors viewed Unani as a challenge. They knew that even though Indians were not averse to the dispensary, it was faith in individual doctors and certain remedies that attracted them to Western medicine; their mindsets was still inclined in favour of Unani. The administration was concerned about their 'prejudices in favour of their own system, especially their theories of hot and cold as applied to the effects of medicine'.[61] They attributed the popular inclination towards Unani to the uneducated and primitive ways of such people. Until this was addressed, lamented men like James McRae, superintendent of the Kanpur dispensary, 'we can never expect to acquire their full confidence

[60] V/24/732, J. Duncan, Supdt, Bareilly Dispensary, to G. Playfair, Supdt, Surgeon Meerut, 1 February 1841, pp. 169–70.

[61] V/24/732. Report on Cawnpore Dispensary by James McRae, Supdt, of Cawnpore dispensary, 1 February 1841, p. 131.

as physicians'.[62] However, others felt that one of the deterrents to the popularity of their medicine was ignorance, and lack of knowledge about its rationale and method of cure. From H.C. Tucker, the magistrate in Azamgarh district, to Inayat Husain, the sub-assistant surgeon at Lucknow, anyone who championed the cause of British medicine felt that the popular ignorance about its medical rationale and ethics was the major obstacle in its success over Unani.[63] The problem was compounded by the relatively greater awareness and understandings of Unani healing.

The dispensary was geared to rectify this problem. It spearheaded the campaign to disseminate scientific knowledge about Western medical science in Indian society. This was necessary to compete with Unani and other indigenous systems. If there was a key player in this imperial project, it was primarily the clinic. It was also a centre where knowledge about British medicine and awareness of its drugs could be professionally acquired. It was a place of instruction where local people could enlist as apprentices, hakims educated in British medicine, and where anyone interested in the British system of medicine could enrol. The dispensary, in short, made medical training accessible to a wider range of society. And in offering easy and quick routes to become a 'native doctor' it aimed to outstrip Unani, which offered no such short cuts.

Much before the dispensary came up, the training of hakims in Western methods had gone on informally. We saw that, in the 1820s, on a long journey from Jwalamukhi to Lahore, William Moorcroft, the British surgeon traveller, performed several cataract surgeries and trained innumerable hakims in the surgical art. He had engaged a hakim from 'Hushourpoor' to accompany him to acquire his surgical style. He promised to train him, bear his expenses, and also set him up with the requisite instruments.[64] Moorcroft entertained many such

[62] Ibid.: 131.

[63] L no. 62, W.C. Tucker, Magistrate Azamgarh district, to E.P. Smith, Officiating Commissioner of Circuit for 5th Benares Div., 2 March, Proceedings, May 1838, NWP, Gov. General's Proceedings, General, January–June 1838, P/214/34. Also see L no. 127, Inayat Husain to I.S. Login, n.d., sent to Medical Board 28 March 1845, Consult, 28 March 1845, India Political & Foreign Consult, 28 March 1845, P/197/10.

[64] Moorcroft Papers MSs EurD. 238. Third Fasciculus of Journal from

requests from the Lamas in the Leh area of the Himalayas for training hakims in cataract surgery.[65] Similarly, at the time of epidemics like smallpox, British compounders trained Indian vaccinators informally. We noted earlier that in the Jungle Tarai region, for instance, where smallpox was particularly virulent, the hill chiefs commanding the Hill Corps had trained as vaccinators.[66]

But it was only in the 1830s that this informal training began to be institutionalized. In 1838, W.C. Tucker, the collector and magistrate at Azamgarh, made a strong plea to the government to introduce formal medical education in government schools alongside general instruction. He suggested the appointment of a teacher from the Calcutta Medical College to give medical instruction in government schools. He proposed lectures on anatomy, surgery, medicines, chemistry, botany, and physics. These, he hoped, would go a long way in disseminating Western medical science.[67] This suggestion remained under the consideration of the government for many years and was used in a modified way at dispensary level.

Governor General Lord Auckland was instrumental in converting the dispensary to a medical training nursery for fresh recruits. Trainees, on completion of their course, were called native doctors. They added to the ranks of native doctors educated at the NMI, which had been abolished in 1835. These new doctors were meant to fill the subordinate positions in the dispensary.[68] In this way the dispensary became one of the early institutions that made access to British medical training

---

Joalamookhee to Lahore the capital of the Sikh dominions from 28 March to May 13, 1820, f. 60.

[65] Moorcroft Papers MSs Eur. D 246, Journey from Leh to Dras, f. 87. The Deputy Superior, Lama Yoarou, the Deba, made such a request. Moorcroft agreed to train and supply with instruments the hakims of his area.

[66] L no. 45, Capt. I. Graham, adjtt. to Hill Rangers, to Capt. Caldwell, ADC to Gov. Gen., 30 August 1822, Extract Bengal Pub. Consult, 3 October 1822, Boards Collection, F/4/761, file No. 20686, pp. 15–19.

[67] L no 62, W.C.Tucker, Magistrate Azamgarh, to E.P. Smith, Officiating Commissioner of Circuit for 5th Benares Div., 2 March 1838, Proceedings, May 1838, NWP Gov. General's Proceedings, General Department, Jan–June 1838, May 1838, P/214/34.

[68] L no. 84, Medical Board, to J. Thomson, Lt. Gov., NWP, 30 October 1851, Proceedings, October 1851, Lt Governor's Proceedings, General, December 1851, P/215/13.

possible at the local level. Trained native doctors were the harbingers of Western medical practice locally; and they projected the benevolent face of British administration.

The dispensary selected its apprentices with care. It was always interested in training hakims. This also ensured the contact with Unani that was essential for acquiring the knowledge of its materia medica. Dispensaries enlisted hakims for administrative and medical help and familiarized them with European medical techniques. But full administrative charge was rarely given to them. They worked under the sub-assistant surgeons—graduates from the Calcutta Medical College—posted at every dispensary. Only in rare cases, in the event of the unavailability of a trained native doctor, was the administration of the dispensary left to a hakim.[69]

The hakim's main job was to write prescriptions for patients. The British, as we saw earlier, showed keen interest in the hakim's prescriptions and pharmacy, though remaining critical of his diagnostics. For instance, in the Moradabad dispensary, two Muslims and a Hindu hakim were permanently employed at decent salaries of Rs 30 per month each. Their main function was to write prescriptions for patients—in which the overseer British civil surgeon showed keen interest. However, in the 1830s the dispensary took responsibility to train them in Western diagnostics and surgery techniques as well. Thus, in the Moradabad dispensary two additional Muslim hakims at Rs 50/month and two Hindu hakims at Rs 20/month were hired for the purpose of vaccine inoculation. Since they were not sufficiently qualified to perform such medical duties, they worked and trained with four native doctors from the Calcutta institution who were specially recruited for the job. As they trained in surgery and inoculation techniques, their salaries, too, were accordingly scaled. One of the Muslim hakim received Rs 25/month, which was to be increased to Rs 30 after five years; and the Hindu hakim received Rs 20 per month, to be increased to Rs 25 after the same period.[70] And of course by the mid nineteenth century

[69] L no. 73, W. I. Conolly, commissioner Rohilcund to Sudder Board of Revenue, NWP, Agra, 7 February 1844, Proceedings, April 1844, NWP, Lt Governor's Proceedings, General, P/214/63. Conolly reported that because of the unavailability of the sub-assistant surgeon at the Shahjahanpur dispensary, it was being taken care of by a native hakim who worked under the supervision of the civil surgeon.

[70] L no. 88, Note on the dispensary at Moradabad, 30 December 1837, sent

the dispensaries of the Punjab region enlisted hakims in large numbers in their rural health programmes, where they were trained as vaccinators and medical assistants, and familiarized with Western medical practices. In 1867 the Punjab government sanctioned a programme proposed by Lt. Col. T.W. Mercer to establish a district-wide scheme for medical relief that depended on the labour of hakims.[71]

Hakims were not the only target group. Dispensaries opened their doors to anyone who wished to train in British medicine. But until 1851 no specific rules existed regarding recruitment, qualifications for further professional training, and administrative jurisdiction. In the absence of rules, confusion prevailed in every dispensary about levels of training and the eventual employment of trained apprentices. In the early nineteenth century no uniformity existed in the dispensaries regarding apprentice training. Many dispensaries did not entertain apprentices at all. Others trained them only as compounders and dressers to fill up such vacancies in their own dispensary, while still others were provided instructions in surgery, inoculation, materia medica, anatomy, and pharmacology.

Not surprisingly, all British doctors did not always approve the idea that the dispensaries be opened up as training centres to the local populace. In 1845 John Balfour, the civil assistant surgeon at Bareilly, cautioned against opening up the dispensary in his area for apprentice trainees. He was of the view that the level of training offered would only provide some basic expertise on medicines to such apprentices. They could practice as native doctors only if they worked under the guidance of qualified doctors educated at the Calcutta Medical College. He also felt that this mass-scale production of semi-qualified native doctors would also generate unemployment problems in the district where no lucrative jobs existed even for the hakims and *jarrahs* (barbers) that lived there. Balfour was against making the dispensary a general

by Jacob L. Hoff, Resident Judicial Dept, NWP, 30 December 1837, Proceedings, April 1838, NWP Gov. General's Proceedings, General, January–June 1838, P/214/34. The reduction affected by scaling down the salaries of hakims was used to send a native doctor to each of the larger cities of Almora, Sambal, and Chandausi. This plan was approved by the government and directed to be carried into effect as an experiment and a report was required after six months.

[71] Hume 1977: 214–31.

factory of British-trained native doctors. He was not however averse to the idea of increasing the number of apprentices marginally from two to four.[72]

His reservation notwithstanding, the dispensary continued to play a role as a centre for medical instruction. In 1851 the publication of the Bengal Medical Regulations systematized the dispensary as a training institute. It laid out that 'two or more respectable Hindoo or Mussulman boys who can read and write the Nagri or Persian character will be attached as apprentices to each dispensary, who will receive from the sub-assistant surgeon such instructions in the subordinate branches and professional education, as may qualify them to serve elsewhere as compounder or third class native doctor.'[73] It also elaborated the role of apprentices as assisting in 'compounding medicines and in carrying on the duties of the dispensary'. Their salaries were fixed at Rs 3/month for the first year, Rs 4/month for the second year, and Rs 5/month for the third year. It was also noted that if they were found to be incapable or unwilling they were to be discharged by sanction of the medical committee. Fresh recruits would be appointed in their place.[74]

The regulation of 1851 systematized the training of recruits as apprentices in dispensaries. It fixed the entry age at around seventeen years. It made knowledge of reading and writing the vernacular language a prerequisite for entrance. Medical officers superintending the dispensary tested their fluency in the vernacular at entry point. The testing committee also included a sub-assistant surgeon. The same committee and officers tested the apprentices every year and decided their eligibility for promotion to a higher salary. Apprentices after the end of two to four years of training passed a credible exam before the divisional medical committee and became class 3 Native Doctors. Three medical officers at the dispensary itself conducted their exam at the end of their

[72] L no. 17, Enclosure letter from John Balfour, civil assistant surgeon, Bareilly, 24 May 1845, para IX, Proceedings, May 1845, NWP, Lt Governor's Proceedings, General, April–July 1845, NWP, n.d., May 1845, P/214/68.

[73] Extract from Bengal medical regulation, chapter XL, para 29, last edition 1851. Enclosed in L no. 85 being circular addressed to the committees superintending government dispensaries, signed by J. Thornton, Sec to Government, NWP, Lt Governor's camp, 20 December 1851, Proceedings NWP, December 1851, Lt Governor's Proceedings, General, October–December 1851, P/215/13.

[74] Ibid.

term. In case three such medical officers were not available at their dispensary, they moved to the nearest station for the exam. Qualified apprentices were issued a certificate.[75] After they qualified it was optional to enlist as native doctors with the army.[76] Depending on their choice, they could get postings either with the army, or at government hospitals and branch dispensaries. They were often also posted with gangs of prisoners.[77] Such native doctors also had the option of joining the medical college at Calcutta for further training.[78] There was also the possibility of the more promising recruits being sent from mofussil dispensaries to the Thomason Hospital at Agra to receive a more intense course of instruction. At the Agra Hospital they were expected to be exposed to a wider range of both indigenous and European pharmacy and pharmacological procedures, and take advantage of a better library and clinical observation. Besides, of course they attended advanced lectures on anatomy, physiology, chemistry, materia medica, pharmacy, surgery, etc.[79]

Hakims and fresh recruits at the dispensary read a range of medical texts.[80] They read important Urdu medical tracts produced by the

[75] Ibid., L no. 84, Medical Board to J. Thomason, Lt Governor NWP, 30 October 1851.

[76] Ibid., L no. 85, J. Thornton, Sec. to Govt, to Sec. of Committees, Supdt, Govt dispensaries, NWP, 20 December 1851.

[77] Ibid., L no. 84, Medical Board to J. Thomason, Lt Governor NWP, 30 October 1851.

[78] L no. 338, James Irving, Sec. to dispensary committee, Allahabad, to William Muir, Sec. to Govt, NWP, Allahabad, 23 March 1855, Proceedings, April 1855, NWP, Lt Governor's Proceedings, General, 20 April–31 May 1855, P/215/43.

[79] L. no. 168, John Murray, Civil Surgeon and Sec. to Thomasan Hospital Agra to W. Muir, Sec. to Govt, NWP, Agra, 1 September 1854, NWP, Lt Governor's Proceedings, General, November–December 1854, P/215/38. The Thomason hospital was envisaged as a nucleus around which a medical school and eventually a college would emerge. The idea was that people from the NWP were reluctant to go to Calcutta because of differences in climate, distance, and culture for medical training. Since a great demand existed for native doctors, the Agra hospital and school were expected to train people locally and thus to some extent meet the demand for native doctors.

[80] L no. 338, James Irving, Sec.to Dispensary Committee, Allahabad, to William Muir, Sec. to Govt, NWP, Allahabad, 23 March 1855, Proceedings,

government lithographic press and authored by British medics. Assistant surgeon Breton's Hindustani translation of the *London Pharmacopoeia*, translated once again in 1841 into Hindustani by George Spilsbury and into the Persian script (1845) by F.J. Monat, was an important text used in dispensary classrooms.[81] Breton's dictionary on human anatomy was also widely used. The latter was presented to the king of Awadh in 1828, who had it printed and distributed amongst the dispensaries in his region.[82] Their reading list also included a range of materia medicas in Hindustani produced by dispensary superintendents; these borrowed from the British pharmacopoeia and attempted to familiarize native doctors at dispensaries with a variety of new drugs. In 1815 the government approved for purchase 500 copies of one such materia medica authored by A.F. Ramsay, assistant surgeon at Barrackpore, so that it could be used by native doctors posted all over the country.[83] Other superintendents wrote the texts in consultation with local pharmacopoeias. These were also used as learning texts. The superintendent of Allahabad dispensary, James Irving, wrote a materia medica precisely for dispensary apprentices. He included in it the knowledge he gained of local drugs and prescriptions while working at the dispensary and interacting with colleagues at other dispensaries.[84] Many tracts on anatomy and surgery were specially prepared for dispensary apprentices by the NMI.[85] Indeed, so great was the demand for medical literature at the dispensary that the medical board always felt the shortage of

---

April 1855, NWP, Lt Governor's Proceedings, General, 20 April–31 May 1855, P/215/43.

[81] Spilsbury & Dutt 1845: preface.

[82] L no. 53, P Breton, Supdt, NMI, to Persian Secretary, n.d., Bengal Pol. Consult, 18 July 1828, P/125/54; see also Breton 1825.

[83] L no. 387, W. Casement, Sec. to Govt in Mil. Dept, to Sec. to Medical Board, 23 June 1821, F/4/655, file no. 18063, F/4/655, file no. 18063, pp. 14–15.

[84] L no. 446, Supdt, Gov. dispensary Allahabad to Sec. to Govt, NWP, Allahabad, 13 February 1856, Proceedings, NWP, February 1856, NWP, Lt Governor's Proceedings, General, 25 February–31 March 1856, P/215/53. See also follow-up letters, nos. 447–50 in the same proceedings.

[85] L no. 84, Medical Board to J. Thomason, Lt Governor NWP, 30 October 1856, Proceedings NWP, December 1851, P/215/13. Native Medical Institution prepared many such texts for the dispensary apprentices. For knowledge of

adequate medical manuals in the 'native language' and of an 'elementary nature' that could be used as reading material by apprentices at the dispensary. In 1854 it recommended that two medical officers be commissioned to prepare such a manual. The government agreed to finance and support such an exercise.[86] The problem of shortage of reading material was believed by some officers to be solved if government colleges were involved in apprentice training programmes. Thornton requested a greater interaction between government colleges and dispensaries in areas like Benares, Agra, Delhi, Bareilly and Ajmer, where both institutions existed. In the event of shortage of learning texts, he suggested the use of college libraries by apprentices where copies of the desired texts were to be lodged.[87]

In fact, Urdu texts were in demand even outside the dispensaries. Many collectors and magistrates felt the need for instructing even hakims who had not enlisted in the dispensary system. They requested the government to dispatch to local libraries 'cheap standard work in the vernacular, simple enough to be read—and yet copious enough to afford all needful education to the native hakims and others—.'[88]

The medical board envisaged the dispensary as a dialogical site for hakims and British doctors. It was keen that training classes for apprentices should also on occasion be used as fora where recruits could teach British surgeons the vernacular languages. The board regretted that very few British doctors knew the local languages as it was an impediment in the spread of their science. It urged them to use dispensary postings to learn vernaculars from the hakims and native

---

dissection, etc., Mr Monat's text was used. And since it was in the Persian script, there was always a demand for a translator to convert it into Devnagari.

[86] L no. 44, E. Spilsbury, Physician Gen., Charles Benny, Surgeon Gen., I.B. Clapperton, Inspector Gen., to J.R. Colom, Lt. Gov. NWP, 10 March 1854; L no. 45, William Muir, Sec. to Govt, NWP, to Medical Board, FW, 5 April 1854, Proceedings NWP, April 1854, Lt Governor's Proceedings, General, April 1854, P/215/32.

[87] L no. 86, J. Thornton, Sec. to Govt, NWP, to medical board, 20 December 1851, Proceedings, December 1851, Lt Governor's Proceedings, General, October–December 1851, P/215/13.

[88] L no. 60, W.C. Tucker, Collector and Magistrate, Azamgarh, to J. Thomason, Sec. to Gov. Gen. in General. Dept, 30 April 1838, Proceedings, May 1838, NWP, Gov. General's Proceedings, General, January–June 1838, P/214/34.

doctor.[89] The board also encouraged doctors to write and print medical vocabulary books in Hindustani and English for the use of medical staff posted in the dispensaries. In 1822 it ordered 500 copies of Dr Swiney's medical vocabulary printed and priced at Rs 5/copy so that the assistant surgeons at the dispensary would be able to 'hold communication with patients in form and symptom of the disease and issue orders to his native doctors in clear and intelligible terms.'[90]

However, the response to apprentice training was never to the satisfaction of the government. Members of the medical board complained that dispensaries were training far less recruits than required. After the initial years of enthusiasm, in 1851, of the 48 dispensaries and relief hospitals distributed over the Lower and Upper divisions of the Bengal Presidency, there was a total of 46 apprentices on record. Out of these, merely 8–10 had over the previous five years passed credible exams and rendered satisfactory service as class 3 native doctors to the government. The rest had not been able to clear the exam. The distribution of apprentices in the North West Provinces, exclusive of the dispensaries in the Sagaur and Nerbada territories and those supported by the native government, were as follows:

Ghazipur—2	Agra—]2	Almora—1
Mathura—2	Srinagar—unknown	
Mirzapur—2	Ajmer—1	Shahjahanpur—2
Benares—none	Naina Nagar—1	Saharanpur—2
Allahabad—3	Delhi—3	Badaun—1
Farrukhabad—2	Bareilly—2	
Kanpur—2	Moradabad—2	Total—30[91]

Part of the reason for the poor response was the fact that recruits were not always keen to move around as part of their training, exam,

[89] L no. 84, Medical Board, to J. Thomason, Lt Governor, NWP, 30 October 1851, Proceeding, December 1851, NWP, Lt Governor's Proceedings, General, October–December 1851, P/215/13.

[90] Bengal Mil. Collection, No. 29. Subscription to Dr Swineys medical vocabulary, Mil. Sec. office 1823, Boards Collection , F/4/737, file no. 20142.

[91] L no. 84, Medical Board to J. Thomason, Lt Governor NWP, 30 October 1851, NWP, Lt Governor's Proceedings, General, October–December 1851, P/215/13.

and then posting. In 1855, when the government began to send promising apprentices from the dispensaries to the Thomason Hospital at Agra for further training and proper education, they met with some opposition and reluctance to travel. Secretaries of dispensary committees at Badaun, Gorakhpur, Allahabad, and other towns reported the unwillingness of their trainees to proceed to Agra.[92] Most of the time the reasons were financial. Most recruits could not afford travel expenses.[93] And dispensary committees were reluctant to foot the bill. In fact, while such dispensary committees took responsibility to arrange their travel to Agra, they expressed inability to continue to pay even their salaries while they were away. Both I.J. Thornton, secretary to the dispensary committee at Mathura, and James Irving, secretary to the dispensary committee at Allahabad, expressed their inability to pay recruits while they trained in Agra.[94]

However, the tardy response to dispensary instruction notwithstanding, by 1855 the idea of a diversified British medical education available easily at the district level had caught on in northern India. Regional elites, who had already pitched in with financial help to set

[92] See L nos 337, Babu Nund Kumar Mitter, Sec. of Dispensary Committee, Budaun, to W. Muir, Sec. to Govt, NWP, 23 March 1855; also L no. 336, T. Atchison, Sec. to Dispensary Committee, Gorakhpur, to William Muir, Sec. to Govt, NWP, 22 May 1855, Proceedings, April 1855, NWP, Lt Governor's Proceedings, General, 20 April–31 May 1855, P/215/43.

[93] L no. 122, G.R. Pemberton, Civil Assistant Surgeon and Secretary to Dispensary Committee to W. Muir, Sec. to Govt, NWP, Mirzapur, 26 March 1855, Proceedings April 1855, NWP, Lt Governor's Proceedings, General, 20 March–20 April 1855, P/215/42. Pemberton wrote that the dresser at his dispensary, who was very efficient, could be sent to Agra for further training, but was reluctant because of want of financial means to travel . Since the dispensary was unable to foot his travel expenses, Pemberton wanted advice from Mr Muir on how to proceed. See also ibid., L no. 124, Sec. Dispensary Committee Aligarh, to W. Muir, Sec. to Govt, NWP, Agra, 27 March 1855. S. Clarke, Secretary of the Committee, expressed his inability to send his best apprentice Roop Singh to Agra because he was 21 years old and thus over age. He recommended him for favourable consideration, and financial help.

[94] L no. 338, James Irving, Sec. to Dispensary Committee, Allahabad, to W. Muir, Sec. to Govt, NWP, Allahabad, 23 March 1855; L no. 339, J. Thornton, Sec. to Dispensary Committee, Mathura, to W. Muir, Sec. to Govt, NWP, 29 March 1855, Proceedings, April 1855, NWP, Lt Governor's Proceedings, General, 20 April–31 May 1855, P/215/43.

up dispensaries, were encouraged by the administration to pool in money for medical instruction schools as well that would be attached to the dispensary. The idea was to make British medicine training easily accessible to a wider range of people and pull it out of the confines of Calcutta and the Bengali population.[95] In 1854 the Rani of 'Vizeanagram' gave her dispensary, hospital, and bungalow at Bhelupur with a monthly allowance of Rs 130 for the use and support of a medical school in connection with the dispensary and hospital. The allowance was to be used for the salary of the assistants, pupils, food, and clothing of poorhouse patients. There was a bungalow also in the enclave for the doctor's residence. An additional Rs 830 was raised by subscription from both European and Indian gentlemen for the purchase of books, chemicals, and furniture.[96] Babu Neel Madhub, the best pupil according to some at the medical college in Calcutta, was posted at Benares as sub-assistant surgeon and placed under the superintendence of Dr Leckie.[97] The government agreed to take over the charge of this dispensary and hospital school as long as the building and the Rs 150 amount continued to be made by the rani. It recommended the establishment of a committee for the control of the dispensary and the establishment. The indents of the civil surgeon were to be complied with in regard to medicines and hospital necessaries. It also underlined that the instructions given to apprentices at the dispensary would be of a practical kind so as to enable them to become subordinate servants in it or in similar institutions.[98]

Contesting Western Medicine in the Dispensary: The Native Doctor and Unani Learning

The clinical testing of Unani drugs and materia medica, pharmacopoeia, and anatomy texts gave them a new scientificity, but also distanced

[95] L no. 364, H.C. Tucker, Commissioner 5th Division, to William Muir, Sec. to Govt, NWP, Agra, Benares, 4 May 1854, Proceedings, May 1854, NWP, Lt Governor's Proceedings, General, 2–28 May 1854, P/215/33.

[96] Ibid., Subscribers were as follows: Babu Rajinder Mitr—Rs 400; Row Deo Narain Singh—Rs 200; Baboo Sobnauth Mittr of Rajshy—Rs 100; Dr Leckie—Rs 30; Babu Raee Narain Dass—Rs 50; Mr H.C. Tucker—Rs 50.

[97] Ibid.

[98] Ibid., L no. 367, W. Muir, Sec. to Govt, NWP to H.C. Tucker, commissioner 5th Division, 26 May 1854.

such texts further from their original contexts. It also moved their knowledge away from the community to the laboratory, making it even more scientistic and distant from healing. However, even though medical knowledge was sanitized of its local flavour, the dispensary appointments and training were structured on the organizational format of Unani as perceived by British doctors. They saw class, religion, and family as the defining features of Unani medicine and the cause of its success. They tried to tailor the dispensary to these referents so as to compete with Unani's popularity. And thus the British administration of the dispensary began to change the social contours of Unani, stamping it with a fresh religious and class profile. This religious profiling of Unani attracted the attention of the Company's Muslim native doctors. They were quick to use it to resist colonial medical authority.

We saw in Chapter 1 that Unani acquired its first layering of a communitarian Muslim profile when it shifted from Persian to Arabic in the late eighteenth century. The British dispensary only underlined further its Muslim profile. By the 1830s most British dispensaries that enlisted hakims began to list them in Muslim and Hindu categories. The Moradabad dispensary, which had patients from all religions even though its funding came from the Hindu shrines at Kasheepur, was administered with a clear division of labour between a Hindu hakim and two Muslim hakims. British religious profiling was most evident when they specifically identified in their records Hakim Dataram, the chief financial collector of funds for the dispensary and a popular Unani physician of Moradabad, as a Hindu hakim.[99]

The religious profiling of Unani was most evident in the British administration of the Lucknow dispensary that had its origins in the 1834 Shahi Shifakhaana (Royal Hospital or Dispensary) of Nawab Nasir-ud din Hyder. This had both a British as well as a Unani section.[100] Indeed, the British obsession with the idea that Unani was an elitist

[99] L no. 88, Note on the dispensary at Moradabad, Jacob L Hoff, Resident Judicial Dept, NWP, Agra, 30 December 1837, Proceedings, April 1838, NWP, Governor General's Proceedings General, January–June 1838. P/214/34.

[100] For the government instruction to convert the Shahi Shifakhaana (Royal Hospital and Dispensary) to the Lucknow dispensary on 1 August 1856, see L no. 404, G. Couper, Sec. to Chief Commissioner of Oudh, Lucknow, to G.F. Edmonstone, Sec. to Govt of India, Calcutta, 1 August 1856, Consult, FW,

Muslim medical system that was in harmony with the Muslim elites of Lucknow, led them to tailor their entire administration of the dispensary to cater to Muslim sensitivities. This was seen as the best way to compete with Unani's popularity. The case of the Lucknow dispensary is interesting also because it became the site where the now religiously stamped Unani was used by British-trained Indian doctors to contest colonial authority. As they foregrounded Unani in their resistance to colonial intrusion they invoked also the authority of the Prophet as a healer. The Prophet's authority was not seen separate from the locale that Unani emphasized as critical to healing.

The Shahi Shifakhaana

According to a deed drawn up between the Oudh nawab and the East India Company in 1836 and executed by Nawab Muhammad Ali Shah in 1840, the nawab invested the sum of Rs 3,40,800 in 4 per cent government securities as endowment for the Royal Hospital and Dispensary.[101] The interest of the sum was fixed by the nawab's successor, Muhammad Ali Shah, for its expenses in perpetuity. As per the deed, the superintendent of the hospital was to be appointed by the Oudh government and the interest amounting to Rs 1208 per month was paid to him. The hospital had a section on the practice and teaching of Western medicine and an identical section for Unani medicine. Rs 450 of the monthly interest was assigned for the maintenance of the European branch, and the balance Rs 758 supported the Unani section. Dr Stevenson was to head the Western medicine part and one Zafar-ud-daulah was made in-charge of the Unani section. The British appointed a sub-assistant surgeon to manage the European side of the hospital's medical care. The Resident extended his aid and assistance in overseeing the management of the hospital.[102]

---

26 September 1856, India Political & Foreign Consult, 26 September–3 October 1856, P/201/68.

[101] L no. 98, Col. W.H. Sleeman, Resident at Lucknow, to I.P. Grant, Sec. to Govt of India, Foreign Dept, FW, 26 April 1854, No Consult, cited, India Political & Foreign Consult, 5–12 May 1854, P/200/60.

[102] L no. 257, J. Couper, Sec. to Chief Commissioner of Awadh, to G.F. Edmonstone, Sec. to Govt of India, Foreign Dept, Calcutta, n.d.; L no. 260,

From the outset, the presence of the Unani section in the Royal Hospital and Dispensary provoked the British to colour the institution with a Muslim hue. They projected the hospital as being located in the 'Mahomedan metropolis' financed by an 'independent Musalman state', and catering to a large population of the 'Musalman faith'.[103] This religious profiling of a city, its patients, and doctors was new to Lucknow. The city had been known for its religious syncretism. Moreover, the list of patients attending the hospital was always an even mix of Hindus and Muslims.[104] The British administration of the Royal Hospital and Dispensary not only gave the city a new cultural stance, it stamped its Unani physicians and their patients with a religious denomination.

In the British mindset so complete was the identification of the Royal Hospital and Dispensary with Muslims that they competed with its Unani component using the Muslim card. Their choice of a sub-assistant surgeon to be posted at Lucknow was guided by considerations of his religious and family background. In November 1840 the government at Calcutta was very clear that if the king of Oudh agreed to the post of a sub-assistant surgeon at the Lucknow hospital and dispensary then only a 'Musalman shall be selected'.[105] The Lucknow Resident was even willing to wait until November of the following year when a 'perfectly qualified' person of a 'Mahomedan persuasion' was likely to be available.[106] Finally, Inayat Husain, at that time the only

---

Enclosure copy of a deed dated 1840 and a letter form the King of Oudh, to Col. Caulfield, acting Resident Lucknow, 26 January 1840, Consult, FW, 27 June 1856, India Political and Foreign Consultation, 27 June–4 July 1856, P/201/60. The deed was never ratified, even though it was executed and followed. After the British annexation of Awadh, the superintendence of the Royal Hospital fell into the hands of people appointed by the British government.

[103] L no. 198, the medical board to Lt Gen. Sir Henry Harding, Gov. Gen. of India, 3 April 1845, Consult, FW 25 April 1845, India Political and Foreign Consultation, 25 April 1845, P/197/12.

[104] L no. 146, list of fatal cases in His Majesty the King of Oude's hospital during the 6 months ending 31 March 1847, Consult, 23 October 1847, India Political and Foreign Consultation, 20–30 October 1847, P/198/2.

[105] L no. 99, Correspondence with Government on the subject of the hospital at Lucknow, Governnment, to Col. Low, British Resident at Lucknow, 23 November 1840, India Political and Foreign Consultation, 5–12 May 1854, P/200/60.

[106] Ibid., Col. Law, Resident at Lucknow, to Government, 28 June 1841.

Muslim graduate of the Calcutta Medical College, was posted at the Royal Hospital and Dispensary precisely because he was a Muslim.[107] Through his respectable family background and his Muslim religion he was expected to strike a bond with the people of rank in the city and thus be a conduit for creating in them a 'taste for European science and medical skill'.[108] And Inayat Husain's appointment just marked the beginning of what soon became an institutionalized practice. Young Muslim boys of 'good family' were instructed in English at the Lucknow Boys School and then encouraged to train with Inayat Husain at the Lucknow hospital. In 1846, one Mirza Ahmad Ali had such a career and I.S. Login, the residency surgeon at Lucknow, was hopeful that in time 'he will do credit to his teacher'.[109] And it is not surprising that when in 1851 Inayat Husain resigned,[110] there was a frantic search for a Muslim sub-assistant surgeon to substitute him at Lucknow. After a wait of several months since no Muslim graduate was ready at the Calcutta Medical College, they finally got one Ameenuddin in 1852 as a suitable substitute.[111]

This religious profiling did not stop with the Muslim stamp. If Lucknow's hakims and doctors were profiled as Muslim, Garhwal and its medical staff were given a Hindu hue. During the British administration of the 1852 plague epidemic at Garhwal, the medical board was

[107] L no. 198, from the Medical Board to the Lt Gen. Sir Henry Harding, Gov. Gen. of India in Council, 3 April 1845, Consult, FW, 25 April 1845, India Political and Foreign Consultation, 25 April 1845, P/197/12.

[108] L no. 81, Major General G. Pollock, Officiating Envoy to King of Oudh to F. Currie, Sec. to Govt of India, 17 June 1844, Consult, FW, 29 June 1844, India Political Consultation, 29 June–13 July 1844, P/196/55. Very often he was accused of having failed in this role.

[109] L no. 2231, I.S. Login, Residency Surgeon, to R. Davidson, Sec. to Govt of India, 3 August 1840, Consult, FW, 17 October1846, India Political and Foreign Consultation, 17–24 October 1846, P/197/46.

[110] L no. 183, A. Bell, Residency Surgeon and Supdt, of Oudh Hospital, to Col. W.H. Sleeman, Resident at Lucknow, 21 December 1851, Consult, 23 January 1852, India Political and Foreign Consultation, 23–30 January 1852, P/199/57.

[111] L no. 90, Lt. Col. W.H. Sleeman, Resident Lucknow, to C. Allen, Officiating Sec. to Govt of India, Lucknow, 3 July 1852, Consult, 20 August 1852, India Political and Foreign Consultations, 13–20 August 1852, P/200/1.

very clear that the religious sensitivities of the people there had to be taken care of if British medical drives were to make a success. The 'low caste' medical establishment had to be embellished also with two Hindu native doctors from the plains.[112] They recommended that 'Hindu sub assistant surgeons' be attached to European medical officers. And 'Brahmins and Hindus of good caste' were preferred to 'endeavour to persuade the diseased to submit to proper medical treatment and exhibit the remedies where they consent to take them'.[113] In the city hospital, dispensary, and 'insane hospital' at Secroh it was believed that the presence of many Chamar (low caste) individuals on duty was always resented. An additional qualification was of course high class, 'respectable birth', good character and the knowledge of English.[114]

Alongside caste, the class and 'respectability' of Indians employed in the Lucknow and other dispensaries were matters of concern for the British. This derived largely from a class profiling of Unani, which was seen to be successful because it based itself on these referents. They identified Unani as catering to the elite class and contrasted it with the relatively class-neutral British medicine that opened its doors to the poor.[115] Inayat Husain was convinced that, in the ultimate analysis, Unani physicians and their medical system were more intelligible to the elites because they could strike bonds with them through class affiliations that centred on proper conduct and deportment.[116] British medicine, he felt, could compete with Unani only if it took the issues

[112] No letter No., Medical Board to J. Thomason, Lt. Gov. NWP, 22 January 1852, Proceedings, NWP, February 1852, NWP, Lt Governors Proceedings, General, January–February 1852, P/215/15.

[113] No letter No. Medical Board to J. Thomason, Lt. Gov. NWP, FW 28 June 1852, NWP, Lt Governors Proceedings, General, July–August 1852, Proceedings NWP July 1852, P/215/18.

[114] L no. 69, D. Butter, Civil Surgeon Benares, to I Halliday, Supdt, Surgeon Benares, n.d., NWP, Lt Governors Proceedings , General, April–June 1842, P/214/53.

[115] L no. 127, Inayat Husain, Sub-Assistant Surgeon, to I.S. Login, Residency Surgeon, in L no. 128, dated 1845, Medical Board to T. Currie, Sec. to Govt of India, India Political and Foreign Consultations, 28 March 1845, P/197/10.

[116] L No. 127, Inayat Husain, Sub-Assistant Surgeon, to I.S. Login, Residency Surgeon, n.d., sent to Medical Board 28 March 1845, Consult, 28 March 1845, India Pol. & Foreign Consult, 28 March 1845, P/197/10.

of class, respectability of family background, and the deportment of its apprentices with some seriousness.

As British medicine focused on the class, family, and individual comportment of its native doctors, the spotlight was once again on Inayat Husain. His elite family background and urbane appearance became matters of immense significance for the British. Special financial benefits were offered to him to make himself 'presentable'. Login vehemently resisted any efforts to cut down the salary of Inayat Husain as this would bring down his standard of living. Indeed he defended also the additional allowance of Rs 50 given to him with the intention of 'affording him more easily the means of keeping up a presentable appearance among the natives at Lucknow, among whom it was expected that he would be able to introduce a taste for European science and medical skill.'[117]

Indeed the British attempt to compete with Unani on its terms backfired when Inayat Husain used its healing techniques to challenge Western medical practice. The highly 'presentable' Inayat Husain struck close intellectual bonds with learned hakims and the literati of Lucknow society. Their influence was palpable in his writings, which subtly challenged colonial authority. He wrote many tracts on diseases and sent them to the medical board for publication and circulation amongst 'interested' Indians and the British. His texts were heavily influenced by Unani medical thinking and reflected his discussions with Unani practitioners. At Lucknow he was in interaction with two or three 'intelligent men' who were good 'Arabic scholars' and who had commenced the study of European-style pathology.[118] Even though Husain was critical of the Unani quacks and precocious patients who claimed they knew the medical system, he did not disagree with Unani's emphasis on the Galenic theory of humours and the role of external medicine to restore their balance. He made a case for the commonalities between Unani and British medicine and argued that Unani drugs differed

[117] L no. 82, I.S. Login, Residency Surgeon, to Major Gen. G. Pollock, official envoy to Oudh, 3 June 1844, India Political Consultation, 29 June–13 July 1844, P/196/55.

[118] L no. 127, Inayat Husain, Sub-Assistant Surgeon, to I.S. Login, Residency Surgeon, report of Husain sent to Medical Board on 28 March 1845, India Political and Foreign Consultation, 28 March 1845, P/197/10.

from European ones only in name. He straddled the two medical worlds with caution and suggested the use of medication to restore humoral balance. However, he was aware of other physiological causes of disease and was convinced that 'they [humours and medicines for them] were not the only agents that produce such innumerable phenomenon in the human system'. Husain's careful balancing derived from the practical reality in which he functioned. He was aware that people were convinced of the benefits and logic of their own medical practices, the ones they comprehended: 'we cannot expect to alter their minds very soon as the difficulty of eradicating ideas early impressed is well known'.[119]

In this he was not alone. In 1875 Edward Balfour, former civil surgeon at Delhi and surgeon general at the Madras medical department, regretted that, 'Indians trained in British Medical schools are also enamoured by ancient medieval theories and talk of demerits of European medicine'.[120] He lamented the fact that even with the spread of Western medical schools the 'correct understanding of human physiology, pathology of diseases, etc.' did not spread.[121]

Balfour's misgivings about British trained Indian doctors were not entirely misleading. Indeed the writings of Inayat Husain showed how the British pandering to family, religion, and class only further entrenched British-trained doctors in the Indian healing traditions. This entangled interest could backfire to the disadvantage of the state. For instance, the *Tracts* that Husain wrote suggest that through his essays on diseases like the chronic lesions of the spleen and the liver he was in fact challenging the established medical principles of Western medicine and arguing closely in line with Unani.

Husain argued in favour of locating medicine in the community—the locale—and critiqued the Western emphasis on rooting it in the clinic-laboratory. He pointed to the element of 'uncertainty' and 'error' in clinically verifiable Western medicine. This grey area, he said, was detrimental to the image of doctors. It opened the space for popular speculations that bordered on the theory of the divine causation of disease. He said these encouraged superstition and the undervaluing

[119] Ibid.

[120] Balfour 1875: 16–17.

[121] Ibid.

of true sciences.[122] Very much like Unani, he made a case for making natural philosophy and natural laws that govern the human body central to understanding disease. He debunked the British idea of applying 'external means' (drugs, surgery) to target the diseased part of the body. Instead, like the hakims, he suggested restoring the natural balance of the body through control of the elements (air, water, earth, and fire). Making a strong argument for the more 'rational and philosophical' mode of medical practice, he said, 'much due credit is gained by an observing and trying habit of the mind, in examining the various functions of the body naturally and pathologically and explicating the too intricate diverse and deceptive connections and distinction of morbid cause and effect before proceeding with the more important and ultimate grand object of obviating them by external means'.[123] Again, in line with Unani reasoning, he argued that Western doctors were often carried away by the urgency of symptoms and did not focus on the original morbid source that caused it. This often proved fatal to the patient and deceived the physician. He underlined that 'a permanent establishment of health and the ensurance of life depend upon the restoration of the primary source of things by removing the first morbid cause.'[124]

Again, Husain's *Tracts* fitted closely with the emerging British idea that health was related to the general well being of society. For him, too, a healthy body was conditional on its locale. And that included not just climate and environment but also the political and social context of being that set certain aspiration and target levels for individuals. However, he used Unani's humoral theory to explain the connection between a healthy body and its environment. Husain explained the physiology of bile as being one of the key humours which could provide dividends only in conjunction with locale 'clothing, diet, national and ceremonial rites'.[125] These contributed to the body balance that defined a healthy

[122] L no. 40. Remarks on some chronic lesions of the liver and spleen by Inayat Husain, Lucknow, 5 January 1840, Consult, FW 22 April 1848, India Political and Foreign Consultation 14–28 April 1848, P/198/24.

[123] Ibid.

[124] Ibid.

[125] Ibid. But the centrality of dress, deportment, and locale in disease meant also that the same disease took different forms in the country. This was only logical, given the cultural plurality of India. Discussing diseases of the liver, he argued that they had different peculiarities in the different parts of India. And this was

person. With such views, Husain was very much in the framework of working closely with Unani.

But this defeated the very purpose of his position and the role of his institution, which was meant to steer British medicine on to an independent trajectory. It was no surprise that the Company did not take too kindly to Husain's critique of Western medicine and leanings towards Unani: it did not encourage his intrusion here. It was of the view that Husain's primary objective was to report on impediments to the spread of European medicine, not to indulge in medical research and causation in consultation with Indians. The medical board returned his report on the diseases of the liver and did not use it in their publications. Login was of the view that Husain was extending his brief in writing such reports. The Company did not want medical research reports from him. It was more keen that he use his 'general intelligence and abilities' to communicate the practical difficulties which occur in adapting European practice to the 'habits and prejudices' of the people.[126] In addition, he was expected to use his contacts and familiarity in the city to supply 'very useful and interesting' information on the state of native medical practice.[127]

Constrained by the demands of his official position, Husain complied with the request. In 1848 he withdrew his earlier report and sent instead some 'cursory observations on the nature of the most prevalent cutaneous diseases, the ideas of natives regarding them and their treatment'.[128] In this five-page report Husain confined himself to his principal objective of being a conduit to the 'useful medical knowledge' of Indians and thereby helping the progress and advancement of Western medicine. However, he could never entirely give up his intellectual

---

because of climatic and dietary difference. In Bengal they were inflammatory in nature with the tendency to get purulent; in Central India by which he meant Awadh, they were not inflammatory or purulent.

[126] L no. 29, I.S. Login, late Supdt, of Oudh hospital and dispensary, to Lt. Col. A.F. Richmond, Resident Lucknow, 28 August 1848, Consult, FW, 4 November 1848, India Political and Foreign Consultation, 28 October–25 November 1848, P/198/36.

[127] Ibid.

[128] L no. 30, Inayat Husain, Sub-Assistant Surgeon, to I.S. Login, Residency Surgeon, 16 August 1848, Consult, FW, 4 November 1848, India Political and Foreign Consultation, 28 October–25 November 1848, P/198/36.

stance in favour of the Unani system. He built a sociology of disease in India that combined Unani's slant on locale with the early-nineteenth-century epidemics-driven concern about the social and cultural causes of illness. But Husain's sociology of disease did not stop at locale, diet, dress, and deportment. It was heavily informed also by the religion, customs, and class profiles of the sick. In effect, he deflected the focus from the physiological causes of disease and reiterated the idea that health was about social well being, and that concern for 'public health' was the answer to many problems.

He listed 421 cutaneous diseases prevalent in the natives of India. There were some like scabies and itches that cut across religion, caste, class, age, and gender. But there were others, like dry ringworm, that were more common in the Hindu population. This disease was painful and itchy and discharged bloody matter on the skin. Husain linked its prevalence in the Hindus to their religious purity- and pollution-related customs. In compliance with their religious rituals they observed daily cleanliness, and baths during which they kept their clothes on. He explained that wet and dirty clothes, when left on the body to dry, created skin infections. The disease appeared in areas where the cloth had clung the most: waist, thighs, and around the groins.[129] Again, he indicated the difference between Hindus and Muslims in their reaction to disease. For instance, Hindus and Muslims reacted differently to the occurrence of fevers. While Hindus starved and left it to nature to cure them, Muslims were prepared to go for medication. They took some mixtures of their own for the first five or seven days and the public at large was prejudiced towards European-style purging in fevers.[130] In another report, on the diseases in the Lucknow area, he linked certain diseases like leprosy to low-class people. He reported that in Lucknow leprosy was rare in the higher class of society.[131]

With his idea of health as part of well being, he tuned into the referents of medical authority embedded in the cultural world of patients. He even challenged the authority of the Company when he highlighted

[129] Ibid.

[130] Ibid.

[131] L no. 127, Inayat Husain, Sub-Assistant Surgeon, to I.S. Login, Residency Surgeon, list of sick treated for last 6 months in Royal Hospital and dispensary, Consult, FW, 28 March 1845, India Political and Foreign Consultation, 28 March 1845, P/197/10.

the Prophet's authority as a healer, embarrassing the Company with his adulation of Prophetic miracles that cured leprosy. The Company was no less appalled when he spoke appreciatively of remedies from native mendicants for the cure of the disease. The medicinal value of *neem* (margosa) was highlighted by him as he described native remedies for the cure of leprosy.[132] And if native medicine was appreciated in his reports, charms and talismans were called 'truly wondrous'.[133] 'I have among one of my patients', he wrote in one of his reports, 'that will sit to read some charm with the rising till the meridian sun—his eyes fixed on this source of heat and light. He is now cured of his leprosy and attributes the cure to this operation.'[134]

Contesting Colonial Medicine Outside the Dispensary: Unani in the Family and Private Libraries

Contestations to colonial authority also went on in the domestic confines of the family and private libraries where Unani learning, in its unalloyed form, continued to be zealously protected and disseminated within a distinct circle of family and associates. Here, even though the unalloyed Unani upheld 'secular' learning like natural philosophy, natural law, and psychology, it emulated the discursive teaching format of the seminary and used the authority of its learned teachers. If in the dispensary the authority of the Prophet provided Inayat Husain a tool with which to counter colonial designs, here it was the authority of seminary learning which emerged as a model to counter colonial authority.

In 1875 Edward Balfour, surgeon general at the Madras medical department, toured North and South India and noted that a small number of families of both hakims and vaids strove to retain their traditional learning by the exclusive reliance on Arabic, Persian, and Sanskrit texts, respectively. They trained apprentices within the family and disseminated their knowledge only to limited circuits of family and

[132] L no. 30, Inayat Husain, Sub-Assistant Surgeon, to I.S. Login, Residency Surgeon, 16 August 1848.Appended his report, Consult, FW, 4 November 1848, India Political and Foreign Consultation, 28 October–25 November 1848, P/198/36.

[133] Ibid.

[134] Ibid.

friends. In Lucknow, he reported, 'there are only 10–12 hakims of any repute all of whom take pupils, generally the sons of friends—medical works are written on paper or are printed; among the latter being that by Avicenna.'[135] He also noted the absence of Unani institutions of medical learning for Unani. The absence of institutions and public libraries was compensated by the exclusive reliance on individual family libraries. For instance, he noted that even though the surgeons in Punjab were quite illiterate, the 'library of the Punjab physician contains all the Arabic and Persian books by rote.'[136] He went on to write that Punjab was a land of the vaids, 'but hakims—have the usual books of the Yunani school and others as the *Tuhfat-ul-Ashaqin* and the *Manazar-ul-Abdal*.'[137] Similarly, he noticed that both vaids and hakims in the province of Gujarat possessed richer medical libraries than any other of the Hindu or Muhammadan physicians of South India. According to him, 'the Muhammadan hakim has no fewer than 18 of the standard Persian works in use by the Unani school of medicine.'[138] He gave details of the books read by the hakims of South India which he was convinced were more than a hundred in Arabic and Persian. In many other family collections, books on Unani existed and were written in India or Iran but had information culled from both the Indian as well as Arabic sources.[139] Balfour noted that so steadfast were the hakims in preserving their knowledge that there was a complete absence of any book on European medicine or any attempt to translate any European text by the hakims.[140]

The Unani knowledge preserved in family libraries was disseminated in a narrow circuit of family and close associates through *darrs* (lectures, lessons or instruction) and *nuskha navisi* (prescription writing). These

[135] Balfour 1857: 5–7.

[136] Ibid.

[137] Ibid.

[138] Ibid.: 8.

[139] Ibid.: 15. The books he lists include *Ikhtiar-i-Badii* written in Persian by Ali ibn Husain Ansari of Baghdad in 1392; the *Tuhfat-ul-Muminin*, by Hakim Muhammad Mumin-ibn-Muhammad Dilimi in 1669; the *Mizan-ul-Tibb*; the book in Arabic *Sharh ul Asbab* by Nafiz bin Iwaz and its transalation the *Tibb-i-Akbari*; the *Ulfaz-ul-Adwiyah* by Muhammad Yakub bin Yusuf; *Makhzan-ul-Adwiyah* by Muhammad Husain Khan in 1769.

[140] Ibid.: 16.

were two discursive techniques adopted to consolidate and disseminate, within a narrow family circuit, unalloyed Unani knowledge. In this inner-enclave instruction it was natural philosophy, the mastery of natural law and psychology, that were preserved. And it was this unalloyed Unani that continued to attract itinerant scholars and medics from the Islamic lands, and continued to keep alive the memory of Dar al-Islam much after Muslims had lost their political hegemony over much of this abode of Islam.

In the late eighteenth century, the *darrs* was a discursive technique borrowed from the Islamic world of theological training to consolidate and safeguard specialized Unani learning within the family and a limited circle of close specialist associates. If Lucknow in the eighteenth century was famous for its Darrs-i-Nizamiya, it was also emerging as the new centre of the Darrs-i-Tibbi (lectures in medicine). The founder of the Darrs-i-Tibbi tradition was the Hakim Sayid Asghar Husain. His son, Mir Muhammad Murtaish, and some other students received their preliminary instruction in Unani medicine at his *darrs*, held in his clinic cum residence.[141] The Darrs-i-Tibbi of the eighteenth century Lucknow Hakim Sayid Asghar Husain who trained young hakims in Unani was modelled on the teaching technique of his city's famous Sunni Muslim seminary of Farangi Mahal. The latter's Darrs-i-Nizamiya offered a feel of very different kinds of texts as compared to the Darrs-i-Tibbi. But they followed similar learning techniques and discipline codes: a premium on memory, oratory, writing skills, perpetuation of the name of the *ustaad* (teacher), and of course that of the family name.

Hakim Murtaish was very fond of holding *darrs* in his house. Of the many students who attended his *darrs*, his favourite was Hakim Yaqub, founder member of the Azizi family of Lucknow hakims. Yaqub was his best student; he called him *Fakhr-i-Ustaad* (pride of the teacher) and hoped he would keep his name alive. He concentrated on disseminating his knowledge to a specialized group of students with whom he built strong teacher–student (*ustaad–shaagird*) bonds. The *darrs* sessions in his family house were his private domain where, through his *khidmat-i-khalq* (service to ordinary pupils and patients), he consolidated and safeguarded both the theory and practice of Unani. Even

[141] Rahman 1978: 38–9.

though he accepted a *wazeefa* (scholarship) from the king (*badshah*) of his age, he refused his offer to become the *tabib-i-khaas* (chief tabib) at the Shahi Shifakhaana in Chauk Lucknow.[142]

He died in the reign of Nawab Nasir-ud din Hyder (1827–37), and in his last years, wrote a specialized book which is considered a masterpiece in *Usool Tashkhees* (Principles of Diagnostics) and *fann nuskha navisi* (art of writing prescriptions). He gifted the book to his student Hakim Yaqub with the idea that he would preserve its knowledge. As he said, 'he had organized it in a way so that it would benefit his study and he hoped that the book would stay with him and continue to perpetuate his memory.'[143] He hoped that the book would be followed by hakims as a seminal text and that they would not waver from it.

Hakim Yaqub honoured his teacher's desire and kept the *darrs* tradition alive in his house. Like his teacher, he shunned public office and continued the service from his family domain. This was true also of most of his contemporaries, like Hakim Mirza Muhammad Ali who came to Lucknow in Nawab Asaf-ud-daula's age and carried on *darrs* and *tadrees* from his residence-cum-clinic. He wrote books and commentaries on *Himayat-i-Qanun* and *nafeesi* (psychology) and refused the nawab's offer to become the chief of the Shahi Shifakhaana in Lucknow. But, like Hakim Murtaish, he did not refuse royal grants and endowments as long as they did not interfere with his undiluted service to Unani. He happily accepted a *khilat* and cash award from Nasir-ud din Hyder, who also granted him a Rs 200/month scholarship for having cured him of a difficult illness. During his long innings in Lucknow until the reign of Wajid Ali Shah, he had amassed considerable wealth through royal grants and favours. He had a pucca house built in Top Darwaza, Lucknow, and also had his own mausoleum built, in which he was buried.[144] Hakim Yaqub's family, as we will see in the concluding chapter, preserved Unani learning closely within the family, which of course extended to include students who were close associates.

Indeed, in the hierarchy of Islamic knowledge traditions, theology and religious instruction and its instructors held top positions. Not

[142] Ali 1978: 18.

[143] Ibid.

[144] Maulvi Syed Nawab Ali, 'Tazkira-i-Atibba Lucknow', *Risaalah-i-Takmil-ut-Tibb*, May 1948, p. 22.

only did their persona command respect in Unani families, their many discursive teaching styles, like the *darrs*, became models for 'secular' instruction. Many of the learned men of religion were role models for hakims because they possessed knowledge of Unani as well. More attractive was their ability to write easily communicative texts for quick memorization. Their services were always honoured. Thus, for instance, Hakim Masih-ud daulah, the royal physician of Nawab Nasir-ud din Hyder,[145] hired a man learned in theology, Maulana Anwar Ali, to tutor his son in both religion and medicine. Ali was honoured by the hakim, who considered it his good luck to have him live with the family and teach his son for as long as he lived.[146] And he was not merely a tutor to his son: the maulana wrote one of the early easy-to-read and memorize texts for Unani—*Anwar-al Hawashi*—at the behest of Masih-ud daulah. This was a commentary (*haashiah*) on psychology (*nafeesi*), a very central part of Unani. His commentary, called *Anwar-al-Hawashi*, is considered exemplary by most ulema and hakims. This text was written with the objective to simplify and make accessible to people the many complicated aspects of Unani, and make it easy for them to memorize its basic cures and principals. He wrote commentaries on the *Qanun* as well and again the idea was to simplify it for ordinary people.[147]

The domestic clinic-classroom of the hakims maintained close ties with the seminary. Most young hakims at Lucknow first trained at the Farangi Mahal seminary. After their basic instruction in both rational sciences (*maaqulaat*) and religion (*manqulaat*) they came to the Tibbi-darrs of the hakims for further education.[148] But students of the seminaries who trained to be theologians were sent to the Tibbi-darrs as well for instruction.[149] The hakims considered their presence there

[145] Ibid.: 23. Hakim Masih-ud daulah, who was a famous *shahi tabib* (royal physician) had the administration of the Darul Shifa, hospital and *khairat-khanah*. He received a monthly *wazeefa* (stipend) of Rs 200 per month. He received another Rs 200 for looking after the health of the royalty (*shaahi ilaaj*). He did not have a clinic at home and saw his patients in the Darul Shifa.

[146] Ibid.

[147] Ibid.: 23.

[148] Robinson 2001.

[149] *Risaalah-i Takmil-ut-Tibb*, August 1948, p. 21–2. The ulema of Farangi Mahal, Lucknow, Maulana Abu al Hasnat Muhammad Hayee, regarded Hakim

a great honour: the attendance of the ulema and their students in the Tibbi-darrs was always noted with pride by hakims. In the late nineteenth century Maulana Lutf Ashad Sahib, renowned ulema of the times, visited Lucknow and was medically treated by Hakim Abd al Aziz—the premier physician who founded the Azizi family (*gharana*) of Lucknow hakims. Upon regaining his health, he expressed a desire to attend the *darrs* of Hakim Abd al Aziz since he had heard a lot about it. Hakim Sahib's response was indicative of the respect that the ulema held in the scholarly society of Muslims. He replied, 'You are cordially invited whenever you wish to visit. I am just an ordinary physician (*tabib*). In the elated circle (*huzoor*) of the Ulema my *darrs* is nothing.'[150] The maulana accepted the invitation and was notably impressed by the *darrs* of Hakim Abd al Aziz, calling it a mixture of medicine and philosophy (*phalsapha aur tibb kaa majoon*).

The *darrs* of Hakim Abd al Aziz, held in a *darr* (corridor) of his father's palatial residence,[151] was particularly popular overseas. The ulema and professionals there attended his *darrs* from Afghanistan, Bukhara, and Hijaz. Indeed, the space that it offered for interconnectedness with other Islamic societies kept the idea of the cultural world of Dar al-Islam alive.

Nuskha Navisi (Prescription-writing)

The premium on memory and learning by rote that characterized seminary training echoed in the medical classroom as well. So did the importance of ornate handwriting skills. The art of writing the medical prescription (*nuskha navisi*) was as important as scholarly knowledge of a hakim. A good teacher was one who could both write beautifully and memorize and repeat the *nuskha* in 'one breath'. This was the art (*fann*) that hakims transmitted to students in their clinics.

Hakim Ibrahim took great pains to teach not just his students but also his family members the art of *nuskha navisi.* He also emphasized that it was equally important that they memorized his *nuskhas* for

---

Sahab as his teacher. Students from the Lucknow seminary Farangi Mahal were sent to attend his *darrs*.

150 Ibid.: 23.

151 Ibid.

future use. Students were then tested in *nuskha navisi.* In his clinic he would communicate to his students his diagnosis and then ask them to write the appropriate *nuskha.* He personally corrected the scripts of those who failed the test. The family tradition of training to write prescriptions continued into the time of Hakim Abd al Aziz. He dictated his *nuskhas* aloud (*ba awaaz-i-buland*) and students would inscribe them in their chapters. After he had approved them, he would hand them over to patients.[152]

Both the content of the *nuskhas* as well as the writing styles were considered family treasures. They embodied family knowledge. As markers of family identity they protected Unani within the embrace of the family. The control of the family was further cemented when books on the art of writing *nuskhas* (*fann nuskha navisi*), as well as compendiums of family *nuskhas,* began to be printed in the late nineteenth century. Such compendiums were dedicated 'for the benefit' of younger family members and, at times, to favourite students—who were seen as part of the family. They ranged from those dictated by father to son, or those that went back to knowledge passed on by family ancestors, to those that claimed to include knowledge passed over from teachers to students not necessarily belonging to the primordial family.

Thus we have a collection of *nuskhas* dictated by Hakim Abd al Aziz to his son Hakim Abd al Hamid in 1902–3.[153] Similarly, the precise medical instructions that Hakim Abd al Wahid (1859–1902) gave to his son Hakim Abd al Moid were codified in the important compendium called *Qanun-i-Wahid-fi-ifaadat-i-Moid* or 'Instructions of Wahid for the Benefit of Moid'.[154] Alongside we have the *bayaaz* (notebook) of Hakim Husain Raza, which has 2523 *nuskhas.* These include his own prescriptions, but also those of the family (*khandaani*) starting from his ancestors—Hakim Muhammad Yaqub, Hakim Ibrahim, Hakim Maseeh—and coming down to his father Hakim Hasan Raza and brother Akbar Hakim Ahmad Raza.[155] Again, the Lucknow Hakim Abd al Wahid compiled *risaalah* of family *nuskhas* called *Mukhtasar Tariqah Ilaaj, Kaali Jamiah Amraaz Abd al Wahid Sahab.* This

[152] Ibid.: 147

[153] *Risaalah-i-Takmil-ut-Tibb,* August 1948, p. 147.

[154] Ibid.: 186–7.

[155] Ibid.: 203.

had the author's prescriptions but it also included prescriptions of his prominent ancestor Hakim Abd al Aziz.

Teachers too immortalized their *nuskhas* by passing them on and dedicating their compendiums to favourite students, both within and outside the family. Thus the master don of the Azizi family, Mir Muhammad Murtaish, dedicated his book on the art of writing *nuskhas* to his student, the founder member of the Azizi family, Hakim Yaqub.[156] Favoured students reciprocated in upholding the passed-on tradition of family healing. They considered themselves part of the professional family of hakims. The prescription notebooks of students trained by a particular family of hakims followed their pattern of writing *nuskha* compendiums. In their books, references always existed to the *nuskha* of their teachers. For instance, a reference to Hakim Abd al Wahid's *risaalah* of family *nuskhas*, mentioned above, figures prominently in the *bayaaz* (notebook) of one of his Sufi students, Hakim Maulvi Abad Ahmad Komandwi.[157] The creation of this 'professional family' is evident once again in the *bayaaz* of Hakim Sayid Wahi-ud-Din Ashraf, the *sajjadanashin* of Khanqah Kachwahasharif. Many of the *nuskhas* he mentions are those he had picked up during his education with Hakim Abd al Aziz. But he was also a student of Hakim Abd al Wahid and, at the end of the *bayaaz*, there is a chapter devoted exclusively to the methods of treatment prescribed by his teacher Abd al Wahid: *Tarikh-i-Ilaji-i-Kulli Muallafah Muratabah Janam Ustaadi Hakim Abd al Wahid Sahib Lucknowi.* (History of the Cures Prescribed by Master Abd al Wahid Sahib Lucknowi.)[158] Such trends suggest that even though the primordial family tradition remained the base of the hakims' professional identity, this could extend to accommodate a professional family as well. It was not rare to find students like Hakim Abd-ul Ahad Kasmandavi, who had in their possession *nuskhas* of their teacher. Hakim Abd al Wahid's compendium, 'A Short Course of the Treatment of All Ailments', was always available with Kasmandavi.[159]

The writing of the *nuskha* was an interactive exercise where the

156 Ibid.: 41.
157 Ibid.: 186–7.
158 Ibid.: Rahman 2002.
159 Ibid.: 186.

hakim entered the cultural world of the patient. And here, even though theoretically Unani upheld natural philosophy and natural law, in practice it tapped many Muslim beliefs about illness. One of them was the relevance of the name of the patient to his illness. In one case, Hakim Shamsuddin of Lucknow, who was particularly keen on rechristening people in case they were named after villains in Islamic history and religion, made a patient change his name when he discovered that his name was Haman—the prime minister of the hated Firon (Pharoah) who was associated with the assassination story of the Prophet's grandchildren. As if to retain the delicate balance, that Unani maintained between religion and medicine, the hakim rechristened the patient—not on a religious figure—but as Harun Rashid, the noted Abbasid caliph known for his patronage of Unani medicine in Baghdad.[160]

[160] Ahmad 1950: 33–4.

5

Urdu Medical Texts in the Late Nineteenth Century

Introduction

THIS CHAPTER DISCUSSES THE LATE-NINETEENTH-CENTURY Unani response to colonial medical drives. It argues that Unani's reaction did not constitute a narrow anti-colonial shrill. Instead, it derived from the plural medical culture that colonialism itself had helped sustain. It shows that Unani healing underwent tremendous internal change in an age characterized both by 'modern' capitalist infrastructure and print culture, as well as the related disturbances of modernity: population explosions, famines, disease, epidemics and related death. This conjuncture was used by Muslims who did not come from traditional hakim families to fashion a new Unani. They challenged both the traditional custodians of Unani knowledge as well as the colonial state. Very much like religious dissent movements of the period, they questioned the received medical knowledge of the Perso-Arabic brand. They used colonial artifacts like print to create a wider medical public sphere. Through their writings they created a culturally larger Unani in the vernacular, Urdu. This mirrored social changes, and the close connections between medicine and *ashrafiyat*—upward social mobility.

The new Urdu Unani literature was a body of knowledge that had a momentum of its own. It remained beyond the control of both the state as well as the traditional Persianate families of healers and older

Arabic-knowing communities. And yet it was truly inspired by the Company-sponsored Urdu medical literature, as well as by the bodily deportment frills of Persian medical texts. It upheld also the humoral theory of Arabic texts preserved by medical families and used by dispensary hakims. It combined these medical pluralities to fashion a new Unani tradition. This fresh Unani doubted the ability of traditional hakims to effectively contest colonial medical intrusions by preserving medicine within the family and via inaccessible Arabic texts. The new authors argued that medicine was not the preserve of an elite that had access to ancient wisdom locked in Arabic texts of antiquity. Rather, it was a gift of God and anyone could learn it through travel, experimentation, and observation.[1] The ancient wisdom of Hippocrates, Aristotle, and Galen was acknowledged. But to master their 'theory of medicine' was not the only way to become a hakim. The new proponents shifted focus to the 'practice of medicine'. They catered to a slowly growing lay literate audience that desired the practical benefits of medical knowledge for their use in the household, barber's shop, and surgery.[2]

Print helped reconstitute Unani as a culturally all-encompassing medical tradition. It did not create an 'imagined community' of Western-influenced healers. Instead, it enabled new hakims to reach out to people and cultural referents hitherto beyond the purview of Unani: the person of the Prophet, Sufi healing, and astrology. It also helped maintain the contact with older Persianate medical traditions that still simmered in *ashraf* society. As Unani embraced new medical localisms, it diversified and made them part of its own world of 'scientific rational' medicine. These medical localisms were not reified,[3] nor were they imbued with

[1] They were very much like the Reformation Protestant physician Paracelsus, who was outside the circuit of university-trained doctors and was all for popularizing and vernacularizing medical knowledge. He wanted to retrieve it from the exclusive hold of elite 'rational and learned' university-trained doctors. He believed that knowledge could be acquired not just from books and libraries but from travel and observation. He argued that the texts of Galen and Hippocrates should be burnt. He represented a direct threat to the university-trained 'learned' doctor'. French 2003: 148.

[2] For a similar pattern in medieval Europe, where the vernacularization of medical knowledge from Latin followed a similar trajectory, see Jones 1990: 1–25.

[3] Dirks 1992.

Western scientific instrumentation.[4] They were incorporated within what had so far been a relatively 'mechanistic' Unani culture. For instance, the new hakim's stress on medical practice brought him closer to the religious and cultural etiology of illness. This meant that well being was not just about the physical and social context of the body, but included the spiritual, moral, and psychological state of being as well.[5] And as Unani incorporated religious beliefs about illness, the imprint of Muslim reformist and revivalist influences on it became more than evident. The new Unani, with its fresh authors and wider clientele, had tremendous potential to challenge both the traditional hakim families as well as the expanding colonial medical system.

Persian Medical Literature Translated into Urdu: The *Zakhirah-i-Khwarzmshahi* (Thesaurus of the Shah of Khwarzm)

The new focus on the cultural etiology of disease brought the Persian rather than the Arabic canonical literature centrestage for translation into Urdu. As we saw in Chapter 1, the Persian medical literature that existed on the fringes of classical Unani had claimed for medicine certain cultural beliefs about illness. As a model for the new Unani it represented an obvious choice for the translator. Thus, medieval Persian texts like the *Zakhirah-i-Khwarzmshahi*, which reflected the Unani shift from canonical materialistic knowledge to demonstrable and culturally diversified knowledge about illness, became the models of Urdu medical literature. So it was that in 1878 the twelfth-century Persian text *Zakhirah-i-Khwarzmshahi* was translated into Urdu.

The original text was written by Muhammad Ahmad-al-Hasani-al Jurjani in the twelfth century. Jurjani was the court physician of Ala-ul Doula Atziz, the son of Qutb-ul-Din. However, he says in the text that he was in the service of Ala-ul Din Muhammad Khwarzmshah (the sultan of the regional kingdom of Khwarzm). He wrote the book at the behest of the sultan, who wanted a comprehensive book to be

[4] Prakash 1999.

[5] For a discussion on the Prophet's medicine and the Islamic notion of well being, see Rahman 1987: 29–58. See also Alavi 2005: 101–29.

available for people to consult at the time of illness.[6] *Zakhirah* was one of the first Unani medical texts to be written in Persian. It represented an important movement at the end of the Abbasid caliphate to pull medicine out of the hold of Arabic-educated scholastic elites and locate it in a popular, non-specialist sphere, a realm where its practice could be easily demonstrated and its ambit extended to encompass popular beliefs about illness. Even though it is in size as big as the *Canon* of Avicenna, its contents are far more concise. It is a relatively easier understood exposition of the whole system of medicine as taught in Jurjani's time. After al Razi and Avicenna, this became the most frequently quoted and consulted textbook of medicine. Its user-friendly and popular style made it far more lively reading than the dull style of the *Canon* of Avicenna.[7]

The *Zakhirah* exists in nine volumes. Collectively, these upheld the humoral theory of health but explained its functioning by constructing a list of popular ideas and responses to illness. According to this work, ecological and dietary factors top the list of beliefs about illness: these factors disrupted humoral balance and caused ill health. The text identified environmental elements (*arkaan*),—hot, cold, dry, and wet—that needed to be in harmony with body humours. It noted specific foods or diet (*ghiza*) that were also essential for the maintenance of equilibrium. Secondary to these ideas was the belief that illness was also caused by the malfunctioning of bodily (*jism*) anatomy. The text advocated *nabz* (pulse) and urine testing as important diagnostic techniques, and offered medicines and surgery as cures to rectify anatomical dislocations. The fourth tier of illness beliefs gives importance to dress and bodily deportment: it highlights the connection of place of residence, peace of mind (*sukoonat*), dress, sleeping habits, and the effect of smells on health. Finally, the text ends with a discussion on religious beliefs about illness. God is believed to control the balance of humours that ensures good health. The *Zakhirah* discusses religious dictums (*tarteeb deeni*) related to health and lists prayers (*dua*) as part of the healing regime. And, very much like the European Protestant idea that endorsed the Hippocratic notion of health and healers as divinely ordained, the emphasis in the *Zakhirah* is that the essence of medicine could not be

[6] Elgood 1951: 215.
[7] Ibid.: 217–18.

acquired from textual authority: it existed in God's natural bounties. A true physician was one who could recognize from God's natural bounties the remedies for cure.

The Persian *Zakhirah* stands apart from the Arabic canonical texts in combining Aristotleian natural philosophy and Galenic anatomical rationality with the social, moral, and religious beliefs about illness prevalent in society. It also incorporates aspects of bodily deportment, etiquette (*akhlaaq*), piety, and spirituality in the healing process. As we saw, in the Arab literature texts that stressed such cultural beliefs about illness remained either marginal or were constructed in opposition to the established Aristotelian natural philosophy dominant canon.[8]

The *Zakhirah* was a text that claimed culture and religious practice for medicine. It thus served as a useful model for the new Unani literature, which was looking for a similar encompassing sphere of authority to contest colonial authority. In 1878 the noted publisher of Lucknow, Munshi Newal Kishore, asked Hakim Haadi Hasan Khan Moradabadi to translate the text. The job was done over a year and a half. The Urdu *Zakhirah* made it possible now for anyone to become a hakim. Newal Kishore clearly stated this objective when he said that the Urdu translation was to benefit 'not only *aalim* [learned] and *faazil* [scholarly] hakims, but all those who were interested and could read Urdu'.[9] The translation into Urdu was to encourage the local medical tradition's popularization so as to counter the fast dissemination of British medical literature in the vernacular. Newal Kishore made this intent even clearer when he said that 'even though people were getting interested in modern medicine the *Zakhirah* would be of general benefit so that (*naazreen*) and those interested in medicine could have access to Unani knowledge as well.'[10] It was for similar reasons that the text had a list of Urdu translations of 33 Arabic and Persian medical books (also published by Munshi Newal Kishore) appended to it for publicity. These included important texts like *Makhzan al Adwiyah* and *Ilaj al Ghuraba*.[11]

In the late nineteenth century, when epidemics ravaged India, the translation of a text like the *Zakhirah* was useful for many reasons. In

[8] Burgel 1998.

[9] Khan 1878: 1922–33.

[10] Ibid.

[11] Ibid., vol. 1 : 2–4.

the first place, this text laid out very clearly that Unani was not mechanistic and abstract but sensitive to popular beliefs about health. It was a dynamic 'cultural system'.[12] This projection of Unani as a cultural system was useful in attracting a wider clientele, specially at a time when the state was seen to have failed to check disease. Secondly, the *Zakhirah's* thrust on the practice of medicine, experience, and observation opened the door to the self-taught hakim. It guaranteed social mobility in a society where, thus far, medicine had been an elite preserve. This was again very useful in an age of competition with British medicine.

Hakim Haadi Khan clearly had these political compulsions in mind, for his work is clearly tailored to the political and social concerns of the day; indeed his text reflects the Islamic reformist and revivalist influences that had begun to imprint knowledge produced by Muslims. Thus the Urdu *Zakhirah* gives a new Islamic orientation to healing. In its preface the hakim deflects attention from the Aristotelian rationality that was the defining feature of the Persian original. Instead, he claims religious practice as its defining rationality. Haadi Khan views God as the ultimate healer, for He created both disease and its cure. The scientia of medicine, according to him, lie buried in the natural habitat that God created. The role of the physician was to identify these correctly and apply them appropriately. Extolling God as the ultimate medical authority, he dedicates his book to Him: 'This book is dedicated to that hakim [God] who created with his *hikmat* (wisdom) vegetation, trees, minerals. And created in each of them a special effect so that people get satisfied.'[13]

The Persian *Zakhirah* had embraced religion as part of healing, but Haadi Khan gives it far greater emphasis, identifying the divine power that he sees as critical to healing. Very much like Muslim reformers of his time, he upholds an Islamic notion—in his case the physical, spiritual, moral, and psychological health of the body, projecting the Prophet as the centre of this Islamic rationality. Unlike the Persian *Zakhirah*, which believes in an abstract God as the author of the rational natural world, the Urdu *Zakhirah* projects the Prophet Muhammad as the

[12] Good, Broadwin & Kleinman 1994.
[13] Khan 1878, vol. 3: 1922.

guide who helps the individual to comprehend a rational medical knowledge which lies embedded in nature. According to Khan, God was indeed the greatest physician who created the boundless sea of the natural world in which the scientia of medicine lay buried, but Muhammad was the most important healer (*tabib*) who gave direction to man to explore the natural remedies that lay scattered in nature, waiting to be discovered. The Prophet was a miraculous healer. He was also the exemplar of moral rectitude and a repository of knowledge about the natural world. He helped end ignorance regarding disease and good health: his 'guidance and instruction and sayings helped in eliminating ignorance [*gumraahi*]) and illiteracy [*jihaalat*], and helped the people by restoring in them wisdom and knowledge. And his intervention gave the knowledge of medicine [*ilm-i-abdaan*] the veneer of knowledge of religion [*ilm-i-adyaan*] and lent to it a certain piety.'[14] Thus the exploration of God's natural world under the guidance of Prophetic sayings was the route to becoming a knowledgeable hakim. Prophetic piety and medicine were central to Unani. Indeed, religion and Unani were connected naturally because physical health was a prerequisite for *ibaadat* (meditation). This religious dimension of Unani made it even more necessary that people consider it mandatory (*waajib*) and important (*muhtamm*).

In short, Urdu translation of the *Zakhirah* brought Islam, as represented by the Prophet, firmly within the ambit of Unani and made the religo-cultural world of people central to the practice of medicine. The *Zakhirah* stressed the cultural contextualization of illness. The hakim was encouraged to enter the world of the ill and construct an illness narrative based on beliefs about the disease. He was expected to combine the oracular ideas of Hippocratic ancient wisdom with those of the Aristotelian as well as Galenic. In this role as medical-ethnographer, the new hakim would be similar to Galen in unifying the Hippocratic divinity that allowed cultural contact with the world of the patient, with Aristotelian rationality. Indeed, Galen was the big hero of Urdu medical literature. Most texts invoked his name and praised their authors as the 'Galen of their times.'[15]

[14] Ibid.: 1922–3.

[15] This was a selective appropriation of Galen as well. For Galen was against

Because the Urdu *Zakhirah* culturally contextualized illness, it took very seriously the role of social conduct and etiquette as markers of health. It addressed the humoral theory also from the social angle of proper conduct, etiquette, and bodily deportment. These too could disrupt humoral balance and cause illness, just as much as ecological factors. The Urdu *Zakhirah* retained the section in the Persian version on the 'purity and decoration' of the body (*jism kee pakeezgee aur aaraastagi*), entitled *Kitab-i-Zeenat*.[16] Here, the purity of the human body was integral to health; and purity was linked to piety, beauty, deportment, and the relaxed and peaceful state of the body. Thus, in the cure of fevers, it also urges alongside medicines that people give importance to baths (*ghusl*), cleanliness, clean clothing, and massages. Health was increasingly associated with the deportment of the ideal body: and the management of body deportment did not require any scholastic training. It was a divinely ordained responsibility that needed prescriptive dictums on conduct, etiquette (*akhlaaq*), purity, and pollution.

Haadi Khan did go on to elaborate that healing was not *only* about the cultural understanding of disease: it also included medicalizing the patient. Medicalizing meant that the hakim drew an empirical remedy into the illness narrative of his patients. In the nineteenth century, empirical details about making *nuskhas* and disease-management procedures, hitherto a guarded preserve of the hakim, began to be published for popular consumption. In this regard too the *Zakhirah* was an appropriate choice for translation into Urdu. It had an entire part devoted to the methods of the making of medical prescriptions (*nuskhas*) and disease management.

The *Zakhirah* has *nuskhas* on diseases of the liver, fevers, intestines, deworming, dysentery, etc.[17] It does not list the sources of prescriptions. No name of any hakim or that of any family of hakims is cited. The *nuskha* begins with detailed descriptions of the ailment. Then the procedural details of disease management are described. Finally, the

---

the popularization of medicine and was critical of Roman doctors who believed in quick shortcuts to become doctors. French 2003: 36, 44.

[16] Khan 1878, vol. 3, part 8, is devoted to matters of purity.

[17] Ibid., vol. II : 1186–1240.

medicines and method of preparing them are detailed. The medical ingredients, as well as the method of making them into effective medical remedies, are listed. The art of making tablets (*golee*), syrups (*sherbets*), powder (*safoof*), and electuary (*majoon*) is also described in detail.[18] A typical *nuskha* detail is listed as follows:

> *Naasoor ka ilaj.* [Remedy for a weeping wound.]
> There is a description of the wound.
>
> The management of disease and treatment procedures follow. This should be treated with strong [*tez*] and dry [*khushk*] medicines. It should be treated in such a way that first it is pressed so that the fluids ooze out. Then medicine should be applied on the wound with cotton wool wrapped on a slim rod. It can be bandaged if desired but first it should be allowed to dry, then medicine should be put on it with cotton wool. When the wound is dry, medical ointment [*marham*] mixed with egg white is put on it. The wound should be washed with water boiled with fruits and herbs like dried pomegranate [*anaar*] and *naagar mootha*. If the wound does not heal, a medicine that accelerates the healing process be applied.

The name of this medicine is listed as *akshareen* and its method of preparation along with proportion of ingredients in detail is listed.[19] The text also has *nuskhas* on removing kidney stones. The meat of several birds is recommended, as also medicines whose details of preparation are listed.

The *Zakhirah* does not stop at *nuskhas*. It also provides copious details about making medical instruments, like a catheter. One such description for the construction of a catheter (*qaashaateer aalah*) reveals the depths of medical precision being made public to dent the authority of the learned rational hakim who prided himself on the theory of medicine. The text lists the usage of the instrument as a device for enabling the passage of urine. It gives details of its manufacture: it can be made of lead (*seesah*), silver, zinc-like metal (*raangah*), or copper. Holes may be punched into it, as per the medical needs of the patient: this helps because if one hole gets blocked, the other can remain open.

[18] Ibid.: 1258.
[19] Ibid.: 1260.

There are detailed descriptions of how to use the instrument, how to make the patient sit on a chair in a urinating position, and how to deploy it effectively.[20] This is accompanied by detailed descriptions of orthopaedic surgeries of the skull, chest, and other bones.[21]

The Ilaj-i-Ghuraba *and the* Mizan-i-Tibb

The *Zakhirah* was not the only text whose translator reached out to the Prophet even as he kept the core humoral theory intact. Another such text was the *Alaijab Mutahsir* of Fakhr-al-Din Ahmad. This was written in 1260 hijri (1844) in the period of Lord Dalhousie, for the 'ordinary people'.[22] Based on Persian documents, the text is replete with empirical details of anatomy, orthopaedics, pulse and urine testing as diagnostic techniques, and so on. But like the *Zakhirah* it interprets the humoral theory via the Prophet's spiritual knowledge, attributing the healing powers of the elements to the Prophet's spiritual touch. Such association with the Prophet makes him call Unani *sharif ilm* (respectable knowledge). The translator underlines the need for a popularization of this system because it ensures a healthy body, which is seen as connecting the material to the spiritual world.[23]

A similar trend can be seen in the Urdu translation of the *Ilaj-i-Ghuraba.* This text by Hakim Ghulam Imam Sahib, which was a compendium of 'simple and difficult' *nuskhas*, was another Persian text chosen for translation into Urdu. In 1864, Hakim Asghar Ali of Gopalmau completed the translation.[24] Like the *Zakhirah Khwarzmshahi*, this text made medical knowledge public through its *nuskhas*. Munshi Newal Kishore got it translated so that ordinary people got the benefits of good healthcare (*gauhar-i-giraan maayah*). The *Ilaj-i-Ghuraba* attributes ill health to physical changes in the identifiable elements (air, water, fire, earth). It has very precise *nuskhas* arranged by the name of the disease: treatment of headache (*Ilaj-i-sar dard*), of constipation,

[20] Ibid.: 1332–3.

[21] Ibid.: 1599.

[22] Ahmad 1260 hijri, 1844. It went to the Lt Governor of Agra for approval, after which it was hoped that it would be made public.

[23] Ibid.: 3.

[24] Ali 1281 hijri, 1864.

throat ailments, etc. It gives copious details for preparing the recommended medicines. For example, it has procedural details for making fig-willow oil (*roghan-baid-anjeer*), powder of dried pomegranate seed (*anaar daana safoof*), and the syrup of berries (*sharbat phalsa*).[25] The *nuskhas* were popularized by the wide circulation of excerpts of their translation through Urdu journals and newspapers: both *Oudh Akhbar* and the journal *Tibbi Ayeena* carried excerpts of the translated text.[26]

The medicalizing drives of the *Ilaj* notwithstanding, it too recognized God as the author of medical empiricism. Thus the translator, Hakim Asghar Ali, like the translator of the *Zakhirah*, salutes the highest hakim, God (*hakim alal Itlaaq*), who 'used his wisdom' (*hikmat*) to produce so many *aalims* (learned men) and made the ordinary human being a prominent and extraordinary creature. And again like the *Zakhirah's* translator, his preface underlines the role of God in medicine much more than in the original Persian text. God he says, 'discovered disease and for the health of man discovered requisite food and medicines. He provided for man's spiritual and physical problems and saved him from many evils by putting an end to ignorance and illiteracy with the spread of his wisdom.'[27]

This general trend of simplifying medical knowledge so as to make it accessible to ordinary people and invocation of the Prophet to popularize medicine is evident also in Urdu translations of the Mughal-period text, the *Mizan-i-Tibb*. As we saw in Chapter 1, the original Persian version of the text written by Hakim Arzani, heralded the trend making Unani prescriptions simple and accessible to the non-specialist. Now in the nineteenth century, when Urdu became the language of the masses, the text was translated into Urdu. In 1852 a Delhi writer, Muhammad Hasan Alijaan, son of Muhammad Mohsin Khan, translated the text with the help of one Syed Qasim Ali Munshi. He was attached to the Delhi madrasa and was a resident of qasba Hindor, Zilla Bijnor.[28] It was published under the supervision of Syed Sharf Ali. Its organization into three sections (*maqaalas*) that deal with

[25] Ibid.: 21–32.

[26] Ibid., foreword, no page number.

[27] Ibid.: 1–2.

[28] Ali 1852.

temperaments, diet, and medicine and diseases and their symptoms, respectively, are similar to the Persian version of Arzani. The translator does not provide elaborations in the margins or in footnotes to explain the Persian text.

An 1872 edition of the text published from Kanpur makes unani even more user-friendly. This version of the *Mizan* was translated by one Muhammad Hasan, who was asked to do the job by Hakim Mannan Muhammad Abdur Rahman bin Haji Muhammad Roshan Khan. This later edition spells out prescriptions in far greater detail. It claims to explain and elaborate on the prescriptions of the Persian text; indeed, not content with elaborations on the text, the translator gives further explanations in the margins. These clarifications appear as footnotes on every page and tease out in simple ways the names of medicines and drugs.[29] A more significant addition that the translator makes to the Persian version is the introduction of the person of the Prophet, whom he eulogizes as the greatest healer who created both man and remedies to keep man healthy.[30] The 'scientific' rationality of the *nuskhas* in the Urdu *Mizan* are wrapped around this powerful Islamic imagery of the Prophet as healer.

Reaching Out to the Prophet: Homegrown Urdu Texts—The *Mazhar-ul-Ulum* and the *Tibb-i-Nabawi*

Urdu texts written in India were stylistically very similar to a genre of medieval Arab medical literature that made a feeble attempt to appropriate religion to silence theological critics.[31] Late-nineteenth-century Unani claimed religious practice with a vengeance. The new Urdu

[29] Hasan 1872. Every page has footnotes in the margin.

[30] Ibid.: Preface. No page numbers.

[31] Rosenthal 1990: 519–32. This was authored by men like Abd-al-Wadud who, over the tenth to twelfth centuries, claimed religion for medicine. Wadud silenced its critics, both religionists and others, who doubted its efficacy. Such writers argued that since the subject (*maudu*) of medicine was the human body—the noblest of all of God's creation—it had to be a sacrosanct discipline that in no way contravened religion. This genre of literature thus glorified the subject of the medical text to earn itself a position within the religious tradition.

medical writers did not come from established families of hakims. They therefore looked for alternative symbols of authority. Their obvious choice was the person of the Prophet. His authority became the ultimate reference point of legitimacy for Unani. And this was practical, since the hakim found readymade ground for such an appropriation. In this period the heightened pilgrim traffic to cities closely associated with the life of the Prophet, Mecca and Medina, had at one level increased the fear of disease because people were seen to carry infection as they travelled across long distances. But it also put the spotlight on his life and teachings and shifted popular attention to his medical experiences and ideas about illness. The shrine at Mecca—the Kaaba—and the fountain of holy water—*aab-i-zam zam*—had a dominant therapeutic narrative closely tied to the person of the Prophet and the history of Islam. Indian pilgrim traffic to Mecca brought these stories home. They further popularized the belief in the Prophet's medicine.

Closer home, nineteenth-century Muslim society was in a state of change with new notions of individuality, personhood, and self being articulated by people as they came to grips with modifications in the political culture around them. In this new milieu, the drift was towards new religious knowledge from revealed Arabic scriptures. As the focus shifted from the more Iranian-inspired knowledge of Sufis, pirs, and rational sciences to knowledge from the Arab lands, the person of the Prophet acquired heightened interest.[32] The Prophet was up for new interpretations by Muslim revivalists and reformers of all hues. They demythologized him, making him more approachable and familiar—in the cast of an exemplary human being. Many emphasized his socio-economic reformist profile and his moral and social conduct as the ideal that would salvage Muslims from their current distress.[33] This intensified emphasis on the life and conduct of the Prophet was reflected also in an increased interest in *mawlid* ceremonies that commemorated

[32] Robinson 2000: 105–21.

[33] Schimmel 1985. This trend began in the late eighteenth century with Shah Waliullah, and was picked up with greater fervour by modernist reformers like Sir Syed Ahmad, who wrote essays on the Prophet's moral and social conduct as a reformer. Print culture aided the proliferation and dissemination of these ideas, and the explosion of the hadith literature on the Prophet's sayings and the biographies on him marked the culmination of the trend.

his birthday These were used as a means to communicate the new religious knowledge inspired by Arabic scriptures from Arabic-speaking lands. This was also evident in the Haj travelogues of Muslims, as well as in the explosion of biographies on him and texts on his tradition—the Hadith.[34] Urdu texts on Unani were affected by this trend and added to the expanding genre of literature on the Prophet. And, like the reformist literature, they used the Prophet's authority to negotiate colonial intrusions into the medical realm.

Unani, more than ever before, now associated itself with all medical knowledge identified with the Prophet. It continued with the idea of health as part of well being, but defined well being as viewed by the Prophet to mean spiritual, moral, psychological, and physical health. Medication as offered by the Prophet was seen as part of this Islamic idea of well being. Belief in medical knowledge associated with the Prophet, and a life patterned on his conduct, with significance attached to prayers (*dua*), diet, Islamic bodily deportment, and moral and spiritual etiquette now became central to Unani. Unani had always seen its rationality as sanctioned by the Prophet, but now it saw no contradiction between the Islamic prescriptive path to well being and rational medical knowledge. Indeed, it saw medication as an essential part of this Islamic notion of well being. From here Unani soon moved out to competing versions of Islam, such as Sufic healing through the control of the soul, the charms of fakirs, and astrology. These rubbed shoulders with aspects of medical knowledge that Unani cannibalized from Western practice.

As we saw in Chapter 1, texts like the *Tibb-i-Nabawi*—the Prophet's medicine—that upheld this Islamic view of well being existed in the Persian medical genre. Those, however, were written by Muslim conservatives who opposed Unani's tilt towards Graeco-Roman philosophy. The notable exception of a qualified hakim writing such literature was the well known eighteenth-century Persian physician of Aurangzeb's court, Arzani. But Arzani lived in exceptional times and had written his text to please his religiously inclined master.[35] And this was not his only work: his better-known contributions to Unani were the *Tibb-i-Akbari* and the *Mizan-i-Tibb*.

[34] Metcalf 1990. See also for Bengali *sirat* literature on the Prophet, Dey 2006.
[35] Elgood 1978: 29.

In the late nineteenth century, texts modelled on Arzani's *Tibb-i-Nabawi* were written by hakims with a view to making the Prophet central to Unani. In 1877 Ilahi Baksh of Kanpur, a practising hakim and teacher in a madrasa, wrote the *Tashreeh al asbaab musammabeh, Mazhar-ul-Ulum.* This was a text written in Urdu in consultation with a range of Persian medical texts to make accessible to ordinary people essential anatomical details. The text was meant to be a useful self-help directory for medical emergencies. The Persian texts consulted included *Hawaashi Qanooncha, Mujizul-Qanun, Nafeesi aur uske hawaashi*, two volumes of *Kulliyaat Qanun, Hamiyaat-al-Qanun Kamil-al-Sinaaitah, Dastoor-al-Ilaj, Zakhirah-i-Khwarzmshahi, Tibb-i-Akbari, Ilaj-i-Ghuraba*, and many other books on science.[36]

The *Mazhar-ul-Ulum* claims religion for Unani. God is as usual projected as the greatest physician (*hakim mutlaq*) and is said to have created wisdom and the many sciences that produce cures. The Prophet is the repository of God's medical wisdom and his house of knowledge is called Darul Shifa (the house of healing). According to the author, the Prophet's association with Unani makes medicine both *afzal* (extraordinary) and *ashraf* (elite).[37] The text upholds the Islamic notion of well being, with its stress on moral and psychological health as crucial to physical strength. However, the text does not entirely dismiss Aristotelian rationality: its dietary and environmental regulations are listed for the maintenance of physical health, and Hippocrates is seen in the text's reliance on dietary regulations as a system of cure. Like most texts of this genre, it is a typical self-help manual which gives copious details on how to use certain medicines so that both the learning and practice of tibb (medicine) become easy.

In 1869 Ikram al Din Hafiz, a practising hakim of Lucknow, wrote the *Tibb-i-Nabawi* and made a strong case for allying Unani with Prophetic medicine: he made it more than clear that health was about moral, spiritual, psychological and physical well being.[38] His 54-page text stresses the critical importance of prayers (dua), piety, and medicine (dawa). It even offers cures for diseases, like anger, combining medicines

[36] Baksh 1877.
[37] Ibid.: 2–3.
[38] Hafiz 1869.

with piety. Forms of piety derive from Quranic injunctions and the recitation of its holy verses. These as well as the Prophet traditions (Hadith) are the ultimate reference point for all suggested cures.

The *Tibb-i-Nabawi* is critical of Greek physicians like Galen who followed Aristotle and frowned on the mixing of medicine with piety: 'such people [Galen] whose wisdom [*hikmat*] is merely empirical experience [*tajurba*] do not consider mere prayers [*dua*] as beneficial for bodily diseases [*amraaz-i-jismaani*]. Instead they criticize Muslims [*ahl-i-Islam*] as foolish and rely on prayers. They argue that prayers are recited from the lips and mouth, but how can something that emanates from the mouth have any effect inside the body?'[39] Hafiz counters such allegations by stating that 'to ignore the effect of what people utter from their mouths on the body is foolishness because everyone is convinced about the effects of such utterances.'[40] He elaborates the link by arguing that if somebody abuses a person, it causes anger in the heart. And if someone praises an individual, it causes happiness in the heart. This reveals the link between utterances and their effect on human anatomy: so, 'a person who does not know the effects of prayers (*dua*) on bodily health reflects his own foolishness' (*himaqat*).

Very much in the tradition of Islamic healing, Ikram al Din Hafiz portrays the best hakim as one who combines the art of *ilaj-i-badan* and *ilaj-i-rooh* (bodily and spiritual healing). The foremost example of this exemplary combination is the Prophet himself. Hafiz reasons that this had to be the case since the health of the world is defined by the physical as well as the spiritual well being of the people. And since the Prophet came as the benefactor (*rahmaan*) of the world, he would have been criticized if he did not contribute to both the physical and spiritual health of the people who constitute the world. He therefore combined the powers of spiritual and physical healing in his person.

Such Urdu medical literature created the new Unani. Hafiz made a distinction between hakims who did not make Prophetic medicine central to Unani and those who did derive from its authority. He defined as Muslim only those who continued the Prophet's tradition. He said that the Prophet taught and trained his community's learned men in both bodily as well as spiritual healing. And thus Muslim hakims

[39] Ibid.: 2.
[40] Ibid.: 2.

who combine dua and dawa are different from those who cure only with medicines (*attiba-i-jismaani*). The text reiterates the importance of such dual healing by arguing that some diseases cannot be cured only with medicines. These include ailments relating to magic, evil spells, and the evil eye. Truly Muslim hakims teach people how to ward off such diseases through Quranic recitation and prayers recited by the Prophet in his moments of distress. Hafiz urges people to treat medicine only as the means to approach God, who is ultimately responsible for their cure. He dubs people infidel who regard medicine as merely the agent of medical redress (*shifa shaafi*).[41]

Hafiz does not completely dismiss the value of medication. He is critical of ulema who regard the taking of medicine as un-Islamic. He projects 'correct religion' (*sahee mazhab*) as that which declares as *sunnat* (in keeping with the Prophet's sayings) the consumption of medicines. The text cites a Hadith to say that people asked the Prophet if it was sinful to take medicines when they were ill. The Prophet replied that they should take medicines freely because God has created both the disease as well as the medicine for its cure.[42] The text then cites the medicinal properties of items such as fruits from the gourd family, and camel's milk, as useful medicines.[43]

The cures in the *Tibb-i-Nabawi* stress not just dua and dawa but also proper deportment for a healthy body. Deportment here derives heavily from the Islamic way of life. Unani's humoral theory is interpreted through the Islamic prescriptive path for proper living. For instance, Islamic matrimony or *nikaah* is recommended as a sure way to ensure the purity of the body's blood. And a specific fasting (*rozah*) schedule is suggested to balance the body's fluids. This includes an entire month of fasting in the month of *shabaan*, 6 fasts after Eid, 2 in Muharram, 9 in *ashurah*. The Hadith says that if this is followed, the fluids of the body do not dry.[44]

Good and evil influences that affect disease cycles are also regulated through proper adherence to the recommended Islamic way of life. The text cites the Hadith and lists a code of rituals that include prayers

[41] Ibid.: 2–3.
[42] Ibid.: 3.
[43] Ibid.: 10.
[44] Ibid.: 4–5.

to be followed when a new bride enters her home. Special prayers get rid of the evil doings of the woman in her house, and spread her goodness. Some ulema are shown to recommend that the feet of the new bride be washed and water sprinkled in corners of the house to increase its prosperity.[45]

The best and most detailed deportment instructions pertain to sexual intercourse so as to ensure the birth of a healthy child. Interestingly, each deportment rule is listed as an *ilaj* or cure for safe childbirth. The first ilaj is a Quranic prayer to be read at the time of copulation, so that Satan runs away and a pious child is born. Second, the preferred hour of copulation is given as late night rather than the early hours of the night. Third, at the time of intercourse men and women need to cover themselves with a cloth, or else a *behayaa* (shameless) child will be born. Fourth, to avoid the birth of a child who is mute or dumb, couples are advised not to converse during intercourse. Fifth, a mad child may be born if someone has *ihtalaam* (wet dream) and copulates without bathing. Sixth, if one copulates on a full stomach, a mentally retarded child is born. Copulation in a standing position makes the body weak. Seventh, to avoid disease in men, it is recommended that they urinate after sexual intercourse, else the child born will be mad. Eighth, a blind boy may be born if the couple happen to see each others' genitals during copulation. Ninth, an ilaj is to tie a talisman (*taweez*) to the left leg of the woman to ease and facilitate childbirth. Tenth, a prayer reiterating the Islamic faith in only one God (*azaan)* should be spoken into the right ear of a child so that he lives his whole life with its blessings. Eleventh, the child should be named on the seventh day after his birth and a good name should be given to him, a bad name may harm him. Lastly, the text recommends various remedies to cast off the spell of the evil eye upon the child.

The Prophet and Beyond: Maqaalat-i-Ihsani, Tibb-i-Ihsani, *and the* Qarabadeen-i-Ihsani

Tibb-i-Nabawi set the trend. Soon, many other hakims used the printing press to incorporate the authority of the Prophet to medical

[45] Ibid.: 4.

experience and medical knowledge. Such hakims emerged as the new authors of medical texts that relied on Prophetic learning. Having invoked the Prophet, they moved on to embrace other dispersed authority referents, such as Sufi and fakir healers, as well. Simultaneously, belief in the humoral theory that justified the critical role of a hakim as healer continued.

In 1878 Ihsan Ali Khan, the hakim (*tabib*) of Mouzah Naarah, Pargana Karra, Zilla Allahabad, former government pleader at Hamirpur and resident of Qasba Salon, wrote the *Maqaalaat-i-Ihsani* in Urdu.[46] This is a 192-page compendium of uncompounded medicines or single drugs (*mufradaat mustamalah*) and *nuskhas* in which the names of medicines are listed in Arabic, Persian, and Hindi. The text very clearly claims the authority of the Prophet for Unani.

The *Maqaalaat* borrows in style from the Arabic medical literature of the medieval Islamic world that claimed religion for medicine. Like the Arab literature, it highlights the fact that the 'subject' (*maudu*) of Unani is the human body—the noblest of God's creation. This makes medical knowledge sacral. Religion is firmly located within the medical framework, making medicine the noblest and most sacral in the hierarchy of crafts.

However, the ultimate seal of approval for Unani comes from Prophetic authority. Ihsan Ali argues that the significance of acquiring knowledge of the body (*ilm al abdaan*) is underlined in the Quran and the Hadith.[47] Both these are intimately linked to the Prophet—the former being revealed to him and the latter being a compendium of his Traditions. Therefore to know Unani is *sunnat*—i.e. to follow the tradition of the Prophet: 'the Prophets Moses and Muhammad declared knowledge of Tibb [*ilm-tibb*] as mandatory [*mafrooz*]. This Unani conforms to the Traditions of the Prophet [*masnoon* or *sunnat*].'[48]

Having established this connection, Ihsan Ali very much like Hafiz, marks out hakims who follow Prophetic dictates in healing. He calls them the 'blessed ones' and says: 'They [Prophets] made man exalted and extraordinary [*ashraful makhlooqaat*] and blessed some men with

[46] Khan 1878: 1–192.

[47] Ibid., title page, no page number.

[48] Ibid.: 1–3.

beneficial knowledge [*ilm-i-waafi*] and perseverance [*himmat-i-kaafi*] so that they could write books and compendiums on the body. Such knowledge was critical for maintaining the body.'[49] He makes a plea for the popularization of the new Unani and the texts authored by its hakims. He concludes that if God's diktat on maintenance of the body is to be obeyed, these books need to be written in the vernacular and popularized to reach as many people as possible.

The *ilm al abdaan* (knowledge of the body) that the text popularizes is in conformity with the idea of Islamic well being. Prescriptions hinge around proper bodily deportment and moral conduct, common good sense, piety, good judgement, and a diet in keeping with the practices of the Prophet. The humoral theory of medicine is diluted with a concept of the human body that is dietetic, moral, and spiritual rather than based on intelligible and easily identifiable physical elements. In this integrationist theory, mastery of the body is more a matter of accumulated experience and bodily practice than scholastic learning about physical change.

Even though the Prophet's authority remains the ultimate reference point for Ihsan Ali, he also encompasses into his Unani other referents of healing. In another text that he authors, *Tibb-i-Ihsani*,[50] published by Munshi Newal Kishore of Lucknow, Unani steers even further away from its 'secular' mechanistic stance, as here he sees illness in terms of the relationship between the body and the soul. Very much like the Sufi healing practices of medieval Islam and the pietist ideas of eighteenth-century Europe, he sees the body and the soul united in health with the soul controlling bodily functions. According to him, the soul is created when food (*ghiza*) reaches the liver and is processed. One part of the end product goes to the heart and produces the soul (*rooh*). This *rooh* is *rooh-i-haiwaani* (animal instincts), and it gets converted to *rooh-i-nafsaani* (human instincts) once it reaches the brain. These instincts include those of smell, hearing, sensation, taste, etc. Blood is also produced by food; in fact it is the food of the body. The ideal clean blood is warm, moderate in consistency, bright red, and sweet. It keeps the bones and other organs moist. Ultimately, the soul preserves blood. And in illness the cognizant functions of the soul become so misdirected

[49] Ibid.
[50] Ali 1864.

that disease takes on a lethal course. In such a view, the hakim's role as the moderator of diet, exercise, regimen, air, water, etc. is minimized, the soul being the chief regulator.[51]

In his discussion on pharmacy and the making of medicine he does not confine himself to the Prophet's dietary regimen but invokes the 'superstitious charms of fakirs' as well. Thus, for instance, *mongaa* (a precious stone), when tied to the forehead (*peshaani*), helps in curing headaches; the wearing of the Christian cross around the neck is also recommended for its healing qualities. He warns the public that such cures are integral to health and should not be dismissed as the superstitious charms of fakirs. And certain medicines, such as rose and vinegar, are effective in both winter and summer.[52]

Ihsan Ali sees no contradiction between Prophetic medicine, religious spirituality of the Sufi kind, and medical rationality as represented in the humoral theory of Aristotle. Thus, despite leanings towards the centrality of the soul in maintaining the body, his text enumerates a long list of dietary regulations for the balance of humours. Unnecessary abstinence from food by a healthy man is forbidden. Food combinations that should be avoided to maintain body humours are listed: fish and milk, sweet porridge with egg, and vinegar with dry rice, curd with radish, etc. He warns that if easily digestible food (*zood hazm*) is mixed with food that is difficult to digest (*deer hazm*), people can get very ill.

The hakim's role is thus important for regulating the environmental and dietry regime of the ill. Indeed, it lays out a medical etiquette for the hakim. The social conduct of the hakim as a sensitive mediator and 'cultural broker' is highlighted as an essential part of treatment. The text underlines the importance of the hakim as a compassionate consoler. He is a person who consoles (*tashaffee*) patients, earns their good will, and maintains good relations and conduct with his attendants as well. Consoling the patient enhances his *quwwat* (stamina), and this in turn fights disease. And by good conduct (*khush akhlaaqi*) an external examination of the patient can be adequately conducted, and his outward condition (*kaifiyat-i-khaarji*) suitably ascertained. This is useful

[51] Ibid.: 7–11; see also French 1990: 88–110. He brings out the dialogue between the Pietist and the Mechanist views on illness in the eighteenth century, when the Enlightenment changed ideas of religion and medicine in Europe.

[52] Ali 1864: 7–11.

for both diagnosis and treatment. The *Tibb-i-Ihsani* lists five elements in an individual as crucial to his being a successful hakim. These include knowledge of the basics of medicine and pharmacy combined with common sense, patience, and proper conduct. According to Ihsan Ali, the essentials for obtaining mastery over the body are knowledge of qanun-i-tibb (rules of medicine); competence to use suitable ingredients to make effective syrups; and most importantly, a sensitive, alert, and caring individual who is large hearted and not a slave of his desires or selfish (*khwahish-i-nafsaani*).[53] Such prescriptive norms gave a hakim far more agency than before. His sensitivity to the 'real world' of his patients rather than his scholarship, became the yardstick of competence, making the hakim a self-styled master of his skill.

This flexibility in the definition of the healer was reflected also in the pharmacy rules prescribed in Urdu texts that upheld the hakim's individual judgement and discretion. Texts like the *Tibb-i-Ihsani* have pharmacy rules that do not uphold the medical precision which doctors were striving to achieve. Instead, they emphasize flexibility and ambivalence in conformity with the cultural world of patients. Thus, for instance, they recommend that if the hakim has not understood the cause of the disease, he should prescribe medicine that can cover both the ailment he has diagnosed as well as the one he suspects. Ihsani's text recommends symptomatic treatments: medicines should be given as per symptoms. It lays out four stages of medicinal effects: the first being when the medicine causes no change in the body; the second when it causes change but the bad effects are not reflected; the third when the bad effects are not fatal; and fourth when the bad effects are fatal. These, along with symptomatic treatment, are the recommended regimen when administering medicine. In a separate section on medicines, the writer stresses the importance of proper regulation of medicines for each individual, as per his body constitution. Unlike the natural philosophy stress on generalized medicines, here the variations in individual body constitution and social context are underlined as important to the regimen.

There were no steadfast universal rules and precise formulae recommended for pharmacies in Urdu texts. Texts that boasted of the

[53] Ibid.

existence of a sophisticated pharmacy tradition in Unani were written, but they emphasized that the making of medicines did not need specialist training. It was sensitivity to the cultural world of the patient that was essential in modulating medicine to individual needs. This was best exemplified in the 168-page *nuskha* compendium called *Qarabadeen Ihsani* written by Ihsan Ali Khan in hijri 1281 (1864).[54]

The *Qarabadeen* gives medical recipes for making syrups (*sharbats*), powder (*safoof*), tablets (*golee*), and *majoon* (electuary). It underlines the need to make compound medicines not necessarily as per standardized global norms, but in accordance with the needs of individual patients. It recommends that substitute constituents be used if the originals are not available.[55] Compound medicines are recommended in case simple ones do not work. Additional compounds are recommended to modulate the smell, taste, and intensity of the compound medicine being prepared, as also to enable it to affect special organs and diseased areas. It notes that complex diseases also require a combination of medicines to cure them.[56]

The *Qarabadeen* does not promote any standardized medicine manufactory or place or even a consultant hakim. It leaves its consumers to choose the place from where they can procure the best medicines. It encourages them to consult self-help texts that discuss the constituents that go into the making of compound medicines.[57] In its descriptions of potions and special medicines like China root (*chob-i-cheeni*), it leaves the composition entirely to the discretion of the hakim. His

[54] Khan hijri 1281, 1864.

[55] Ibid.: 9.

[56] Ibid. The text however classifies three types of medicine: *Jamaadaat* (made from non-living constituents), *banataat* (made from barks of trees), and *haiwaanaat* (made from animal matter). *Jamaadaat* requires no time schedule and can be purchased from any where and at any time. *Banataat* comes from trees that flower annually and either have a life of a year or a few years. Different parts of the flowers and fruits of the tree need to be plucked at the requisite time for the preparation of these medicines. And *haiwaanaat* is made from various parts of the animal body that differ in their medicinal propensities. Such medicines need to be kept in a dry place and are fit for consumption until they retain their smell, taste and colour.

[57] Ibid.: 12–14

discretion and judgement in keeping with the specific cases he handles is central to all such preparations.[58]

The text stresses a proper code of conduct—*adab*—for the dispensation of medicines as well. Ihsan Ali offered a code which is comprehensible and interactive with the patient, and which includes him in the healing process. It lays out norms for the hakim and patient's interaction. He highlights the importance of writing medical prescriptions (*nuskhas*) in a legible manner so that patients do not have difficulty discerning the names of the medicine and can participate in the healing process. Keeping a finger on the sensitive pulse of society, he insists that the rich not be prescribed cheap medicines because they perceive them as being less effective. Further, compound medicines (*dawa-i-murakkab*) must not to be doled out to just anyone. They should be sold in public, with witnesses around, so that if there is a problem the pharmacists are not accused of highhandedness.

Scientificity, Astrology, and the New Unani: The Risaalah-i-Qarura

The new hakims embraced the authority of astrology as well. Astrology had been linked closely in the Arabic sources to medicine and even to Aristotelian natural philosophy. But when it was used in twelfth-century Europe its Aristotelianism meant very little. Medical academies excluded it from natural philosophy.[59] It continued as a system of diagnosis outside the academies. Unani medicine, heavily influenced by the Aristotelian tradition and eager to develop its 'rationality', had excluded it from scholastic learning. However, in the late nineteenth century, as Unani became more culturally widespread, it incorporated astrology. This was combined with scientific techniques of urine testing.

Astrology was now used by hakims in their practice, even though it remained beyond the purview of its literature. In the late nineteenth

[58] Ibid.

[59] French 2003: 132–3. Aristotle held that numbers that are the basis of astrologers' impressive calculations could not reveal the essence of things.The personalities of the planets and their influences on things in the world were antithetical to Aristotle.

century, specilaized texts on uroscopy that combined astrological techniques to offer prognostics began to be written. In 1877 Hakim Ghulam Yahya published an article on urine testing called '*Risaalah-i-Qarura*'. This was published in the *Risaalah Naadir Ahwal* that was brought out by Munshi Newal Kishore's press. Hakim Ghulam Yahya wrote the text in a question–answer form at the behest of his friend and close associate Hakim Sayid Muhammad Jaan sahib.[60]

The text has details of how urine is produced and where and how it circulates in the body, and how it exits the human system. The focus is on urine testing and it offers details of related diagnostic techniques. Copious details about the instruments required, and tips for making diagnostic judgements, render this text an easy self-help manual for those not formally trained in uroscopy. The instructions include urine preferences, choice of bottles for storage, the kind of light in which hakims should examine urine, and the duration it should be allowed to stay in the specimen bottle. Perhaps the most important instruction is the uroscopy etiquette to be followed, which gives due regard to astrological influences in diagnostic techniques.

According to Hakim Ghulam Yahya, the best urine for testing is that which is fresh and has been kept for some time so that its constituents have mixed well. After 6-7 hours, the urine is not fit for testing because impurities have set in it. Etiquette requires that the person who examines the urine (*qarura*) should hold the bottle in his left hand. It is desirable and wise for the hakim to keep the bottle at a distance from himself so that its colour does not change. He should examine the urine in daylight, allowing the sun to fall on it from the north. This is because when sunlight reflects on the bottle, the urine shines and can be examined best.[61]

The colour, smell, and viscosity of urine as gauged through astrological configurations are then used to interpret the humoral theory. The text offers the hakim critical clues to diagnosis. For instance, red and yellow urine indicate hot temperaments and anger in patients. The text lists a series of diseases that can be identified from the viscosity and density of urine. It also provides clues to the anatomy of the human body and the gender of the patient. Thus, the urine of non-adults is

60 Yahya 1877.

61 Ibid.: 5–6.

more viscous than that of adults. That of women is dirtier and whiter and less luminous than that of men, and it does not become translucent when shaken.[62] And interestingly, the temperament (*mizaaj*) of a patient can be discerned from his urine. Whitish urine indicates a cold temperament, and so on.

Risaalah-i-Qarura was not the only text on uroscopy written in the nineteenth century. In 1877 Hakim Ali Hasan Saiyid wrote the *Zubdat al Mufradaat*, in Lucknow. This is an important text of 72 pages that has sections on how to examine the pulse, diagnose diseases through urine, and follow astrological configurations while testing urine. And if the *Risaalah* combined uroscopy with astrology, the *Mufradaat* followed the poetic writing style of *akhlaaq* literature. The entire text is written in prose. The section on urine testing has different poetic sections that reveal the range of possible deductions to be drawn from the various colours of the urine and the diseases to be identified from its turbidity and density.[63]

Unani as Persianate Social Etiquette:
The Akhlaaq-i-Kashi

In the late nineteenth century, Unani reached out to the Persian norms of medical civility that still simmered in society. Urdu texts modelled themselves in part on the Persian Mughal manuscripts which, as we saw in Chapter 1, stressed proper bodily deportment and aristocratic virtue as central to good health. However, these texts, unlike the Mughal elite literature, were accessible to ordinary people. Pandit Kashi Nath's *Akhlaaq-i-Kashi* written in 1870 for 'ordinary people', was one such text. Published from Agra, it was priced at Rs 12 and was written primarily to be taught to Hindus and Muslims in their schools and homes. Later, it was included in the list of books prescribed by government schools.

Akhlaaq-i-Kashi belonged to the genre of polite literature on conduct and etiquette that characterized Persian literary culture. Indeed, Kashi Nath wrote it in consultation with three well-known Persian texts: *Akhlaaq-i-Nasiri, Akhlaaq-i-Jalili*, and *Akhlaaq-i-Mohsini*.[64] The text

[62] Ibid.: 37–8.
[63] Saiyid 1877.
[64] Nath 1870: 11.

defines a perfect man (*kamaal-i-insaani*) as someone who is not only healthy but is complete with humility (*sharf-i-insaani*) and has knowledge of the polite culture of moral self improvement, etiquette, and conduct (*ilm-i- tehzeeb-i-Akhlaaq*). *Akhlaaq* is an acquired virtue. Its value lies in the fact that, through this trait—learnt via the intellect, talent, and advice of elders—a man can move from his natural state as the most extraordinary of all created beings to the self-acquired state of a complete and perfect human being.

Very much in the manner of Islamic medical ethics, the text projects the hakim as one such perfect man who can combine medical knowledge or learning (*quwwat-i-ilmi*) with piety and moral conduct (*akhlaaq*), and experience and practice (*quwwat-i-amali*). Through the former he can know the medical condition or disease, and through the latter he can build the illness narrative of the people and 'govern' health (*hukmaaran-i-sihat*). The hakim can intrude into the moral and cultural world of his patients and change what is injurious to what is beneficial. But this ability he can only acquire if, through his conduct and manners (*akhlaaq* and *tehzeeb*), he can make his every action praiseworthy.[65] Kashi Nath argues that if, for the hakim, *tabaabat* (the practice of medicine) is the most privileged learned tradition because it deals with the body of man, knowledge of *akhlaaq* is the most noble, for only through it can he reach out to the moral and cultural world of his patients. Indeed, he defines a hakim as one who combines the two traits of *amal* (experience) and *ilm* (learning and knowledge of living beings) and who cements them with good conduct and etiquette (*akhlaaq*).[66]

Kashi Nath presented this Islamic medical ethics as a general code for all hakims, irrespective of religion. His text describes the manner in which one can learn to use one's intellect, exhibit good manners, and respond to the advice of elders to become the 'perfect gentleman' and then a good hakim. The social etiquette ladder through which one becomes a perfect man (*insaan-i- kaamil*) with characteristics of humility (*sharf-i-insaani*) is called *tehzeeb-i-akhlaaq*—the culture of etiquette. Kashi Nath sees this *tehzeeb-i-akhlaaq* as the characteristic feature of a good hakim.

[65] Ibid.: 4
[66] Ibid.: 10.

Again, very much like Prophetic medicine which saw health linked to moral and psychological well being, Kashi Nath connects disease to psyche (*nafs*), arguing that the health of the nafs is central to the health of the body. Indeed, *nafs* can be regulated through *akhlaaq*. Kashi Nath underlines *akhlaaq* as key to the control of the *nafs*—meaning desires. He highlights the fact that the knowledge of medicine and knowledge of *akhlaaq* complement each other; together they create the 'perfect' healer. The former forbids certain food items as injurious to health, and the latter identifies types of social company as detrimental to the health of the *nafs*. Since a healthy *nafs* is critical to a healthy body, *akhlaaq* becomes critical to the overall health of the individual, since it controls *nafs* or desires.[67]

The text urges individuals to have good friends who can tell them about their behavioral defects. It exhorts its clientele to observe carefully the *akhlaaq* of their friends and enemies and use it as their mirror. It cites the example of Yaqub Kindi, an Islamic wise man who said that students of *hifz-i-sihat* (healthcare) should try and avoid social traits that they see as defects in the *akhlaaq* of their friends.[68] It also elaborates the nafs-*akhlaaq* connection by citing the example of certain diseases that are cured by proper *akhlaaq*. One important example of how this works is the case of the disease of ignorance and illiteracy (*jahl-i-baseet*). Its cure is to keep the illiterate man in the company of the literate. This enables the ill (illiterate) to develop an interest in education, and be thereby cured.[69] Finally, Kashi Nath gives a four-point prescription for becoming the master of one's psyche (*sihat-i-nafs ka hafiz*). According to him it is imperative that the master of the 'health of psyche' does four things: keep good company and listen to wise men; read and acquire knowledge; follow *amal* (practice) in accordance with *ilm*; know the merits and flaws of his action as seen through the eyes of God.[70] In sum, he underlines the combination of *ilm* (learning), *amal* (practice), and *akhlaaq* (conduct) as key to a perfect and complete healthy man.

The link between *tibb-i-badani* (medicine of the body) and *tibb-i-nafsaanee* (medicine of the psyche) is further elaborated when Kashi

67 Ibid., part III: 1–2.
68 Ibid.: 3.
69 Ibid.: 9–14.
70 Ibid.: 4.

Nath compares its logic to that of the theory of humours. He says that just as in *tibb-i-badani* every ailment is cured by giving its antidote, similarly in *tibb-i-nafsanee* every social conduct has an antidote that can cure the disease. So, for instance, if you have a habit of being a miser, then acquire the habit of being a spendthrift. And this similarity of treatment is because temperament (mizaaj), which has to be balanced for a healthy body, is constituted by both *jismaani* (bodily) and *nafsaanee* (desires and soul-related) traits. The body and the *nafs* have a natural link and rapport, and they cannot be separated from each other without the order (*hukm*) of God. Changes in one realm cause changes in the other. So, if nafs has some tension or crisis, or if some melancholy envelops it, then the body also gets a trembling sensation and lethargy takes over. And if a body gets diseased, it causes tension and produces injury in the *nafs*. Kashi Nath argues that for perfect health the specialist of nafs should work in consultation with the specialist of the body. Diseases should be identified as originating from the body or the *nafs* and treated together by specialists in the two fields.[71]

A dead body is one which is rid of the *nafs* and experiences no sensation. The ignorance of people with regards to nafs as central to their health can be removed by knowledge and education. This is crucial to maintain their good health.[72]

Self-help Primers—Zubdat-al Hikmat, *Or the* Sihat Ayeena

The fusion of religion, spirituality, and bodily deportment with Aristotleian medical rationality also meant that Unani oscillated from upholding the relevance of the hakim to diluting his significance by providing easy self-help medical texts to people. Qamar Ali's prescriptions on health, suggestively called the *Sihat Ayeena* or Mirror of Health, and also titled *Zubdat-al hikmat* or Rules of Hikmat (medicine), constructed a medical ethic and urged every individual to become the 'master of his health' (*haafizan-i-sihat*).

This book is a compendium of about seventy-two *nuskhas* collected from a range of physicians. It was written at the behest of Lala Bhagwandas Seth, resident of Suba Mathura, so that 'every ordinary person

[71] Ibid.: 5.
[72] Ibid.: 14.

becomes the master of his health [*haafizan-i-sihat*].'[73] Its prescriptions combine the Hippocratic emphasis on proper diet and deportment with the Aristotelian stress on proper environment, body pathology, and medication as critical for good health. In combining the two it offers a foolproof regimen of preferred diets for different seasons so as to maintain good health. It promotes the idea of a dietary regime, deportment, and a regimen of emotional and physical etiquette as the basis of good health. Qamar Ali is against medicalization. He categorically states—'we do not need medicine. For medicine melts away the body. And if you are not feeling well do not go for medicine.'[74]

Instead, he puts his finger on diet and deportment as the best healers. A healthy person is advised to stay away from extremes of weather and polluting environments—e.g. hot and strong winds, graveyards, abattoirs, the foul air of latrines, drains, ghats, etc. Regions like Kashmir are also to be avoided. Fragrances that keep the house and body warm, such as *lobaan* (camphor), attars, and amber incense are recommended.

Extreme weather conditions require a change in dietary preferences as well. The rules prescribed here suggest different diets in different seasons. In winters (*mausam-i-sarda*), hot meals need to be eaten at midday. Pure wheat bread, to be eaten with mint, ginger, and fenugreek, is advocated. Almonds, fresh pistachio, nutmeg (*jaifal*), and mace (*jawatari*) are to be consumed morning and night. The sweet items preferred are puddings (*halwa-i-tarkhub*) and scented sugar, and musk and cloves added in betel leaves. A balm (*roghan*) of jasmine (*chameli*) is to be rubbed on the body at night. Relaxation with friends is also part of the recommended health regime. In summers (*mausam-i-garm*) an alternative diet is recommended: pure wheat bread soaked in grape juice, tamarind, or limejuice is preferred. Apple, pomegranate drinks, and vegetables like pumpkin, spinach, and gourd—that are cooling—are suggested. The diet for the monsoon season (*mausam-i-barsaat*) includes stir-fried items. If the monsoon is hot, then a cooling scented essence (*kewra*) and unrefined sugar (*misee*) are recommended. For the health of the heart and mind, mixed drinks of rose, pomegranate, and oranges are useful. The use of tobacco in betel leaves is prohibited.[75]

[73] Ali 1867: 3–4.
[74] Ibid.: 7.
[75] Ibid.: 4–5.

Alongside this, etiquettes of eating are outlined. One should eat only when hungry and stop eating when thirteen parts of hunger remain. Similarly, water is to be had only when one is thirsty. And one should never follow dietary abstinence when healthy. One meal a day is the preferred diet. Chewing the food slowly and not gulping down large portions quickly are key to relax and stabilize the liver. Too many varieties of food should not be eaten together. Sweet dishes should be partaken at the end as they help in digestion.[76] Easily digestible (*lateef*) and difficult to digest (*kaseef*) food combinations are to be avoided. Sour foods are to be avoided since they age the mind and make one old before one's time. Too much sweet weakens the liver and reduces the appetite. Too much salt should be avoided since it dries the body.[77] Food combinations such as radish and curd, milk and sour things, milk and spinach, and melon and honey are to be avoided.

An entire regimen of etiquette for drinking water is also recommended. In between meals, water is not to be consumed. Neither is it to be had immediately after meals. The best time is 4–5 *gharis* after thirst. The preferred way to drink water is in three goes. The text also offers a hierarchy of preferred waters: river water is better than well water, and water that flows from the mountains is best. The water of the river Ganges is privileged over all other waters. Well water can be mixed with river water for drinking. Hot water (*aab-i-garm*) eases constipation; cold water (*aab-i-sard*) gives stamina (*quwwat*).

The text also offers a regimen of exercise for the stability of certain organs like the liver. It suggests fasting (*faaqah*), sleep, and rose water as the remedy for a defective liver. It recommends an exercise regimen of lying on the right side, and turning to the left after a while, and then going for a walk before sleeping at night. Some form of exercise (*riyazat*) is suggested for every morning and evening. Emotional well being—keeping sadness out of the system and keeping the heart happy—are the best route to good health.[78] Food cooked in silver and gold pots is said to add to the energy of the heart and mind (*dil* and *dimaagh*). Hakims do not recommend copper and brass vessels as cooking pots.

Most of the *nuskhas* in this compendium reflect a sophisticated understanding of the anatomy of the ear, head, nose, teeth, and heart.

[76] Ibid.: 6–7.
[77] Ibid.: 5.
[78] Ibid.: 7.

But instead of suggesting medicines for their health, they highlight the importance of relaxation, cleanliness, and a regimen of emotional and physical decorum as the crux of care.[79] The compendium discusses the medicinal qualities of a range of materia medica without going into their chemical compositions or their medicinal forms. Thus, *chana* or gram is cited as possessing properties that purify blood and, when eaten soaked in water, give energy. If eaten continuously, channa reduces phlegm and improves the voice, reduces toothache, urine flow, and constipation. Its paste softens the body and is good for skin ailments like ringworm (*daad*).[80] Similarly, maize (*jau*) removes thirst, is good for the liver, and removes phlegm when mixed with fig. Its paste removes pain. A kind of vetch-ling (*moong*) is beneficial for colds and sore throats.[81] The herb *til* is recommended for removing phlegm, clearing the chest, improving voice quality, etc.[82]

The Unani Defence: Competing with British Medical Literature

The Urdu medical literature discussed above socially constructed a new culturally widespread and totalitarian Unani. It created a new hakim. It thus brought to the forefront fresh ideas about the causes and treatment of diseases that were neither reducible biologically or pathologically, nor ecologically identifiable. Instead they were seen as lying embedded in a diverse cultural matrix. This medical literature was used both for creating an awareness of Unani in society, as well as propagating primers and self-learners among those who wanted to use it as a profession.

The new Unani created a firmer cultural niche for itself than the old Unani. Indeed, it remained beyond the control of older medical communities and competed with them. It joined the race for popularity with the British medical literature in the vernacular as well. Like government-sponsored materia medicas, Urdu texts were low priced and

[79] Ibid.: 17–18, 23, 28, 30–2.
[80] Ibid.: 48.
[81] Ibid.: 49.
[82] Ibid.: 51.

easily available.[83] Indeed through this competition Unani underlined its difference with British medicine. Unani saw its scientific rationality as deriving from a global Islamic tradition, including both pre-Islamic Aristotelian strands as well as the person of the Prophet. It also struck local roots and reached out to the fakirs and Sufis of India. This futuristic, rather than preservationist, tradition underlined the specific modernity of Unani.[84] Such a Unani was more suited to the temperament or mizaaj of the Indian people, a mizaaj deriving from the four elements that constituted the locale of an individual. Culture—in its moral, spiritual, and psychological dimensions—was seen as part of the locale. Hakims who authored Unani texts emphasized that they and their patients shared this locale, and were thus compatible with each other. They projected the British as external to this shared world of the healer and the healed.

In Unani, as we saw above, health was all about the balance of humours; but by the end of the nineteenth century, humours were interpreted through many ways: ecology, diet, bodily deportment, and Prophetic piety. These factors rarely had universal referents. They were bound in locales. And thus health was also about locality. The best healer was one who shared the locale with the patient. Such definitions pushed British medics outside the shared locale and therefore projected them, even if as good and competent, as also unsuitable for Indians.

[83] *Oudh Akhbar*, 1 January 1879. This issue carried a price list of the Urdu texts as an advertisement. The list cited the following prices: *Risalah Zabtah al mufradat*—2 annas; *Zubtadah al hikmat*—6 annas; *Mufid al ajsaam fawwaid ajibah*—3 annas; *Ilaj-al ghuraba*—8 annas; *Makhzan al adwiyah*—Rs 5; *Tarjummah Tibb-i-Akbari*—Rs 4 and 4 annas; *Qanun Itrat*—4 annas; *Tohfatul Atibba*—1 anna; *Qarabdeen Shifai*—6 annas; *Qarabdeen Zakai*—12 annas; *Mujarrabaat Akbari*—5 annas; *Tibb-i-Hindi*—2 annas; *Ramooz al hikmat*—2 annas; *Mualajat Ihsaani*—4 annas; *Tarjummah Urdu Ilaj al Amraaz*—Re 1 and 10 annas; *Risaalah Qarura*—1 paisa; *Markiyat Ihsani*—5 annas; *Ilaj Ihsani*—2 paise; *Mizan-i-Tibb*—8 paise; *Tashreeh-al ajsaam*—7 annas; *Majmuah al Bahrein*—Rs 3; *Maqalaat Ihsani*—6 annas; *Ilaj al mawaish*—3 annas; *Tarjummah Khwarzmshahi*— Rs 10.

[84] This particularistic modernity of Unani is discussed in further detail in the next chapter. For this futuristic notion of tradition, see Gadamer 2004; Geyer & Paulmann 2001; Goswami 2004.

If the locale shared by the hakim and the patient drew the line of difference between Unani and Western medicine, the stylistic and writing techniques of Urdu texts highlighted this difference even more. As we saw above, almost every Urdu medical tract, despite its propensity for Prophetic piety, responded to the colonial 'narrative of progress' by retaining as its core the humoral diagnostic techniques and awareness of advanced techniques of orthopaedics, pharmacology, and pathology. At the same time, stylistically as well as in their definition of health and disease, these tracts contrasted themselves with Western medical science. Stylistically, such texts were couched in the powerful Mughal-style polite literature of moral self-improvement—*akhlaaq*—as well as popular wisdom literature available in a gnomic style. It is significant that many of these texts were categorized or named with *akhlaaq*. And many of the journals were called *Tibbi Ayeena*, in line with the title of the genre of polite literature on conduct and etiquette published for princes in Mughal India. Their content and language were not different either. For instance, as we saw above, a text like the *Akhlaaq-i-Kashi*, written by Munshi Kashi Nath in consultation with three well-known *akhlaaq* texts, is written in Persianized Urdu, and is hyperbolic in style, has a section on etiquette, and a long section on health heavily influenced by the Persianate Islamic way of life. Good health is about proper governance of the body (*hukm sihat*); a good hakim is one who combines *ilm* (knowledge of the body constituents) with *amal* (experience and implementation) and turns what is injurious into beneficial.[85] As a concept, this is very similar to the role of the all-knowing monarch in *akhlaaq* literature. The recommended deportment for good health—sleep, clothes, lifestyle—is also comparable to those recommended for the ashraf and the ajlaf in *akhlaaq* literature.

This trend can be seen in several other texts, such as the *Zubdat al Hikmat* or the *Sihat Ayeena*. In these, health is about the maintenance of appropriate decorum in dress, cooking, and domestic comforts.[86] Such Mughal-style texts mingled with those like the *Tibb-i-Nabawi*, which used the Prophet's authority to legitimate Unani. The argument about the divine origin of medical knowledge and its transmission to

[85] Nath 1870, part II: 10.
[86] Ali 1867: 4–5.

special people in society projected Unani practitioners as *afzal makhloo-qaat*—extraordinary and noble. This extraordinariness underlined the difference between them and others and such a noble profession could not be allowed to dissipate. Almost every Urdu text underlined the extraordinariness of Unani as a consequence of its divine origin. While Hafiz's *Tibb-i-Nabawi*, as we saw above, openly exhorted hakims to combine Prophetic learning and piety with empirical experience and wisdom, other authors were happy to use Prophetic authority even as they reached out to Sufi and other referents.[87]

Underlining the difference with British medicine was only one aspect of the Unani defence. Cannibalizing parts of English medicine was another aspect of survival. Professionals trained in the English system of medicine were familiarized with Unani. At the same time, hakims were made aware of some of the useful techniques and medicines of the English doctors as a way to combine the two systems.

An important illustration of this assimilative strategy, which brought the two systems together even as it underlined their difference, was the publication of a monthly Urdu journal called *Mirror of Health* (*Tibbi Ayeena*) whose editor, Imad al Din Ahmad Tabaish, was the curator of the Agra Medical School. He published the journal of his own volition. Both hospital personnel as well as hakims read this journal. It served as a critical bridge between Unani and British medicine, even though its clientele was primarily Indian native doctors who did not know English and needed a journal in Urdu for their regular medical update. This journal aimed at bringing Hindustani cures to the notice of professionals trained in the British system, and to familiarize hakims with the latest treatments and cures known to British medics.

The introductory pages of the journal list its benefits and state that it is primarily aimed at teaching the modern experiences of surgeons to those hospital assistants who do not know English. It also aimed to publish photographs of modern medical instruments that came from Europe and list their characteristics so that people could familiarize themselves with modern instrumentation. Hospital staff and Unani professionals were both on its list of people who needed to be educated about Western medical treatments and techniques. Its aim was also to

[87] This is most glaring in all the texts of Ihsan Ali Khan discussed above.

make Unani knowledge and Hindustani cures popular amongst hospital staff.[88] The *Tibbi Ayeena* was a monthly publication of approximately thirty-two pages. It reproduced useful articles in Urdu from English newspapers and journals like the *Indian Medical Gazette*. Its prices were reasonable. It could be subscribed to on a monthly or annual basis by writing and sending a money-order to the editor.

Apart from lifting articles on scientific instrumentation from other English journals, it had editorials on medicines for fevers, and details of their preparation, including the locales from where their raw forms could be acquired. In an important editorial on the medicine '*sarkari sankuna*' or medicine of fevers, details of its preparation are deliberately listed for the benefits of Unani hakims. As the editor says, 'because this risaalah is read by attiba-I-Unani and Hindustanis, therefore it is hoped that we describe this medicine's reality and method of use'. It goes on to describe its place of cultivation—Darjeeling—apparently it was available in government gardens, in the form of a tree bark with ingredients relevant to Tibb-i-Unani. Well-known doctors of the Calcutta Medical College as well as learned people (*sahaban-i-mausuf*) had deemed it useful as a substitute for quinine. It could be had in mixture form, or as a syrup. The editorial gives details on how to prepare tablets out of this medicine and ends with the announcement that the medicine can be had from both English as well as Hindustani medical shops in Calcutta.[89]

One of the important sections of the journal was the section called 'Letters to the Editor'. Here, a cross-section of people from all over the country, which included both ordinary patients as well as hakims, shared their medical experiences with readers. In this column both intellectual arguments on aspects of medicine as well as effectiveness of Unani *vis-à-vis* Western medicine were passionately debated. For instance, one Najmul Haq sent to the editor two *nuskhas* for publication from Zillah Basti. The *nuskhas* were for diseases like swellings and used medicines like carbonate of potash, ammonium chloride, boric acid, etc. The editor published this with the note that he had never seen anyone use them, and hoping they would be useful.[90]

[88] Ahmad 1879: 2.
[89] Ibid.: 5.
[90] Ibid.: 8.

The journal published other letters too, from hospital superintendents and assistants who shared successful remedies with readers. Thus, Syed Ghulam Husain, hospital superintendent, sent a *nuskha* for acute bronchitis that was passed to him by Maulana Baksh Sahib, hospital assistant. This included the use of English medicines like spirit ammonia, spirit chloroform, brandy, ordinary water, etc.[91] Similarly, Lala Bhagwan Das, a hospital superintendent at the Sitapur Cantonment, not only wrote to the editor about shortcomings in the *risaalah*, but also sent an article on diet and physique. The write-up was based on the findings published in an article by one Dr Line Sahib, who was interested in the diets of prisoners.[92] Doctors and hospital compounders and staff from as far as Darjeeling used the forum of the risaalah to publicize *nuskhas* that they had observed or experienced as being successful. Mohammad Muneer Khan, compounder at the charitable hospital in Darjiling, sent a *nuskha* for diarrhoea and dysentery, and one for dog bite. The editor, who could comment on these, cautioned readers that people with *baarid* (cold) *mizaaj* (temperament) should not have this *nuskha*.

A section of the journal had essays by contributors who delved into topical concerns, such as the causes of diseases like cholera, smallpox, etc. Fakhir Ghulam Husain's article, *Asbaab al Amraaz* or 'Causes of Diseases', was a case in point. The article attributed most diseases to air pollution and made a case for cleaning the environment of Indian cities. This was, in the author's view, a sure preventive against disease.

[91] Ibid.: 10.
[92] Ibid.: 11.

6

Argumentative Hakims: Debates in the *Oudh Akhbar*

Introduction

URDU MEDICAL TEXTS WERE DEBATED IN THE PUBLIC SPHERE created by Urdu newspapers.[1] These newspapers mirrored the ambiguities of society as it came to grips with the plurality of Unani traditions: they carried viewpoints on both the new futuristic as well as the old preservationist Unani, And of course, the print capitalist Munshi Newal Kishore (1836–95), editor and proprietor of Lucknow's Urdu newspaper the *Oudh Akhbar*, gave voice to the range of Unani traditions in his columns.

[1] For the idea of the public sphere in eighteenth-century Europe, see Habermas 1989: 27. His definition of 'public sphere' stems from his analysis of the development of bourgeois society in seventeenth-eighteenth-century Europe, in which 'private citizens [came] together as a public' to discuss matters of 'public concern' or 'common interest' and to criticize and put pressure on the absolutist state. The public sphere of Habermas has discursive and institutional spaces, a common language, and a set of procedural principles, and activism and engagement with a general public. It thus provides a space for debate and discussion between different voices. See also Fraser 1992: 109–42. For the creation of a public sphere in colonial Lucknow through the production of Urdu newspapers and journals that were produced by the middle class and which in turn shaped their politics and aspirations, see Joshi 2001: 23–58.

However, since his paper's core readership were the higher echelons of society, he allowed the voice of the elite who supported traditional Unani learning, to loom large over all others.[2] Indeed, Munshi Newal Kishore, the most notable entrepreneur of Lucknow, himself belonged to this upper crust of society, and he used his newspaper to strengthen his connections with the landed aristocracy, the service gentry, and the high minded in Awadh's British society. Their support was critical for his own professional growth. The *Oudh Akhbar* thus became an important advocate of the old canonical Unani which had the support of the local elite.[3]

The *Oudh Akhbar* created for Unani, to use Juan Cole's phrase, a new 'textuality' and 'authoriality'. This meant that through publishing the views of hakims on the future of Unani, it created a new concept of an author as public intellectual.[4] He was a professional who voiced not just medical but larger public concerns relating to Unani's survival in a colonial setting. In the columns of the *Oudh Akhbar* the Muslim public intellectual reworked the global ideas of professionalization, historicism, and modernity from within his own tradition. He challenged the Eurocentric nature of these concepts via those that were

[2] In 1881 the *Outh Akhbar* was reported to have a circulation of 715 copies. L/R/5/58, selections from the Vernacular Newspapers, received up to 18 June 1881, p. 348.

[3] Stark 2003: 68, 73, 75. She shows that the *Oudh Akhbar* had British patronage in the form of subscriptions and overall support extended to Munshi Newal Kishore. He had a business collaboration with the colonial administration, and had hogged the lion's share , about 75 per cent of printing commissions in 1882, for his press. The *Akhbar* supported the progressive agenda of Sir Syed Ahmad Khan, and its financial and moral supporters included the upper crust of Awadh society: Lucknow's printer publisher Maulana Haji, Harmain Sharifain, CSI, deputy collector, Raja Kishan Das, CSI, former secretary of Syed Ahmad Khan's British Indian Association. It had a printrun of 700–800 annually, of which only 20 per cent went to government. The rest was distributed around Awadh and the North West Provinces, and even Punjab. The readership far exceeded these circulation figures, since the newspaper was recirculated and read out to groups of non-literate people in public arenas and homes.

[4] I borrow this idea from Juan R.I. Cole, who uses it to illustrate the many different implications of print on the Islamic reform movements in nineteenth-century Egypt. See Cole 2002: 354–5.

rooted within the Unani tradition. The attempt was to balance the tensions between gentlemanly comportment, spiritual healing, and forms of knowledge beyond the control of family and state, with medical wisdom or science that lay within the purview of state and family control. For instance, the professionalization of Unani meant freeing society from the clutches of ill-trained Urdu-read hakims and the dispensaries that employed them. Professionalization meant proper state control of Unani: accountability, identification of 'proper' practitioners, formalization of the terms and conditions of their service, their representation in municipal boards, sanitary commissions and other government bodies, and adequate financial inputs for Unani by the state.[5] Professionalization was a strategy to streamline Unani and bring its core Arabic canon and new scientific experiments to the notice of English medical practice. It was a strategy used by well-heeled hakims to withdraw Unani into the family fold and place it under the surveillance of the state.

The Muslim public intellectual grounded his professionalization demand in a critique of Unani practice. Knowledges that had spilt out of the control of the family and state were the obvious targets of attack. These knowledges were represented in the practice of the new-style hakims. It was not the system of medicine that was at fault; it was the practitioner who needed to be reformed. Very much like the Arab writers of medieval medical deontology who complained about the quacks and charlatans of their times, the Muslim intellectual now complained about the fall in the standards of professional ethics and the incompetence of the new hakim.[6] Professionalization thus meant tighter state regulation of Unani's public face: practitioners, pharmacists, and clinics. This was the only way the public could be best served. The *Oudh Akhbar* argued that the government needed to intervene to pull Unani out of the hands of the new hakims so that it should serve the people better: government's intervention was urgently required because the dispensaries had failed to deliver. Indeed, these needed to be reformed as much as Unani practice.

[5] Haynes 2006: 130–56. This was different from the European concept of professionalization.

[6] Rosenthal 1990: 484.

Second, the public intellectual made a case for cannibalizing aspects of Western medicine, such as surgery. Yet he did not see this borrowing as emulating the scientific instrumentation of Western medicine. He viewed it as a way of embellishing Unani's own robust scientific tradition. He saw the scientificity of Unani as more relevant because it had a history in Hindustan that went back to antiquity. Indeed, he saw the unique modernity and scientificity of Unani as located firmly within the geographical and territorial imaginary of Hindustan as carved out by the colonial state. However, he reconstituted within it a national space for Unani by rewriting its history. This national space was constituted by a cultural imagery. The new history of Unani located its core in Hindustan, which was now seen as a distinct cultural space or locale. As the public intellectual mapped Unani onto the imagined cultural space of the nation, he projected it as part of Indian tradition and culture. He contrasted this lofty status of Unani with the mundane status of English medicine, whose 'modernity' was located outside this cultural space of the nation. He argued that Unani's distinctiveness lay in the fact that it had historical roots, whereas Western modernity had no history or territoriality in Hindustan.

Third, the public intellectual used the novel technique of advertisements to balance the tension between the colonial medical space and the national space newly carved out for Unani. Both the colonial as well as Unani medical pharmacies and practitioners were advertised. A vernacular public was created, and the two systems of medicine made interactive. Newspaper print culture thus located Unani in a lively and vibrant public sphere where medical debates had the potential of spilling into the political domain.

The Critique of the Dispensary

The *Oudh Akhbar* critiqued the everyday functioning of the dispensary, not the Western medical tradition that the dispensary upheld. Dispensaries were reported to be expensive, short of medical supplies, and racially discriminatory. The *Oudh Akhbar* reported that their doctors lacked *akhlaaq* (manners), provided no home visits, and did not take into account the class difference between the rich and poor patients who visited them. In addition, English medicine was seen as inappropriate for Indian constitutions.

Editorials in the *Oudh Akhbar* commented on reasons for the unpopularity of government dispensaries and enumerated why some people were not enthusiastic to visit them. The fact that these dispensaries were insensitive to the class consciousness of Indian elites was identified as a major inconvenience: they were neither sensitive to the *sharaafat* (gentlemanliness) of high-class people, nor to the *razaalat* (pettiness) of patients. They clubbed all patients together, irrespective of class. The paper suggested it would be better if every dispensary had separate branches for the gentleman (*shareef*) and the poor man (*razeel*). The editor pointed out that the changing times had reduced many wealthy gentlemen to paupers, and they could not afford home visits by doctors. At the same time, they found it below their dignity to attend dispensaries alongside the indigent.[7]

If class was an issue, so was race. The *Oudh Akhbar* regretted that the dispensaries discriminated between white and dark skins (*kaala* and *gora*) when dispensing medicine. The editor admited that, in view of a visible drop in the availability of good hakims and their medicines, government dispensaries were the only way out. While he was happy that they were proliferating, he lamented that their benefits were not being fully utilized by the people. This was because they discriminated on racial lines and the dispensations supplied by them were few and poor in quality. Good and expensive medicines were reserved for the Europeans. This, he said, was unfortunate since many dispensaries owed their origin to financial contributions from Indians.[8]

The dispensaries were found to be insensitive to caste taboos as well. A report in the *Oudh Akhbar* suggested that many Hindus avoided the dispensary because native doctors were chosen largely from the lower castes. The correspondent added that, 'although the Hindu population are quite willing to submit to treatment at the hands of a Musalman or a Christian of a different race from their own, yet they cannot be expected to look with the same respect on a low caste man of their own race.'[9]

[7] *Oudh Akhbar*, 1 October 1876, p. 1823.

[8] *Oudh Akhbar*, 15 October 1876.

[9] L/R/5/47. Selections from the Vernacular Press, received upto 22 July 1870, pp. 264, 263–5.

Added to this was the complete lack of personal involvement of doctors with their patients. According to the *Oudh Akhbar* both elites and the poor were disenchanted with the dispensary administration. Yet they had no recourse except to knock at their doors for help since they could not afford the home visits of doctors, these being priced Rs 4–5 on average. One editorial appealed to the government to encourage doctors to make home visits, treat patients with politeness (*shafqat*), and allow them an additional allowance if necessary—at least over periods of seasonal disease.[10] An anonymous long essay in the *Oudh Akhbar* by a reader reflected similar concerns and noted that too many patients had made personal care of the sick questionable. Added to this was the expense of buying prescribed medicines from chemists (*attars*) or paying a hefty fee to doctors so as to receive medicines. Such problems created popular apathy towards dispensaries. The writer also noted the lack of compassion and etiquette as further deterrents to the patients. Indeed he indicted the dispensaries, saying that a doctor lacking in *akhlaaq* was not a doctor at all, for the *akhlaaq* of a doctor is as important as his medicines and diagnostic skills.[11] He suggested that the dispensaries needed to integrate *akhlaaq* into the profession, especially since doctors obtained handsome government salaries. And if *akhlaaq* was important for doctors, it was even more essential for compounders and other staff in the dispensaries. The writer regretted the fact that dispensary staff lacked etiquette and was steeped in corruption.[12]

Other newspapers too wrote about the blatant arrogance and criminal behaviour of many native doctors posted at the dispensaries. In 1882 the native doctor at the government dispensary in Amroha was the butt of a storm as people accused him of 'extortion' and 'avarice'. He allegedly poisoned a patient to death since he could not meet the doctor's financial demands. The newspapers of Amroha took sides on the issue. While the *Ahsanu-'l Akhbar*, the Urdu weekly published by Ali Husain Khan, pressurized the government to take action against the doctor, the *Ainu-l-Akhbar*, the Urdu weekly published from neighbouring Moradabad

[10] Ibid.

[11] *Oudh Akhbar*, 27 September 1879, p. 3034.

[12] Ibid.

by Dilawar Ali, supported the doctor. The former published a petition which the people of Amroha had submitted to the government elaborating upon his corrupt activities. The weekly stated that the doctor had in the past been fined and transferred by the late civil surgeon of Moradabad, Mr Deane. However, once Mr Deane left, the doctor used his influence to return to the city. The war of words between Urdu periodicals brought the issue of corruption in dispensaries to the public gaze.[13] The correspondents of the *Oudh Akhbar* did not cease to report on innumerable other cases of extortion and corruption by native doctors in the mofussil towns of North India. Most cases of high-handedness arose because the dispensaries were short of supply with medicines and the doctor's fee, at Rs 2, was very high. It requested the government to look into these problem zones.[14]

Ordinary people too voiced their concerns about the efficacy of English medicines and doctors in their letters to the editor. They exhorted the newly formed Indian National Congress to take up these issues and pressurize the government to act.[15] Many letters also cautioned the public against rumours about Western medicines and doctors that were often spread by jealous and competitive hakims. In a letter that urged the government to pay attention to the plight of Unani hakims, the writer cited the anecdote of an ailing wife of a *sahukar* in Ajmer who visited a hakim after British medicines could not cure her. The hakim asked her for the medicines she had been having. He smelt them and declared they contained alcohol. When the sahukar insisted this was not the case, the hakim offered Rs 1000 as a bet if he was proved wrong. The hakim's men smelt it, and they too declared it had alcohol, so the sahukar threw away the medicines. However, the English doctor swore that the medicine had no alcohol since it was not required to cure the patient. The writer refers to such mindsets as deterrents to

[13] L/R/5/59. Selections from the Vernacular Press, received upto 19 April 1882, p. 257; also L/R/5/59. Selections from the Vernacular Press, received upto 17 May 1882, p. 324; and L/R/5/59. Selections from the Vernacular Press, received upto 27 May 1882, p. 342.

[14] L/R/5/60. Selections from the Vernacular Press, received upto 19 September 1883, p. 764.

[15] L/R/5/66. Selections from the Vernacular Press, received upto 9 December 1889, p. 759.

the popularity of Western medicine. Such impediments could be removed only if the condition of hakims was improved. He pleaded to the government to improve the condition of the Unani practitioners, towards whom people were naturally inclined, and establish institutions for Unani learning.[16]

In 1870 the *Oudh Akhbar* reviewed the annual report of the Awadh dispensaries for 1869. The statistics confirmed many of its observations on the functioning of dispensaries. The number of patients for 1869 was less by 8739 than in the previous year. It also appeared from the report that a large number of dispensaries were badly managed because of inattention by officials. People did not visit these dispensaries not only for fear of English medicine, but also because of the rude and arrogant behaviour of native doctors and staff. According to the *Oudh Akhbar* the general view in society was that British trained doctors were good in surgery and skin diseases, but that in epidemics their diagnostics failed. Dispensaries had to cope with such prejudices.[17]

The Critique of Urdu Unani

The critique of government dispensaries was matched by an equally scathing critique of the new Urdu Unani in the columns of the same newspaper. Here too it was the self-read, home-bred, Urdu-knowing hakims and their incompetence that were targeted. The diffuse and uncontrollable realms of knowledge that they represented were attacked. Unani as a system of medicine, identifiable in family-controlled and state-sanctioned texts, was glorified. Indeed, the critique brought to the forefront Unani's robust scientific core—its practical wisdom identifiable in surgery, pharmacology, and chemistry. In 1876, in a long article in the *Oudh Akhbar* titled '*Tibb-i-Unani kei fawaaid ba muqaabalah doctry*' (The Benefits of Tibb-i-Unani as Compared to 'Doctry'), the anonymous writer enumerates the virtues of Unani medicine but laments the degradation of its practitioners and the incompetence of self-read hakims, and bemoans the decline in standards of learning. According to him, it is the lack of government interest that has created

[16] *Oudh Akhbar*, 18 November 1879, p. 3567.

[17] L/R/5/47. Selections from the Vernacular Press, received upto 22 July 1870, pp. 263–4.

a situation where untrained and uneducated people have become self-styled hakims: 'today such *tabibs* [physicians] remain who have read one or two Tibbi risalahs of Phaarsi and Urdu as lessons from a *ustaad* [teacher], or by themselves and have become hakims.'[18] He also regrets that even though the concept of anatomy (*tashreeh*) exists in Unani, incompetent and ill-read hakims do not practise it as a diagnostic tool. The same is the case with surgery (*jarrahat*), which is very much present in Unani, but not practised by the hakims. Chemistry (*fann-i-keemiyaa*), which results in the making of acids and solutions (*arraqs*), also exists in Unani, but again is not practised in as advanced a form as in the English system of medicine.[19]

In yet another article entitled '*Kamzor aadmi kei jism mein khoon kei zariyei sei quwwat pahuchanei kaa zariyaah*' (Injecting stamina through blood transfusion in a weak man) the anonymous writer regrets that, despite experimentations in the technique of blood transfusion by English doctors, no foolproof success story has emerged. He laments that in the past Unani specialists have written about this technique in their books, but the present generation has not been keen to research the technique any further. He says one of the main reasons for this is the slump in the educational and professional standards of those who claim to be hakims: 'they are not even well versed in their own subject, and whatever texts they know like *Qanun*, *Nafeesi*, etc. they have forgotten.' He critiques the corruption in the family-defined Unani educational system where medical knowledge was traditionally passed from father to son. Educational standards have fallen so low that 'if you are son of hakim sahib you too for nothing will become hakim, and without any qualms will practice—If you ask them have you read anything they say we know everything—why are you asking us about our *ilm* [knowledge], see our *amal* [experience, practice].'[20] Another writer observed that the amal were such that they saw several patients at a time and provided no personalized care. Lambasting the new Urdu-read hakims, he says: 'they are feeling the *nabz* [pulse] of some patient, and at the same time analysing the *qarura* [urine] of others, and listening

[18] *Oudh Akhbar*, 11 August 1876, p. 1341.

[19] Ibid.: 1341–1442.

[20] *Oudh Akhbar*, 18 February 1877, p. 294.

to a third in fear so that they don't lose a single patient.'[21] He noted that their knowledge was equally shocking; most had read no Arabic text; they had read only one or two Urdu medical texts like *Tibb-i-Akbari* and *Tibb-i-Ihsani*. They had spent a few months with some hakim or other and learnt how to write prescriptions, and thus claimed the status of hakims. They experimented on patients and prescribed expensive medicines. Such was the pitiable state of Unani.[22]

In 1881 another contributor, Muhammad Ali Fatehpuri, lashed out at Urdu self-taught hakims. He regretted that the present generation of hakims knew neither the Arabic medical canon, nor had any experience. They had no diagnostic or pharmacological skills, no knowledge of family *nuskhas*, and were therefore inept in figuring out the temperament *mizaaj* of their patients. They just read, he pointed out, 'a few Urdu medical texts and trained in local clinics and called themselves hakims.'[23] Muhammad Ali defined a good hakim as one who had not merely read books but had a certain mindset (*zahnee isteqaamat*), aptitude (*munaasbat*), who offered sober advice (*mataanat-i-rai*) and had the ability to evaluate his data (*quwwat-i-fauqiaat*).[24] In 1883 the *Mihr-i-Nimroz*, a Urdu bi-weekly from Bijnor, complained that native physicians generally know little of medicine and kill hundreds of people every year through their ignorance. It regretted that the government was not taking any action against them.[25]

In its 30 January 1900 issue the same forum took up the matter once more since the problem had only intensified. It complained that native physicians received no regular medical education and had only read one or two elementary medical books. They were responsible for many deaths in the city. The writer suggested the government set up a proper board to monitor such characters.[26] Not just the Urdu press launched this critique: the Hindi weekly *Bharat Bandhu*, from Aligarh, was

[21] *Oudh Akhbar*, 20 October 1879. p. 3283.

[22] Ibid.

[23] *Oudh Akhbar*, 20 May 1881, pp. 1639–40.

[24] Ibid.

[25] L/R/5/60. Selections from the Vernacular Press, received upto 24 October 1883, p. 864.

[26] L/R/5/77. Selections from the Vernacular Press, received upto 30 January 1900, p. 22.

critical of both ignorant vaids and the hakims who practised without any qualifications or certificates or degrees. It too was critical of government apathy in the matter and demanded the proper regulation of such native medics.[27]

The *Oudh Akhbar* made it a practice to report all cases where illiterate Urdu-read hakims had messed with the lives of patients. It referred to such hakims as *malikul maut* (angel of death). It went out of its way, in its daily columns, to create public opinion about malpractice in Unani practitioners and urged the government to act. The newspaper underlined the seriousness of the issue when it called the proliferation of such hakims an epidemic (*wabaa*).[28] In 1880 a news item, poetically titled, '*neem hakim khatraah-i-jaan wa neem mullaa khtaraah-i-imaan*' (spurious hakims who are life-threatening and spurious priests who are faith-threatening) once again reminded the public of the absence of any law that forbade illiterate and untrained men to practice Unani medicine. It noted that most hakims 'neither study Arabic Tibb books, nor even Persian or Urdu ones. Ideally, they should practise with an established hakim for at least 5 years. But they do not do so for even 6 months or a year and begin their own practice. They are like *malikul maut* [angel of death].'[29] It reported that in Saran, in the district of Chapra, a Muslim youth from Patna was given three doses of medicine by a hakim; he died after eating the first dose. The dead body, as well as the medicine, had been sent to Calcutta for investigation. The newspaper hoped the government would wake up to the crisis and appealed to it to take action against such hakims and pharmacists that infested almost all cities of India.[30]

The Urdu press did not end its critique of new Unani practitioners by commenting on their lack of knowledge in the basics of their practice; it also noted that they were equally inept in pharmacology and pharmacy administration (*attari* and *dawa faroshi*). It considered both these as critical trades that needed to be conducted with care for the success of Unani. In 1879, in an essay titled, '*Tibb-i-Unani wa Angrezi*' (Tibb-i-Unani and English Medicine), published in the *Oudh Akhbar*, yet

[27] L/R/5/56. Selections from the Vernacular Press, receieved upto 10 July 1879, p. 537.

[28] *Oudh Akhbar*, 21 February 1880, p. 622.

[29] Ibid.

[30] Ibid.

another anonymous writer regretted the falling standards of pharmacy. He lamented that a profession in which hakims who had written copiously on the art of making medicines had come to such a pass that its practitioners were illiterate in relation to such knowledge: 'they have no knowledge of how to regulate the quantity and proportion of medicinal constituents so as to cater to different ages and sexes.'[31] Hakims, he said, lacked the expertise to extract medicinal essence from the range of herbs and flora that was available to them. Unani medicines were difficult to consume as hakims prescribed elaborate procedures for patients before they could be used. Unani had to be pulled up on all these fronts. Added to this, of course, was the general lack of a spirit of research and experimentation that enveloped practitioners. He contrasted the lack of medical and pharmacological research and the absence of medical print culture in Hindustan to the vibrancy of medical research in Europe where journals, newspapers, and periodicals kept the issue of medical research alive in the public domain.[32]

In October the same year, in another essay entitled '*Beemari aur ilaj ka tareeqah*' (Diseases and the method of treatment), another writer commented on the unprofessional way pharmacists functioned in India: 'Each has ten to twenty pots [*haandis*] which have turmeric [*haldi*], tamarind paste [*sonth*], black pepper [*kaali mirch*] and they begin dispensing medicines.'[33] He was appalled at their lack of knowledge and their ignorance of languages like Persian and Sanskrit, in which the hakims and vaids often wrote their prescriptions. Their linguistic deficiencies did not stop such pharmacists from dispensing medicines. They took help from people who knew these languages to decipher prescriptions and then went ahead with their sales: 'they do not bother about weight nor anything, they pack every rubbish [*ghaans phoos*] and sell it. Such is their pathetic state of affairs that they dole out from one container [*qarrabah*] all kinds of solutions [*arraq*]; and all kinds of syrups [*sharbats*] from one bottle.'[34] The corruption of the attars was a theme that found mention in other Urdu periodicals as well. In 1889 a local correspondent of the *Azad*, an Urdu weekly from Lucknow, highlighted the corrupt practices of Lucknow's attars, who

31 *Oudh Akhbar*, 27 September 1879, p. 3034.

32 Ibid.

33 *Oudh Akhbar*, 20 October 1879, p. 3283.

34 Ibid.

dispensed the root of the rosebay, or oleander, which was poisonous, in place of medicinal marshmallow.[35]

Some contributors urged the people (*qaum*) to become themselves medically aware and knowledgeable about medicinal matters to protect themselves from illiterate hakims and corrupt pharmacists. In 1880, in an article entitled '*Desi Dawa khaanah*' (Indigenous Pharmacy), the anonymous writer lamented the corruption in the ranks of the pharmacists, which had made the prescriptions of even good hakims defunct. But he regretted even more the popular apathy to the problem: 'But people do not even know the names of medicines they are buying leave alone test their purity; and whatever attar gives them they obediently and blindly give to patient. If people (*qaum*) themselves do not become alert to problems of fraudulent attars, government cannot do anything. If they complain, government will act. But the people are not bothered'.[36] This writer attributed the spread of epidemics in India to the lax attitude of people towards health and medical practitioners. The combination of uneducated hakims and corrupt and inept pharmacists had facilitated epidemics. The situation, he pointed out, was so bad that 'you could say they [people] go to attars to buy *nuskhas* but buy *amraaz* [diseases] instead.'[37]

The Demand for Professionalization

The only way out of this mess was to professionalize Unani. Professionalization meant protecting the scientific core of Unani from the new hakim. He was dangerous because he represented knowledges beyond the control of state and family. Professionalization meant restoring the hold of the older families on Unani. It meant asking for state patronage and control of Unani. The *Oudh Akhbar* was clear that the government had to step in as the new patron of Unani.

The *Oudh Akhbar* made a case for government patronage on the grounds that Unani was the most popular medical system. It had to be protected from the new hakims who had driven it beyond the control of family and state. From its editor to the hakim contributors, everyone vouched for the popularity of Unani. For instance, an anonymous

[35] L/R/5/66. Selections from the Vernacular Press, received upto 21 October 1889, p. 655.

[36] *Oudh Akhbar*, 4 May 1880, pp. 1546–7.

[37] Ibid.

writer from Rajasthan argued that he did not doubt the efficiency of government dispensaries and their medicines, but the people preferred hakims even if they had to pay more to consult them. The people in Rajasthan, he argued, went to a hakim who charged them 4 annas fee for the cure of seasonal fevers and ignored the charitable and cheaper government dispensaries. He said that in Hindustan 25 crore people flocked to Unani healing and hakims cured lakhs of people. Along with many others, he felt that the government was thus duty bound to step in as a patron for this most popular system.[38]

The letters to the editor and articles received from the people reminded the government of its responsibility to protect Unani from the new illiterate hakims. In 1876 a reader sent an article penned by a hakim who lamented the new hakims' infiltration of Unani ranks. The reader wanted the article published because it highlighted the latter's dismal lack of professional knowledge. As a solution to the menace, it recommended the formalization of Unani training; it was the responsibility (*farz*) of the government to save the people from spurious hakims (*neem hakim*). It went on to outline the responsibilities of government in streamlining Unani education: state patronage should extend to the teaching of Unani, as was being done for all other sciences (*uloom* and *funoon*). Students successful in the final exam needed to be given a certificate that would allow them to practice. And as with British-trained doctors, the government should patronize renowned hakims in special ways: they should be checked out by government officials, given certificates (*sanads*), be registered, and only then allowed to practice. The government should also regulate hakims employed by the princely states. They too should be registered and given qualifying certificates once they cleared exams conducted by reputed hakims (*atibba-i-haaziq*).[39] The article reiterated the excellent diagnostic techniques of Unani and its unmatched cure of diseases in which British doctors had failed. The only problem with Unani, it said, was the lack of a professional culture. The writer hoped that the government would provide the lead to cultivate such professionalization.[40]

The *Oudh Akhbar* supported this definition of professionalization. In its editorial of 15 April 1877 it urged the people to focus attention

[38] *Oudh Akhbar*, 18 November 1879, p. 3567.

[39] *Oudh Akhbar*, 11 August 1876, pp. 1341–1442.

[40] Ibid.

on the miserable plight of the Unani medical profession. It urged the government to protect the scientific core of Unani from the new hakim. It pointed out that even though the 'Ionian' or Unani system of medicine was the most prevalent in the country, very few qualified and knowledgeable hakims existed. The worst hit was therefore the patient, who risked his life under the treatment of a half-educated practitioner. It was the duty of people to appeal to the government to improve this sad state of affairs. To regulate against newcomers and restore the hold of the old order, the editor argued that government should constitute a committee to consider the means by which able physicians well versed in their field were made available to people. The newspaper was optimistic of a positive response from the government: in the past the government had heeded its opinion columns. It had referred the matter to the Chief Commissioner of Awadh and the municipal committee of Lucknow. The editor regretted that, on this occasion, the government's initiative failed since neither the members of the municipal committee nor the inhabitants of Lucknow took the matter seriously.[41]

As the demand for the restoration of the hold of the older Unani families picked up, a lively discussion went on between readers on the modalities of this brand of professionalization. While some wanted a formal exam as a precondition for medical practice, others wanted the government to also pass a regulation requiring all native physicians, apothecaries, and druggists to be examined in the prescribed texts. Only those who cleared the exam based on such texts would get the requisite certificate to practice. Writers laid out the modalities of conducting the exam. Many recommended that a committee at the district level should oversee its operations.[42] The *Oudh Akhbar* wanted these overseer committees to be filled with learned *tabibs* (physicians) from traditional families. All aspirants to indigenous medicine were to take exams under the supervision of such committees. It also delineated the subjects in which exams were to be held: diagnostics (*ilm-i-taskhees*),

[41] L/R/5/54. Selections from the Vernacular Press, received upto 21 April 1877; *Oudh Akhbar*, 15 April 1877, pp. 282–3.

[42] L/R/5/48. Selections from the Vernacular Newspapers, received upto 3 June 1871, pp. 260–1. The committee members would have the additional role of keeping an eye on druggists and apothecaries, and inspect their shops. The writer also recommended a tax to be levied on druggists and shopkeepers and the money to be used to appoint an officer for supervising them.

principles of physiology (*usool-i-tabee'at*), botany (*ilm-i-nabataat*), chemistry (*ilm-i-keemiyaa*), anatomy (*ilm-i-tashreeh*), surgery (*ilm-i-jarrahi*), pharmacology (*ilm-i-adwiyah*). It urged desi tabibs to consult translations of English books to enhance their knowledge of some of these disciplines.[43]

Other Urdu weeklies like the *Shula-i-Tur*, a periodical published by Tasawwar Husain from Kanpur, wanted hakims from traditional families as examiners and not examinees. It felt that the idea of exams was irrelevant in cities like Lucknow where competent hakims existed who needed no certificates. However, it could be implemented in other cities like Kanpur, which were not endowed with such professionals.[44]

Almost every month an article or letter appeared in the *Oudh Akhbar* to pressurize the government to act. In these write-ups the government was reminded of its responsibility to formalize Unani education, regulate practitioners, and systematize their service codes, fees, and salary—as it had done in the case of its British-trained doctors. To set the ball rolling, the *Oudh Akhbar*, in 1879, made a request for instituting Unani lessons in existing medical schools.[45] Other contributors requested the government to establish institutions (*madaaris*) of Unani learning in a few big cities; or else include the study of Unani and hold exams for its students in those institutions where Arabic and Persian were taught.[46] Other newspapers like the *Azad*, an Urdu weekly published by Ahmad Ali from Lucknow, went so far as to suggest that the proposed Allahabad University could hold Unani exams as well, and issue certificates to successful hakims.[47]

The *Oudh Akhbar* reassured the government that it would have the support of well-heeled traditional hakim families in its efforts to administer Unani learning. To support its claim it published a petition, which the Rifaah-i-aam (a peoples' social club) committee of Lucknow had sent to the government. This petition from the notables of the city,

[43] *Oudh Akhbar*, 2 July 1875, no page number.

[44] L/R/5/60. Selections from the Vernacular Newspapers, received upto 19 April 1883, p. 323.

[45] L/R/5/56. Selections from the Vernacular Newspapers, received upto 27 September 1879, p. 753.

[46] *Oudh Akhbar*, 18 November 1879, p. 3567.

[47] L/R/5/64. Selections from the Vernacular Newspapers, received upto 19 April 1887, p. 255.

again, urged the government to formalize both Unani education and practice and thus save the people from spurious practitioners. It suggested a committee of reputed hakims to be set up as a screening body that would have the authority to grant certificates of qualification to other hakims. The petition clarified that the Rifaah-i-aam committee had already consulted prominent hakims of the city and was making its suggestions in consultation with them.[48]

The campaign for the professionalization of Unani begun by the *Oudh Akhbar* was picked up by other Urdu and Hindi newspapers and periodicals. In 1881 *Kavivachan Sudha*, the Hindi weekly from Benares which had a circulation of 350 copies, commented on the pathetic state of public health in villages where people were at the mercy of ill-trained hakims and vaids. It asked the government to institute a set of exams for hakims under the supervision of a committee of leading hakims and vaids at the headquarters of every district. The civil surgeon of the district could head and control the activities of the committee. Like the *Oudh Akhbar* it also suggested that arrangements be made for hakims and vaids to receive instruction in existing medical schools.[49] Similar demands were made by the *Bharat Bandhu*, a Hindi and English weekly from Aligarh. Its publisher Tota Ram was totally in line with the *Oudh Akhbar* position on this issue.[50] *The Municipal Guide*, published from Agra, followed a similar line.[51]

Thus, the Urdu press, alongside Hindi and English periodicals, pressurized the government rather hard to legislate against spurious hakims.[52] It used this argument for professionalization to bargain for greater government investment and control over Unani learning. For instance, it made government help to Unani education a prerequisite

[48] L/R/5/57. Selections from the Vernacular Newspapers, received upto 3 June 1880, p. 378.

[49] L/R/5/58. Selections from the Vernacular Newspapers, published in the Punjab, NW Provinces, Oudh, Central Provinces and Berar, received upto 6 December 1881.

[50] L/R/5/56. Selections from the Vernacular Newspapers, received upto 10 July 1879, p.537.

[51] L/R/5/60. Selections from the Vernacular Newspapers, received upto 21 July 1883, p. 603.

[52] *Oudh Akhbar*, 21 February 1880, p. 622.

for its support to any regulation. Thus it argued that legislation was urgently required to provide for special schools for the teaching of all native systems of medicine.[53] Such an education would produce qualified hakims and was a prerequisite to make meaningful any registration act that would draw the line between qualified and unqualified practitioners.

Thus in 1887, when the government proposed to introduce the Medical Registration Act to curb the menace of quacks, the vernacular press was quick to react. No such act was acceptable to it without adequate government patronage to Unani learning. A correspondent of the newspaper *Hindustan* argued that this regulation would harm genuine, well-educated native practitioners, both hakims and vaids. This was particularly true if such an act were introduced before the government laid out adequate norms that defined a qualified hakim and vaid as distinct from a spurious one.[54] It then went on to delineate what needed to be done to formalize Unani and Vedic education: books needed to be prescribed in both subjects; arrangements for holding exams had to be made; certificates had to be granted from time to time; and municipal boards had to be involved in passing on information about syllabus and exam dates to the concerned professionals. Once all this was done, the Medical Registration Act would be able to serve its desired purpose.

Other newpapers joined the debate on the Registration Act as well. *The English Journal of the Anjuman-i-Punjab*, Lahore, cautioned that the proposed bill would be an impediment to the skeletal public health services of small mofussil towns. It pointed out that here a range of ill-trained professionals were the only healthcare providers. In these towns, government dispensaries were scarce and inept when handling public health. People here relied on men who would be disqualified from practice if the bill came into force.[55]

[53] L/R/5/57. Selections from the Vernacular Newspapers, received upto 22 April 1880, p. 378.

[54] L/R/5/64. Selections from the Vernacular Newspapers, received upto 19 April 1887, pp. 565–6.

[55] L/R/5/60. Selections from the Vernacular Newspapers, received upto 25 January 1883, pp. 80–1.

In 1904, when plague raged in most parts of Awadh and many hakims performed remarkable feats, the issue of government apathy towards them once again became the focus in the Urdu press. The correspondents of the *Oudh Akhbar* reported on the dismal state of Unani education in North India. Government apathy was particularly shocking in view of the services the hakims had rendered in treating plague victims. The correspondents pressurized the government to arrange for Unani learning in at least some important places in the country. They were, however, sceptical that the government would yield to their demands as this would adversely affect the European system of practice and would be commercially suicidal *vis-à-vis* the sale of Western drugs in India.[56]

Professionalization was used also to remind government about the rights and privileges (*huqooq*) of hakims. One writer appealed to the government to define their rights in the same manner as it had done for other functionaries, such as *patwaris*. It clarified that it was not asking for set incomes for them from the government treasury, nor was it demanding hereditary scholarships (*wazeefahs*), as was the practice in pre-colonial times. But it was keen that their rights as hakims be laid out by the government. Once people were clear about such matters, they would take to Unani learning and contribute towards the establishment of educational institutions for such learning. The writer exhorted the elites to also pressurize the government to see to it that such an ancient medical system was protected and did not die out.[57]

And of course one of the most important rights the newspaper identified was employment. The *Oudh Akhbar* pointed out that lack of employment was an impediment in the education of hakims. In the absence of regular jobs, hakims remained preoccupied with putting together resources for their basic monthly expenses. This resulted in the neglect of medical research.[58] The government was pressurized to provide employment for hakims in the municipalities. A distraught contributor asked: 'Isn't it the job of municipalities that they should open new *dawakhaanahs* [clinics] with donations? And employ in

[56] L/R/5/80. Selections from the Native Newspapers, published in UP, received upto 22 October 1904, p. 363.

[57] *Oudh Akhbar*, 16 July 1880, p. 2378.

[58] Ibid.

them *desi tabibs* [hakims] who were respected by the *khaas* and *aam* [elites and ordinary people]?'[59] In 1904 a letter to the editor in the *Rohilkhand Gazette*, a Bareilly-based newspaper, lamented the lack of government recognition to the splendid services provided by hakims. In some towns of Rohilkhand, the municipal board had fixed monthly grants for some hakims. The writer wondered why this practice could not be expanded in a more systematic way all over Bareilly.[60] Another contributor lamented the lack of government patronage for Unani and attributed its dismal state to this lack. He asked the government to look at the notable medical feats some hakims in Lucknow performed despite hardships. He requested the government to fix a nominal monthly salary for hakims, ranging from Rs 30 to Rs 60 per month, depending upon their abilities.[61]

Professionalization also meant protecting pharmacists, the attar, as well from the ranks of their nonconformist colleagues. In 1881 Asghar Husain, a practising hakim of Lucknow, suggested that registers be placed in the shops of attars so that they maintained a record of the patient and his hakim. This implied that attars had to be literate. By this he suggested that the government should ensure their basic literacy. Attars should pass qualifying exams so that they could read *nuskhas* and not sell wrong medicines on account of illiteracy. They were also expected to mark their medicine bottles with labels for the benefit of patients. Hakims were expected to keep an eye on attars and ensure that they dispensed the proper medicines.[62] In another set of suggestions geared to create better coordination between attars and hakims, Husain said the latter should publish a list of medicines and their characteristics and sell it to attars for their consultation. Hakims were to oversee their activities and ensure that they threw away all the medicines that had crossed their expiry date. He also desired a law prohibiting the cuts and commissions fixed between hakims and attars. An *iqraar naamah* signed by attars prohibiting such cuts was offered as a suggestion. And finally, the medicines that attars prepared were to be sent to the

59 *Oudh Akhbar*, 4 May 1880, pp. 1546–7.

60 L/R/5/80. Selections from the Native Newspapers, published in UP, received upto 10 September 1904, p. 309.

61 *Oudh Akhbar*, 10 September 1875, no page number.

62 *Oudh Akhbar*, 14 June 1881, pp. 1918–19.

government to check their authenticity before being sold or even advertised in the press.[63] Husain's appeal was to the municipal committees and the government to curb the corruption in the ranks of the attars. Attars, he argued, sullied the reputation of hakims by preparing their medicines in the most unprofessional way. Thus, if Unani was to be preserved, the attars too had to be regulated.[64]

The *Oudh Akhbar* offered concrete suggestions for protecting genuine attars from the activities of spurious ones. It suggested that they be monitored by municipality officials. The municipal committee could give licenses for the sale of drugs. The sale of drugs without the license would be prohibited. It laid out that the licence fee would not exceed 4–5 annas a year. Municipal hakims were asked to inspect the shops of druggists twice or thrice every month and hand over defaulters to the magistrate. It also advised people to be more vigilant and report the matter to the police whenever the need arose.[65] These regulatory suggestions made sense, since the proposed Medical Practitioners Bill was beset with dangers if introduced in Awadh. The Medical Practitioners Bill being discussed in Bomaby and Calcutta in the 1880s to register all qualified native pharmacists and practitioners was seen as impractical in the small mofussil towns of the Awadh region, where medical help was still dispensed largely by unqualified practitioners. It was widely believed that the passage of the bill would extinguish all such men and thus imperil the public.[66]

By the turn of the century, the editor and contributors of the *Oudh Akhbar* emerged as the public intellectuals who were vigorously pursuing a campaign to professionalize Unani. This centrally meant cutting out uncontrollable and unidentifiable forms of knowledges from the Unani core. Once social control over the core was established, the rights and privileges of its practitioners could be delineated. The *Akhbar* argued that such a professionalized Unani, set within the hold of the family and the state, was better suited to the Hindustani habitat than

[63] *Oudh Akhbar*, 19 September 1881, pp. 2988–9.

[64] *Oudh Akhbar*, 19 August 1881, pp. 2647–8.

[65] Ibid.: 287.

[66] L/R/5/60. Selections from the Native Press, received upto 25 January 1883, *The English Journal of the Anjuman-i-Punjab* (Lahore), 17 January 1883 , pp. 80–1.

Western medicine. They expanded their campaign by connecting themselves with other newspapers, journals, and scientific societies. The *Oudh Akhbar* not only reacted to articles on Unani that were published in other newspapers and fora, it also urged them to reproduce its own articles and give them wide circulation. The *Akhbar* considered this campaign for Unani an act of public welfare and urged all the newspapers of Hindustan to join this mass campaign. It was grateful to the editor of the newspaper brought out by the Scientific Society, Aligarh, which readily reproduced its articles in its issues. Munshi Newal Kishore not only gave wide circulation to articles from other newspapers by publishing them in the *Oudh Akhbar*, he also used his influence to bring them to the notice of government.

The *Oudh Akhbar* and the vernacular press that it influenced thus acted as a pressure group to force the government to act. Their influence was evident when, on many occasions, its initiative brought the issue of Unani to the government table. An important article, provocatively titled, '*Atibba-i-Unani aur government kee qadrdaani*' (Unani physicians and government respect) published in the *Oudh Akhbar* of 1 September 1875, was presented through the help of Sir John Inglis, head of the local government in Awadh and resident commissioner, to the city's municipality for necessary action. Members were asked by government to send in their views on how to improve Unani's state of affairs. The editor hoped that this would help set the ball rolling as far as government intervention was concerned.[67]

The Urdu Press and the Making of Unani 'Modernity'

Even though a range of people contributed to the *Oudh Akhbar* on the feasibility of cannibalizing aspects of Western medical training for Unani and giving it a distinct modern profile, there were three major contributors: Hakim Asghar Husain, a native of Farrukhabad who served as assistant surgeon in many government dispensaries and was trained both in the English system as well as in Unani;[68] Mukhlis al

[67] Ibid.

[68] *Oudh Akhbar*, 3 December 1880, p. 3839. See also an entry on him in *Physicians–Authors of Graeco-Arab Medicine in India*, p. 50. Here he is listed as

Daulah, a hakim; and Abdul Muhammad Syed Jamaluddin, Farrukhabadi, a doctor in the dispensary at Ghori, Zillah Sagar. These three men came from modest backgrounds, with no links to traditional hakim families. Yet, they shared the views of the older families on professionalization. Their prominence as writers and the lively responses to their pieces in the *Oudh Akhbar* reflected the critical role of the newspaper in creating a new concept of the author as public intellectual.

All three wanted to emulate the educational model of the Muslim reformer Sir Syed Ahmad Khan and combine Unani with British medical education. They were agreed that this was possible because Unani was as scientifically inclined and advanced as Western medicine. Indeed, according to them a professionalized Unani, purged of the illiterate hakim, was a natural ally of Western medicine. The scientific core of a cleansed Unani could move parallel with such medicine. But it also had to develop in contrast to it, and eventually carve out for itself a 'national' space within the discursive space it shared with colonial medicine. The reconstitution of a national space for Unani from within the colonial framework was achieved by constructing a distinct cultural profile for it in Hindustan. Thus, intellectuals underlined the compatibility of Unani with Hindustan's locale (*aab-o-hawa*) as they set about unpacking their cultural project. They wrapped the cultural compatibility narrative with a hurriedly constructed local history of Unani in Hindustan. This sought to prove that Unani was the ancient (*qadeemi*) medical tradition of the region, which had to be in synch with Hindustan's culture and therefore loftier than those who were outside of it. Unani was for them not just a medical system but a cultural system that had an ancient past in Hindustan. In contrast, they defined English medicine as modern (*jadeed*), which meant that it did not have territoriality and a history in Hindustan. The contributors argued that Unani, being ancient, remained rooted in the time–space

---

having written a series of essays and texts, many of which were printed over the period 1850–71. Notable ones on the theme of bringing ancient and modern medicine together were *Al-Bahr al-Muhit fi'l Tibb al Qadim wa'l Hadith'*, or, Circumventing Sea on Ancient and Modern Medicine; and *Dastur al-Najat Masa'ib al-Hummayat f'il Qadim wa'l Hadith*, or, Constitution of Cure from Affections of Fevers in Ancient and Modern Medicines. In 1871 he wrote the famous tract called *Risaalah dar Hal-i-Tibb*, or, Treatise on the Position of the Medicine. Besides these, he wrote tracts on cholera and fevers.

confines of locality and was sensitive to localized relations; whereas the modernity of Western medicine, destroyed the primacy of place, the locale, in healing as it was hooked on global and universal medical norms.[69] Unani was said to share with other Indian medical traditions, like Ayurveda, the concept of an unobjectified body that was porous and conditioned by locale. This was the point of departure between it and culture-deficient Western medicine, which upheld an objectified notion of the body.[70] This medicine's modernity, defined in terms of it being external to Hindustan's culture, became its critical point of difference with Unani. Unani was projected as being far more in tune both with the place and space in which it operated.

The construction of Unani as the ancient medical tradition of Hindustan was a significant shift from its larger Graeco–Roman–Arabic tradition in the past. And thus the project to read Unani as culture was fraught with tension. A niche had to be carved out for it by trimming the larger Dar al-Islam tradition, which had been its main locus thus far. The Muslim public intellectual balanced the two worlds of Unani, just as he managed the Islamic universal and the local territorial strands of his own identity.[71]

Unani and Colonial Medicine: Sharing a Discursive Space

In 1880 the editor of the *Oudh Akhbar* published a note titled '*Tibb-i-Unani wa Doctory*' (Tibb-i-Unani and English Medicine) which favoured the fusion of Unani with English medicine. He acknowledged that the task was easier said than done. Both doctors as well as the

[69] Giddens 2003: 18–19. This is similar to the concept of modernity as defined by Giddens to mean that, in conditions of modernity, place becomes increasingly phantasmagoric. Which means that locales are thoroughly penetrated by and shaped in terms of social influences quite distant from them. What structures the locale is not simply that which is present on the scene; the visible form of the locale conceals the distanciated relations which determine its nature.

[70] Langford 2001: 1–24. Langford talks about a similar process of constructing Ayurveda as a cultural system to mark its difference from English medicine in the early twentieth century. It became the nationalist project to read Ayurveda as culture.

[71] Jalal 2001.

hakims were ignorant of each others' knowledge systems and the modalities of interaction had to be put in place. The editor piloted this dialogue and opened up the columns of the newspaper for suggestions from its readers: 'We appeal for this debate to continue for some more days in our columns. So that we get more and more ideas and advice.'[72]

His appeal only accelerated the growing curiosity amongst public intellectuals about English medical practices. Indeed, this curiosity was inspired by Muslim reformers like Sir Syed Ahmad Khan, who steadfastly created a favourable public opinion for the mapping of traditional learning onto English education. In a long article in the *Oudh Akhbar* of 27 September 1880, Hakim Asghar Husain made a strong case for Indians being educated in both the English and the Unani system of medicine. He said that, in his own clinic, he used the best of both systems for cure and it worked very successfully.[73] In an opinion piece of 16 October 1880 he urged the people to set up a school to combine the teaching of the two systems of medicine. He appealed to the public to donate money, and asked for financial help from the government as well.[74] He said that his model of medical fusion was inspired by the educational reforms of Sir Syed Ahmad Khan, founder of the Aligarh Muslim University. He cited the successful educational experiment of Sir Syed in several other essays as well. He never failed to mention that Aligarh's success was due to both the emotional and the financial support of the people. He hoped that people would contribute to his school, as they had done for Aligarh, and that their initiative would stir government officials to extend a hand of support for the proposed medical school.[75] He felt that the English and Unani fusion would benefit if books of modern medicine from Egypt were brought into India and translated into Urdu. He also urged government to circulate more widely Urdu medical periodicals published from the government medical schools in Agra and Lahore. He was careful to add

[72] *Oudh Akhbar*, 8 December 1880, p. 3893.

[73] *Oudh Akhbar*, 27 September 1880, pp. 3123–9.

[74] *Oudh Akhbar*, 16 October 1880, pp. 3294–6.

[75] *Oudh Akhbar*, 22 October 1880, pp. 3400–1.The public enthusiasm for the Aligarh Movement he contrasted to the relative lack of popular effort for the setting up of the religious seminary at Deoband. And this lack of popular enthusiasm he said explained why Deoband does not possess that influence (*quwwat*) which Aligarh today holds.

that the dissemination of Western medical knowledge was in no way a substitute for Unani: it was only to supplement Unani learning, which already had a scientific core. He cited many cases where the combination of Unani and Western medical treatment had been successful. These were stomach problems, for which both surgery and Unani purgatories helped. He listed different kinds of surgeries and anaesthetic techniques, and the use of instruments like the microscope and the thermometer, which were compatible with Unani diagnostics.[76]

Asghar Husain's repeated pleas made a difference and contributions soon started pouring in.[77] But more than the donations, his success lay in the public awareness that he generated about the implications of Unani's encounter with English medicine. Soon, people from all over India responded to his views in the columns of the *Oudh Akhbar*. New authors and writers from ordinary backgrounds emerged as they found in the paper a forum to voice opinions they had long held. Even though the editor of the *Oudh Akhbar* was never fully satisfied with the range of popular response, it was enough to generate a discussion on the theme.

Most people were in agreement with Asghar Husain and their responses encouraged him to write more persuasively on the issue. On 3 December 1880 he responded to the article '*Tibb-i Unani wa Doctory*' (Tibb-i-Unani and English Medicine) written in response to one of his pieces by Abdul Muhammad Jamaluddin, Farrukhabadi. Abdul Muhammad was broadly in agreement with Asghar Husain, and the latter used the opportunity to reiterate that his long interaction with Western doctors and intellectuals had convinced him that they were as responsive and attentive to his lectures on Unani as he was curious about their medical expertise. He wanted to transfer his personal experience of interaction to a broader institutional level by the establishment of a medical school that would combine the learning of the two systems of medicine. He was convinced that there were many advantages in this approach.[78]

[76] *Oudh Akhbar*, 27 September 1880, pp. 3123–9.

[77] *Oudh Akhbar*, 22 October 1880, pp. 3400–1. Shaikh Riazuddin Ahmad, the headmaster of *risaalah* no. 18 Chavni, promised a donation soon after reading Asghar Husain's convincing arguments.

[78] *Oudh Akhbar*, 3 December 1880, pp. 3839–42.

Asghar Husain's idea of a Unani–British medical fusion was not confined to educating the medics. It covered pharmacists or attars as well. In an article titled '*Tibb-i-Qadeem*' (Ancient Medicine) he urged attars to learn how to make British medicines and sell them in their shops. The demand for such medicines, he said, was great, because many hakims increasingly prescribed them. And the fusion of Unani and Western medical practice would only increase their demand. He was of the view that since most British medicines were manufactured in medical halls in Allahabad, Kanpur, Calcutta, and Bombay it would not be difficult for attars to learn the art of making medicines the Western way. Moreover, he pointed out that most English pharmacopoeias, or *qarabadeen*, were now available in Hindustani translations. This would make it easy for attars to learn how to prepare British medicines.[79]

More importantly, he pointed out that the concept of the pharmacopoeia and formulae to make compound medicines were not entirely novel to the attar. This was because such texts existed in Unani as well. We had, he writes, 'books and *risaalahs* [magazines] on *qarabadeen* [pharmacy], and texts on chemistry translated into Arabic in our medical tradition. And we have books to make compounds like acetates, sugrates, carbonates, compounds, in which names have often been translated from Latin words. There might be mistakes because of the translations, but such texts do exist. And so we should not shy away from the progress of our tibb.'[80] Husain was of the view that it would prove very lucrative for attars if they educated themselves in pharmacological skills. They could benefit by brisk sales since the medical halls where such medicines were currently sold had very high prices.[81]

From Colonial Frame to National Space: Constructing a Cultural Sphere for Unani

The advocacy of Unani's scientific commonalities with Western medicine did not mean that the advocates ceased to underline its difference with the latter. Indeed, they reconstituted the colonial space as the national space of Unani. Asghar Husain, himself the chief protagonist of the fusion thesis, stressed that Unani was the ancient medicine of

[79] *Oudh Akhbar*, 19 September 1881, pp. 2988–9.
[80] *Oudh Akhbar*, 29 March 1881, pp. 1038–9.
[81] *Oudh Akhbar*, 19 September 1881, pp. 2988–9.

Hindustan and gave a definite territorial profile and history to Unani. But territory was not merely bounded geographical contours, as delineated by colonial medicine. It represented a cultural entity as well—the locale.

The Urdu word for locale is *aab-o-hawa*, which literally translates as 'water and air'. But its cultural connotation was explained by an anonymous author in a write-up titled '*Aab-o-hawa*'. He defined locale as a geographical-cultural entity. He said that in Unani *aab-o-hawa* meant a collection of geographical and cultural influences that makes a place distinct and different from all others. He identified these distinctions as the quality of air (light or heavy, humid or dry, hot or cold), the quality of water (sweet or hard, *khaara*, light or heavy), the quality of soil (fertile or infertile, jungle or desert, its productivity), food that people consume, the rituals and customs they follow, whether they live close to the sea or far from it, etc. All these influences collectively constitute *aab-o-hawa*, or locale. According to the writer, all these referents shape the body and define its health. When a hakim recommends change in *aab-o-hawa*, he means change of locale for betterment of health (*tabi'at*).[82]

The author constructs a hierarchy of health-inducing places that corresponded to a range of locales. Places where there were no fluctuations of weather produced least illnesses. Places with both hot and cold climates produced more disease. Cold countries had more instances of coughs and colds. Places with both dry and humid conditions were most prone to disease. Places with dry conditions were good for certain temperaments, but not for others. Thus, change of place, diet, and clothes was a must to counter these influences and remain healthy.[83] A further extension of the argument was to classify people as well on the basis of locale. Thus, he argued that people of dry climates were clever, and those of humid areas dull.[84]

According to Husain the centrality of this personalized locale in Unani marked its point of difference with Western medicine.[85] Many other respondents to Husain's pieces agreed with him. In 1876 an

[82] *Oudh Akhbar*, 31 July 1880, p. 2546.
[83] Ibid.
[84] Ibid.
[85] Ibid.

anonymous contributor to the newspaper put this difference bluntly: 'doctors often mis-diagnose because the principles of their medicine make no concessions for weather [*mausam*], water [*aab*], air [*hawa*] and temperament [*mizaaj*] of patients. But principles [*usool*] of Unani take all this into account and therefore benefit us Hindustanis'.[86]

Indeed, many other contributors to the newspaper took the argument further when they mapped the contours of the nation onto this imagined cultural entity. They challenged the universal claims of medicine and showed how each nation, if viewed as a cultural construct, would logically have its own distinct medicine. In 1879, in an article titled 'Tibb-i-Unani wa Angrezi' (Unani and British Medicine), one Raqim Mustafser argued that every country had its own specific medicines because their locale was different. Different locales produced different temperaments, varied diseases, and the requisite flora and fauna for cures. Thus, the people of Europe, he argued, lived in cold climates and needed hot items (*ajzaa*) for treatment. But in the East people lived in hot climates, and for them cold items were a must for good health.[87] There could be no universal medicine equally suited to all.

According to the author, the locale had subdivisions as well. For instance, the people of the East used a range of different medicines because of the variety of locales that shaped Eastern societies. He argued that people living in village societies found it more useful to use herbs and flora within their locale for treatments rather than the medicine of *tabibs* (medics). And the people of the jungles were similarly more in tune with the medicinal plants of their locale. Birth in a certain locale defined temperament and it was the produce of that locale which alone balanced temperament.[88]

All these contributors argued that Unani took care of such principles of healing and did not disrupt the time–space balance in the locality. It was therefore the natural choice of the sick. This cultural space chalked out for Unani was projected as the national space. Unani was seen as ancient and thus different from colonial territorial space, which was modern. And thus it is not surprising that almost every essay of

[86] *Oudh Akhbar*, 11 August 1876, pp. 1341–1442.
[87] *Oudh Akhbar*, 27 March 1879, pp. 994–6.
[88] Ibid.

Husain and other contributors in the *Oudh Akhbar* was titled using the word *qadeem* (ancient) for Unani; quite clearly this connoted rootedness in the locale. The same authors called Western medicine *jadeed* (modern), meaning it was medicine located external to the Hindustani locale.

Sensitivity to locale also meant compatibility with the Hindustani temperament (*mizaaj*). Temperament was embedded in locale, and Unani's sensitivity to Hindustani temperament became the other key word to define its ancient and thus lofty status, and to contrast it with the 'modern' and thus mundane nature of Western medicine. More than one contributor to the *Oudh Akhbar* argued that the 'ancient' Unani was compatible in diagnostics, surgery, and chemistry to British medicine. But it was still preferred over the latter because of its rootedness in the Hindustani locale and compatibility with the temperament of the people. In a long essay, '*Hindustaniyon kei liyei Tibb-i-Unani banisbat doctori kei zyaadaa mufeed hai*' (Unani medicine is more beneficial for Hindustanis than British medicine), an anonymous writer gave reasons for his medical preference: an important reason was the sensitivity of Unani to the temperament of the Hindustani people. As he put it: 'its principles and Quliyat have been tailored to the mizaaj of Hindustanis.' He provided a scathing critique of the lack of such rootedness in British medicine, arguing that it may be suitable for Europeans (*ahl-i-Europe*) but was absolutely inappropriate (*naa muwaafiq*) for Hindustanis. European medicine, he said, is 'not tailored to the temperament of Hindustanis whose distinct locale makes them of a more warmer and weaker disposition than the Europeans.' He argued that quinine was the only exception, but even this was beneficial for only some kind of fevers.[89] The writer did not deny the advanced diagnostic techniques of British medicine but said that they were tuned to the temperaments of Europeans. And this was natural, for English medicine had a history and territorial identity in England: it was the ancient (*qadeemi*) medicine of that country. But in Hindustan the same medicine was modern as it had neither a territorial nor historical affiliation to the region. Thus, despite their technological advancements, Western medics could not correctly diagnose the ailments of Hindustanis. Their medical training

[89] *Oudh Akhbar*, 10 September 1875, no page cited.

was insufficient for understanding Hindustani temperament; there could be no universal medicine as locale was central to healing.[90]

He cited numerous cases of incompatibility between Western medicines and Indian patients. In most of these cases, faulty diagnosis had been made by English doctors and wrong medicines prescribed to Hindustani patients. The lives of many such patients were saved because they approached a hakim on time.[91] The writer alluded to Unani as representative of the personalized care that structured itself on the faith and trust on an individual rather than abstract knowledge systems. He argued that people preferred Unani because they were unaccustomed to the 'faces and company of English doctors' (*soorat aur sohbat*).[92] The personality of the healer was very critical in the locality-centred medical belief system that Hindustanis identified with.

This freshly constructed antiquity of Unani that made it most suited for Hindustanis continued to be defined in the *Oudh Akhbar* in terms of its rootedness in locale. Most essays referred to Unani in their titles as *qadeem* or ancient and argued that its antiquity made it so popular in Hindustan. In 1881, one Muhammad Ali responded to the issue of ancient *vs* modern in an article titled '*Qadeem Tibb*' (Ancient Medicine). By ancient he meant sensitivity to locale and temperament; these were also considerations that went into the making of Unani medicine and its prescriptive cures. Thus the method (*tarkeeb*) of making and consuming medicines was as important as the medicine itself. Due consideration had to be given to water (*aab*) and air (*hawa*), activity (*harkat*), and relaxation (*sukoon*), alongside the medicine regime (*tadbeer-i-dawa*).[93]

The same year, Hakim Asghar Husain reinforced this argument when he said that those medicines of the English doctors that relied on local content were more successful as compared to those that were based on universally applicable formulae. For instance alum (*phitkari*) mixed in water prescribed by English doctors as a cure for eye pain proved very successful, since it was in tune with the temperaments of

[90] Ibid.

[91] Ibid.

[92] Ibid. The other reasons cited for Unani's popularity was the easy availability of its drugs at low prices. For the elites it was a preferred system because they found it demeaning to visit government dispensaries.

[93] *Oudh Akhbar*, 13 June 1881, pp. 1881–3.

the patients.[94] Indeed, he praised Western medicines and said they would be most effective if they were prescribed taking into account temprement, stamina, and local context. These he identified as the water and air of the place where the patient lived, and the seasons that he had to endure in his particular locality. For instance, if he was a washerman his professional hazards—being in water for long hours—needed to be taken into account in both diagnosis and prescription. Equally important was the role of the seasons that he withstood, and his age and temprement. Unani took into account all these factors; Western medicine based on abstract medical notions did not. Except quinine, which had proved successful in the cure of fevers, no other English medicine suited the Hindustanis.[95]

The issue of locality and its related influence on healing was central also to health care (*hifzan-i-sihat*). Asghar Husain wrote a long essay called '*Tareeqah-i-Hifzan-i-sihat*' (method of health care) to suggest how town planning should be done keeping in mind the quality and nature of air in the locality. Dry and moist airs were related to varying ailments. This required appropriate housing, town planning, and clothing. For instance, dead bodies and crowds polluted the air and caused disease. Thus houses, he argued, should be spacious and built on dry ground, not close to graveyards or densely populated areas.[96]

The *Oudh Akhbar* took the issue of public health very seriously and tackled it as a problem that was entirely local, and which could be sorted out by regulating the environmental influences which shaped the locality. These influences were air, water, diet, and appropriate clothes, as per the seasons of the locality. The regulation of these was a matter of personal hygiene (*hifz-i-sihat*). A daily bath, preferably with fresh water, was recommended for good health. Wearing clean and fresh clothes after a bath kept disease at bay. Wearing fragrant odours and keeping the house fragrant and freshly painted also kept the germs of disease away. Morning and evening walks and exercise were all recommended for a healthy body. So was a diet that included fresh vegetables, vinegar, pickles, and chutneys.[97] Diet or *ghiza* was an important constituent in maintaining good health. Cold and easily digestible diets

[94] *Oudh Akhbar*, 11 June 1881, pp. 1868–9.
[95] *Oudh Akhbar*, 1 September 1881, p. 2787.
[96] *Oudh Akhbar*, 11 October 1881, pp. 3255–7.
[97] *Oudh Akhbar*, 19 April 1881, p. 1810.

were always recommended. Heavy and oily foods were said to cause cholera (*haiza*). And this haiza, related to diet, could take an epidemic form (*wabai haiza*).[98] News items titled '*Hifzan-i-sihat*' (Health Care) appeared very frequently in the *Oudh Akhbar* and stressed the importance of suitable clothing for health, citing diseases that could occur if one was not appropriately attired.[99]

Writing a History of the 'Ancient' Unani in Hindustan

The 'modernity' of English medicine was not just about being external to Hindustani locale, but about having no history there as well. Asghar Husain constructed a Hindustani history for Unani to reinforce its putative ancient status, and in order to define Western medicine as one with no history in Hindustan.

He pointed out that Unani and European medicine owed their origin to a common Graeco-Arab intellectual tradition and that both were thus at par in their scholarly advancements. Unani incorporated the Arabic translations of many of the Greek philosophic works of Aristotle, Socrates, and Galen under the patronage of the Ummayad caliphs in Damascus. Later, the Greek works of the physician philosopher Avicenna and the surgeon Al Farabi too were translated into Arabic and became integral to Unani. This wide knowledge collective ensured that Unani not only had a philosophic strand but also a component of pharmacology and the fine art of surgery. Husain argued that European medicine shared this intellectual legacy with Unani. Thus, both were equally scientifically inclined and advanced. If being scientific was a criterion for progress and advancement, then Unani could be said to have surpassed European medicine, because 'it borrowed from Greece when it was at its zenith, and firangis learnt from it when it was in decline. So they could not have the advantage of Ibn Sina, Farabi etc. who taught how to make medicine.'[100]

However, if their shared scientific legacy brought together Unani and colonial medicine, the former's rootedness in Hindustan's locale made it different from European medicine. Husain constructed a

[98] *Oudh Akhbar*, 8 April 1881, p. 1160.
[99] *Oudh Akhbar*, 31 July 1880, p. 2545.
[100] *Oudh Akhbar*, 1 March 1881, p. 736.

fascinating history of Unani which located its core in Hindustan. He traced its Hindustani links back to its Arab phase of expansion. He argued that medical knowledge gathered from Hindustan by Arab traders from the seventh century went into the making of many of the ancient Arab medical texts. He clarified that even though the philosophic texts of Unani were compiled in pre-Arab times, the prescriptive compendiums on practical wisdom owed their origins to the medical prescriptions (*nuskha-i-amraaz*) the Arabs had collected from the countries they traded with. Hindustani prescriptions, such as *attriphal ka nuskha* (prescriptions made from three fruits), were very much part of these compendiums. The inclusion of Hindustani prescriptions also meant that Unani's remedies were most effective in the specific dietetic and environmental context of Hindustan. Thus, Unani prescriptions gave due regard to Hindustan's water (*aab*), air (*hawa*), and seasons (*mausam*) in the making and consumption of medicines. This sensitivity was reflected in the importance Unani gave to seasonal fruits and vegetables in a patient's diet. Even the non-vegetarian food recommended for a patient took into account what the animals had eaten before being slaughtered.[101] According to Husain, Unani's stress on individual well being reflected its greater sensitivity to the comportment and cultural practices of people in Hindustan. Its choice of diet, clothes, and environment was heavily influenced by the religious cosmologies of the locality. Health was intimately related to religion as certain foods were forbidden for reasons of health. *Halaal* (sacrificed in the name of Allah) food was healthy for Islamic people, but *haraam* (not sacrificed in the name of Allah) food was injurious. Similarly, many religions prescribed vegetarianism as a sure route to health. And matters of religious prescription extended to attire as well. Unani was sensitive to these issues. Indeed certain hakims linked clothes (*poshaak*) with their medical prescriptions. Tight and loose clothes were recommended for different kinds of patients, keeping their health requirements in mind.[102] Husain concluded that Unani's long history in Hindustan enabled it to incorporate age-old knowledges on practical wisdom that circulated within families and identifiable texts. This made it the ancient medicine of Hindustan. He contrasted this to the relative historical

[101] Ibid.

[102] *Oudh Akhbar*, 16 October 1880, pp. 3294–6.

shallowness of Western medicine because it did not care for 'ancient books that were sensitive to the preservation of the practical wisdom that made the individual compatible with his locale.'[103]

The *Oudh Akhbar* endorsed Asghar Husain's categorization of Unani as ancient, agreeing that its long history in Hindustan was reflected in the knowledge it had of the locale. It agreed that Unani engaged with the cultural etiquette and rituals of the people. In its editiorial of 10 October 1878 titled '*Beemar aur Teemar*' (The Sick and their Attendants) it enumerated its prescription for good health, upholding the Unani idea of health being about individual well being and prescribing a proper relationship between the individual and his locale for good health. It emphasized morning and evening walks for fresh air, cleanliness of house and attire, exercise, ventilated homes, and the right medicines. It pointed out Unani's prescriptive norms about proper etiquette among not just hakims but also the patient's attendants. It stressed that Unani respected the patient's belief in prayer (*dua*) and charity (*khairaat*) as part of the healing regime.[104] Again, in its January 1880 issue the *Oudh Akhbar* endorsed the publication of a health primer by one Mirza Nissar Husain that had been received in response to a health department advertisement inviting people to submit manuscripts of this nature. The primer made a strong connection between religious prescriptions and personal hygiene for public health. It urged Muslims to take care of the cleanliness and purity of their air, water, surroundings, houses, and clothes so as to both remain healthy and please God, who recommended such behaviour.[105]

From Culture and History to a 'Nation' (Mulk) for Unani

In the disease-ravaged late nineteenth century, medical discussions and debates in the Urdu press created a distinct literary space. In this space, both directly through their contributions and indirectly as consumers of news, people from all walks of life joined to move medicine

[103] Ibid. He stressed that to tailor Unani even more closely to the Hindustani locale its texts, like *Qanun*, should be edited. Sections inapplicable to Indian problems and contexts should slowly be deleted from the text as they were taught in the Unani syllabus.

[104] *Oudh Akhbar*, 10 October 1878, no page cited.

[105] *Oudh Akhbar*, 22 January 1880, pp. 256–7.

beyond the pale of elites, and used it to politically and culturally educate the public. In this 'political' education the stress on locale, temperament, and history became a way also of mapping the fight for Unani onto the fight for one's country, *mulk*. In 1881 Mukhlis al Daulah, a public intellectual from Mattiaburj near Calcutta, and contributor to the *Oudh Akhbar* who was familiar with Unani practice, argued that the fight for the cause of Unani mirrored the fight for mulk. He praised the efforts of the *Oudh Akhbar* to keep the issue of Unani alive and called it *khairkhwaah-i-mulk*, or welfare of the country. Like many others, he exhorted the *rausa*—the elites—to take up the cause of Unani.[106] Asghar Husain went one step ahead of this when he argued that the progress of Unani was a responsibility not just of the elites but also of all people because it concerned physical health, which was critical to the profile of the country (*raunaq-i-mulk*). The struggle for the advancement of Unani needed to become a popular struggle as it was connected to the well being of the country.[107] At this point, for Unani professionals, mulk was about a geographically defined bounded territory framed by colonialism. But it was also a cultural entity—the locale—that conditioned the health of its people. It was a national space that they had reconstituted from within the discursive colonial space.

The correspondence of this mulk with the national cultural space critical for well being was evident in pieces in the *Oudh Akhbar* that made a similar case with respect to Ayurveda. In an item dated 18 March 1879, one Babu Rajendranath made a case for a medical college at Calcutta for *hindi adwiyah* (local medicines). This was imperative because, he said, '*Hindi adwiyah* (medicines) yields better results as compared to those of foreign lands (*ghair mulk*).' He saw the greater efficacy of 'Hindi' medicines because they were more in tune with the cultural cosmologies that conditioned the Indian body. He linked this to the temperament and constitution of the human body of Indians, which had been conditioned as per things Hindustani. The body, he argued, 'responds more to those medicines that are produced in one's own country'. He warned that 'it was going against nature if we use the medicines of foreign countries.[108]

[106] *Oudh Akhbar*, 28 July 1881, p. 2394.
[107] *Oudh Akhbar*, 29 March 1881, pp. 1038–9.
[108] *Oudh Akhbar*, 18 March 1879, p. 882.

The dialectics of colonial *vs* national space as carved out by Unani were clear in the way the contributors criticized the neglect of Unani. They were critical of elites who had been vested with administrative powers by the government but did not perform their duties towards their mulk—by which they meant the national space carved out by Unani. And this neglect of mulk meant a lack of interest in Unani. Mukhlis al Daulah used this general apathy of the elites to show also the limited role of the press in Hindustan. He was shocked that, despite the concerted campaign for Unani taken up by the *Oudh Akhbar*, the response of the elites was in particular poor. He compared this dismal situation to Europe, where the press shaped national opinion because everyone read newspapers. In contrast, he said, 'in our country those interested in newspapers will be 1–2 people out of 2–4 thousand.'[109] In these circumstances, 'how can the country progress?'[110]

As the Unani *vs* Western medicine debate proceeded, the definition of the hakim came to include a new referent—*mulki hamdardi* (sympathy/feelings for the country). Here, sympathy basically meant sensitivity to a geographical-cultural construct, the locale (*aab-o-hawa*) that was being projected as the national space. Hakims, as opposed to British medics, were valorized as they were said to have mulki hamdardi, which made them understand the social problems of Hindustan better; and this social context was critical to healing. As one contributor put it, with a twist he seemed oblivious of: 'They have mulki hamdardi. They often visit patients at home and don't charge fee.'[111] The extension of such arguments was the demonization of English doctors, who never visited patients at home; and whose 'fees is high and prescriptions expensive.' A contributor to the *Oudh Akhbar* thought that was reason enough for such doctors being unsuitable among Hindustanis.[112]

Newspaper Advertisements: The National–Colonial Dialectic

The newspaper did not just create a new concept of the author; it also created a new concept of the public. The core readership of the *Oudh*

[109] *Oudh Akhbar*, 21 April 1881, p. 1315.
[110] Ibid.
[111] *Oudh Akhbar*, 25 May 1881, pp. 1671–2.
[112] Ibid.

Akhbar may have been only within the limited circles of literate elites, but the stories it carried and the views it shared had a life beyond that circuit. And this was particularly true of news and views relating to medicine, which was of paramount interest to both elites and the populace in mofussils and villages as well as big towns. Ideas about the greater suitability of Unani to Indian temperaments, its distinct territoriality and historicity in a culturally imagined nation, and the benefits to be derived by Unani's advancements along Western lines were transmitted from readers to a range of people with whom they interacted. This created a public opinion that cut across class and bridged the urban–rural divide. Ideas and news about medicine travelled particularly fast at a time when cholera, plague, and fevers ravaged the country and made medicine and medical care a hot topic of social discussion. Indeed, the thrust on medical issues in the Urdu press created a wide 'vernacular public' that comprised both readers and listeners.

A new development, advertising, closely connected the new author and the new public. The newspaper provided not only ideas about the suitability of different systems of medicines, it also carried advertisements on useful medical knowledge: new medicines, new scientific experiments by lesser-known hakims, the medical advice of doctors, and employment news for medics. Indeed, advertising was the new technique that both cemented the vernacular public as well as used it as a viable commercial market for the sale of drugs, medical texts, and medical services. The vernacular press thus invigorated the medical public sphere by keeping medical news and views of both the Western and Unani kind alive for actual experimentation and discussion. Both doctors and hakims used the Urdu press to keep the vernacular public agog, both for larger politics and as a viable commercial market for their medicines and medical practices.

British Doctors, Medicines, and Advertisements

The daily *Oudh Akhbar* provided advertisement space in every issue. Indeed as Ulrike Stark has shown, it ran on sound commercial lines and partly financed itself through advertisements. A typical issue of the *Akhbar* in 1871 had for instance sixteen pages. Of these, on average four were given over to advertisements. In 1879 the rate obtaining for occasional small advertisements was 2 annas per line and column.

Rates for regular commercial advertisements depended on how often an advertisement was placed in the paper and ranged from Rs 12 for insertion once a month to Rs 60 for insertion six times a month.[113]

But more importantly, the *Oudh Akhbar* also provided its own advisory also in these advertisements. This made the advertisements 'advertorials' i.e. news items that carried the legitimacy and support of the patron-proprietor of the newspaper. Thereby in the late nineteenth century, when the vernacular public became a commercial market for Hindustani and British pharmacies and medical practitioners, it was not a free for all. Through the introduction of the advertorial, the market was still controlled by proactive proprietors and editors of newspapers who shaped public opinion.

At one level, advertisements reflected the Hindustani support for Western medicine. Proprietors like Newal Kishore, who encouraged Unani, also lent their seal of approval to Western medicines in their advertorials. The most frequently appearing advertisement, and one that occupied most space in the *Oudh Akhbar*, was that of the English pharmaceuticals firm Johnson Company, based in London. It advertised medical prescription books and the medicines of Dr De rooz, MD, resident of Hallford, London. The medicines could be obtained directly from London, and Dr De rooz provided free consultation for patients through correspondence. A book by Dr De rooz, 'medical adviser', called '*Tibbi-Nassah*' (Directory of Advice on Medicine), which contained his prescriptions for a variety of diseases, was advertised on 7 July 1875. Prescriptions included those for diseases of the mind and body, including sexual prowess. As with most such advertisements, Newal Kishore gave his seal of approval to the medicines advertised: 'the medicines of this book do not have the dangerous and bad effects which many dangerous medicines have. Clearly this kind of Qulliyat on such diseases can come only from someone who is experienced and thus Dr De rooz is proud of the fact that he can refer to many people in his preface who get cured under his treatment, and they too can vouch for their good health.'[114]

[113] Stark 2003: 69. Also see *Fihrist e-Kutub-e matba-e Munshi Newal Kishore*, Lucknow: Newal Kishore Press, 1879, p. 4.

[114] *Oudh Akhbar*, 7 July 1875, pp. 978–9.

The advertorial went on to detail the different kinds of medicines of Dr De rooz, their prices and ways of obtaining them from London. De rooz's provocatively titled medicine, '*Dawa-i-Qatraat-i-Hayaat*' or 'Life-giving Drops' was praised as a great energizer for diseases of the mind and body that produced lethargy. The medicine was priced at 4 shillings and 6 pence per bottle, or Rs 2 and 4 annas; and for a double size bottle 11 shillings or Rs 5 and 50 paise; a bottle four times as large was priced at 33 shillings, or Rs 100 and 50 paise. The proprietor announced that the medicines could be obtained by sending a hundi or bill of exchange to Dr De rooz, encashable in London.[115]

The other popular medicines of Dr De rooz were tablets to remove pains caused during urinating, bladder problems, back pain, kidney afflictions, and so on. The advertorial promised quick results—in three days. It also differed advice on matters of purchase. Each box of these tablets was priced at Re 1 or 21 pence; Rs 9 or 2 shillings and 9 pence; Rs 16 and 50 paise or 33 shillings.[116]

Through his detailed advisory in the advertorial Newal Kishore exercised substantial control over the drugs market. He gave his own recommendations and also published popular responses to some of these drugs. Thus in an advertisement for the 'Life Drops' of Dr De rooz, he clearly wrote that he was promoting this medicine 'with confidence [*itmenaan sei*] to those people who feel weakness or to those who have blood deficiency and feel lethargic'; this confidence stemmed from the fact that 'many people have written back praises of this medicine'.[117] Again in his advertisement for the 'Renal Pills' of Dr De rooz, he made it a point to add that 'people have been buying these a lot.'[118] Newal Kishore's important mediation in the market was reflected in a special column called 'Zaroori Itilaa Awaam Ko' (Important Message/News/Information to the Public) that often accompanied advertisements in the *Oudh Akhbar*. This was an interactive column where he listed the various effective qualities of the medicines and published the reaction of doctors as they modified the chemical

[115] Ibid.

[116] Ibid.

[117] *Oudh Akhbar*, 3 May 1876, p. 800.

[118] Ibid.

compositions of their drugs in response to public reactions published in the *Oudh Akhbar*. Newal Kishore's role as mediator between the British doctor and his medicines and the Indian public was critical to the legitimacy of the doctor, as well as to the effectiveness of the vernacular public as a laboratory for testing such medicines. In 1875 he noted, in one such column, that since the Indian public had privileged Dr De rooz's drugs over all other medicines, the doctor had been encouraged to enhance their energy potential four times. He was also confident of disseminating it with equal effectiveness all over the world.[119] Such mediation was accompanied by a word of caution to the public. In a parallel column titled 'Zaroori Ihteyaat Awaam Ko' (Important Caution to the Public), he warned the public against buying spurious drugs that did the rounds in the market. He asked people to check the government seal before buying these medicines, and to report any case of fraud that they saw.[120]

The *Oudh Akhbar* did not stop at mere pharmaceutical advertisements and advisories. It was also the translator-interlocutor between the doctor and the public. Drugs like Halloway's Balm or tablets advertised regularly in the *Oudh Akhbar* had a detailed prescription in English with the method of use packed into each box. The public needed a translator to ensure the effective use of the medicine. And the *Akhbar* stepped into this role willingly. Newal Kishore was of the view that translation of the material that came with the medicines would increase its popularity. He not only translated into Urdu the prescriptions enclosed in a medicine like Halloway's Balm but published the translation in his newspaper for greater circulation. Apart from prescriptions that detailed methods of use, books that elaborated on the nature of the medicine were translated into Urdu and serialized in the *Oudh Akhbar*.[121] For instance, Newal Kishore published in two parts an important book authored by Professor Thomas Halloway on the usage of his medicines. The translation was titled *Tareeqa-i-istimaal adwiyaat-i-Halloway, jiss-se beshtar amraaz ko faaydah hotaa hai; jo amraaz jism-i-insaan sei taaluq rakhtei hain* (Method of use of Halloway's

[119] Ibid.

[120] Ibid.

[121] *Oudh Akhbar*, 29 August 1879, pp. 2665–7.

medicines which are beneficial for most diseases related to the human body); it appeared in Urdu in the *Oudh Akhbar* and indicated the significant role that the Urdu press played between Western medicines and the Indian public.[122] Professor Halloway also used the *Oudh Akhbar* to communicate with the public: in a write-up that he sent from London, and which appeared in the *Oudh Akhbar* in August 1879, he cautioned people against fraudulent duplication of his medicines. People needed moreover to be careful of being misled by cheats who diverted them for use within other medicines.[123] A month later he advertised in the *Oudh Akhbar* that his medicines were very popular in different parts of Europe as well, and that respectable (*mu'azzaz*) people both in India and Europe had taken to them in a big way.[124]

Newal Kishore often issued advertorials for Western drugs without the permission of the respective governments. Thus, in the 14 July 1875 issue of the newspaper he published, without permission of the French government, a modern French treatment based on the use of turpentine—'*Tehrapeen mualaje aur mualajaat*' (Treatment with turpentine). As with all his other advisories, he praised the effectiveness of the drug as an energizer for piles and other growths. It was to be used in different ways for varying ailments. Three kinds of turpentine remedies were offered. The instructions for use were detailed on the packet. Their prices were Re 1 and 4 annas, Rs 2 and 4 annas, Rs 5 and 8 annas, and stretched upto Rs 16 and 8 annas, depending upon the size. The big packet of £5 or Rs 50 could be got directly from London. Once again he cautioned users to check the seal of the commissioner before using the medicine.[125]

French doctors also used the *Oudh Akhbar* for publicity. In March 1875 an advertisement in the newspaper announced that Dr Lali Mond had 'arrived in Lucknow from Paris, France. He was lodged in Noor Muhammad's hotel in Husaingunj.' He had come not only equipped with special medicines to treat paralysis, pains, tooth problems, indigestion, eye diseases, and leprosy, but had also brought sophisticated

[122] Ibid. The booklet is published on pp. 2666–70; and the second part is in *Oudh Akhbar*, 2 September 1879, pp. 2753–6.

[123] Ibid., pp. 2665–7.

[124] *Oudh Akhbar*, 2 September 1879, p. 2753.

[125] *Oudh Akhbar*, 14 July 1875, p. 1018.

electro-magnetic devices (*taarbarqi*) and instruments for diagnosis and treatment. These equipments could treat cholera and many stamina-related ailments, the advertisement added.[126]

The *Oudh Akhbar* also acted as a dealer for Western medicines that it advertised.[127] Indeed, the *Akhbar* office was one spoke in the larger wheel of commerce that Western medicine in India generated. The consumer market for these medicines in Hindustan also reflected the trading networks that constituted the commercial wheels of the British–Unani encounter. If lucrative commerce in India made it imperative to introduce Western medicine in India, then the spin-off of that trade provided the profits of trade to a lobby of Indian traders and dealers as well. Some doctors had deputed specific individuals as their agents in the cities. Thus, in Lucknow Dr Abdul al Rahman Khan was the agent of Dr De rooz. He had his chemist shop (*dawakhaana*), called Medical Hall, in the locality of Aminabad. Large pharmaceutical companies also operated through their Indian dealers. Medicines could be obtained from Indian dealers who had spread out to Calcutta, Hyderabad, Karachi, Pune, Delhi, Lucknow, Madras, and Lahore. Almost every big city of India had an agent attached to a pharmaceutical company. Some important names were Barkat Ali and Brothers, located on No. 95 Farrington Street, London. Fida Husain Khan in the office of the newspaper *Darwesh* was their agent. Some other important agents were Hilton Company on Wesley Street, Calcutta, and Nooshi Company, Lal Bazaar, Calcutta. Other names included Sibtan wa Jeevanji Company; Military and Insurance Company on Apollo Street; and Fasdeemji Nawsherwaanjee, Forbes Street.[128]

Medicines, like those of Dr De rooz, or the French medicine based on turpentine, could also be obtained directly from London through the post. Barkat Ali and Brothers in London were big agents of Dr De rooz's medicines in London.[129] And a range of commercial companies that included Barclay Company and Newbery Company in London were listed as dealers for the French turpentine-based drugs.[130] Their addresses were provided in the advertisements for these drugs.

[126] *Oudh Akhbar*, 7 March 1875, no page cited.
[127] *Oudh Akhbar*, 3 May 1876, p. 800.
[128] *Oudh Akhbar*, 7 July 1875, pp. 978–9.
[129] Ibid.
[130] *Oudh Akhbar*, 14 July 1875, p. 1018.

Advertisements: Hakims and their Nuskhas

British medicines and doctors shared this discursive space with Indian doctors and hakims, who also advertised their expertise and drugs. For the hakims the advertisement provided a welcome forum to carry forward the battle for survival against Western medicine. Advertising was also a mode of popularizing their medical texts and advisories with an eye on commercial profits and employment. Munshi Newal Kishore, himself a canny businessman, saw no contradiction in advertising Urdu texts and the prescriptions and services of new hakims, even as he critiqued the new Unani in his columns. Advertisements brought him commercial profits.

The *Oudh Akhbar* was always happy to entertain advertisements of prescriptions sent by hakims from places as far as Hyderabad. In an age of tough medical competition, hakims were happy to go public with their medical knowledge. Detailed prescriptions for common ailments like diarrhoea and dysentery were published in its columns under the names of the sender and were titled 'Announcements'. In 1876, Maulana Abdul Aleem Nasrulla Khan, a trained hakim and pensioner (*wazeefah khwar*) of the Hyderabad nizam, detailed his remedy for loose motions. He listed the ingredients for making the medicines and detailed the method of consuming the preparation, and the variations on it that were necessary for adults and children. At least three *nuskhas* of the Maulana were published in the advertisement.[131]

Hakims advertised their services as well. In 1876 one Syed Bahadurshah, a noted hakim in the Junagarh estate of Saurashtra and a resident of Bareilly, advertised in the *Oudh Akhbar* his expertise in cures like leucoderma. Many feats of the hakim were cited in the news item.[132] In 1880, in a similar kind of advertisement, the editor advertised the services of Hakim Bandah Hasan Sahib of Lucknow, who was employed by Chaudhury Muhammad Khuslat Husain, the talluqdar and honorary magistrate of Sandila. He was available in Lucknow and looking for employment. The *Oudh Akhbar* announced that he could be contacted in Lucknow in case any *riyaasat* (estate) required his services.[133] And

[131] *Oudh Akhbar*, 9 June 1876, p. 917. See also another *nuskha* of his in the *Oudh Akhbar*, 21 June 1876, p. 1009.

[132] *Oudh Akhbar*, 4 August 1876, p. 1288.

[133] *Oudh Akhbar*, 4 February 1880, p. 442.

riyasats did look out in the Urdu press for the perfect hakim. In 1879, Rai Munshi Parmesher Das, superintendent of the riyasat of Bownee, Sandila, and Jagni advertised for a hakim for his estate. The advertisement specified that the hakim should be 'clever, experienced [*tajurbe-kaar*], competent [*laaiq*] and friendly [*khaleeq*]'. A salary ranging between Rs 25 and Rs 40 was promised.[134]

Many times, unusual and miraculous cures performed by hakims appeared as news items/advertisements for them. In December 1878 the *Oudh Akhbar* carried one such advertisement that elaborated the miraculous curing powers of a Muslim hakim from Delhi. He cured a *seth* who was brought to him in an old and sick state: his complaint was frogs in his stomach. The news item gave vivid procedural descriptions of the cure and the medicines used, and exhorted the people to contact the hakim if required.[135]

Finally, most importantly hakims used the *Oudh Akhbar* to popularize their popular texts and *akhlaaq*-driven definitions of health. Indeed, the emphasis on *akhlaaq* in everyday behaviour was so central that even rulers often asked for advice on proper conduct. Such royal recognition was considered an honour worthy enough to be publicized through the *Oudh Akhbar*. In July 1876, in a column titled 'Advice of hakim Abu al Hasan Arabi', the hakim advised his students to always live in a place where there was a generous ruler and a reputed and pleasant (*haaziq*) hakim. The laudable *akhlaaq* of both were elaborated and stated to be critical to the discharge of their respective duties.[136]

And if *akhlaaq* was stressed in the advisories of hakims, Urdu medical texts that underlined it as crucial to the definition of a hakim were also advertised in the columns of newspapers. Urdu medical literature that we discussed in the previous chapter, which underlined deportment and *akhlaaq* as crucial to healing, was advertised in the columns of the *Oudh Akhbar*. Most of these Persian and Arabic medical texts were translated into Urdu at the behest of Munshi Newal Kishore and published by his press. The *Oudh Akhbar*, the newspaper of this press, was

[134] *Oudh Akhbar*, 27 November 1879, p. 3648.
[135] *Oudh Akhbar*, 7 December 1878, p. 3882.
[136] *Oudh Akhbar*, 12 July 1876, p. 1171.

thus the ideal forum for their advertisements.[137] The advertisements came in the form of summary news items from the publisher-proprietor himself. Very often, the references used by the author were also listed, which indicated the range of texts consulted by him. Thus, the Urdu medical text *Mazhar ul Uloom* by Qazi Ilahi Baksh was advertised in the *Oudh Akhbar* of 20 August 1876. The newspaper proprietor said that the public needed an abridged and simple book of medical knowledge, and he had published the book on popular demand. He listed a range of books that included *Dastur-al Ilaj, Zakhirah-i-Khwarzmshahi, Tibb-i-Akbari, Tashreeh-al-a falak* etc. that went into the making of this text.[138] Most of these, as we saw, projected the ideal hakim as cultured and well mannered apart from being professionally efficient. Many of these reference texts were themselves advertised separately. Thus, in the *Oudh Akhbar* of 8 April 1878, one of the above-mentioned texts, the Urdu translation of the *Zakhirah-i-Khwarzmshahi*, was advertised. The advertisement stated that Hadi Hasan Khan Moradabadi translated the original Persian version into Urdu, gave its price, enumerated its usefulness, and exhorted buyers to get it by post from the press.[139]

In keeping with the spirit of medical fusion that the Urdu press encouraged, journals like *Risaalah Ilm Keemiyaa* by Maulvi Zakaullah, professor of science, Muir Central College, Allahabad, which printed articles on new themes like the significance of chemistry in pharmacies, were also advertised in the *Oudh Akhbar*.[140] It was this spirit that led the Urdu press, in particular, and the vernacular press in general, to encourage translations of not just Unani medical literature into local languages, but even of the English medical journals and periodicals that carried articles on modern techniques of pharmacy and healing. In December 1879 the newspaper *Anjuman-i-Punjab* obtained a copy of a sanitary health journal of Dr Cunninghan—the *Hifzan-i-Sihat*—

[137] *Oudh Akhbar*, 8 April 1878, p. 960. The proprietor stated this connection between the texts advertised and the newspaper in the advertisement of the *Zakhirah-i-Khwarzmshahi*.

[138] *Oudh Akhbar*, 20 August 1876, p. 1399.

[139] *Oudh Akhbar*, 8 April 1878, p. 960.

[140] *Oudh Akhbar*, 1 May 1880, p. 1520.

for translation. The *Oudh Akhbar* appreciated the move and announced the news in the most appreciative and encouraging manner. In a news item titled 'Risaalah Sihat kaa Tarjumaah' (Translation of Risaalah Sihat), it said: '[We are] pleased that they [*Anjuman-i-Punjab*] have undertaken its translation' and [we agree] with the advice of the editor of *Anjuman-i-Punjab* that *risaalah* be translated in Urdu and Gurmukhi'.[141] The same spirit of open-mindedness also made Munshi Newal Kishore the foremost publisher of translations also of Vedic and 'doctory' medical texts. As with the Unani translations, here again the idea behind such translation was to make medical knowledge accessible. In August 1880, in an advertisement, the *Oudh Akhbar* announced the publication of a *Majmuah Tibb*, a Compendium of Medicine, in Urdu. This collection, published by the Newal Kishore Press, combined Unani, Vedic, and 'doctory' treatments. The advertisement made the objective of the publisher very clear: it had been published for the public (*rifaah-i-aam*), which needed access to medical knowledge because, 'they have suffered often at hands of inexperienced *nuskhas*. People in villages, qasbas, etc. will benefit with this since they will now no longer be deprived of cures.'[142]

The editor and proprietor added their usual advice on the items and services advertised. Thus in the case of Hakim Syed Bahadurshah, the hakim who we saw advertised his services, the editor added that 'he was a well wisher of the people and this had inspired him to inform us to spread the news of his expertise far and wide.' He hoped that the readers would spread this news and cited the case of a patient, Munshi Khairat Ali Khan Bangash, whose kidney pain had been cured by the said hakim.[143] The editor often had to mediate more emphatically when the dialogue that the advertisements provoked between the hakim and the consumers tended to become bitter. In an interesting incident, when a letter to the editor complained that the medicines of one Hakim Abul Hasan Sahib Banarsi advertised in the *Oudh Akhbar* were fraudulent (*dhokei kee goliyaan*), the editor intervened to argue that the medicines did not work perhaps because they were not taken in the prescribed manner or did not suit the temperament of the

[141] *Oudh Akhbar*, 12 December 1879, p. 3820.
[142] *Oudh Akhbar*, 5 August 1880, p. 3598.
[143] *Oudh Akhbar*, 4 August 1876, p. 1288.

patient concerned. And this lack of full information he attributed to the method of advertising, which he said was a novel mode to reach out to the public. He offered to provide an advertisement format to hakims which would ensure that all aspects of caution and care in the consumption of the medicine were clearly indicated in future. Such an advisory was sent to Hakim Abul Hasan so that his future advertisements took care of these aspects.[144]

Indian doctors, Hindu and Muslim, both also used the *Oudh Akhbar* to reach out to the public. Dr Harish Chandar Sarma advertised his medicine for sexual-weakness-related diseases (*zof-i-baah*). The medicines could be procured personally from the doctor or in the mail by writing to him, the advertisement added. The price of this medicine and a price list of other medicines available with him were mentioned. Thus Dr Sarma's hair preserver for turning grey hair black was for Rs 10; Dr Sarma's Hem Sagar Tel (oil) was for Rs 7; Dr Sarma's *bawaseer dawa* (piles medicine) was available, and so was his special toothpaste.[145] In another provocatively titled advertisement called 'Royal Alternative', a doctor from Patna, Dr Muqaddar Husain, advertised a special *arraq* (syrup) he had made combining Unani and Western medical potions. This was not only said to be good for aches and pains but also for general health and prowess of the brain. It was priced at 6 annas per bottle.[146] Pharmacists too put their advertisements in the *Oudh Akhbar*. In an advertisement entitled '*Waaste rifaah-i-aam kei*' (For the Ordinary Public) medicines for asthma (*dama*) and from a pharmacist, Lala Nakamal of Lalitpur, Calcutta, were advertised. One who guaranteed cures for toothaches and the swellings of gums also advertised special tooth powders. The Lala's full address was listed so that people could obtain his medicines via the mail.[147] Advertisements for local mixtures (*chooran*) for indigestion were also advertised under the names of local drug companies. Thus, the Pakalan Company of Lucknow had several such advertisements in the columns of the paper.[148] Here too the office of the *Oudh Akhbar* acted as the counter from which many such local

[144] *Oudh Akhbar*, 15 May 1879, pp. 1545–6.
[145] *Oudh Akhbar*, 21 July 1875, no page given.
[146] *Oudh Akhbar*, 20 November 1878, p. 3718.
[147] Ibid.
[148] *Oudh Akhbar*, 13 September 1879, p. 2896.

medicines could be obtained. The *Oudh Akhbar* of 19 September 1879 advertised that *arraq baid-i-mashk* (syrup) could be obtained by post by writing to its office in Lucknow, or from its printing press (*chhapakhana*) at Kanpur.[149]

[149] *Oudh Akhbar*, 19 September 1879, p. 2956.

7

From Jhawain-tola to Takmil-ut-Tibb, Lucknow

Introduction

IN THE LAST CHAPTER WE SAW THAT THE *OUDH AKHBAR* created a public sphere wherein the new Unani was critiqued by hakims who supported the older scientistic Unani of the canonical texts. These hakims writing on the corruptions in Unani emerged as the new Muslim public intellectuals. They strengthened the case of the canonical Unani, demanded its professionalization, and gave it a distinct cultural and historical profile in Hindustan. And they constituted its new profile within the colonial medical framework rather than in opposition to it, even as they never stopped at contrasting its antiquity to that of colonial medicine. They highlighted Unani's cultural roots in Hindustan. They argued that the modernity of colonial medicine could not make any claims to a history or intellectual legacy in their country.

In 1901 the issues of professionalism, historicity, and modernity as popularized by the Urdu press were further reworked to their own advantage by the traditional hakim families. They established institutions of Unani learning and took the professionalization process under their exclusive control. They wrote a history for Unani that made it dig its roots deeper in Hindustani society. They defined its modernity by locating it in a national space culled out of the colonial framework. And yet the 'national' Unani remained tied to strands of religious communitarianism and Islamic universalism. It could never author a narrow anti-colonial territorial nationalism.

This chapter will look at the private records of one established family of hakims who lived in the Jhawain-tola locality of what is now the old city of Lucknow—the Azizi family. It will analyse the working of their institution, the Takmil-ut-Tibb college, to understand Unani's development in the early twentieth century as it engaged with Western medicine. It will show how the changed political context of the twentieth century shaped the predicaments of the Azizi family as they balanced Unani's 'national' profile with a Muslim communitarian and Islamic universalist image.

In 1902 the Azizi family established the first institute of Unani learning in Lucknow. It was named the Takmil-ut-Tibb college. The name of the college, Takmil-ut-Tibb, meaning completion of education in medicine, indicated that it was a finishing school for medical training. But for our narrative the word *takmil* very appropriately summed up also the completion of the task that the public intellectuals had initiated in the late nineteenth century: professionalization of Unani and carving for it a territorial profile and history in Hindustan. The former meant bringing Unani under tighter family control. The latter strove to create for it a national space from within the colonial medical framework.

At the Takmil-ut-Tibb college the Azizi family began to professionalize Unani education. This meant purging it of its uncontrollable knowledge frills as represented by Urdu-read hakims. It amounted to shoring up Unani's Arabic textual core: Arabic texts were to be the kernel of the reformed Unani. Formal apprenticeship with a teacher at the institute was its defining feature. Most teachers belonged to the Azizi family, so Unani came firmly under their control. However, the family threw its net wider to define itself also as a professional family, with members who were not only linked by blood, religion, ethnicity, and dynasty, but by professional training as well. The institution also wrote a history for Unani which located it in a Hindustani past, giving their profession a national space.

This national Unani never was completely separated from colonial medicine. It cannibalized aspects of Western medical practice while protecting its scientistic core from the new Urdu-read hakims. It trimmed its sacral frills. At a time when Hindus and Muslims were consolidating separate educational traditions, the medical reforms of the Azizi family, which opened the door to talent and scholarship irrespective of religion, attracted immense popular curiosity. By the 1920s, under popular pressure, medical debates often spilt into the communalized political sphere. And politics often coloured Unani with a Muslim hue. Yet a sophisticated balance between the national and the Muslim communitarian defined the politics of the Azizi family.[1]

The Azizi Family of Lucknow Hakims: A Profile

The Azizi family of Lucknow hakims traces its origin to Hakim Muhammad Yaqub (1790–1870). His ancestors lived in Kashmir and migrated to Delhi at the time of the Afghan invasions in 1750. His father, Maulvi Ghulam Nabi, was a disciple of Shah Abd al Aziz, the Delhi-based Sufi. After the political eclipse of Delhi, the family moved to Faizabad and then to Lucknow. It was in Lucknow in 1790 that Hakim Yaqub was born.

Like most sons of hakim families, he trained at the local Farangi Mahal madrasa and then completed his apprenticeship in the family school of the Lucknow hakim, Mir Muhammad Murtaish. He set up his clinic in the Jhawain-tola quarter of his natal town, Lucknow. He divided his time between his students, his clinic, and attending to various members of the Awadh court. The family earned fame in the period of his grandson, Hakim Abd al Aziz (1855–1911). It called itself the Azizi family as a tribute to his contributions.

Hakim Abd al Aziz was the son of Hakim Mohammad Ismail (1822–86), who was the court physician of the Nawab of Rampur and the Nawab of Awadh. He became famous for his view that Unani medical education needed to protect its Arabic canon and be standardized and institutionalized. He saw control of the family over Unani knowledge

[1] The Azizi family was very much part of the Lucknow–Delhi sharif culture to which Sir Syed Ahmad Khan, Maulvi Zakaullah, and Nazir Ahmad of Delhi College belonged. All wished to enagage with colonial 'new learning' and translate British rule to their best advantage. The language and the constituency to be catered to created in-house differences between these Muslim intellectuals. For Syed Ahmad it was the landed Muslim elite, for Zakaullah it was the Muslim Urdu-reading community, and for the Azizi family it was the Arabic-canon-reading Muslim professionals into which Hindus were unproblematically included. See Hasan 2006: 261–98.

a guarantee of its safety. In 1870 he set up his clinic in Jhawain-tola in Lucknow. At the clinic he taught medicine (tibb) to students who had the basic madrasa degree from the city's Sunni Muslim seminary of Farangi Mahal. The medical training ran like a family apprenticeship, and most members of the Hakim Sahib's extended family were enrolled as students. Teaching and management were at the clinic an exclusively family-run enterprise.

The dream of standardization and institutionalization of this medical education was fulfilled in 1902 when Hakim Abd al Aziz founded the Takmil-ut-Tibb college at Lucknow. This was a family-run institution. Hakim Abd al Aziz provided the finances and infrastructure for the college. Other family members contributed by teaching for free. The main emphasis in the college was the teaching of Ibn Sina's *Qanun*. This was complemented by the introduction of Western-style surgery and anatomy in its curriculum. The college offered a three-year course, with theoretical and practical parts, which included surgery, anatomy, physiology, and chemistry. A teaching hospital was also attached to the college. Teaching in all departments was controlled by the family. Hakim Abd al Aziz lectured on most topics. After his death his two eldest sons, Hakim Abd al Rashid and Hakim Abd al Hamid, took over the running of Takmil-ut-Tibb. Hakim Abd al Rashid, the elder son, set up an executive committee for the school in 1911 and became its first secretary. The college expanded in this period and depended on financial contributions from the family and many other Indian landed aristocracies.

The Azizi family was important also because its members were the archetype Muslim public intellectuals, very active in the local- and national-level organization of Unani practitioners. Very like medical professionals in Europe, who organized themselves in associations in the late nineteenth century to demand rights and protect privileges, the Azizi family also established professional associations of hakims. In 1911 the hakims of Lucknow founded the Anjuman-i-Tibia, the medical society whose general aim was the advancement of Unani and protection of the rights of hakims as medical professionals. Hakim Abd al Rashid became its secretary. It was due to the efforts of such associations that, later, other members of the family became part of government-run bodies like the UP Board of Indian Medicine. Some members, like

Hakim Abd al Hamid, also became members of the All India Sanitary Conference (1914) and the Provincial Board for the Prevention of Epidemics (1929). Other family members were part of government boards as well as various advisory committees.

Members of the Azizi family did not remain confined to their college at Lucknow. Many moved beyond the Takmil-ut-Tibb institution in Lucknow and were involved in other medical schools and colleges. For instance, after setting up his own medical school, Hakim Abd al Halim (1905–54), the fourth son of Hakim Abd al Aziz, became principal of another medical institution in Lucknow over whose development and subsequent merger he presided until 1953. Similarly, other members of the family were involved in the Tibbiya College of Aligarh Muslim University. Family members like Hakim Abd al Latif were on advisory committees of Tibbi institutions from Patna to Hyderabad.

Hakim Abd al Aziz and the Challenge of Colonial Medicine: The Takmil-ut-Tibb College at Lucknow

By the early twentieth century the colonial state, far from establishing its cultural hegemony, stood overwhelmed by the problems created by the new Unani and its diversified knowledge base. The floodgates of Unani knowledge, difficult to control, had been opened by Urdu print culture. There was public pressure on the state to monitor and regulate the activities of the so-called non-conformist, 'illiterate' hakims. As the British looked for allies to check the further growth of the new Unani, their interests coalesced with those of traditional Unani families who, in the earlier century, had lost out to the new hakim.

These established families as well as the colonial state were eager to control the new hakim. The families, with the support of the British, set up Unani institutions in the early twentieth century as a way to reassert their hold over their medical tradition, produce qualified professionals, and displace the new hakim. The British supported the institutions as they saw in it the end of the menace of ill-trained hakims and the creation instead of a pool of trained hakims to assist their medical service. This shared interest made families like those of the Azizi at Lucknow work with the British government, even as they underlined their indispensability as unique healers to their in-house clientele. The *Oudh Akhbar* had already created a favourable public opinion for the families, and for the coming together of Unani and British medical practice. All this made it easy for the traditional families to work with the British. And yet of course the institution provided a space for competition with the colonial state as well, the national space for Unani being forged through this process of cooperation and competition. This unique relationship ensured that the colonial state's political and cultural hegemony remained questionable, as did the notion of an exclusive elite national hegemony.[2]

The Takmil-ut-Tibb: Co-operation and Competition with the Colonial State

The opening minute of the first executive committee meeting of the Takmil-ut-Tibb (1902), which details the idea behind its establishment, shows that the family had the support of the British administration. It was titled, 'On the occasion of the coronation of Emperor Edward VII the establishment of the Qaiser-i-Baitulshifa' (the king's hospital). This title revealed that Hakim Abd al Aziz, the founder of the college, acknowledged and appreciated the support of Edward VII to medical philanthropy both in England and India. He inaugurated the college on the day of his coronation to strike bonds of co-operation with the benign emperor.

The inaugural speech of Hakim Abd al Aziz made it even more clear that his college would work in tandem with the British. He praised the benevolent and philanthropic character of British rule in India and

[2] David Arnold shows that the urban elites encouraged the popular colonial theme of medical philanthropy. This idea was actively popularized and encouraged by the Indian middle classes. Donations to existing funds and setting up of new charitable hospitals by wealthy Indians followed. Arnold argues that to a large extent this was to buy peace, status, and privilege from the colonial state; often, it was for personal reasons to perpetuate the memory of a dead relative. But it did make Western medicine more familiar and discernible to the people. At the same time, this also dented its hegemony as middle-class participation came with its own agenda. The colonial state needed to accept their authority and leadership before it could carry on its colonizing medical drives. Arnold shows that this was best reflected in the anti-plague riots and the policy of conciliation towards the Indian middle-class leadership that followed. Arnold 1993: 240–89.

highlighted the colonial government's sanity in instituting innumerable hospitals, dispensaries, medical arenas of learning, and *hifzan-i-sihat* (hygiene) for the welfare of the people of India.[3] According to the Hakim Sahib, the most remarkable act of state benevolence was evident in the realm of women's health. He detailed very positively the colonial initiative, in particular the efforts of the vicerene, Lady Dufferin, to establish separate hospitals for women. The training and education of Indian *dais* in midwifery was a worthwhile project that he said needed to be appreciated.[4]

This appreciation of colonial medical drives in India was followed by some interesting logic—he was sure the colonial philanthropic state would encourage similar initiatives by Indians. Thus, he continued, 'in view of all this [state philanthropy] we suggest that on the occasion of the coronation [of Emperor Edward VII] a hospital [*baitful shifa*] should be established by the hakims in Lucknow after the name of the Emperor.'[5]

The Azizi family and the colonial state were thus partners in bringing Unani under tighter control, with the element of competition between them carefully crafted into their cooperation. The hakims were meant to slowly intrude into domains monopolized by the colonial doctor. The institution's desire to compete with the colonial state in its fast-expanding medical drives was evident when it laid out its objectives. Interestingly, even though this Hakim Sahib's institute was to be set up to 'impart the best education in Unani medicine', its aims corresponded closely with those laid out by the colonial state in the establishment of its hospitals, dispensaries, and medical learning centres. It sought to produce hakims who were as well equipped as doctors to serve in government medical service. To fulfil this objective, several steps were urgently required.

First, the syllabus was formalized. Arabic medical texts were retained as the core of Unani learning, and practical training in Western-style surgery was introduced as an additional course. For madrasa-trained fresh recruits, a three-year course was offered. This was divided into

[3] Minutes of the Executive Committee of Takmil-ut-Tibb College, Lucknow: 1902–9, pp. 1–4. Henceforth Minutes.

[4] Ibid.

[5] Ibid.

practical and theoretical halves. Surgery was introduced as part of practical training. Anatomy, physiology, and chemistry were included in the syllabus as compulsory theoretical subjects. The central text of the course was Ibn Sina's (Avicenna) *Qanun-fi-al-Tibb*. The curriculum had specialized papers based on this text. It was organized as follows:

Hakim Abd al Aziz taught the general principles of Tibb as enunciated in Avicenna's text.
Hakim Abd al Hafiz taught *Hummayat-i-Qanun* (the volume of *al Qanun* on fevers).
Hakim Abd al Rashid taught *Sharh-i-Asbab* (diseases, their causes, signs and symptoms, vol. I).
Hakim Hafiz Abd al Majid taught *Sharh-i-Asbab*, vol. II.
Hakim Abd al Hamid taught *Nafisi* and *Kamil ul Sana*.
Hakim Abd al Moid taught *Qanuncha*.
Emphasis was laid on the teaching of *al Qanun* and its commentaries. Later, other special tracts penned by family hakims were added.[6]

Second, the college gave formal education in midwifery, in particular to 'literate' women. Special training in gynaecology was offered to trainees. So far, such training had been imparted only by the British. Hakim Sahib said that the aim was to train 'properly such women who are at present involved in *daigiri*' (nursing).

Third, like British hospitals, the new Unani college highlighted its charitable face. In both medical training and out-patient treatment, the institute was meant to cater not just to elite sections of society but to the masses as well. Hakim Sahib hoped to have 'proper funding for those students or trainees who are not financially well off.'[7] He emphasized that poor and destitute patients would be treated as an act of charity (*Allah-ki wasté*).

The revised curriculum made the Azizi family churn out hakims equipped for government jobs and helped it strengthen its hold over

[6] Rahim 2001: 34. Hakim Abd al Rashid wrote his tract on plague, *Taun*, and Hakim Abd al Aziz wrote a tract in Persian called *Risaalah Tuhfa i Azizi* (1888). Many other tracts were penned by hakims of the Azizi family. In order to make such books easily available, Hakim Abd al Aziz persuaded Nami and Newal Kishore, the two leading printing houses of Lucknow, to undertake the printing of Unani books and supply them at reasonable prices.

[7] Minutes.

Unani. Those whose learning skills did not correspond to this textual regimen were no more seen as hakims. The axe fell first on Urdu-reading self-taught and self-styled hakims who had proliferated in the nineteenth century. The professional class of jarrahs (barbers who acted as surgeons), who got away as hakims, was also affected. Hakim Abd al Aziz clarified that jarrahs were illiterate (jahil), and only hakims with certificates from his college were skilled surgeons. Similarly, he was critical of *jahil dais* (illiterate midwives) and urged women to train at his college.[8]

These objectives, which carefully matched those of the colonial state, were justified by using a terminology that derived heavily from Islamic notions of charity and philanthropy. Thus, Hakim Sahib clarified that no personal gains were achieved through the establishment of his Unani medicine institution. He said he needed no personal gain because he came from a family 'where charity and service of Allah dominate'. Relating the Islamic notion of charity to himself, he said he had '*waqf* [gifted] himself and his two sons to this mission'.[9] Underlining his family's charitable instincts, he pointed out that he taught students and trained them in his clinic free of cost. He also encouraged his *rais* (rich) friends to help students with donations. The system needed to be institutionalized because charity satisfied the donor but often embarrassed the receiver. Moreover, the informal practice was cumbersome for students, often humiliating, and cut into their study time. Hakim Sahib was of the view that an institution, where all their daily needs were taken care of, without personal adhocism and obligations, provided them more time to concentrate on their studies.[10]

He laid out a list of infrastructure and financial requirements to achieve the objective of systematizing medical learning and building his institution on a reasonable footing, like British hospitals. This reflected a clear attempt to move beyond the hitherto informal family-apprenticeship-based system of medical training. The two urgent requirements of this changed medical system were a big library and teaching facilities for students with adequate equipment for surgery and demonstration. It was also proposed to start a students' hostel in the fashion of a residential campus, with scholarships offered to those

[8] Ibid.: 1–4, 53.

[9] Ibid.: 28.

[10] Ibid.

coming from economically weak backgrounds. The library, hostel, and residential campus were meant to break the cumbersome and often humiliating exercise of obtaining food and facilities for students via personal charitable initiatives.

Through its revised curriculum and new infrastructure, the institute aimed to produce hakims who could effectively enter medical realms, such as sanitary commissions and municipalities, hitherto regarded an exclusive preserve of Western-style doctors. Surgery and midwifery were realms where such doctors prevailed. As David Arnold has shown, even though surgery was not so fiercely resisted because of the quick relief from pain that it offered, the issue of colonial intervention in women's health was fraught with tensions.[11] Hakim Abd al Aziz hoped that, with the setting up of his college, both these medical fields would have Unani practitioners competing with doctors on an equal footing.

Early signs of the success of this policy of intruding terrains, which the colonial state wanted to monopolize, was revealed when plague broke out in Lucknow in 1903. At a time when, as Arnold has shown, there was considerable resistance to the colonial state's anti-plague drives,[12] the hakims of Lucknow became the alternative medical practitioners to whom people flocked for the surgical care that they resisted in colonial hospitals and dispensaries. In his annual report of the Takmil-ut-Tibb, Hakim Abd al Aziz writes that the training and certificates in surgery that his two sons, Hakim Abd al Rashid and Hakim Abd al Hamid, had received from the Agra Medical College came as a boon when tackling the disease. These two young hakims used their newly-acquired surgery skills to remove plague glands from the armpits of scores of patients. Whereas patients in state dispensaries resisted these bodily intrusions, once in the hands of the hakims they were less worried about such surgeries. The popularity of the hakims in these difficult times, he said, could be gauged from the fact that whereas in the big cities of India the casualty rate of plague was 80–90 per cent, in Lucknow the cure rate (*shifayaabi*) was 80–90 per cent. This was due to the successful techniques of Unani practitioners. Through their

[11] Arnold 1993: 256–60.

[12] Ibid.: 200–40. People objected to the inspections of their bodies for diagnosis, to restrictions on their mobility that were imposed to prevent the spread of disease and to the attempt to control their private spheres.

reformed medical system they largely succeeded in controlling plague, and thus proved that their 'dying system of medicine has still some life in it.'[13]

By his emphasis on intervening in domains that colonial medical institutions had dominated, Hakim Sahib's argument had an interesting edge. At one level he highlighted the similarity of colonial and Unani medical interests and objectives in controlling medical knowledge. At another level, this compatibility also invigorated and made the medical market more competitive. Indeed, it challenged the monopoly of the British-trained doctor over the medical market of North India.

Hakim Abd al Aziz alluded to this remarkable change on the occasion of a farewell function for Col. J. Anderson, the civil surgeon in Lucknow. In his speech he appreciated the colonial state's contribution in training hakims to effectively participate in the medical market. He praised the state's role in curbing the activities of the non-conformist new hakims, whom he referred to as *jahil* or illiterate. He was of the view that such encouragements had helped restore the scholar-hakim to his rightful place:

> I thank you [Col. Anderson] for training my sons. The young physicians were formally also adept in the sister science of surgery, but for a long time they have given up entirely that branch of their profession and as a consequence that knowledge has been exclusively confined to illiterate and insufficiently trained practitioners, the *jarrahs*, in whose skill the public has no confidence.—If the knowledge of surgery could be imported to hakims it would confer a great boon on the inhabitants of remote country side villages, where no government hospitals exist and no doctors are available.[14]

Underlining once again the vibrancy that the Unani–British cooperation had lent to the medical market, he said:

> We attribute all the success of the institute to the patronage and attention which you have bestowed on it. You have also given pecuniary assistance and have laid the present and future generation of Unani physicians

[13] Rahman 1978: 239–41. Hakim Rashid also wrote a tract on plague called *Almaun fi al Taun* in Urdu, which was published in 1897 at Matba Anwari i Muhammadi, Lucknow.

[14] Ibid.: 154.

under a deep debt of gratitude. You have also brought the professors of Unani and European systems of treatment into an amicable relationship and have bridged the gap between these two which had hitherto existed and caused them to co-operate in the noble task of alleviating pain and palliating human sufferings.[15]

Col. Anderson, in his reply, was equally appreciative of the medical dynamism that the co-operation had triggered:

> Personally I think that there should be no jealousies between practitioners of different systems. We are all fellow workers in a common cause namely the alleviation of human suffering. Our motto should be 'live and let live'. A man who discovers a new remedy or improved method of treatment should spread his knowledge far and wide. He who keeps such a discovery to himself with a view to pecuniary gains is an unworthy member of the whole profession which we represent.[16]

This co-operative yet competitive work culture that connected the Muslim professional class and the colonial state also led to some intimate personal friendships across the divide. This only added to the medical ferment of the times. The personal friendship between the civil surgeon of Lucknow, Col. Anderson, and Hakim Abd al Aziz, is revealing. Their friendship was structured on the shared premise that Unani medicine needed to purge itself of the new hakim, shield its Arabic core, and adapt to Western biomedicine techniques like surgery. Also, that this was in the interest of 'public welfare'. To join this colonial project, Hakim Sahib made a point by sending his two sons, Hakim Abd al Hamid and Hakim Abd al Rashid, to be trained in surgery by Col. Anderson. The Colonel took them with him to the Agra Medical College once he was transferred there, and the boys obtained a certificate in surgery from Agra. While with Col. Anderson, they were no ordinary students. They were treated with the utmost paternal care by him and his wife and they lived with the couple. On their return, they performed a successful eye surgery on their father, Hakim Abd al Aziz.[17] Later, the two nephews of Hakim Sahib, Hakim Abd al Majid and Hakim Abd

[15] Ibid.: 153.

[16] Ibid.: 155.

[17] Minutes, pp. 10–11.

al Moid, were similarly trained.[18] Col. Anderson and his team occasionally visited the Takmil-ut-Tibb for inspections and invariably expressed their satisfaction at the surgery operations of trainees.[19] They also noted in the inspection records their appreciation for the 'public welfare' agenda that the college pursued.[20] Their appreciation was followed by generous grants of surgical equipment to the college, much of which was transported there from England.[21]

Hakim Sahib slowly worked out a 'national' character for Unani, even as his institution functioned within the discursive framework of colonial medicine. He was of the view that Unani could compete best with Western medicine if it had a national character. His objective was to establish Unani institutions all over Hindustan. In the first few years, he mobilized students as well as donors from all over the country. After completion of their course, the students were encouraged to go back to their respective regions and establish Tibbi schools there. The donors to the college were both Hindu and Muslim zamindars and taluqdars of Awadh, namely:

Raja Muhammad Siddiq Khan of Nanpara
Rani Srat Kunwar, taluqdar Khirigarh
Chaudhury Muhammad Jaan, taluqdar Sandila
Raja Bhup Inder Vikram Singh, taluqdar Prayagpur
Raja Prithipal Singh, taluqdar Surajpur, district Barabanki
Nihaluddin Ahmad, registrar judicial commissioner, Awadh
Khan Bahadur, Munshi Mumtaz Ali, former manager of Balrampur state, and finance minister of Bhopal state
Munshi Ihtesham Ali, taluqdar Kakori, zilla Lucknow
Munshi Prayag Narain, proprietor Newal Kishore Press, Lucknow
Raja Udit Narayan Singh, taluqdar Ramnagar, district Barabanki
Chaudhury Nusrat Ali, Asstt. Sec. Taluqdaran-i-Awadh
Captain Rameshwar Datt Singh, adviser Balrampur state[22]

[18] Ibid.: 17.

[19] Ibid.: 18; and Rahim 2001: 34. This is a revised version of a booklet: Zillur Rahman and T. Siddiqui, 'Studies in the History of medicine'.

[20] Minutes, p. 17.

[21] Ibid.

[22] Ibid.: 20.

Later additions were

Ganga Baksh Singh, of the estate of Balrampur
Babu Ganga Singh, taluqdar Tikaari, zilla Rae Bareilly
Sher Bahadur Singh, tehsildar Bahraich
Lala Baboo, tehsildar Jahan Bar, zilla Gonda
Tej Bahadur Singh, tehsildar Tulsipur, Gonda.[23]

This list of patrons and donors within UP grew larger in the years of Hakim Abd al Hamid and Hakim Abd al Rashid, the two sons of Hakim Abd al Aziz. It came to include the Raja of Nanpara, the Raja of Prayagpur, the Raja of Singhai (Kheri), the Raja of Chandapura, the Raja of Kasmanda, the Raja of Khajurgaon, and the Raja of Samrauta.[24] The Raja of Mahmudabad, the Raja of Jahangirabad Sir Tasadduq Rasul Khan, and his successor Raja Ijaz Rasul Khan, donated large sums of money to the college. Outside UP supporters of the college included the Maharaja of Gwalior, Mir Usman Ali Khan (the ruler of Hyderabad), and the Maharaja of Baroda. Dignitaries like Sir Aga Khan also encouraged the college initiative and remained closely associated with it.[25]

Like the patrons and donors, the students also initially came from UP and Bihar. Later they came from other states as well:

Maulvi Muhammad Husain Khan (Sitaput, UP)
Maulvi Altaf Ali Khan (Moradabad, UP)
Maulvi Shamsul Haq (district Muzaffarpur, Bihar)
Maulvi Muhammad Sayid (Distt. Azamgarh, UP)
Maulvi Muhammad Abdur Rahman (Junagarh, Kathiawar)
Hafiz Muhammad Ishaq Khan (Patna, Bihar)
Maulvi Hidayat-ul-Hasan (Unnao, UP)
Maulvi Mazhar Nabi (Gaya, Bihar)
Maulvi Fakhruddin (Rae Bareli, UP)
Maulvi Hizb-ul-Rahman (Patna, Bihar)
Maulvi Muhammad Naim (Patna, Bihar)
Maulvi Abul Hasan (Patna, Bihar)

[23] Ibid.: 57.
[24] Rahim 2001: 49.
[25] Ibid.: 54.

Maulvi Muhammad Bashar (Patna, Bihar)
Maulvi Bashir-ul-Haq (Patna, Bihar)
Maulvi Abd ul Aziz (Azimabad, Bihar)
Maulvi Muhammad Raza (Biharsharif, Bihar)
Maulvi Abd al Hai (Patna, Bihar)
Maulvi Zahoor Ahmad (Moradabad, UP)
Maulvi Abdur Rahim (Arra, Bihar)
Maulvi Abdullah (Sitapur, UP)
Maulvi Wasimuddin Ahmad (Unnao, UP)
Maulvi Sirajuddin Ahmad (Monghyr, Bihar)
Maulvi Muhammad Sayid Askari (Saran, Bihar)
Maulvi Muhammad Usman (Monghyr, Bihar)
Maulvi Wali Muhammad Khan (Pratapgarh, UP)
Maulvi Shahud-ul Hai (Monghyr, Bihar)
Maulvi Manzoor Ahmad (Patna, Bihar)
Maulvi Muhammad Idris (Gaya, Bihar).[26]

Over 1903–19 their number had increased from 9 to 73 per year. And the total number of students who completed the course by 1919 was 1154. Out of these, only 256 were found competent enough for the award of certificates.[27] After completing their education at Lucknow these students, normally forty in a batch, either went back to establish their own clinics in their hometowns, or else got jobs in other princely states like Bhopal and Baroda. The regional spread of the students of Takmil-ut-Tibb only expanded in the years that followed. In the period of Hakim Aziz's sons—Hakim Abd al Hamid and Hakim Abd al Rashid—Unani medicine spread in a systematic manner all over North and Central India. Hakim Abd al Hamid's students acquired fame as physicians not only in the princely states but as far east as Calcutta and as far west as Lahore.[28] Similarly Hakim Abd al Rashid also strove to

[26] Minutes, pp. 76–9.

[27] UP Proceedings in Local Self Government, Allahabad 1916, Progs no. 19, sr. no. 37. Medical Dept UP, September 1916, p. 110. The total number of patients over 1902–18 had swelled from 105 to 27,887. By 1919 the total number of eye surgeries performed was 49,475. Surgeries of other kinds: 54, 723; and other diseases treated were 207,286.

[28] In Calcutta, Shifa ul Mulk Hakim Nisar Ahmad; and in Lahore Hakim

find jobs for his students all over India. Many of his students taught and practised in the princely state of Gwalior.[29] A prominent student of the college, Hakim Sayid Wazir Ali practised medicine in its Lucknow version as far away as in Mecca.[30]

Creating a National Profile for Unani

The Azizi family worked within the colonial medical space—hospitals, municipalities, sanitary commissions, etc.—projecting Unani as the medicine of Hindustan that was open for all, irrespective of community. But their task was far more difficult than had been the case in the late nineteenth century. In the communally surcharged early twentieth century, the balancing of Unani's local territorial and professional profile with the Islamic universal strand of its Muslim identity was not easy.

In a period when seminaries like Deoband and Nadwat-ul-Ulama (Lucknow), and 'secular' institutions like the Muhammadan Anglo-Oriental College were hiving off Muslims into separate compartments, the non-communitarian profile of the Takmil-ut-Tibb evoked immense curiosity. Muslims wanted to know how the proposed reforms would benefit both the community (*qaum*) as well as the nation (*mulk*). In this surcharged context every reformer had to engage with both *qaum* and *mulk*, even if these were not of direct concern.

The Azizi Muslim professional perspective on Western medical learning was not as narrowly focused on 'Muslim' interests as that of Muslim intellectuals like Sir Syed Ahmad Khan. Indeed, the connection of Unani with Muslims figures reluctantly in the family literature, and only in response to queries from critics who wanted to know how the reforms would benefit the community.[31] Instead, Hakim Abd al Aziz saw his initiative as generally benefiting the people of Hindustan (*ahl-i-Hind*). He stated that his proposed reforms were based on his understanding that Hindustanis appreciated Unani medicine the most,

---

Nayyar Wasti, who was also an eminent historian of Tibb, set up clinics and schools for the spread of Unani medicine. Rahim 2001: 60.

[29] Ibid.: 50. His student, Hakim Khwaja Shams ud-Din, was sent to Gwalior to treat the maharaja, who had been a patient of Hakim Rashid.

[30] Ibid.

[31] The questionnare of Anjuman-i-Rifaah-i-Aam Club, Rahman 1978: 119.

and that it had roots in the Indian locale. He tapped into the locally defined notion of nation that contributors to the *Oudh Akhbar* had articulated to underline the exclusivity of Unani: like nineteenth-century writers, he highlighted the compatibility of Unani medicine and Hindustan. He alluded to its long existence and rootedness in the country by highlighting its ready acceptance and easy availability in Hindustan. Its popularity, he stressed, derived from the fact that 'a good number [*mo'tabdeh girroh*] of Hindustanis consider it easily available [*sahlul wasool*], harmless [*be khatar*] and suited to their temperament [*mizaaj*].'[32] As such, he continued, 'we must promote it, but also include in it as and when necessary researches made in the allopathic system; and also get acquainted with equipment in surgery which in certain cases are required.'[33]

Interestingly, if the people whose interests were being talked about were Hindustanis, the word used to describe their groupings was *girroh*, indicating a more expansive and encompassing social grouping than say *qaum* or *jamaat*, which are more exclusionary in their connotation. Hakim Sahib's proposed medical reforms that leaned towards Western medical techniques were thus justified in terms of their benefits to not just the Muslim community but to the larger girroh of Hindustanis of the mulk, or *ahl-i-watan*.

But who exactly constituted his girroh, his watan, his mulk? Mulk was not just a geographical-territorial frame as created by the colonial medical bureaucracy. For Hakim Abd al Aziz it was also a cultural entity defined by diet, climate, flora, and fauna. It was also defined by its plural cultural mix, which included all religions and classes, and both genders—the Hindus and Muslims, the elderly (*buzurg*), women, (who were normally excluded when Muslim intellectuals like Syed Ahmad Khan defined Muslim 'interests' in response to colonial modernity), the poor and the aristocratic class of the rich (*umaraa* and *ruassa*), intellectuals and ordinary people. This composite mix constituted the nation for Unani.

Like public intellectuals of the *Oudh Akhbar*, he gave this Unani a long history in the country. He said that for centuries it was appreciated and admired (*mu'taqid-i-dildaada*) by the large majority of people

[32] Minutes, p. 4.

[33] Ibid.: 1–4.

(*girroh*),[34] making it more popular than Western medicine. As proof of its longer roots in Hindustan he pointed out that 'in this mulk patrons [*sarparast*] of Tibb still exist. And highly intellectual, experienced and elderly [*buzurg*] people of the country are of the view that in certain internal ailments Unani practitioners play the role of *maseeha* [saviour]'; also the *umaraa* and *shurafa* he said have faith in the system.[35] This long history of patronage, faith, and appreciation contrasted with the reception of Western medicine in India.

However, he had to balance Unani's Hindustani history with its Islamic legacy. Hakim Abd al Aziz saw no contradiction in balancing his micro narrative of Hindustani Unani with the mega narrative of its Islamic past. Indeed, the Graeco-Arabic legacy of Islam remained the highest reference point of legitimacy for his proposed reforms aimed at emulating and incorporating aspects of Western medical practice. Indeed he reached out to them to prove that colonial medicine only embellished the scientistic core of Unani. For instance, he argued that the proposed introduction of anatomy and surgery into Unani was not entirely revolutionary. These had already existed in unadulterated Unani—as it existed in the pre-Islamic Graeco-Roman period. To prove the scientistic compatibility of Unani with colonial medicine he located Unani in a very eclectic history that he constructed for Islam. He traced the beginning of Unani medicine to its pre-Islamic Greek origins, to the period of 'Isqaliboos' I (Aesculapius), who he claims was the originator of Unani (the first hakim), born 1697 years after Adam.[36] This first hakim introduced the system of *tajurba (*experience) and practice on which Unani medicine is based, and it remained intact in this form for 1480 years. However, in the subsequent years the tension

[34] Ibid.: 28.

[35] Ibid.

[36] He was a scientist and astronomer, identified with the biblical Enoch. He is credited with introducing both the use of the pen and woven garments to humankind. See Nagrami 1989: 64, 70–4. He mentions that 'Isqaliboos' was a Greek hakim and a disciple of Hazrat Idris, who lived for 90 years. He was well known for his extraordinary powers of prophecy and miraculous cures and revelations. He was so popular as a healer that people worshipped him. There was a difference of 5560 years between him and Galen, and it is said that he was the latter's ideal.

between *tajurba* and *qiyaas* (speculative philosophy) surfaced and remained an issue handled by later Greek philosopher-physicians like Ghoras, Manis, etc. The matter was settled in favour of *qiyaas* by the Greek philosopher Aflatoon (Aristotle), who relied more on the imagination and philosophical dimensions for cure. But after the death of Aflatoon his students drifted away from *qiyaas* and the old tensions resurfaced. Many turned towards *tajurba* again. Some introduced in their treatments different kinds of surgeries: puncture treatments, treatment of boils and wounds, ophthalmology, and orthopedic surgery. But later hakims followed Aristotle and re-established *qiyaas* so firmly that no possibility was left for any further critique. This ensured that surgery remained out of Unani medicine.[37]

It was, he says, 'the Greek philosophers whose intervention changed medical science and made it a different Tibb altogether.' He regretted that in the Arab period surgery and anatomy remained peripheral to Unani since Islamic law was against dissection of the human body. The translations of medical texts from Greek to Arabic underplayed the sections on surgery. The discarding of surgery ensured that it fell into the hands of *jahils* (illiterates). This made Unani medicine almost dead (*neemjaan*).[38]

He identified precedents supportive of his reforms from this same legacy. He pulled out examples of surgeons from this history to justify his reforms. Thus, Hakim Sahib mentions some important hakims in the 'ancient' period who were excellent surgeons, and so knowledgeable that they authored illustrated books on surgery: Khalf-bin Abbas Zuhrawi, who wrote *Kitaab-i-Zuhrawi*, which is an illustrated history of medicine; Abul Mansur Abdullah bin Shaikh us Sadid Abul Hasan Ali, who died in Cairo in 592 hijri; Rashid-ud-din Ali bin Khalifa, who died in 1611 hijri; and the famous Hakim al Razi.[39]

[37] Minutes, pp. 7–12.

[38] Ibid.

[39] Abu Bakr Muhammad bin Zakariya al Razi (865–925), whose position in Unani medicine has been compared to that of a second Galen. Born in Raiy, he spent his youth studying music, philosophy, literature, and alchemy, and did not embark on medicine until the age of 30. He wrote fifty-six medical texts. The most well known were *Kitab al Mansuri, Kitab al Jadari* on smallpox and measles, and the *Kitab al Tibb al Ruhani*, which examined the psychological dimensions

These examples from the Islamic history of Unani had convinced him that surgery was a part of the training of former hakims and that its reintroduction was nothing novel. All that was needed was the weeding away of 'adulterating' influences and making it the streamlined system that it originally was.[40] But he was quick to add that if Unani was rooted in its accretive Islamic legacy, it was also an integral part of Hindustan. It was connected to the mulk, its patrons were here, and people here preferred it to state medicine because of their greater familiarity with a system that suited their temperament.[41]

This language, which emphasized society's relatively greater familiarity with Unani and alluded to the 'alien' character of English medicine, was intensified in the speeches of the two sons of Hakim Abd al Aziz. For them, Unani represented the 'ancient tradition' of Hindustan. They offered a more strident definition of Unani's national contours that were said to be lacking in Western medicine. In fact they appealed to people for funds and patronage to preserve Unani's 'ancient and indigenous' forms.

Indeed, Hakim Abd al Rashid compared Unani to the ancient monuments of the country. He insisted that the preservation of traditional medicine as well as other traditional arts and sciences demanded a treatment at least equal to that meted out to ancient monuments:

> On the preservation of ancient monuments, which are made of bricks and mortar, vast sums of money are spent, while the garden which our ancestors built and nurtured with their sweat, their mental prowess, and incessant labour, which benefited a large number of people and which is still rendering a distinguished service, is being neglected. This garden has to be kept in proper shape, occasionally pruned, irrigated and guarded. All this needs funds which the government should release freely.[42]

---

of illness. Although al Razi relied heavily on his Greek predecessors, especially Galen, he was not afraid to criticize or correct the Greek texts wherever he found them wanting. Liebeskind 1995: 48.

[40] Minutes, pp. 7–12.

[41] Ibid.: 4.

[42] Rahim 2001: 47–8.

This conflation of Unani with the nation, as represented by its syncretic architecture, was a significant shift. But the mapping was not entirely foolproof. For, at the same time, Unani was reluctant to acknowledge its connections with other indigenous medical traditions like Ayurveda, aspects of which it had appropriated in its earlier phase in pre-colonial Hindustan. Indeed in the early twentieth century it vehemently denied any engagement with Ayurveda. At the 1906 All India Unani-Tibbi Conference (organized by the Delhi-based Hakim Ajmal Khan family) held at Lucknow, the Azizi family fiercely resisted any suggestion of engaging with Ayurveda.[43] And at the 1910 All India Ayurvedic and Unani-Tibbi Conference, Hakim Abd al Aziz and his son Hakim Abd al Rashid resisted any combination of the Ayurveda and Unani systems. Such a combination, they argued, would lead to the death of the Unani system. Later, in 1935, while formulating the course for the Tibbi College at Hyderabad, Hakim Abd al Hamid was of the view that the syllabus should confine itself to Unani and exclude all those theories and concepts which were not compatible with it. In a comprehensive note to the college committee, he and his group said: 'if theories and fundamental principles borrowed from other medical systems were allowed to make inroads into the well defined territory of Tibb, it would lose its individuality.'[44]

The Hakim as a Professional: Balancing the National Local with the Subcontinental Professional

This aversion to Ayurveda was the first inkling of Unani's effort to carve out for itself an exclusive professional identity as a community of hakims firmly rooted in Hindustan. Indeed, a heightened sense of being foremost a professional class, and of the safeguarding of professional interests, are central to the Azizi family discussion on reforms.

[43] Rahman 1978: 143–4. At the 1906 All India Unani Tibbi Conference, Hakim Abd al Aziz and Hakim Abd al Rashid resisted the combination of Ayurveda and Tibb. Indeed they said it would lead to the downfall of Tibb. This anti-Ayurveda stand of the Azizi family was the major point of difference also between them and the Delhi-based family of Hakim Ajmal Khan.

[44] Rahim 2001: 57, Rahman 1978: 143–4. See also *Raudad-i-Takmil-ut-Tibb*, 1908–11.

Hakim Aziz hoped that the reforms would blot out the taint of ineffectiveness (*be-kamali*) that had stained his profession.[45] He hoped that a reformed medical system would enable hakims to pursue their trade more professionally (*peshe-ke-bataur pesha*). That the Takmil-ut-Tibb was established to not just serve the mulk but also to enhance the profession of tabibs was most evident when Hakim Aziz says, 'I established the Takmil-ut-Tibb institution for *mulki* and *fanni* khidmat.' (service of the nation and the profession).[46] The obsessive concern with this overpowering professional identity was not just typical of Lucknow hakims. Sir Syed Ahmad reported its prevalence in early-twentieth-century Delhi as well. He said that the hakims of the city were very conscious of their responsibilities as medical professionals. They rarely took fees from patients and considered it their responsibility (*farz*) to cure people.[47]

This professional identity often spilt beyond the territorial and cultural borders of the imagined country, and its balance with the national profile had to be carefully engineered. The two key referents central to the making of this professional identity were the art of writing medical prescriptions (*nuskha navisi*), and the art of disseminating medical knowledge to large gatherings (*darrs*). Both these professional attributes give a premium to knowledge embellished by memory, oratory, and ornate writing skills. These were specialized skills that had to be professionally learnt. The process of acquiring them from specialists located in traditional hakim families only reinforced further the latter's control over knowledge. These professional specialities also provided a connecting bridge that brought together co-professionals of the subcontinent. They were welded into a large family of professionals tied not by blood but by intellectual debts to a traditional family.

Both *nuskha navisi* as well as *darrs* were carefully guarded family treasures. The profession was thus deeply enmeshed in a familial primordial identity. Yet these family-knowledge traditions could be transmitted and exchanged across families spread around the world, weaving hakims into a large professional family. Later, in the twentieth century, the professional family was further strengthened as families codified

[45] Rahman 1978: 118.
[46] Ibid.: 132–3.
[47] Lahori 1996: 317.

medical knowledge in printed books. Older members of the family wrote notebooks and compendiums for the benefit of the younger ones. As such books could be bought and sold in the market, they further encouraged the dissemination of family knowledge and brought into its ambit professionals not necessarily connected to the originating family by blood. Thus *darrs* and *nuskha navisi* connected hakims of one family to those of others both within and outside the country. Indeed these skills connected them to other professional groups like the ulema as well, who were themselves organized around similar prescriptive and disciplinarian learning traditions. This gave the community of hakims a region-specific family rooting alongside an identity which, professionally at least, extended into the subcontinent.

We saw earlier that the Darrs-i-Nizamiya and the Darrs-i-Tibbi offered a feel of very different kinds of texts. But they followed similar learning techniques and discipline codes: a premium on memory, oratory, writing skills, and perpetuation of the family name. The Takmil-ut-Tibb borrowed heavily from seminary-style learning, the *darrs*. This seemed to be the best option, given the fact that the local Farangi Mahal madrasa at Lucknow was the main catchment area for student trainees at the Takmil-ut-Tibb.

The Tibbi *darrs* was generally organized in the hakim's house. Even though it was meant for students of Unani medicine, everyone was welcome. Though most of the students were Muslims, non-Muslims also displayed curiosity and interest in such gatherings. For instance, the Hindu magistrate of Lucknow, Hakim Rai Deen Dayal, who was proficient in Arabic, was one of the prominent participants in the *darrs* of Hakim Abd al Rashid.[48] The Tibbi *darrs* drew large and mixed crowds because it was essentially a forum where the theoretical-philosophical issues in Unani medicine were discussed. More importantly, one of its objectives was to familiarize not just specialists but society at large about the treatments offered by Unani. It was hoped that the creation of this familiarity would attract people to hakims, since their system contrasted with enigmatic and alien colonial biomedicine.

By the middle of the nineteenth century Hakim Abd al Aziz reported at least forty to fifty students attending his *darrs*. Aziz organized the *darrs* in his house daily. He lectured on theories of anatomy, causes of

[48] Rahman 1978: 146.

ailments and their symptoms, and the relevant medicines for their cure. 'I acquaint them with researches and experiences of past hakims and also of my own experiences and studies', he says. When he was satisfied with their theoretical knowledge, he invited them to his clinic to learn the art of *nuskha navisi*.[49]

By the end of the nineteenth century, as hakims increasingly specialized in specific branches of medicine, the *darrs* also became specialized. Lectures by specialists on specific themes became a characteristic feature of such gatherings. For instance, Hakim Abd al Aziz was known as someone who had memorized the *Qanun*; he was considered exemplary in the *darrs* of *Kulliyat-i-Qanun*.[50] His son Hakim Abd al Rashid also held his own *Kulliyat-i-Qanun darrs*. Hakim Hafiz Abd al Majeed, the famous hakim of Delhi, who attended his *darrs* continuously for fourteen years, said that he never heard the same lecture repeated. Each time it was a new perspective on the *Kulliyat*.[51]

The *darrs* retained its centrality in medical training until the early twentieth century. But the organizational principles of lecturing and the style of interaction with students did not remain stagnant. From the late eighteenth to the early twentieth centuries the format of lectures shifted from a monologue to a stress on dialogue with students. Hakim Abd al Hafiz, the brother of Hakim Aziz, initiated the practice of introducing debates between students on passages of prescribed medical books.[52] These debates, even though an advance from the monologues of the earlier period, remained very teacher-centric. They were couched in a vocabulary that heaped praises on the *ustaad* rather than one which addressed issues. It was Hakim Hafiz Abd al Wali who broke this tradition and initiated scholarly debates on Tibbi issues. His new ideas of organizing the discussions in the *darrs* became so popular that he also established an *anjuman* (organization) for discussion only of medical controversies. He also organized weekly meetings in which he, along with his students, read *tibbi risaalahs* (medical magazines).[53]

The seminary students' knowledge of Arabic, and their familiarity with *darrs*-style teaching, made it easy for them to train in a medical

[49] Ibid.: 121.
[50] Ibid.: 108–9.
[51] Ibid.: 145.
[52] Ibid.: 179.
[53] Ibid.: 207–8.

education that used similar learning techniques and languages of instruction. Indeed, well into the twentieth century, both the Shia and the Sunni ulema sent their students to the Tibbi *darrs* of Hakim Abd al Aziz. And not just students; established ulema like *Ustaad-ul-Ulema* Maulana Lutf-ullah Aligarhi, Maulana Abd al Haq Khairabadi, and Shibli Nomani, among others, attended the Tibbi *darrs* of Hakim Abd al Aziz. They appreciated it for the knowledge they gained in philosophy and Tibb. Maulana Abd al Hai, Nazim of the famous Sunni Muslim Lucknow seminary Nadwat-ul-Ulama, attended the *darrs* and was so impressed that he composed an Arabic *qasida* (encomium) in praise of the sessions.[54] Ulema from all over Awadh attended the *darrs* of Hakim Abd al Hamid and Abd al Rashid.[55]

Darrs and *tadrees* (instruction) in Tibbi philosophy connected the Azizi hakims not only to hakims and ulema within India but also to similar professionals outside the country. In the mid nineteenth century Hakim Ismael (1822–86), the father of Hakim Abd al Aziz, had students from Kabul and Najarah in his *darrs*.[56] Later, the Tibbi *darrs* of Hakim Abd al Aziz attracted professionals from Afghanistan, Bukhara, Herat, and Hijaz.[57] Similarly the *darrs* of his son, Hakim Rashid, was attended by hakims from Kabul and Herat, such as Maulana Hakim Habibullah Khan Kabuli, Maulana Hakim Mozamil Khan Herati, and Hakim Yusuf Kabuli.[58] The attendance of people from outside the country and their active participation in the debates and discussions in these *darrs* reflected how a family-specific learning tradition was open to knowledge exchanges with hakim families across the subcontinent and beyond.

If scholarly knowledge and oratory underlined the popularity of the Tibbi *darrs* and made it the critical link in knowledge dissemination, the art of writing the medical prescription (*nuskha navisi*) grounded this circulating family knowledge in a written tradition. As in the *darrs*, here also the premium was on memory. A good teacher was one

[54] Ibid.: 110–11.

[55] Ibid.: 145.

[56] Ibid.: 76. See also Robinson 2001: 211–39. He discusses the shared knowledge and connective systems of the ulema across the subcontinent.

[57] Rahman 1978: 108–9.

[58] Ibid.: 146. Hakim Rai Deen Dayal, Magistrate, Lucknow, who was proficient in Arabic also attended his *darrs*.

who could memorize and repeat in 'one breath' a *nuskha* and inscribe it beautifully on paper as well. This was also a family art (*fann*) and the hakims transmitted it to their students in their clinics. As with the *darrs* here too the students came from all over, including Afghanistan and Central Asia.

Hakim Ibrahim (*d.* 1882) took great pains to teach his students and family members the art of *nuskha navisi*. As we saw earlier, he was emphatic that they memorized his *nuskhas* for future use. Students were then tested in *nuskha navisi*. In his clinic he would communicate to his students his diagnosis and then ask them to write the appropriate *nuskha*. He personally corrected the scripts of those who failed the test. The family tradition of training to write prescriptions continued into the time of Hakim Abd al Aziz. He dictated his *nuskhas* aloud (*ba awaaz-i-buland*) and his students would inscribe them in their notepads. After he had approved them, he would hand them over to the patients.[59]

Both the content of the *nuskhas* as well as the writing styles were considered family treasures, markers of family identity. The professional class that they defined rested firmly on primordial family identity. This professional-familial identity was further cemented when books on the art of writing *nuskhas* (*fann-i-nuskha navisi*), as well as compendiums of family *nuskhas*, began to be printed. The trend began in the late eighteenth century, but picked up in the late nineteenth century with the expansion of print culture. In this period, older family members dedicated compendiums 'for the benefit' of the younger generation. But dedications sometimes moved out of the primordial family and into the domain of the 'professional family of hakims'. Most of these compilations of individual hakims had family remedies passed on for 'the benefit' of sons or favourite students. Thus we have a collection of family *nuskhas* dictated by Hakim Abd al Aziz to his son Hakim Abd al Hamid in 1902–3.[60] Or the precise medical instructions that Hakim Wahid gave to his son Hakim Abd al Moid that were codified in the important compendium called *Qanun-i-Wahid-fi-ifaadat-i-Moid* (Instructions of Wahid for the Benefit of Moid).[61] Or the instance of

[59] Ibid.: 147.
[60] Ibid.
[61] Ibid.: 186–7.

the master don of the Azizi family, Mir Muhammad Murtaish, who dedicated his book on the art of writing *nuskhas* to his student and the founder member of the Azizi family, Hakim Yaqub.[62]

It was not just family members who followed this tradition. Favoured and promising students were steadfast in upholding the family tradition of healing. They considered themselves part of the professional family of hakims. The prescription notebooks of students trained by a particular family of hakims followed their master's pattern of writing *nuskha* compendiums. In such books, references always existed to the *nuskha* of their teachers. For instance, a reference to Hakim Abd al Wahid's *risaalah* of family *nuskhas* mentioned above figures prominently in the *bayaaz* (notebook) of one of his Sufi students, Hakim Maulvi Abad Ahmad Komandwi.[63] The creation of this 'professional family' is evident once again in the *bayaaz* of Hakim Sayid Wahi-ud-Din Ashraf, *sajjadanashin* of *khanqah* Kachwahasharif. Many of the *nuskhas* he mentions are those he had picked up from his teacher Hakim Abd al Aziz. At the end of the bayaaz there is a chapter devoted exclusively to the methods of treatment he learnt from his other teacher Hakim Abd al Wahid: *'Tarikh-i-Ilaji-i-Kulli Muallafah muratabah Janam Ustaadi Hakim Abd al Wahid Sahib Lucknowi Madda Zilluhul Aali'* (History of the Cures Prescribed by Master Abd al Wahid Sahib Lucknowi).[64] Such trends suggest that the primordial family tradition could extend to accommodate a professional family as well. Together, they constituted the hakims professional identity. Students like Hakim Abd-ul-Ahad Kasmandavi who had in their possession *nuskhas* of their teacher were not rare to find. Hakim Abd al Wahid's 'A short course of treatment of all ailments', was always available with Kasmandavi.[65]

We also have instances of Hakim Aziz's *nuskhas* being repeated in later compendiums of the younger generation of his family.[66] For instance, Hakim Abd al Wahid compiled a risalah of family *nuskhas* called *'Mukhtasar Tariqah-i-Ilaj, Kaali Jamiah amraaz Abd al Wahid*

[62] Ibid.: 41.
[63] Ibid.: 186–7.
[64] Ibid.; and Rahman 2002.
[65] Ibid.: 186.
[66] Ibid.: 41.

Sahab' (Brief Methods of Cure). This had the author's prescriptions but it also included prescriptions of his prominent eldest brother, Hakim Abd al Aziz. Alongside, we have the *bayaaz* of Hakim Husain Raza, which has 2523 *nuskhas*. These include his own prescriptions, but also those of the family (*khaandaani*), starting from his ancestors, Hakim Muhammad Yaqub, Hakim Ibrahim, Hakim Maseeh, and coming down to his own father, Hakim Hasan Raza, and brother Hakim Ahmad Raza.[67]

The Professional Hakim and the Colonial Medical Space

The professional hakim in the 'national' Unani tradition had a particular dialectics with colonial medical space. He was linked to its discursive framework as represented in institutions of control: the municipality, sanitary commissions, hospitals, asylums, and legal bodies. At the same time, he was distant from and distinct from it in terms of the specific cultural space he carved out for himself in Hindustan. Until 1912 this delicate balance created little problem, since the law did not discriminate between doctors and hakims as far as government appointments and privileges were concerned. The national Unani could nestle within the colonial frame.

However, over 1912–17, the government passed a series of medical acts at the level of provinces that began to register doctors with qualified degrees so as to segregate them from hakims and vaids who remained beyond its purview. The objects of these acts were to define a recognized professional as one who came with an allopathic medical degree. Those who had such a degree were registered as recognized medical practitioners. The acts also proposed to institute authoritative bodies that would supervise medical education and exercise control over and promote the interest of the profession.[68] These bodies were constituted by registered practitioners. Hakims and vaids being unregistered by law, were left out of these fora. These legal introductions disturbed the national–colonial balance.

[67] Ibid.: 203.

[68] UP Proceedings in Local Self Government Dept, 1916, Allahabad, Progs no. 14, Medical Dept, Sr. no. 25, September 1916, p. 35, and Progs no. 12, Sr. no. 23, September 1916, P/9934.

The government emphasized that this legal innovation did not adversely target the hakims and vaids, for they were allowed to practice as before. But the registration acts were seen by the hakims and the vaids as a de-recognition of Indian medical traditions. The Usman Committee set up to report on the indigenous system of medicine remarked that the acts had reduced even the distinguished hakims and vaids to the 'position of unrecognized, unqualified laymen, whose professional opinion is at a discount in courts of law and in the eyes of the government.'[69] It regretted that the hakims and vaids had been completely eliminated from the proposed medical councils that would oversee the registration of practitioners and the grant of degrees and medical education. It warned that the acts separated the colonial from the indigenous medical system.

Both the Lucknow and Delhi families reacted equally strongly to the severing of their ties from the discursive colonial space. They argued that the 1916 United Provinces Medical Act was objectionable as it would leave out the hakims and vaids from registering as qualified practitioners.[70] They feared that it would de-recognize the hakims and vaids and eliminate them from government employment. Hakim Abd al Rashid of the Azizi family submitted a memorandum of amendments to the government. He regretted that in its present form the act was unacceptable. He pointed out that the sanitary commissions, hospitals, infirmaries, asylums, municipal boards, and medical committees where they had so far raised their voices would henceforth shut their doors to hakims. The UP Registration Act authorized such fora to enlist only registered practitioners.[71]

The reactions of the hakims against the registration act revealed clearly that their 'national' space thrived in conjunction with colonial medical space, and that they regretted the move to sever this critical link. It proved that even in the high period of Indian nationalism they upheld an outward-looking, robust Unani tradition, not a reified

[69] Khan Bahadur Muhammad Usman. The report of the committee on the indigenous systems of medicine, Madras, part 1, 1923, p. 23, V/26/850/13.

[70] UP Proceedings in Local Self Government Dept, 1916, Allahabad, Progs no. 14, Medical Dept, Sr. no. 25, September 1916, p. 35, and Progs no. 19, Sr. no. 37, September 1916, p. 4042, P/9934.

[71] UP Proceedings in Local Self Government Dept, 1916, Allahabad, Progs no. 14, Medical Dept, Sr. no. 25, September 1916, p. 35, P/9934.

preservationist one. In 1925, as a result of hakim reactions to the registration acts, the government decided to set up a committee to advise on the best means of improving the traditional system of medicine. The committee, headed by Justice Gokaran Nath Misra of the Oudh court, consisted of prominent hakims and vaids. These included two hakims of the Takmil-ut-Tibb: Hakim Abd al Hamid and Hakim Abd al Hasib. The committee's report reiterated the idea that Unani and Ayurveda both needed the patronage and investments of the colonial state. They did not want a severing of relations with the state. Indeed they recommended a greater presence of culturally distinct indigenous medical traditions within the discursive medical framework of the colonial state. Thus they argued for greater control of Unani and Ayurveda knowledge, both by the familes and the state so as to keep these shielded from the menace of medical knowledges that were difficult to control. They proposed the following: a state-aided Unani and Ayurveda school and college; provision of degrees and certificates to candidates passing from these state-aided Ayurvedic and Unani institutions; increased grant to private institutions and dispensaries; and the arrangement of special research on Ayurvedic and Unani drugs.[72] In addition the committee recommended the establishment of a Board of Indian Medicine consisting of twenty-two members and a president. These would include hakims and vaids and would have supervisory powers as far as indigenous medicine was concerned. The most important of these were seen as supervising education, conferring degrees, and registering medical practitioners.[73]

The government accepted their recommendation and a representative body of hakims and vaids was formed with eighteen members. This was called the Board of Indian Medicine and was headed by Syed Wajid Husain, deputy collector and secretary Husainabad, Lucknow.[74] In 1931 it also agreed to establish a state-run Unani school in Lucknow.[75]

[72] Medical Dept, UP, December 1930, Progs. no. 11, Sr. no. 4, 995/V-234, p. 12, P/11889.

[73] Ibid.

[74] Ibid.: 15.

[75] For details of this Unani school, see Proceedings of UP in Medical Dept, 1931, pp. 8–25, P/11957.

National *vs* Communitarian: Unani Gets a Muslim Hue

But in the late nineteenth century was it at all possible for Muslim professionals to articulate such professional communities and make much headway in their bargains with the state without engaging with the dominant communitarian discourse? The manner in which even the 'liberal' ulema, represented by people like Shibli Nomani, reacted to the Takmil-ut-Tibb experiment indicates the dominance of their discourse. The son of a wealthy vakil and landholder of Azamgarh, Shibli Nomani was one of the leading intellectual lights of Aligarh's formative phase. He was also one of the founders of the Nadwat-ul-Ulama at Lucknow (1897).[76] Being a leading Muslim intellectual of the time, it is not surprising that he commented on the formation of the Takmil-ut-Tibb as well. In a speech that he delivered on 14 October 1906 during the annual function of the college, he appreciated and encouraged Hakim Abd al Aziz's effort to reform Unani medicine on Western lines. However, unlike Hakim Abd al Aziz, he did not confine himself to endorsing and pushing for reforms only in the larger interests of country or profession. The 'national' interest was mentioned, but the need for making a special mention of the 'Muslim community' interests that such reforms would serve was never lost track of. Indeed, the relevance of such reforms for the 'Muslim community' was the central issue of Nomani's intervention: 'who does not know that in India modern medicine, Unani medicine and the Vedic system of medicines are in vogue. The point at issue is that both *for our country and for our qaum* [emphasis mine] the conditions are such that we have to find out what system we should adopt.'[77] Clearly the interests of the mulk in Nomani's view did not necessarily encompass those of the Muslim qaum. That was why the two interests were separately mentioned.

This separation of mulk and qaum was followed by the decentring of Unani from the history and territory that the Muslim public intellectual and the Azizi family had striven to carve out for it in Hindustan.

[76] Azmi 1994: 190–205; Troll 1993: 1–19. For an informative and exhaustive biography, see Suleiman Nadwi 1943.

[77] Minutes, pp. 109–12.

Unani was given a new history where its accretive strands, in particular its pre-Islamic Graeco-Hellenic past, and its development in Hindustan, was marginalized and the achievements made in the Islamic classical empires highlighted.[78] The rewriting of the history of Unani medicine was followed by the predictable plea that it be rechristened 'Islamic medicine' instead of Unani medicine. In an emotional speech Nomani said, 'Our Tibb has not originally originated from Greece. Those who know history are well aware that our ulema and *atibba* [scholars of medicine] have contributed such a lot by their researches that this system can hardly be called Unani Tibb. This system of medicine can now rightly be called Islami Tibb.'[79]

While Hakim Abd al Aziz, as we saw in the last section, in his history of Unani medicine, acknowledged the contributions of pre-Islamic Greek philosophers and made a plea to combine their methods with those of later Muslim surgeons, Nomani pits them against the so-called 'Islamic' or 'ulema' contributors. In a clear effort to tear apart the synthetic history of this medicine, Nomani mentions how, in face of the ulema's contributions to Unani medicine, 'Greek philosophy could not stand before Islamic philosophy'.[80] Underlining not only the difference of Greek and Islamic contributors but emphasizing the superiority of the latter, he adds, 'Our *atibba* admit that though this system of medicine originated from Greece, they have made such changes, additions, and deletions that now to call Islami Tibb Unani is like calling the Jam Meh's hotel the Aminabad shop of Misraji.'[81]

But if the thrust was to shift the contours of Unani medicine to a narrow Islamic domain and talk about the benefits of the proposed reforms in terms of the interest of qaum, the interest of the mulk was not entirely lost either. In fact Nomani too balances the contours of the Islamic transnational Unani with its national profile.[82] Thus, for Nomani Unani was Islami Tibb; its development took place alongside

[78] Aziz also streamlined the medicine and its history but never negated its Greek past. His argument was that Greek philosophers had underplayed the centrality of surgery in Unani medicine.

[79] Minutes, pp. 109–12.

[80] Ibid.

[81] Ibid.

[82] Jalal 2001.

Islamic expansion. It reached Hindustan with the Muslims and entrenched itself in the country even as its connections with its natal home remained intact.

Thus, in support of the Takmil-ut-Tibb experiment, Nomani too argues that Unani medicines are in accord with the nation's *aab-o-hawa* (locale) and thus need to be encouraged. He also supports the spread of the Unani experiment on grounds of it being a medicine whose practitioners and their language the people are more familiar with. The language referred to here is clearly Urdu which had, by the late nineteenth century, become the popular language of North India. By highlighting Urdu as the language of Unani, or as he called it Islami Tibb, once again the Hindustani strand was being hooked on to its larger history in the Islamic world. As he says, 'The major problem with doctori ilaj is that we cannot read the prescription and the medicines they have written—in contrast in Tibb-i-Unani our *atibba* write their prescription in a language that everyone knows.'[83] Nomani's intervention in the annual convocation of the Takmil-ut-Tibb steered the Muslim hakim's discourse on colonial learning to a different terrain.

The Muhammadan Educational Conference, the Anjuman-i-Tibia, and the Unani–Vedic Conference

In the changing political climate of early-twentieth-century communitarian politics, represented by people like Nomani, the trans-community professional identity that the Azizi family strove to create for Unani was threatened. The religious community was now an important pressure group. The state yielded to demands with relative ease if they came from community platforms. At the same time, social approval was more forthcoming if the community's interests were highlighted in discussions of medical reform.

Takmil-ut-Tibb college too used the community card to further its interests. Even though it saw Unani as constituting a professional rather than a religious community, it did not hesitate in asking support from the Muhammadan Educational Conference—a body that had been set up to deliberate upon the affairs of the Aligarh experiment.

[83] Minutes, pp. 109–12.

The 1904 session of the conference considered the college's request, and said that the 'Takmil-ut-Tibb college particularly and Muslims in Lucknow in general want that in the next session of the Muhammadan Educational Conference, a resolution should be presented that urges it to endorse the aims and objectives of Takmil-ut-Tibb Lucknow and extend full cooperation to it for the promotion of Tibb-i-Unani.'[84] The resolution pleaded that the Muhammadan Educational Conference, in its capacity as a national body, should include Unani in its educational basket and support it because the tradition is also 'favourable to Muslims and beneficial for their welfare, religion, culture, and civilization.' In another resolution submitted to the conference, the college cleared any misgivings about Unani being inconsistent with Islamic principles. These efforts brought the desired results. The 28 December meeting of the Muhammadan Educational Conference endorsed the college resolution and made the spread of Unani part of its educational agenda. This led to the relatively faster acquisition of funding and the nationwide spread of colleges and hospitals of Unani medicine. This was just the beginning of what became a tension-prone drift towards the consolidation of communitarian identities around medicine.

Its identification with the Muhammadan Educational Conference was only the beginning of the tension between the professional and the communitarian pulls on Unani. Soon after the endorsement of their cause at the Muhammadan Educational Conference, the Lucknow hakims reacted with a vengeance. They formed their exclusive association, where they closed ranks as a community of professionals who were connected by learning and not religion. Formed in 1911, this association was called Anjuman-i-Tibia (association of medics), and it had Hakim Abd al Rashid as its secretary. It sought to protect the rights of hakims as medical professionals, demanded jobs for hakims in government service, and piloted the professionalization process. However, it remained a family-based association and excluded hakims outside the family ambit. In order to sharply define its professional identity as an association only of hakims, it distanced itself from the Delhi hakims who wanted a merger with other medical systems such as Ayurveda. The clash of egos and personalities of the Delhi family of Ajmal Khan

[84] Ibid.: 83.

and the Lucknow family of Abd al Aziz widened the rift between these two prominent Unani families and further consolidated the Anjuman-i-Tibia (also known as Anjuman-i-Tibb) as a forum for only those hakims who supported an unalloyed professional Unani.[85] In 1911 the Anjuman opposed Ajmal Khan's All India Vedic and Unani Conference, which represented the coming together of Unani and Ayurveda.

In a letter presented in Delhi before the All India Unani-Vedic Conference of 27 November 1910, Hakim Abd al Aziz argued that he would join the organization only if it did not dabble with Ayurveda and retained the unalloyed Unani as its core.[86] Some hakims of Lucknow attended this meeting to push their case. But they felt slighted as the conference completely ignored their proposals. Relations soured further when, in March 1911, Ajmal Khan arrived in Lucknow to establish a local committee of the All India Unani-Vedic Conference. Ajmal Khan went ahead with his plan despite requests by the Lucknow hakims to postpone this event until Abd al Aziz returned from his Haj pilgrimage. On his return from the pilgrimage Abd al Aziz again attempted a rapprochement: he requisitioned a meeting where the issue could be discussed by a wider cross-section of hakims from the different provinces. Even this request was turned down on grounds that such a discussion could be held at the next annual meeting of the All India Unani-Vedic Conference, to be held at Lucknow.

The differences soon echoed in larger Lucknow society and lines were sharply drawn around the two associations. This made the hold of individual families over Unani even stronger. It ensured that professionalization in India remained a family enterprise and did not transcend such referents, as had been the case in Europe. In fact, as Unani professionalized, it began to develop a regional identity defined loosely around the location of its family patron. Thus, Abid Ali Khan, the owner and

[85] 'All India Ayurvedic and All India Unani Tibbi Conference, Delhi, Ek Munsafanah Nazar', by Abid Ali Khan, Lucknow, undated pamphlet. Hereafter Abid Ali Khan. The clash of personalities between the two leading Unani families ensured that on 24 June 1911, at a public meeting presided over by Raja Mehmudabad, Sir Ali Muhammad Khan, the Lucknow family established their separate organization called Anjuman-i-Tibia.

[86] Pamphlet called 'Reasons for Remaining Away from the All India Vedic and Unani Conference, Delhi', Lucknow: Nami Press, 1911.

proprietor of the Nami Press at Lucknow and a supporter of the Azizi family, whipped up a 'Lucknowi' patriotism for the local hakims. He exhorted people to boycott the All India Unani-Vedic Conference because the hakims of Lucknow were under-represented in its fora—it had just one member from Lucknow in the organizational committee, whereas Delhi had twenty. This, he thought, was ironic since Lucknow was the centre of medical learning. He chided those hakims of the city who, despite this neglect of their city and mulk, had joined the conference.[87] He went on to link the defence of the Lucknow Unani tradition with the defence of the city of Lucknow. He said the city represented the mulk for Lucknow hakims, and joining the All India Unani-Vedic Conference by the hakims of the city amounted to their lack of interest in defending their part of the mulk of Hindustan (*hissa* Hindustan). He argued that if they could not protect their part of Hindustan (the city of Lucknow), how could they ever be expected to defend all of Hindustan. Khan clearly mapped the contours of Unani to those of a narrowly redefined notion of the nation.[88]

In its 15 October 1911 meeting, the Anjuman-i-Tibia made its stand very clear: 'The Anjuman-i-Tibia Lucknow does not agree with it [All India Unani-Vedic Conference] and does not consider it as the representative conference of all of Hindustan. It endorses all the published literature which appeals to people to stay away from the conference.'[89] In 1911 the Anjuman-i-Tibia laid out its objectives, which aimed at the consolidation of 'like-minded' hakims under the aegis of the Azizi family. It pledged to work for the advancement of Unani alone, and promote the research and publications of its practitioners. It had the following agenda:

To publish Tibbi books and promote its teaching.

To publish the historical circumstances in which practitioners lived and disseminated their healing skills.

To help in the establishment of Tibbi institutions and provide in them adequate infrastructure for the teaching of surgery and midwifery.

To assist in the establishment of charitable Tibbi hospitals and dispensaries.

[87] Abid Ali Khan.
[88] Ibid.
[89] *Raudad-i-Takmil-ut-Tibb*, Lucknow: 1911, p. 5.

To make arrangements for the supply of simple drugs and compound medicines and ensure their quality at every stage.

To promote co-operation among doctors and practitioners of other systems and protect their legitimate interests.[90]

In its 14 November 1911 session, the Lucknow family emphasized its wide regional reach when it resolved to call the Anjuman-i-Tibia the All India Unani-Tibbi Conference.[91]

Detailed by-laws (*dastur amal*) of the conference were soon laid out to systematize its functioning. The conference had office bearers elected for a period of one year: President, Vice President, Secretary, Assistant Secretary, and Treasurer. It was decided that the managing committee would meet every two months. Initially, it had a membership of 40 people and they were given permission to increase its membership to 100. The managing committee decided that the conference would have two kinds of members: life members who paid Rs 100 at one go, and ordinary members who paid Rs 3 per annum. Alongside, there were 'visitors' who paid Re 1 per year. Only members had voting rights in the sessions of the conference. In order to expand its ambit to bring like-minded people together under its banner, it also had a provision for co-opted members. These paid Rs 10 in addition to their membership fee.[92]

The logistics of membership was as follows: members of the managing committee paid Rs 6 per year once, or in instalments. They were approved by the open session of the conference after every three years on the recommendation of the managing committee. Election of members would be on the basis of proposals made by existing members and these were to be seconded by another colleague. The Azizi family always had its members on the managing committee, and it ensured that Lucknow remained the headquarters of the Anjuman. Professionalization in terms of organizing to fight for rights and privileges remained very much a family-controlled enterprise.

[90] Ibid.: 2.

[91] *Raudad Ijlaas Awwal All India Unani Tibbi Conference*, 14 November 1911, Resolution no. 4, in *Raudad-i-Takmil-ut-Tibb*, p. 15.

[92] *Dastur-al Amal*, Managing Committee, All India Unani Tibbi Conference, on request of secretary Hakim Syed Muhammad Taqi, in Report on 3rd All India Unani Tibbi Conference, 13 April 1914, Lucknow, pp. 4–14.

Tensions within the Anjuman-i-Tibia: The Drift to a Muslim Unani

At the time of its formation the Anjuman-i-Tibia projected itself as a professional body of hakims organized by the Azizi family which stood for an unalloyed Unani. Its professional rather than communitarian identity was reflected also in the composition of its members, who were a religious and sectarian mix of the local notables of Lucknow, including Sunnis, Shias, and Hindus. Some of the important members were Rai Deen Dayal, the (Hindu) magistrate in Lucknow, and the prominent Shia barrister of Lucknow Sir Wazir Hasan, both supporters of the Azizi family and patrons of Unani learning.[93]

But the minutes of the Anjuman-i-Tibia, which came to be known as the All India Unani-Tibbi Conference over 1911–26, reflect tensions within the organization on the issue of an exclusive professional identity that ignored the community strands. On 14 November 1911 Mirza Samiullah Beg, a vakil at the Lucknow High Court, objected to the effort of some hakims to link Unani with Islam. He tabled a resolution at the Anjuman-i-Tibia meeting that said that 'Tibb-i-Unani is not linked to any particular religion [*mazhab*], community [*qaum*] or group [*girroh*].' He argued that though this was stating the obvious, it needed saying because an advertisement tabled at the Conference linked Tibb-i-Unani to Islam. Indeed, it linked the progress of Islam to that of medicine and rechristened Ilm-i-Tibb as Islamic Tibb. Samiullah Beg felt this was objectionable and ahistorical since the origin of Unani medicine pre-dated that of Islam.[94] Reacting to the same advertisement Rai Burj Mohan Dayal, a Hindu vakil at the high court in Lucknow whose father had studied Unani at the Takmil-ut-Tibb college, stated that one of the reason why people erroneously associated Unani with Islam was because its texts were in Arabic and thus it was more accessible to Muslims who were proficient in that language. This, he argued, created the impression of Unani as linked to Islam. He argued that the misunderstanding could be corrected if Unani was taught in Urdu, that being familiar to both Muslims and Hindus.[95]

[93] *Raudad-i-Takmil-ut-Tibb*, pp. 2–3.

[94] Resolution no. 3 in *Raudad Ijlaas awwal*, All India Unani Tibbi Conference, 14 November 1911, *Raudad-i-Takmil-ut-Tibb*, pp. 9–10.

[95] Ibid.: 10.

The Azizi family and patrons of the All India Unani-Tibbi Conference tried hard to extinguish the communitarian slant within their professional group. Indeed, many of them argued that they had distanced themselves from the All India Unani-Vedic Conference precisely because it saw Unani and Ayurveda as distinct Muslim and Hindu medicines: Hakim Abd al Aziz himself argued precisely this, emphasizing that 'no system of medicine can be called exclusively for a particular community'. He said 'this was particularly true of Unani whose practitioners had been of different religions, regions and countries.'[96] And he was not the only one to object to the All India Unani-Vedic Conference for such reasons. Abid Ali Khan, owner and proprietor of the Shahi Press in Lucknow, reacted similarly to an article by Hakim Abd al Wali, who was both a member of the Muhammedan Conference as well as an advocate of the All India Unani-Vedic Conference. Khan alleged that Wali's invitation to the hakims of Lucknow to join the All India Unani-Vedic Conference was unacceptable because it was based on the assumption that Unani was the medicine of Muslims and Ayurveda of Hindus, and the conference would bring the two communities together to fight for their professional interests. Like Hakim Abd al Aziz, he argued that this was far from the truth and that his All India Unani-Tibbi Conference attempted to make Unani open to every community, region, and nation—in the same manner that Western medicine had transcended religion and community issues. Khan said that in case the view of Wali was to be accepted that Unani was Muslim medicine, there was all the more reason that it should progress separately from Ayurveda since the political and educational agendas of Muslims were being considered by organizations like the Muhammedan Conference as separate from those of the Hindus.[97]

Khan also indicated the dangers in the All India Unani-Vedic Conference's claim that certification from it would facilitate the entry of vaids and hakims to municipality jobs; such people otherwise would find their appointments opposed by their Muslim and Hindu counterparts. According to Wali, such suggestions of the All India Unani-Vedic Conference could create a divide between hakims and the vaids on communal lines, and that would be detrimental to the long-term

[96] Pamphlet, 'Reasons for remaining away from the All India Vedic and Unani Conference, Delhi'.

[97] Abid Ali Khan.

interests of both. He was of the view that at a time when the government was showing interest in matters of indigenous medicine, Hindus and Muslims should come together under the banner of an unalloyed Unani and protect it. He thus made a strong plea to open up unani even more than before to both Hindus as well as Muslims—in the past this had benefited both communities. A hakim, he said, had to be promoted 'irrespective of whether he was a Hindu or a Muslim'.[98]

Rai Burj Mohan Dayal could not agree more. He said that Ayurveda was already cast as a Hindu medicine, and 'the Hindu University is already involved in the progress of the Ayurveda. The 1911 meeting [*jalsah*] of the Conference also shows that Hindus are trying to make it progress.' There was therefore no need for Unani to get involved in communitarian politics around medicine.[99] Other hakims, like Shah Muhammad Fakhir Allahabadi, articulated their opposition on similar lines. In a speech at the 14 November session of the All India Unani-Tibbi Conference he said that at a public meeting in Allahabad organized by the Hindus (*ahl-i-Hindus*) to discuss issues of medicine, he was not allowed to speak because he was a Muslim and therefore seen as an advocate of Unani. He got permission to speak on condition that he would speak on Ayurveda. This unfortunate incident convinced him that the Hindus had consolidated their religious identity around Ayurveda, and there was therefore no need to combine Unani with a medicine so stamped. In fact Fakhir Allahabadi pointed out that so aggressive was the communitarian profile of Ayurveda that some vaids were even opposed to the formation of the All India Unani-Vedic Conference as they thought that their alliance with Unani diluted their communitarian agenda. Such vaids broke away from the All India Unani-Vedic Conference and formed a parallel Vedic Sabha in Lucknow to promote the cause of their medicine. This prompted Fakhir Allahabadi to make a strong case against the merger of Ayurveda and Unani under one public platform. Fakhir Allahabadi contrasted Ayurveda's Hindu profile to the non-communitarian openmindedness of his own discipline, which remained open to patients and learners of all communities.[100]

[98] Ibid.: 9.

[99] Resolution no. 3 in *Raudad Ijlaas awwal*, All India Unani Tibbi Conference, 14 November 1911, *Raudad-i-Takmil-ut-Tibb*, p. 10.

[100] Ibid.: 12–13.

The encompassing and non-communitarian professional identity of Unani was given a history by many delegates who penned an accretive and eclectic history of Unani which ran from its Graeco-Roman past, to Arabia, to Asia, Africa, and Europe. They underlined that many of its renowned hakims, like al-Razi, Avicenna, Ali bin Abbas, and others belonged to a nice mix of Jewish, Muslim, Christian, and Zoroastrian descent. And thus to call this system of medicine exclusively Islamic was ahistorical.[101] The advocates of Unani eclecticism, like Maulana Shah Muhammad Suleiman Qadri Chishti, argued that, despite its wide embrace, Unani retained its core humoral theory.[102]

However, by the second All India Unani-Tibbi Conference in 1913, the communitarian lobby had made considerable headway. Both in the kind of reconstructed histories of Unani that were tabled, as well as in the symbolism used and exhortations made by speakers to mobilize financial support and pressurize the government to offer help, the link of Unani with the Muslim community was more clearly laid out than ever before. The president of the second All India Unani-Tibbi Conference, a renowned taluqdar of Sandila—Chaudhury Muhammad Altafat Rasool—presented a reconstructed history of Unani that highlighted its Islamic core and ignored its Graeco-Roman philosophic past. At best, the pre-Arab past was seen as the period of Unani's 'infancy'. Like Nomani, Rasool linked the evolution and progress of Unani to Islamic expansion. The high points of its progress corresponded to those of Islamic expansion. Predictably, in the period of the Abbasids, Unani was seen to flourish under royal patronage. Surgery, the discovery of treatment for smallpox, the introduction of chemistry in Unani, and the setting up of charitable hospitals (*shifakhaana* and *dawakhaana*) were all located in this period of high Islamic expansion.[103] In yet another speech by Hakim Muhammad Abd al Moid Lucknawi of the Takmil-ut-Tibb College, Muslim contributions to Unani in the period of Ummayad rule in Andalusia, and the advancements in Baghdad under the aegis of the Abbasids, were highlighted. Major discoveries

[101] Ibid., Resolution no. 5, *Raudad-i-Ijlaas Awwal*, All India Unani Tibbi Conference, 14 November 1911, p. 15.

[102] Ibid.: 15.

[103] Karawai Jalsah Awwal. Presidential Address by Chaudhury Muhammad Altafat Rasool, Report, *Ijlaas Doawam*, 2nd All India Unani Tibbi Conference, 4 February 1913, Lucknow: Nami Press, July 1913, pp. 6–11.

of the period, like surgical instruments (*aalaat-i-surgery*), were attributed to Abu-al Qasim-ibn Abbas Andalusi, who wrote many books on these as well.[104]

In a discussion at the second Unani-Tibbi meeting at Lucknow, the Muslim profile of Unani was linked to the linguistic identity of Muslims as well. One hakim, Mukhlis al Daulah Jaipuri, saw the formation of the All India Unani-Tibbi Conference as a mark of separation from the Hindus. He underlined Unani's religious profile, using Islamic symbolism. He said that Unani's proximity to Islam was proved by the fact that Arabic, a language associated with Islam, was Unani's canonical language, whereas Sanskrit, the language of the Hindus, was the language in which the Ayurveda texts were compiled. Unani and Vedic texts thus represented two different cultural worlds and could never come together on the same platform. He argued an even greater exclusivist cultural identity for Unani when he critiqued the blowing of the conch shell, a Hindu auspicious ritual, along with the chanting of Allah-o-Akbar, the Muslim way of greeting, at the sessions of the All India Unani-Vedic Conference. He contrasted this objectionable practice to the sessions of the All India Unani-Tibbi Conference where, he said, of late, rituals were tailored to narrow Islamic idioms like the reciting of *tauhid* (belief in one God and Muhammad as his Prophet), and Allah-o-Akbar. Again, he made a plea to learn from the way exclusivist organizations like the Hindu University or the Muslim University or even the Muslim League operated to further the interests of their narrow constituencies. Mukhlis-ud-Daulah used Quranic verses to provoke people to engage in Unani knowledge.[105]

By the third annual meeting of the conference in April 1914, the secretary made it very clear that the objective of the All India Unani-Tibbi Conference was to work for the community. It was for this objective that he wanted to bring all the hakims of the country together onto one platform. He regretted that the conference had not achieved an all-India character, like that of the Anglo-Muhammedan Conference

[104] *Taqreer* (speech), Hakim Muhammad Abdul Moid Sahib Lucknowi, ibid., pp. 28–31.

[105] Hakim Mukhlis-ul-Daulah Jaipuri, Report; Ijlaas Doawam, 2nd All India Unani Tibbi Conference, 4 February 1913, Lucknow: Nami Press, July 1913, pp. 18, 21.

of Sir Syed Ahmad Khan. Indeed, so much did the conference work like the Anglo-Muhammedan Conference that the secretary continuously referred to its organizational and fund-raising strategies as the model to be followed. And like Sir Syed Ahmad, the secretary too saw mobilization of the qaum as the key to its success, since community pressure would alone budge the government to act and take interest in Unani. By the 1920s, Unani had drifted towards a fuzzy Muslim profile. And yet its entanglements with a professional identity were never entirely severed.

Thirty years later, in 1950, Hakim Abd al Latif, a leading hakim of the Azizi family, upheld the value of logic and metaphysics to underline Unani's advanced nature as a science. He contrasted Unani's scientificity to the relatively lower reliability of observation and laboratory testing that Western medicine claimed as its marker of scientificity. But even as Hakim Abd al Latif championed the cause of logic and metaphysics that he had inherited from the Graeco-Arab philosophical tradition, he introduced God as a new element in his critique of Western claims to authenticity of observation. He noted that things had causal properties that were hidden from direct observation. In addition God, in an Islamic understanding of the world, could intervene at any time to alter the behaviour of objects.[106] This was indeed a far cry from the efforts of Unani, through its long history, to grow separately from religion. But even now, Latif's predominant plea was to underline the unique scientific core of Unani and not highlight its particularistic Muslim aspect. And this careful balancing act between the scientific, secular-national-territorial and Muslim communitarian was once again noteworthy because none of the members of the family participated directly in either the anti-colonial national movement or in the separatist politics of the time. It is of considerable significance that no one from the Azizi family participated in the Muslim League politics of the time, or migrated to Pakistan in 1947. Such Muslim professionals walked a unique social space which is seldom recognized in the binaries of nationalist and separatist that define studies on Muslim politics.[107]

[106] Liebeskind 2002: 66.

[107] A notable exception being the scholarly work of Ayesha Jalal, which has revolutionized our thinking on Muslim politics in the nineteenth and twentieth centuries. See Jalal 2001.

Conclusion

THIS BOOK HAS TRACED THE TRAJECTORIES OF INDO-MUSLIM—Unani—medicine and its largely Muslim practitioners from the Mughal period to the early twentieth century. It has sought to capture the social history of the period as it unfolded in the Gangetic valley, away from the glare of the colonial state. It suggests that in this vital and considerable terrain of life, political oppression and economic exploitation notwithstanding, colonial subjects substantially made their own history. This was possible because they had in Unani their own robust intellectual tradition, which was characterized by ingenuity and adaptability.

My analysis shows the making of a medical tradition in pre-modern India. It reveals the functioning of an 'archaic globalization' in which universal religion and the humoral understanding of body and mind offered space for the accommodation of new ideas. Muslim rulers and their subjects were receptive to the late-eighteenth-century notion of Muslim universalism inspired by the new Arabic learning, which was itself imbued with Hellenism. This intellectual legacy encouraged the emergence of knowledgeable communities, both autonomous and under state patronage, that energized the public welfare sphere in pre-colonial India.

This book has mapped the transformation of this tradition in the course of India's transition to colonialism, arguing that change in immediate pre-colonial India was triggered by churnings within the tradition caused by extra-European influences. Rather than seeing the European expansion into India as the propeller of change, the discussions here have highlighted the role of Arabic scientistic ideas which defined the contours of Unani's colonial encounter. I suggest that the colonial intervention in the early nineteenth century only reinforced the Arabic-driven idea of medicine as science and made it more popular. I have also tried to expand the pre-modern idea of well being, showing how it shifted its focus from the individual to society as a whole. These interactions created new referents of medical status and opened up a rudimentary notion of public service. Finally, it was vernacular print culture, rather than any anglicist push, that played a critical role in sketching out this particularistic 'public' and 'civil society' in early-nineteenth-century India.

In the postmodern scholarship of the last two decades, the narrative of a formidable state that monopolized 'knowledge solidus power' has dominated the history of the nineteenth century. This has been true also of the way colonial medical drives and issues of public health have been understood—as callous imports from Western-style models of civil society. These are either said to have colonized the indigenous medical traditions or to have reified them as cultural practices. In contrast, this book has shown that people were not willing pawns on the colonial chessboard, waiting to be pushed around. They translated the colonial context to their own advantage and put their medical tradition to new uses despite a politically oppressive state. Of course this was possible because they had a rich intellectual and political legacy at hand, and because the colonial state was conspicuously less oppressive in the medical sphere than in the political.

Indeed, the colonial context led to far greater in-house introspection: indigenous intellectual traditions were put to greater scrutiny, a literature of dissent both in religion and the sciences emerged, and the plurality of opinions became more audible. The monopoly enjoyed by some on the knowledge of medicine, science, and religion was questioned. And colonial consequences such as print were used to diversify intellectual reservoirs. These generated fresh contestation over knowledge within the Unani tradition, which then expanded older public arenas and offered greater space for both in-house debates as well as opportunities of dialogue with the British. An energized medical public sphere became the site where British public health reforms were perfected and at times exported to the metropolis.

It is of course true that colonial rule created new contesters and triggered contestations over public roles and control of civil society. But pre-modern arenas continued to provide the frame within which colonial public health drives were shaped. For instance, the king or elites as welfare managers had always been central to the pre-colonial public arena. This book has shown that the colonial state remained dependent on them and the communities of knowledge they commanded. It had to carry on a protracted dialogic contest with local elites to monopolize the welfare space. Public health, far from being a Western-derived import into India, evolved out of this interaction between state and traditional elites.

Indeed, I argue that it was public health reforms perfected during the cholera epidemic of 1818–19 in India that were imported into the metropolis when cholera ravaged London in the 1830s. The colony held other benefits as well. While recent researches have highlighted the revolutionary impact of print in changing the mentalities of the colonized, this book has focused on how it advanced British careers in the metropolis. It has argued that in the early nineteenth century the printing press in the colony was used foremost by British doctors to claim authorship and authority over the medical knowledge of the colonized. Print enabled this hitherto informally shared knowledge to enter British-authored pharmacopoeias and materia medicas. This was of course used to further British experimentation on pharmacy and therapeutics and help the development of global science. More significantly, British doctors flaunted their authorship of pharmacopoeias and other texts printed in the colony to further their careers in the medical academies of Britain. This often created tensions between the colony and the metropolis.

The analysis of this Indo-Muslim healing tradition has also enabled a review of issues of modernity, professionalization, and scientific rationality over which Europe has claimed monopoly. I have argued that people used print to articulate dissonances in their tradition and give them wider publicity. This process saw the emergence of notions of modernity and professionalism that were different from the ones which originated in Europe. People derived these definitions from their own freshly invigorated medical tradition. The discussions herein have shown how the specific conjuncture of the nineteenth century, where both global capitalism and its unsettling effects coincided, provided opportunities for new contenders to challenge traditional communities of knowledge. This enabled Unani to spill out of its 'scientistic' scholarly mould and reconstitute itself as an accessible and culturally encompassing medical tradition. It enabled new hakims to reach out to people and cultural referents that hitherto lay beyond the purview of traditional Unani: the person of the Prophet, Sufi healing, and astrology. As Unani embraced new medical localisms, it diversified and made them part of its own world of 'scientific rational' medicine.

The new Unani threatened traditional medical families. Even as these longstanding custodians of specialist knowledge ducked for cover, they leaned simultaneously on the colonial state, territorial nativism, Muslim communitarianism, and Islamic universalism to brave the onslaught of pressures for change from within. In the process they imbued modernity and universalism with fresh meanings. Thus, the demands for public health reforms, modernization, and professionalization voiced by traditional families did not have an anti-colonial tone. Yet they also had meanings very different to European claims. These derived from the efforts of elites to use such control mechanisms to hit out at new in-house contenders of knowledge. Far from any correspondence with Western ideas of professionalism or modernity, these were seen as colonial artifacts to combat the new Urdu Unani. And support of the colonial state was always welcome.

The book discusses, finally, the Azizi family of Lucknow hakims to show how their articulations of modernity were at times embedded in a national space that was culled out of the colonial discursive framework. This exemplary family gave their modernity a historicity within the territorial confines of the nation. This definition contrasts with the modernity of Western medicine, which was said to have neither historicity nor territoriality in India. Yet, at least in the case of the Muslim physicians like those of the Azizi, family, a 'national' medicine did not automatically translate into the vigour of 'secular' territorial nationalism. It remained riveted within competing ideas of Muslim communitarianism and Islamic universalism; 'national' Unani had to contend with strands of religious communitarianism and Islamic universalism. It could never author a narrow anti-colonial territorial nationalism. I suggest that the nature of Unani's complex encounter with colonialism shaped the predicaments of the Azizi family as they balanced Unani's 'national' profile with the new Muslim communitarian and Islamic universalist frame.

Thus my book, even as it traces the trajectories of the practitioners of Unani medicine in colonial India, does not have a story of anti-colonial political activism to offer. It does justice to the story of how Muslim physicians and their followers translated for themselves the new situation, came to terms with pulls towards Muslim universalism and communitarianism, and arrived at an anti-colonialism that blurred the boundaries of the secular and the religious, and the national and universal.[1]

In this sense and others, the book concludes in agreement with recent understandings of the modern nation state offered by C.A. Bayly, Ayesha Jalal, and Sugata Bose. It shows that the modern nation state, if examined on the basis of its medical encounter with Unani, was nowhere near its postmodern understanding of knowledge as power. In contrast, it stood entrenched in networks of formal and informal knowledge resources that were represented by communities of Arabic, Persian, and Urdu medical literature, and also by the healing powers of Sufis and fakirs. Its 'new' institutions of public welfare, like the dispensary, exemplified this marriage of older networks of practitioners, notions of individual well being, scholastic learning, and elite medical patronage with new medical knowledge as exemplified in British texts on anatomy, notions of public service, and social well being. The colonial context, exemplified foremost by the expansion of print capitalism, made it easier for older communities to put their resources to new uses. This of course caused immense change within their own learning traditions and created tensions and contestations, as new contenders for knowledge emerged. Ultimately, it was this in-house revolution that expanded the medical public sphere. It shaped both British public health reforms as well as Indian understandings of modernity and professionalism.

Such being the nature of the colonial encounter within the medical terrain, I have argued that Muslim public intellectuals of the nineteenth century straddled two worlds—a pre-colonial medical ethos, and Western traditions represented by the British—with ease. They carved out a national space for their medicine from within the colonial frame. And yet, as Jalal and Bose argue, there was no freeway for territorial nationalism in nineteenth-century India. Unani as national medicine and the potential for territorial nationalism that it encoded rubbed

[1] Bose: 2006.

shoulders and had to contend with strands of Muslim communitarianism and Islamic universalism. Its journey to the status of a national medicine was far too complex to appropriate into some neat political anti-colonial activism, or to show simplistically as an aspect of Muslim separatism.

Glossary

aab-o-hawa	locale; climate
aab-i-garm	hot water
aab-i-sard	cold water
aab-i-zam zam	holy water; water of the sacred Meccan well, Zam Zam
aalaat	instruments
aalim	learned; Muslim scholar
aam	ordinary people
aaraastagi	decoration
adab	proper code of conduct
adl	justice
afzal	extraordinary; excellent; supreme; most prominent
ahl-i-Europe	people of Europe
ahl-i-Hind	people of India
ahl-i-Islam	Muslims; people of Islam
ahl-i-watan	people of the country;
ajlaaf	ignoble people; the scum of society
ajzaa	ingredients; items; parts
akhlaaq	etiquette, conduct, moral behaviour
amal	practice

amali tibb	practical medicine
amraaz	diseases
amraaz-i-jismaani	bodily diseases
anaar	pomegranate
anaar daana	dried pomegranate seed
anjuman	organization; association
arraq	solution; distilled water
arkaan	elements
arq baid mashk	syrup; distilled water of musk willow
asbaab	reasons; causes; factors
ashraf	exalted; noble
ashraful makhlooqaat	exalted and extraordinary person (man as the most eminent of created beings)
ashrafiyat	nobleness
attar	chemist; seller of perfume
atibba-i-haaziq	reputed hakims
azaan	call for prayer
baarid	cold
ba awaaz-i-buland	in a loud voice; loudly
baitul shifa	house of recovery or cure; hospital
bandagi	piety; servility; servitude
bayaaz	commonplace-book; fair copy
behayaa	shameless
be kamaali	ineffectiveness
be khatar	harmless
bihishti	water carrier
bismillah	in the name of Allah
buzurg	aged; noble

chana	gram
chhapakhaana	printing press
chataank	sixteenth part of a seer; weight equivalent to about two ounces
cherayta	plant (a kind of a herb)
chob-i-cheeni	China root; Chinese wood
chooran	digestive powder; aromatic powder
chutsal	Hindi school
daad	ringworm
daayagiri	midwifery
dai	midwife
dama	asthma
darr	corridor
daroga	inspector; police station officer
darrs	lesson
darul shifa	the house of healing; hospital
dawa faroshi	the profession of druggists/ chemists
dawakhaana	dispensary; pharmacy
dawa-i-murakkab	compound medicines
deer hazm	difficult to digest
desi	indigenous; local
dil	heart
dimaagh	mind
dua	prayer
faaqah	fasting
faazil	scholarly person
fakhr-i-ustaad	pride of the teacher
fann	art; profession
fann-i-keemiyaa	chemistry

fanni khidmat	professional service; technical service
fann-i-jauhar	medicinal essence
fann-i-nuskha navisi	art of writing prescription
farz	duty; obligation
faujdari	post of military commandant
fawaaid	benefits
funoon	sciences; branches of knowledge; arts
gaanjah	hemp
gauhar-i-giraan maayah	of great value; valuable
ghaans phoos	grass; straw
ghair mulk	foreign land; foreign country; abroad
gharana	family
ghari	span of time equal to 24 minutes
ghazab	shock; anger
ghiza	diet; food
ghuraba	the poor
ghusl	bath
girroh	group
golee	tablet
gora	white
gumraahi	ignorance about religion; deviation from the right path
haandis	pots
haaziq	reputed; expert
haiza	cholera
hakim	physician; doctor
hakim mutlaq	greatest physician

halaal	legal; lawful; permissible
haldi	turmeric
halwa-i-tarkhub	pudding
haqeeqat	reality
haraam	illegal; unlawful; not permissible
hararat	heat
hareer	silken clothes
harkat	movement; motion; activity
hawa	air
hayaat	life
hifzan-i-sihat	health care; hygiene
hifz-i-sihat	personal hygiene; health care
hikmat	medical practice; wisdom; philosophy
himaqat	folly; absurdity; foolishness
himmat-i- kaafi	perseverance; enough courage
hindi adwiyah	Indian medicines
hukm	order
hukmaran-i-sihat	to govern health
hummayaat	psychology; fevers
huqooq	rights; privileges
huzoor	sir; Your Honour; presence
ibaadat	meditation; worship
ihtalaam	wet dream; nocturnal pollution
ilaj	cure
ilaj-i-badan	treatment of the body
ilaj-i-rooh	treatment of the soul
ilaj-i-sar dard	treatment of headache
ilm	knowledge; awareness
ilm-i-abdaan/ilm al abdaan	knowledge of the body

ilm-i-adyaan/ilm al adyaan	knowledge of religions
ilm-i-adwiyah	pharmacology
ilm-i-jarrahi	surgery
ilm-i-keemiyaa	chemistry
ilm-i-nabataat	botany
ilm-i-tashreeh	anatomy
ilm-i-taskhees	diagnostics
ilm-i-tehzeeb-i-akhlaaq	etiquette and conduct
ilm-i-tibb	theoretical medicine
ilm waafi	beneficial knowledge
insaan	human being
insaan-i-kaamil	perfect man; perfect human being
insha	correspondence
iqraar namah	agreement; contract; written bond
itmenaan sei	calmly; satisfactorily; with confidence
jadeed	new; modern
jahl-i-baseet	illiteracy
jahil	illiterate
jaifal	nutmeg
jalsah	sitting; public gathering; meeting
jamaat	group; congregation
jarrahat	surgery
jarrah	surgeon
jau	maize
jawatari	mace
jihaalat	ignorance
jism	body
jismaani	bodily; physical

kaala	black
kaala bhangra	a kind of herb
kaala daana	indigo seed used as purgative
kaali mirch	black pepper
kahar	palanquin bearer
kaifiyat	temperament; condition
kaifiyat-i-khaarji	outward condition
kamaal-i-insaani	human perfection
kaseef	difficult to digest
kataan	clothes
kewra	cooling scented essence/a sweet fragrant plant
khaara	brackish water
khairaat	charity
khairkhwaah-i-mulk	well-wishing for the country; welfare of the country
khaleeq	courteous; polite; amiable; friendly
khatimah	conclusion
khandaani	ancestral; hereditary; family
khidmat-i-khalq	service to the human being
khilat	robe of honour as princely award or investiture
khushk	dry
khush akhlaaqi	good conduct
khwahish-i-nafsaani	carnal desire; sensual desire
kutcha	raw
laaiq	worthy; capable; suitable
lateef	easily digestible
lobaan	camphor
maaqulaat	rational sciences

madaaris	institutions
madrasa	institution; Islamic school
mafrooz	mandatory
majoon	electuary
maktabkhaana	primary school; small madrasa
malikul maut	angel of death
manqulaat	transmitted sciences
maqaalahs/maqaalaat	section; treatise
marham	ointment
maseeha	healer; saviour
mataanat-i-raai	sobriety of advice
maudu	subject
mausam	weather; season
mausam-i-barsaat	rainy season
mausam-i-garma	summer
mausam-i-sarda	winter
mawlid/mauwlood/milaad	birth anniversary of the Prophet
mazhab	religion
misee	unrefined sugar
mizaaj	temperament; disposition
mongaa	a precious stone
moong	vetchling
mualajaat/mualaje	treatment; cure
mu'azzaz	respectable
mufradaat musta'malah	single drugs
muhtamm	important
mulk	country
mulki hamdardi	patriotism; sympathy/good feelings for the country

munaasbat	aptitude; suitability; appropriateness
munaazra	polemics
muqaddimah	preface
mu'tadbeh girroh	sufficient, substantial group; a good number
mu'taqid-i-dildaada	sincere devotee
naagar motha	a kind of fragrant grass
naa muwaafiq	unsuitable; inappropriate
naasoor	weeping wound; running sore; ulcer
naazreen	spectators; onlookers; readers
nabz	pulse
nafs	psyche; soul
nafsaani	carnal; sensual
natijah	result; conclusion; consequence
nautch	dance
neem	margosa; margo tree
neemjan	half-dead
neem hakim	spurious hakims; quack
nikaah	matrimony; marriage
nuskha	prescription
nuskha navisi	prescription writing
nuskha-i-amraaz	medical prescriptions
pakizgee	purity; cleanliness
patwaris	revenue official; village registrar
peshaani	forehead
phitkari	alum
poshaak	clothes
putsal	Sanskrit school

qaashaateer aalah	catheter
qadeemi/qadeem	ancient
qadardaani	respect; appreciation
qaidah	formula; rule
qanun-i-tibb	rules of medicine
qarabadeen	pharmacopoeia; materia medica
qarrabah	container
qaroora	urine
qasba	town
qasida	encomium
qatraat	drops
qaum	community; nation; people
qazi	Muslim judge
qiyaas	speculative philosophy (analogy)
quolaj	colic pains
quwwat	stamina; power; strength
quwwat-i- amali	experience and practice
quwwat-i-fauqiaat	ability
quwwat-i-ilmi	academic learning
raangah	silver; zinc-like metal
rahmaan	benefactor
rais	landed class
rausa	landed class; elite; the rich
raunaq-i-mulk	grace for the country
razaalat	pettiness; ignobleness
razeel	ignoble; petty
rifaah-i-aam	public welfare
risaalah	article; treatise; magazine
riyaasat	state

riyaazat	exercise
roghan-i-baid anjeer	fig-willow oil
roghan-i-chameli	balm of jasmine; jasmine oil
rooh	soul
rooh-i-haiwaani	animal instincts
rooh-i-nafsaani	human instincts
rozah	fasting
sadarat	presidency; chairmanship
safoof	powder
sahaiban-i-mausuf	learned people
sahlul wasool	easily available
sahukar	moneylender
sanad	certificate
sardi	cold
sarparast	patron
seesah	lead
seth	tycoon; business magnate
shafqat	politeness; affection
sharaafat	gentlemanliness
sharbat/sherbet	syrup/drink
sharbat-i-kadar	syrup
sharf-i-insaani	human dignity
shareef ilm	respectable knowledge
shifakhaana	hospital
shifayaabi	recovery
shurafa	the gentry
sihat-i-nafs ka hafiz	master of one's psyche/health
sonth	dry ginger
soorat aur sohbat	appearance/look and company

sukoon	peace; tranquillity; relief; relaxation
sukoonat	peace of mind; calmness
sunnat	practice of the Prophet; according to the Traditions of the Prophet
taarbarqi	telegraph; telegram
tabaabat	practice of medicine
tabib	physician; healer
tabib-i-khaas	special physician
tabi'at	temperament; disposition
tadbeer-i- dawa	use of medicine; application of medicine
tadrees	to teach
tajurba	experience
tajurbekaar	experienced
takmil	to complete; to finish
tarkeeb	method
tarteeb-i-deeni	religious dictums
tashaffee	console; solace; satisfaction
tashreeh	anatomy
tauhid	oneness of Allah; monotheism
taweez	talisman
tehzeeb	culture; good manner; civilization
tez	strong
tibb-i-badani	medicine of the body
tibb-i-nafsanee	medicine of the psyche
tibb-i-qadeem	ancient medicine
til	sesamum seed; sesamum
uloom	sciences
umaraa	nobles; aristocratic class

ummat	community
usool	principles
usool-i-tabee'at	physiology
usool-i-tashkhees	principles of diagnosis
ustaad	teacher
ustaad –shaagird	teacher–pupil
vakil	counsel; lawyer; advocate
waajib	mandatory
wabaa	epidemic
waqf	endowment
wazeefah	scholarships; stipend; pension
wazeefah khwar	pensioner
wizarat	ministry
zaat-ul-janb	pleurisy
zehnee isteqaamat	mindset; mental perseverance
zenaana	feminine; female
zof-i-baah	sexual weakness; impotency
zood hazm	digestible food

Bibliography

Published Secondary Literature

Ahmed, J. & A. Qadeer, *Unani the Science of Greco-Arabic Medicine*, Delhi, nd

Alam, M., & Subrahmanyam, S., eds, *The Mughal State, 1526–1750*, New Delhi, 1998

——, 'Shari'a, Akhlaq and Governance', in M. Alam, *The Languages of Political Islam—India 1200–1800*, Chicago, Delhi, 2004, pp. 26–80

——, 'The Pursuit of Persian: Language in Mughal Politics', *Modern Asian Studies* (*MAS*), 32, 2, May 1998, pp. 317–49

Alavi, S., *Sepoys and the Company. Tradition and Transition in North India*, Delhi, 1995

——, 'Unani Medicine in the Nineteenth Century Public Sphere: Urdu texts and the *Oudh Akhbar*', *Indian Economic and Social History Review* (*IESHR*), 42, 1, 2005, pp. 101–29

Arnold, D., ed., *Imperial Medicine and Indigenous Societies*, Manchester, 1998

——, *Colonising the Body. State Medicine and Epidemic Disease in Nineteenth Century India*, London, 1993

——, *Science, Technology and Medicine in Colonial India.* The New Cambridge History of India, Cambridge, 2000

——, 'Cholera and Colonialism in British India', *Past and Present*, 113, 1986, pp. 118–57

——, 'Touching the Body: Perspective on the Indian Plague, 1896–1900, in R. Guha, ed., *Subaltern Studies V*, Delhi, 1987, pp. 55–90

Askari, S.H., 'Medicines and Hospitals in Muslim India, *Journal of Bihar Research Society*, 43, 1957, pp. 7–21

Attewell, G., *Refiguring Unani Tibb: Plural Healing in Late Colonial India*, Delhi, 2007

Azmi, A.A., *History of Unani Medicine in India*, Delhi, 2004

Azmi, Neyaz A., 'Shibli on Muslim Education and Politics', in *Islam and the Modern Age*, 25:3, 1994, pp. 190–205

Bayly, C.A., *Empire and Information. Intelligence Gathering and Social Communication in India, 1780–1870*, Cambridge, 1996

Brass, P., *Language, Religion and Politics in North India*, Cambridge, 1974

Bose, S., *A Hundred Horizons. The Indian Ocean in the Age of Global Empire*, Delhi, 2006

Bose, D.M., S.N. Sen & B.V. Subbarayappa, *A Concise History of Science in India*, Delhi, 1971

Brown, Cynthia, J., 'Text, Image, and Authorial Self-consciousness in Late Medieval Paris', in Sandra Hindman, ed., *Printing the Written Word. The Social History of Books, 1450–1520*, Ithaca, 1991, pp. 103–42

Burgel, Jan. C., 'Secular and Religious Features in Medieval Arabic Medicine', in C. Leslie, ed., *Asian Medical System—A Comparative Study*, reprint Delhi, 1998, pp. 44–62

Bynum, W.F., *Science and the Practice of Medicine in the Nineteenth Century*, Cambridge, 1994

Catanach, I., 'Plague and the Indian Village, 1896–1914', in P. Robb, ed., *Rural India: Land, Power and Society under British Rule*, Delhi, 1986, pp. 216–43

Chakrabarty, D., *Provincialising Europe. Postcolonial Thought and Historical Difference*, Princeton, 2000

Chakrabarty, P., *Western Science in Modern India: Metropolitan Methods, Colonial Practices*, Delhi, 2004

Chandavarkar, R., 'Plague, Panic and Epidemic Politics in India, 1896–1914', in T. Ranger and P. Slack, *Epidemics and Ideas: Essays on the Historical Perception of Pestilence*, Cambridge, 1992, pp. 203–40

Chatterji, P., *The Nation and its Fragments: Colonial and Postcolonial Histories*, Delhi, 1993

Cohn, B., *Colonialism and its Forms of Knowledge: The British in India*, Princeton, 1996

Cole, J.R.I., *Comparing Muslim Societies: Knowledge and the State in a World Civilization*, Ann Arbor, 1992

——, 'Printing and Urban Islam in the Mediterranean World, 1890–1920', in Leila Tarazi Fawaz and C.A. Bayly, eds, *Modernity and Culture. From the Mediterranean to the Indian Ocean*, Columbia, 2002, pp. 344–64

Davis, Natalie Z., 'Printing and the People', in idem, ed., *Society and Culture in Early Modern France*, Stanford, 1975, pp. 189–226

Dey, A., *The Image of the Prophet in Bengal Muslim Piety 1850–1947*, Calcutta, 2006

Dirks, N.B., ed., *Colonialism and Culture*, Ann Arbor, 1992, pp. 1–25

Dols, Michael W., and Adil S. Gaman, *Medieval Islamic Medicine: Ibn Ridwan's Treatise 'On the Preservation of Bodily Ills in Egypt'*, London, 1984

Dunn, F.L., 'Traditional Asian Medicine and Cosmopolitan Medicine as Adaptive Systems', in Charles Leslie, *Asian Medical System—A Comparative Study*, Berkeley, 1976, pp. 133–58

Eaton, R., *The Rise of Islam and the Bengal Frontier, 1204–1760*, Berkeley, 1993

——, 'Sufi Folk Literature and the Expansion of Indian Islam', in R. Eaton, ed., *Essays on Islam and Indian History*, Delhi, 2001, pp. 189–202

Ebrahimnejad, H., *Medicine, Public Health and the Qajar State: Patterns of Medical Modernization in Nineteenth Century Iran*, Leiden, 2004

Eisenstein, E.L., *The Printing Press as an Agent of Change: Communication and Cultural Transformation in Early Modern Europe*, vol. II, Cambridge, 1979

Elgood, C., *Medicine in Persia*, New York, 1934, reprint 1978

———, *A Medical History of Persia and the Eastern Caliphate from the Earliest Times until the Year AD 1932*, Cambridge, 1951

——, *Analecta Medico-Historica: Safavid Surgery*, London, 1966

Fraser, N., 'Rethinking the Public Sphere: A Contribution to the Critique of Actually Existing Democracy', in C. Calhoun, ed.,

Habermas and the Public Sphere, Cambridge, Massachusetts, 1992, pp. 109–42

French, R., *Medicine before Science. The Business of Medicine from the Middle Ages to the Enlightenment*, Cambridge, 2003

——, 'Science and the Soul: Stahl, Hoffman and Sauvages on Pathology', in A. Cunningham and R. French, eds, *The Medical Enlightenment of the Eighteenth Century*, Cambridge, 1990, pp. 88–110

Gadamer, H.G., *Truth and Method*, reprint New York, 2004

Galdston, I., 'Medicine and Culture', in F.N.L. Poynter, *Medicine and Culture*, London, 1969, pp. 15–25

Geyer, M.H. & J. Paulmann, eds, *The Mechanics of Internationalism: Culture, Society and Politics from the 1840s to the First World War*, Oxford, 2001

Giddens, A., *The Consequences of Modernity*, London, 2003

Goswami, M., *Producing India: From Colonial Economy to National Space*, Chicago, 2004

Good, M., B. Brodwin, B. Good, & A. Kleinman, eds, *Pain as Human Experience: An Anthropological Perspective*, Berkeley, 1994

Good, B. and Mary-Jo DelVecchio Good, 'The Comparative Study of Greco-Islamic Medicine: The Integration of Medical Knowledge into Local Symbolic Contexts', in Charles Leslie and Allan Young, eds, *Paths to Asian Medical Knowledge*, Berkeley, 1992, pp. 259–65

Greaves, M.A., *Education in British India 1698–1947*, London, 1967

Guha, S., 'Early Propaganda for Vaccination in India: Thomas Coats' Gaichya Devichi Utpatti va Guna Tyacha Tapshil', in Sumit Guha, *Health and Population in South Asia from Earliest Times to the Present*, London, 2001, pp. 140–55

Gupta, B., 'Indigenous Medicine in Nineteenth and Twentieth Century Bengal', in C. Leslie, ed., *Asian Medical Systems: A Comparative Study*, Berkeley, 1976, pp. 368–78

Habermas, J., *The Structural Transformation of the Public Sphere*. Cambridge MIT, 1989

Habib, I., 'Akbar and Technology', in I. Habib, ed., *Akbar and His India*, Delhi, 1997, pp. 129–48

——, 'The Technology and Economy of Mughal India', *Indian Economic and Social History Review* (*IESHR*), 27, 1980, pp. 1–34

Harrison, M., *Public Health in British India: Anglo-Indian Preventive Medicine, 1859–1914*, Cambridge, 1994

Hasan, F., *State and Locality in Mughal India: Power Relations in Western India, c. 1572–1730*, Cambridge, 2006

Hasan, M., *Nationalism and Communal Politics in India 1916–28*, 2nd edition, Delhi, 1979

——, *A Nationalist Conscience: M.A. Ansari, the Congress and the Raj*, Delhi, 1987

——, 'Maulawi Zaka Ullah: Sharif Culture and Colonial Rule', in M. Pernau, ed., *The Delhi College. Traditional Elites, the Colonial State, and Education before 1857*, Delhi, 2006, pp. 261–98

Haynes, Douglas, M., 'Victorian Imperialism in the Making of the British Medical Profession: An Argument', in D. Ghosh & D. Kennedy, eds, *Decentering Empire: Britain, India and the Transcolonial World*, Delhi, 2006, pp. 130–56

Hobsbawm, E. & T. Ranger, eds, *The Invention of Tradition*, Cambridge, 1983

Hume, John C., 'Rival Traditions: Western Medicine and Yunan-i-Tibb in the Punjab, 1849–1889', *Bulletin of the History of Medicine*, vol. 51, 1977, pp. 214–31

Iliffe, J., *East African Doctors: A History of the Modern Profession*, Cambridge, 1998

Jalal, A., *Self and Sovereignty. Individual and Community in South Asian Islam since 1850*, Delhi, 2001

Jones, Peter M., 'Medical Books before the Invention of Printing', in Alain Besson, ed., *Thornton's Medical Books, Libraries and Collectors: A Study of Bibliography and the Book Trade in Relation to the Medical Sciences*, 3rd edition, Hants, England, 1990, pp. 1–29

Joshi, S., *Fractured Modernity: Making of a Middle Class in Colonial North India*, Delhi, 2001

Johnson, T., 'Imperialism and the Professions: Notes on the Development of Professional Occupations in Britain's Colonies and the New States', in P. Hamos, ed., *Professionalisation and Social Change*, Keel, 1973

Keshavarz, F., *A Descriptive and Analytical Catalogue of Persian Manuscripts in the Library of the Wellcome Institute of Medicine*, London, 1986.

Khan, I.G., 'Medieval Theories of Vision and the Introduction of Spectacles in India, *c.*1200–1750', in D. Kumar, ed., *Disease and Medicine in India: A Historical Overview*, Delhi, 2001, pp. 26–39

Khan, Muhammad, S., *Islamic Medicine*, London, 1986

Kolff, D.H.A., *Naukar, Rajput and Sepoy: The Ethnohistory of the Military Labour Market in Hindustan, 1450–1850*, Cambridge, 1990

Kopf, D., *British Orientalism and the Bengal Renaissance: The Dynamics of Indian Modernization, 1773–1835*, Berkeley, 1969

Kumar, A., *Medicine and the Raj: British Medical Policy in India, 1835–1911*, Delhi, 1998

Kumar, D., *Science and the Raj: A Study of the British in India*, Delhi, 1995

——, ed., *Disease and Medicine in India: A Historical Overview*, New Delhi, 2001

——, 'Unequal Contenders, Uneven Ground: Medical Encounters in British India, 1820–1920, in A. Cunningham & B. Andrews, eds, *Western Medicine as Contested Knowledge*, Manchester, 1992, pp. 172–90

Lane, J., 'The Role of Apprenticeship in Eighteenth Century Medical Education in England', in W.F. Bynum and R. Porter, eds, *William Hunter and the Eighteenth Century Medical World*, Cambridge, 1985, pp. 57–103

Langford, J., *Fluent Bodies. Ayurvedic Remedies for Postcolonial Imbalance*, Duke, 2002

Law, N.N., *Promotion of Learning in India during the Muhammadan Rule by Muhammadans*, Calcutta, 1916

Lelyveld, D., *Aligarh's First Generation: Muslim Solidarity in British India*, Lahore, 1991

Liebeskind, C., 'Arguing Science: Unani Tibb, Hakims and Biomedicine in India, 1900–50', in W. Ernst, ed., *Plural Medicine, Tradition and Modernity, 1800–2000*, London &New York, 2002, pp. 58–75

———, 'Unani Medicine of the Subcontinent', in *Oriental Medicine. An Illustrated Guide to the Asian Arts of Healing*, ed. J.V. Alphen & A. Aris, London, 1995, pp. 39–64

Max, M., 'New Light on Hunain Ibn Ishaq and his Period', in Penelope Johnstone, ed., *Studies in Medieval Arabic Medicine. Theories and Practice*, Variorum, reprint 1984, pp. 685–724

Metcalf, B.D., *Islamic Revival in British India: Deoband 1860–1900*, Princeton, 1982

——, 'The Pilgrimage Remembered: South Asian Accounts of the Hajj', in Dale F. Eickelman and James Piscatori, eds, *Muslim Travellers. Pilgrimage, Migration and the Religious Imagination*, Routledge, 1990, pp. 85–107

Moir, M. & Zastoupil, L., *The Great Education Debate: Documents Relating to the Orientalist-Anglicist Controversy, 1781–1843*, Surrey, 1999

Moosvi, S., 'A Sixteenth Century Code for Physicians', in D. Kumar, ed., *Disease and Medicine in India: A Historical Overview*, Delhi, 2001, pp. 66–70

Palit, C., *Scientific Bengal. Science, Technology, Medicine and Environment under the Raj*, Delhi, 2006

Pannikar, K.N., 'Indigenous Medicine and Cultural Hegemony', in K.N. Pannikar, *Culture, Ideology, Hegemony: Intelligence and Social Consciousness in Colonial India*, Delhi, 1995, pp. 145–75

Pati, B. & M. Harrison, eds, *Health, Medicine and Empire. Perspectives on Colonial India*, Delhi, 2001

Patton, A., Jr., *Physicians, Colonial Racism and Diaspora in West Africa*, Florida, 1996

Pernau, M., ed., *The Delhi College. Traditional Elites, the Colonial State, and Education before 1857*, Delhi, 2006

Pollock, S., ed., *Literary Cultures in History. Reconstructions from South Asia*, Berkeley, 2003

Porter, R., 'William Hunter: A Surgeon and a Gentleman', in W.F. Bynum & R. Porter, eds, *William Hunter and the Eighteenth Century Medical World*, Cambridge, 1985

——, 'The Gift Relation: Philanthropy and Provincial Hospitals in

Eighteenth Century England', L. Granshaw & Roy Porter, eds, *The Hospital in History*, London, 1989, pp. 149–78

Prakash, G., *Another Reason. Science and the Imagination of Modern India,* Princeton, 1999

Qaiser, A.J., *The Indian Response to European Technology and Culture, 1498–1707,* Delhi, 1982

Qaiser, N., 'Politics, Culture and Colonialism: Unani's Debate with Doctors', in B. Pati & M. Harrison, eds, *Health, Medicine, and Empire: Perspectives on Colonial India*, Delhi, 2001, pp. 317–55

Rahim, A., *The Azizi Family of Physicians*, Lucknow, 2001

Rahman, F., *Health and Medicine in the Islamic Tradition: Change and Identity*, New York, 1987

Raj, K., *Relocating Modern Science. Circulation and the Construction of Scientific Knowledge in South Asia and Europe, Seventeenth to Nineteenth Centuries*, Delhi, 2006

Ramanna, M., 'Gauging Indian Response to Western Medicine: Hospitals and Dispensaries, Bombay Presidency, 1900–1920', in D. Kumar, ed., *Disease and Medicine in India: A Historical Overview*, Delhi, 2001, pp. 233–48

Ramasubban, R., 'Imperial Health in British India, 1857–1900', in E. MacLeod and L. Milton, eds, *Disease, Medicine and Empire: Perspectives on Western Medicine and the Experience of European Expansion*, London, 1988, pp. 38–60

Reddy, S., 'Dar-ul-Shifa Built by Sultan Muhammad Quli: The First Unani Teaching Hospital in Deccan', in *Indian Journal of History of Medicine*, II, 1957, pp. 102–5

Rezavi, S.A.N., 'Physicians as Professionals in Medieval India', D. Kumar, ed., *Disease and Medicine in India: A Historical Overview,* Delhi, 2001 pp. 40–65

Rizvi, S.A.A., *Religious and Intellectual History of the Muslims in Akbar's Reign, with Special Reference to Abul Fazl (1556–1605*), Delhi, 1975

Robinson, F., *The Ulema of Farangi Mahall and Islamic Culture in South Asia*, Delhi, 2001

——, 'Islam and Muslim Separatism', in F. Robinson, ed., *Islam and Muslim History in South Asia*, Delhi, 2001, pp. 177–209

——, 'Religious Change and the Self in Muslim South Asia since

1800', in F. Robinson, ed., *Islam and Muslim History in South Asia*, Delhi, 2000, pp. 105–21

Rosenthal, F., 'The Defence of Medicine in the Medieval Muslim World', in Franz Rosenthal, ed., *Science and Medicine in Islam. A Collection of Essays*, Hampshire, UK, 1990, pp. 519–32

——, 'The Physician in Medieval Muslim Society', in Franz Rosenthal, ed., *Science and Medicine in Islam: A Collection of Essays*, Hampshire, UK, 1990, pp. 474–91

Roy, A., *The Islamic Syncretistic Tradition in Bengal*, Princeton, 1983

Schimmel, A.M., *And Muhammad is His Messenger. The Veneration of the Prophet in Islamic Piety*, Berkeley, 1985

Singh, D.K., 'Clouds of Cholera and Clouds around Cholera, 1817–70', in D. Kumar, ed., *Disease and Medicine in India. A Historical Overview*, Delhi, 2001, pp. 144–65

Stark, U., 'Politics, Public Issues and the Promotion of Urdu Literature: *Awadh Akhbar*, the First Urdu Daily in Northern India', *Annual of Urdu Studies*, 18.1, 2003, pp. 66–94

Storey, C.A., *Persian Literature. A Bio-bibliographical Survey*, vol. II, part II, London, 1971

Siddiqui, M.Z., *Studies in Arabic and Persian Medical Literature*, Calcutta, 1959

Sivaramakrishnan, K., *Old Potions, New Bottles: Recasting Indigenous Medicine in Colonial Punjab 1850–1945*, Delhi, 2006

Trawick, M., 'Writing the Body and Ruling the Land: Western Reflections on Chinese and Indian Medicine', in D. Bates, ed., *Knowledge and the Scholarly Medical Traditions*, Cambridge, 1995, pp. 279–96

Troll, C.W., *Sayyid Ahmad Khan. A Reinterpretation of Muslim Theology*, Delhi, 1978

——, 'Muhammad Shibli Nu'mani (1857–1914) and the Reform of Muslim Religious Education', *Islam and the Modern Age*, 24, 1993, pp. 1–19

Vaughn, M., *Curing their Ills: Colonial Power and African Illness*, Berkeley, 1991

Verma, R.L. & N.H. Keswani, 'Unani Medicine in Medieval India. Its Teachers and Texts', in N.H. Keswani, ed., *The Science of Medicine in Ancient and Medieval India*, Delhi, 1974, pp. 127–42

Voll, John O., 'Islam as a Special World System', *Journal of World History*, 5, 1994, pp. 213–26

Wallis, F., 'The Experience of the Book: Manuscripts, Texts, and the Role of Epistemology in Early Medieval Medicine', in D. Bates, ed., *Knowledge and the Scholarly Medical Traditions*, Cambridge, 1995, pp. 101–26

Unpublished Secondary Literature

Attewell, G., 'Authority, Knowledge and Practice in Unani Tibb in India, *c.*1890–1930', SOAS PhD, 2004

Prior, K.H., Introduction to C.E. Trevelyan's 'On the Education of the People of India (1838)', unpublished Ms. 2004

European MSs In the British Library (BL) & Wellcome Library (WL)

Moorcroft papers, BL:

MSS Eur. D. 236

MSS Eur. D. 238

MSS Eur. D. 239

MSS Eur. D. 240

MSS Eur. D. 243. Eighth Fascicilus of a journal from the 17 Aug. to Sep. 16 and Ritanka ke joth in Koolloo to Toong-loong ghath in Tartary

MSS Eur. D244. Ninth Fascicilus of a journal from Sept. 16 to Oct. 21, 1828 in the country of Ladakh. Memorandum Respecting Rhurbarb' ff 684–92

MSS Eur. D.246

W.B. O'Shaughnessy Papers, BL

MSS Eur. E 301/4

Monckton Papers, WL

MSS 3596–3611. Edward Henry Cradock Monckton, collections of chemical, alchemical, medical and miscellaneous recipes, Indian in origin 1842–53

Law, J.A., 'Essay on Cholera Founded on Observations of the Disease in various parts of India and in Sunderland, Newcastle and Gateshead, with Notices of the Treatment and of the Civil and Hospital Police Adopted in the Towns', T.923

Printed Books in English, Eighteenth-Nineteenth Centuries

Adam, W., *Reports on the state of education in Bengal (1835–8) including some account of the state of education in Bihar and a consideration of the means adapted to the improvement and extension of public instruction in both provinces*, ed. A. Basu, Calcutta, 1941

Adam, W., *Third report on the state of education in Bengal including some account of the state of education in Bihar, and a consideration of the means adapted to the improvement and extension of public instruction in both provinces*, Calcutta, 1838

Ainslie, W.H., *Materia Medica or some account of those articles which are employed by the Hindoos and other eastern nations in their medicine, arts, and agriculture*, 2 vols, Calcutta, 1813, second edition, London, 1826

——, *Observations on the cholera morbus of India: a letter addressed to the Court of directors of the English East India Company*, London, 1825

Annesley, J., *Researches into the causes, nature and treatment of the more prevalent diseases of India and of warm climates generally illustrated with cases, postmortem examinations and numerous coloured engravings of morbid structures*, 2 vols, London, 1828

——, *Sketches of the most prevalent diseases of India, comprising a treatise on the epidemic cholera of the East. Statistical and topographical reports of the disease in the different divisions of the army under the Madras presidency; embracing also the annual rate of mortality etc of European troops and practical observations on the effects of calomel on the alimentary canal, and on the diseases most prevalent in India*, London, 1825

Balfour, E., *Medical hints to the people of India. The Vydian and the Hakim, what do they know of medicine?*, Madras 1857

Breton, P., *A vocabulary of the names of the various parts of the human body and of medical and technical terms in English, Arabic, Persian, Hindee, Sanskrit for the use of the medical department in India*, Calcutta, 1825

Broughton, T.D., *Letters written in a Mahratta camp during the year 1809, descriptive of the character, manners, domestic habits and religious ceremonies of the Mahrattas*, London, 1813

Corbyn, F., *A treatise on the epidemic cholera as it prevailed in India. Together with reports of the medical officers made to the Medical Boards of the presidencies of Bengal, Madras, Bombay, for the purpose of ascertaining a successful mode of treating that destructive disease and a critical examination of all the works which have hitherto appeared on the subject*, Calcutta, 1832

Davies, S., Report of govt. dispensary, Patna. Prepared with inputs sent by Ram Eshur Awushtee, incharge govt. dispensary, January 1840

Gladwin, F., Tr., *Ulfaz Udwiyeh or the materia medica in the Arabic, Persian and Hindevy languages compiled by Noureddeen Muhammad Abdullah Shirazy*, Calcutta, 1793

Hooper, R., *The anis ul Musharrahin or the Anatomist's Vade-Mecum*, tr. into Arabic by J. Tytler, Calcutta, 1830

Holwell, J.Z., *An account of the manner of inoculating for the small pox in the east Indies. With some observations on the practice and mode of treating that disease in those parts*, London, 1769

Hutchinson, J., *Code of regulations for medical deptt. of the Bengal establishment compiled by order of government under Supdtt. of Medical Board*, Calcutta, 1838

——, *Observations on some of the most important points connected with the consideration and treatment of cholera asphyxia*, Calcutta, 1832

Irvine, R.H., *A short account of the materia medica of Patna*, Calcutta, 1848

Jameson, J., *Report on the epidemic cholera morbus as it visited the territories subject to the presidency of Bengal, in the years 1817, 1818 & 1819*, Calcutta, 1820

Kerr, J., *A review of public instruction in the Bengal presidency, from 1835–1851*, Calcutta, 1852, part I

——, *A review of public instruction in the Bengal presidency, from 1835–1851*, London, 1853, part II

Leitner, G.W., *Reports on the state of education in the Punjab since annexation and in 1882*, Lahore, 1882

——, *History of indigenous education in the Punjab since annexation and in 1882*, Calcutta, 1882

McRae, J., *Report on Cawnpur Dispensary*, 1 February 1841

Memorial of Surgeon John Tytler of the Bengal establishment, addressed to the court of directors of the EIC. L. No. ix, J Tytler's reply to the committee of native medical education as to the best language to be used as a medium of instruction, Tracts 15.

Moore, J., *The history of the small pox*, London, 1815

O'Shaughnessy, W.B., *The Bengal Pharmacopoeia and general conspectus of medicinal plants arranged according to the natural and therapeutical systems*, Calcutta, MDCCCXLIV

——, *The Bengal Dispensatory and Companion to the Pharmacopoeia*, Calcutta, MDCCCXLII

Playfair, G., tr., *The Taleef Shereef or Indian Materia Medica*, Calcutta, 1793

Playfair, G. & A. Ross, *Report on Delhi city dispensary for 1840*, V/24/732

Selections from the records of the Bengal government, no. xiv, papers relating to the establishment of the Presidency college of Bengal, Calcutta, 1854, MF 1/871

Shoolbred, J., *Report on the progress of vaccine inoculation in Bengal from the period of its introduction in Nov. 1802 to the end of the year1803 with an appendix submitted to the medical board*, Calcutta, 1804

Sprenger, A., *Report of the researches into the Muhammadan libraries of Lucknow, by Alois Sprenger*, Calcutta, 1896

Spilsbury G.G. and S. Dutt, *Hindustani Version of the London Pharmacopoeia, edition 1836, 1841. Done into the Persian from the Nagri character with the addition of an appendix. Contains an abstract of the actions, uses and doses of all the preparations contained in the pharmacopoeia, by Fred Mouat*, Calcutta, 1845

Stewart, C., *A Descriptive catalogue of the oriental library of the late Tipu Sultan of Mysore*, Cambridge, 1809

Thackrah, C.T., *Cholera its character and treatment with remarks on the identity of the Indian and English and a particular reference to the disease as now existent at New Castle and its neighbourhood*, Leeds, 1832. Wellcome Library, T 923

Tytler, J., tr. into Arabic, *A short anatomical description of the heart extracted from the Edinburgh medical dictionary*, Calcutta 1828, p. v

——, tr. into Arabic, *Hooper, R: The anis ul Musharrahin or the Anatomist's Vade-Mecum*, Calcutta, 1830
——, Tracts 1881 a 3, *The Calcutta Literary Gazette*, 14 June 1834, essay no. VII
——, ed., *Arabic tr of the Aphorism of Hippocrates or the Fusool-i-Abkrat by Hoanin Ben Motawukkue*, Calcutta, 1832
Usman, Khan Bahadur Muhammad, *The Report of the committee on the indigenous systems of medicine*, Madras, part I, 1923

Proceedings of Governments and Other Bodies, Oriental and India Office Collections, British Library

Board's Collection
Bengal Political Consultations 1790–1850
Bengal Public Consultations 1770–1850
Bengal Judicial Consultations 1780–1850
Bengal Military Consultations 1780–1836
Indian Military Consultations 1840–50
Indian Public Consultation 1840–50
Proceedings of the Governor General NWP 1838–1849
Lt. Gov. NWP Proceedings General 1835–50
UP Proceedings in local self government, Allahabad 1916
Proceedings of UP in Medical Department 1930–1931
Indian Political and Foreign Consultations
Selections from the vernacular press, received from 1870–1904

Persian Manuscripts

WMS. Per. 199 (D), 3 folios, Wellcome Library, London
WMS. Per. 580, Wellcome Library, London
Bibliotheque Nationale Paris, no. 342, supplement
Ilajat-i-Dara Shikuhi, 342A, vol 2, Paris
Ilajat-i-Dara Shikuhi, 857–9. Supplement Persan 342, 342A, 342B
Mizan-i-Tibb, Add Mss. 17949, British Library, London
Egerton 1006 & 1007 being vols 1 & 2 of *Tibb-i-Akbari*

Urdu Books, Newspapers, and Articles

Ahmad, Hakim Iftekhaar, 'Matba-i-Ustaad Kei Chand Fakahaat', *Risaalah Takmil ut Tibb*, Lucknow, May 1950

Ahmad, Imad al Din, ed., *Tibbi Ayeenaa*, 1879

Ahmad, Munshi Sayeed Fakhr-al-Din Ahmad, *Alaijab Mutahsir*, 1260 hijri, 1844

All India Ayurvedic and Unani Tibbi Conference, Delhi, Ek Munsafanah Nazar, by Abid Ali Khan, Lucknow, undated pamphlet

Ali, Hakim Ihsan , Tibb-i-Ihsani, 3rd edition, Lucknow, 1281 hijri, 1864

Ali, Qamar, *Kitaab Lajwaab wa kimyaab nuskha naushadarwee sihat ayeena*, Lucknow, 1867

Ali, Maulvi Syed Nawab, 'Tazkira Attiba-i-Lucknow', *Risala-i-Takmil ut Tibb*, March 1948

Ali, Muhammad Hasan,tr., *Mizan-i-Tibb*, Delhi, 1852

——, *Mizan-i-Tibb*, Kanpur, 1872

Ali, Asghar, *Ilaj al Ghuraba, Lucknow*, 1281 hijri, 1864

Aziz, Hakim Abd al, *Risaalah Tuhfa-i-Azizi*, Lucknow, 1888

Baksh, Ilahi, *Tashreeh al asbaab musammabeh Mazhar al ulum*, Kanpur, 1877

Hafiz, Ikram al Din, *Tibb-i-Nabawi*, Lucknow, 1869

Khan, Hakim Hadi Husain, tr., *Zakhirah-I-Khwarzmshahi*, Lucknow, 1878, 3 vols, 10 parts

Khan, Ihsan Ali, *Maqaalat-i-Ihsani*, Lucknow, 1878

——, *Qarabdeen Ihsaani*, Kanpur, hijri 1281, 1864

Karawai Jalsah Awwal. (Proceedings of the first convention.) Presidential address by Chaudhury Muhammad Altafat Rasool, Report Ijlaas Doawwam (2nd). All India Unani Tibbi Conference, 4 February 1913, Lucknow, Nami Press Lucknow, July 1913

Lahori, Zia ud Din, *Khud Nawisht. Hayat-i-Sir Syed*, Karachi, 1996

Minutes of the Executive Committee of Takmil-ut-Tibb College, Lucknow, 1902–9

Nadwi, Maulana Suleiman Syed, *Hayat-i-Shibli*, Azamgarh, 1943, reprint, 1999

Nagrami, Hakim Syed Muhammad Hasan, *Tarikh Tibb*, Delhi, 1989
Nath, Pandit Kashi, *Akhlaaq-i-Kashi*, Agra, 1870
Oudh Akhbar, 1876–81
Pamphlet on 'Reasons for remaining away from the All India Vedic and Unani conference, Delhi', Nami Press Lucknow, 1911
Raudad-i-Takmil-ut-Tibb (Proceedings of Takmil-ut-Tibb) 1908–11
Risaalah-i-Takmil-ut-Tibb (Magazine of Takmil-ut-Tibb). All issues for the year 1948–50
Rahman, Hakim Zillur, *Tazkira-i-Khandan-i-Azizi*, Aligarh, 1978
Rahman, Hakim Zillur, ed., *Bayaaz-i-Wahidi*, Aligarh, 2002
Rashid, Hakim, *Almaun fi al Taun,* Matba Anwari Muhammadi, Lucknow, 1897
Saiyid, Hakim Ali Hasan, *Zubdat al Mufradat*, Lucknow, 1877
Tahreer Hakim Mukhlis-ul-Daulah Jaipuri (Speech of Hakim Mukhlis-ul Daulah Jaipuri), Report-Ijlaas Doawwam (2nd) All India Unani Tibbi Conference 4 February 1913, Lucknow, Nami Press Lucknow, July 1913
Yahya, Hakim Ghulam 'Risalah-i-Qarurah', in *Risalah Naadir Ahwal*, Munshi Newal Kishore Press, Lucknow, 1877

Index